Rapid Methods and Automation in Microbiology and Immunology

Edited by K.-O. Habermehl

With 212 Figures and 159 Tables

Springer-Verlag
Berlin Heidelberg New York Tokyo

Professor Dr. K.-O. Habermehl
Institut für Klinische und Experimentelle Virologie
Freie Universität Berlin
Hindenburgdamm 27, D-1000 Berlin 45

Fourth International Symposium on Rapid Methods and Automation in Microbiology and Immunology, Berlin, June 7–10, 1984

ISBN 3-540-13695-9 Springer-Verlag Berlin Heidelberg New York Tokyo
ISBN 0-387-13695-9 Springer-Verlag New York Heidelberg Berlin Tokyo

Library of Congress Cataloging in Publication Data. International Symposium on Rapid Methods and Automation in Microbiology and Immunology (4th: 1984: Berlin, Germany) Rapid methods and automation in microbiology and immunology. "Fourth International Symposium on Rapid Methods and Automation in Microbiology and Immunology, Berlin, June 7–10, 1984" – T.p. verso. Bibliography: p. 1. Diagnostic microbiology – Automation – Congresses. 2. Immunology – Automation – Congresses. 3. Hybridomas – Congresses. 4. Antibodies, Monoclonal – Congresses. I. Habermehl, K.-O. (Karl-Otto), 1927– . II. Title. [DNLM: 1. Automation – congresses. 2. Immunologic Technics – congresses. 3. Microbiological Technics – congresses. W3 IN922RM 4th 1984r/QW 25 I6104 1984r] QR67.I573 1984 616.07'56 85-10031
ISBN 0-387-13695-9 (U.S.)

Printed in Germany

Typesetting, printing and bookbinding: Brühlsche Universitätsdruckerei, Giessen
2127/3130-543210

Preface

Rapid progress in molecular biology, genetic engineering, and basic research in immunology has opened up new possibilities for application to diagnostic procedures and to clinical research. In a short period a new era of diagnosis dawned, covering nearly all fields of microbiology, immunology, and food technology. In consequence of this rapid development, scientists of many disciplines are involved studying infections of humans, animals, and plants or working in technical microbiology.

The application of the newest findings of basic research to diagnostic work and to clinical research covers nearly all fields of microbiology and immunology. Moreover, it underlines the close relationship between diagnosis, therapy, and epidemiology. An outstanding example of these connections is given by the recent development of hepatitis B vaccine. The discovery and identification of a non-cultivable agent by physicochemical and immunological methods were the heralds of a new era in the prevention of infectious diseases.

This book provides an up-to-date, comprehensive review of developments and future aspects in various fields. I am convinced that the authors have succeeded in furnishing a large variety of new ideas and possibilities.

K.-O. Habermehl

Contents

Physical Detection of Microorganisms

Chromatographic Analysis of Microbial Components

Isolation of Microorganisms

Antigen Detection and Identification

Fluoroimmunoassays, Time-Resolved Fluorometry, and Bioluminescence

List of Contributors

You will find the addresses at the beginning of the respective contribution

The Evaluation of Vaccines

M. R. Hilleman [1]

Introduction

It seems somewhat incongruous to speak of the evaluation of vaccines in a symposium dedicated to rapid methods and automation. The evaluation of vaccines for man involves the conduct of tests in human subjects, and there speed cannot be a dominant attribute. Instead, it involves the slow and gradual accumulation of data, since safety must be the prime concern. The investigator may, however, be greatly aided in the conduct of the needed laboratory support tests by the application of the most efficient and rapid methods available, and here is where speed and automation may be of very substantial indirect value.

That aside, there is no subject of greater importance in preventive medicine than that of prophylactic immunization. Vaccines are a cornerstone of modern infectious disease control, and an increase in the use of vaccines for this purpose is the inevitable wave of the future, since so much social and economic benefit can be achieved at so little cost. Of added importance are the new and remarkable technological breakthroughs in biochemistry, especially in the production of polypeptide and protein antigens in unnatural host cells by DNA recombinant technology, and the production of monoclonal antibodies in hybridoma cells that increase the ability to define the significant immunologic determinants of infectious agents and to facilitate their purification. What we are seeing now is both technical possibility and economic feasibility for vaccines that would have been technically and economically impossible just a few years ago. In fact, it is now possible to develop vaccines against agents that cannot be grown in the laboratory.

The development of a new or improved vaccine for general application in man follows no chartable course. The primary aim is to establish safety and protective efficacy. The rigor of the evidence required to establish safety and efficacy varies from country to country and may be increased or diminished with the importance of the disease that the vaccine is intended to prevent. However, judgmental guidelines based on prior experience and precedent may be followed, and most countries impose legal requirements. Among the countries most engaged in vaccine development, the USA, I believe, imposes the most rigid legal requirements [1–4] on vaccine development and licensure, and I shall use them as a guide in my further presentation.

---

1 Merck Sharp & Dohme Research Laboratories, West Point, PA 19486, USA

Concepts and Feasibility

The development of a vaccine begins with conceptual and feasibility considerations. First and foremost, perhaps, is a determination of need for the vaccine. Does the disease for which the vaccine is intended occur in a large enough segment of the population, and is it serious enough to justify the time and effort required to develop it? Assuming that a real need can be established and documented, the important further considerations are technical feasibility and practicality, i.e., the state of the art.

For live virus vaccines, it is necessary to consider the source of the virus. Can it be grown in cells acceptable for human use, free of contaminating agents and potential carcinogenicity? Can attentuation for man be achieved and stabilized genetically so as to preclude either reversion to virulence in the host or possible contagious transmission to susceptible persons who are in contact with the vaccinee? These are very important considerations being applied to the varicella-zoster vaccine now in clinical evaluation. New genetic technology may facilitate the development of safe and stable vaccines through directed rather than random genetic mutation, selection, and stabilization. Most of the same principles that are applied to live virus vaccines apply also to live attentuated bacterial vaccines, except that bacterial cells are directly cultivable.

For killed whole-organism or subunit antigens, the question is whether the key immunodeterminants can be identified, produced, purified to remove toxic and allergenic factors, rendered free of possible living contaminants, and presented to the host in a way that will elicit an immune response. This was a most important consideration in the development of human hepatitis B vaccine, prepared by using surface antigen from the plasma of human carriers of hepatitis B virus infection.

There is then the question of whether the vaccine can be reproducibly made in adequate amount at a cost that is acceptable. Finally, and perhaps of greatest importance, is the matter of whether, or to what extent, the vacine would actually be used in the population for which it is intended. These considerations bear on the basic economics of the whole vaccine enterprise, since vaccines that sit on the shelf and are not used give no beneficial effect to those who should receive them and provide no incentive to the producer to invent new vaccines. Simplified combined vaccine, such as one that might include antigens of herpes simplex 1 and 2, Epstein-Barr virus, cytomegalovirus, and varicella-zoster virus, would go far toward making immunization more beneficial to the user.

Development

Once technical feasibility has been established and the background research and development have been carried out, the next step is to prepare one or more lots of vaccine for clinical test. The necessary research and development that precede vaccine preparation require substantial expenditures of time and funds, and success is not likely to be achieved in the absence of a critical mass of scientific workers of diverse disciplines working in adequate facilities. The target must be a safe and effective vaccine capable of manufacture in adequate amount by a reproducible process. A vaccine that can be made only once is simply not a vaccine.

In preparing vaccine, the common seed lot system is employed, and all possible markers of attentuation are applied to vaccines containing live, attenuated organisms. Precise measurement of the applicable physical and chemical attributes is required for all vaccines. Greatest possible purity must be sought.

Very extensive tests need to be carried out in animals, in cell cultures, and in artificial media to assure freedom from contaminating organisms. Tests are also carried out in animals by the best possible procedures to give the greatest assurance of safety and immunogenicity. Few animal models have direct comparability for man, and animal tests can only provide guidance in averting problems. The safety and efficacy of a vaccine need to be proved in the target species for which it is intended, and laboratory test data coupled with seasoned judgment have usually been quite effective in the safe conduct of tests of vaccines for man.

In the USA, there is a requirement that all vaccines for investigative use be made employing good manufacturing [1–2] and good laboratory practices [3]. Procedures and findings in vaccine preparation are carefully documented. As a result, vaccines for experimental use are as carefully prepared and controlled as are lots of vaccine for commercial distribution. A detailed review of procedures and findings is given by the U.S. Food & Drug Administration in the form of Investigational New Drug regulations [4] before first tests in man may be started.

Clinical Evaluation

The third and most critical step in vaccine development is that of clinical evaluation to measure safety and protective efficacy in the short and long terms. Such tests are carried out by qualified clinical investigators having adequate laboratory and epidemiologic support capability.

First tests of a new vaccine intended to measure safety are necessarily conducted on a small scale under carefully controlled conditions in persons who are least likely to encounter natural infection in the course of the test and for whom any adverse effect of the vaccine would be least injurious. In essence, they are normal, healthy persons. For this reason, it is common practice to employ adults in initial tests of vaccines intended for children. However, while useful in ruling out nonspecific vaccine effects, adults are immune to most childhood infections, and safety needs to be established in the population for which the vaccine is intended. Observations are made for adverse reactions of any sort and for contagious spread to susceptible persons who are in contact with vaccinees when living agent vaccines are being tested.

Efficacy trials must be well planned to provide definitive answers. In such trials, laboratory tests are carried out to measure immune responses, and protection is measured in the field. Placebo-controlled field studies are commonly carried out in populations in which there is the highest possible attack rate, thereby providing the greatest opportunity of demonstrating significant protection in manageable numbers of subjects. Though desirable, it is not always possible to measure efficacy in the target populations for whom annual attack rates are very low, as was the case for the pneumococcal vaccine that was proved effective in young African gold miners.

Studies to measure the duration of protection may involve many years of follow-up, during which the indirect benefit derived from the reduction in the natural spread of the infectious agent through the establishment of herd immunity is also observed. Long-term surveillance for possible adverse effects is carried out, based on both field reports of vaccine in general use following licensure and clinical follow-up of subjects who were in the pre-licensure studies.

Ethical Considerations in Clinical Testing

The conduct of research, development, and clinical testing of vaccines is for the most part a purely scientific activity reasonably free from emotional overtones. The application of ethical standards to the selection and treatment of participants, however, is an issue of considerable controversy, influenced by religious beliefs, philosophy, cultural background, and even the political persuasions of those who engage in it. This problem has been dealt with by many committees in many national and international forums without resulting in the adoption of any international standards or procedures, though the controversy has cooled considerably in the last 2 or 3 years.

Minimally, however, there is general agreement that there must be full disclosure of information concerning potential risks and benefits and that meaningful, informed consent must be obtained from participants in clinical trials. In the case of children and mentally handicapped persons, consent may be given by parents, guardians, or other responsible individuals or institutions. Inducement to participate in clinical trials by monetary or other reward, especially as applied to prisoners, is commonly deemed to be coercive and ethically suspect, but reward can be justified in many instances as a payment for service or inconvenience. Wherever possible, effort is made to assure that the vaccine being tested can provide a potential benefit to the recipient. This may not always be possible, as with the testing of child vaccines in adults who may already be immune to the infection. The right to participate and the potential for societal gain are overriding considerations in such situations. Provision is made in most trials for compensation for injury to a participant who may have taken part in the test of a vaccine that is defective and causative. The safety record for investigative vaccines has been exemplary, and I know of no instance of damage by experimental vaccine in the last two decades.

To be meaningful, efficacy trials, must be controlled, and a group that receives placebo is used to match the group receiving vaccine. This means that a person may be injected with a substance providing no benefit. This problem is sometimes circumvented by testing two different vaccines, each of which serves as a control for the other. It is also common practice to terminate clinical trials as soon as definitive results are obtained and to vaccinate those persons who were in the control group so as to give them protection against the disease as well.

Governmental regulation commonly governs the conduct of clinical trials. By such procedure, background information and test protocols are presented to an institutional review board made up of health professionals and lay persons such as lawyers, clergy, or other members of the community. Such committees determine whether a trial can be allowed to take place and monitor the progress of the trials. All clinical testing in the USA is carried out under the provisions of the Investigational New Drug laws [4], which provide for mandatory governmental review and monitoring of the vaccine trials.

Candidate vaccines that provide adequate evidence of safety and efficacy and that have been shown capable of routine and consistent production may be considered for licensure for commercial distribution.

Three Vaccine Examples

The three speakers who are to follow are experts in the clinical evaluation of vaccines. They have chosen to discuss three vaccines representative of the tedious and long-term collection of data essential to prove safety and efficacy. Hepatitis B virus vaccine, licensed in 1981 and now finding widespread application throughout the world, was chosen as an example of a killed subunit virus vaccine. Varicella-zoster vaccine, now in the final stages of development and possibly to be licensed within the next 2 years, was chosen to represent an attenuated live virus vaccine. Finally, herpes simplex 2 vaccine, currently being tested to determine the feasibility of vaccination against herpes, was chosen to represent a killed subunit vaccine derived from infected chick embryo cells in culture. To be efficacious such a vaccine will need to be produced by a more economic procedure, such as DNA recombinant technology, and availability for general use may not come for at least 5 years. These three examples are highly instructive, indicating the time span required in the evaluation of vaccines for safety and efficacy.

References

1. 21 Code of Federal Regulations (21CFR), Food & Drugs (1983) Office of the Federal Register National Archives and Records Service, General Services Administration, U.S. Government Printing Office. Current good manufacturing practives in manufacturing, processing, packing, or holding of drugs: general, part 210, April 1, 1983, pp 73–74
2. 21 Code of Federal Regulation (21CFR), Food & Drugs (1983) Office of the Federal Register National Archives and Records Service, General Services administration, U.S. Government Printing Office. Current good manufacturing practices for finished pharmaceuticals, part 211, April 1, 1983, pp 75–95
3. 21 Code of Federal Regulations (21CFR), Food & Drugs (1983) Office of the Federal Register National Archives and Records Service, General Services Administration, U.S. Government Printing Office. Good laboratory practice for non-laboratory studies, part 58, April 1, 1983, pp 209–223
4. 21 Code of Federal Regulations (21CFR), Food & Drugs (1977) Office of the Federal Register National Archives and Records Service, General Services Administration, U.S. Government Printing Office. New drugs for investigational use, part 312, April, 1977, p 72

Evaluation of Hepatitis B Vaccine

S. KRUGMAN [1]

The development of rapid methods to identify the etiology of various infectious diseases is an important objective. However, in the development and evaluation of vaccines, caution and prudence rather than rapidity are essential. The acceleration of vaccine development by so-called "crash programs" can prove to be disastrous.

The history of the development of polio, measles, and rubella vaccines reveals that the time interval between identification and cultivation of the agents and subsequent licensure of the vaccines ranged between 7 and 16 years. Inactivated poliovaccine was licensed for use in 1956 and live attenuated poliovaccine in 1961, 7 and 11 years respectively after successful cultivation of the virus in 1949. Measles vaccine was licensed in 1963, 9 years after the growth of the virus in cell culture. The time interval between propagation of rubella virus in 1962 and subsequent licensure of the attenuated vaccine in 1969 was 7 years. The development of these viral vaccines required the successful propagation of large quantities of virus for inactivation or attenuation.

Unlike polio, measles, and rubella viruses, the successful cultivation of hepatitis B virus (HBV) has not been achieved. Nevertheless, it has been possible to develop, evaluate, and license various hepatitis B vaccines. The sequence of events that culminated in this important development occurred over a span of 16 years. The many milestones that led to this achievement stemmed from the contributions of various investigators during the period 1965–1981.

History of the Development of the Hepatitis B Vaccine

The first and most important milestone was the discovery of Australia antigen by Blumberg et al. (1965). The antigen that they detected in the blood of an Australian aborigine subsequently proved to be associated with hepatitis B infection (Prince 1968). In 1970, Dane et al. (1970) identified 42-nm, double-shelled particles in the blood of patients with Australia antigen-associated hepatitis. They detected these particles by immune electronmicroscopy. Later, it became apparent that the so-called Dane particle was HBV. It also became apparent that Australia antigen was immunologically identical with the envelope or surface component of HBV and immunologically distinct from the core component of HBV.

1 New York University Medical Center, 550 First Avenue, New York, NY 10016, USA

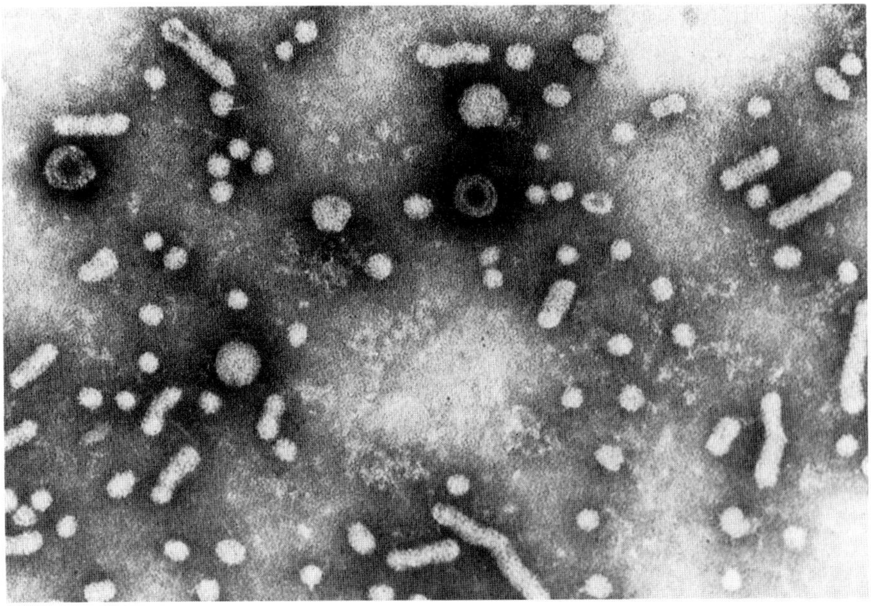

Fig. 1. Electronmicrograph of serum containing HBV particles, 42 nm in diameter, and HBsAg particles that are either 20-nm spheres or tubular structures of similar diameter and variable length

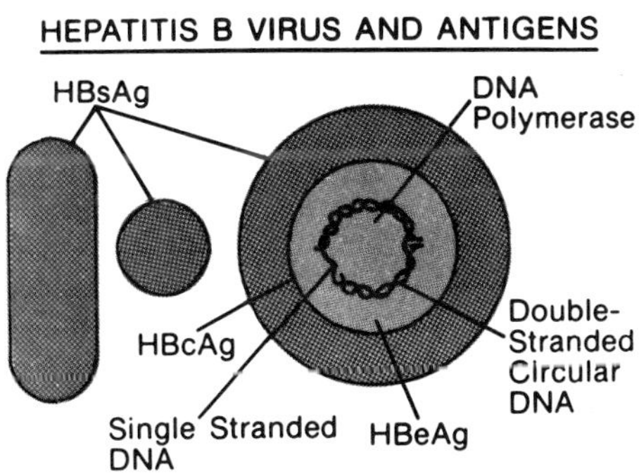

Fig. 2. Schematic illustration of hepatitis B virus (HBV) and its antigens, HBsAg, HBcAg, and HBeAg

An electronmicrograph of HBV infected blood reveals three morphologically distinct particles: 20-nm spheres, tubular structures of similar diameter and variable length, and 42-nm HBV (Dane) particles (Fig. 1). The spherical and tubular particles are immunologically identical with the surface component of HBV and are now designated hepatitis B surface antigen (HBsAg). The core component contains two immunologically distinct antigens, hepatitis B core antigen (HBcAG) and hepatitis B e antigen (HBeAg) (Fig. 2). The respective antibodies of these antigens have been designated anti-HBs, anti-HBc, and anti-HBe.

In the early 1970s, the development of highly sensitive and specific tests to detect such hepatitis B markers as HBsAg, anti-HBs, and anti-HBc provided the technology required to evaluate safety, immunogenicity, and efficacy of hepatitis B vaccine. Studies on the natural history of hepatitis B had revealed that induction of anti-HBs and/or anti-HBc was indicative of immunity.

Another milestone was reached in 1971 when my colleagues and I demonstrated that heat-inactivated MS-2 hepatitis serum was no longer infectious but remained antigenic (Krugman et al. 1971). Thus, the heat that inactivated HBV did not affect the antigenic potential of HBsAg. The detection of anti-HBs in the serum of recipients of this heat-inactivated preparation indicated that the noninfectious HBsAg particle was the antigen needed for vaccine production. Subsequent studies revealed that the heat-inactivated serum was not only immunogenic but also partially protective.

The results of our studies with heat-inactivated MS-2 hepatitis serum demonstrated that hepatitis B vaccine development was an achievable objective. These findings stimulated Hilleman and his colleagues and other investigators to accelerate their efforts to extract and purify antigen from HBsAg-positive plasma for subsequent vaccine development.

Vaccine development gathered momentum in 1973 when HBV infection was transmitted successfully to nonhuman primates (Barker et al. 1973). Seronegative chimpanzees proved to be excellent animal models for the study of hepatitis B infection and the evaluation of vaccines for safety and efficacy.

A third milestone was the development of safe, practical, and effective vaccines for routine application in man. Starting in 1968, Hilleman et al. (1975) developed a highly purified hepatitis B surface antigen vaccine that was first tested clinically in 1975 and proved highly effective not only in stimulating antibody but also in conferring protective immunity. Procedures for development of other hepatitis B vaccines have been described by Purcell and Gerin (1975), by Maupas et al. (1976), and by Reesink et al. (1978). The following discussion of the manufacturing process for the hepatitis B vaccine will summarize the procedures employed by Hilleman et al. to yield a highly purified product which, by incorporating various measures, assures safety from possible adventitious agents. The objective underlying the steps used in the process is to extract and purify plasma-derived HBsAg by utilizing biophysical technology capable of eliminating all possible living agents that may be present in human blood.

Manufacturing Process

Plasma obtained from hepatitis B carriers contains enormous quantities of HBsAg and small quantities of HBV. The ratio may be as high as 10,000 HBsAg particles to one HBV particle. The manufacturing process that was developed by Hilleman et al. (1975) involves a series of complex physical and chemical procedures to isolate the 20-nm spherical, noninfectious particles. The procedures include concentration of HBsAg by ammonium sulfate precipitation, isopyknic banding in sodium bromide, and rate zonal centrifugation in sucrose. The partially purified HBsAg particles are then treated with pepsin at pH 2, followed by treatment with 8*M* urea to remove extraneous blood and human liver proteins. After gel filtration, the antigen is treated with formalin in a 1:4,000 dilution. The final step is formulation of the vaccine to contain 20 μg of HBsAg and 0.5 mg of Al^{3-} as aluminum hydroxide in each 1.0-ml dose. Thiomersal, in a concentration of 1:20,000, is added as a preservative.

Each of these three steps, (pepsin, urea, and formalin) has been shown to inactivate HBV in chimpanzee studies. In addition, pepsin has been shown to inactivate viruses from every known group, including rhabdoviruses (vesicular stomatitis virus), poxviruses (vaccinia), togavirus (sindbis), herpesviruses (herpes simplex, type 1), coronaviruses (infectious bronchitis virus), and reovirus. The urea treatment inactivates myxoviruses (Newcastle disease virus), picornaviruses (mengovirus), and slow viruses (scrapie agent), as well as the viruses inactivated by pepsin. Formalin inactivates a wide variety of viruses, including non-A, non-B hepatitis viruses, parvoviruses, retroviruses, and the delta agent. However, formalin does not inactivate such viruses as SV40 virus, and inactivation can be inhibited by the presence of impurities. Therefore, the use of multiple purification and inactivation procedures provides assurance of inactivation of any agent that might be carried over from human blood, including the putative AIDS agent.

The putative AIDS agent was recently described in the U.S.A. (Gallo et al. 1984) and in France (Vilmer et al. 1984). This agent, though still unproved as the cause of AIDS, seems a likely candidate. It is of some interest that this agent is a retrovirus of the human T cell leukemia group of viruses (HTLV) and that representatives of the group are known to be inactivated by formaldehyde. Additionally, the surface glycoproteins of retroviruses are at least partially destroyed by proteolytic enzymes. This gives added evidential support for confidence that the process serves to destroy all known life forms.

Electronmicrographs of the original plasma and the purified HBsAg are shown in Fig. 3. All of the protein in the final product is accounted for as HBsAg,

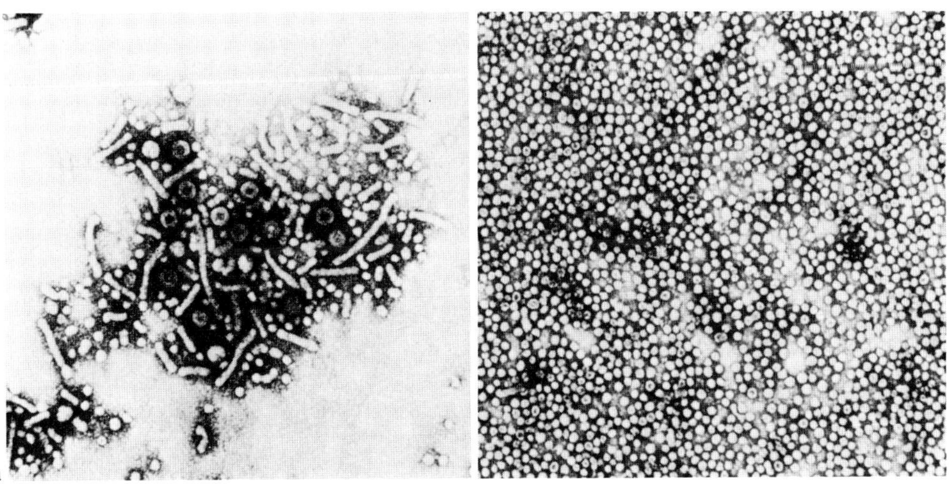

a b

Fig. 3a, b. Electronmicrographs **a** of plasma containing infectious HBV particles and excess HBsAg particles and **b** of hepatitis B vaccine (purified HBsAg particles) thereby decreasing the potential risk of an autoimmune response to extraneous protein.

Tests for safety of the vaccine include in vitro and in vivo assays for viral and microbial sterility of the bulk plasma, the purified antigen, and the final product. Each lot of vaccine is tested in chimpanzees to detect possible residual infectivity to hepatitis B virus or other viral agents.

The initial studies in chimpanzees confirmed the safety, immunogenicity, and efficacy of the vaccine for these animals. Vaccine-induced anti-HBs protected immunized chimpanzees who were subsequently challenged with live HBV. In contrast, unimmunized, seronegative chimpanzees contracted hepatitis B infection after a comparable challenge.

Evaluation of the Hepatitis B Vaccine

Safety – Evaluation Before Licensure

The first human trials with the vaccine prepared by Hilleman et al. were begun in November 1975. These initial studies involved 11 seronegative adult volunteers who received one dose of vaccine. Follow-up observation for 6 months revealed no evidence of HBV infection. Serial samples of blood at 1-month intervals revealed no detectable HBsAg or anti-HBc, and the serum aminotransferase (ALT) values were normal. Subsequent studies by various investigators from 1976 to 1978 confirmed the safety of the vaccine.

Immunogenicity

The inactivated hepatitis B vaccine has proved to be highly immunogenic when given intramuscularly in a three-dose schedule at 0, 1, and 6 months. The present recommended dose ist 10 µg (0.5 ml) for children under 10 years of age and 20 µg (1.0 ml) for older children and adults. Immunocompromised persons, such as hemodialysis patients, should receive a 40-µg dose (2.0 ml) because they have not responded as well as healthy adults. After three 40-µg doses of vaccine, the anti-HBs response may range between 60% and 70%. Hepatitis B vaccine will induce an anti-HBs response in 80%–90% of immunocompetent recipients by 1 month after the second dose. The third (booster) dose, given at 6 months, will increase the antibody response to over 95%, as well as increasing the anti-HBs titers of those who responded previously. The immune response may be sluggish in individuals older than 40 years, and males do not respond as well as females. To date, vaccine-induced anti-HBs has persisted for at least 4 years in about 90% of vaccinees. Since antibody titers would be expected to decline, it is likely that an additional booster dose may be required 5 or more years after the third dose, unless it can be shown that exposure to HBV will induce an anamnestic response that will afford protection against this "long incubation disease."

In one reported study by Francis et al. (1982) involving 1,402 homosexual men, the antibody response after three doses of vaccine was 85% instead of 95%. Subsequent investigation revealed that many vials of vaccine used in this study had been inadvertently frozen. It has been shown that freezing decreases the potency. The optimum temperature for shipment and storage of inactivated hepatitis B vaccine is 2°–8 °C. It should not be frozen.

Efficacy

The efficacy of hepatitis B vaccine has been evaluated in three randomized, placebo-controlled, double-blind trials among 1,083 homosexual men in New York (Szmuness et al. 1980), 1,402 homosexual men in five additional cities (Francis et al. 1982), and 865 staff members of 43 hemodialysis units in the United States (Szmuness et al. 1982). The vaccine proved to be highly effective in preventing HBV infection and disease. The difference in cumulative attack rates between the vaccine and placebo groups was highly significant. The vaccine recipients were protected against acute hepatitis B, asymptomatic infection, and chronic HBsAg carrier state. All vaccine recipients who had an anti-HBs response were protected against disease. The only cases of hepatitis B disease in the vaccine group occurred in those who did not respond and in those who were already infected at the time of entry in the study. However, the vaccine appeared to be partially effective when given after exposure. In the hemodialysis study, the vaccine which contained HBsAg subtype ad protected medical staff exposed to patients suffering from HBsAg subtype ay infection.

Safety – Evaluation Since Licensure

Observation of the 3,350 persons who participated in the double-blind, placebo-controlled trials revealed that the only increased side effect in those who received vaccine was transient, local soreness at the site of inoculation. The frequency of fever and other complaints was the same in placebo and vaccine groups, and no serious vaccine-related reactions were observed. In September 1982, a Public Health Service Inter-Agency Group reported that 19,000 persons were immunized in studies prior to licensure between October 1975 and December 1981 (United States Public Health Service Center for Disease Control 1982). No serious side effects were reported among members of this group.

A subsequent report in March (MMWR 1983) reviewed the reactions of about 200,000 individuals who received the vaccine after its licensure in 1981. Based on previous double-blind, placebo-controlled studies, the reported side effects and illnesses that occurred within 6 weeks of immunization did not appear to be causally related to hepatitis B vaccine. The various manifestations reported did not exceed the number of such events that would be expected to have occurred by chance alone within 6 weeks of vaccinating 200,000 people.

As of June 1984, more than 500,000 persons in the United States and an additional 500,000 persons worldwide have been immunized with the plasma-derived hepatitis B vaccine. To date, no serious side effects have been attributed to the use of vaccine. For example, the speculation and concern about a possible association between acquired immune deficiency disease (AIDS) and hepatitis B vaccine have proved to be groundless. The lack of association is based on three findings: (a) no cases of AIDS have been detected among the many thousands of health professionals who have been immunized during the past 2 years; (b) no increased incidence of AIDS has been observed among immunized high-risk male homosexuals when compared with unimmunized high-risk homosexuals, and (c) it would be incredible that the putative AIDS agent could survive the inactivating procedures involving pepsin at pH 2, 8*M* urea, and formalin. The physical and biochemical procedures involved in the preparation of hepatitis B vaccine are capable of eliminating all of the agents proposed as possible causes of AIDS.

Conclusion

The first-generation, plasma-derived hepatitis B vaccines have proved to be safe, immunogenic, and effective. These licensed products are recommended for persons of all ages who are at high risk of contracting hepatitis B infection. Studies currently in progress indicate that future second-generation vaccines will be prepared by DNA recombinant technology. In the meantime, extensive experience during the past 2 years has revealed that the demonstrated benefits of the currently licensed plasma-derived vaccines far exceed any theoretical risks.

References

Barker LF, Chisari FV, McGrath PP, et al. (1973) Transmission of type B viral hepatitis to chimpanzees. J Infect Dis 127:648–662

Blumberg BS, Alter HJ, Visnich S (1965) A "new" antigen in leukemia sera. JAMA 191:541–546

Dane DS, Cameron CH, Briggs M (1970) Virus-like particles in serum of patients with Australia antigen-associated hepatitis. Lancet 1:695–698

Francis DP, Hadler SC, Thompson SE, et al. (1982) The prevention of hepatitis B with vaccine. Ann Intern Med 97:362–366

Gallo RC, Salahuddin SZ, Popovic M, et al. (1984) Frequent detection and isolation of cytopathic retroviruses (HTLV-III) from patients with AIDS and at risk for AIDS. Science 224:550–552

Hilleman MR, Buynak EB, Roehm RR, et al. (1975) Purified and inactivated human hepatitis B vaccine: progress report. Am J Med Sci 270:401–403

Krugman S, Giles JP, Hammond J (1971) Viral hepatitis, type B (Ms-2 strain): studies on active immunization. JAMA 217:41–45

Maupas P, Goudeau A, Coursaget P, et al. (1976) Immunization against hepatitis B in man. Lancet 1:1367–1370

Prince AM (1968) An antigen detected in the blood during the incubation period of serum hepatitis. Proc Natl Acad Sci USA 60:814–821

Purcell RH, Gerin JL (1975) Hepatitis B subunit vaccine: a preliminary report of safety and efficacy tests in chimpansees. Am J Med Sci 270:395–399

Reesink HW, Brummelhuis EH, van Elven LJT, et al. (1978) The preparation and evaluation of a hepatitis B vaccine in the Netherlands. In: Vyas GN, Cohen SN, Schmidt R (eds) Franklin Institute, Philadelphia, p 714

Szmuness W, Stevens CE, Harley EJ, et al. (1980) Hepatitis B vaccine: demonstration of efficacy in a controlled clinical trial in a high risk population in the United States. N Engl J Med 303:833–841

United States Public Health Service Centers for Disease Control (1982) Hepatitis B vaccine safety: report of an inter-agency group. MMWR 31:465–467

United States Public Health Service Centers for Disease Control (1983) The safety of hepatitis B vaccine. MMWR 32:134–136

Vilmer E, Barre-Sinoussi F, Rouzioux C, et al. (1984) Isolation of new lymphotropic retrovirus from two siblings with hemophilia B, one with AIDS. Lancet 1:753–757

Concepts in the Development of Vaccines Against Genital Herpes Simplex Virus Infections

L. COREY [1] and G. MERTZ [2]

Introduction

Genital herpes simplex (HSV) infection is a disease of public health importance. In the United States the prevalence of genital herpes has increased ten fold between the years 1965 through 1980 [1]. Concomitant with this increase in genital herpes has been a parallel increase in the incidence of neonatal HSV infections. The incidence of neonatal HSV infection has increased from 2.5 cases per 100,000 live births to King County residents in the years 1966–1969 to 13.4 cases per 100,000 live births in the years 1982–1983 [2]. Many mothers who deliver infants with neonatal herpes have acquired disease asymptomatically [2, 3]. Approximately 50% of the source contacts who transmit genital herpes are asymptomatic at the time of transmission and an additional 40% of the symptomatic source contacts have clinically undiagnosed or atypical symptoms [4]. As such, any treatment strategy designed to interrupt the transmission of genital herpes must be directed at preventing infection in exposed sexual partners and/or infants of asymptomatic source contacts.

Table 1. Approaches to the prevention of genital herpes

	Advantages	Disadvantages
Vaccines	Generally most cost-effective approach to prevention if effective and long lasting	Immunization of person in low-risk categories
Chemotherapeutics	Can be targed and used in high-risk populations	Need to be given after all sexual exposures. Effective or safe? Long-term compliance may be difficult.
Barrier forms of contraceptive	Inexpensive – readily available	Often not used. Reduced efficacy in skin-to-skin transmission?

Several approaches to the prevention of genital herpes simplex virus infections have been proposed (Table 1). The use of vaccines, antiviral agents after intercourse, and barrier forms of contraceptives are the most commonly advocated ap-

1 Departments of Laboratory Medicine, Microbiology and Medicine, University of Washington, Children's Orthopedic Hospital and Medical Center, 4800 Sand Point Way N.E., Seattle, WA 98105, USA

2 Presently Department of Medicine, Division of Infectious Diseases, University of New Mexico, USA

proaches. Barrier forms of contraception such as condoms have, in vitro, been shown to prevent the passage of HSV [5]. As such, they may decrease transmission from virus asymptomatically shed from the cervix or in semen. However, transmission of diseases spread by skin-to-skin transmission (i.e., herpes and syphilis) despite the use of condoms has been reported [6]. While antiviral drugs such as acyclovir may be effective in reducing the severity of episodes of genital herpes, to date no antiviral medication tested has eradicated an established latent HSV infection [6–8]. In order to prevent acquisition of genital herpes, an antiviral drug such as acyclovir would have to be taken systemically for an extended period of time, for example after intercourse. Patients compliance is likely to wane over time. In addition, the medications are not likely to be available to partners of persons with asymptomatic infection. As such, an HSV vaccine is the most likely method of effectively reducing the transmission of genital herpes. However, several formidable problems currently exist in the development of effective, safe vaccines against genital herpes. This paper will review some of these issues as well as discuss our current trial evaluating a subunit inactivated HSV-2 glycoprotein vaccine.

Problems in the Development of HSV Vaccines

In Vitro Markers of Vaccine Efficacy

One of the most difficult problems in the development of HSV vaccines is the lack of in vitro and animal models predictive of protection from infection or resolution of clinical disease. Hyperimmune antiserum to HSV-1 and HSV-2 can, in animal models, prevent acquisition of disease, decrease spread of virus into the central and peripheral nervous system and decrease the number of latently infected dorsal root sensory ganglia [9–11]. Recently monoclonal antibodies to the HSV surface glycoproteins have been shown to protect animals against subsequent viral challenge [13]. Antibodies to the viral glycoproteins will neutralize HSV [13]. However, in humans the possession of neutralizing antibodies has not been associated with resolution of clinical disease, prevention of dissemination of disease in immunosuppressed patients and/or transmission of disease to the infant [14–16]. In fact, genital herpes patients with high titers of neutralizing antibodies have been shown to have more frequent recurrences of disease than those with low neutralizing antibody titers, probably reflecting the fact that antibody may be a "marker" of frequent and/or intense prior antigenic stimulation and/or a large number of latently infected cells [15]. While other antibodies such as those involved in antibody-dependent cell-mediated cytotoxicity (ADCC) are measures of HSV immunity, these immune responses have not been shown to be associated with protection against subsequent reactivation and/or resolution of disease [17, 18]. Patients with depressed cellular immune responses have more severe and frequent HSV infections [14, 19]. However, no specific defect in in vitro cell-mediated response has been associated with resolution of recurrences or resistance to acquisition of genital or oral-labial HSV [20–24]. In addition to the in vitro production of humoral and cellular immune responses, vaccines for the prevention of genital herpes must be able to sustain immunity for an extended (10–30 year) period. Differences in life-style and sexual and social habits vary greatly from person to person and within individuals by decade. A vaccine requiring frequent booster doses may have a markedly reduced impact on the public health aspect of disease control.

Animal models of primary genital HSV infection appear to be reasonably similar to that of primary human genital HSV [25]. While recurrences of disease do occur in the guinea pig model of genital herpes, viral shedding appears to be persistent, not intermittent as in man, and recurrences of disease in the guinea pig model are quite different in pattern and nature than that seen in human disease [25]. As such, established predictive in vitro markers to guide the development and economically assess the potential efficacy of candidate HSV-2 vaccines are not currently available. To date evidence suggests that vaccines against genital herpes should produce neutralizing and antibody dependent cell-mediated cytotoxicity antibody responses of equal quality and quantity to those produced by naturally acquired infection. In addition, some parameters of cell-mediated immunity, such as in vitro lymphoproliferative responses to HSV antigens, production of HSV specific gamma interferon, or HSV natural killer cell responses, seem desirable.

Design of Clinical Trials to Evaluate Candidate HSV Vaccines

Genital HSV infections pose segeral new challenges for designing clinical trials to assess their efficacy. Little is known about the frequency of transmission of genital HSV infection, and hence, calculations of the number of patients and duration of follow-up needed to asses efficacy may at first be difficult. However, as several groups are currently engaged in some form of vaccine evaluation, data should soon be forthcoming concerning the transmission rates of disease among the populations being studied [26]. Parameters for evaluating the clinical efficacy of a genital herpes vaccine are outlined in Table 2. Classically vaccine efficacy has been measured by the prevention of clinical disease and vaccine trials have been designed to determine the frequency of acquisition of clinical disease between vaccines and placebo recipients and to measure disease severity among those who "break through" vaccination.

Primary genital herpes is a disease of considerable morbidity; painful genital lesions last an average of 3 weeks, as many as 10% patients will require hospital-

Table 2. Goals for the development of a genital HSV vaccine

1. Prevention of acquisition of clinical episodes of genital herpes
2. Amelioration of first episode disease in those who have vaccine "breakthrough", i.e., prevention of aseptic meningitis, neurologic complications
3. Reduction of subsequent recurrence rate of disease in those who have vaccine breakthrough
4. Prevention of the development of ganglionic latency and subsequent reactivation and/or transmission
5. Prevention of infection – (asymptomatic and symptomatic)

ization for aseptic meningitis, and autoinoculation of HSV to distant body sites is common [27]. As such, a vaccine that would reduce the frequency as well as ameliorate the severity of primary first episode genital herpes would in itself be clinically useful. However, a truly effective HSV vaccine should also prevent infection with HSV, or if infection occurs, prevent the development of ganglionic latency and subsequent reactivation and transmission of disease.

Many persons with naturally acquired genital herpes are asymptomatic [14]. Persons who acquire the disease asymptomatically may subsequently develop clinical reactivation of symptomatic disease and transmit infection both symptomatically and asymptomatically years later [3, 4, 6, 14]. In addition, if immunosuppressive therapy or an immunosuppressive disease occurs years later, reactivation of latent infection may result in severe mucocutaneous as well as disseminated disease [19, 28]. An HSV vaccine which does not prevent the development of ganglionic latency but only increases the frequency of asymptomatic infection may actually be "harmful", if it increases the percentage of persons with undiagnosed and/or atypical disease. If this occurs, subsequent sexual partners or infants of these persons may, years later, be at risk of developing disease. As this type of reactivation and transmission may occur years after enrollment, it may be very difficult to design appropriate clinical trials to measure these long-term outcomes. One potential approach to address this concern would be to develop in vitro markers which separate naturally acquired HSV infection from vaccination, similar to the use of anti-HBc testing in trials evaluating HBsAg subunit vaccines [29]. Recent evidence indicates that this type of approach may also be adapted to the testing of HSV vaccines. Several specific viral polypeptides that occur invariably after naturally acquired HSV-2 infection have been identified [30–32]. Two non-glycosylated polypeptides seen after naturally acquired infection have been used to distinguish asymptomatic acquisition of wild-type HSV-2 infection from vaccination with an HSV glycoprotein subunit vaccine [33]. If serological markers to distinguish between live attenuated and naturally acquired HSV-3 disease are difficult to develop, vaccine trials of live HSV vaccines may need to be conducted over an extended period and include enrollment and follow-up of subsequent sexual partners and infants of vaccinees.

Types of Genital HSV Vaccines

Several types of HSV vaccines have been proposed. These include formalinized inactivated whole virus products, solubilized whole virus preparations, subunit vaccines (both inactivated and those expressed in vectors such as vaccinia virus), and attenuated whole virus preparations (Table 3). Inactivated as well as cloned fragments of HSV-2 DNA may cause cell transformation and the subsequent development of tumors in animals [14, 34, 35]. In addition, there are numerous seroepidemiologic studies delineating an increased risk of genitourinary tumors among persons with HSV-2 infection [14, 36–38]. The recent demonstration of HSV-2 DNA and RNA in specimens obtained from cervical dysplasia has strengthened this association [39, 40]. As such, it appears prudent that HSV vaccines should be free of viral DNA and/or the potential for inducing cell transformation.

Table 3. Herpesvirus vaccines

	Advantages	Disadvantages
Inactivated vaccines		
Formalized whole virus vaccines	Already available	No proven efficacy. Contains viral DNA – capable of causing cell transformation. No way to document asymptomatic acquisition
Solubilized whole viral antigens	More defined components. Theoretically can remove viral DNA	Deletion of important antigens or present antigens in "nonphysiologic" manner. Need to develop marker of wild vs. vaccine immunity
Subunit vaccines: Mixed glycoprotein vaccine	Important immunogenic proteins	Will all important "protective" antigens be present in vaccine? For glycoprotein vaccine are there nonglycosylated polypeptides important in preventing infection?
Purified specific glycoprotein	Cloned – defined molecule. Easily made in large quantities. Can be incorporated into "replicating vectors"	Only one antigen. Will in vitro translation products be similar to in vivo product?
Synthetic polypeptides	Lower cost-defined protein	Too specific – not as immunogenic as "natural molecule"
Live virus vaccine		
Attenuated virus	Potentially should produce longer lasting immune response and a local mucosal immunity	Animal models have not proven to be predictor of attenuation. Question of latency and reactivation years later
Genetically engineered live virus vaccine	Stable, attenuated	Transmissable revertants? Relationship to genitourinary tumors? Necessity to exactly define factors associated with cell transformation

Traditionally live virus vaccines offer the potential for producing a quicker and more sustained immune response, as well as for inducing local immune responses. However, live HSV viral vaccines also pose several significant safety issues which must be addressed: (1) At present there are no defined animal or in vitro markers of viral attenuation, cell transformation, and/or tumor-producing potential; (2) there is no animal model of recurrent genital herpes to assess the stability of the attenuated strain to reactivate; and (3) if future immunosuppression occurs to the host, e.g., pregnancy and/or thc subsequent use of cytotoxic agents, will reactivation, dissemination, and/or transmission of the vaccine virus occur? Because of the formidable problems discussed above, inactivated HSV vaccines lacking viral DNA at present provide the potentially safest approach for the development of effective genital HSV vaccines.

Current Status of a Trial of a Subunit HSV Glycoprotein Vaccine for the Prevention of Genital HSV Infection

As mentioned earlier, inactivated subunit vaccines offer several advantages in the prevention of HSV infection. Among them is the fact that serologic markers separating naturally acquired HSV infection from immunity due to vaccination can be developed. Because of the importance of th HSV surface gylcoproteins in the cellular and humoral immune responses to HSV infection, several candidate glycoprotein vaccines have been developed. One of these vaccines, an inactivated glycoprotein-enriched vaccine developed by Hilleman et al. is immunogenic in mice, guinea pigs, and cebus monkeys [41]. To evaluate the immunogenicity, reactogenicity, and efficacy of this vaccine, a series of clinical trials evaluating its use in immunocompetent patients seronegative to HSV but at risk of acquiring genital herpes have been initiated at the University of Washington.

Immunogenicity Study Design

Twenty-three healthy volunteers between the ages of 21 and 39 who were without a history of genital herpes, who denied having a sexual partner with genital herpes, and who lacked HSV neutralizing antibody prior to vaccine administration, were enrolled in a clinical immunogenicity trial. At 0, 1, and 5 months, 0.5 ml (50-μg dose) of HSV-2 vaccine was administered by intramuscular injection to 13 subjects an 1 ml (100-μg dose) of HSV-2 vaccine was administered to an additional 10 subjects. Blood was obtained weekly from week 0 through 4 and at weeks 8, 22, 23, and 52. Subjects recorded their temperatures and any adverse effects daily for 5 days and were examined 48 h arter each injection.

Vaccine

HSV-2 glycoprotein subunit vaccine, lot 806 was supplied by Merck, Sharp, and Dohme Research Laboratories. Curtis Strain E-304 of HSV-2 was grown in chicken embryo fibroblasts, and the infected monolayer was harvested after treatment with 2% Triton X-100 at pH 7.5. This extract was clarified by high-speed centrifugation, and the supernatant was digested with DNAase. The glycoproteins were then absorbed from the extract onto a lectin affinity column, eluted with methyl-D-mannoside, and purified by affinity chromatography with *Lens culinaris* hemaglutanin (LcH)-Sepharose. The glycoprotein-enriched fraction was treated with formalin and combined with an alum adjuvant. The vaccine extract was free of detectable nucleic acid (<0.05 μg/200 μg protein), contained no virus particles by electron microscopy, and was free of live virus by isolation attempts on susceptible cell monolayers.

Neutralizing Antibody

HSV-1 and -2 neutralizing antibody was determined in a complement-independent microneutralization assay [44]. HSV-1, Strain 2,931, and HSV-2, Strain MS-2, were employed. Antibody specificity to HSV-1 or -2 was determined by calculating the potency of neutralization (pN) of HSV-1 minus the pN of HSV-2 (pN_1-pN_2). Anti-HSV-1 specificity was defined as pN_1-pN_2 of >0.5 and HSV-2 specificity as $pN_1-pN_2 < 0.05$. Antibody with pN_1-pN_2 between 0.5 and 0.05 was classified as indeterminate. Prior to vaccine administration, all seronegative subjects had titers < 1:4 to HSV-1 and -2. In subsequent neutralizing antibody determinations, HSV neutralizing antibody was considered present if the titer to HSV-1 or -2 was ≧ 1:8 or if the titer to both HSV-1 and HSV-2 was ≧ 1:4.

Lymphocyte Transformation Response

In vitro lymphocyte transformation (LT) responses to the mitogen concanavalin A and the soluble antigens *Candida albicans* and inactivated HSV-1 and -2 antigen (HSV-1, Strain E115, and HSV-2, Strain 333) were performed as previously described [18]. The lymphocyte transformation response to the HSV-2 glycoprotein subunit vaccine, lot 804, was also measured. The vaccine was first dialysed to remove the alum adjuvant. Preliminary testing indicated that a 1:10 dilution of vaccine gave an optimal blastogenic response. Lymphocyte transformation responses were calculated using the stimulation index. The stimulation index for HSV antigens was calculated as the mean counts per minute in the HSV antigen stimulation wells, divided by the mean counts per minute in unstimulated cell control wells [18].

Vaccine Tolerance

Vaccine was well tolerated with transient local symptoms being reported after one-third of the doses. Sixty percent of subjects reported no local or systemic reactions Local reactions consisted of local pain, erythema, and itching, noted in 40% of subjects after vaccination. The median duration of local pain was 1 day. Systemic symptoms were reported by less than 10% of vaccinees after vaccination. No subject discontinued vaccination or missed school or work owing to local or systemic reactions. The frequency of local reactions was similar among those who received 50 μg and those who received 100-μg doses.

Neutralizing Antibody Response in Initially Seronegative Subjects

Among the 23 seronegative individuals, HSV neutralizing antibody was first detected in five subjects 3 weeks after the initial dose of the vaccine and was present in 91% 3 weeks after the 1-month booster (Table 4). Neutralizing antibody

Table 4. Presence of HSV-neutralizing antibody in seronegative recipients of an HSV-2 subunit glycoprotein vaccine

Days after vaccination	0↓	7	14	21	28↓	35	56	140↓	147	352
50-μg dose	0/13	0/13	0/13	3/13	4/13	8/13	12/13	5/12 $p = 0.04$	10/12	7/10
100-μg dose	0/10	0/10	0/10	2/10	2/10	6/9	9/10	8/9	8/9	7/9
Total ($n = 23$)	0/23	0/23	0/23	5/23 22%	6/23 26%	14/22 64%	21/23 (91%)	13/21 62%	18/21 80%	4/19 (74%)

* Fisher's exact test (one failed)
→, day of vaccine administration

waned after the second booster and was detected in only 13 of 21 vaccinees (62%) at day 140, just prior to the second booster. Recipients of the 100-µg dose vaccine had a more sustained neutralizing antibody response after the second injection of vaccine than those who received the 50-µ dose. Eight of nine recipients of the 100-µg dose who had serum drawn at day 140 possessed HSV neutralizing antibody compared with 5 of 12 recipients of the 50-µg dose ($P = 0.04$, Fisher's Exact Test). All 23 recipients developed detectable HSV neutralizing antibody sometime during the course of follow-up; and all had HSV neutralizing antibody with HSV-2 specificity. One year after the initial dose of vaccine, 14 (74%) of 19 subjects had detectable HSV neutralizing antibodies.

Lymphocyte Transformation Responses in Initially Seronegative Individuals

The mean stimulation index to inactivated HSV-2 glycoprotein extract (dialyzed vaccine) and inactivated whole virus (HSV-2) antigen is shown in Table 5. A stimulation index to inactivated HSV-2 glycoprotein extract of >3.0 was detected in 80% of recipients by week 4 after vaccination, and was present in all recipients by week 5. The mean stimulation index to the HSV-2 glycoprotein extract was similar in those seronegative patients who received 50 µg doses of vaccine and those who received 100-µg doses. The in vitro lymphocyte transformation response to inactivated HSV antigens was less in both 50- and 100-µg recipients than that to the dialyzed HSV-2 glycoprotein extract (Table 5). At the 1-year follow-up, 55% of 50-µg recipients and 50% of 100-µg recipients had an SI >3.0 to the glycoprotein extract, while 91% of 50-µg recipients and 70% of 100-µg recipients had an SI of >3.0 to the inactivated whole virus preparation. During the course of the study, an SI >3.0 to whole HSV-2 antigen was present in at least two assays in 22 of the 23 initially seronegative individuals. An SI >3.0 to the HSV-2-glycoprotein extract was detected in at least two assays in all 23 individuals.

Table 5. Lymphocyte transformation response (stimulation index) among seronegative recipients of HSV glycoprotein vaccine

Days after vaccination	0	7	14	21	28	35	56	140	147	356
50 µg-dose recipients ($n = 13$)										
Mean stimulation index (SI) to inactivated HSV-2 antigens	0.9 (0.2)	4.2 (2.1)	5.8 (3.5)	3.3 (1.4)	3.5 (1.5)	4.9 (1.3)	5.9 (2.0)	8.1 (2.7)	16.7 (9.7)	13.2 (3.8)
Mean SI to HSV-2 glycoprotein extract	1.6 (0.6)	1.3 (0.3)	19.1 (7.6)	12.7 (4.2)	28.0 (18.8)	18.8 (3.8)	30.0 (13.6)	15.6 (7.5)	10.9 (5.8)	10.5 (4.6)
100 µg-dose recipients ($n = 10$)										
Mean stimulation index (SI) to inactivated HSV-2 vaccine	1.6 (0.4)	3.0 (0.7)	7.9 (2.6)	5.6 (2.2)	3.7 (1.7)	7.3 (2.0)	9.4 (2.2)	9.7 (3.0)	15.1 (3.7)	3.7 (0.9)
Mean SI to HSV-2 glycoprotein extract	1.8 (0.6)	2.5 (0.8)	7.2 (2.3)	14.5 (4.6)	28.3 (9.7)	18.4 (5.8)	15.4 (4.3)	20.6 (5.8)	26.8 (4.7)	133.7 (6.0)

( ) = standard error of the mean

Discussion

Development of safe, immunogenic, and effective HSV vaccines afford the best opportunity to alter the increasing prevalence of genital herpes. Several formidable problems remain in achieving prevention of this disease:

1. It is not currently known which viral antigens are important in preventing acquisition and/or resolution of disease; and effective in vitro and animal models to predict safety, attenuation, and efficacy of candidate vaccine are unavailable.
2. Both HSV-1 and HSV-2 can cause genital infection and little is known about the transmission of disease in various at-risk populations.
3. HSV vaccines will be required to be effective for many years after initial vaccination.
4. Efficacy evaluation for HSV vaccines must measure not only a reduction in clinical disease and amelioration in the severity of first episode genital herpes but also a reduction in asymptomatic acquisition of infection.

Asymptomatic HSV infections can lead to subsequent ganglionic latency and subsequent transmission months to years later to sexual partners and/or infants. HSV vaccines must prevent this latter form of infection and disease. As such, clinical trials of candidate HSV vaccine must include sensitive serologic markers to determine when naturally acquired infection occurs, as well as markers to distinguish between vaccination and natural infection. At present, subunit vaccines directed at the viral glycoproteins appear to offer one of the most promising approaches for the development of effective HSV vaccines. One glycoprotein vaccine containing a mixture of glycoproteins B, C, D, E, and G developed by Hilleman and associates at the Merck, Sharp, & Dohme Research Laboratories, U.S.A., is currently under clinical trial. Both 50- and 100-μg doses of this vaccine given in three injections are well tolerated and produce neutralizing and cellular immune responses. Neutralizing antibody responses to the 100 μg per dose vaccine appear more sustained than those seen with the 50-μg doses. Efficacy trials evaluating the protective effect of both these dose ranges are currently underway. Whatever the results of these ongoing trials, they will provide valuable information for the development of future candidate HSV vaccines.

In many respects, the current vaccines under development for the prevention of HSV may be considered to be a first generation approach in the development of an effective means of prevention of genital herpes. New HSV polypeptides continue to be described, and studies of their importance in disease pathogenesis are just being initiated. In addition, new techniques to utilize recombinant DNA technology to produce purified defined glycoproteins in large quantities with subsequent higher antibody responses than are achieved with previously utilized purification techniques have emerged [43]. Several groups have suggested that antibodies to nonglycosylated polypeptides may also be important in the resolution of HSV infections [32]. Further studies defining the human immune response to HSV may allow the incorporation of several highly immunogenic viral polypeptides into a preparation that will provide quicker and more sustained immune responses than has been previously achieved.

Acknowledgment. This work was supported by NIH grant AI-20381.

References

1. Anonymous (1982) Estimated rate of patient consultations with private physicians for genital herpes: United States 1966–1979. Morbidity and Mortality Weekly Report 31:138–9
2. Sullivan-Bolyai J, Hull HF, Wilson C, Corey L (1983) Neonatal herpes simplex virus infection in King County, Washington: increasing incidence and epidemiologic correlates. JAMA 250:3059–62
3. Whitley RJ, Nahmias AJ, Visintine AM, Fleming CL, Alford CA (1980) The natural history of herpes simplex virus infection of mother and newborn. Pediatrics 66:489–94
4. Mertz GJ, Schmidt O, Jourden JL, Guinan ME, Remington ML, Fahnlander A, Winter C, Holmes KK, Corey L (to be published) Frequency of acquisition of first episode genital herpes simplex virus infection from symptomatic and asymptomatic source contacts. Sex Trans Dis
5. Conant MA, Delano W, Spicer DW, Smith CD (1984) Herpes simplex virus transmission: condom studies. Sex Trans Dis 11(2):94–5
6. Corey L, Holmes KK (1983) Genital herpes simplex virus infection: Current concepts in diagnosis, therapy and prevention. Ann Intern Med 98:973–78
7. Reichman RC, Badger GJ, Mertz GJ, Corey L, Richmann DD, Connor JD, Redfield D, Savoia MC, Oxman MN, Bryson Y, Tyrrell L, Portnoy J, Creogh-Kirk T, Keeney RE, Ashikaga T, Dolin R (1984) Treatment of recurrent herpes simplex infections with oral acyclovir: a controlled trial. JAMA 251(16):2103–8
8. Corey L, Fife KH, Benedetti JK, et al. (1983) Intravenous acyclovir for the treatment of primary genital herpes. Ann Intern Med 98:914–21
9. Oaks J, Rosemond-Hornbeak H (1978) Antibody-mediated recovery from subcutaneous herpes simplex virus type 2 infection. Infect Immun 21:489–95
10. McKendall R, Klassen T, Baringer J (1979) Host defenses in herpes simplex infections of the nervous system; effect of antibody on disease and viral spread. Infect Immun 23:305–11
11. Stevens J, Cook M (1974) Maintenance of latent herpetic infection: an apparent role für antiviral IgG. J Immunol 113:1685–93
12. Kristensson K, Vahlne A, Person LA, Lycke E (1978) Neural spread of herpes simplex virus type 1 and 2 in mice after corneal or subcutaneous (footpad) inoculation. J Neurol Sci 35:331–40
13. Dix R, Pereira L, Baringer J (1981) Use of monoclonal antibodies directed against herpes simplex using glycoproteins to protext mice against acute virus-induced neurological disease. Infect Immun 34:192–99
14. Nahmias AJ, Roizman D (1973) Infection with herpes simplex virus 1 and 2. N Engl J Med 299:667–74, 781–89
15. Reeves W, Corey L, Adams H, Vontver L, Holmes K (1981) Risk of recurrence after first episodes of genital herpes: relation to HSV type and antibody response. N Engl J Med 305:315–19
16. Yeager AS, Arvin AM, Urbani LJ, Kemp JA (1980) Relationship of antibody to outcome in neonatal herpes simplex virus infections. Infect Immun 29:532–8
17. Kohl S, Loo LS (1982) Protection of neonatal mice against herpes simplex virus infection: probable in vivo antibody dependent cellular cytotoxicity. J Immunol 129:370–76
18. Moller-Larsen A, Haarh S, Black FT (1978) Cellular and humoral immune response to herpes simplex virus during and after primary gingivostomatitis. Infect Immun 22:445–57
19. Meyers JD, Flournoy N, Thomas ED (1980) Infection with herpes simplex virus and cell mediated immunity after marrow transplant. J Infect Dis 142:338–46
20. Corey L, Reeves WC, Holmes KK (1978) Cellular immune response in genital herpes simplex virus infection. N Engl J Med 299:986–91
21. Sheridan JF, Donnenberg AD, Aurelian L, Elpern DJ (1983) Immunity to herpes simplex virus type 2. J Immunol 129:326
22. O'Reilly RJ, Chibbaro A, Anger E, et al (1977) Cell-mediated immune response in patients with recurrent herpes simplex virus infections. II. Infection-associated deficiency of lymphokine production in patients with recurrent herpes labialis or herpes progenitalis. J Immunol 1089:1095–1102
23. Cunningham AL, Mergian TC (1983) Interferon production appears to predict time of recurrence of herpes labialis. Am Assoc Immunol 130:2397–2400
24. Ching C, Lopez (1979) Natural killing of herpes simplex virus type 1-infected target cells: normal human responses and influence of antiviral antibody. Infect Immun 26:49–56

25. Stanbury LR, Kern ER, Richards JT, Abbott T, Overall JC Jr (to be published) Genital herpes in guinea pigs, pathogenesis of primary infection and description of recurrent disease. J Infect Dis
26. Mertz GJ, Peterman G, Ashley RA, Jourden J, Salter D, Morrison L, McLean A, Cores L (to be published) Herpes simplex virus type-2 glycoprotein subunit vaccine: Tolerance and humoral and cellular immune responses in man. J Infect Dis
27. Corey L, Adams HG, Brown ZA, Holmes KK (1983) Genital herpes simplex virus infection: Clinical manifestations, course and complications. Ann Intern Med 98:958–72
28. Ramsey PG, Fife K, Hackman R, Meyers JD, Corey L (1982) Herpes simplex virus pneumonia: clinical presentation and pathogenesis. Ann Intern Med 97:813–20
29. Szmuness W, Stevens CE, Harley EJ, Zang EA, Oleszko WR, William DC, Sadovsky R, Morrison JM, Kellner A (1980) Hepatitis B vaccine: demonstration of efficacy in a controlled clinical trial in a high-risk population in the United States. N Engl J Med 303:833–841
30. Zweerink HJ, Corey L (1983) Virus-specific antibodies in sera from patients with genital herpes simplex virus infection. Infect Immun 37:413–21
31. Ashley R, Corey L 1984 Association of herpes simplex virus polypeptide specific antibodies and the natural history of genital herpes infections. In: *Herpesvirus*, vol 21. Liss, New York, pp 37–54
32. Ashley RL, Corey L (1984) Effect of acyclovir treatment of primary genital herpes on the antibody response to herpes simplex virus. J Clin Invest 73(3):681–688
33. Ashley RL, Corey L (to be published) Development of antibodies to a herpes simplex virus type 2 (HSV-2) subunit vaccine in seronegative and seropositive volunteers. Proc. of International Herpes Virus Workshop, Oxford, England
34. Galloway DA, McDougall JK (1981) Transformation of rodent cells by a cloned DNA fragment of herpes simplex virus type 2. J Virol 38:749–60
35. Jariwalla RJ, Aurelian L, Ts'O PO (1980) Tumorigenic transformation induced by a specific fragment of DNA from herpes simplex virus type 2. Microbiol 77:2279–83
36. Royston I, Aurelian L (1970) The association of genital herpes virus with cervical atypia and carcinoma in situ. Am J Epidemiol 91:531–38
37. Nahmias AJ, Naib ZM, Josey WE (1973) Prospective studies of the association of genital herpes simplex virus infection and cervical anaplasia. Cancer Res 33:1491–97
38. Kaufman RH, Dreesman GR, Burek J, Korhonen MO, Matson DO, Melnick JL, Powell KL, Purifoy DJM, Courtney RJ, Adam E (1981) Herpesvirus-induced antigens in squamous-cell carcinoma in situ of the vulva. N Engl J Med 305:483–88
39. Eglin RP, Sharp F, MacLean AB, Macnab JCM, Clements JB, Wilkie NM (1981) Detection of RNA complementary to herpes simplex virus DNA in human cervical squamous cell neoplasms. Cancer Res 41:3597–3603
30. McDougall JK, Galloway DA, Fenoglio CM (1980) Cervical carcinoma: detection of herpes simplex virus RNA in cells undergoing neoplastic change. Int J Cancer 25:1–8
41. Hilleman MR, Larson VM, Lehman ED et al. (1981) Subunit herpes simplex 2 vaccine. In: Nahmias AJ, Dowdle WR, Schinazi RF (eds) The human herpes-viruses. Elsevier, New York 503–6
42. Chan W (1983) Protective immunization of mice with specific HSV-1 glycoproteins. Immunol 39:343–52
43. Berman PW, Dowbenko D, Lasky LA (1983) Detection of antibodies to herpes simplex virus with a continuous cell line expressing cloned glycoprotein D. Science 222:524–27
44. Rawls WE, Iwamoto K, Adam E, Melnick JL (1970) Measurement of antibodies to herpesvirus types 1 and 2 in human sera. J Immunol 104:599–606

Development of a Live Attenuated Vaccine Against Varicella

A. Gershon [1]

Varicella zoster (VZ) virus was first isolated in tissue culture by Weller (1953), about 30 years ago. As Krugman (1984) has pointed out, vaccines against polio, measles, and rubella became licensed at the most within 16 years after it was possible to propagate the respective agents in the laboratory. The hepatitis B virus vaccine was developed even before that agent could be propagated. It might therefore seem surprising that there is, after 30 years, no vaccine licensed for prevention of disease caused by VZ virus. Fortunately, however, despite this long lag period a vaccine has been developed and it is being tested. This vaccine will probably become commercially available within the next few years.

How can the long lag period between virus isolation and vaccine development be accounted for? One explanation is that until rather recently, varicella was not considered a disease serious enough to warrant prevention. Most children contract varicella before they are 10 years old, and the disease is usually mild. Varicella in normal children lasts 5–7 days, with only low grade to moderate temperature elevation; complications are rather unusual. Among those reported are, in order of frequency: bacterial superinfection, postinfectious encephalitis, Reye's syndrome, arthritis, purpura fulminans, and glomerulonephritis.

Individuals at High Risk to Develop Severe Varicella

In contrast to normal children, however, complications of varicella are frequent and serious in immunocompromised children, as first recognized in 1956 (Cheatham et al. 1956). These complications include primary viral pneumonia, severe hepatitis, encephalitis, and death. In one retrospective study of 60 children with an underlying malignant disease who were receiving maintenance chemotherapy, 19 (32%) had severe disseminated disease and four (7%) died (Feldman et al. 1975). Children recognized to be at risk to develop severe, disseminated varicella include those with an underlying malignancy receiving chemotherapy and/or radiotherapy, those on high doses of steroids (greater than 1.5 mg/kg/day) for any reason, patients with deficiencies in cellular immunity, and newborn infants whose mothers have acute varicella at the time of delivery. Obviously, as advances in medicine have improved the life span for children with many of the diseases mentioned, there are more and more individuals at risk to develop severe varicella.

1 New York University Medical Center, 550 First Avenue, New York, NY 10016, USA

Normal adults who have never had varicella are also at some risk to develop severe infection. Fatalities due to varicella have also been reported in pregnant women with the infection. In addition, varicella-susceptible adults who are employed in hospitals are potentially vectors for transmission of nosocomial varicella.

Thus while varicella has traditionally been considered a rather mild disease, currently this would seem to be an oversimplification of the potential seriousness of the infection.

Latency and VZ Virus

Another important deterrant to the development of a vaccine against varicella was the well recognized potential for VZ virus to cause not one but two diseases, varicella and zoster. VZ virus becomes latent in the human host after varicella, residing in sensory ganglia. It appears that most if not all individuals harbor the virus in latent form after primary infection. Reactivation of latent VZ virus results in zoster; this was first proposed by Garland (1943). Bokay (1909) previously had demonstrated in human volunteers that the virus causing zoster also causes varicella (see Table 1). It was postulated by early opponents to the use of varicella vaccine (Brunell 1977) that zoster might be either increased or more severe in vaccinees than in those who had experienced the natural infection. While opinion was divided (Plotkin 1977), fear of the possible outcome of reactivation of latent infection slowed progress in testing of the live attenuated varicella vaccine after its development.

Table 1. Major events in the history of varicella zoster (VZ) virus and development of live attenuated vaccine

Date	Person	Event
1909	Bokay	Realization that varicella and zoster are caused by the same agent
1943	Garland	Hypothesis that zoster results from reactivation of latent VZ virus
1953	Weller	Isolation of VZ virus in tissue culture
1956	Cheatham et al.	First report of severe varicella in children with underlying malignant disease
1969	Brunell et al.	Reported efficacy of zoster immune globulin (ZIG) for prevention of varicella in normal children
1974	Gershon et al.	Efficacy of ZIG shown in children with underlying malignancy
1974	Takalashi et al.	Report on development of live attenuated varicella vaccine
1977	Izawa et al.	Reported safety of varicella vaccine in children with leukemia
1980		First studies on varicella vaccine in United States began
1984	Gershon et al.	Efficacy of varicella vaccine for prevention of severe varicella in leukemic children demontrated

Development of Varicella Vaccine

Takahashi and his colleagues, cognizant of the potential severity of at least some cases of varicella, and recognizing the potential for significant benefit to patients, reported the development of a live attenuated varicella vaccine in 1974 (reviewed by Gershon 1980). These investigators had successfully attenuated the virus by passage in animal cells (a parallel with the naturally occurring bovine passage of cowpox virus) and by temperature adaptation. The vaccine virus was passaged 11 times in human embryonic lung fibroblasts at 34 °C, 12 times in guinea pig lung cells at 37 °C, and finally 1–21 times in WI-38 human embryonic lung fibroblasts at 37 °C.

Because no practical animal model for this virus infection existed (or exists), all testing had to be performed in humans. First the vaccine was given to normal children. When there were no significant side-effects and an immune response was found to occur, the vaccine was given to abnormal hosts at increasing risk to develop severe chickenpox. Then children on low to moderate doses of steroids (considered to be at moderate risk) were vaccinated; again no significant side-effects occurred but there was an immune response (Takahashi et al. 1974). Finally children with underlying malignant diseases were immunized successfully. Side-effects of immunization were mild and an immune response occurred although the vaccinees were immunocompromised (Izawa et al. 1977). This was the first time in medical history that a live virus vaccine had been successfully administered to immunocompromised children. Prior to this time, immunization of such children with live virus vaccines was contraindicated.

Clinical Trials of Varicella Vaccine

Because of the success of Takahashi and his colleagues with immunization against varicella, investigators in the United States and Europe began to study the immunogenicity and safety of the vaccine. Studies in the US began around 1980. Some groups chose to immunize normal children (Arbeter et al. 1980; Weibel et al. 1984), and others concentrated on immunocompromised children (Brunell et al. 1981; Gershon et al. 1984). In general the studies of Americans corroborated those of Japanese workers.

Studies in normal children confirmed the immunogenicity and safety of this vaccine for these children (Arbeter et al. 1982). Efficacy studies in families in which one child received vaccine and one received placebo revealed that those vaccinated were totally protected from clinical illness after close exposure while controls who received placebo were not (Weibel et al. 1984).

Studies in immunocompromised children, most of whom had leukemia in remission, revealed that the vaccine significantly protected against clinical illness after exposure (Gershon et al. 1984). The following findings of the Japanese investigators were also confirmed. The live vaccine was clearly safe to administer to immunocompromised children, and the vaccine was immunogenic for them, although two doses were required to induce a seroconversion rate of greater than 90% in leukemics. In addition, no increase in the incidence of zoster (thus far less than 1%) was found in vaccinees in comparison to leukemics with past natural varicella (6% in this study). While further numbers of vaccinees need to be studied for longer periods of time, it is possible that after vaccination the incidence of zoster will be even less than after the natural infection.

There was also no increase in the incidence of relapse of leukemia (about 5% per year) as a result of immunization. The latter was a concern because to potentiate the immune response to the vaccine, antileukemic chemotherapy was suspended for 2 weeks, 1 week before and 1 week after vaccination. Apparently this suspension of anticancer therapy was not accompanied by ill-effects.

Studies in the United States, however, revealed some unexpected findings that had not been previously reported. One was that the vaccine was potentially contagious to others although at a much lower rate than the natural illness. It was found that occasional varicella-susceptible normal siblings seroconverted against VZ virus after exposure to a vaccinee with rash (Brunell et al. 1982; Gershon et al. 1984). Rare cases of mild varicella were also noted in siblings about 2 weeks after some of the leukemic vaccinees had experienced a vaccine-associated rash (Gershon et al. 1984). No viral spread was noted, however, if the vaccinee had no rash after immunization. Possibly because rash was so rare in normal children who were vaccinated, spread of vaccine virus to family members exposed to normal vaccinees has not been described.

New Concepts Concerning Varicella that Were Recognized During Vaccine Trials

At the time the vaccine began to be studied extensively in the US, a marker that made it possible to distinguish between vaccine type and wild type virus was developed (Martin et al. 1982). By treatment of viral DNA with restriction endonuclease enzymes, differences in wild and vaccine virus DNA are clearly discernible. In addition, differences in temperature sensitivity of wild and vaccine type virus were apparent, vaccine type being thermolabile at 39 °C (Takahashi et al. 1983). Finally the ability of vaccine type virus to grow in guinea pig tissue culture cells distinguished it from wild type virus strains (Takahashi et al. 1983). Thus for vaccinees with rash occurring at any time after immunization, it has been possible to determine whether the infecting virus was wild or vaccine type. It has also become clear that there are a number of minor strain differences in VZ viruses (Straus et al. 1984).

Another surprise concerning the natural history of varicella was found by American investigators. Until vaccine studies were performed, it was always presumed that the presence of detectable antibody to VZ virus in the blood would protect a person exposed to the virus from varicella. This, however, was clearly not necessarily always so. The first observed instance occurred in 1982 when a child vaccinated 10 months previously developed a very mild attack of varicella despite both antibody and cellular immunity to VZ virus in her blood 2 weeks prior to her illness. Virus isolated from her lesions was found to be wild type, indicating that she had been reinfected with wild virus despite her apparent immunity. Clearly she had not experienced reactivation of a latent infection with vaccine virus.

At that time it was not known how often this phenomenon might occur in vaccinees. Occasional attacks of mild varicella had been noted in Japanese vaccinees (Sakurai et al. 1982), but it was not known whether they had detectable antibody at the time of exposure or whether the illness was caused by wild virus or vaccine type virus. It was presumed they might have had atypical disseminated zoster rather than varicella.

Continued follow up of vaccinated immunocompromised children in the US revealed that the attack of clinical varicella in leukemic vaccinees with household exposures was about 20% (Gershon et al. 1984). Thus the vaccine does not always confer absolute immunity to varicella. This attack rate is, however, significantly lower than that seen in varicella-susceptible children with household exposures (80%–90%) (Ross et al. 1962). Those leukemic children who have developed varicella despite immunization have had an extremely MILD disease with no more than 70 vesicles. This is in contrast to normal children with varicella who develop on the average 250 to 500 vesicles (Ross et al. 1962). At present in the United States and Japan out of approximately 600 leukemic children who have been vaccinated, about 40 have developed varicella. In only three in Japan was the illness considered severe, requiring antiviral chemotherapy (Takahashi, personal communication). In the Collaborative Study in the United States (Gershon et al. 1984) all cases in leukemics have been mild, although there have been 13 breakthrough cases. Perhaps this is the best record that can be expected since passive immunization with varicella zoster immune globulin does not always prevent disease either. Nevertheless, it is clear that the vaccine is highly protective for both leukemic and normal children. One possible significant difference between vaccine studies in the United States and those in Japan is that in the US Collaborative Study, two doses of vaccine at least 3 months apart have been given to most vaccinees. In Japan only one dose has usually been used. Further study will be necessary to determine whether use of two doses makes a significant difference in the outcome of immunocompromised patients.

The Future

Since the live attenuated varicella vaccine appears to be safe and effective even for leukemic children, there is now renewed interest in the use of the vaccine on a routine basis for normal children. Ideally the vaccine might be given simultaneously with measles, mumps, and rubella vaccines to children when they are 15 months of age. Several studies of the feasibility of this strategy are currently being undertaken in the USA. In Japan it is expected that the vaccine will be licensed within this year for use in high risk children. It is likely that when the time comes to apply for licensure in the USA, recommended use for all children will be sought.

While it has taken longer from the time of isolation of VZ virus "in vitro" to the licensure of a vaccine than it did for measles, mumps, and rubella, the eventual outcome should be similar: the elimination of a great deal of suffering from infection caused by this virus.

References

Arbeter AM, Starr SE, Weibel RE, et al. (1982) Live attenuated varicella vaccine: immunization of healthy children with the OKA strain. J Pediatr 100:886–893

Bokay J (1909) Über den ätiologischen Zusammenhang der Varizellen mit gewissen Fällen von Herpes zoster. Wein Klin Wochenschr 22:1323–1326

Brunell P (1977) Brunell's brush-off. Pediatrics 59:954

Brunell PA, Geiser C, Shehab Z, et al. (1982) Administration of live varicella vaccine to children with leukemia. Lancet 2:1069–1072

Cheatham WJ, Weller TH, Dolan TF, et al. (1956) Varicella: report of two fatal cases with necropsy, virus, isolation, and serologic studies. Am J Pathol 32:1015–1028

Feldman S, Hughes WT, Daniel CB (1975) Varicella in children with cancer. Pediatrics 56:388–397

Garland J (1943) Varicella following exposure to herpes zoster. N Engl J Med 228:336–337

Gershon AA (1980) Live attenuated varicella-zoster vaccine. Rev Infect Dis 2:393–407

Gershon AA, Steinberg S, Gelb L (1984) The National Institute of Allergy and Infectious Diseases Collaborative Varicella Vaccine Study Group. Efficacy of varicella vaccine in children with leukemia in remission. JAMA 252:355-362

Izawa T, Ihara T, Hattori A, et al. (1977) Application of a live varicella vaccine in children with acute leukemia or other malignant diseases. Pediatrics 60:805–809

Krugman S (1984) Evaluation of hepatitis B vaccine. Presented at 4th international Symposium on rapid methods and automation in microbiology and immunology, Berlin

Martin JH, Dohner DE, Wellinghof WJ, et al. (1982) Restriction endonuclease analysis of varicella-zoster vaccine virus and wild-type DNAs. J Med Virol 9:69–76

Plotkin S (1977) Varicella vendetta: Plotkin's plug. Pediatrics 59:953–954

Ross AH, Lenchner E, Reitman G (1962) Modification of chickenpox in family contacts by administration of gamma globulin. N Engl J Med 267:369–376

Sakurai M, Ihara T, Ito M, et al. (1982) Application of a live varicella vaccine in children with acute leukemia. In: Herpes virus: clinical pharmacological and basic aspects. Shiota, Cheng, Prusoff (eds) Exerpta Medica, Amsterdam, pp 87–93

Straus SE, Hay J, Smith H, et al. (1983) Genome differences among varicella-zoster virus isolates. J Gen Virol 64:1031–1041

Takahashi M (1983) Chickenpox virus. Adv Virus Res 28:285–356

Takahashi M, Otsuka T, Okuno Y, et al. (1974) Live vaccine used to prevent the spread of varicella in children in hospital. Lancet 2:1288–1290

Weibel R, Kuter B, McLean A, et al. (1984) Efficacy of varicella vaccine in normal children. N Engl J Med (to be published)

The Use of DNA Hybridization for the Identification of Pathogenic Bacteria

S. FALKOW [1]

Introduction

The applications of the newer methods of gene cloning and DNA biochemistry have provided those of us interested in infectious diseases with powerful new alternatives for the study of microbial pathogenicity. The availability of cloned genetic sequences specific for a particular microbial pathogen can be exploited in a number of ways. Genes from pathogens cloned into a harmless microorganism, such as *Escherichia coli* K-12, can be amplified, and the gene product may be produced many times more than under usual cultivation conditions. The impact of recombinant DNA methods upon the study of viruses and noncultivable bacteria is obvious. We are just beginning to see the practical application of gene cloning methods as a means to study the fundamental pathogenic mechanisms of medically important microorganisms and as a rationale for the production of a new generation of safe and effective vaccines. The availability of cloned genetic sequences specific for a particular microbial pathogen can also be exploited as a diagnostic reagent. Hence, nucleic acid hybridization, so long the research tool of the molecular biologist, may now find its way into the diagnostic laboratory.

The utility of nucleic acid hybridization for diagnostic purposes is based on the specificity of the hybridization reaction. Nucleic acid hybridization based on the detection of the specific genes of an infecting organism will pretty well unfailingly identify the agent, if present in reasonable numbers, within a clinical sample. However, while nucleic acid hybridization is accurate, it is not necessarily a straightforward procedure. Numerous variations of the nucleic hybridization method now exist and are commonly employed in recombinant DNA research. In general, these procedures are complex and time consuming. Until quite recently, they all required the use of radioisotopes, typically ^{32}P, and therefore have the inherent difficulty of handling and disposal. Because of the time factor(s) and the use of isotopes, DNA and RNA hybridization assays have remained a research tool. They have not become part of the diagnostic armory available to the clinician. It is clear, nonetheless, that hybridization may be ideally suited for important diagnostic applications, such as viral and bacterial identification. One can be reasonably confident that new and innovative methods will be found to simplify nucleic acid hybridization methods so that they may be used in the diagnostic laboratory. Indeed, Professor Ward (this volume) has a number of strategies to achieve this end. In the following sections my role will be to review some

1 Department of Medical Microbiology, Stanford University, Stanford, CA 94305, USA

of the research and epidemiologic studies which have exploited nucleic acid hybridization and which stand as the foundation for the practical application of nucleic acid probes.

Methods and the Criteria for the Development of DNA Probes

DNA probe development has the aim of selecting and identifying a unique nucleotide sequence which will only hybridize with a specific gene(s) of interest. If the sequence is chosen carefully, its detection within the genome of a microorganism will constitute an accurate means of identification. Our laboratory had not deliberately sought to identify and isolate DNA probes for the detection of a specific organism. Rather, our interest in understanding specific virulence determinants in pathogenic bacteria by the study of cloned genetic determinants has led to the identification of specific groups of pathogenic microorganisms by nucleic acid hybridization. For example, our isolation of the specific genes governing enterotoxin biosynthesis in *E. coli* [1–3], could be exploited to identify the presence of toxigenic organisms directly in clinical material [4–6]. Yet, in this instance, the DNA probe does not identify a specific species but rather a group of strains within a species which have specific pathogenic potential. In theory, any microorganism possessing the same enterotoxin gene(s) will be detected by the hybridization reaction. In practice, the base majority of strains detected by the probe are, indeed, toxigenic *E. coli.*

Other DNA probes which have been utilized as a result of our work includes two plasmid species, one found in most, but not all, *Neisseria gonorrhoeae,* and the other found exlusively in all tested isolates of *Chlamydia trachomatis*. Both of these probes have been fruitfully used to identify infecting microorganisms directly in clinical material [7, 8]. Their predictive value has been 100% in the sense that a positive hybridization signal within a sample indicates that the pathogen is present. Hence, no obvious cross-reactivity with other bacterial species has been detected with either DNA probe within clinical material thus far. Interestingly enough, both of the plasmids are cryptic in the sense that we are unaware of their contribution to the gene pool of their host organism. In rather a similar manner, "minicircle" kinetoplast DNA of *Leishmania* may be utilized to identify *Leishmania* species [9]. In this case as well as species of DNA, cryptic in the phenotypic sense, serves as a means for the rapid identification of a microorganism which has been difficult to identify by the more classical procedures of microbiology.

The utilization of DNA probes defined by genes which contribute to bacterial virulence may be of practical importance, but this approach may not be the best from the standpoint of the clinical microbiologist or epidemiologist. For example, it is possible to utilize the identical genes involved in epithelial cell invasion by invasive. *E. coli* and *Shigella* species as a means for detecting these microorganisms in the feces of patients with diarrhea. From the clinician's standpoint, it is largely irrelevant whether an individual patient is suffering from an infection caused by an *Escherichia* or a member of the *Shigellae*. The management of the patient would be the same and the cause of the patient's dysentery as a established as a bacterial infection. Of course, the microbiologist, and in particular the epidemiologist, would prefer to know precisely the cause of the patient's infection. Thus, ultimately, it may be preferable to develop DNA probes which can detect only a particular species or even a particular strain of a species. I have found it of some interest that the most straightforward approach for the development of DNA probes has yet to be reported. That is, it is perfectly apparent that one can select a unique sequence of DNA by reacting the DNA from one species with that of another closely related organism. For example, virtually any fragment of DNA from a *Shigella* species will cross-react only with *Escherichia* species or with *Salmonella* species. Hence, selecting for *any* sequence of *Shigella* DNA which does not hybridize with *E. coli* results in the isolation of a specific nucleotide sequence for the identification of members of the genus *Shigellae*. The hybridization reaction itself is exploited to select for the specific sequence. Thus, if one has *E. coli* DNA covalently bound to a specific supporting matrix, it is clear that hybridization with *Shigellae* DNA will result in the binding of related sequences to the matrix while the unreacted DNA, the *Shigella*-specific sequences, will remain unbound. Any variation of this theme could surely be utilized profitably to isolate the desired nucleotide sequence with the desired specificity. In my judgement, therefore, the extraordinary promise of DNA probes for the detection of microorganisms arises from the ease with which one may isolate specific sequences and the relative freedom of such sequences from the variability that one sometimes experiences with important antigenic determinants of microorganisms used in antibody detection methods.

Recent Utilization of Nucleic Acid Hybridization for Diagnostic and Epidemiologic Applications

Currently, hybridization methods are based on radioautography of radioisotope-labeled nucleic acid probes bound to immobilized specimen DNA. Despite this serious limitation, a surprising number of publications encompassing a wide variety of applications for the detection of specific pathogens by hybridization have appeared over the last few years. While it is not my intent to provide an exhaustive review of these investigations. Table 1 provides some indication of the

Table 1. Application of DNA hybridization for the diagnosis or epidemiologic investigation of infectious disease agents

Pathogen	Source of DNA	Specimen examined
Neisseria gonorrheae	Plasmid or IgA protease gene	Urethral pus
Chlamydia trachomatis	Plasmid	Cervical swab
Enterotoxigenic *E. coli*	Cloned toxin genes	Stool isolates, stool samples, food
Cytomegalovirus	Cloned fragments	Urine and buffy coat
Epstein-Barr virus	Cloned viral fragments	Infected tissue
Adenovirus	Isolated DNA fragment	Nasopharyngeal swabs
Hepatitis B virus	Isolated viral DNA	Liver serum
Leishmania species	Cloned kinetoplast DNA	Patient isolates, cutaneous lesions

specific applications for which nucleic acid hybridization has been found to be worthwhile. Not surprisingly, most of these studies deal with microorganisms which are difficult to detect by ordinary bacteriological means and for which no generally applicable serological detection method has been available. While there is some variation in results, by and large I believe there is reasonable unanimity of opinion that using ^{32}P-labeled nucleotides one can readily detect between 1,000 and 10,000 bacteria or viral particles directly in a clinical sample, or about 0.1 pg of nucleic acid or higher. In those cases where there was some attempt to amplify the nucleic acid signal by providing for some growth period of the organism prior to hybridization, the sensitivity was not significantly different from that of the conventional detection means, and in some cases it appeared that the hybridization method was, indeed, more sensitive. Even with this favorable sensitivity relative to culture or to some other recognized laboratory standard, the sensitivity of the hybridization method in which infectious agents were directly looked for in direct patient material, was generally between 85% and 98%. The sensitivity of the assay in these studies do not necessarily reflect that there were too few organisms to detect by hybridization. Rather, Tompkins (this meeting) has pointed out that it appears that a significant amount of development will be required to find the optimal conditions to perform nucleic acid hybridization within patients' specimens, particularly fecal material and thick mucopurulent material.

The diagnosis of infectious diseases still largely depends upon traditional culture methods and serological techniques. It is, of course, the goal of those concerned with disease diagnosis to provide the most accurate data in the shortest period of time in order to effect treatment of the patient's illness. Recent developments in immunology, immunochemistry, biochemistry, genetics, and recombinant DNA technology have provided new means for the development of highly specific reagents that permit a more rapid and precise clinical diagnosis than ever before. Although I can report at this time only the first steps in the contribution of nucleic acid hybridization to infectious disease diagnosis, it seems likely that the revolution now taking place in the clinical laboratory will include this specific, accurate methodology.

Microbial Diagnosis by Nucleic Acid Sandwich Hybridization

M. RANKI [1], M. LAAKSONEN [1], A. PALVA [1], A.-C. SYVÄNEN [1], M. VIRTANEN [1], and H. SÖDERLUND [2]

Introduction

Nucleic acid hybridization is a highly specific method to identify a given DNA sequence and, thus, the microbe(s) containing it. The specificity depends on two basic properties of the hybridization reaction: (a) The double-stranded hybrid which is formed between the probe and the sample strand is stable only if the sequences of these DNA strands share a high degree of homology. (b) There is no chance that a nucleotide sequence in the probe which is long enough to form a stable hybrid would occur at random. There are, however, conserved genes that are preserved during evolution. They have to be taken into account when selecting the probe. They have to be avoided if strictly species-specific identification is required (Kingsbury et al. 1969; Totten et al. 1983), or they can used advantageously when setting up group-specific diagnosis (see results) (Fitts et al. 1983; Palva 1983; Palva et al. 1984).

Hybridization fulfills the generally accepted criteria for a good diagnostic method. It is becoming applicable to routine microbial diagnosis due to recombinant DNA techniques. Probes with the desired level of specificity can be selected from genomic libraries. Furthermore, genes coding for certain properties, e.g. virulence, can be identified and used as a probe. Characterization of the plasmid encoded toxin genes of the enterotoxigenic *Escherichia coli* strains (So et al. 1978, 1981) and the subsequent use of them as probes in diagnosis of diarrhea (Moseley et al. 1982) present an illustrative example (see Falkow, this volume, pp. 30-33).

To be applicable to routine use a test must be reliable, reproducible and simple to perform. With that aim we have developed a sandwich hybridization method (Ranki et al. 1983a) which has the following advantages: In this method the sample is kept in solution as opposed to most hybridization methods, in which the sample DNA is fixed onto a solid carrier. Thus, sample pretreatment can be kept simple, and fairly crude samples can be studied without causing unspecific hybridization background (Ranki et al. 1983b). The second advantage concerns the use of recombinant DNA as probe. Specific hybridization background due to vector sequences hybridizing to homologous sequences possibly present in the sample does not occur. The method uses two separate DNA reagents derived from the microbial nucleic acid to be identified. One of the DNA fragments is fixed onto a solid carrier (nitrocellulose filter) and the other is radiolabelled and

---

1 Orion Genetic Engineering Laboratory, Valimotie 7, SF-00380 Helsinki

2 Recombinant DNA Laboratory, University of Helsinki, Valimotie 7, SF-00380 Helsinki

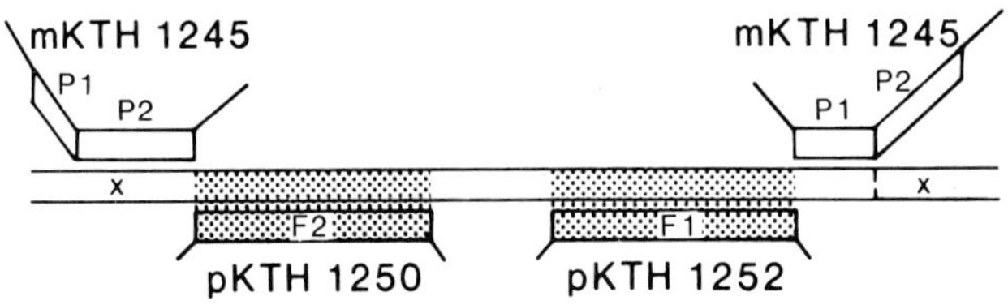

Fig. 1. Sandwich hybridization with two reagent pairs. *pKTH1250* and *pKTH1252* (F_2, F_1) are two recombinant plasmids, each containing a different 3 kb long *C. trachomatis* DNA insert. *mKTH1245* (P_1P_2) is a recombinant phage with a 2 kb long *C. trachomatis* insert. This insert is a composite of two DNA fragments (0.8 kb and 1.2 kb) which in the genome are situated next to the DNA fragments used as filter reagents. *X, C. trachomatis* DNA hybridizing to both filter-bound and probe DNA

functions as probe. When present in the sample, a DNA molecule homologous to the two reagents will hybridize both to the filter-bound DNA and to the labelled probe and thus mediate labelling of the filter (Ranki et al. 1983a). The recent modification of the method using multiple reagent pairs is shown in Fig. 1. In the following the applicability of the method to viral and bacterial diagnosis will be described.

Methods and Specimens

Sandwich Hybridization Reagents

All nucleic acid reagents were prepared using recombinant DNA techniques as described previously. For each microbe the identifying reagent pair consisted of a recombinant plasmid attached to a nitrocellulose filter ($4–10 \times 10^{10}$ molec/10-mm filter) and a single-stranded recombinant phage DNA which was chemically iodinated to a specific activity $10^7–10^8$ cpm/µg DNA (Palva 1983; Ranki et al. 1983a). The multiple identifying reagent pairs for *C. trachomatis* are depicted in Fig. 1 (Palva et al. 1984).

Sandwich Hybridization

Each hybridization contained one filter with microbe-specific DNA and one or two control filters with calf thymus or no DNA, respectively. The reaction was carried out in 400 µl containing the denatured sample, the labelled probe (200,000–400,000 cpm), 0.6 *M* NaCl, 0.06 *M* sodium citrate (4 × SSC), 0.02% each of Ficoll, polyvinylpyrrolidone and occasionally bovine serum albumin, 200 µg/ml denatured herring sperm DNA, 0.25% SDS. Incubations were for 19–24 h at 65 °C, whereafter the filters were carefully washed in 0.1 × SSC, 0.2% SDS at 37 °C or 50 °C and finally counted in a gamma counter. For details of the assay, see Ranki et al. (1983b).

Treatment of Specimens

Virus, Infected Cells and Bacteria. DNA from virus or animal cells was released by treatment with proteinase-K (1 mg/ml, 37 °C, 1 h) in 1% SDS. If cells were present, viscosity was reduced by sucking through a hypodermic needle. Gram-negative bacteria were lysed by lysozyme (500 µg/ml), EDTA (70 m*M*) treatment (37 °C, 30 min) followed by addition of SDS (0.25%). DNA was denatured by boiling for 5 min in 14 m*M* NaOH (bacteria) or 7 m*M* NaOH (CMV), cooled, neutralized and subjected to the assay.

Concentration of CMV DNA from Urine. Urine specimens were supplemented with 200 µg herring sperm DNA as carrier and 40 µg/ml of gentamycin and treated with sarkosyl (2%) and EDTA (5 m*M*). An equal volume of isopropanol was then added and DNA precipitated at room temperature for 10 min. The DNA was collected by centrifugation (27,000 g, 30 min, 4 °C) and the DNA dissolved in TE (10 m*M* Tris-HCl pH 7.5, 1 m*M* EDTA).

Clinical Specimens. Nasopharyngeal mucus aspirates were collected from children with acute respiratory infection. They were diluted into the radioimmunoassay dilution buffer, sonicated, treated with SDS and proteinase-K and finally boiled for 5 min (Virtanen et al. 1983).

The lung biopsies were collected at the University Central Hospital, Helsinki, from immunesuppressed patients suffering of severe pneumonitis. Virus isolation was performed from a tiny tissue specimen. A separate piece (a few mm^3) was frozen for the hybridization test. The specimen in TE was minsed with scissors, 1% SDS added and the homogenate digested with proteinase-K (1 mg/ml, 1 h, 37 °C) and finally homogenized by sucking through a hypodermic needle. DNA denaturation was as above.

Female endocervical specimens were selected for this study on the basis of results of chlamydial isolation from the collection of the Department of Virology, University of Helsinki. The cotton-tipped or dacron swabs were removed from the transport medium and washed with 200 µl of distilled water. The medium and the wash were combined and 40 µl of 2% SDS, 150 m*M* NaOH and 60 m*M* EDTA were added. The sample was lyophilized, dissolved in 400 µl of distilled water, denatured, neutralized and subjected to the assay (500 µl).

Results and Discussion

Group-Specific Reagents for Bacterial Detection

The possibility of creating a test with broad but defined specificity was studied using a gene conserved among the Enterobacteriaceae as a model. Sandwich hybridization reagents were prepared from the *omp* A gene of *Escherichia coli* (Palva 1983). Table 1 shows that this assay detected all the tested members of the Enterobacteriaceae family even if the detection level for the distantly related members (*Serratia marcescens* and *Proteus mirabilis*) was decreased about 100-fold as compared to homologous DNA. The specificity of the test was controlled with 13 bac-

Table 1. Specificity of *Escherichia coli omp* A reagents

Organism[a]	Sandwich hybridization (cpm)
E. coli K-12	561
E. coli (urine isolate)	462
Salmonella typhimurium	475
Enterobacter cloacae	253
Yersinia enterocolitica	173
Serratia marcescens	113
Proteus mirabilis	111
Control bacteria	65[b]

[a] 10^8 lysed cells

[b] Mean for 13 different bacteria unrelated to the Enterobacteriaceae (variation range 41–81 cpm). Background hybridization without sample 83 cpm (Palva 1983)

terial strains representing gram-negative and gram-positive aerobic and anaerobic bacteria. They were all negative in the test even if present in 10-fold higher amounts. The experiment suggests that group-specific tests can be constructed but that the entire enterobacterial family is probably too large to be reliably diagnosed with a single reagent.

Recombinant DNA techniques also allow selection of specific reagents for micro-organisms that have been poorly characterized. It is not necessary to know the gene content of a probe fragment nor do the genes need to be expressed (Salyers et al. 1983). This, however, means that a wide range of control microbes should be included in the specificity testing because of unpredictable occurrence of conserved properties among the microbes. Selection of DNA-reagents specific for the detection of *Chlamydia trachomatis* serves as an example of the process outlined above. *C. trachomatis* L2 DNA was shot gun cloned into the pBR322 plasmid vector after restriction enzyme digestion. The specificity of some of the clones was tested by hybridizing them with DNA isolated from several *C. trachomatis* serotypes (positive controls) and from unrelated bacteria (negative controls). Sandwich hybridization reagents were constructed from a clone with the desired properties and their specificity subsequently confirmed by hybridizing with 41 different microbes. These included *Neisseria gonorrhoeae* and *Gardnerella vaginalis, Candida albicans, Trichomonas vaginalis* and herpes simplex virus (Palva et al. 1984). The selected reagents detected all *C. trachomatis* serotypes but none of the unrelated organisms even if the latter were present at a molar concentration 5,000-fold the detection limit for the homologous DNA.

Specificity of Viral Reagents

Adenoviruses are structurally and antigenically well conserved, but the DNA homology between subgroups is only in the order of 10%–20% whereas within the subgroups it is close to 90% (Green et al. 1979). Table 2 demonstrates that

Table 2. Specificity of adenovirus reagents

Specimen	Sandwich hybridization with reagents from	
	AD2 (cpm)	AD3 (cpm)
Subgroup C-virus (40 ng DNA) (AD1, AD2 or AD5)	4,892	213
Subgroup B-virus (15 ng DNA) (AD3 or AD7)	472	4,473
No DNA	107–159	140–390

Specimen were clinical nasopharyngeal mucus aspirates. The DNA content was calculated according to the standard curve. For details see Virtanen et al. (1983)

subgroup-specific sandwich hybridization tests could be constructed (Virtanen et al. 1983). No cross hybridization with herpes simplex virus or human DNA was demonstrated.

More elaborate specificity controls are required for DNA fragments derived from large, poorly characterized viruses such as cytomegalovirus (CMV). During evolution these viruses apparently have picked up sequences from their host cell genomes (Heller et al. 1982; Peden et al. 1982; Spector and Vacquier 1983) which obviously should not occur in virus-specific DNA-reagents. When selecting CMV-specific sandwich hybridization reagents we therefore tested them with human and calf DNA as well as herpes simplex virus DNA (Virtanen et al. 1984). A pair of *Bam*HI fragments derived from the *Hind*III L-fragment of the CMV AD169 strain (Oram et al. 1982) fulfilled the specificity criteria.

Sensitivity of Sandwich Hybridization

The adenovirus and CMV tests detect about $4–5 \times 10^6$ DNA molecules in an overnight test (Ranki et al. 1983a; Virtanen et al. 1984). For detection of *C. trachomatis* two identifying reagent pairs were constructed (Fig. 1), because we postulated that this should increase the sensitivity of the assay. This was indeed the case, because now 10^6 molecules of *C. trachomatis* DNA were detected (Palva et al. 1984).

Clinical Applicability of the Sandwich Hybridization Method

Detection of Adenovirus from Nasopharyngeal Secretions. Nasopharyngeal secretions had been collected from children with acute respiratory infection. They were subjected to the sandwich hybridization in a crude form. The results were compared with those obtained from a radioimmunoassay (RIA) measuring the major adenovirus structural antigen (hexon). Table 3 shows that the sensitivity of the hybridization test was comparable to that of the highly sensitive RIA and no false positives were detected.

Table 3. Detection of adenovirus from nasopharyngeal mucus specimens by sandwich hybridization. For details see Ranki et al. (1983b) and Virtanen et al. (1983)

	No. of specimens	Sandwich hybridization	RIA (hexon antigen)	Virus isolation
Exp. 1	5	+	+	+
	3	–	–	–
	2	–	–	+
Exp. 2	22	+	+	ND
	3	–	+	+
	23	–	–	ND

Table 4. Detection of CMV DNA from urine after concentration

Addition of CMV to 18 ml urine (particles $\times 10^7$)	Yield of DNA after concentration		
	cpm hybridized	No. of DNA molec	
		$\times 10^7$	%
20	1,522	16	80
8	908	6	75
4	539	2.7	68
2	387	1.8	90
0.8	193	0.7	88
–	54	–	

Viral DNA was released by sarkosyl, EDTA treatment and DNA concentrated by precipitation with isopropanol. Recovery was estimated according to standard curve

Detection of CMV from Urine and Lung Biopsies. Urine is the most common clinical substance studied for CMV. The sensitivity of a hybridization test is, however, too low to allow detection without prior concentration. We decided to concentrate the DNA instead of virus. This was achieved by precipitation with isopropanol from sarkosyl-treated urine. Table 4 demonstrates that the virus DNA yield after such a process as measured by quantitative sandwich hybridization is about 80%. The results also indicate that no impurities harmful to the test were concentrated with the DNA. The method has been successfully applied to detection of CMV from infant urine (Virtanen et al. 1984).

We had previously shown that viral DNA could be demonstrated from a crude cell lysate (Ranki et al. 1983a). We therefore applied an analogous approach to studying lung biopsies for the presence of CMV. Table 5 shows the results of testing all the specimens suspected to contain CMV that were taken during the period from July 1982 to end of 1983 at the University Central Hospital, Helsinki. Two separate samples of the patient 1 were positive by sandwich hybridization, but the other two specimens taken from patients with CMV disease re-

Table 5. Detection of cytomegalovirus from lung biopsies by sandwich hybridization

Sample	Virus isolation	Hybridization[b]	
		cpm	Interpretation
1a[a]	CMV	512	+
1b[a]	CMV	280	+
2	CMV	67	–
3	CMV	39	–
4	HSV	48	–
5	–	36	–
6	–	47	–
7	–	28	–
8	–	57	–

[a] 1a and b are two independent specimens from the same patient
[b] CMV DNA standards gave 626 cpm (4×10^7 molec), 121 cpm (4×10^6 molec), and 60 cpm (no DNA)

Table 6. Detection of *C. trachomatis* from endocervical specimens by sandwich hybridization

Specimen[a]	Chlamydia isolation	Sandwich hybridization	
		cpm[b]	Interpretation
1	+	140	+
2	+	159	+
3	+	412	+
4	+	144	+
5	+	150	+
6	+	170	+
7	+	211	+
8	–	144	+
8–20	–	79–144 ($\bar{x}_{12}$ 101)	

[a] For treatment, see Methods and Specimens
[b] cpm values higher than the mean +2 SD (=137 cpm) of the culture negative specimens were considered positive. The empty control filters in the tests gave 39–89 cpm, mean 61 cpm. Chlamydia DNA standards gave 1,043 cpm and 250 cpm for 5×10^7 and 5×10^6 DNA molec, respectively

mained negative. This is perhaps not surprising, because Volpi et al. (1983) have shown, by using monoclonal antibodies, that very few infected cells were present in their lung biopsy specimens. No false positives were detected in our test.

Detection of C. trachomatis from Endocervical Specimens. For this study we collected a series of endocervical specimens, part of which were positive and part negative by chlamydial isolation. Table 6 demonstrates that the number of elementary bodies in the specimens varied between about 10^6–10^7. One of the culture negative specimens was weakly positive by the hybridization test. We believe that this is a true positive, because otherwise the variation in the results with the isolation negative specimens was low. It should be noted that several of the specimens which scored positive in the hybridization test contained only slightly more chlamydial DNA than required for detection (10^6 molecules). We have also tested samples that were isolation positive but negative in the hybridization test. Thus the sensitivity of the test should preferably be improved. A larger clinical study is required to evaluate the usefulness of the sandwich hybridization test in chlamydial diagnosis.

Conclusions

Sandwich hybridization is a three component reaction which allows keeping the specimen in solution. Two microbe-specific recombinant DNA reagents are used to catch the homologous sample DNA to nitrocellulose filter. Overnight tests specific for adenovirus, cytomegalovirus and *Chlamydia trachomatis* have been constructed.

Because specimens are tested in a soluble form minimum pretreatment is required. Usually the microbial nucleic acid is released by a combined enzyme and detergent treatment, and the DNA is denatured by boiling in mild alkali and added to the test. Several types of crude specimens have been studied without background problems: specimens contaminated with a large number of unrelated bacteria (Palva 1983), cell lysates (Ranki et al. 1983a), homogenized lung biopsies, endocervical swabs (here), nasopharyngeal mucus (Virtanen et al. 1983) and urine (Virtanen et al. 1984). The clinical applications of the method have indicated that sandwich hybridization is reliable and very specific with no false positive reactions. It is well suited to diagnosis of infections from which samples with moderate or high microbe content can be obtained.

Acknowledgements. We wish to thank Prof. Pekka Halonen (University of Turku) for providing us with the nasopharyngeal mucus specimens and performing the radioimmunoassays; Dr. Jon Oram (PHLS Center for Applied Microbiology and Research, Porton Down) for cytomegalovirus DNA clones; the Department of Virology, University of Helsinki, for virus isolations; the University Central Hospital, Helsinki, for lung biopsy specimens; and The National Public Health Laboratory, Helsinki, for various bacterial strains.

References

Fitts R, Diamond M, Hamilton C, Neri M (1983) DNA-DNA hybridization assay for detection of *Salmonella* spp. in foods. Appl Environ Microbiol 46:1146–1151

Green M, Mackey JK, Wold WSM, Rigden P (1979) Thirty-one human adenovirus serotypes (Ad1–Ad31) form five groups (A–E) based upon DNA genome homologies. Virology 93:481–492

Heller M, Henderson A, Kief E (1982) Repeat array in Epstein-Barr virus DNA is related to cell DNA sequences interspersed on human chromosomes. Proc Natl Acad Sci USA 79:5916–5920

Kingsbury DT, Fanning GR, Johnson KE, Brenner DJ (1969) Thermal stability of interspecies *Neisseria* DNA duplexes. J Gen Virol 55:201–208

Moseley SL, Echeverria P, Seriwatana J, Tirapat C, Chaicumpa W, Sakuldaipeara T, Falkow S (1982) Identification of enterotoxigenic *Escherichia coli* by colony hybridization using three enterotoxin gene probes. J Infect Dis 145:863–869

Oram JD, Downing RG, Akrigg A, Dollery AA, Duggleby CJ, Wilkinson WG, Greenaway PJ (1982) Use of recombinant plasmids to investigate the structure of the human cytomegalovirus genome. J Gen Virol 59:111–129

Palva A (1983) *OmpA* gene in detection of *Escherichia coli* and other *Enterobacteriaceae* by nucleic acid sandwich hybridization. J Clin Microbiol 18:92–100

Palva A, Jousimies-Somer H, Saikku P, Väänänen P, Söderlund H, Ranki M (1984) Detection of *Chlamydia trachomatis* by nucleic acid sandwich hybridization. FEMS Microbiol Lett 23:83–89

Peden K, Mounts P, Hayward GS (1982) Homology between mammalian cell DNA sequences and human herpesvirus genomes detected by a hybridization procedure with high-complexity probe. Cell 31:71–80

Ranki M, Palva A, Virtanen M, Laaksonen M, Söderlund H (1983 a) Sandwich hybridization as a convenient method for the detection of nucleic acids in crude samples. Gene 21:77–85

Ranki M, Virtanen M, Palva A, Laaksonen M, Pettersson RF, Kääriäinen L, Halonen PE, Söderlund H (1983 b) Nucleic acid sandwich hybridization in adenovirus diagnostics. Curr Top Microbiol Immunol 104:309–318

Salyers AA, Lynn SP, Gardner JF (1983) Use of randomly cloned DNA fragments for identification of *Bacteroides thetaiotaomicron*. J Bacteriol 154:287–293

So M, Dallas WS, Falkow S (1978) Characterization of an *Escherichia coli* plasmid encoding for synthesis of heat-labile toxin: molecular cloning of the toxin determinant. Infect Immun 21:405–411

So M, Atchison S, Falkow S, Moseley S, McCarthy BJ (1981) A study of the dissemination of Tn 1681: a bacterial transposon encoding a heat-stable toxin among enterotoxigenic *Escherichia coli* isolates. Cold Sring Harbor Symp Quant Biol 45:53–58

Spector DH, Vacquier JP (1983) Human cytomegalovirus (strain AD169) contains sequences related to the avian retrovirus oncogene v-myc. Proc Natl Acad Sci USA 80:3889–3893

Totten PA, Holmes KK, Handsfield HH, Knapp JS, Perine PL, Falkow S (1983) DNA hybridization technique for the detection of *Neisseria gonorrhoeae* in men with urethritis. J Infect Dis 148:462–471

Virtanen M, Palva A, Laaksonen M, Halonen P, Söderlund H, Ranki M (1983) Novel test for rapid viral diagnosis: detection of adenovirus in nasopharyngeal mucus aspirates by means of nucleic acid sandwich hybridization. Lancet I:381–383

Virtanen M, Syvänen A-C, Oram J, Söderlund H, Ranki M (1984) Cytomegalovirus in urine: Detection of viral DNA by sandwich hybridization. J Clin Microbiol 20:1083–1088

Volpi A, Whitley RJ, Ceballos R, Stagno S, Pereira L (1983) Rapid diagnosis of pneumonia due to cytomegalovirus with specific monoclonal antibodies. J Infect Dis 147:1119–1120

Diagnosis of Papillomavirus Infection by Nucleic Acid Hybridization

L. GISSMANN [1] and A. SCHNEIDER [2]

The human papillomaviruses (HPV) represent a heterogeneous group of agents, each of which is associated with specific epithelial proliferations (Table 1) [1]. Whereas most of these lesions are usually typical benign warts, others clearly represent premalignant tumors or bear the potential for conversion into malignant growth. As shown in Table 1, the carcinomas themselves harbor only particular types of HPV. Therefore virus-typing might be of diagnostic value in evaluating the risk of a given lesion to develop into a malignant tumor.

Table 1. Association of HPV types with human tumors

Clinical diagnosis	HPV type[a]	Malignant conversion	HPV type in the invasive carcinoma
Common wart	1, 2, 4, 7	No	–
Epidermodysplasia verruciformis	3, 5, 8, 9, 10, 12, 14, 15, 17, 19–24	Yes	5,8
Morbus Heck	13	?	–
Laryngeal papilloma	6, 11	Yes	?
Condyloma acuminatum	6, 11	Rare	16
Bowenoid papulosis, Bowen disease	16	Yes	16
Cervical lesions	6, 11, 16, 18	Yes	16, 18

[a] Only the commonly prevalent HPV types are mentioned; data taken from [1] and [9]

Since an in vitro system for virus replication is thus for not available, serological methods are hardly applicable in determining the HPV type. The different viruses have therefore been defined on the basis of molecular hybridization of their DNA. By definition, a new type of papillomavirus shares less than 50% sequence homology with the other types, as measured under high stringent conditions of hybridization [2]. An unknown isolate is first tested by Southern blot hybridization against all the various papillomaviruses. In the case of a positive reaction with another HPV DNA but a different restriction enzyme pattern, both genomes are compared by the kinetics of their reassociation. Molecular cloning of the DNA is necessary prior to hybridization in order to correlate the speed of reassociation with definite DNA concentrations.

1 Deutsches Krebsforschungszentrum, Institut für Virusforschung, Im Neuenheimer Feld 280, D-6900 Heidelberg

2 Frauenklinik, Universität Ulm, Prittwitzstr. 43, D-7900 Ulm

As stated before, there is a correlation between the clinical and histological appearance of particular papillomas and the virus types present. Therefore, only a limited number of HPV DNAs have to be tested as candidates within a given lesion. Southern blot hybridization analysis is the method of choice, using the individual HPV DNAs as probes, and the cellular DNA blotted in parallel on different nitrocellulose filters. Digestion with restriction enzymes, which cut several times within the HPV genomes, thus resulting in a typical fragment pattern seen after hybridization, easily allows the differentiation between closely related virus types, e.g., HPV 3 and 10 or HPV 6 and 11 [3,4]. In the case of a limited amount of cellular DNA available, hybridization with the single HPV DNAs can be performed sequentially onto the same nitrocellulose filter. This procedure, however, usually takes a considerable amount of time until the final result is obtained. Thus, a different approach can be applied which may provide the direct information by a single experiment. Labeling 1 μg of the total cellular DNA and hybridization with approximately 0.1 μg of the individual HPV DNAs blotted onto nitrocellulose allows the detection of ten genome equivalents per cell within the DNA of interest [5]. Differentiation between the closely related papillomavirus types (see above) is feasible since these DNAs will react with a different intensity.

The latter method is applicable as a routine screening test since the nitrocellulose filters can be prepared in advance. However, like the conventional Southern blotting procedure, it is rather time comsuming since it requires regular DNA extraction from the tumor material. Moreover, as stated above, the sensitivity of not much less than ten genome copies per cell might be insufficient when in the case of very small lesions the tumor cells are contaminated with normal tissue devoid of papillomavirus DNA. Such a situation usually occurs when cervical neoplasias have to be analyzed for the presence of HPV sequences.

As has been shown in previous studies, HPV types 6 or 11 can be detected in the majority of benign genital warts, whereas HPV 16 and 18 are mainly associated with malignant genital tumors, especially carcinomas of the cervix uteri [6]. Cervical dysplasias, which are considered putative preneoplastic lesions with a certain probability of progressing into more advanced cellular atypia, carcinoma in situ, and invasive cervical carcinoma proved to contain either HPV 6/11 or HPV 16/18 in approximately 60% of the cases tested by Southern blot analysis [6]. Typing of the HPV DNA might therefore be of diagnostic importance in order to evaluate – in combination with cytological and histological criteria – the potential of a given lesion to develop into a more malignant state.

A hybridization technique was established using epithelial cells gained from a routine cervical swab. The material is fixed onto a nitrocellulose filter, hybridized with ^{32}P-labeled HPV DNA, and exposed to an X-ray film as outlined in Fig. 1. The reliability of the test can be shown when parallel swabs taken subsequently from the same patient were hybridized with radiolabeled HPV DNA (Fig. 2). The results appear to be similar even in those materials showing only a slightly positive reaction (filters 59a–c in Fig. 2), arguing against unspecific binding of the radioactivity onto the filter. In order to eliminate the possibility of hybridization with pBR 322-like sequences, which might be present in the vaginal flora, the HPV DNA was cut off from the vector and purified by gel electropho-

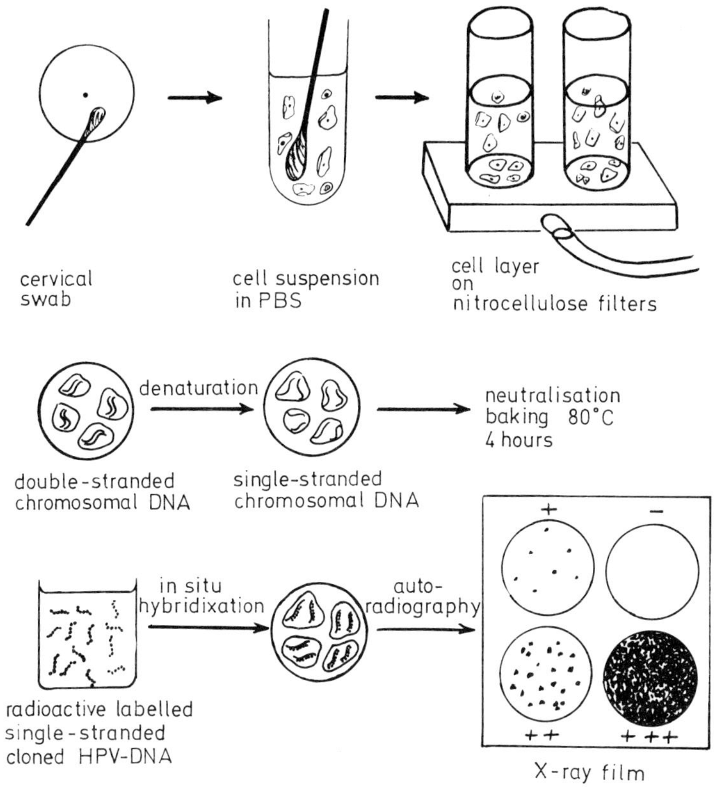

Fig. 1. In situ hybridization of cervical epithelial cells with radioactively labeled HPV DNA. The procedure is described in detail be Gissmann et al. and Wagner et al. [7, 8]

resis prior to labeling. This procedure usually gives rise to a more than 99% pure probe, as a 100- to 1,000-fold excess of pBR 322 compared with the respective HPV DNA is required to give an equal signal in a Southern blot hybridization. Retesting of positive filters with labeled pBR 322 DNA (data not shown) clearly indicated that no vector sequences were responsible for the reaction. The sensitivity of this test is hardly to determine. One can speculate, however, that the spots seen on filters 59a–c in Fig. 2 represent single cells containing papillomavirus DNA, since 10^5 molecules of HPV DNA can easily be detected in a spot hybridization test (data not shown) when distributed over nitrocellulose, corresponding to at least one hundred times the area of a cell. In addition, under the conditions of hybridization used in this test, the virus-specific RNA does participate in the reaction.

Screening a number of materials indicated that HPV 16 or 18 were present in approximately two-thirds of the cases which were diagnosed cytologically from parallel swabs as severe dysplasia or carcinoma in situ. These viruses, however, were never found in normal epithelial cells. The results are described elsewhere in more detail [7, 8]. In general, the same frequency of association between partic-

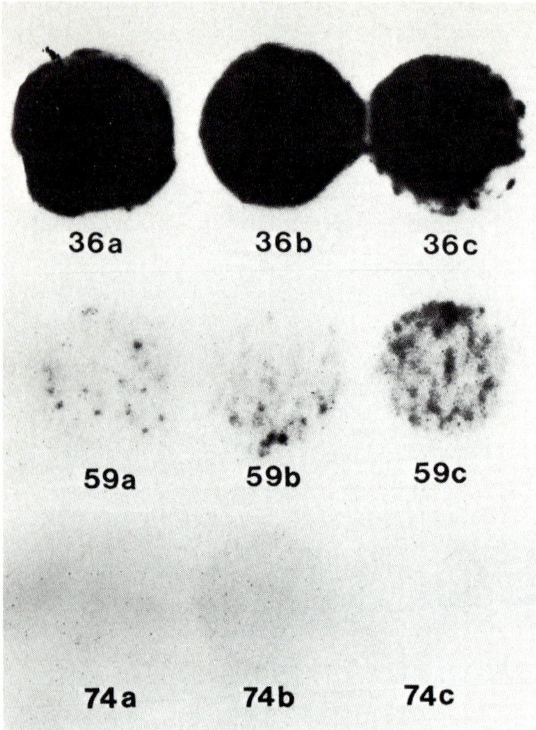

Fig. 2. Autoradiograph of cervical cells after hybridization with ^{32}P-labeled HPV 11 DNA. The cells were obtained and processed as outlined in Fig. 1. Parallel swabs (*a–c*) from the same patients show a similar reaction

ular HPV types and more or less malignant lesions was found as in Southern blot analysis with DNA extracted from biopsy materials. Therefore, in situ filter hybridization on epithelial cells may be used as a routine test by many laboratories, especially if the technology of DNA hybridization, involving nonradioactively labeled DNA is available.

References

1. Pfister H (1984) Biology and biochemistry of papillomaviruses. Rev Physiol Biochem Pharmacol 99:111–181
2. Coggin JR, zur Hausen H (1979) Workshop on papillomaviruses and cancer. Cancer Res 39:545–546
3. Kremsdorf D, Jablonska S, Favre M, Orth G (1983) Human papillomaviruses associated with epidermodysplasia verruciformis. II. Molecular cloning and biochemical characterization of human papillomavirus 3a, 8, 10 and 12 genomes. J Virol 48:340–351
4. Gissmann L, Wolnik L, Ikenberg H, Koldovsky U, Schnürch H-G, zur Hausen H (1983) Human papillomavirus types 6 and 11 sequences in genital and laryngeal papillomas and in some cancer biopsies. Proc Natl Acad Sci 80:560–563
5. Gissmann L, Schwarz E (1984) Cloning of papillomavirus DNA. In: Becker Y (ed) Development on molecular virology, vol 5. Recombinant DNA. Nijhoff, Hingham
6. Gissmann L (1984) Papillomaviruses and their association to cancer in animals and in man. Cancer Surveys (to be published)
7. Gissmann L, Boshart M, Dürst M, Ikenberg H, Wagner D, zur Hausen H (1984) Presence of human papillomavirus (HPV) DNA in genital tumors. J Invest Dermatol [Suppl] 83

8. Wagner D, Ikenberg H, Böhm N, Gissmann L (1984) Type specific identification of human papillomavirus in cells obtained from cervical swabs by DNA in situ hybridization. Obstet Gynecol 64:767–772
9. Kremsdorf D, Favre M, Jablonska S, Obalek S, Rueda LA, Lutzner MA, Blanchet-Bardon C, van Voorst Vader PC, Orth G (1984) Molecular cloning and characterization of the genome of nine newly recognized HPV types associated with epidermodysplasia verruciformis. J Virol 1013–1018

Detection of Epstein-Barr Virus DNA by Nucleic Acid Hybridization

G. W. Bornkamm [1], S. Wiest [1], and C. Desgranges [2]

Introduction

The progress made in modern biology during the last two decades is closely associated with the possibility of identifying specific sequences or genes by means of nucleic acid hybridization. Up to now the use of nucleic acid hybridization has been limited to experienced researchers usually working in well equipped laboratories on questions of basic research rather than diagnostics applied directly to the patient. As will be discussed below, this situation is now changing slowly and the possibilities of using nucleic acid hybridization as a widely applicable diagnostic tool is apparently emerging. Obviously the diagnostic application of nucleic acid hybridization is not limited to questions of clinical virology and may also be used in other fields of microbiology to identify specific infectious agents and their epidemiologic routes. In addition, nucleic acid hybridization is presently becoming a most important diagnostic tool for the prenatal diagnosis of hereditary diseases in man. The principles used for the diagnosis of hereditary diseases are essentially identical to those applied in the study of the epidemiologic traits of virus infections. Since we have recently reviewed the different types of hybridization technique (Bornkamm et al. 1983) we will concentrate here on some new aspects, particularly in the preparation of the labeled probe and the application of nucleic acid hybridization to EBV-associated diseases.

Preparation of the Labeled Probe

The detection of specific sequences by nucleic acid hybridization requires that a labeled DNA or RNA probe is available. Besides the widely used nick translation we will describe here some new developments in preparation of the labeled probe, some of which make use of specific properties of modern cloning vehicles.

1 Institut für Virologie, Zentrum für Hygiene der Universität Freiburg, Hermann-Herder-Str. 11, D-7800 Freiburg i. Br.

2 Laboratoire d'Epidemiologie et Immunovirologie des Tumeurs, Faculté Alexis Carrel, F-69372 Lyon Cedex 2

Nick Translation

DNA can be labeled to a specific activity of 5×10^7 to 2×10^8 cpm/μg making use of the so-called nick translation reaction of *E. coli* DNA polymerase I. If single-stranded nicks are introduced into the DNA by DNase, the *E. coli* DNA polymerase I starts at these nicks to remove nucleotides by its 5′-3′ exonuclease activity, while the polymerase activity replaces the removed nucleotides and thus shifts the single-stranded nick from 5′ to 3′ (nick translation). In the presence of labeled deoxyribonucleotide triphosphates the DNA is thus labeled to high specific activity (Rigby et al. 1977).

Primer Extension

If a probe is cloned into a vector, an oligonucleotide complementary to vector sequences close to the cloning site can be hybridized to the chimeric single-stranded molecule. Using this oligonucleotide as a primer, the second strand of the cloned insert may be synthesized by the large fragments of *E. coli* DNA polymerase I (Klenow enzyme) or by reverse transcriptase using labeled deoxyribonucleotide triphosphates. This technique allows not only the preparation of probes of very high specific activity (ca. 10^9 cpm/μg) but also allows the generation of probes which are specific for one or the other of both strands of the DNA (Hu and Messing 1982).

Synthesis of Complementary RNA from the SP6 Promoter

This technique combines several features, each of which may be advantageous in specific cases. Here the probe has to be cloned into a vector which carries the promoter used by phage SP6 RNA polymerase close to the cloning site. Using SP6 RNA polymerase and labeled ribonucleotide triphosphates, RNA complementary to one or the other of both strands of the cloned probe is synthesized at a high rate starting at the SP6 promoter (Butler and Chamberlin 1982). Since the RNA polymerase produces many copies of the same template, a large quantity of labeled probe may be prepared. The melting temperature of RNA–RNA hybrids is significantly higher than that of DNA–RNA hybrids, and more stringent and thus more specific hybridization conditions may therefore be used. This may be important particularly for in situ hybridization.

Biotin Labeling

All of these labeling techniques may be used with biotin-labeled deoxy- or ribonucleotide triphosphates instead of radioactively labeled nucleotides (Leary et al. 1983). Biotin-labeled DNA or RNA can then be used as a probe for nucleic acid hybridization. The hybridization reaction may be monitored using a complex of avidin and biotinylated polymers of alkaline phosphatase. Today it is still easier to achieve a high sensitivity with ^{32}P-labeled probes. However, since the technique is still being improved, it is possible that the same or an even higher sensitivity will be achieved routinely with biotin-labeled probes in the future. There would then apparently be little or no further need to use radioactively labeled hybridization probes. Another great advantage would obviously be the stability of the probe. The high stability would in addition facilitate the standardization of the hybridization procedure and enable the probes to be made available commercially.

Sensitive Detection of EBV DNA

Nowadays, hybridization reactions of high sensitivity are usually monitored by binding of a radioactively labeled probe to a nitrocellulose membrane followed by autoradiography using intensifying screens. Because of its higher sensitivity, autoradiography has almost completely replaced liquid scintillation counting to monitor the hybridization.

EBV DNA can be detected most sensitively by "Southern blot hybridization" (Southern 1975). DNA is extracted from the clinical specimens, digested with a restriction endonuclease, and the fragments separated by agarose gel electrophoresis. The separated fragments are denatured by alkali within the gel and then transferred onto a nitrocellulose filter. After hybridization with a labeled probe the filter is washed to remove the unbound probe and exposed to an X-ray film. For highest sensitivity we do not use the total viral genome as a probe, but rather a cloned 3.1-kb fragment, the large internal repeat, which is represented in the viral genome in many copies. Two assumptions are made if only a fragment of the viral genome is used as a hybridization probe: first, that EBV viruses do not exist lacking these sequences, and secondly, that the remainder of the viral sequences will also be present if the large internal repeat can be detected. Our knowledge of the structural organization of the EBV genome justifies these assumptions. Figure 1 shows schematically the structure of the viral genome, while Fig. 2 gives an example of a blot hybridization using the 3.1-kb repeat as a probe. The limit of detection is in the range of 0.1–1 pg of the large internal repeat and corresponds to an absolute number of 60,000–600,000 molecules (6,000–60,000 copies of the viral genome). Since the EBV genome carries about ten copies of the large internal repeat, and one tumor cell usually contains 5–50 copies of the viral genome, one EBV genome carrying tumor cell may thus be detected within 100–1,000 normal cells.

Filter hybridization can also be used to search for viral RNA sequences if RNA is separated on agarose-formaldehyde gels (Seed 1982) and transferred to nitrocellulose filters under appropriate conditions (Thomas 1980). For detection of RNA the term "Northern" is used, reminiscent of the work of Southern (1975).

The advantage of Southern and Northern blot hybridization is based on the fact that the binding of the labeled probe to the filter is confined to specific bands and thus provides an internal control for the high specificity of the binding. In the "dot" hybridization technique the advantage of specificity is abandoned for

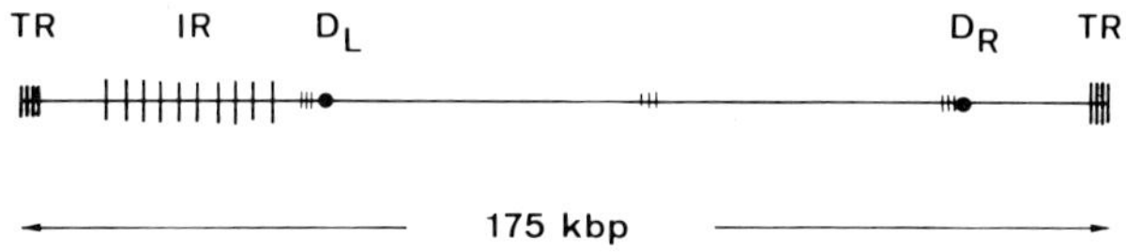

Fig. 1. Schematic diagram of EBV DNA. *TR*, terminal repeats; *IR*, internal repeats; D_L and D_R, left and right duplication. The genome contains additional arrays of tandem repeats

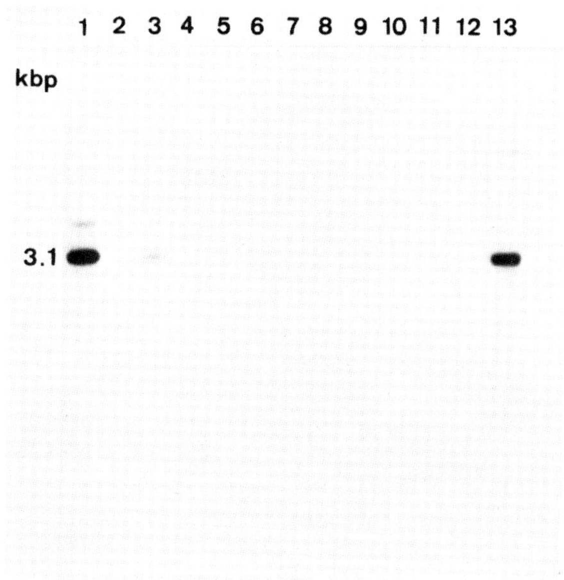

Fig. 2. Hybridization of ^{32}P-labeled large internal EBV repeat to tumor DNAs (lanes 5–13). The DNAs were digested with BglII, the fragments separated on a 0.6% agarose gel, denatured, and transferred to nitrocellulose, BJAB DNA served as negative control (lane 4). Lanes 1, 2, and 3 contained 50 pg, 0.5 pg, and 5 pg pSL9 DNA (B95-8 BamHI-W), respectively

the sake of simplicity in handling and processing a large number of specimens. For the detection of specific DNA or RNA sequences cells may be lysed and denatured directly on nitrocellulose filters before hybridization. Alternatively, cells may be lysed and denatured in suspension before being applied to the filter membrane. It has been reported that messenger RNAs can be specifically identified in this way if the conditions of cell lysis and adsorption to nitrocellulose are well controlled (Bresser et al. 1983). Because they are so simple these techniques are becoming more and more important for diagnostic purposes, also for viruses other than EBV.

In situ hybridization using labeled DNA or RNA probes followed by autoradiography may be used to search for viral sequences in touch smears or thin sections. This technique provides the advantage that the histology is simultaneously evaluated and that the sequences of interest can be assigned to specific target cells. In situ hybridization has been used to search for the site of EBV replication in the oropharynx (Wolf et al. 1981). Improvements of the techniques have also allowed to study the expression of specific genes in histologic sections (Hafen et al. 1983). If only a few grains can be observed, however, it may be difficult to be sure of the specificity of the in situ hybridization. These types of problems have been extensively faced by those who searched for HSV specific sequences in human tumors (for a review, see Galloway and Mc Dougall 1983).

Association of EBV with Malignant Diseases

Burkitt's Lymphoma

Since its original description (zur Hausen et al. 1970), the presence of EBV DNA in Burkitt's lymphoma and nasopharyngeal carcinoma cells has been widely confirmed (Nonoyama et al. 1973; Lindahl et al. 1974; Olweny et al. 1977; Geser et al. 1983). Up to now a total number of 125 African Burkitt's lymphoma cases have been carefully analysed by nucleic acid hybridization (Table 1), 120 (96%) of which were EBV DNA positive (Geser et al. 1983). The five EBV-negative cases were also well documented cases of Burkitt's lymphoma.

Burkitt's lymphoma occurs with a low incidence all over the world. Because of a different clinical presentation the incidence of the tumor outside Africa has certainly been underestimated. The non-African cases are usually characterized by abdominal manifestation (also a frequent symptom in African patients), while involvement of the jaws is only rarely observed outside Africa. In contrast to African Burkitt's lymphoma, only a small portion of the non-African cases are associated with EBV.

A compilation of all reports of Burkitt's lymphoma cases from outside Africa revealed that only 15%–20% of cases in low incidence areas are associated with EBV (Lenoir et al. 1983).

A more constant finding is the occurrence of characteristic chromosomal translocations in Burkitt lymphoma cells, regardless of whether the tumor developed in or outside Africa. Taking into account the incidence of Burkitt's lymphoma in and outside Africa, it becomes obvious that the EBV-negative cases have about the same incidence all over the world, while the high incidence of the tumor in Africa is somehow associated with the virus.

Table 1. Presence of EBV DNA in cases of Burkitt's lymphoma from Africa

Reference	No. of cases studied	No. of EBV-positive cases	Percentage of EBV-positive cases
zur Hausen et al. (1970)	10	10	100
Nonoyama et al. (1973)	20	19	95
Lindahl et al. (1974)	27	26	96
Olweny et al. (1977)	15	14	93
Geser et al. (1983)	53	51	96
Total	125	120	96

Association of EBV with Nasopharyngeal Carcinoma (NPC)

In contrast to Burkitt's lymphoma, where the virus appears to be an important risk factor (de Thé et al. 1978) but no conditio sine qua non for the development of the tumor, the importance of EBV for the occurrence of NPC has not been questioned.

So far all cases of nasopharangeal carcinoma of the undifferentiated type as well as some of the nonkeratinizing more differentiated tumors have been shown to carry the viral genome in many copies (zur Hausen et al. 1970; Nonoyama et al. 1973; Desgranges et al. 1975; Lanier et al. 1981; Andersson-Anvret et al. 1977, 1979).

Association of EBV with Other Malignant Diseases

In Alaska, which is also a high incidence area for NPC (Lanier et al. 1981), a number of cases of parotid tumors have been reported which histologically resemble NPC (Lanier et al. 1981; Krishnamurthy et al. 1984). In none of these cases was a primary tumor in the nasopharynx found. Using nuclear acid hybridization, EBV DNA was detected in these tumors but not in control tissue, suggesting that this type of parotid tumor is a new EBV-associated tumor.

The association of EBV with another malignant condition is now well established. Purtilo described an X-linked immunodeficiency syndrome which leads to fatal lymphoproliferation after primary infection with EBV. A similar fatal lymphoproliferative disorder has been observed in renal transplant recipients (for a review, see Purtilo 1981). Again, the tumor cells harbor the EBV genome, as revealed by nucleic acid hybridization, and express the viral nuclear antigen EBNA (Saemundsen et al. 1981; Hanto et al. 1981). In contrast to Burkitt's lymphoma, the tumor cells do not show the characteristic chromosomal translocations and may be polyclonal in origin. The fatal lymphoproliferation observed in immunodeficient conditions is highly reminiscent of the fatal lymphoproliferative disease induced in new world monkeys by EBV.

Identification of Epidemiologic Traits by Nucleic Acid Hybridization

The degeneration of the genetic code allows for some changes in the nucleotide sequence without changing the primary structure of the coded proteins. Since there is no selection against such mutations, they will remain in the population and will be inherited as a Mendelian trait once they have appeared. Using such mutations as genetic markers it is thus easily possible to trace genetic descendance. These mutations can be visualized if the recognition site of a restriction endonuclease is altered and the size of the restriction enzyme fragment is changed (restriction enzyme length polymorphism). If a molecularly cloned probe for the fragment is available, the change in the size of the fragment can be visualized by

Southern blot hybridization using the cloned fragment as a hybridization probe. This technique is beginning to be widely used for the prenatal diagnosis of hereditary diseases in man. Similarly, it is possible to discriminate any wild type virus from others and to follow epidemiologic routes.

We have used this approach to investigate the question of whether the EBV in Burkitt's lymphoma tumor cells is identical or different from that found in the nonmalignant cells of the same patients. Studying three patients we were able to show that the virus in the tumor cells is identical to the one in the nonmalignant cells of the same patients (Bornkamm et al. 1984).

Nucleic Acid Hybridization as a Tool for Early Diagnosis of NPC?

Patients with NPC usually develop anti-EBV IgA titers in their sera before the clinical onset of the disease. The search for anti-EBV IgA antibodies may thus be used to identify persons with high risk for NPC in the high incidence area in South China. In some cases in which biopsies from anti-VCA IgA positive individuals were examined, the detection of EBV DNA has helped to establish the diagnosis of NPC early in the course of the disease (Desgranges et al. 1982).

We have therefore attempted to evaluate nucleic acid hybridization as a diagnostic tool in patients at high risk for NPC. Specimens with a sufficient quantity of cells can easily be collected from the oropharynx of the patient using a negative pressure suction apparatus described by the Zhangiang Medical College (1976) and Zeng et al. (1980). Cells are suctioned from the surface of the mucosa and collected on a nylon net. The cells may then be examined cytologically, stained for EBNA or other antigens, and finally extracted for nucleic acid hybridization.

Results obtained on exfoliated cells of anti-IgA positive and negative individuals studied over a period of 1 year by dot hybridization were not, however, too promising, since there was no good correlation between detectability of EBV sequences and serology (Desgranges et al. 1983). A final evaluation nevertheless requires careful follow-up of a sufficient number of patients over a period of several years.

Concluding Remarks

Nucleic acid hybridization has been of great importance in associating EBV with specific malignant diseases, even though the role of the virus in these diseases is not understood. In the past the use of nucleic acid hybridization for the detection of EBV DNA sequences has been limited to a few specialized institutions which have the experience and the equipment to carry out the experiments. Provided that nucleic acid hybridization turns out to be useful for early diagnosis of NPC, it will be most important to adapt nucleic acid hybridization to the field in the high incidence area in China. The availability of a biotin-labeled hybridization probe will be a prerequisite for this endeavor. Provided a well standardized protocol has been worked out, it should be possible to carry out dot hybridizations with a biotin-labeled probe in the laboratory of any hospital of the world.

Acknowledgement. Work reported here was supported by Die Deutsche Krebshilfe (Diagnostik und Differenzierung EBV-assoziierter Tumoren) and by Die Deutsche Forschungsgemeinschaft (SFB 31).

References

Andersson-Anvret M, Forsby N, Klein G, Henle W (1977) Relationship between the Epstein-Barr virus and undifferentiated nasopharyngeal carcinoma: correlation between nucleic acid hybridization and histopathological examination. Int J Cancer 20:486–494

Andersson-Anvret M, Forsby N, Klein G, Henle W, Biörklund A (1979) Relationship between the Epstein-Barr virus genome and nasopharyngeal carcinoma in Caucasian patients. Int J Cancer 23:762–767

Bornkamm GW, Desgranges C, Gissmann L (1983) Nucleic acid hybridization for the detection of viral genomes. Curr Top Microbiol Immunol 104:287–298

Bornkamm GW, von Knebel Doeberitz M, Lenoir G (1984) No evidence for differences in the Epstein-Barr virus genome carried in Burkitt lymphoma cells and nonmalignant lymphoblastoid cells from the same patients. Proc Natl Acad Sci USA (in press)

Bresser J, Hubbell HR, Gillespie D (1983) Biological activity of m-RNA immobilized on nitrocellulose in NaI. Proc Natl Acad Sci USA 80:6523–6527

Butler ET, Chamberlin MJ (1982) Bacteriophage SP6-specific RNA polymerase: isolation and characterization of the enzyme. J Biol Chem 257:5772–5778

Desgranges C, Wolf H, de Thé G, Shanmugaratnam K, Cammoun N, Ellouz R, Klein G, Lennert K, Munoz N, zur Hausen H (1975) Nasopharyngeal carcinoma. X. Presence of Epstein-Barr genomes in separated epithelial cells of tumours in patients from Singapore, Tunisia and Kenya. Int J Cancer 16:7–15

Desgranges C, Bornkamm GW, Zeng Y, Wang PC, Zhu JS, Shang M, de Thé G (1982) Detection of Epstein-Barr viral DNA internal repeats in the nasopharyngeal mucosa of Chinese with IgA/EBV-specific antibodies. Int J Cancer 29:87–91

Desgranges C, Pi GH, Bornkamm GW, Legrand C, Zeng Y, de Thé G (1983) Presence of EBV-DNA sequences in nasopharyngeal cells of individuals without IgA-VCA antibodies. Int J Cancer 32:543–545

de Thé G, Geser A, Day NE, Tukei PM, Williams EH, Beri DP, Smith PG, Dean AG, Bornkamm GW, Feorino P, Henle W (1978) Epidemiological evidence for causal relationship between Epstein-Barr virus and Burkitt's lymphoma from Ugandan prospective study. Nature 274:756–761

Galloway D, Mc Dougall JK (1983) The oncogenic potential of herpes simplex viruses: evidence for a "hit and run" mechanism. Nature 302:21–24

Geser A, Lenoir GM, Anvret M, Bornkamm GW, Klein G, Williams EH, Wright DH, de Thé G (1983) Epstein-Barr virus markers in a series of Burkitt's lymphomas from the West Nile district, Uganda. Eur J Cancer Clin Oncol 19:1393–1404

Hafen E, Levine M, Garber RL, Gehring WJ (1983) An improved in situ hybridization method for the detection of cellular RNAs in *Drosophila* tissue sections and its application for localizing transcripts of the homeotic Antenna-pedia gene complex. EMBO J 2:617–623

Hanto DW, Frizzera G, Purtilo D, Sakamoto K, Sullivan JL, Saemundsen AK, Klein G, Simmons RL, Najarian JS (1981) Clinical spectrum of lymphoproliferative disorders in renal transplant recipients and evidence for the role of Epstein-Barr virus. Cancer Res 41:4253–4261

Hu N, Messing J (1982) The making of strand-specific M 13 probes. Gene 17:271–277

Krishnamurthy S, Lanier AP, Clift SE, Kline KT, Bornkamm GW, Henle W, Gown A, Thorning D (1984) Malignant lymphoepithelial lesion of the parotid gland: pathologic and Epstein-Barr viral studies. (submitted for publication)

Lanier AP, Bornkamm GW, Henle W, Henle G, Bender TR, Talbot ML, Dohan P (1981) Association of Epstein-Barr virus with nasopharyngeal carcinoma in Alaskan native patients: serum antibodies and tissue EBNA and DNA. Int J Cancer 28:301–305

Leary JJ, Brigati DJ, Ward DC (1983) Rapid and sensitive calorimetric method for visualizing biotin-labeled DNA probes hybridized to DNA or RNA immobilized on nitrocellulose: Bio-blots. Proc Natl Acad Sci USA 80:4045–4049

Lenoir GM, Philip T, Sohier R (1983) Burkitt's type lymphoma: EBV association and cytogenetic markers in cases from various geographic origins. In: Magrath IT, O'Conor GT, Ramot B (eds) Environmental influences in the pathogenesis of leukemias and lymphomas. Raven, New York, pp 283–295

Lindahl T, Klein G, Reedman BB, Johansson B, Singh S (1974) Relationship between Epstein-Barr virus (EBV) DNA and the EBV-determined nuclear antigen (EBNA) in Burkitt lymphoma biopsies and other lymphoproliferative malignancies. Int J Cancer 13:764–772

Nonoyama M, Huang CH, Pagano JS, Klein G, Singh S (1983) DNA of Epstein-Barr virus detected in tissue of Burkitt's lymphoma and nasopharyngeal carcinoma. Proc Natl Acad Sci USA 70:3265–3268

Olweny CL, Atine I, Kaddu-Muleasa A, Owor R, Andersson-Anvret M, Klein G, Henle W, de Thé G (1977) Epstein-Barr virus genome studies in Burkitt's and non-Burkitt's lymphomas in Uganda. J Natl Cancer Inst 58:1191–1196

Purtilo D (1981) Malignant lymphoproliferative diseases induced by Epstein-Barr virus in immunodeficient patients, including X-linked cytogenetic and familial syndromes. Cancer Genet Cytogenet 4:251–268

Rigby RW, Dieckmann M, Rhodes C, Berg P (1977) Labelling deoxyribonucleic acid to high specific activity in vitro by nick translation with DNA polymerase I. J Mol Biol 113:237–251

Saemundsen AK, Purtilo D, Sakemoto K, Sullivan JL, Synnerholm AC, Hanto D, Simmons R, Anvret M, Collins R, Klein G (1981) Documentation of Epstein-Barr virus infection in immunodeficient patients with life-threatening lymphoproliferative diseases by Epstein-Barr virus complementary RNA/DNA and viral DNA/DNA hybridization. Cancer Res 41:4237–4242

Seed B (1982) Diazotizable arylamine cellulose papers for the coupling and hybridization of nucleic acids. Nucleic Acids Res 10:1799–1810

Southern E (1975) Detection of specific sequences among DNA fragments separated by gel electrophoresis. J Mol Biol 98:503–517

Thomas PS (1980) Hybridization of denatured RNA and small DNA fragments transferred to nitrocellulose. Proc Natl Acad Sci USA 77:5201–5205

Wolf H, Bayliss GJ, Wilmes E (1981) Biological properties of Epstein-Barr virus. In: Grundmann E, Krueger GRF, Ablashi DV (eds) Nasopharyngeal carcinoma. Cancer campaign, vol 5. Fischer, Stuttgart, pp 101–109

Zeng Y, Liu UX, Liu CR, Chen SW, Wei JN, Zhu JS, Zai HJ (1980) Application of immunoenzymatic method and immuno-autoradiographic method for the mass survey of nasopharyngeal carcinoma. Intervirology 13:162–168

Zhangiang Medical College (1976) Diagnosis of nasopharyngeal carcinoma by cytological examination of exfoliated cells taken by negative pressure suction. Chin Med J 1:45–47

zur Hausen H, Schulte-Holthausen H, Klein G, Henle W, Henle G, Clifford P, Santesson L (1970) EBV-DNA in biopsies of Burkitt's tumors and anaplastic carcinomas of the nasopharynx. Nature 228:1056–1058

Diagnosis of Viroid-Caused Plant Diseases by Nucleic Acid Hybridization

T. O. DIENER and R. A. OWENS [1]

Introduction

In agricultural practice, control of viroid (or virus)-incited plant diseases relies mainly on prevention rather than cure. An economically feasible and environmentally safe chemical treatment has not yet been devised for the elimination of viroids (or viruses) from infected plants. Preventive measures include the growing of crops from viroid (or virus)-free seeds or stocks and prevention of these pathogens from entering and spreading through crops (Gibbs and Harrison 1976). In these procedures suitable diagnostic tests for the rapid and reliable detection of viroids (and viruses) are of paramount importance.

In recent years, diagnostic tests based on enzyme-linked immunosorbent assays (ELISA) have been developed for the detection of a number of important plant viruses and several of these tests are now in widespread use. For example, a sensitive automated ELISA test for the simultaneous detection in potato tubers of several important viruses has been developed (Gugerli 1979). This procedure permits sampling of up to 300 tubers per hour.

Because viroids lack the antigenic protein coat characteristic of viruses, assays for viroids that are based on immunological principles, such as ELISA, have not been reported. Two types of diagnostic test have been used in the past to detect the potato spindle tuber viroid (PSTV): bioassay on suitable tomato cultivars (Raymer and O'Brien 1962; Fernow 1967) and polyacrylamide gel electrophoresis (PAGE) of extracted nucleic acids (e.g., Morris and Wright 1975). Although Fernow et al. (1969) have demonstrated that a double-inoculation technique can be used to detect both mild and severe strains of PSTV in potato seed stocks, bioassays on tomato are slow and are unreliable unless specific environmental conditions are maintained. Detection by PAGE, on the other hand, is laborious and expensive. Therefore neither method is suitable for the rapid screening of thousands of seed potato tubers that is so necessary in maintaining suitable seed stocks.

One promising alternative to the above methods is hybridization of highly radioactive DNA complementary to PSTV (PSTV cDNA) with PSTV bound to a solid support and with autoradiographic detection of the resulting DNA-RNA hybrids. Hybridization in solution has been used to detect the presence of viroids in purified RNA preparations (Owens et al. 1978; Palukaitis et al. 1979). Two

1 Plant Virology Laboratory, Agricultural Research Service, U.S. Department of Agriculture, Beltsville, MD 20705, USA

conditions had to be fulfilled before hybridization with PSTV cDNA could be developed into a practical diagnostic test for the detection of PSTV in numerous samples: (1) PSTV cDNA had to be available in unlimited quantities and at high specific radioactivity, requiring cloning of the cDNA by recombinant DNA technology and nick translation of the DNA; and (2) to expedite and simplify sample preparation, clarified sap rather than purified nucleic acid had to be suitable as the viroid source. Both conditions have been fulfilled and a practical diagnostic test for the detection of PSTV based on nucleic acid hybridization has been developed (Owens and Diener 1981).

In this chapter we first describe the nucleic acid dot hybridization method, then present evidence of its sensitivity, specificity, and reliability, and finally examine the potential impact of this novel diagnostic method on practical agriculture – on commercial potato production and in its application to viroid-incited diseases of other crops. Finally, we consider possible uses of similar dot hybridization procedures for purposes other than viroid detection.

Molecular Cloning of PSTV cDNA

Determination of the 359 nucleotide sequence of PSTV established viroids as the first naturally occurring examples of covalently closed circular RNA (Gross et al. 1978). Knowledge of the RNA sequence facilitated the molecular cloning and characterization of double-stranded (ds) PSTV cDNA (Owens and Cress 1980).

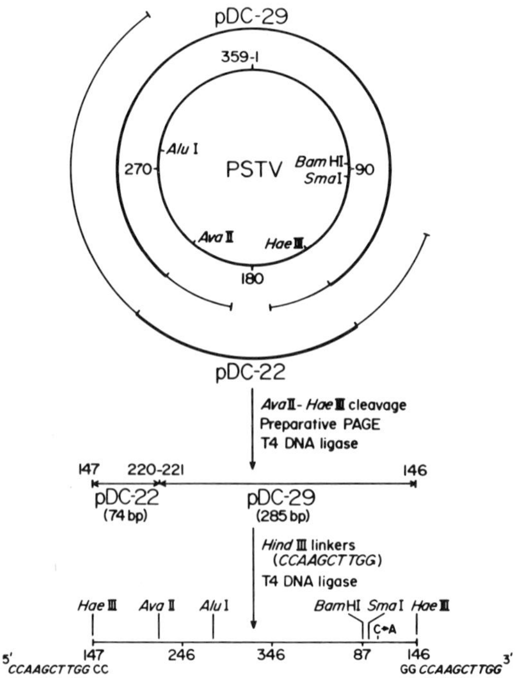

Fig. 1. Construction and molecular cloning of full-length dsPSTV cDNA

PSTV Detection by Nucleic Acid Hybridization

Tissue Sample

↓ Homogenize in 200 mM K_2HPO_4
10 mM DIECA
5 mM DTT
0.1% Triton X-100
Clinical centrifugation

Clarified Sap

↓ Transfer 3–5 μl aliquot to nitrocellulose membrane
Bake 2 hrs 80 °C *in vacuo*

PSTV Bound to Nitrocellulose Membrane

↓ Hybridization with [^{32}P] recombinat DNA
Wash 55 °C

[^{32}P] DNA · PSTV Hybrids Bound to Nitrocellulose Membrane

↓ X-ray film
Intensifying screen

Detection of Hybrids by Autoradiography

Fig. 2. Outline of nucleic acid hybridization procedure. *DIECA*, diethyldithiocarbamate; *DTT*, dithiothreitol

In the latter work, ds PSTV cDNA was synthesized from a polyadenylated linear PSTV template and inserted in the *Pst*I endonuclease site of plasmid pBR322 using the oligo (dC) · oligo(dG) tailing procedure. Although one recombinant clone (pDC-29) contained a larger than expected 460 base-pair (bp) insert, restriction endonuclease mapping and nucleotide sequence determinations showed that all recombinants contained less than a complete copy of PSTV.

Figure 1 shows the strategy used to construct clones containing full-length dsPSTV cDNA. Restriction analysis of the PSTV-specific inserts from two of these clones, pDC-29 and pDC-22, suggested that these clones were partially overlapping and could be used to construct a full-length clone. The 285 bp *Ava*II-*Hae*III fragment from pDC-22 was ligated at the *Ava*II site to the contiguous 74 bp *AvaII-Hae*III fragment from pDC-22. *Hind*III oligodeoxy-nucleotide linkers were added to the *Hae*III blunt ends of this linear ligation product and, following digestion with *Hind*III, this fragment was cloned in the *Hind*III site of pBR322. *Hind*III digestion of the resulting clones released the expected 365 bp PSTV-specific fragment (Cress et al. 1983).

Because use of this particular *Hind*III decanucleotide linker reconstructed the terminal *Hae*III sites of the PSTV insert, digestion of the recombinant DNA with *Hae*III releases a 359 bp fragment. This fragment is a full-length dsPSTV cDNA. Determination of the complete nucleotide sequence of this DNA by a combination of the Maxam-Gilbert chemical degradation and M13 dideoxy chain termination methodologies has shown the cloned DNA sequence to be identical with that predicted by the published RNA sequence (Cress et al. 1983).

PSTV Diagnosis by Nucleic Acid Dot Hybridization

Figure 2 summarizes the sample preparation and test procedure used. A relatively high ionic strength and diethyldithiocarbamate concentration of the extraction buffer serve to release PSTV from nuclei and inhibit enzymatic polyphenol oxidation, respectively. Each sprout or seed sample (0.1–0.2 g) is homogenized at room temperature in a small conical ground-glass homogenizer in the extraction buffer shown (1.5 ml per gram of tissue or 0.2 ml per 16 true seeds). Tissue debris is removed by brief low-speed centrifugation, and the sap is used immediately. Three- to five-microliter portions of the sap samples to be tested are pipetted onto a treated nitrocellulose membrane (Thomas 1980) and then baked as indicated in Fig. 2. Although RNA covalently binds to diazobenzyloxymethyl (DBM) paper (Alwine et al. 1979), this support is expensive and requires a number of carefully controlled steps to prepare and use. Because RNA also stably binds to a nitrocellulose membrane (Thomas 1980), and because this method is simple, inexpensive, and reproducible, nitrocellulose rather than DBM paper is used.

Hybridization reactions are performed essentially as described (Wahl et al. 1979), but we have found that the prehybridization reaction can be omitted. The same buffer [40% (v/v) formamide, 0.18 M NaCl, 10 mM sodium cacodylate. 1 mM EDTA, 0.1% sodium dodecyl sulfate, and yeast transfer RNA (400 μg/ml) at pH 7.0] is used for both hybridization and briefly wetting the membrane (5–10 min at 20°–25 °C) before hybridization. The hybridization reaction is performed for 24 h at 55 °C in the presence of 10% dextran sulfate and ^{32}P-labeled nick-translated recombinant PSTV cDNA ($1–2.5 \times 10^6$ cpm/ml). The ratio of buffer volume to membrane area is at least 1 ml per 35 cm^2. Nick-translated DNA is denatured by heating for 2 min at 100 °C in the presence of 50% formamide before addition to the hybridization reaction. Nitrocellulose membranes are washed at 55 °C with five changes of 0.36 M NaCl, 10 mM tris-HCl (ph 7.5), and 0.1% sodium dodecyl sulfate and then with two changes of the same buffer diluted tenfold. Autoradiography is carried out for 24–48 h at −70 °C with Kodak X-Omat film [2] and Dupont Cronex Lightning-Plus Intensifying Screens.

Evaluation of Nucleic Acid Dot Hybridization Test

Tuber Sprouts from Commercial Varieties

The feasibility and reliability of this method for the rapid and sensitive detection of PSTV by nucleic acid hybridization has been demonstrated with tubers harvested from commercially important U.S. potato varieties that had been intentionally infected with PSTV (Owens and Diener 1981).

PSTV was shown to bind stably to either DBM paper or nitrocellulose membrane (Fig. 3). Comparison of relative autoradiographic intensities showed that the presence of sap from uninfected tuber sprouts reduces the binding approxi-

2 Mention of a commercial company or specific equipment does not constitute its endorsement by the U.S. Department of Agriculture over similiar equipment or companies not named

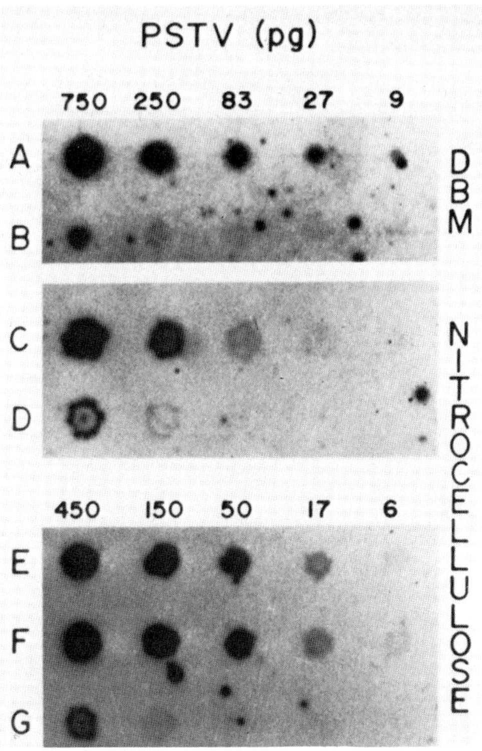

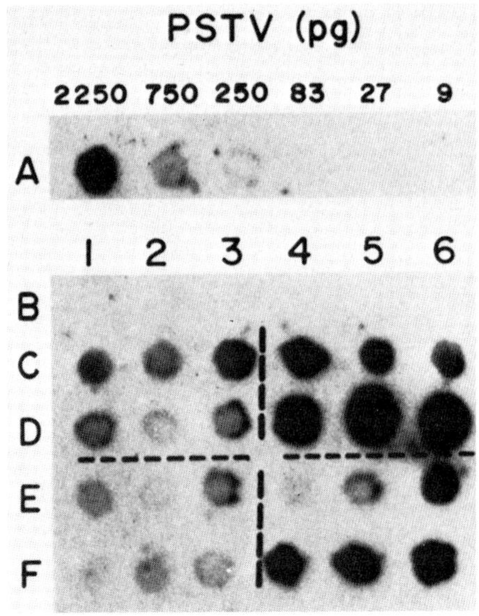

Fig. 4.

◄ **Fig. 3**

Fig. 3. Sensitivity of PSTV detection by hybridization and autoradiography after binding to DBM paper and nitrocellulose membranes. Low molecular weight RNA containing 0.4–0.5% PSTV was isolated from PSTV-infected tomato seedlings and diluted with a mixture of 0.15 *M* sodium acetate and 0.85 *M* acetic acid (*rows A and C*), sap prepared from healthy Katahdin sprouts (*rows B, D, and G*), water (*row E*), or homogenization buffer (*row F*). Portions of successive threefold dilutions were then transferred to DBM paper (*rows A and B*) or nitrocellulose membranes (*rows C–G*)

Fig. 4. Reliability of PSTV detection by hybridization and autoradiography after binding to nitrocellulose membrane. *Row A:* Successive threefold dilutions of low molecular weight RNA containing PSTV that was diluted with sap prepared from healthy Katahdin sprouts; the sample in the first column did not contain sap. *Row B, left to right:* Composite sap samples prepared from sprouts taken from several healthy Chieftain, Irish Cobbler, Green Mountain, Katahdin, Kennebec, and Saco tubers. *Rows C and D:* Sap samples prepared from sprouts taken from individual PSTV-infected tubers; *row C, columns 1–3* from Katahdin, *columns 4–6* from Kennebec; *row D, columns 1–3* from Saco, *columns 4–6* from Katahdin line 235. *Rows E and F:* Composite sap samples, each prepared from tissue taken from three individual PSTV-infected tubers; *columns 1 and 4,* epidermis from between eyes; *columns 2 and 5,* barely sprouted eyes (≦2 mm); *columns 3 and 6,* sprouts (≧5 mm). *Row E, columns 1–3* from Katahdin, *columns 4–6* from Kennebec; *row F, columns 1–3* from Saco, columns 4–6 from Katahdin line 235

mately tenfold, but 83–250 pg PSTV are still easily detected after hybridization with radioactive recombinant DNA (Fig. 3). Additional experience with the methodology has shown, however, that the limit of detection is approximately tenfold lower (< 10 pg PSTV/spot). This amount is equivalent to a concentration of about 0.005 μg PSTV per gram of tuber sprouts. Actively growing potato tissue contains ≥ 0.5 µg PSTV per gram of tissue (Morris and Wright 1975). Our hybridization method is therefore adequate for detecting PSTV in potato tissue. The relatively high ionic strength and diethyldithiocarbamate concentration of the extraction buffer required to release PSTV from nuclei and inhibit enzymatic polyphenol oxidation do not interfere with PSTV binding to nitrocellulose (Owens and Diener 1981). No reaction was detected with sap prepared from healthy tubers of six commercial varieties (Fig. 4).

Tuber Sprouts from CIP Potato Clones

Recently, we have extended our studies in collaboration with Dr. L. F. Salazar of the Centro International de la Papa (CIP), Lima, Peru.

Tubers from 20 potato clones currently under investigation at CIP were selected for testing by nucleic acid hybridization at Beltsville. Most of the selections had been tested at CIP by PAGE, but the results of these analyses were not revealed to us at the Beltsville Agricultural Research Center until the spot hybridization tests had been completed.

Of 20 CIP potato clones selected, PAGE analysis of nucleic acid extracts from tuber sprouts and/or foliage had disclosed PSTV in ten. One clone had not been tested at CIP, and the remaining clones had given negative results. In contrast to PAGE analysis, nucleic acid hybridization with tuber sprout extracts gave positive results with 17 clones, a weakly positive result with one clone, and negative results with only two clones.

As expected, all clones that had tested positive by PAGE, also tested positive by nucleic acid hybridization. However, eight additional clones tested positive in the dot hybridization test. The question thus arose of whether these additional positive results were due to false-positive reactions or to increased sensitivity of the dot hybridization test as compared with PAGE.

Reevaluation of these initial results by bioassay on tomato, followed by dot hybridization with tissue from the inoculated tomato plants, confirmed the presence of PSTV in all but one of the plants that had tested negative by PAGE analysis but positive in the dot hybridization assay (Salazar et al. 1983).

True Potato Seed

Major efforts are underway to propagate potatoes by true seed (Upadhya 1979). Because PSTV is both seed and pollen transmitted (Fernow et al. 1970), it is important to ascertain that seed is free of PSTV. Ideally, a test for this purpose should be able to detect PSTV in a composite seed sample even if only one or a few percent of the seeds in the sample are infected. To determine whether the nucleic acid dot hybridization test is suitable for this purpose, true seeds were collected from potato plants known to be free of PSTV and from others known to be infected. Seeds from healthy and infected plants were allowed to germinate for 5 days in the dark, were mixed in various proportions, and were then homogenized in extraction buffer.

Conventional diagnosis using tomato bioassay failed to detect PSTV in any of the composite samples, including one sample that consisted of seeds from infected plants only. Although none of the inoculated tomato plants showed symptoms, PAGE analysis of extracted nucleic acid demonstrated that PSTV was present in those tomato plants that had been inoculated with seed extracts of samples containing at least 25% seeds from PSTV-infected potato plants (Salazar et al. 1983).

Other portions of the composite seed samples were analyzed by the nucleic acid dot hybridization test. Positive results were obtained with all but one of the samples containing seed from infected potato plants, including a sample in which the ratio of seed from infected plants to seed from healthy plants was 1:15 (Salazar et al. 1983).

Conclusions

Our data demonstrate that the nucleic acid dot hybridization test can reliably detect both severe and mild strains of PSTV in crude extracts from potato tuber sprouts. The test is specific and, in our hands, never results in false-positive indications.

The dot hybridization test is also suitable for detecting PSTV in true potato seed. In composite samples consisting of seeds from both PSTV-infected and healthy potato plants, PSTV could be detected when as few as one in 16 seeds had originated from an infected plant. In view of our limited efforts with true seed testing it appears likely that further work may result in detection of still lower levels of PSTV. Although transmission of PSTV through the seed from open-pollinated female parents occurs frequently, it may vary in individual collections from zero to 100 percent (Fernow et al. 1969).

The nucleic acid dot hybridization test is at least 10 times more sensitive than PAGE. Furthermore, results indicate that the test may also be more sensitive than bioassay on tomato. This fact is most clearly evident in the case of the composite seed samples. Apparently, PSTV concentration in these samples, although sufficient to infect some of the assay plants (as shown by subsequent PAGE analysis), was not high enough to lead to recognizable symptoms in the test plants. The dot hybridization test, on the other hand, readily detected PSTV in all but one of the very same composite seed samples.

One note of caution is indicated. Our results demonstrate that it may not be sufficient to test sprouts from only one tuber of a clone. With two clones one tuber tested negative in the hybridization test (in seeming confirmation of earlier results obtained by PAGE), but when sprouts from a second tuber of each clone were analyzed, positive reactions were obtained (Salazar et al. 1983). We conclude that PSTV may be unevenly distributed in plants, giving rise to both PSTV-free and PSTV-infected tubers or tuber portions. Conceivably, such plants have become infected recently, and the viroid has neither accumulated to any large extent nor spread evenly throughout the plant. At this stage, PAGE analysis might not detect PSTV, but, because of its higher sensitivity, dot hybridization might.

Although the dot hybridization test takes 4 days to complete, no manipulations are required during much of this time. Thus, many samples may be applied to one membrane and be processed simultaneously and multiple membranes may be run sequentially at brief intervals. Also, automation of sample preparation (Gugerli 1979) may further reduce the amount of labor required. The nucleic acid dot hybridization test is consequently suitable for the reliable assay of large numbers of samples for the presence of PSTV.

Impact on Agriculture

Practical Consequences of PSTV Infection

The potato spindle tuber disease poses a potentially serious threat in seed production, germ plasm collections, and cultivar development (reviewed by Diener 1979). PSTV is transmitted through vegetative propagation, foliar contact, and true seed and pollen.

Tomato has been used as a diagnostic host for PSTV because symptoms in some potato cultivars are indistinct. A number of independently isolated PSTV strains have been described and can be classified as either severe or mild strains on the basis of the symptoms produced in Rutgers tomato. Severe strains cause extreme shortening of the internodes, severe epinasty, shortening of petioles and midribs, and necrosis of stems, petioles and midribs in tomatoes. Symptoms of mild strains, on the other hand, are slow to develop and are often so mild that they are easily overlooked (Fernow 1967).

Depending on the potato cultivar, the strain of PSTV present, and environmental conditions, symptoms of PSTV infection in potato may vary considerably. Foliage symptoms are often obscure, and the severity of the characteristic tuber symptoms – elongation with the appearance of prominent bud scales and growth cracks – depends upon temperature and length of infection. While the disease causes neither total crop destruction nor storage losses, it can cause a serious reduction in total production. As long as control is effective in keeping the incidence of disease low, yield losses are of little consequence in temperate growing areas (1%–2%). Inadequate control, however, can lead to catastrophic losses (~20% for mild strains or ~60% for severe strains) in a relatively short time (reviewed by Diener 1979).

When potatoes are grown at high temperatures, PSTV causes severe damage and total crop loss may ensue. Because of its importance as a food source, major efforts are underway to supplement the world's food supply by adapting the potato to growth in subtropical and tropical climates, and a lowland tropical potato cultivar has already been developed (Sawyer 1979). Under these environmental conditions, PSTV poses a potentially serious threat and the importance of a sensitive, rapid, and reliable method for PSTV diagnosis is readily apparent. This is true regardless of whether potatoes are propagated vegetatively or by true seed.

Because, at present, potatoes are vegetatively propagated, virus diseases are a major problem in potato production, and practical means for the exclusion of viruses from "seed" potatoes are a necessity. A sensitive automated method for the simultaneous detection in potato tubers of several important viruses has been developed (Gugerli 1979). This method is based on an enzyme-linked immunosorbent assay; it permits sampling of up to 300 tubers per hour. With the availability of the nucleic acid dot hybridization test, "seed" potato tubers can now be tested simultaneously for the presence of important viruses and for PSTV. Thus, the danger of introducing PSTV into commercial potato fields should become drastically reduced.

Application to Other Viroid Diseases

Diagnostic tests paralleling the one developed for the detection of PSTV in potato may be developed for the diagnosis of economically important diseases caused by other viroids. Because of the considerable sequence homologies of certain viroids with PSTV (Visvader et al. 1982), it may even be possible to use the PSTV probe for their detection also. This applies particularly to the detection of chrysanthemum stunt and citrus exocortis viroids, but the development of specific probes will probably yield more reliable results.

Diagnostic tests are most urgently needed with vegetatively propagated crops that are susceptible to viroid-induced diseases, such as chrysanthemum (stunt and chlorotic mottle viroids), hops (stunt viroid), citrus (exocortis viroid), and avocado (sunblotch viroid). In each case, development of a nucleic acid dot hybridization test requires isolation and purification of the respective viroid, preparation of specific cDNA, and cloning of the cDNA. Also, use of crude extracts from the various hosts as the viroid source may require modification of the extraction buffer, because these host plants, in contrast to tomato and potato, may contain substances that interfere with the hybridization reaction.

Development of specific dot hybridization tests for these viroid diseases would be advantageous, because such tests could replace presently used but far less satisfactory diagnostic procedures that are based on bioassay in suitable indicator plants or analysis by PAGE. As with PSTV, bioassay often is exceedingly slow and requires extensive greenhouse space, whereas PAGE analysis is prohibitively expensive for most applications.

Nucleic acid dot hybridization tests may also become useful for the early detection of viroid infection in young seedlings. One case in point is the coconut cadang-cadang disease. Because of the great sensitivity of the hybridization test, it is possible that routine testing of seedlings in new plantations could identify infected palm trees long before symptoms appear or before the cadang-cadang viroid would become detectable by PAGE analysis. Early roguing of infected palm trees might materially diminish tree-to-tree spread of the viroid in these plantations. Similar considerations may apply to avocado plantations and their protection from the avocado sunblotch viroid.

Other Potential Applications

Nucleic acid dot hybridization tests may prove useful for the diagnosis of infection by certain conventional viruses. Immunological assays may not be practical with viruses that produce labile virions or whose coat proteins are only weakly antigenic; in such cases tests based on hybridization of the viral genome with specific cloned recombinant DNA may be more advantageous than conventional tests, although preparation of the tissue extracts would probably be more cumbersome than with viroids. Presumably, the viral genome, to be reactive in the test, would have to be liberated from virions which would require some sort of nucleic acid isolation procedure. Conceivably, however, nonencapsidated viral nucleic acid, such as replicative form (RF) or replicative intermediate (RI), might be present in sufficient quantity to react with the cDNA probe.

Nucleic acid dot hybridization tests are excellently suited for the identification of cryptic virus infections and of integrated viral genomes. Indeed, a conceptually similar nucleic acid dot hybridization procedure has been described to screen eukaryotic cell lines for their Epstein-Barr virus DNA content (Brandsma and Miller 1980), and Maule et al. (1983) have developed a similar technique that allows rapid screening of plant leaf tissue for a range of DNA and RNA viruses.

Another possible application for nucleic acid dot hybridization tests involves the rapid screening of cell lines for the presence of genes that have been inserted into the cells' genome by recombinant DNA procedures. Undoubtedly, nucleic acid dot hybridization tests will be useful in many applications in addition to those listed here.

References

Alwine JC, Kemp DJ, Parker BA, Reiser J, Renart J, Stark GR, Wahl GM (1979) Detection of specific RNAs or specific fragments of DNA by fractionation in gels and transfer to diazobenzyloxymethyl paper. Methods Enzymol 68:220–242

Brandsma J, Miller G (1980) Nucleic acid spot hybridization: Rapid quantitative screening of lymphoid cell lines for Epstein-Barr viral DNA. Proc Natl Acad Sci USA 77:6851–6855

Cress DE, Kiefer MC, Owens RA (1983) Construction of infectious potato spindle tuber viroid cDNA clones. Nucl Acids Res 11:6821–6835

Diener TO (1979) Viroids and viroid diseases. Wiley, New York

Fernow KH (1967) Tomato as a test plant for detecting mild strains of potato spindle tuber virus. Phytopathology 57:1347–1352

Fernow KH, Peterson LC, Plaisted RL (1969) The tomato test for eliminating spindle tuber from potato planting stock. Am Potato J 46:424–429

Fernow KH, Peterson LC, Plaisted RL (1970) Spindle tuber virus in seeds and pollen of infected potato plants. Am Potato J 47:75–80

Gibbs A, Harrison B (1976) Plant virology. The principles. Wiley, New York

Gross HJ. Domdey H, Lossow C, Jank P, Raba M, Alberty H, Sänger HL (1978) Nucleotide sequence and secondary structure of potato spindle tuber viroid. Nature 273:203–208

Gugerli P (1979) Le test immuno-enzymatique (ELISA) et son application pour le diagnostic rapide des viroses de la pomme de terre. Rev Suisse Agric 11:253–260

Maule AJ, Hull R, Donson J (1983) The application of spot hybridization to the detection of DNA and RNA viruses in plant tissue. J Virol Methods 6:215–224

Morris TJ, Wright NS (1975) Detection on polyacrylamide gel of a diagnostic nucleic acid from tissue infected with potato spindle tuber viroid. Am Potato J 52:57–63

Owens RA, Cress DE (1980) Molecular cloning and characterization of potato spindle tuber viroid cDNA sequences. Proc Natl Acad Sci USA 77:5302–5306

Owens RA, Diener TO (1981) Sensitive and rapid diagnosis of potato spindle tuber viroid disease by nucleic acid hybridization. Science 213:670–672

Owens RA, Smith DR, Diener TO (1978) Measurement of viroid sequence homology by hybridization with complementary DNA prepared in vitro. Virology 89:388–394

Palukaitis P, Hatta T, Alexander D McE, Symons RH (1979) Characterization of a viroid associated with avocado sunblotch disease. Virology 99:145–151

Raymer WB, O'Brien MJ (1962) Transmission of potato spindle tuber virus to tomato. Am Potato J 39:401–408

Salazar LF, Owens RA, Smith DR, Diener TO (1983) Detection of potato spindle tuber viroid by nucleic acid hybridization: Evaluation with tuber sprouts and true potato seed. Am Potato J 60:587–597

Sawyer RL (1979) Annual report 1978. International Potato Center, Lima, p iii

Thomas PS (1980) Hybridization of denatured RNA and small DNA fragments transferred to nitrocellulose. Proc Natl Acad Sci USA 77:5201–5205

Upadhya MD (1979) Potential for potato production from true seed under developing country conditions. In: CIP report of the planning conference on the production of potatoes from true seed. Manila, p 12–20

Visvader JE, Gould AR, Bruening GE, Symons RH (1982) Citrus exocortis viroid: Nucleotide sequence and secondary structure of an Australian isolate. FEBS Lett 137:288–292

Wahl GM, Stern M, Stark GR (1979) Efficient transfer of large DNA fragments from agarose gels to diazobenzyloxymethyl-paper and rapid hybridization by using dextran sulfate. Proc Natl Acad Sci USA 76:3683–3687

DNA Technology Applied to the Detection and Epidemiology of Enteric Pathogens

L. S. TOMPKINS, P. A. MICKELSEN, and N. TROUP [1]

New techniques to isolate, clone, and manipulate DNA can now be applied to the detection of enteric pathogens in clinical materials, using specific gene sequences as probes. Although no DNA probes are yet commercially available, several have been developed and applied in research laboratories. In presenting our work on the development of a DNA probe for *Campylobacter jejuni* and its application to fecal samples, we will address the strategies which can be used in probe development and will outline the technical limitations which now exist in probing samples directly. In addition to using DNA as a diagnostic reagent, the plasmid or chromosomal DNA of enteric pathogens, including *Campylobacter* and *Salmonella*, can be used as an epidemiological tool to differentiate strains of the same species. Preliminary data from experiments on *Salmonella enteritidis* will be presented.

DNA probe molecules detect microorganisms by combining with genetic sequences which are homologous with the DNA of microbes contained in the clinical sample, forming duplexes (double stranded hybrids) which can then be detected in some fashion. The colony hybridization technique of Grunstein and Hogness [1] was first applied to bacterial colonies growing on the surface of nitrocellulose paper. Colonies to be screened for the gene of interest, for example *Escherichia coli* colonies containing the gene encoding for LT enterotoxin, are grown on a filter, lysed, and the bacterial DNA denatured into single stranded molecules. Using a labeled probe consisting of a portion of the gene encoding for LT enterotoxin, the unlabeled DNA on the filter is reacted with the probe and hybrids are allowed to form between complementary strands. Unreacted probe DNA is then removed and the filter is then scored for the presence of labeled hybrid molecules.

Moseley and colleagues were the first to apply DNA probes to clinical materials in order to study the epidemiology of enterotoxigenic *E. coli* in Bangladesh and Thailand [2]. Initially, *E. coli* colonies recovered from diarrheal stool samples were reacted with radiolabeled probes specific for LT and two different ST enterotoxins. Later, they observed that the sensitivity and specificity of colony hybridization was equally great if the probe was applied to a "macrocolony" of stool spotted onto nitrocellulose and grown overnight, thus speeding up the time to completion of the test by 24 h. Subsequently, Totten et al. have developed a probe to detect *Neisseria gonorrhoeae* in clinical specimens directly without allowing bacterial growth to occur [3]. A study of more than 50 patients with culture-

1 Stanford University Medical Center S-156 Stanford, CA 94305, USA

proven gonococcal urethritis showed that the sensitivity of the probe applied to the clinical sample directly was greater than 90%. In addition, the probe which was employed reacted only with *N. gonorrhoeae* and virtually no other Neisserial species.

These studies demonstrate two important concepts: (a) DNA probes can be applied to clinical samples directly (with some technical limitations) and (b) the selection of a relevant, specific DNA sequence is extremely important. The essence of DNA probes are their extreme specificity, since under the appropriate hybridization conditions only molecules which are highly homologous will form duplexes. If one can select and isolate a gene which is found in only one strain, species, or genus, then this DNA should detect only microorganisms containing an identical sequence. Whereas Moseley's work is an example of the use of a known virulence determinant to detect virulent strains, Totten's strategy of probe development epitomizes a second approach, that of selecting a gene which one can prove to be encoded in the DNA of one species, irrespective of its contribution to virulence. In this case, the cryptic plasmid, carried in the cytoplasm of more than 90% of gonococcal strains in North America, was selected to detect only one *Neisseria* species. Similarly, one can employ the cryptic plasmid carried by all serovars of *Chlamydia trachomatis* to detect the presence of *Chlamydia* in infected tissue culture cells and in endocervical secretions (Palmer and Falkow, personal communication).

Our approach to developing a specific *C. jejuni* probe took into account the G+C content of this species. We reasoned that the low ratio (34%–38%) essentially precluded homology with other enteric pathogens and other members of the fecal flora. Therefore, a probe consisting of random length fragments of chromosomal DNA was employed to measure sensitivity and specificity. Specificity was determined by reacting the probe with colonies of other enteric pathogens, including *Campylobacter fetus*, *C. coli*, *Campylobacter*-like organisms, *Vibrio cholerae*, *Salmonella*, *Shigella*, enterotoxigenic *E. coli* and a variety of *Bacteroides* species. Of these, only *C. coli* reacted weakly with the *C. jejuni* probe, while 7 different *C. jejuni* isolates reacted strongly. Sensitivity of the chromosomal probe was initially measured by seeding stool slurry preparations with varying concentrations of *C. jejuni* and probing a 10 μl sample spotted onto nitrocellulose. Preliminary experiments suggested that 1×10^4 organisms per spot could be detected.

A field trial of 299 patient stool samples spotted onto nitrocellulose and hybridized without allowing for bacterial growth suggested that direct hybridization was less sensitive than expected. Using the chromosomal probe, the sensitivity of the probe, compared with culture, was only 44%, with a specificity of 96%. Currently, the probe detects between 2×10^4 and 8×10^5 organisms per spot. The probe was able to detect several logs fewer bacteria suspended in broth than suspended in feces, indicating that substances in stool interfere with the detection of specific DNA sequences. Attempts to improve sensitivity by eliminating mucous or other proteins from stool by treatment with a variety of agents, including dithiothreotol, chloroform, and phenol, were not successful. Other approaches are now being taken to extract DNA from fecal samples prior to hybridization to improve the sensitivity. Thus, although DNA probes appear to be capable of detecting microorganisms in some samples, such as tissue imprints or pus, preparation of other specimens remains a major stumbling block to the commercial application of DNA as a diagnostic reagent.

An additional hurdle which must be leapt in applying probes to specimens is that of labeling. Whereas radiolabeled probes have been used extensively in genetic research, they are unsuitable for the diagnostic laboratory because of the problems of handling of radioactive materials and short shelf-life. Recently, Ward and colleagues have synthesized biotin-labeled probes containing nucleotides linked to biotin, a water soluble vitamin [4]. Biotinylated DNA can then be detected by reacting it with complexes composed of avidin, a protein which has an extremely high affinity constant for biotin, linked to biotin and enzyme polymers. Addition of specific substrate produces a color reaction on nitrocellulose which can be measured visually.

In addition to employing DNA probes as detection reagents which could potentially hasten the identification of an enteric pathogen by several days, possibly within an hour, we have also exploited the DNA content of enteric organisms as a means to differentiate between strains of the same species. We discovered several years ago that many enteric species contain a large variety of plasmids, most of which are cryptic, since no function can be assigned to them. After studying the plasmid content of many isolates of several nosocomial epidemic strains, we discerned that isolates which were phenotypically identical, by biotyping, colicin typing, phage typing or serotyping, had identical plasmid fingerprint patterns whereas isolates of the same species, collected at the same time and location, which had different phenotypes could also be distinguished on the basis of their plasmid fingerprints [5]. This type of analysis has also proven to be quite a useful epidemiologic tool for gram positive organisms, especially staphylococci.

However, some strains and species are mostly plasmid-free, and thus cannot be differentiated by plasmid profile. Therefore, we set about to determine whether the chromosomal DNA, specifically the nucleotide sequences comprising cleavage sites for several restriction endonucleases, differed from strain to strain. Virologists were the first to note that viral strains could be profiled by reacting the genome with restriction endonucleases, enzymes which cleave DNA sequences at precise sites determined by the sequence of 4–6 nucleotides. For example, the enzyme Eco RI recognizes the sequence GAATC (reading from 5′ to 3′) on one strand of DNA and cleaves each strand between the G and A, producing a series of different sized linear fragments, the number and size of which depend on the number and position of Eco RI cleavage sites. These fragments can be separated and enumerated by electrophoresis in agarose gels. Each molecularly unique replicon, either plasmid or chromosome, contains a unique restriction digest pattern. Therefore, DNA molecules which have identical restriction digest fingerprints are considered to be homologous. This idea was first applied to distinguish bacterial strains by Australian investigators who showed that two serovars of *Leptospira interrogans* could be differentiated by restriction-endonuclease analysis [6]. Subsequently, Kaper et al. demonstrated that different strains of *Vibrio cholerae* isolated from the U.S. Gulf Coast waters had different digestion patterns that could be easily seen [7]. Bradbury has also shown that *C. jejuni* isolates which are epidemiologically linked and have similar serogroup antigens have identical chromosomal digestion patterns while isolates which were not associated with an outbreak of disease, and are serologically different from the epidemic strain, have distinctive patterns [7].

Very recently we have applied this technique to several *Salmonella enteridis* isolates sent to us by Dr. Mitchell Cohen at the Center for Disease Control, Atlanta, Ga. These isolates, and one recently recovered at Stanford, all have a common serogroup and cannot be differentiated serologically. The isolates from the CDC were collected over a 15-year period in several New England states and a common source was considered to be a possibility in several instances. Our analysis of the isolates by agarose gel electrophoresis confirms the studies done in Atlanta, disclosing that all isolates, except for one, contain a 36 megadalton plasmid. Several others had additional plasmids of lower molecular weight. One the basis of plasmid fingerprinting it is difficult to establish whether these isolates represent a single clone of *S. enteritidis* which has remained relatively stable in several geographic locations, except for the addition or deletion of nonessential plasmids. Total DNA digests from these isolates, including the Stanford isolate, show that 7 of 8 appear to have similar fragment patterns, using two different endonucleases which cleave at different sites. The eighth isolate contains at least one band which is not observed in the others and also lacks a larger molecular weight fragment. These changes are subtle, but reproducible.

As an additional way of analyzing the distribution of chromosomal restriction sites of these *Salmonella* isolates, we employed the technique of Southern hybridization, probing the chromosomal DNA fragments which had been digested, separated by gel electrophoresis, and transferred from the gel to nitrocellulose paper. In this instance, we employed as probes *Salmonella* chromosomal gene sequences which were derived by cosmid-cloning, employing size-fractionated, restricted fragments. The source of the DNA was one of the isolates containing only the 36 megadalton plasmid.

Previous studies by Kaper et al. on the distribution of the enterotoxin gene in *Vibrio cholerae* isolates suggested that different strains of toxigenic isolates all contained a chromosomal gene which hybridized with the cloned probe, consisting of a portion of the LT enterotoxin gene from *E. coli*, but that the size of the chromosomal fragments encoding this gene were strain dependent [7]. Isolates of the same strain all contained the gene on the same sized fragment, and thus demonstrated identical Southern digest fingerprints, while toxigenic isolates of different *V. cholerae* strains contained the gene on one or more fragments of different sizes. Since the length of the chromosomal fragment, produced by endonuclease cleavage, depends on the position of the specific sites bounding genetic sequences, Southern blot analysis should reflect the heterogeneity of restriction sites. Thus, we reasoned that if two isolates contained different restriction sites for a particular enzyme, each should have a distinctive probe pattern. It also seemed logical to assume that any *Salmonella*-specific gene sequence, or pool of sequences, should be able to serve as a specific probe to identify the location of the gene sequence.

The analysis of the Southern blot hybridization patterns of several of the *S. enteritidis* isolates revealed that those which had identical chromosomal digestion patterns also had identical Southern blot patterns, whereas the one isolate which appeared to be different by chromosomal digest analysis also lacked a fragment homologous to the cosmid clones used as probes. In addition, control salmonellae, consisting of isolates of different *Salmonella* species, appear to have probe patterns which are quite dissimilar to the *S. enteritidis* isolates. This analysis is quite preliminary, however, and must be repeated using different restriction endonucleases and a larger number of control *Salmonella* species.

It would appear, therefore, that chromosomal digestion patterns, and possibly cloned chromosomal fragment probes, can distinguish between isolates of the same species and can be employed in epidemiologic investigations of infections due to enteric pathogens.

References

1. Grunstein M, Hogness DS (1975) Colony hybridization: a method for the isolation of cloned DNAs that contain a specific gene. Proc Natl Acad Sci USA 72:3961–3965
2. Moseley SL, Huq I, Alim ARMA, So M, Samadpour-Motalebi, Falkow S (1980) Detection of enterotoxigenic *E. coli* by DNA colony hybridization. J Infect Dis 142:892–898
3. Totten PA, Holmes KK, Handsfield HH, Knapp JS, Perine P, Falkow S (1983) DNA hybridization technique for the detection of *Neisseria gonorrhoeae* in men with urethritis. J Infect Dis 148:462–471
4. Langer PR, Waldrop AA, Ward DC (1981) Enzymatic synthesis of biotin-labeled polynucleotides: novel nucleic acid affinity probes. Proc Natl Acad Sci USA 78:66–6337
5. Schaberg DR, Tompkins LS, Falkow S (1981) Use of agarose gel electrophoresis of plasmid deoxyribonucleic acid to fingerprint gram-negative bacilli. J Clin Microbiol 13:1105–1108
6. Marshall RB, Wilton BE, Robinson AJ (1981) Identification of leptospira serovars by restriction-endonuclease analysis. Med Microbiol 14:163–166
7. Kaper JB, Bradford HB, Roberts NC, Falkow S (1982) Molecular epidemiology of *Vibrio cholerae* in the U.S. Gulf Coast. J Clin Microbiol 16:129–134
8. Bradbury WC, Pearson AD, Marko MA, Congi RV, Penner JL (1984) Investigation of a *Campylobacter jejuni* outbreak by serotyping and chromosomal restriction endonuclease analysis. J Clin Microbiol 19:342–346

New Developments in Nucleic Acid Hybridization

H. WOLF[1], M. HAUS[1], U. LESER[1], S. MODROW[1], M. MOTZ[1], S.-Y. GU[2]
N. FALSER[3], I. BANDLOW[3], W. RICHTER[4], and R. PATHMANATHAN[5]

Introduction

Nucleic acid hybridization is based on the tendency of nucleic acids to form double-stranded hydrogen-bonded complexes if strands of complementary sequences are incubated under appropriate salt and temperature conditions. The technique has been widely applied to study mainly the following questions:

Presence of genes
State of genes (free linear, circular, integrated)
Localization of genes in specific tissue or cell types or in subcellular structures
Transcriptional activity of genes.

Formulae Used to Describe Hybridization Conditions

The hydrogen bonds between two complementary strands of nucleic acids can be reversibly separated. This can be achieved at neutral pH and low ionic strength by the application of heat or, in case of DNA, by treatment with alkali. Neutralization and rapid chilling as well as dilution reduce self-annealing. The addition of likewise single-stranded nucleic acids from other sources will lead to the formation of hybrid molecules, which is the characteristic name-giving event in this technique. The formation of double-stranded nucleic acid molecules occurs best at about 18°–32 °C below the melting point of th product (Gillis et al. 1970). The melting point of the product is identical to that of the parental species if these are completely homologous. Random sequence differences of the hybrid strands will lower the T_m of the product (see formula). T_m is also influenced by the base composition, the monovalent ion concentration, and dielectric constant of the solvent. The latter effect is frequently used to lower the T_m by addition of formamide, and, as a consequence, the hybridization temperature decreases (McConaughy et al. 1969). The same effect of lowered T_m can be observed if a significant fraction of nucleotides is chemically modified. This can be seen, for example, with iodi-

1 Max von Pettenkofer-Institut, Universität München, Pettenkofer Str. 9a, D-8000 München 2
2 Institute of Virology, Beijing, Peoples Republic of China
3 Hals-Nasen-Ohren-Klinik, Universität Innsbruck, A-6020 Innsbruck
4 Hals-Nasen-Ohren-Klinik, Universität Würzburg, Josef-Schneider-Straße, D-8700 Würzburg
5 Institute of Pathology, University of Malaya, Kuala Lumpur, Malaysia

nated probes and may be compensated by lowering the hybridization temperature.

For DNA-DNA (Howley et al. 1979):

$$T_m = 81.5 + 16.6 \log(\mathrm{Na}^+) + 0.41(\text{percentage GC}) \\ - 0.7\ (\text{percentage formamide}) - 1.4\ (\text{percentage mismatch}).$$

The relative hybridization (to 0.18 M Na^+) rate is:

$$F = K^{(0.24/K)^{0.42}},$$

where K = monovalent ion concentration and F = acceleration factor (Britten and Smith 1970).

It should be noted that the molar concentration of DNA molecules influences the rate of renaturation, not the concentration of nucleotides. Consequently, DNA from smaller viruses hybridizes faster than DNA from larger viruses when similar optical densities of DNA are used in the test. The relative hybridization rates are therefore directly correlated to the genetic complexity if repetitive sequences are absent.

Commonly Used Hybridization Techniques and Their Objectives

Detection of Certain Genes in a Mixture with Unrelated Nucleic Acids

This application is widely used to detect viral nucleic acid sequences in clinical specimens or specific plasmids in bacterial isolates. The specific advantage of nucleic acid hybridization is its independence of viral replication, which allows detection of viral genes in infected cells. This is of particular importance in tumor virology, where virus production is not likely to occur. Hybridization was used to correlate Epstein-Barr Virus (EBV), then the first candidate for a human tumor virus, with neoplastic disease of man (zur Hausen et al. 1970; Wolf et al. 1973).

The other important application of this technique is based on the possibility of detecting viral nucleic acids where biological activity may be too labile to be preserved or too time consuming to be detected, as is the case for cytomegaloviral DNA in urine (Chou and Merigan 1983). The presence of hepatitis B- viral DNA in HBS-antigen-positive sera in the absence of detectable e-antigen, as it occurs in persons recovering from hepatitis B infection, seems to be the best marker for infectivity and is another important clinical application of nucleic acid hybridization (Scotto et al. 1983).

Some enteroviruses (eg. Rota) and viruses infecting the respiratory organs have been detected with nucleic acid hybridization and may soon allow fast characterization of the isolates (Flores et al. 1983). Another application which may be of considerable value could be the type-specific detection of human papilloma viruses in laryngeal and cervical materials (see this volume).

The detection of EBV DNA aspirated cells from a group defined by serological pretesting as high risk in developing nasopharyngeal carcinoma is under investigation for its diagnostic and prognostic value (S. Gu, Y. Zeng, H. Wolf, in press) and may be one of the first mass screening tests for neoplastic and preneoplastic conditions is man based on nucleic acid hybridization.

Filter Hybridization Techniques

The most frequently used technique is based on the immobilization of one hybridization partner on DNA binding membranes (Gillespie and Spiegelman 1965; Alwine et al. 1977). Extracted nucleic acid is denatured and fixed to membranes, so as to avoid self-annealing. Labeled single-stranded nucleic acid is then added and its binding is monitored. The labeled probe can be removed by heat treatment under low salt conditions and the filters which retained the sample nucleic acid can be rehybridized with probes of different specifity (Thomas 1980).

Colony Hybridization Test

A very helpful modification of this technique was introduced by Grunstein and Hogness (Grunstein and Hogness 1975), who circumvented the need for extraction of nucleic acids by directly lysing cells on the membrane. This technique has been successfully adapted to eukaryotic cells (Brandsma and Miller 1980; Wolf 1981).

Spot Hybridization

The large number of specimens that have to be tested for clinical purposes require an economic means of hybridization. Multichannel microfiltration units are now widely used for the application of series of samples on membranes (Schleicher and Schuell, Keene NH, USA). These can then be hybridized using standard procedures. Bresser succeeded in using suspensions of biological materials without phenol extraction and could selectively bind DNA or RNA with concentrated NaI (Bresser et al. 1983). The limitations of this technique are the binding capacity of the membrane for nucleic acid per area. This limits the detection of very rare sequences in a vast excess of unrelated sequences. Digestion of DNA with restriction enzymes, separation of the fragments, and transfer to membranes has been used with excellent success to overcome these problems (southern blot) (Southern 1975; Desgranges et al. 1982). The same technique has been used in a most elegant way to describe areas of higher and lower homology between related viruses by adding, at a constant hybridization temperature to replicate blots, varying amounts of formamide to control the melting point and thus the stringency of hybridization conditions (Howley et al. 1979).

Reassociation Kinetics

This technique has been widely used to detect sequences of low abundance, to detect partial homologies, and to determine relative concentrations of certain sequences. The principle of this technique is that both hybrid-forming components are kept in solution. The labeled sequence has to be present in a very low concentration to minimize self-annealing. The conversion rate of the labeled sequence into double-stranded nucleic acid is measured and compared with the rate of self-annealing in a control experiment. This technique requires the determination of the amount of label converted into double-stranded nucleic acid by chromatographic procedures (Kohne and Britten 1971) or by digestion of the single-stranded fraction with the specific nuclease S1 and determination of the activities found in the fractions. Large amounts of DNA (up to 5 mg) were necessary for the detection of 0.1–1 single-copy genes. (For application and limitations of the method and further references see Frenkel et al. 1976.) For this reason and because of the time-consuming experimental protocol, this procedure is now frequently replaced by other methods.

Detection of the State of Genes

The use of appropriate restriction enzymes in connection with blot hybridization allows the determination of the free linear or circular or integrated status of genes under investigation. When DNA from tissue is digested with restriction enzymes and the fragments are separated, blotted on membranes, and hybridized with nucleic acid sequences from the ends of the gene under investigation, fragments of expected molecular weight, derived from the free linear form, indicate free sequences. If the bands appear at different molecular weight regions, this is likely to be due to circularization or integration (Botchan et al. 1976).

Localization of Sequences in Specific Cells

Frequently, certain genes or genomes have to be localized in certain cell types or even subcellular structures, such as chromosomes. Although tissue fractionation has been used successfully to locate viral sequences in specific cell types (Desgranges et al. 1975), in situ hybridization is the method of choice in answering these questions (Gall and Pardue 1969). The first application of in situ hybridization in human medical virology was the successful detection of EBV DNA in the epithelial tumor cell fraction of nasopharyngeal carcinoma (Wolf et al. 1973). Since then, it has found widespread application in cytological and virological problems. Recently, in situ hybridization has been used to identify cells in the lumen of parotid glands as sites of lifelong production of EBV following primary infection (Wolf et al. 1984). The powerful technique has also been used for localizing viral genes on specific chromosomes (Wolf et al. 1975; Henderson et al. 1983; Gerhard et al. 1981). Due to the presence of proteins, this technique can produce signals unrelated to specific hybridization. Carefully controlled experiments, experience in evaluation, and, ideally, confirmation by independent techniques are therefore necessary.

Detection of Transcription

Separation of mRNA by molecular weight and hybridization of the fractions with fragments of specific genomic areas have been widely used to study transcription (Thomas 1980). The availability of single-stranded hybridization probes in the direction of transcription and in reverse orientation allow, after a comparison of the hybridization results, the detection of transcribed genes directly by in situ hybridization. This approach has been used with clones from hepatitis B virus in M13. These clones span the HBs and HBc antigen-encoding regions, respectively. Transcription of viral genes (HBs) has been detected in a human hepatoma cell line. Hybridization with the complementary strand of viral DNA was negative at the resolution obtained in this test (Modrow et al., in preparation).

Preparation of Labeled Hybridization Probes

In Vivo Labeled Probes

The most important requirement for nucleic acid hybridization concerns the preparation of hybridization probes. In vivo labeling of nucleic acids by metabolic pathways has been used to obtain probes. The main difficulties are the usually limiting low specific activity (Frenkel et al. 1976), the difficulty in obtaining enough labeled material, and the problem of unspecific or cross-reacting sequences. These sequences may appear due to incomplete purification of the desired gene, for example from cell lysates, or can be due to sequences within the genome used as probe. Cytomegaloviral DNA is a typical example, as several areas of the viral genome cross-hybridize with DNA from uninfected human cells (Peden et al. 1982).

In Vitro Labeled Probes

Various procedures have been used to introduce labels into nucleic acids. DNA can be transcribed in vitro in the presence of radioactive ribonucleoside triphosphates into cRNA using RNA polymerases. Similarly, RNA can be reverse-transcribed into cDNA. The original template can be destroyed with appropriate nucleases. Although of great importance in earlier studies, these techniques have been largely replaced by other methods.

The most frequently used protocol is based on the use of *Escherichia coli* DNA polymerase I. This enzyme acts on DNAase-introduced nicks in double-stranded DNA as a 5′–3′ exonuclease and a 3′–5′ synthetase, which allows the efficient and random introduction of label. With P^{32} specific activities between 10^8 and 10^9 cpm/μg, DNA can be obtained.

Single-stranded nucleic acids and, with reduced efficiency, also double-stranded nucleic acids can be labeled by catalytic introduction of I^{125} into the cytosines of DNA or RNA (Commerford 1971; Prensky 1976; Gu et al. 1983; Han and Harding 1983). The specific advantage of chemical modification is that it can be easily scaled up for greater production of labeled probes.

Problems with the template specificity of probes are effectively controlled through recombinant DNA techniques. Sequences without the cross-reacting portions can be cloned in plasmids and prepared with good yields. Wherever the plasmid part introduces a danger of unwanted signals, it can be replaced by another vector or removed by restriction enzyme digestion and electrophoretic separation. The desired sequence can be eluted from the gel using special procedures (Langridge et al. 1980; Vogelstein and Gillespie 1979), which abolish otherwise observed inhibitory effects of remaining contaminants from the gel matrix. Recently the single-stranded DNA phage m13 has been used as a cloning vehicle for probes (Messing 1983). This approach is very helpful as m13 sequences hardly occur in natural specimens and do not have to be removed; the recombinant phage DNA can be purified in large amounts without using on ultracentrifuge or other specific equipment. Single-stranded DNA can be labeled by primer-directed synthesis of a second strand, which spans all or part of the m13 sequences (Hu and Messing 1982; Han and Harding 1983; Gu et al. 1983), or by chemical reactions.

Indirect "Sandwich" Hybridization

It was only recently (H. Wolf, in preparation) that a new approach was developed which overcomes the need to introduce label into each specific hybridization probe. The technique takes advantage of the recombinant DNA technique and links the specific sequence 1 to another sequence 2. Sequence 2 can conveniently be M13 DNA. This readily available probe is used in a first hybridization step unlabeled and in high concentrations, which favors fast and complete hybridization. After removal of excess probe, a second probe is added which is homologous to sequence 2 and ideally contains both orientations of the DNA strands. This second probe can be universally applied and can, under appropriate conditions, form a network on top of sequence 2, which leads to an amplification of the signal by up to 100-fold. We have used this technique to detect a fragment of EBV DNA in southern blots and used the P^{32}-labeled the double-stranded replicative form of M13 for second-step hybridization. We were able to detect 0.3 pg of the fragment of EBV.

Selection of Label

Tritium has been widely used for most labeling procedures. Due to its weak radiation, the obtainable resolution in the commonly used autoradiographic systems is rather low. However, for in situ hybridization this label gives well-locatable signals and good resolution if the samples are exposed after dipping in photoemulsion. P^{32}-labeled nucleotides give the highest specific activity. P^{32} is still the label of choice if highest resolution is required in all variations of filter hybridization. It is not suitable for in situ hybridization, as the signals are relatively weak and diffuse. The main disadvantage of P^{32} is its short half-life of 2 weeks.

S^{35} nucleoside triphosphates may be used to replace P^{32} where longer shelf-life is required (half-life 87 days) and where a slightly reduced sensitivity is acceptable. Probes labeled with this isotope should be acceptable for in situ hybridization (Vincent et al. 1982). I^{125} labeling can be efficiently achieved with simple chemical procedures, especially if single-stranded probes are used. The sensitivity of labeled samples is comparable with P^{32}-labeled samples when used in filter assays. I^{125}-labeled probes can also be used for in situ hybridization with slightly less satisfying results than with H^3 due to higher background (Prensky 1976; Gu et al. 1983; Commerford 1971).

Because of the simple labeling procedure, the acceptable half-life of 60 days, and the applicability for in situ and filter hybridization tests, I^{125}-labeled probes appear very attractive for routine use. The modification in the cytosine may lower the melting point by a few degrees Celsius.

Nonradioactive labels have been used to detect hybridization (Langer et al. 1981). At present, the use of biotin covalently linked with nucleic acids via a molecular spacer is the most popular nonradioactive label. It is currently introduced into DNA by using appropriately altered nucleoside triphosphates in the nick translation reaction. The biotin is detected in further steps with standard procedures, using either avidin or antibiotin antibodies covalently linked to fluorochoromes or enzymes used in ELISA tests (β-galactosidase or horseradish peroxidase).

These labels are the big hope of routine laboratories, as they would avoid the handling of radioactive materials with all its decontamination problems and because of the absence of nuclear decay the keeping of usable stocks of probes would be facilitated. This advantage becomes even stronger if only one labeled probe for the sandwich hybridization technique is kept.

The sensitivity of the reaction was found by many groups, including ours, to be not quite as high as with radiolabeled probes in direct comparisons, especially in the filter hybridization test. It will be interesting to see whether sandwich hybridization can increase the sensitivity to fully acceptable levels.

Chemical introduction of the label would further facilitate the mass production of probes, which becomes possible if only one labeled sequence is used in the sandwich technique. This procedure should reduce the cost per test significantly.

Preparation of Test Materials

In many instances, the collection of samples is theoretically easy because the lesions which yield the test materials are accessible without operation. However, the lack of specialized equipment for sample collection enforces compromises in sample quality or quantity. Using a recently designed set of equipment we were able to try out a new probe (Richter et al. 1983) (also available from Haselmeier, Stuttgart, FRG). The probe has a slim head and a cylinder connected via parallel tubing to a buffer reservoir and a vacuum source. Pressing the head on to a suspected area creates a vacuum, which leads to aspiration of buffer and a continuous passage of liquid over the mucosal area. The force of the passing fluid is the only force used in collecting the cells. This procedure is painless in most cases. The stream of isotonic buffer will exfoliate cells from the mucosa and transport them to a membrane filter, composed of nitrocellulose ester, where the cells are concentrated from a large volume.

Alternatively, the cells can be collected with the buffer in a trap and pelleted by centrifugation. The maintenance of isotonic conditions minimizes cell damage and allows the specimen to be used for cytological examination as well as for nucleic acid hybridization. Both techniques may be carried out directly on the filter membranes (Richter et al. 1983; Wolf 1981).

Where histological conditions are important, a section of the material under investigation has to be made. Frozen sectioning has been successfully used to locate viral genomes in tumor cells (Wolf et al. 1983) or to identify "virus factories" in biopsy material (Wolf 1984, in press).

Frequently, fresh samples are not available or fixation is important for better results in the cytological and histological examination by light or electron microscopy. Glutaraldehyde-fixed, paraffin-embedded materials have been successfully used to detect hepatitis B viral DNA (Brigati et al. 1983) in liver tissue and *Pseudorabies* virus DNA in brain tissue (Falser et al., in preparation) after removal of paraffin and in situ hybridization.

Paraffin blocks are not suitable for sectioning when cartilage is involved. We tested several alternative embedding media and found that sections from the acrylate-based K4M (Balzers, Frankfurt, FRG) embedded materials gave acceptable hybridization signals. K4M has the advantage that it still polymerizes at temperatures as low as -20 °C. The use of osmium tetroxide treatment following glutaraldehyde fixation did not further reduce the hybridization signal. Formalin fixation for 20 min or more drastically reduced the hybridization efficiency (Faber et al., in preparation).

Discussion

Nucleic acid hybridization was long considered a technique usable only in highly specialized laboratories. More efficient labeling procedures for nucleic acids have increased the sensitivity of detection to about 10^{-19} mol of a specific sequence under optimal conditions and the signal can be associated with cellular structures. Still, this powerful technique has not found widespread use in clinical testing. The reasons may have been in part a lack of information, but the main issue seems to be the apparent need for specialized and expensive equipment and for highly qualified personnel to perform the complex experimental protocols. This review gives some approaches, some of them new, which should help to establish nucleic acid hybridization in routine microbiological or pathological laboratories. The main issues are seen as follows:

1. Use of single-stranded phages as vectors for specific hybridization probes. This allows the production of highly characterized specific probes without expensive equipment.

2. Labeling procedures which can be scaled up and give probes of long shelf-life are available through chemical iodination and similar procedures involving nonradioactive labels may soon be available.

3. Indirect or "sandwich" hybridization can be used to increase the sensitivity and avoid expensive labeling of specific probes for each sequence to be traced.

4. Solid embedding media allow the detection of specific sequences with high resolution in large sections which may include bone or cartilage.

References

Alwine J, Kemp D, Stark G (1977) Method for detection of specific RNAs in agarose gels by transfer to diazobenzyloxymethyl-paper and hybridization with DNA probes. Proc Natl Acad Sci USA 74/12:5350–5354

Botchan M, Topp W, Sambrook J (1976) The arrangement of simian virus 40 sequences in the DNA of transformed cell. 9:269

Brandsma J, Müller G (1980) Nucleic acid spot hybridization: rapid quantitative screening of lymphoid cell lines for Epstein-Barr viral DNA. Proc Natl Acad Sci USA 77:6851–5

Bresser J. Doering J. Gillespie D (1983) Quick-blot: Selective mRNA or DNA immobilization from whole cells. DNA 2:3

Brigati D, Myerson D, Leary J, Spalholz B, Travis S, Fong C, Hsiung G, Ward D (1983) Detection of viral genomes in cultured cells and paraffin-embedded tissue sections using biotin-labeled hybridization probes. Virology 126:32–50

Britten R, Smith J (1970) A bovine genome. Carnegie Inst Washington Yearb 68:378:386

Chou S, Merigan T (1983) Rapid detection and quantitation of human cytomegalovirus in urine through DNA hybridization. N Engl J Med 308/16:921

Commerford S (1971) Iodination of nucleic acids in vitro. Biochemistry 10/10:1993–1999

Desgranges C, Wolf H, De-The G, Shanmugaratnam K, Cammoun N, Ellouz R, Klein G, Lennert K, Munez N, Zurhausen H (1975) Nasopharyngeal carcinoma, X. Presence of Epstein-Barr genomes in separated epithelial cells of tumors in patients from Singapore, Tunisia and Kenya. Int J Cancer 16:7–15

Desgranges C, Bornkamm G, Zeng Y, Wang P, Zhu J, Shang M, De-The G (1982) Detection of Epstein-Barr viral DNA internal repeats in the nasopharyngeal mucosa of chinese with IgA/EBV-specific antibodies. Int J Cancer 29:87–91

Flores J, Purcell R, Perez I, Wyatt R, Boeggeman E, Sereno M, White L, Chanock R, Kapikian A (1983) A dot hybridisation assay for detection of rotavirus. Lancet I:555–558

Frenkel N, Locker H, Cox B, Roizman B, Rapp F (1976) Herpes simplex virus DNA in transformed cells: Sequence complexity in five hamster cell lines and one derived hamster tumor. J Virol 18/3:885–893

Gall J, Pardue J (1969) Formation and detection of RNA-DNA hybrid molecules in cytological preparations. DNAS 63:378–383

Gerhard D, Kawasaki E, Bancroft F, Szabo P (1981) Localization of a unique gene by direct hybridization in situ. Proc Natl Acad Sci USA 78/6:3755–3759

Gillespie D, Spiegelman S (1965) A quantitative assay for DNA/RNA hybrids with DNA immobilized on a membrane. J Mol Biol 12:829

Gillis M, De Ley J, De Cleene M (1970) The determination of molecular weight of bacterial genome DNA from renaturation rates. Eur J Biochem 12:143 153

Grunstein M, Hogness D (1975) Colony hybridization: A method for the isolation of cloned DNAs that contain a specific gene. Proc Natl Acad Sci 72/10:3961–3965

Gu S, Wolf H, Yi Z (1983) Cloning fragments of EBV-DNA in single-stranded phage M13 mp^8 I preparation and identification of cloned DNA. Cancer (China) 129–135

Han J, Harding J (1983) Using iodinated single-stranded M13 probes to facilitate rapid DNA sequence analysis-nucleotide sequence of a mouse lysine tRNA gene. Nucleic Acids Res 11:7

Henderson A, Ripley S, Heller M, Kieff E (1983) Chromosome site for Epstein-Barr virus DNA in a Burkitt tumor cell line and in lymphocytes growth-transformed in vitro. Proc Natl Acad Sci USA 80:1987–1991

Howley P, Israel M, Law M, Martin M (1979) Rapid methods for detecting and mapping homology between heterologous DNAs. J Biol Chem 254/10:4876–4883

Hu N, Messing J (1982) The making of strand-specific M13 probes. Gene 17:271–277

Kohne D, Britten R (1971) Hydroxyapatit techniques for nucleic acid reassociations. In: Davies (ed) Procedures in nucleic Acid Research, vol 2. Harper and Rowe, New York, pp 500–512

Langer P, Waldrop A, Ward D (1981) Enzymatic synthesis of biotin-labeled polynucleotides: Novel nucleic acid affinity probes. Proc Natl Acad Sci USA 78/11:6633–6637

Langridge J, Langridge P, Bergquist P (1980) Extraction of nucleic acids from agarose gels. Anal Biochem 103:264–271

McConaughy L, Laird C, McCarthy B (1969) Nucleic acid reassociation in formamide. Biochemistry 8:8

Messing J (1983) New M13 vectors for cloning. Methods Enzymol 101:20

Peden K, Mounts P, Hayward G (1982) Homology between mammalian cell DNA sequences and human herpesvirus genomes detected by a hybridization procedure with high-complexity probe. Cell 31:71–80

Prensky W (1976) The radioiodination of RNA and DNA to high specific activities. Methods Cell Biol 13:121–152

Richter W, Gu S, Seibl R, Wolf H (1983) A new method for examination of carcinomas of the nasopharynx. In: Prasad U et al. (eds) Nasopharyngeal carcinoma: Current concepts. University of Malaya Press, pp 25–33

Scotto J, Hadchouel M, Hery C, Yvart J, Tiollais P, Brechot C (1983) Detection of hepatitis B virus DNA in serum by a simple spot hybridization technique: Comparison with results for other viral markers. Hepatology 3/3:279:284

Southern E (1975) Detection of specific sequences among DNA fragments separated by gel electrophoresis. J Mol Biol 98:503–517

Thomas P (1980) Hybridization of denatured RNA and small DNA fragments transferred to nitrocellulose. Proc Natl Acad Sci USA 77/9:5201–5205

Vincent M, Beltz W, Ashton S (1982) Preparation of DNA labeled with high specific activity S^{35}-deoxyadenosine 5'- (a-thio) triphosphate; the use of ^{35}S-Labeled nucleic acids as molecular hybridization probes. Miami Winter Symposium

Vogelstein B, Gillespie D (1979) Preparative and analytical purification of DNA from agarose. Proc Natl Acad Sci USA 76/2:615:619

Wolf H (1981) Die Verwendung verschiedener Nukleinsäure-Hybridisierungstechniken am Beispiel von Epstein-Barr-Virus korrelierter Erkrankungen. Verh Dtsch Ges Pathol 65:47–57

Wolf H, zurHausen H, Becker V (1973) EB viral genomes in epithelial nasopharyngeal carcinoma cells. Nature New Biol 138:245–247

Wolf H, zurHausen H, Klein G, Becker V, Henle G, Henle W (1975) Attempts to detect virus-specific DNA sequences in human tumors, III. Epstein-Barr viral DNA in non-lymphoid nasopharyngeal carcinoma cells. Med Microbiol Immunol (Berl) 161:15–21

Wolf H, Haus M, Wilmes E (1984) EBV persists in the parotid gland. J Virol 51:795–798

zurHausen H, Diehl V, Wolf H, Schulte-Holthausen H (1974) EB virus associated macromolecules in cells derived from human tumors. Mol Studies Viral Neoplasia 517–530

zurHausen H, Schulte-Holthausen H, Klein G, Henle W, Henle G, Clifford P, Santesson L (1970) EBV DNA in biopsies of Burkitt tumors and anaplastic carcinomas of the nasopharynx. Nature 228:1056–1058

Variability of Antigen Epitopes of Monotypic Viruses

E. NORRBY [1]

Introduction

The definition of a virus species has been extensively debated. The pragmatic conclusion has been reached that at our present state of knowledge a species should be synonymous with a serotype of a virus. The serotype of a virus is defined by use of polyclonal convalescent or hyperimmune sera in neutralization assays. Since the process of neutralization involves only certain virion surface structures (usually in minority), the type identification (speciation) is determined by only a small fraction of the total genome. The consequences of this are obvious and well known. These are the facts that the same serotype can be carried by viruses (a) with varying pathogenetic capacity and (b) with varying virulence. Variations in virulence is exploited in the design of live virus vaccines, in which case an attenuated strain of virus is used to provide immune protection against the antigenically identical (or closely related) virulent strain of virus.

Monotypic and Antigenically Unstable Viruses

The surface structures of virions generally are effective immunogenes. As a consequence viruses causing a generalized infection extensively influence the clonal composition of the immune system for the remaining life span. The antigenic stability of a virus, therefore, is of decisive importance for the effect of subsequent infections with the same serotype of virus. Thus after infection with a monotypic virus, e.g., measles virus, a lifelong protection against disease is established. In contrast, a virus capable of showing antigenic drift and/or shift can cause recurrent clinical infections. Somewhat inconsistently, strains of influenza A viruses which have altered surface antigens allowing circumvention of immunity in an individual are referred to as subtypes.

Interestingly most viruses are antigenically stable (monotypic). This must reflect some inherent restriction in the degree of variation allowed in surface structures. It seems likely that in monotype viruses there is at least a partial overlap between structures carrying dominating surface antigens and structures carrying important functions such as receptor seeking activity. The advent of hybridoma technology, however, has allowed elucidation of the fact that a monotypic virus permits the continuous emergence of variants although it is functionally stable.

1 Department of Virology, Karolinska Institute, School of Medicine, S-10521 Stockholm

It has been realized that the product of virus replication is a partly heterogeneous population of virions and that, e.g., negative single-stranded RNA viruses show similar particle population dynamics, be they antigenically stable or unstable. Thus it was shown in studies with monoclonal antibodies that the frequency of emergence of nonneutralizable variants was about $10^{4 \cdot 6}$ for both influenza, Sendai, and vesicular stomatitis virus (Portner et al. 1981). A similar frequency of mutation was demonstrated for rabies (Wiktor and Koprowski 1980; Lafon et al. 1983), polio (Emini et al. 1983; Blondel et al. 1983), and coxsackieviruses (Prabhakar et al. 1982).

Structural-Functional Characteristics of Virus Components Determined by Monoclonal Antibodies

Each structural protein component of a virus carries some different antigenic domains and within each domain generally more than one epitope can be identified. A monoclonal antibody only reacts with one epitope, and since the antigenic structure is complex a single structural component can react with a large number of monoclonal antibodies of different specificities. An epitope may be represented by a stretch of 5–6 or more amino acids, but it may also be formed by noncontinuous parts of a polypeptide chain or of more than one polypeptide chain. Furthermore, allosteric phenomena apply to polypeptides, and consequently a focal change by, e.g., a mutation or the action of a monoclonal antibody may cause conformational changes of an epitope located elsewhere. Since in addition the avidity and isotype of a monoclonal antibody also influence its capacity to react with a given epitope, a detailed characterization of a set of monoclonal antibodies specific for a certain structural component is a complex matter. The primary screening of hybridoma products usually is made by some general test for antibodies, e.g., enzyme-linked immunosorbent assay (ELISA) or radioimmunoassay (RIA). These tests may be supplemented by immune fluorescence analyses or the more specific neutralization and hemagglutination-inhibition (HI) tests. The best general method to get precise information about the virus polypeptide, which is the target for a monoclonal antibody is the radioimmune precipitation assay (RIPA). The results obtained in comparative serological analyses with a certain hybridoma product are not always concordant. This is due to the fact that the various tests are influenced differently by the nature of the epitope involved and by the properties of the monoclonal antibody concerned. A particular epitope may show varying capacity to withstand, e.g., drying, acetone fixation, and treatment with a detergent-containing lysis buffer. A certain monoclonal antibody may therefore show a high antibody titer in ELISA and immunofluorescence tests but not display any reactivity in RIPA. As a consequence, the specificity of some monoclonal antibodies may be difficult to determine. Occasionally the Western immunoblotting technique is a useful supplementary test. Another problem in RIPA can be the identification of polypeptides, which are highly sensitive to proteolysis. Even in the presence of proteolytic inhibitors difficulties may be encountered in the demonstration of, e.g., the polymerase (P) protein of some paramyxoviruses. Short time labelling (less than 2 h) or exchange of cell lysates for purified virions as antigen preparation may be helpful.

Separate identification of different epitopes on a single structural protein is usually attempted by competition ELISA or RIA. This approach is successful in some cases, but in others the control homologous competition does not work. Even in a situation when the assay appears to work, ambiguous results may be obtained. Partial inhibition or even enhancement of attachment of the labelled antibody can be encountered. These findings frequently are difficult to interprete, and possible explanations are steric or allosteric phenomena.

For pragmatic reasons the following terminology will be adopted to describe antigenic properties (cf. Yewdell and Gerhard 1981). An *epitope* represents a defined combination of amino acid residues on the surface of a protein antigen. By operational analysis, e.g., competition experiments or induction of mutational changes (see below), individual or groups of epitopes can be defined to represent an *antigenic site*. These sites are interpreted to be topographically (functionally) nonoverlapping.

Individual structural components can carry different biological activities. In each case the effect of monoclonal antibodies on the activity can be evaluated. By this approach it was shown that the hemagglutinating and neuraminidase activity of Sendai virus is carried by different parts of the same polypeptide (Portner 1981; Yewdell and Gerhard 1982; Örvell and Grandien 1982). Other biological activities that can be exploited in antibody-inhibition assays are the hemolytic (cell fusion) activity, different kinds of polymerase activities and, of particular interest, virus infectively.

Characterization of a group of antibodies with capacity to neutralize the virus can be approached by selection of variants capable of growing in the presence of individual monoclonal antibodies. The capacity of virus variants to react with heterologous antibodies is then determinated. In one study nine epitopes were identified on the surface of the glycoprotein of rabies virus (Flamand et al. 1980 b). Later it was shown that these epitopes could be grouped into three antigenic sites (Lafon et al. 1983). Analysis of nonneutralizable mutants as a supplement to competition assays was required to functionally separate two of these sites. Similarly in another study (van Wyke and Murphy, personal communication) a group of monoclonal antibodies against the hemagglutinin of parainfluenza 3 virus was shown by competition assays to have the same antigen site specificity, but when nonneutralizable variants with hemagglutinins that did not react with certain monoclonal antibodies were examined the antibodies could be segregated into five separate subgroups.

Studies with neutralizing antibodies carry a certain significance since they may allow identification of the part of a molecule responsible for induction of a potentially protective immune response. This site(s) is often referred to as the neutralizing epitope.

Emeni et al. (1983) used a panel of 12 neutralizing antibodies for selection of nonneutralizable variants of poliovirus type 1. In this study seven functionally distinct epitopes involved in neutralization was identified. Similarly selection of nonneutralizable variants of coxsackievirus type B4 by use of nine neutralizing antibodies permitted identification of five different epitopes (Cao et al. 1984). In more detailed studies, examination of antigenic mutant viruses of poliovirus type 3 selected for resistance to individual monoclonal antibodies showed point mutations concentrated in the region 277–294 bases downstream from the start of the region of viral RNA coding for the peptide VP1 (Evans et al. 1983; Minor et al. 1983).

Demonstration of Spontaneously Occurring Intratypic Variation in Different Monotypic Viruses by Use of Monoclonal Antibodies

The availability of monoclonal antibodies has allowed identification of naturally occurring antigenic variation in structural components of monotypic viruses which hitherto have remained undiscovered. These variations have been found to concern epitopes on both surface and internal structural components.

Certain logistic considerations have to be made in comparison of virus strains by monoclonal antibodies. Obviously, the chances of seeing variation increases with the number of monoclonal antibodies or rather the number of epitopes identified by a given set of antibodies. However, for practical reasons only a moderate number of monoclonal antibodies can be employed in a given study and, further, the comparative analyses usually have to be performed one way, i.e., antibodies against one strain of virus are tested against other strains. Some examples of strain comparison with monoclonal antibodies will be given in the following.

The group B coxsackievirus can be divided into six serotypes by use of polyclonal antisera. The prototype strain and 15 consecutive isolates of type B4 coxsackievirus were characterized in neutralization tests with a set of 18 monoclonal antibodies (Prabhakar et al. 1982). Thirteen spontaneously occurring variants were identified. In a subsequent study (Cao et al. 1984) five different variants were found among the prototype strain and eight field strains in tests with nine neutralizing monoclonal antibodies.

Strains of rabies virus have for a long time been considered antigenically similar, although variation in pathogenicity of strains has been noticed to occur. By use of polyclonal antisera the lyssa virus genus has been shown to encompass at least four different serotypes, rabies and the three rabies-related viruses Lagos bat, Mokola, and Duvenhage. By use of monoclonal antibodies differences between strains of the various serotypes have been detected (Wiktor et al. 1980; Wiktor and Koprowski 1980; Flamand et al. 1980a, b). Both antibodies to the nucleocapsid protein (Flamand et al. 1980a) and to the glycoprotein (Flamand et al. 1980b) were used in these comparisons of strains. Twenty-one monoclonal antibodies against the nucleocapsid of either of four rabies virus strains or the rabies-related Mokola virus were employed. On the basis of their reactivity with five rabies virus strains and three rabies-related viruses they were divided into eight groups. All antibodies from hybridomas induced by fixed strains of rabies virus showed complete cross-reactions, except in tests with the Flury-HEP strain. Some antibodies against rabies virus strains also reacted with rabies-related viruses, reflecting intertype specificities. A set of 25 hybridomas producing antibodies with specificity for the glycoprotein allowed further differentiation of strains within serotypes and allowed a separation into 14 groups. The four rabies virus strains could be distinguished by differences in a few epitopes. Some antibodies to glycoprotein epitopes of rabies viruses showed complete cross-reaction between all

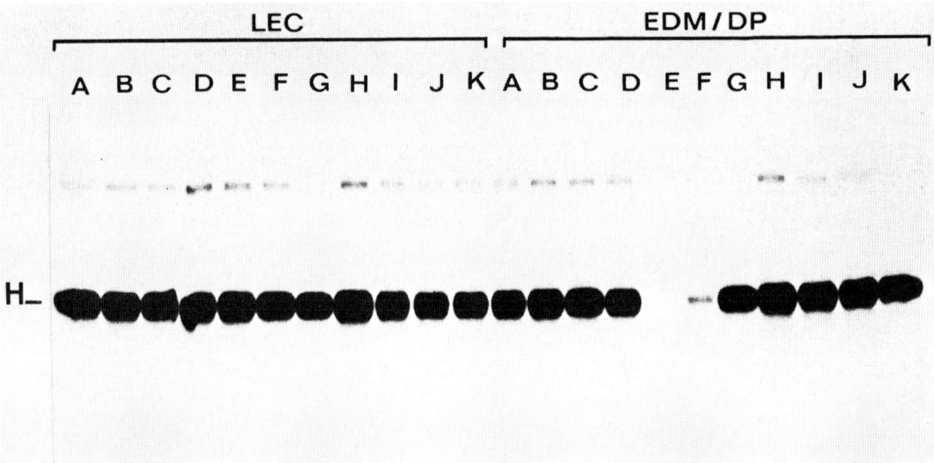

Fig. 1. Reactivity of the hemagglutinin of the LEC and EDM/DP measles virus strains with 11 monoclonal antibodies in RIPA (cf. Sheshberadaran et al. 1983)

rabies virus strains but also reacted with one, two, or three of the rabies-related viruses.

Measles virus is the prototype for a ubiquitous monotypic virus. Spontaneous variation in the occurrence of epitopes on the hemagglutinin has now been demonstrated by use of monoclonal antibodies by several groups (Birrer et al. 1981; Giraudon and Wild 1981; ter Meulen et al. 1981; Trudgett et al. 1981; Sheshberadaran et al. 1983). In studies employing a larger number of monoclonal antibodies it was noted that the differences concerned only a few epitopes and that the majority of monoclonal antibodies reacted equally well with different strains. Thus in the study by Sheshberadaran et al. (1983) a set of 11 different monoclonal antibodies identifying nine different epitopes was tested against nine different strains of measles virus. Three epitope variations were demonstrated (cf. Fig. 1).

In the case of rabies virus strains discussed above, studies with monoclonal antibodies to the nucleocapsid showed only limited intratypic variations. In contrast, they allowed demonstration of intertypic cross-reactions. Similarly it was found that in the case of measles (Sheshberadaran et al. 1983) monoclonal antibodies to the nucleocapsid did not show any variation in the occurrence of epitopes on strains, but a variable capacity to cross-react with other morbillivirus serotypes, distemper, and rinderpest virus were found (Sheshberadaran et al., to be published). A priori no prediction about antigenic stability of internal components can be made. Although they are not subjected to the selective pressure by the immune system, the variation allowed, as in the case of surface components, is limited by the requirements for fulfillment of structural and metabolic functions. In this context it is interesting that a pronounced variation in the occurrence of epitopes on the matrix (M) protein of measles virus was found (Sheshberadaran et al. 1983). This original investigation included nine strains but in further studies another nine strains have been analyzed (Sheshberadaran and

Table 1. Reaction of lytically replicating measles virus strains in RIPA with nine monoclonal antibodies identifying six epitopes on the M protein (cf. Sheshberadaran et al. 1983)

Monoclonal antibody		LEC–KI, Woodfolk, Schwarz, Moraten	Hu2, J. M.	MVO, MVP, Hallé, Mc Clellan, Dean, Zistev, Mantooth	LEC–WI	Edmonston (DP, NIH), Fischer	Edmonston (V7), McI
Group no.	Clone no.						
1	10 EF 10	+	+	+	+	+	+
2	16 BB 2	+	+	+	+	+	+
3	19 AG 10	+	+	+	−	+	+
4	19 CG 6 19 GF 6	+	+	+ (weak)	+	−	−
5	19 DC 5	+	−	−	+	−	−
6	19 DF 10 19 HC 5 19 HF 6	+	−	+	−	+	−

The character and source of measles virus strains was the following: Fresh isolates from cases of acute measles – Woodfolk, MVO, MVP, Hu2 (cf. Sheshberadaran et al., 1983), Edmonston V7 (six passages in HEK cells and seven passages in vero cells; from Dr. P. Albrecht, Div. Virology, Natl. Center Drugs and Biologies, Bethesda, Md.), McI (2 passages in RhMK cells and 6 passages in vero cells; isolated by Dr. K. McIntosh and provided by Dr. P. Albrecht), JM (11 passages in vero cells; from Dr. P. Albrecht); vaccine strains Schwarz, Moraten; SSPE strains – Hallé, Mc Clelland, Dean, Zistev, Mantooth, Fischer (from Dr. J. Sever, Natl. Inst. Health, Bethesda, Md.), LEC–WI (eight passages in vero cells; from Dr. H. Koprowski, Wistar Institute, Philadelphia, Penn.); laboratory strains – Edmonston NIH (from Dr. R. Lazzarini, Natl. Inst. Health, Bethesda, Md.), Edmonston DP and LEC–KI (passaged repeatedly in vero cells in this department).

Norrby, unpublished). These 18 strains displayed six different reaction patterns in tests with a panel of nine monoclonal antibodies identifying six different epitopes (Table 1). Attempts were made to correlate M protein reaction pattern with virulence or pathogenicity features. Five out of seven lytically replicating strains from patients with subacute sclerosing panencephalitis (SSPE) showed the same pattern of M epitopes. A similar pattern was also found in two fresh isolates, MVO and MVP. Since all five SSPE strains that showed the same M protein epitope pattern derived from the same laboratory, further SSPE strains from other laboratories should be studied before any generalizing conclusion can be made. In a separate study (Norrby, unpublished) it was found that 30 consecutive limiting dilution passages of four strains with unique M epitope patterns did not lead to any detectable alteration in antigen properties of the M antigen. This observation together with the experience that strains with the same designation obtained from different laboratories generally display the same M protein epitope pattern indicates that this pattern is relatively stable and therefore may be used for strain identification.

As concerns population stability of virus strains, note should be taken of the potential importance of host cells used for propagation of the virus. In studies of monoclonal antibodies against the hemagglutinin-neuraminidase of mumps virus it was found (Örvell, unpublished) that a subgroup of antibodies reacted with egg-grown but not with vero cell produced virus. Since the same group of monoclonal antibodies also reacted with influenza A virus grown in eggs, they were interpreted to be directed against carbohydrate antigen of unique egg cell origin. Although differences in glycosylation represent one source of host cell induced modification of antigenic properties of virus components also other phenomena may be influential as indicated by studies of influenza B virus (Schild et al. 1983). This study emphasized the importance of different host cell tropisms of viral subpopulations. Passage of virus in different cell types gave a selection of virus populations with different antigenic properties as demonstrated with monoclonal antibodies.

Change of Viral Pathogenicity by Selection Against Neutralizing Epitopes

The surface of a virion plays an important role in the reaction with neutralizing antibody, as already mentioned, and it also mediates the reaction with cellular receptors in virus adsorption. Interaction with receptors on the cytoplasmic membrane has been postulated to represent a factor of major importance in determining viral tropism. In cases when the antigenically dominating region and the receptor seeking region overlap it can be anticipated that the two properties may be altered in parallel. The occurrence of phenomena of this kind was suggested in some of the studies of spontaneous variants discussed above, but they emerge more clearly from studies of variants selected in the laboratory by use of neutralizing monoclonal antibodies.

In the case of monotypic viruses no spontaneous selection of antigenically modified variants occurs in vivo. This stability of a virus population in the presence of a polyclonal antibody response contrasts with the ease with which non-neutralizable variants can be generated by propagation in the presence of monoclonal antibodies. Studies of such variants in four different virus systems with regard to pathogenetic properties of the viruses will be discussed in the following.

The S1 protein (hemagglutinin) of reovirus type 3, which specifies a different cell tropism from that of reovirus type 1, is also the major determinant of neurovirulence (cf. Fields 1982). The critical role of this viral hemagglutinin was demonstrated by selecting virus variants with an altered neutralizing epitope identified with a certain monoclonal antibody. These variants showed a markedly reduced neuropathogenicity (Spriggs et al. 1983). Similarly it was shown that the pathogenicity of rabies virus in adult mice was influenced by the antigenic character of the viral glycoprotein (Coulon et al. 1982; Dietzschold et al. 1983). A rabies strain with a high neuropathogenicity was propagated in the presence of a selected neutralizing monoclonal Uantibody. In one study (Coulon et al. 1982) 13 neutralization-resistent mutants were isolated. Two of these showed a loss of neuropathogenicity for adult mice. These two mutants could be segregated out by their lack of capacity to react with a certain second neutralizing monoclonal

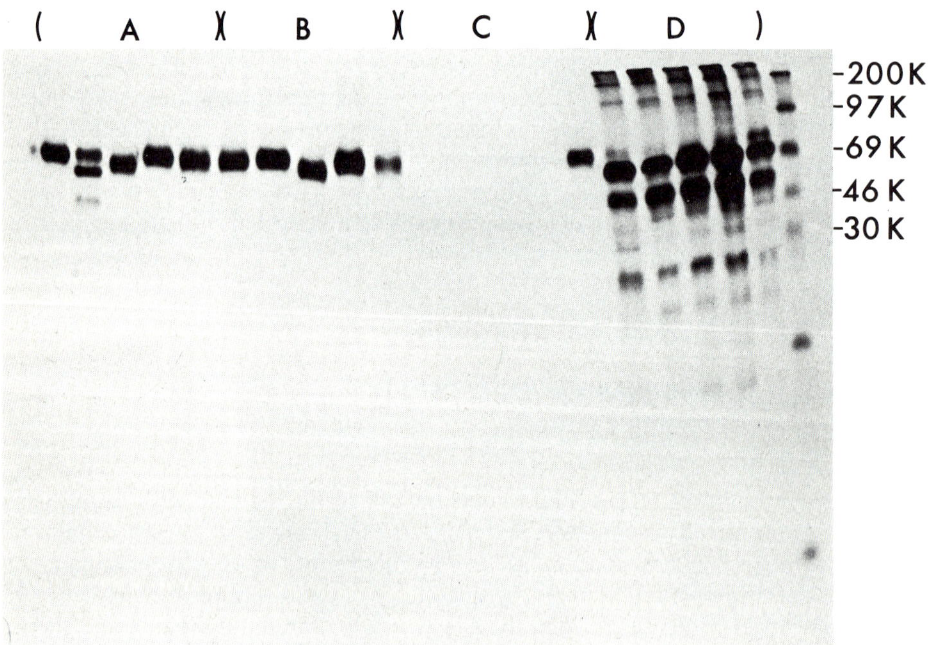

Fig. 2. Characterization of four mutants of the Kilham mumps virus strain selected by propagation of virus in the presence of monoclonal antibody C and the parental virus by RIPA. The RIPA was performed with three different monoclonal antibodies against the hemagglutinin-neuraminidase (*A, B, C*) and a polyclonal hyperimmune serum (*D*). The column on the right side contains molecular weight markers (cf. Löve et al. 1985)

antibody. In a subsequent study (Dietzschold et al. 1983) it was confirmed that all variants that had lost their ability to react with both monoclonal antibodies were nonpathogenic for adult mice. Analysis of tryptic peptides of the glycoproteins of the parental virus and the mutants revealed that the change in pathogenicity corresponded to a single amino acid substitution at position 333 of the glycoprotein molecule.

Finally, a change in neuropathogenicity has also been achieved by selection of nonneutralizable mumps virus variants (Löve et al. 1985). The neurovirulent Kilham strain of mumps virus was passaged in the presence of a neutralizing monoclonal antibody. Four mutants were selected for further characterization. All four mutants showed an increased neuraminidase activity, a property carried by a topographically distinct part of the glycoprotein. Two mutants had lost their capacity to agglutinate erythrocytes under standard conditions, and a third mutant showed an increased rate of migration of the hemagglutinin-neuraminidase polypeptide in SDS-polyacrylamide gel electrophoresis (Fig. 2). All three of these mutants showed an unaltered neuropathogenicity. In contrast, the fourth mutant, which did not display alteration in hemagglutinating property or molecular size, showed a markedly reduced neuropathogenicity.

Five groups of different mutants of coxsackievirus type 4 were identified by use of nine monoclonal antibodies for selection of nonneutralizable variants (Cao et al. 1984). These variants were found to differ from the parental strain and from one another in their myocarditic and cardiotropic properties in a murine model. Two of the variants produced more extensive cardiac pathology and one strain was notable for extensive production of necrotic lesions in the myocardium. The interesting observation was made that a field strain showing accentuated capacity to give myocarditis with extensive necrosis in mice was resistant to neutralization by the same monoclonal antibody used for selection of a variant with similar properties.

Implications for Clinical Virological Analysis of Intratypic Variation of Epitopes

The potential use of monoclonal antibodies in clinical virological analysis requires separate evaluation in different virus systems. A certain guidance in selection of reagents can be obtained from the accumulated experience of their application for epitope characterization of different viruses. The antigen properties of both surface and internal components of lytically replicating viruses appear relatively stable, but in the presence of neutralizing monoclonal antibodies resistant variants with modified surface antigenicity readily emerge. Selection of monoclonal antibodies to be used for direct identification of viral antigen or virions in clinical specimens depends on the purpose of the test. If the aim is to identify a strain of virus a strain-specific monoclonal antibody should be used. If, however, type-specific identification is the goal, a monoclonal antibody against a conserved epitope should be employed. In some cases it may be recommendable to use a mixture of monoclonal antibodies. Finally, there may be situations when intertype-specific monoclonal antibodies are most useful. In selection of the appropriate antibody for a certain clinical laboratory investigation, caution should be applied to avoid antibodies which show molecular mimicry (cf. Örvell and Norrby 1985). A considerable number of different virus-specific monoclonal antibodies have been shown not only to react with virus structural components but also with an epitope on host cell protein(s). In many cases the cross-reaction has been found to concern cytoskeletal components and in one case stress proteins.

The antigen target specificity of the antibody should be chosen to best serve the kind of test used. Thus, for example, if an immune fluorescence test with cells from the respiratory tract is to be examined, it may be recommendable to use a monoclonal antibody against the NP component since this occurs in the largest abundance. In an indirect immune fluorescence test for rapid detection of RS virus a monoclonal antibody against the nucleocapsid was chosen (Bell et al. 1983).

In some situations monoclonal antibodies may be considered for anchoring a selected antigen to the bottom of plastic wells for the purpose of quantification of antigens or for the identification of specific antibodies against this antigen in sera. The efficacy of antigen anchoring by hybridoma antibodies varies depending upon their isotype and avidity. If there is a need for a polyclonal hyperimmune serum, it should be kept in mind that monoclonal antibodies frequently can be effectively used for high degree of purification of individual virus structural components by affinity chromatography. Components purified by this method can then be used for preparation of highly specific hyperimmune sera. This approach was successfully employed for characterization of measles virus H, F, and M components (Varsanyi et al. 1984).

Acknowledgement. The author's work cited in this review was supported by a grant from the Swedish Medical Research Council (Project no. B85-16X-00116-21B).

References

Bell DM, Walsh EE, Hruska JF, Schnabel KC, Hall CB (1983) Rapid detection of respiratory syncytial virus with a monoclonal antibody. J Clin Microbiol 17:1099–1101

Birrer MJ, Bloom BR, Udem S (1981) Characterization of measles polypeptides by monoclonal antibodies. Virology 108:381–390

Blondel B, Akacem O, Crainic R, Couillin P, Horodniceanu F (1983) Detection by monoclonal antibodies of an antigenic determinant critical for poliovirus neutralization present on VP1 and on heat inactivated virions. Virology 126:707–710

Cao Y, Schnurr DP, Schmidt NJ (1984) Monoclonal antibodies for study of antigenic variation in Coxsackie virus type B4: association of antigenic determinants with myocarditic properties of the virus. J Gen Virol 65:925–932

Coulon P, Rollin P, Aubert M, Flamand A (1982) Molecular basis of rabies virus virulence. I. Selection of avirulent mutants of the CVS strain with anti-G monoclonal antibodies. J Gen Virol 61:97–100

Dietzschold B, Wunner WH, Wiktor TJ, Lopes AD, Lafon M, Smith CL, Koprowski H (1983) Characterization of an antigenic determinant of the glycoprotein that correlates with pathogenicity of rabies virus. Proc Natl Acad Sci USA 80:70–74

Emini EA, Kao SY, Lewis AJ, Crainic R, Wimmer E (1983) Functional basis of poliovirus neutralization determined with monospecific neutralizing antibodies. J Virol 46:466–474

Evans DMA, Minor PD, Schild GC, Almond JW (1983) Critical role of an eight amino acid sequence of VP1 in the neutralization of poliovirus type 3. Nature 304:459–462

Fields BN (1982) Molecular basis of reovirus virulence. Arch Virol 71:95–107
Flamand A, Wiktor TJ, Koprowski H (1980a) Use of hybridoma monoclonal antibodies in the detection of antigenic differences between rabies and rabies-related virus proteins. I. The nucleocapsid protein. J Gen Virol 48:97–104
Flamand A, Wiktor TJ, Koprowski H (1980b) Use of hybridoma monoclonal antibodies in the detection of antigenic differences between rabies and rabies-related virus proteins. II. The glycoprotein. J Gen Virol 48:105–109
Giraudon P, Wild TF (1981) Differentiation of measles virus strains and a strain of canine distemper by monoclonal antibodies. J Gen Virol 57:179–183
Lafon M, Wiktor TJ, MacFarlan RI (1983) Antigenic sites on the CVS rabies virus glycoprotein: analysis with monoclonal antibodies. J Gen Virol 64:843–851
Löve A, Rydbeck R, Kristensson K, Örvell C, Norrby E (1985) Hemagglutinin-neuraminidase glycoprotein as a determinant of pathogenicity in mumps virus hamster-encephalitis: analysis of mutants selected with monoclonal antibodies. J Virol 53:67–74
Minor PD, Schild GC, Bootman J, Evans DMA, Ferguson M, Reeve P, Spitz M, Stanway G, Cahn AJ, Hauptmann R, Clarke LD, Mountford RC, Almond JW (1983) Location and primary structure of a major antigenic site for poliovirus neutralization. Nature 301:674–679
Örvell C, Grandien M (1982) The effects of monoclonal antibodies on biologic activities of structural proteins of Sendai virus. J Immunol 129:2779–2787
Örvell C, Norrby E (1985) Antigenic structure of paramyxoviruses. In: Neurath AR, van Regenmortel MHV (eds) Immunochemistry of viruses – the basis for serodiagnosis and vaccines. Elsevier Biomedical, Amsterdam, pp 241–264
Portner A (1981) The HN glucoprotein of Sendai virus: analysis of site(s) involved in hemagglutinating and neuraminidase activities. Virology 115:375–384
Portner A, Webster RG, Bean WJ (1980) Similar frequencies of antigenic variants in Sendai, vesicular stomatitis and influenza A viruses. Virology 104:235–238
Prabhakar BS, Haspel MV, McClintock PR, Notkins AL (1982) High frequency of antigenic variants among naturally occurring human Coxsackie B4 virus isolates identified by monoclonal antibodies. Nature 300:374–376
Schild GC, Oxford JS, deJong JC, Webster RG (1983) Evidence for host-cell selection of influenza virus antigenic variants. Nature 303:706–708
Sheshberadaran H, Chen S-H, Norrby E (1983) Monoclonal antibodies against five structural components of measles virus. I. Characterization of antigenic determinants on nine strains of measles virus, Virology 128:341–353
Spriggs DR, Bronson RT, Fields BN (1983) Hemagglutinin variants of reovirus type 3 have altered central nervous system tropism. Science 220:505–507
ter Meulen V, Löffler S, Carter MJ, Stephenson JR (1981) Antigenic characterization of measles and SSPE virus hemagglutinin by monoclonal antibodies. J Gen Virol 57:357–364
Trudgett A, Gould EA, Armstrong M, Mingioli ES, McFarlin DF (1981) Antigenic difference in the hemagglutinin of measles and related viruses. Virology 109:180–182
Varsanyi TM, Utter G, Norrby E (1984) Purification, morphology and antigenic characterization of measles virus envelope components. J Gen Virol 65:355–366
Wiktor TJ, Koprowski H (1980) Antigenic variants of rabies virus. J Exp Med 152:99–112
Wiktor TJ, Flamand A, Koprowski H (1980) Use of monoclonal antibodies in diagnosis of rabies virus infection and differentiation of rabies and rabies-related viruses. J Virol Methods 1:33–46
Yewdell JW, Gerhard W (1981) Antigenic characterization of viruses by monoclonal antibodies. Annu Rev Microbiol 35:185–206
Yewdell J, Gerhard W (1982) Delineation of four antigenic sites on a paramyxovirus glycoprotein via which monoclonal antibodies mediate distinct antiviral activities. J Immunol 128:2670–2675

Immunoblotting with Monoclonal Antibodies: A Highly Specific System for Detection and Identification of Bacterial Outer Membrane Proteins

M. JÜRS [1], H. PETERS [1], K. N. TIMMIS [2], and D. BITTER-SUERMANN [1]

Abbreviations

ABTS, 2,2′-azino-di-(3-ethylbenzthiazoline sulfonic acid); DTT, DL-dithiothreitol; IncF, incompatibility group F; PAGE, polyacrylamide gel electrophoresis; MoAb, monoclonal antibody; M_r, relative molecular weight; SDS, sodium dodecyl sulfate; TSB, Tryptone Soya Broth; TBS, TRIS-buffered saline; UTI, urinary tract infection

Introduction

We are studying the *E. coli* outer membrane proteins traTp, ompAp, and "X", a major outher membrane protein with a molecular weight of about 42000 daltons that has not yet been definitively identified, as potential targets for diagnostic reagents and vaccines.

In this report, we describe our experiments with traTp as an example of an approach towards the detection of such proteins by means of immunoblotting with monoclonal antibodies.

TraTp is an IncF group plasmid-encoded protein that mediates surface exclusion, the phenomenon which inhibits plasmid-carrying bacteria from acting as recipients in crosses with other bacteria carrying a related plasmid, and bacterial resistance to the lethal effects of complement (serum). Certain IncF group plasmids also code for the production of colicins, hemolysins, adhesion antigens, enterotoxins, and resistance to a variety of antibiotics [17]. Thus, in addition to its intrinsic interest, traTp may also serve as a marker for bacterial isolates carrying one or other member of this important group of plasmids.

Not only does traTp mediate bacterial resistance to serum [14] but it also increases bacterial resistance to phagocytosis by macrophages [2a]. Because of the potential importance of these properties for the virulence of pathogenic bacteria, we screened a large number of clinical isolates of *E. coli*, salmonellae, and *Enterobacter cloacae* for the presence of traTp, and analyzed its association with the K1 capsular antigen [2], a well-known virulence factor of some *E. coli* isolates.

1 Institut für Medizinische Mikrobiologie, Johannes-Gutenberg-Universität, Augustusplatz, D-6500 Mainz, FRG

2 Department of Medical Biochemistry, University of Geneva, CH-Geneva

Material and Methods

Bacterial Strains and Plasmids

The traTp-positive control bacteria were derivatives of the laboratory strain of *E. coli* K-12 C600 rif^r [3] carrying either plasmid pKT146 or plasmid pKT107 [14]. The traTp-negative strain was a derivative carrying the pACYC184 vector plasmid [6]. The ompAp-positive strain JC6650 and its ompAp-negative mutant JC6650rh [5] were kindly provided by U. Henning, Tübingen, FRG.

Preparation of Bacteria for SDS-PAGE

E. coli K-12 C600 rif^r bacteria carrying either plasmid pKT146 or pACYC184 were cultivated overnight in TSB (Oxoid, Wesel, FRG) containing tetracycline 5 μg/ml (Sigma), whereas JC6650 und JC6650rh were grown on blood agar plates. TSB cultures were centrifuged at 4,000 rpm for 10 min and the pellets were resuspended in physiological saline. Colonies of strains grown on blood agar plates were scraped from the plates and resuspended in physiological saline. Cells grown by either method were then washed twice by centrifugation and resuspension in saline, and adjusted to a final concentration of 5×10^8 cells/ml. One-milliliter quantities of cell suspensions were then centrifuged (Eppendorf minifuge) and the pellets obtained were resuspended in 100 μl volumes of SDS sample buffer (0.0625 *M* TRIS-HCl, pH 6,8; 5% 2-mercaptoethanol, Merck-Schuchardt, Munich, FRG; 2% SDS; 12.5% glycerol). The suspensions were then held at 100 °C for 10 min.

SDS-PAGE

The method of preparing the sample and stacking gel has already been described [11]. Gradient slab gels were obtained by means of a gradient former. Sample wells were formed in the stacking gel by means of a gel former (Desaga). Protein samples (12 μl each) were applied to the gel with a multichannel pipette (Desaga), and electrophoresis was carried out in a Desaphor electrophoresis apparatus (Desaga, Heidelberg, FRG) containing electrophoresis buffer (5 m*M* TRIS; 0.38 *M* glycine; 0.1% SDS) at 100 V/150 mA for 1 h, then 200 V/250 mA until the bromophenol blue tracking dye had run off the gel. Protein molecular weights were estimated from the migration distances of standard proteins (SDS Marker Kit from Biorad Labs, Munich, FRG) co-electrophoresed in the same gel. Gels were either blotted or stained with Coomassie brilliant blue G250 (Sigma).

Immunoblotting

Proteins fractionated on SDS gels were transferred electrophoretically to nitrocellulose sheets [16] (BA85, Schleicher & Schüll, Dassel, FRG) for 2 h at 25 V/0.2 A followed by 2 h at 50 V/0.4 A using a Transblot apparatus (Biorad, Munich, FRG), as desribed previously [4]. Blotted nitrocellulose sheets were incubated overnight in blocking solution (2% gelatine in buffer A: 10 m*M* K_2HPO_4/KH_2PO_4; HCl pH 7.5; 15 m*M* NaCl), after which they were incubated for 2 h with anti-traTp, anti-ompAp, or anti-"X" monoclonal antibodies. They were subsequently washed six times in buffer A over a 15-min period before incubation for 2 h with peroxidase-conjugated anti-mouse immunoglobulins diluted 1 : 1,000 in blocking solution. The washing procedure was then repeated. Bound conjugated antibody was detected by incubating nitrocellulose sheets with a solution containing 60 mg 4-chloro-1-naphthol (Sigma) dissolved in 20 ml methanol (Merck, Darmstadt, FRG) and diluted in 100 ml TBS (20 m*M* TRIS-HCl pH 7.5, 500 m*M* NaCl) and 100 µl 30% H_2O_2 (Merck), as described previously [7]. The reaction was terminated by washing sheets in distilled water, after which they were dried carefully and stored in the dark. Blue bands indicated positive reactions.

ELISA

Microtiter plates were coated with membrane preparations of traTp-positive or -negative, or ompAp-positive or -negative bacteria (1×10^7 bacterial equivalents/well). Overnight cultures of bacteria were centrifuged and resuspended in physiological saline to a final concentration of 2.5×10^{10}/ml, and then solubilized in detergent solution (0.01 *M* TRIS-HCl pH 8.0; 5 m*M* EDTA, Merck; 0.1 *M* DTT, Sigma; 2% SDS, Serva, Heidelberg, FRG), heated for 10 min, incubated for 1 h at 37 °C, and then treated as described above.

Production of MoAbs

For the production of anti-traTp antibodies, 12-week-old BALB/c mice (Zentralinstitut für Versuchstierkunde, Hannover) were injected intraperitoneally with a partially purified preparation of traTp (20 µg) [4]. Injections were repeated weekly for 5 weeks. In the 6th week, the mice were boosted daily for 3 days. Immune serum was obtained by heart puncture and the spleenocytes were fused [10] with hybridoma line X-63-Ag8.653 [9].

For anti-ompAp and anti-"X" antibodies, BALB/c mice were immunized with 2.5×10^8 bacteria of strain *E. coli* K-12 C600 rif^r (pKT146) according to the immunization regime described above.

The screening of supernatant fluids for anti-traTp antibody was performed in an ELISA with partially purified traTp fixed to poly-L-lysine (M_r 80 000, Sigma, Munich, FRG) coated microtiter plates (Costar, Tecnomara, Fernwald, FRG), crosslinked with glutaraldehyde (Serva, Heidelberg, FRG; 25% stock solution diluted 1 : 2,000 in buffer A) and then blocked with 2% gelatine (Merck, Darmstadt, FRG) in buffer A. Peroxidase-conjugated anti-mouse immunoglobulins (Dakopatts, Hamburg, FRG) were diluted 1 : 500 in blocking solution und ABTS was used as substrate (Sigma, Munich, FRG). Membrane preparations of *E. coli*

K-12 C600 rif^r derivatives carrying pKT146, pKT107, or pACYC184 served as traTp-positive or traTp-negative controls.

Anti-ompAp and anti-"X" antibodies reacted with both traTp-positive and traTp-negative strains. Their reaction pattern was characterized and the proteins with which they reacted were identified by means if immunoblotting and, in case of anti-ompAp MoAbs also by an ELISA using as antigen ompAp-positive (JC6650) and ompAp-negative (JC6650rh) bacteria.

Anti-traTp, -ompAp, and -"X" positive hybridoma cells were isolated, incubated with feeder cells and cloned by limiting dilution [15]. Positive clones were injected intraperitoneally into Pristan (Aldrich Chem. Company, USA) primed BALB/c mice and ascites obtained 2 weeks later.

Results

The immunoblotting of SDS-PAGE-fractionated whole cells with anti-traTp, anti-ompAp, and anti-"X" MoAbs readily detected the traTp, ompAp, and "X" outer membrane proteins in isolates of *E. coli* (Fig. 1). We have screened 353 clinical isolates of *E. coli* for the presence of traTp and found 38% of strains from healthy donors to be positive, 51% of those from patients with UTI infections to be positive, and 56% of those from patients suffering from septicemia to be positive (Table 1). By means of the chi-square statistical test, these differences were found to be significant (χ^2:$p<0.01$).

Examination of the subpopulation of isolates that produced the K1 capsular antigen revealed that 66% of *E. coli* strains from healthy donors were traTp positive, whereas 88% of the isolates from cases of sepsis were traTp positive. The dif-

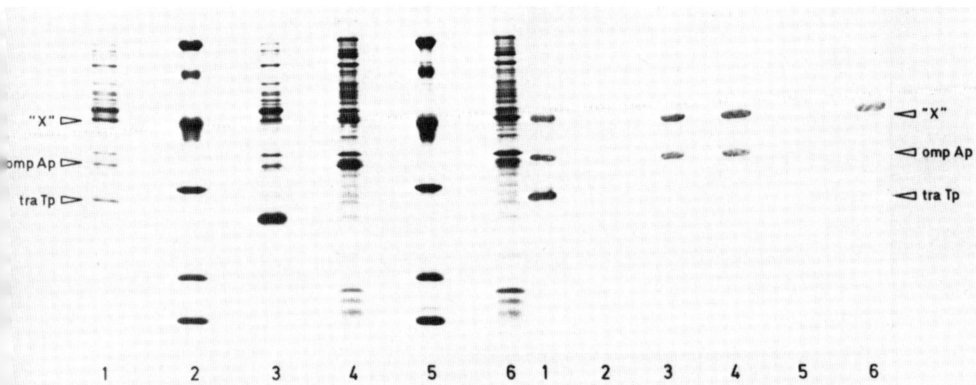

Fig. 1. The immunoblotting procedure. *Left:* Coomassie blue stained SDS, 12%–15% gradient slab gel run under reducing conditions. *Right:* immunoblot with a mixture of MoAbs (anti-traTp MoAb 414-B9, anti-ompAp MoAb 759-G9, anti-"X" MoAb 900E). *Lane 1,* 5×10^7 solubilized bacteria of *E. coli* K-12 C600 rif^r (pKT146), traTp$^+$ ompAp$^+$ X$^+$; *Lanes 2,5,* molecular weight markers; *Lane 3,* solubilized bacteria of *E. coli* K-12 C600 rif^r (pACYC184), traTp$^-$ ompAp$^+$ X$^+$; *Lane 4,* solubilized bacteria of *E. coli* K-12 JC6650, traTp$^-$ ompAp$^+$ X$^+$, *Lane 6,* solubilized bacteria of *E. coli* K-12 JC 6650 rh, traTp$^-$ ompAp$^-$ X$^-$

Table 1. Incidence of TraTp in clinical isolates of Escherichia coli

Type	Total no.	TraTp	
		+(%)	−(%)
Fecal	100	38	62
UTI	100	51	49
Sepsis	153	56	44
χ^2:$p<0.01$			

Table 2. Incidence of TraTp in Enterobacteriaceae

S. typhimurium		Other salmonellae (groups B, C1, C2, D1)		*Enterobacter cloacae*	
TraTp		TraTp		TraTp	
+	−	+	−	+	−
38	6	12	27	33	3
86%	14%	31%	69%	91%	9%

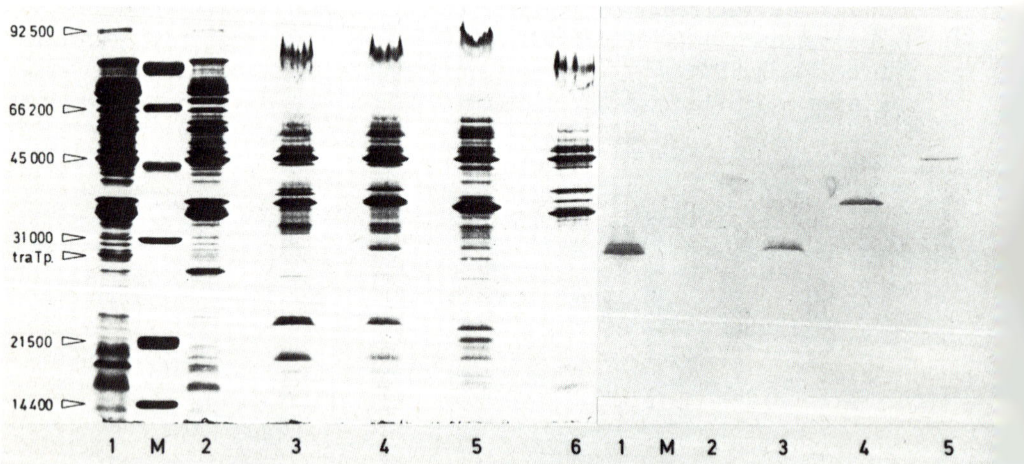

Fig. 2. Immunoblot of non-*E. coli* isolates with anti-traTp MoAb 414-B9. *Left:* stained gel run under reducing conditions. *Right:* immunoblot with MoAb 414-B9. *Lane 1,* 5×10^7 solubilized bacteria of strain *E. coli* K-12 C600 rif (pKT146), traTp$^+$; *Lane 2,* 5×10^7 solubilized bacteria of strain *E. coli* K-12 C600 rif^r (pACYC184), traTp$^-$ (*M,* molecular weight markers); *Lane 3,* 5×10^7 solubilized bacteria of strain *Salmonella typhimurium*; *Lane 4,* 5×10^7 solubilized bacteria of strain *Salmonella derby*; *Lane 5,* 5×10^7 solubilized bacteria of strain *Enterobacter cloacae*; *Lane 6,* 5×10^7 solubilized bacteria of strain *E. coli*

ference in incidence of traTp between K1$^+$ and 1$^-$ isolates was significant (χ^2:$p<0.01$) by the Le Roy omnibus test.

Upon screening other Enterobacteriaceae with the anti-traTp antibody, we detected reacting proteins having the same molecular weight as traTp in 86% of isolates of *Salmonella typhimurium*. Surprisingly, anti-traTp antibody reacting proteins with higher molecular weights were detected in 31% of isolates of salmonellae of serogroups B, C, and D and in 91% of isolates of *Enterobacter cloacae* (Fig. 2; Table 2). Whether or not these traTp-like proteins in non-*E. coli* isolates are plasmid-encoded remains to be determined. The ompAp protein (35000 daltons) exhibitis the phenomenon of heat modifiability [8], i.e., changes in banding pattern of a protein effected by different periods of heating in SDS sample buffer prior to SDS-PAGE. This phenomenon was clearly demonstrated in an immunoblot with anti-ompAp monoclonal antibody 759-G9 (Fig. 3). The same antibody failed to react in an immunoblot or an ELISA in which the antigen was prepared from the ompAp-negative strain JC6650rh, which demonstrates the specificity of this monoclonal (Fig. 4).

Protein "X" (≈45000 daltons) was detected by anti-"X" moAb 900-E in all isolates of bacteria tested that belong to the family of Enterobacteriaceae (i.e., *Klebsiella pneumoniae, Serratia marcescens, Citrobacter freundii, Proteus mirabilis, Salmonella typhimurium, Salmonella bovis* morbif., *Enterobacter cloacae*, and *Enterobacter aerogenes*). Studies are in progress to identify this protein.

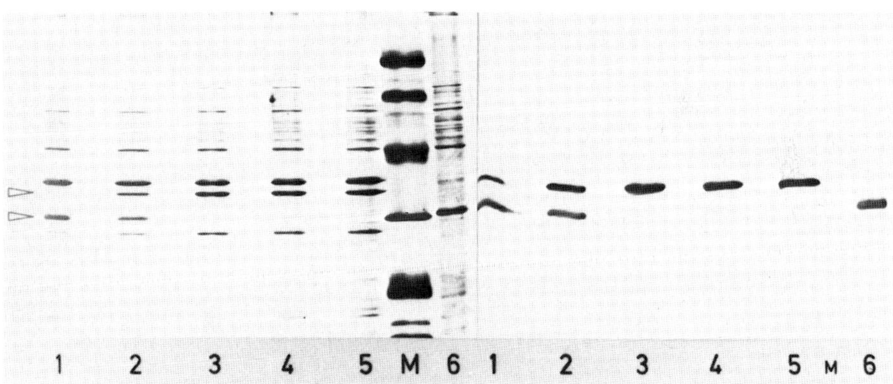

Fig. 3. Heat modifiability of ompAp detected by immunoblotting. *Left:* stained gel. *Right:* immunoblot with anti ompAp MoAb 759-G9. *E. coli* K-12 C600 rif^r (pKT146) bacteria were boiled at 100 °C in SDS sample buffer for 1 min (*Lane 1*), 2 min (*Lane 2*), 5 min (*Lane 3*), 10 min (*Lane 4*), or 20 min (*Lane 5*). Bacteria in *lane 6* were incubated in SDS sample buffer at 37 °C (i.e., without boiling) for 1 h. Note that anti-ompAp antibody detects both forms of ompAp and that the lower molecular weight band decreases, while the higher one becomes more intense, with increasing length of the boiling period

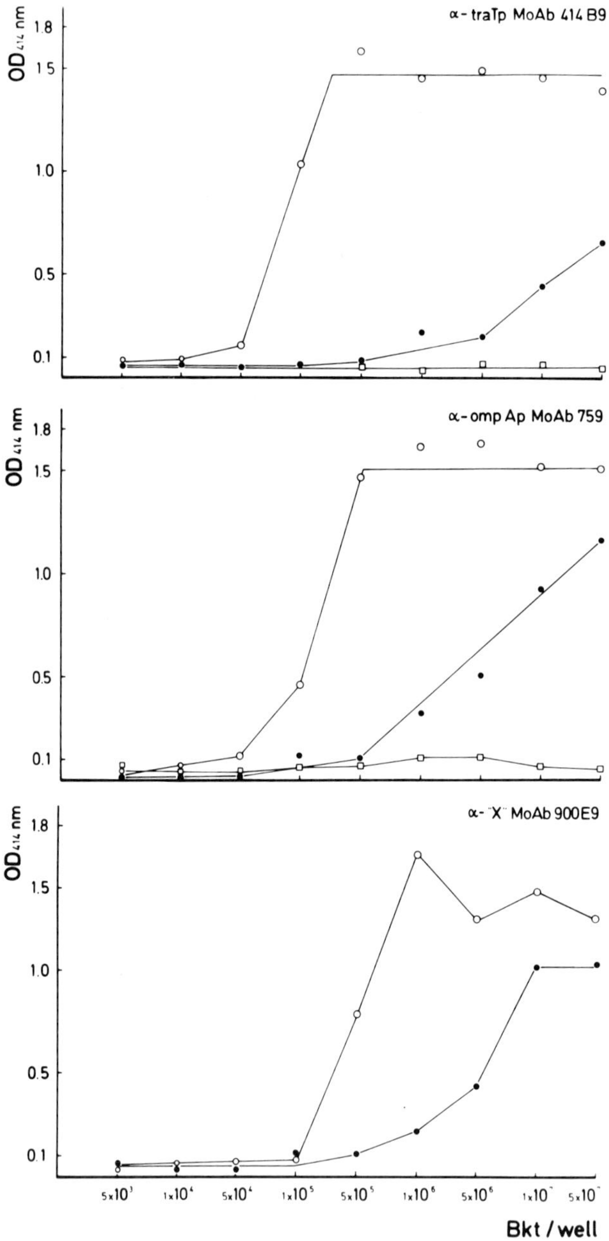

Fig. 4. ELISA with solubilized and whole bacteria. Anti-traTp, anti-ompAp, and anti-"X" MoAbs were used. The *abscissa* shows the absolute number of solubilized or whole bacteria used per well. *Open circles* represent C600 rif^r (pKT146) bacteria (traTp$^+$, ompAp$^+$, X$^+$) solubilized in detergent solution, whereas *filled circles* represent whole bacteria. *Open squares* represent *E. coli* K-12 JC6650 rh bacteria (traTp$^-$, ompAp$^-$, X$^+$) solubilized in detergent solution. The *ordinate* shows the absorbance at 414 nm produced by the peroxidase reaction

Discussion

We have shown here that the presence of specific major outer membrane proteins of *E. coli* is readily demonstrated with MoAbs. Because such proteins may serve as reliable indicators of certain groups of isolates, the MoAbs described here constitute useful reagents for diagnosis and epidemiological studies. For all the proteins studied, solubilized bacteria were better sources of antigen than were whole bacteria, which indicates that the epitopes with which the antibodies react are not exposed on the cell surface. The antibodies can be used either for immunoblotting or in an ELISA.

The ELISA is more sensitive than the immunoblot (limits of detection: 1×10^5 bacteria vs 1×10^7 bacteria), but the immunoblot not only detects the protein, but also gives additional information relating to its molecular weight: In our study, the existence of traTp-like proteins having molecular weights greater than that of traTp and the heat modifiability of ompAp were demonstrated.

In a survey of 353 isolates of *E. coli*, traTp was found with increasing frequencies according to the severity of the disease caused: i.e., 38% of bacterial isolates from healthy donors were traTp positive, compared with 51% of isolates from patients with UTI and 56% of isolates from patients with septicemia.

Moreover, anti-traTp antibody reacting proteins were found in most strains of *S. typhimurium* and *Enterobacter cloacae* and in some strains of salmonellae of groups B, C, and D.

The MoAbs described here, and others in preparation, have a number of other potential uses. The ompA protein or proteins that cross-react with ompAp antibodies are reported to be widespread in strains of *E. coli* and other Enterobacteriaceae and in *Aeromonas salmonicida* [12]. The anti-ompAp MoAb 759-G9 will permit clinical implications of the production of ompAp by these strains to be analyzed. The antibodies can also be used to estimate the degree of exposure of these outer membrane proteins on the cell surface, a crucial aspect for the development of one or more these proteins as an effective vaccine. Finally, the MoAbs can be used to analyze the biological functions of the outer membrane proteins, for example the surface exclusion property of traTp [1, 13].

Summary

MoAbs have been raised against the *E. coli* major outer membrane proteins traTp (28,000 daltons), ompAp (35,000 daltons), and a third one, as yet unidentified, that we refer to as "X" (45,000 daltons). In ELISA reactions, all MoAbs reacted better with solubilized bacteria than with whole cells.

The traT protein, which is encoded by IncF group plasmids, mediates the phenomenon of surface exclusion and confers resistance to serum upon bacteria that produce it. We have screened 353 isolates of *E. coli* obtained from healthy donors, or from patients suffering from UTI or septicemia, for the presence of traTp by means of an anti-traTp monoclonal antibody and Western blotting of total cellular protein fractionated on SDS-polyacrylamide gels. Thirty-eight per cent of fecal isolates of *E. coli* obtained from healthy donors were traTp positive, compared with 51% in the case of UTI strains and 56% in the case of sepsis strains.

When a subgroup of *E. coli* isolates that produce the K1 capsule were selected from the sample, 66% of the fecal isolates from healthy donors and 88% of the sepsis isolates were found to be traTp positive.

In non-*E. coli* clinical isolates, we found a traTp-like protein in *S. typhimurium*, and an anti-traTp antibody reacting protein with a M_r greater than 28,000 daltons in salmonellae belonging to serogroups B, C, and D and in *Enterobacter cloacae*. The anti-traTp antibody would appear to be a useful reagent for detecting traTp-positive and/or IncF group plasmid-carrying isolates.

Acknowledgement. The work reported here was supported by the Bundesministerium für Forschung und Technologie, Bonn, grant no. 01ZR0420.

References

1. Achtman M, Kennedy N, Skurray R (1977) Cell-cell interactions in conjugating *Escherichia coli*: role of traT protein in surface exclusion. Proc Natl Acad Sci 74:5104–5108
2. Agüero ME, Capello FC (1983) Relative contribution of ColV plasmid and K1 antigen to the pathogenicity of *Escherichia coli*. Infect Immun 40:359–368

2a. Agüero ME, Aron L, DeLuca AG, Timmis KN, Cabello FC (1984) A plasmid-encoded outer membrane protein, TraT, enhances resistance of Escherichia coli to phagocytosis. Infect Immun 46:740–746

3. Bagdasarian M, Lenz R, Rückert B, Franklin FCH, Bagdasarian MM, Frey J, Timmis KN (1981) Specific purpose plasmid cloning vectors. II. Broad host range, high copy number RSF 1010-derived vectors, and a host-vector system for gene cloning in pseudomonas. Gene 16:237–247
4. Bitter-Suermann D, Peters H, Jürs M, Nehrbass R, Montenegro M, Timmis KN (1984) Monoclonal antibody detection of IncF group plasmid-encoded TraT protein in clinical isolates of Escherichia coli. Infect Immun 46:308–313
5. Bremer E, Cole ST, Hindenach I, Henning U, Beck E, Kurz C, Schaller H (1982) Export of a protein into the outer membrane of *Escherichia coli* K12. Eur J Biochem 122:223–231
6. Chang ACY, Cohen SN (1978) Construction and characterization of amplifiable multicopy DNA cloning vehicles derived from the p15A cryptic mini plasmid. J Bacteriol 134:1141–1156
7. Hawkes R, Niday E, Gordon J (1982) A Dot-immunobinding assay for monoclonoal and other antibodies. Anal Biochem 119:142–147
8. Hindenach I, Henning U (1975) The major proteins of the *Escherichia coli* outer cell envelope membrane. Preparative isolation of all major proteins. Eur J Biochem 59:207–213
9. Kearney JF, Radbruch A, Liesegang B, Rajewsky K (1979) A new mouse myeloma cell line that has lost immunoglobulin expression but permits the construction of antibody-secreting hybrid cell lines. J Immunol 123:1548–1550
10. Kohler G, Milstein C (1975) Continuous cultures of fused cells secreting antibody of predefined specifity. Nature 256:495–497
11. Laemmli UK (1970) Cleavage of structural proteins during assembly of the head of bacteriophage T4. Nature 227:680–685
12. Lugtenberg B, van Alphen L (1983) Molecular architecture and functioning of the outer membrane of *Eschericha coli* and other gram-negative bacteria. Biochem Biophys Acta 737:51–115
13. Manning PA, Bentin L, Achtman M (1980) Outer membrane of *Escherichia coli*: properties of the F sex factor traT protein which is involved in surface exclusion. J Bacteriol 142:285–294
14. Moll A, Manning PA, Timmis KN (1980) Plasmid-determined resistance to serum bactericidal activity: a major outer membrane protein, the traT gene product, is responsible for plasmid-specified serum resistance in *Escherichia coli*. Infect Immun 28:359–367
15. Oi VT, Herzenberg LA (1980) In: Mishell BB, Shiigi SM (eds) Selected methods in cellular immunology. Freemann, San Francisco, pp 351–371
16. Towbin H, Staekelin T, Gordon J (1979) Electrophoretic transfer of proteins from polyacrylamide gels to nitrocellulose sheets: procedure and some applications. Proc Natl Acad Sci 76:4350–4354
17. Willets N, Skurray R (1980) The conjugation system of F-like plasmids. Ann Rev Genet 14:41–76

Immunoblotting and Dot Immunobinding

J. Gordon [1,2], Ch. Rordorf [3], M. Rosenthal [4], and Y. Z. Sun [2]

Current Immunoassay Technique

The great advances in immunoassay technique in recent years have resulted from the advent of solid-phase methodology. The use of reactants bound to a solid phase permits the easy removal of excess reagents in simple, rapid washing procedures. The resulting methodology is rapid, simple, and quantitative. However, this methodology does not as such give directly qualitative information concerning homogeneity, purity, or identity of immunoreacting species. Classic immunodiffusion and immunoelectrophoresis, depending on diffusion and precipitation in a gel phase, did indeed yield such qualitative information. However, these methods are slow, cumbersome, and insensitive. With the advent of immunoblotting (Towbin et al. 1979), it has been possible to combine the power of the classic methodology for qualitative analysis with the speed, simplicity, and sensitivity of solid-phase methodology.

The drive for the development of immunoblotting came from the need to obtain more powerful analytical procedures for screening and characterization of monoclonal antibodies. In turn, the methodology now provides the vehicle for innumerable applications of monoclonal antibodies (Towbin and Gordon 1984).

Prior to the advent of the hybridoma technique, it was necessary to prepare immunogens in sufficient quantity and of sufficient purity in order to obtain antibodies for use in specific immunoassays. The advent of hybridoma methodology resulted in the substitution of these problems by problems of screening, in order to obtain hybridomas secreting antibodies of the desired specificity. This need provided the impetus for the development of the immunoblotting technique. One consequence of the specificity of monoclonal antibodies is that they will fulfil the criteria of the original assay conditions used for the screening, but may not fulfil the criteria of the end use. With the dot immunobinding assay (Hawkes et al. 982), many specificities may be screened for simultaneously, so that the risk of inappropriate specificity of the selected hybridoma may be minimized.

Since dot immunobinding methodology permits the performance of multiple simultaneous immunoassays on the same sample in a very simple procedure, it also has great potential for applications in routine clinical analysis, utilizing monoclonal or polyclonal antibodies for specific immunoassays, or analysis of pathological antibodies in patient sera. We have, therefore, begun exploring these possibilities.

1 Present Address: Dept. 99H, Abbott Laboratories, North Chicago, IL 60064, USA
2 Friedrich Miescher Institut, Postfach 2543, CH-4002 Basel
3 Pharmaceuticals Division, CIBA-GEIGY AG, Postfach 273, CH–4002 Basel
4 Departement für Medizinische Forschung, Kantonspital Basel, Steinengraben 67, CH-4051 Basel

Immunoblotting Methodology

The technique depends on initial separation of the protein antigens of interest by a suitable gel electrophoresis procedure, followed by transverse electrophoresis onto a solid phase, where the proteins are trapped in a form suitable for subsequent immunoassay. The proteins bind to the solid phase – usually nitrocellulose – in the form of a replica of the original gel pattern. The remaining protein-binding sites are then blocked with some nonimmunoreacting substance such as bovine serum albumin (BSA) or whole serum ("blocking"). The distribution of the immunoreacting species is then determined in an immunodetection step. The immunodetection may be any of the methods used in solid-phase immunoassays provided the signal remains localized. We described protocols for immunodetection with enzyme-, fluorescent-, or [^{125}I]-labeled antiimmunoglobulin (Towbin et al. 1979). Subsequently, Burnette (1981) described immunodetection with [^{125}I]-labeled protein A. This has been widely adopted because of the wide availability of protein A and autoradiography. However, protein A does not recognize all immunoglobulins, and radioactivity is undesirable. Burnette (1981) also introduced the expression "western blotting," which has become widely accepted. We prefer retaining "immunoblotting" as it is more directly descriptive.

Numerous variations of the protocols have been described and are summarized elsewhere (Towbin and Gordon 1984). Two variants were described in the original publication (Towbin et al. 1979). One was for ribosomal protein gels containing urea as a denaturant, and the other for Laemmli (1970) gels containing sodium dodecyl sulfate (SDS). In the former case, transfer was in a medium of low pH, where all proteins were negatively charged, and there was quantitative transfer to the nitrocellulose. In the case of SDS gels, the situation is more complicated. The SDS may be required to maintain the solubility and net negative charge of the proteins, but may also prevent the binding to nitrocellulose. A compromise may therefore have to be struck. This is also discussed in detail elsewhere (Towbin and Gordon 1984). The conditions will have to be tailored to the particular problem. We have described optimal conditions for transfer of ribosomal protein S6 (Nielsen et al. 1982).

Variations of blocking procedures have also been summarized (Towbin and Gordon 1984). Our preferred procedure is to use 10% horse serum, both for blocking and in antibody diluents. This both blocks the binding capacity of the nitrocellulose and minimizes nonspecific protein-protein interactions. The large excess of irrelevant immunoglobulins is important, since we find horse serum is more effective than fetal calf serum. In the case of anti-nucleic acid antibodies, nuclease degradation may occur. Then we use 1 mg/ml bovine serum albumin (BSA) and 0.01% Nonidet P40.

Immunodetection with peroxidase-labeled antibody is the most convenient and sensitive method. Commercially available reagents are used and the chromogenic reaction products remain at the site of reaction. The most convenient chromogenic substrate is 4-chloronaphthol, which is highly sensitive and not known to be carcinogenic (Hawkes et al. 1982).

Dot Immunobinding Methodology

Here, the same methodology as in immunoblotting is used except for the replacement of the electrophoretic transfer step by direct application of the antigens (or antibodies) to the solid support. Any number of samples can be applied in any desired format. As originally described (Hawkes et al. 1982), either single dots were applied to nitrocellulose squares for the screening of hybridoma supernatants, or multiple dots of various viral antigens were applied on nitrocellulose strips.

Gridded nitrocellulose is most suitable as it permits the definition of the location of the dots and facilitates cutting into strips. The application may be with a micropipet or microdispenser. The Drummond (Broomall, Pennsylvania, USA) or Brand (Wertheim, West Germany) Microdispenser, which delivers 5 µl in 1-µl aliquots, was modified to deliver 0.5-µl aliquots (Hawkes et al. 1982; Derer et al. 1984). A microprocessor-controlled syringe-driven device (Microlab M, Hamilton, Bonaduz, Switzerland) delivering dots of down to 0.2 µl is optimal when quantitation is required (Derer et al. 1984). When these small volumes are used, the dot dries spontaneously. The drying might be important for the immobilization of some antigens. The dotted sheets may be stored in a dry state, either before or after blocking.

Quantitative Aspects

Densitometry

The immunoperoxidase stain bears a quantitative relation to the amount of bound antibody, so that immunoblotting and dot immunobinding can be used in a quantitative fashion. We have used a thin-layer chromatogram densitometer (CAMAG, Muttenz, Switzerland) in the reflectance mode. Scans of immunoblots were published for illustrative purposes (Towbin et al. 1982; Rordorf et al. 1983). Strips of dots may be glued to cards with rubber cement for scanning purposes. The optical properties of the nitrocellulose are unchanged, and the strips remain flat without buckling. Parallel tracks are automatically scanned successively.

Integrator Program

The CAMAG densitometer can be provided with an interface for a Spectra Physics (Santa Clara, California, USA) computing integrator, which is BASIC-programmable. With strips containing many different antigen dots, handling the data presents difficulties due to the amount of raw data and the spurious peaks arising from the grid lines of the nitrocellulose. We have written a program which uses the grid lines as a reference frame and selectively prints out the data for the dots only, and not for the grid lines. The program begins with a dialog, where the user is invited to input file data concerning the project, dates of materials and tests, codes for successive sera or samples, names of individual dots, and the de-

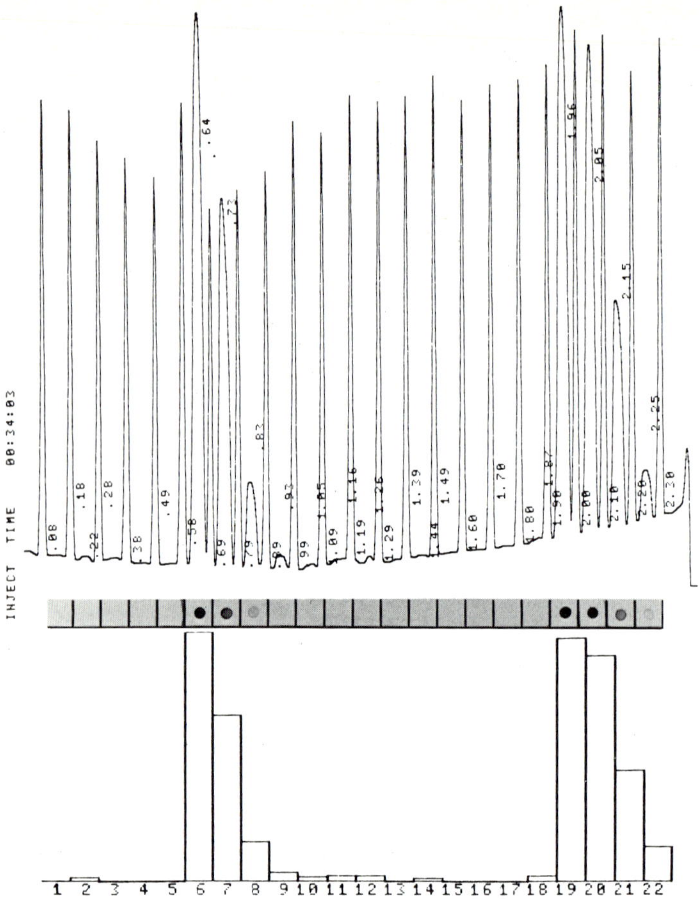


```
PROJECT TITLE:DEMONSTRATION OF SCANNER PROGRAM
DATE OF KIT:NOV.17,83
DATE OF TEST:JAN.06,84
STRIP NR. IS: 1.
SERUM CODE IS:ASC.73C3
******DATA FOR NAMED PEAKS******

NAME          PEAK HEIGHT    RT
 1.BSA             0.         0.18
 2.H2A             1.4        0.22
 3.H2A/5           0.         0.38
 4.H2A/25          0.         0.49
 5.H2A/125         0.         0.58
 6.H2B           130.7        0.64
 7.H2B/5          87.2        0.73
 8.H2B/25         20.4        0.83
 9.H2B/125         3.7        0.93
10.H2B/625         1.5        1.05
11.H3              2.1        1.16
12.H3/5            2.2        1.26
13.H3/25           0.         1.39
14.H3/125          0.8        1.44
15.H4              0.         1.6
16.H4/5            0.         1.7
17.H4/25           0.         1.8
18.H4/125          2.1        1.87
19.TOTAL         127.5        1.96
20.TOTAL/5       118.4        2.05
21.TOTAL/25       58.1        2.15
22.TOTAL/125      17.8        2.25
```

Fig. 1. Program for displaying data from dot immunobinding assays. *Top,* analog trace from Spectra Physic Integrator above actual dots being scanned. *Middle,* histogram output showing selective plot of dots only. *Bottom,* selective output showing actual data. The *numbers* of the dots correspond to the *numbers* in the histogram plot. Copies of the program may be obtained from the authors sired output format. Any combination of the following output options can be obtained: the "Full report" which gives data on all peaks recognized by the Spectra Physics ROM, "Selective report" which gives only data corresponding to the named dots, and "Histogram report" which gives the same data as the selective report, but in a graphic format. The analog trace in real time can also be suppressed if desired. The graphic format permits a rapid visual analysis of the quantitative data. An example of such a report for an analysis of the specificity of a monoclonal antibody against histone (Mueller, Couppez, Briand, Gordon, Sautière, and Van Regenmortel, in preparation) is given in Fig. 1.

Curve Fitting

The instrumental response of the densitometer to the amount of bound immunoglobulin is not linear (Hawkes et al. 1982). The nonlinearity may be made up of several components. When Coomassie blue is applied, the response is also not linear with respect to the amount of dye added. It appears that as increasing surface sites are saturated, the dye binds to sites deeper below the surface. Similarly, when the dot is overloaded with antigen, immunoperoxidase stain may be seen on the reverse side. In addition, there can be nonlinearity in the amount of antibody bound as a function of concentration and nonlinearity in the peroxidase reaction. We have attempted to describe the shape of the curves obtained in order to transform the data into standardizable units.

In the case of a sandwich assays for total IgE, the assay was standardized with reference samples with known amounts of IgE. A fit was obtained with a parabolic function, which permitted the transformation of data into international units (Derer et al. 1984). When a sufficient amount of pure immunoglobulin is available, it is possible to set up an internal standard series with dots of known amounts of immunoglobulin applied directly to the strip. The use of such a standard series is attractive visually and permits the individual correction of samples for a number of experimental errors. However, the method depends on the assumption that the same amount of anti-immunoglobulin will bind regardless of whether the immunoglobulin is bound to its respective antigen or directly to nitrocellulose. This hypothesis is difficult to test rigorously, but the approach nevertheless provides an empirical procedure for standardization of data.

Another weakness of this method of data reduction is that imprecision in the standard values may result in imprecision in the transformed data, which will outweigh the advantage of the standardization. We have adopted a compromise procedure, whereby all of the standards of a particular run are averaged and used to transform the data of that run.

We have found empirically that an equation of the form:

$$1/y = A/x + B/x^2 + C$$

provides the best fit to the data data over a wide range of immunoglobulin amounts, as indicated in the legend to Fig. 2. This figure shows an example of a standard curve with flags indicating the standard deviation of the experimental error. Figure 3 shows the reliability of the transformation by the above function

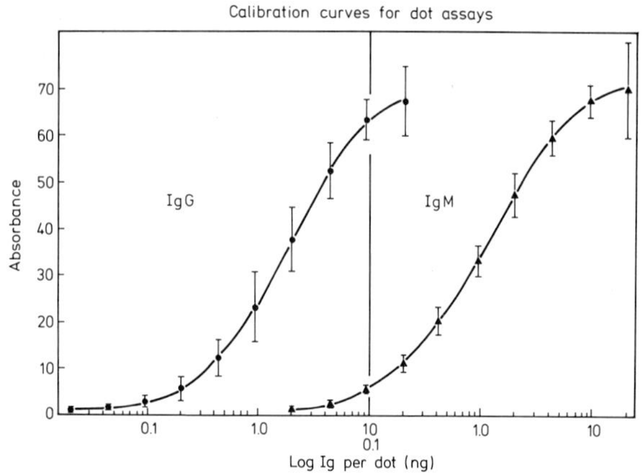

Fig. 2. Plot of calibration dots with amounts of Ig in the dot as indicated on the *abscissa*. The *flags* represent the standard deviations from ten determinations

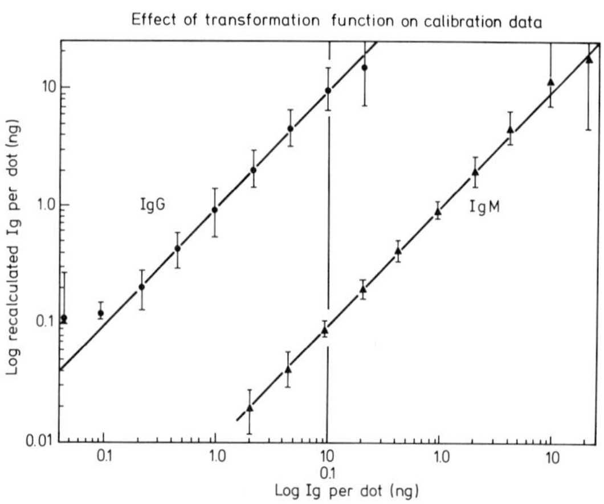

Fig. 3. Test of transformation function. The absorbancy values for the dots of the standards (see Fig. 2) were transformed back to ng/dot by the function: $y = A + B/x + C/x^2$, with the constants A, B, and C being the best fit to equation, y is the amount of Ig per dot, and x is the absorbance value of the dot. The values were also calculated at the extremes of the ranges of the standard deviations of Fig. 1 and are shown here as the *flags*

by transforming the absorbance values for the standards and comparing with the original amounts of immunoglobulin. The limits of the standard deviations have also been transformed to show the effect of error through this transformation. It can be seen that the method permits the quantitation of immunoglobulin over a 1,000-fold range with a confidence limit of approximately a factor of 2.

Applications in Autoimmunity

Immunoblotting of Ribosomal Proteins

In spite of their evolutionary conservativeness (Ramjoue and Gordon 1977), it is possible to raise antibodies which are immunologically cross-reactive with the immunized species. We originally applied immunoblotting to the screening and characterization of monoclonal antibodies against ribosomal proteins (Towbin et al. 1979). We immunized mice with whole eukaryotic ribosomes or preparations of total ribosomal proteins. This was more realistic than obtaining a set of purified ribosomal proteins (approximately 70 in all. McConkey et al. 1978) and performing individual immunizations, a formidable task which has not been achieved with eukaryotic ribosomal proteins. In practice, the frequency of positive clones was such that we obtained only a limited set (Towbin et al. 1982). These were used to demonstrate the existence of a previously unrecognized ribosomal protein, immunologically related to proteins P1/P2 (Towbin et al. 1982). Also, that against S6 was used as part of an assay for the phosphorylation of S6 (Nielsen et al. 1982). While the antibody binding was not dependent on the phosphorylation, it was possible to use the antibody as a selective immunostain for S6 on blots of one-dimensional gels (Nielsen et al. 1982). The shift in electrophoretic mobility was used as an index of the phosphorylation under conditions where the protein was not resolved from a large number of other proteins. Thus, phosphorylation assays were possible on crude material from a variety of cells and tissues which otherwise did not lend themselves to analysis. Ambiguities due to the use of isotopic labeling of biological systems were thus also eliminated. This was used to demonstrate in vivo changes in phosphorylation under various physiological conditions.

It has been known for some years (Schur et al. 1969) that antibodies against ribosomal proteins are sometimes present in the sera of patients with systemic lupus erythematosus (SLE). We therefore proposed that immunoblot methodology would find diagnostic applications in autoimmune diseases (Towbin et al. 1979). Subsequently, we used immunoblots of ribosomal proteins to screen the sera of patients with SLE and other connective tissue diseases (Gordon et al. 1982). We demonstrated the presence of antibodies against the protein moiety of ribosomes in sera or patients with SLE and other connective tissue diseases (Table 1). Further, in spite of the known evolutionary conservativeness of ribosomal proteins

Table 1. Antibodies to ribosomal proteins in various diseases (after Gordon et al., 1982)

Disease	Number positive on blot[a]
SLE	11/47
RA	33/92
MCTD	2/12
SS	0/ 4
PSS	3/ 7
RSDS	1/ 2
Controls	1/61

Abbreviations: SLE, systemic lupus erythematosus; RA, rheumatoid arthritis; MCTD, mixed connective tissue disease; SS, Sjogrens syndrome; PSS, progressive systemic sclerosis; RSDS, reflexive sympathetic dystrophy syndrome

[a] Numbers found giving reactivity with human ribosomal protein on a blot

(Ramjoue et al. 1977), we found a significant number of patient sera containing human-specific antiribosomal protein antibodies. This is in contrast to the nuclear antigens widely used in the study of connective tissue diseases, where antigens of nonhuman origin are routinely used.

Immunoblotting of Nuclear Antigens

Nuclear antigens have in fact received more attention than ribosomes as markers for connective tissue diseases, and the antigen referred to as RNP is in fact diagnostic of mixed connective tissue disease (MCTD). Patient sera showing a reaction with an apparently unique antigen have been used as the basis for a variety of markers. Connective tissue diseases have a complex etiology and a variety of tissue and organ systems may be affected. The diagnosis may often be late. Efforts are therefore being made to develop easy screening tests for such diseases. Several groups have used immunoblot methodology in the characterization of the marker antigens. The molecular weights of the peptides found in this way are summarized in Towbin et al. (1984). There is so far no consensus on the molecular weights of the relevant immunoreactive peptides. More work needs to be done to characterize these antigens completely and to establish their importance in understanding the etiology and their use in the diagnosis of connective tissue diseases.

Dot Immunobinding with Mouse Antigens

While immunoblotting can yield detailed information concerning the nature of antigenic markers for autoimmune disease, discrete information concerning specific classes of antigen are of more immediate applicability and can be provided by the dot immunobinding assay. We therefore used well-characterized sera from autoimmune mice, MRL 1pr/1pr, which are a model for the human SLE, to evaluate the dot immunobinding assay for autoimmunity. We set up assays for single-stranded DNA, double-stranded DNA, a soluble nucleoprotein antigen preparation (SNP), ribosomes, circulating immune complex, and endogenous retroviral antigens (Rordorf et al. 1983).

Endogenous retroviral immune complexes are characteristic of autoimmune strains of mice, but have not been reported for human SLE. We found by immunoblotting that the serum of the mice contained anti-P30. Only anti-gp70 was previously known to be present in circulating immune complexes in MRL mice.

The dot immunobinding assays for all of the above parameters were shown to yield values which increased with age in the autoimmune mice and were specific for the autoimmune strain (Rordorf et al. 1983), confirming data from conventional enzyme immunoassays on the same sera.

Dot Immunobinding with Human Antigens

The work with the mouse sera was extended to human patient sera (Gordon and Rosenthal 1985). In order to make the study as general as possible, we used antigens of human origin (except the DNA and rheumatoid factor assays). We also avoided purified human antigens by using subcellular fractionation cuts which would contain all possible representative antigens. We have found that human tissue material is not suitable as even the purified cuts still contain sufficient contaminating human immunoglobulins to give high background reaction. We therefore used HeLa cells as readily available material which can be taken as a generalized human cell. A possible disadvantage of this approach might be the absence of tissue-specific antigens.

Strips of nitrocellulose were prepared with the following antigens: single-stranded and double-stranded DNA, rabbit IgG (for rheumatoid factor), whole HeLa cells, and subcellular freactions from HeLa cells (nuclei, mitochondria, cytosol, ribosomes, nucleoli, and soluble nuclear ribonucleoprotein). Sera from patients with connective tissue diseases, healthy individuals, and nonautoimmune individuals were screened with these strips. The assays for antinuclear antibodies and rheumatoid factor showed good qualitative correlation with those from standard clinical laboratory assays (Table 2) (Gordon and Rosenthal 1985). The normal sera and irrelevant diseases were used to establish the nonpathological ranges of values for these assays, and the pathological ranges were determined. The rheumatoid factor values were selectively elevated in rheumatoid arthritis patient sera, while the other parameters were elevated in the other connective tissue disease patient sera. Further, each strip yielded a profile which was recognizably characteristic for sera taken at different times from an individual patient. Systematic changes in individual dots in the profile were then readily recognizable (Gordon and Rosenthal 1985).

Table 2. Accordance of assays from autoimmunity dot profile with standard assays

Group	Assay for	No. of agreements		No. of discrepancies		Agreement (%)
		+/+[a, b]	−/−	+/−	−/+	
Healthy	Antinuclei	0	99	3	0	97
	Rheumatoid factor	0	101	1	0	99
Miscellaneous	Antinuclei	4	53	1	1	97
	Rheumatoid factor	0	54	0	5	92
Rheumatoid arthritis	Antinuclei	9	32	6	3	82
	Rheumatoid factor	22	20	5	3	85
Other connective tissue diseases	Antinuclei	10	10	0	2	91
	Rheumatoid factor	5	13	2	2	82

[a] +/+, both assays positive; −/−, both assays negative; +/−, dot assay positive and clinical laboratory negative; −/+, dot assay negative and clinical laboratory positive

[b] Dot assays were taken as positive when 20 or more units of absorbance were obtained. In the case of antinuclear antibodies, agreement was scored +/+ if either IgG or IgM or both tests were positive. Immunofluorescence was taken as positive if a titer of 1:20 was obtained. For rheumatoid factor, only IgM-specific dot assays were taken into account

Table 3. Day-to-day reproducibility of dot immunobinding assays for connective tissue disease

Antigen in dot	IgG[a]		IgM	
	Serum 1	Serum 2	Serum 3	Serum 2
ssDNA	0	0	0.15±0.01	0
dsDNA	0	0	0.44±0.05	0
RF	–	–	0.05±0.008	0.20±0.03
HeLa	2.8 ±0.5	4.6 ±1.8	0.47±0.14	1.9 ±0.3
Nuclei	2.2 ±0.7	4.7 ±0.8	0.39±0.03	2.7 ±1.0
Mitochondria	0.86 ±0.12	2.4 ±0.9	0.1 ±0.02	0.93±0.2
Ribosomes	1.1 ±0.1	3.7 ±2	0.12±0.02	1.7 ±0.6
Cytosol	0.29 ±0.04	0.33±0.06	0	0.10±0.03
Nucleoli	0.087±0.014	0.44±0.1	0.03±0.008	0.18±0.05
Nucleosol a	2.50 ±0.9	3.51±1	0.07±0.02	1.6 ±0.5
Nucleosol b	2.57 ±0.4	4.68±1.4	0.07±0.01	2.78±1.9

[a] Assays were based on dots of subcellular fractions from HeLa cells (details to be published in Gordon and Rosenthal 1984b) and reference sera from the Center for Diseases Control, Atlanta, Georgia. Serum 1 is a reference serum for Sm and RNP, serum 2 for Sm, and serum 3 dsDNA. The data are means and standard deviations on assays repeated on 5 successive days and transformed as described in the legend to Fig. 2 to ngIg bound per dot

With this data on several hundred pathological and nonpathological sera, we attempted to confirm the well-known increase in spontaneous occurrence of autoantibodies with aging. We found in fact no relationship to age (Gordon and Rosenthal 1984). It is not clear whether this is a result of the assay system being used or whether the explanation lies elsewhere. Nevertheless, no age-dependent correction need be applied to the normal ranges of the titers.

Internationally available reference sera can be used for standardization of the assay conditions. These are reference sera for anti-double-stranded DNA, and the antigens Sm and RNP. The presence of other antibodies from these sera cannot be excluded, and some might be present. The results in Table 3 show values obtained with these sera and the transformation function of the type is shown in Fig. 3. The day-to-day variation is also shown.

Hybridoma Banking

Normally, a specific monoclonal antibody is sought following a fusion of spleen cells from immunized mice. Since the MRL mice contain such a variety of spontaneously occurring antibodies, it is reasonable to establish a bank of hybridomas from them, without any prior immunization. The dot immunobinding assay permits the screening of such a bank with multiple assays, thus avoiding repetitive screening of the same supernatants. In this way, we have been able to establish a bank of 650 hybridomas, and have screened them for a variety of specificities. Hybridomas have been obtained with anti-DNA, antihistone, and several antinuclear antibodies with as yet undetermined specificities.

The anti-DNA antibodies were rescreened with synthetic polynucleotides with different structures (Sun and Gordon, in preparation). No clear-cut pattern emerged. Each antibody had a characteristic fingerprint of relative reactivities to double- and single-stranded DNA and to other polynucleotides. In order to complete the characterization of the specificities, we also set up a dot immunobinding assay for cardiolipin, since it has been claimed that the phosphodiester backbone of cardiolipin is sufficiently similar to that of a nucleic acid that extensive cross-reactivities occur (Lafer et al. 1981; Koike et al. 1982; Schoenfeld et al. 1983). We found no reactivity of any of our 20 monoclonal anti-DNA antibodies with cardiolipin. The assay itself was validated with human syphilitic sera, which are diagnosed by anticardiolipin (the classic test). It is not clear whether our failure to find cross reactivity with cardiolipin is due to the assay conditions or due to the selection conditions for the monoclonal antibodies. It is possible that the conditions of the dot immunobinding assay have a higher degree of selectivity than do classic solid-phase assays.

Applications to Allergy

Total IgE

Allergy diagnosis is an attractive application for immunoblot and immunodot technique, because of the potential for multiple assays and high sensitivity. The IgE levels are approximately one millionth of those of the common immunoglobulins. We have established a dot immunobinding assay for total IgE (Derer et al. 1984). A sandwich format was used, with the same antibody preparation applied as dots to the nitrocellulose as that used for the immunodetection. When different antibody preparations were used, backgrounds due to antibody-antibody interactions were found. The antibody preparation was of commercial origin and is of sufficiently high titer that one batch can be used for hundreds of thousands of assays. A sandwich technique was preferable to a direct assay with samples applied as dots, since it is more convenient for the user to perform assays on nitrocellulose already prepared, and the sensitivity is far higher in the sandwich format. Data correlated well with total IgE determined with a commercially available radioimmunoassay kit and had $\pm 3\%$–5% reproducibility.

Specific IgE

Since it is necessary to assay for multiple allergen-specific IgEs during the course of allergy diagnosis, the use of strips with multiple allergen-specific assays is attractive. In order to explore the feasibility of this approach, we set up such assays with bee venom and dust mites as allergens (Derer et al. 1984). The detection steps were the same as for the total IgE, so that it was possible to combine total and specific IgE tests in one strip. We compared the results with a limited number of positive and negative sera, those obtained with the commercially available RAST assays (Pharmacia, Uppsala, Sweden). While the results did not correlate well quantitatively with the RAST results, there was enough qualitative correlation to indicate that dot immunobinding assays for allergen-specific IgE are feasible. Walsh et al. (1984) adopted a similar approach independently and demonstrated the feasibility of such assays as an all-or-none test for bakers' allergies.

References

Burnette WN (1981) "Western blotting": electrophoretic transfer of proteins from sodium dodecyl sulfate gels to unmodified nitrocellulose and radiographic detection with antibodies and radioiodinated protein. Anal Biochem 112:195–203

Derer M, Miescher S, Johanssen B, Frost H, Gordon J (1984) Applications of the dot immunobinding assay to allergy diagnosis. J Allergy Clin Immunol 74:85–92

Gordon J, Rosenthal M (1984) Failure to detect age-related increase of non-pathological autoantibodies. Lancet 1:231

Gordon J, Rosenthal M (1985) A multi-dot immunobinding assay for auto-immune parameters: evaluation in the diagnosis and management of connective tissue disease. J Rheumatol 12:257–264

Gordon J, Towbin H, Rosenthal M (1982) Antibodies directed against ribosomal protein determinants in the sera of patients with connective tissue diseases. J Rheumatol 9:247–252

Hawkes R, Niday E, Gordon J (1982) A dot immunobinding assay for monoclonal and other antibodies. Anal Biochem 119:142–147

Koike T, Tomioka K, Kumakai A (1982) Antibodies cross-reactive with DNA and cardiolipin in patients with systemic lupus erythematosus. Clin Exp Immunol 50:298–302

Laemmli U (1970) Cleavage of structural proteins during the assembly of the head of bacteriophage T4. Nature 227:680–685

Lafer EM, Rauch J, Andrzejeski C, Mudd D, Furie B, Furie B, Schwartz RS, Stollar B (1981). Polyspecific monoclonal antibodies reactive with both polynucleotides and phospholipids. J Exp Med 153:897–909

McConkey EH, Bielka H, Gordon J, Lastick SM, Lin A, Ogata K, Reboud JP, Traugh JA, Traut RR, Warner JR, Welfle H, Wool IG (1979) A new uniform nomenclature for eukaryotic ribosomal proteins. Mol Genetics 169:1–6

Nielsen PJ, Manchester KL, Towbin H, Gordon J, Thomas G (1982) Physiological variations on the phosphorylation of the rat small ribosomal subunit protein S6. J Biol Chem 257:12316–12321

Ramjoue HPR, Gordon J (1977) Evolutionary microdivergence of eukaryotic ribosomal proteins. J Biol Chem 252:9065–9070

Rordorf C, Gambke C, Gordon J (1983) A multidot immunobinding assay for autoimmunity and the demonstration of novel antibodies against retroviral antigens in the sera of MRL mice. J Immunol Methods 59:105–112

Schoenfeld Y, Rauch J, Massicotte H, Datta S, Andre-Schwartz J, Stollar BD, Schwarz RS (1983) Polyspecificity of monoclonal lupus autoantibodies produced by human-human hybridomas. N Engl J Med 308:414–420

Towbin H, Gordon J (1984) Immunoblotting and dot immunobinding – current status and outlook. J Immunol Methods 72:313–340

Towbin H, Staehlin T, Gordon J (1979) Electrophoretic transfer of proteins from polyacrylamide gels to nitrocellulose sheets: procedure and some applications. Proc Natl Acad Sci USA 76:4350–4354

Towbin H, Ramjoue HP, Kuster H, Liverani D, Gordon J (1982) Monoclonal antibodies against eukaryotic ribosomal proteins. Use to characterize a ribosomal protein not previously identified and antigenically related to acidic phosphorproteins P1/P2. J Biol Chem 257:12709–12715

Walsh BJ, Wrigley CW, Baldo BA (1984) Simultaneous detection of IgE binding to several allergens using a nitrocellulose "polydisc." J Immunol Methods 66:99–102

Fluorescence-Activated Cell Sorter and Monoclonal Antibodies: Complementary Tools in Immunodiagnosis and Immunotherapy

T. J. KIPPS and L. A. HERZENBERG [1]

Introduction

The Fluorescence-Activated Cell Sorter and Monoclonal Antibodies as Complementary Tools

The fluorescence-activated cell sorter (FACS) and monoclonal antibodies are complementary tools in cell biology. Monoclonal antibody technology has made generally available many well-characterized, fluorochrome-conjugated antibodies to cell surface antigens that are required for discriminating FACS analysis and sorting. The FACS, on the other hand, has greatly aided the production, characterization, and utilization of monoclonal-producing hybridomas. A recent extension of this has been the use of the FACS to select from existing hybridomas ones producing monoclonal proteins of different antibody isotype. This paper will focus on this most recent application, discussing hybridoma isotype switch variant selection using the FACS.

Monoclonal Antibody Isotype and the Need for Change

The ability to change the isotype of monoclonal antibodies may greatly enhance their application to cell biology and medicine. Most screening techniques for newly developed hybridomas focus on selecting those producing monoclonal antibodies of desired antigen-binding specificity. It is the immunoglobulin isotype, however, that largely directs the functional activity of the selected monoclonal antibody [1]. Thus, while a monoclonal antibody may have the desired binding specificity, it may still fail to satisfy the requirements for a given biologic assay or application. Many monoclonal antibodies to cell surface antigens, for example, are of the IgG_1, isotype, a class of antibody generally unable to fix complement through the conventional pathway. Because of this, many of these antibodies are not suited for complement-mediated cell depletion experiments or standard tissue-typing assays. The ability to change the isotype of these monoclonal antibodies to either IgG_{2a} or IgG_{2b}, antibody isotypes that are very efficient in fixing complement, may enhance their application to these fields [2].

---

1 Department of Genetics, Stanford University, Stanford, CA 94305, USA

Isotype Switch Variants

As first demonstrated by Rajewsky and his associates, the FACS can isolate rare isotype switch variant cells from large cultures of hybridomas [3–6]. The cell surface immunoglobulin of hybridomas is of identical isotype and antigen-binding specificity as the secreted monoclonal antibody. Using fluorescence-conjugated antibody specific for an immunoglobulin isotype not expressed by the parent hybridoma cell line, one can stain and subsequently sort rare switch variant cells that produce antibodies of desired isotypes, while retaining varibale region idiotype(s), binding specificity, and light chain identical to antibodies produced by the parent hybridoma line [5–8]. Switch variants thus apparently switch from the expression of the gene encoding the constant part of one heavy chain to the expression of another gene, while using the same heavy-chain variable region gene and L-chain gene which together encode the antibody's antigen-combining site.

Techniques for Rare Cell Sorting

Improvements in Switch Variant Selection

Since the first variants were isolated with the FACS, we have made several improvements in the sorting of rare cells which greatly facilitate switch variant selection. The first switch variants were isolated only after repeated rounds of FACS sorting, involving months of tissue culture and many hours of FACS time. Currently, it is possible to detect and, in many cases, directly clone switch variant cells with only one round of FACS selection.

Staining Reagents

The specificity and brightness of the staining reagent are critical for variant isotype sorting. Conventional polyclonal anti-isotype antibodies must be absorbed extensively. After affinity purification and subsequent conjugation to fluorescein, such preparations may still nonspecifically stain hybridomas, making them unacceptable for switch variant selection. Many recently developed monoclonal anti-mouse allotype antibodies, however, react with only one isotype [9, 10], making them useful reagents for variant sorting [11, 12]. Staining with two such monoclonal antibodies labeled with fluorescein that bind to different sites (epitopes) on the variant isotype specifically labels variant cells more brightly than many fluorochrome conjugated polyclonal antibody preparations.

Exclusion of Dead Cells

When selecting for rare cells (i.e., at frequencies of $1–10/10^6$ cells), it is important to exclude dead and dying cells. Such cells nonspecifically absorb the fluorescein-labeled antibody, decreasing the reagent's effective staining concentration and increasing background fluorescence. For these reasons, cell populations should have a greater than 90% viability. Even with optimal viability, however, a given population will have many more dead cells than true variants. Therefore, we use a dye, propidium iodide (PI), in our cell-staining experiments. This dye intercalates with RNA and DNA but is excluded from living cells. Because the membranes of dead and dying cells are freely permeable to PI, this dye specifically labels these cells brightly red fluorescent (above 580 nm) when excited by the argon dye laser at 488 nm. We also monitor foward-angle ligth scatter measured from 3°–12° on the FACS, since the scattered ligth of most dead cells is less than that of living cells. Thus, by electronically excluding (gating out) cells with bright red fluorescence and low foward-angle light scatter, dead or dying cells can be largely eliminated.

Cell Flow Rates

To select rare cells such as isotype variants, we use flow rates of 2,500–3,500 cells per second. To minimize cell loss due to "coincidence" gating, in which the drop containing the desired cell is not deflected because of the possible inclusion of unwanded cells, we deflect an average of 1.5 drops per sorted cell. Using an 80-μ nozzle driven at 28–32 kHz, excellent recovery of sorted cells can be obtained even at these high flow rates. Setting the number of deflected drops to average 1.5, however, provides only a ± 0.25 drop cycle latitude for error in the drop deflection. Therefore, it is particularly important to determine carefully the optimal intrinsic sort delay setting for each particular nozzle and FACS apparatus. A recent review on fluorescence-activated cell sorting discusses the problem of optimization [13].

Limiting Dilution Analysis of Sorted Populations

In order to detect directly rare switch variant cells, we devised "pauci-population sorting." With this technique, we sort 5, 25, 50, or 100 selected cells into each well of 96-well microtiter plate. After 1–2 weeks' culture, we test the supernates for antibodies of a desired isotype, using a sensitive, solid-phase radioimmunoassay. The cells from positively identified wells are then expanded in culture and subjected to another round of FACS sorting for individual cell cloning.

This technique has several advantages over sorting into a single large well. First it allows us to assess directly the number of variants present in a given hybridoma population. After establishing the number of wells positive for the variant isotype on the microtiter plate, we use limiting dilution analysis based on the Poisson distribution to calculate the number of true switch variants that were actually sorted onto the plate. Dividing this by the total number of cells sorted onto the plate gives the fraction of cells sorted that were truly switch variants. Multiplying this fraction by the actual fraction of cells sorted with the FACS yields a good approximation of the proportion of variants present in the original population.

A second advantage is the speed at which switch variants can be isolated. One to two weeks after the first round of FACS selection, positive wells will have switch variant cells at frequencies equal to or greater than the reciprocal of the number of cells sorted per well. As the FACS can accurately sort individual cells that have been enriched to fractions greater than 10^{-3}–10^{-4}, switch variants can be readily cloned from positive wells during a second round of FACS sorting.

Switch Variant Frequency Analysis

Mutagenesis Studies

Using pauci-population sorting to quantitate the proportion of switch variants within a given hybridoma population, we tested several mutagens for their abilities to increase the rate of isotype switching. Exponentially growing hybridoma cells were maintained in culture in the presence of a mutagen for one cell cycle period of 18 h. A concentration of the drug which reduced the viability of the treated population by 30%–40% was used. Treatment with higher drug levels resulted in populations of cells unacceptable for variant sorting because of excessive debris and poor cell cloning efficiencies. After washing cells free of the mutagen, the cells were cultured for 72–96 h to allow for cell recovery and cell surface expression of variant immunoglobulin.

Table 1 lists a few early results from this investigation. Since we expect variants to accumulate with the length of time elapsed since cloning, we tabulate the approximate time in weeks elapsed since the subcloning of the IgG_1-producing hybridoma cell lines. The accumulation of variants that occurs as time passes limits the sensitivity of mutagen testing for switch variants that arise acutely. Mitomycin C, a potent inducer of sister chromatid exchange [14], appears not to in-

Table 1. Accumulation of switch variants over time after subcloning of IgG_1-producing hybridoma cell lines

Treatment	Frequency of IgG_2 Variants	
	Experiment 1 3–4 weeks after cloning	Experiment 2 4–8 weeks after cloning
Mitomycin C (50 ng/ml)	1.2×10^{-5}	7.0×10^{-5}
Untreated	8.4×10^{-6}	5.0×10^{-5}
	Experiment 3 2–3 weeks after cloning	Experiment 4 8–10 weeks after cloning
Hydroxyurea (1 µg/ml)	8.3×10^{-6}	8.4×10^{-5}
5-Azacytidine (25 ng/ml)	nt	1.2×10^{-4}
Untreated	1.3×10^{-6}	8.3×10^{-5}

nt, not tested

crease the proportion of switch variants with the limitations stated. With hydroxyurea, however, the frequency of variants increased more than sixfold (Table 1, Experiment 3). However, on the same cell line that had been maintained in culture for several weeks, we did not find a similar proportionate increase in variants compared to the untreated population (Table 1, Experiment 4). Similarly, 5-azacytidine appears to cause only a modest increase in the generation of variants. Note that the frequency of variants found in the untreated population greatly increased after a few weeks in tissue culture.

The Kinetics of Spontaneous Isotype Switching

The increases in the proportion of variants in control populations stimulated investigation into the rate at which isotype switch variants emerge in a logarithmically growing culture. A newly subcloned population of an IgG_1-producing hybridoma was maintained in continuous culture, and at various times, aliquots were assayed for IgG_2 variants by pauci-population sorting. The fraction of variants increased with time in culture (Table 2). Although increases in the proportions of both IgG_{2b} and IgG_{2b} occurred, IgG_{2a} variants were more frequent.

Switch variants should accumulate with time, since they arise by spontaneous events (DNA or RNA rearrangement) similar to spontaneous mutations. As initially described in microbial cells by Luria and Delbruck [15], the proportion of mutant-type cells increases linearly with the number of generations through which the culture is maintained, provided the growth rates of normal and mutant cells are equal, back-mutation is negligible, and the proportion of mutant-type cells remains small. These same conditions apply to many of the switch variant families that we have analyzed.

Catcheside derived a formula to calculate the actual mutation rate from the increase in the proportion of mutants noted with time in a logarithmically growing bulk culture assayed prior to achieving mutational equilibrium [16].

$$M = 2(\ln 2)\left[\frac{\left[\frac{M_2}{N_2}\right] - \left[\frac{M_1}{N_1}\right]}{G}\right].$$

Where m is the mutation rate, M_1 and M_2 are the numbers of mutants at times 1 and 2, N_1 and N_2 are the numbers of cells at these two time points, and G is the number of generations elapsed between times 1 and 2.

Table 2. Increase in variants compared to length of time in culture

Cell generations since cloning	Variants 10^6 IgG_1 Cells (n)		
	IgG_{2b}	IgG_{2a}	Total
25	0.8	0.2	1
36	13	3	16
50	30	90	120
125	50	350	400

The natural logarithm of 2 appears in the formula because the mutation rate is per cell per generation rather than the rate per cell division. Since we are dealing with unsynchronized, exponentially growing populations, an event observed to occur within one generation to produce n number of cells will actually be the result of $n/(2\ln 2)$ divisions. For a mutation rate of m, the proportion of mutant cells within a culture will increase by a fixed increment, $m/(2\ln 2)$ per generation.

Using the formula described above and the data presented in Table 2, we calculated the switch frequency of IgG_1 to IgG_2 to be approximately 6×10^{-6} per cell per generation.

These results have important implications for switch variant isolation. Obviously, the selective pressure applied by the FACS can not induce isotype switching but only isolate variants after they have occurred spontaneously. Thus, prior to attempting switch variant selection on a particular cell line, the line should be maintained in bulk culture for several weeks to increase the proportion of variants. If the frequency of variants within the culture drifts up to 10^{-4}, then these cells may be cloned directly during the first round of FACS sorting.

Biologic Implications of Switch Variants for Human Immunotherapy

Switch Variant Families

Using the techniques outlined above, we have generated switch variant families from many different hybridomas producing monoclonal antibodies to polymorphic human leukocyte antigen (HLA) determinants, tumor-specific antigens, human T cell differentiation antigens, or idiotypic determinants of surface immunoglobulin of human B cell lymphomas. Because of the increase in the proportion of switch variants after propagation in culture, cells were grown for at least 3–4 weeks prior to variant selection. The proportions of variants then found in several hybridoma populations were high enough to permit direct cloning of switch variants during the first round of sorting. Many hybridoma cell populations, however, still required pauci-population sorting enrichment before cloning, perhaps reflecting lower intrinsic isotype switch rates for these cell lines.

Comparative Studies Using Switch Variants

Because the immunoglobulins produced by individual members of a switch variant family share identical light chains and heavy-chain variable regions, valid comparative studies can be performed on the biologic activities of the various murine immunoglobulin isotypes. For example, we recently compared the IgG_{2a}, IgG_{2b}, and IgG_1 antibodies of a switch variant family derived from the IgG_1 producing hybridoma, ME-1, in directing antibody-dependent cell-mediated cytotoxicity (ADCC) mediated by human K cells [17]. Although each antibody has identical binding specificity for the HLA B7 surface antigen expressed by the B lymphoblastoid target cell, the IgG_{2a} antibody is most efficient at directing

ADCC by human effector cells. The IgG_{2b} directs intermediate levels of target cell cytolysis, and the IgG_1 antibody is ineffective.

Switch variant immunoglobulins may resolve the question of whether antibody isotype influences the effectiveness of murine monoclonal antibodies in passive immunotherapy. Mouse immunoglobulin isotypes of antitumor antibodies apparently differ in their ability to influence tumor growth in experimental animal model systems [18–20]. Moreover, it is hypothesized that the isotype of a murine monoclonal antibody may largely determine how effective a reagent it is in the treatment of human malignancy [21]. In collaboration with Drs. Paul Abrams, Kenneth Foon, and Robert Oldham of the Biological Response Modifiers Program, Fredricksburg, VA; with Dr. Benedict Cosimi of the Massachusetts General Hospital, Boston, MA; and Dr. Noel L. Warner of the Becton Dickinson Monoclonal Antibody Center, Mtn. View, CA, clinical trials are in progress with several newly generated variant families, testing monoclonal antibodies of several different isotypes that have identical binding specificities for cell surface differentiation antigens, idiotypic determinants on human B cell lymphomas, or tumor-specific antigens. Determination of any different biologic activities of the murine immunoglublin isotypes through these studies may allow for a more effective application of murine monoclonal antibodies in the immunotherapy of human disease.

Summary

Using improved methods for rare cell detection with the fluorescence activated cell sorter (FACS), we have been able to detect and in several cases directly clone rare switch variant cells after only one round of selection. These variant cells have switched from expressing one heavy-chain constant region gene to another, while maintaining expression of the same light-chain and heavy-chain variable region gene which encodes the antibody-combining site. We have determined that various hybridomas spontaneously switch isotypes at rates ranging from 10^{-5} to 10^{-7} per cell per generation. We have also assessed the ability of mutagens to increase this rate.

We have developed several families of switch variant hybridomas producing monoclonal antibodies to polymorphic HLA determinants, tumor-specific antigens, human T cell differentiation antigens, or idiotypic determinants of surface immunoglobulin of human B cell lymphomas. These families have allowed us to compare directly the biologic effector functions of different immunoglobulin isotypes, including their ability to activate complement, to mediate antibody-dependent cellular cytotoxity, and to influence tumor growth in experimental animals and humans. Such studies underscore the value of carrying out in vitro isotype variant selection in such a way that a hybridoma clone originally selected for the binding specificity of its secreted monoclonal antibody may be modified to produce specific immunoglobulins with isotypes of desired biologic activity.

Acknowledgements. This research was supported by grants GM17367 and CAO4681. T.J. Kipps is a former scholar of the National Leukemia Association and current special fellow of the Leukemia Society of America.

References

1. Dorrington KJ, Klein MH (1983) The CH2 region plays a major role in mediating the biological activities of immunoglobulin G. In: Yamamura Y, Tada T (eds) Progress in immunology V. Fifth international congress of immunology. Academic, Tokyo, pp 37–46
2. Parham P, Kipps TJ, Ward FE, Herzenberg LA (1983) Isolation of heavy chain class switch variants of a monoclonal anti-DC1 hybridoma cell line: effective conversion of non-cytotoxic IgG1 antibodies to cytotoxic IgG2 antibodies. Hum Immunol 8:141–151
3. Liesegang B, Radbruch A, Rajewsky K (1978) Isolation of myeloma variants with predefined variant surface immunoglobulin by cell sorting. Proc Natl Acad Sci USA 75:3901–3905
4. Radbruch A, Liesegang B, Rajewsky K (1980) Isolation of variants of mouse myeloma X63 that express changed immunoglobulin class. Proc Natl Acad Sci USA 77:2909–2913
5. Baumhaeckl H, Liesegang B, Radbruch B, Rajewsky K, Sablitzky F (1982) Switch from NIP specific IgG3 to IgG1 in the mouse hybridoma cell line S24/63/63. J Immunol 128:1217–1220
6. Beyreuther K, Bovens J, Dildrop R, Dorff H, Geske T, Liesegang B, Muller C, Neuberger M, Radbruch A, Rajewsky K, Sablitzky F, Schreier PH, Zaiss S (1981) Isolation and characterisation of class switch variants of myeloma and hybridoma cells. In: Janeway C, Sercarz EE, Wigzel H (eds) Immunoglobulin idiotypes. ICN-UCLA symposia on molecular and cellular biology, vol XX. Academic, New York, pp 229–239
7. Dangl JL, Parks DR, Oi VT, Herzenberg LA (1982) Rapid isolation of cloned isotype switch variants using fluorescence activated cell sorting. Cytometry 2:395–401
8. Oi VT, Vuong TM, Hardy R, Reidler J, Dangl J, Herzenberg LA, Stryer L (1984) Segmental flexibility and effector function of monoclonal antibodies containing identical combining sites joined to different heavy chain C-regions. Nature 307:136–140
9. Huang CC, Parson M, Oi VT, Huang H-J S, Herzenberg LA (1984) Genetic characterization of mouse immunoglobulin allotypic determinants (allotypes) defined by monoclonal antibodies. Immunogenetics 18:311–321
10. Parsons M, Herzenberg LA, Stall AM, Herzenberg LA (1984) Mouse immunoglobulin allotypes: description and special methodology. In: Weir DM, Herzenberg LA, Blackwell CC, Herzenberg LA (eds) The handbook of experimental immunology, 4th Edn. Blackwell Edinburgh
11. Dangl JL, Herzenberg LA (1982) Selection of hybridomas and hybridoma variants using the fluorescence activated cell sorter. J Immunol Methods 52:1–14
12. Kipps TJ, Herzenberg LA (1984) Hybridoma immunoglobulin isotype switch variant selection using the fluorescence activated cell sorter. In: Weir DM, Herzenberg LA, Blackwell CC, Herzenberg LA (eds) Handbook of experimental immunology 4th Edn. Blackwell, Edinburgh (to be published)
13. Parks DR, Herzenberg LA (1984) Fluorescence activated cell sorting: theory, experimental optimization and applications in lymphoid cell biology. In: Methods in enzymology (to be published)
14. Takehisha S (1982) Induction of sister chromatoid exchanges by chemical agents. In: Wolff S (ed) Sister chromatid exchange. Wiley, New York, pp 87–147
15. Luria SE, Delbruck M (1943) Mutation of bacteria from virus sensitivity to virus resistance. Genetics 28:491–511
16. Catcheside DG (1949) In: The genetics of micro-organisms. Pitman, London, pp 158–162
17. Kipps TJ, Parham P, Punt J, Herzenberg LA (1985) Importance of isotype in directing human antibody dependent cellular cytotoxicity by murine monoclonal antibodies. JEM 161:1–17
18. Wright PW, Bernstein ID (1980) Serotherapy of malignant disease. Prog Exp Tumor Res 25:140–162
19. Bernstein ID, Nowinski RC, Tam MR, McMaster B, Houston LL, Clark EA (1980) Monoclonal antibody therapy of mouse leukemia. In: Kennett TJ, McKearn TJ, Bechtol KB (eds) Monoclonal antibodies. Plenum, New York, pp 275–291
20. Matthews TJ, Collins JJ, Roloson G, Thiel H-J, Bolognesi DP (1981) Immunologic control of the ascites form of murine adenocarcinoma 755. IV. Characterization of the protective antibody in hyperimmune serum. J Immunol 126:2332–2336
21. Steplewski Z, Herlyn D, Maul G, Koprowski H (1983) Hypothesis: macrophages as effector cells for human tumor destruction mediated by monoclonal antibody. Hybridoma 2:1–5

A Microfilter Assay for the Detection of Antibody-Producing Cells In Vitro

S. A. Möller and C. A. K. Borrebaeck [1]

Introduction

The haemolytic plaque assay introduced by Jerne and Nordin (1963) has been the most widely used test system for enumeration of specific antibody-secreting cells (ASC). This system was originally developed to detect cells secreting complement binding antibodies against erythrocytes which eventually resulted in lysis of the red target cells. When this assay was applied to soluble antigens several inherent limitations became obvious, such as instability of the conjugated erythrocytes, inconsistent lysability of different batches of antigen-coated red cells, uncertainty about the antibody isotype involved in formation of direct plaque, and effort and time required by this assay procedure, making it unsuited for a large number of samples.

To overcome these limitations and to introduce several new advantages we have developed a rapid filter immuno-plaque assay (FIPA), based on the ELISA principles, for the detection of antigen-specific antibody-producing cells. This assay may also be easily adapted for the measurement of isotype-specific antibody-secreting cells.

Materials and Methods

Antigen

Ovalbumin (OA) (BDH Chemicals Ltd., Poole, England) and bovine serum albumin (BSA) (Miles Laboratories, Inc., Elkhart, IN, USA) were dissolved in 50 m*M* phosphate buffer, pH 7.2 containing 0.15 *M* sodium chloride (PBS) and stirred for 3 h. The solutions were centrifuged (15,000 × g for 30 min) to remove aggregates and the protein content was determined spectrophotometrically at 280 nm.

Animals and Immunizations

Female BALB/c mice (Gl. Bomholtgård Ltd., Ry, Denmark) (4–6 weeks old) were primed with an intraperitoneal injection of 50 μg OA in 0.5 ml Freunds com-

1 Department of Pure and Applied Biochemistry, University of Lund, P.O. Box 124, S-22 100 Lund

plete adjuvant. One week later the mice were given an injection with 50 μg OA in 0.5 ml PBS.

Culture Medium

Dulbecco's modification of Eagles medium was supplemented with streptomycin (50 μg/ml), penicillin (50 IU/ml), 4 m*M* L-glutamine, and 1% (v/v) ×100 non-essential amino acids (Flow Laboratories, Inc., Solna, Sweden) (s-DMEM). Fetal calf serum (FCS) (1% or 10%) was added to s-DMEM before the cells were re-suspended in this medium.

Spleen Cell Preparation

A spleen cell suspension was prepared as recently described (Borrebaeck 1983 a) by removing 2–3 spleens 5–6 days after the last immunization. Briefly, the organs were pressed through a sterile 50 mesh stainless steel screen, using a plunger from a 1 ml syringe, into s-DMEM containing 10% FCS. Clumps were removed and the cells were washed twice by contrifugation (300×g for 10 min) in s-DMEM containing 10% FSC. Viability, determined by trypan blue dye exclusion, was usually more than 95%.

Production of Mouse Hybridomas

Several hybridoma cell lines producing monoclonal antibodies to OA were obtained by fusing Sp 2/0-Ag14 plasmacytoma cells with immunized mouse splenocytes as recently described (Borrebaeck 1983 a). The hybridoma cells were cloned three times after the fusion and screened for antigen specificity using a solid phase enzyme immunoassay (Borrebaeck and Etzler 1981).

Microfilter Assay

The assay utilizes the Millititer Filtration System (Millipore Co., Bedford, MA, USA) with a microtiter plate having the plastic bottom replaced by a nitrocellulose membrane, type HATF (pore size 0.45 μm). The nitrocellulose filter was coated with antigen by the addition of 100 μl antigen solution. The solution was sucked into the filter using the Millititer Vacuum Filtration Holder (Millipore Co.) and incubated at room temperature for 1–2 h. The plate was then washed twice by sucking 200 μl of sterile PBS through the membrane filter.

The bottom of the membranes were blotted using soft tissue paper and the cells, suspended in s-DMEM containing 1% FCS, was added (100 μl/well) and cultivated in a 37 °C humidified incubator for 1–16 h. The cells were discarded by flicking the plate and the wells were then washed four times with PBS and ten times with PBS containing 0.05% Tween 20. 50 μl horseradish peroxidase (HRP) conjugated to rabbit anti-mouse immunoglobulins (Dako A/S, Copenhagen, Denmark) and diluted 1:300 in 10 m*M* sodium phosphate buffer, pH 8.0 containing 0.5 *M* sodium chloride and 0.1% Tween 20, was added to each well. After a 1-h incubation at 37 °C, the plates were washed twice with PBS containing 0.05% Tween 20 and then eight times by sucking the washing solution through the membranes as described above.

Finally, 100 μl of freshly prepared enzyme substrate was added to each well and the enzymatic reaction was terminated after 15–30 min by washing the plate three times with distilled water. The substrate, 4-chloro-1-naphtol (Sigma Chemical Co., St. Louis, MO, USA), was prepared from 2 ml of a stock solution (3 mg/ml methanol) by adding 10 ml of 50 m*M* Tris/HCl, pH 7.8 containing 0.2 *M* sodium chloride, and 40 μl of 3% hydrogen peroxide.

The filter membranes were dried and the blue plaques were counted using 30 × magnification.

Results and Discussion

The microtiter plate that was used in the filter immuno-plaque assay had the conventional plastic bottom replaced by a 0.45 μm nitrocellulose membrane. This ensured a rapid and reproducable coating of the antigen since the porous nitrocellulose membrane had a hundred-fold higher protein adsorption capacity than

Table 1. The effect of different amounts of antigen, adsorbed onto the nitrocellulose membrane, on the outcome of FIPA. Values are expressed as mean ±S.D. for quadruplicate cultures

Spleen cells/well	Palque		
	0.1	1	5 (mg OA/ml)
10^5	9± 2	15± 6	19± 6
10^6	104±12	136± 6	170± 3
10^7	135±15	728±15	60±18

Table 2. The number of specific plaque formed by the hybridoma line (1D5–8B) as a function of different cell incubation times and cells seeded per well. Values are expressed as mean ±S.D. for quadruplicate culture

Incubation time (h)	Plaque		
	10^3	10^4	10^5 (cells/well)
1	44±4	536±42	4035± 54
3	53±9	464±36	4690±230
16	58±5	505±43	3527±363

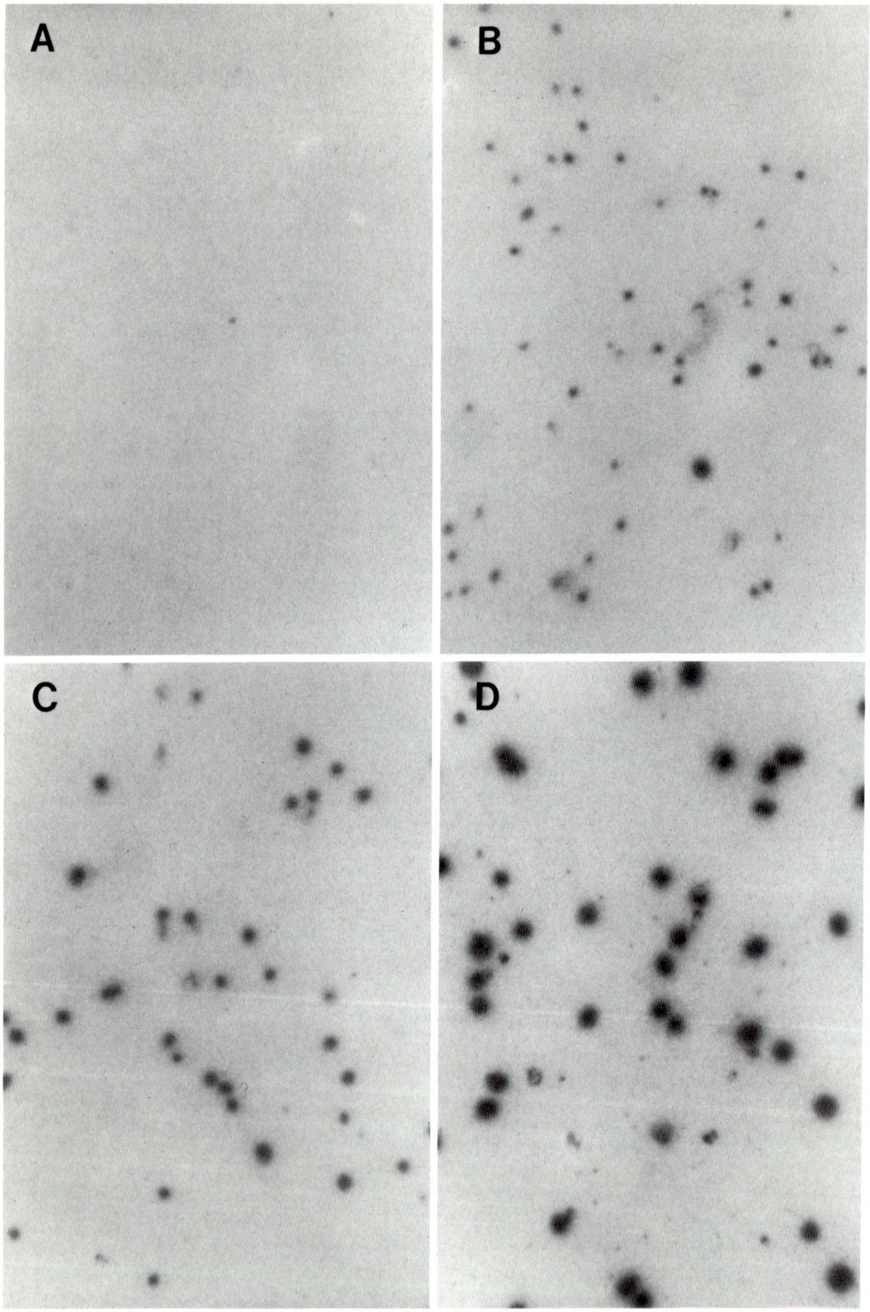

Fig. 1 A–D. Filter immuno-plaque assay using a hybridoma cell line (1D5-8B). **A** In the control panel the nitrocellulose membrane was coated with 1 mg BSA/ml PBS and the cell incubation time was 3 h. In the test panels the nitrocellulose membranes were coated with 1 mg OA/ml PBS and the cell incubation times were **B** 1 h, **C** 3 h, and **D** 16 h

polystyrene wells. Initially, the outcome of the FIPA using different cell and antigen concentrations were tested (Table 1). In the following experiments we chose to use 1 mg antigen/ml of coating buffer. The optimal antigen concentration, however, always has be tested for every individual antigen. A hybridoma cell line (1D5-8B) producing monoclonal antibodies against OA was then used to estimate the optimal incubation time for the cells in antigen-coated microtiter wells. Tests were conducted after 1, 3, and 16 h using 10^3, 10^4, and 10^5 cells/well, and the number of plaque obtained is shown in Table 2. The relation between incubation time and size of the plaque is also clearly illustrated in Fig. 1. The optimal incubation time to obtain easily enumerated plaque seemed to be 1–3 h for both hybridoma cells and in vivo immunized splenocytes, whereas 16-h incubation time sometimes gave a blurred picture since the plaque became too large and diffuse. It should, however, be noted that only the size of the formed plaque increased due to the prolonged incubation time, and not the total number of plaque. BSA (1 mg/ml) was used as control in this test and, as can be seen (Fig. 1 A), the formation of plaque was antigen specific, since no plaque was obtained using this unrelated antigen in the assay.

In vivo immunized mouse splenocytes were also tested using the FIPA, and Fig. 2 shows the obtained plaque. It was obvious that the immune spleen cells gave a greater variation of the plaque size than hybridoma cells producing monoclonal antibodies against OA. This most probably represents the in vivo variation

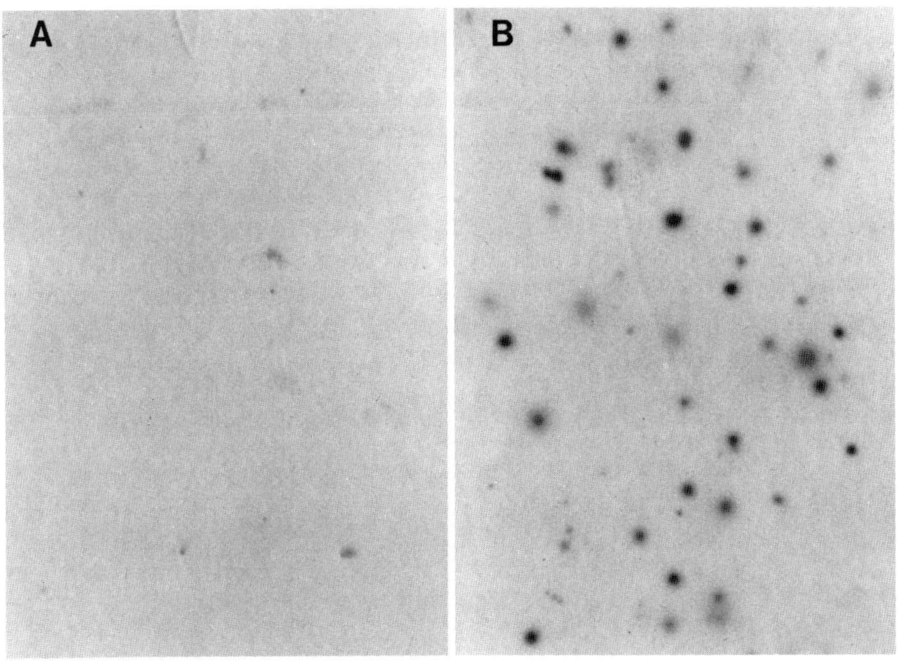

Fig. 2 A, B. Filter immuno-plaque assay using in vivo immunized mouse splenocytes. The nitrocellulose membrane was coated with **A** 1 mg BSA or **B** OA/ml PBS and the cell incubation time was 3 h. 10^5 cells were seeded per well

of antibody producing cells in regard to the amount and rate of produced antibodies. The number of unspecific plaque, when an unrelated antigen was used as control, could also be higher when spleen cells were used. This background was, however, easily reduced (Fig. 2 A) by using the described washing procedure.

The filter immuno-plaque assay offers several advantages compared to existing test systems (Jerne and Nordin 1963; Sedgwick and Holt 1983; Czerkinsky et al. 1983): (a) the nitrocellulose membrane rapidly adsorbs high amounts of a variety of soluble proteins; (b) the adsorption properties of the nitrocellulose membrane increase the reproducability of the assay compared to when polystyrene is utilized as solid phase for the antigen adsorption; (c) the washing procedure is simple due to the Millititer Filtration System, and several hundred samples can be run in a day; and (d) the assay is rapid (4 h) and easy to perform.

FIPA has been developed as a tool to evaluate the requirement for various lymphokines that are necessary to support an antigen-specific in vitro activation of B-lymphocytes. This assay can also be used to monitor the progression of in vitro immunization (Borrebaeck 1983 a, b; Borrebaeck 1984) of mouse and human lymphocytes before the immunized cells are used as parental cells in somatic cell hybridization experiments.

Acknowledgement. This investigation was supported by a grant from the Swedish Medical Research Council, the Medical Faculty (University of Lund) and the John and Augusta Persson Foundation.

References

Borrebaeck CAK (1983 a) In vitro immunization for the production of antigen-specific lymphocyte hybridomas. Scand J Immunol 18:9–12

Borrebaeck CAK (1983 b) In vitro immunization of mouse spleen cells and the production of monoclonal antibodies. Acta Chem Scand B37 7:647–648

Borrebaeck CAK (1984) Dependence on T cell replacing factor and immunogenic dose for the production of monoclonal antibodies using the in vitro immunization technique. Mol Immunol 21:841–845

Borrebaeck CAK, Etzler ME (1981) Production and characterization of a monoclonal antibody against the seed lectin of the Dolichos biflorus plant. J Biol Chem 256:4723–4725

Czerkinsky CC, Nilsson LÅ, Nygren H, Ouchterlony Ö, Tarkowski A (1983) A solid-phase enzyme linked immunospot (ELISPOT) assay for enumeration of antibody-secreting cells. J Immunol Methods 65:109–121

Jerne NK, Nordin AA (1963) Plaque formation in agar by single antibody producing cells. Science 140:405

Sedgwick JD, Holt PG (1983) A solid-phase immunoenzymatic technique for the enumeration of specific antibody-secreting cells. J Immunol Methods 57:301–309

A Biotin Avidin Amplified ELISA Using Exclusively Monoclonal Antibodies for the Detection of Respiratory Syncytial Viral Antigens in Nasopharyngeal Secretions

G. OBERT and C. BEYER [1]

Introduction

Respiratory syncytial virus (RSV) is the most important respiratory viral pathogen in children, particularly in early infancy [3]. The diagnosis of RSV infections by cell culture is time consuming and expensive. Therefore rapid methods for antigen detection in nasopharyngeal secretions have gained popularity. The immunofluorescence assay (IFA) employing either polyclonal [9, 12, 14, 15] or monoclonal antibodies [2, 11] is the most frequently used technique for rapid diagnosis of RSV infections. However, although commercial antisera are available, the need for intact cells containing viral antigens has caused some limitations to the use of this technique since cellular preparations are rapidly destroyed during the specimen transportation. Furthermore, this method also requires a highly skilled microscopist to evaluate the test, and it is not easly automated.

Solid phase immunoassays such as the radioimmunoassay (RIA) [16] or the enzyme immunoassay (ELISA) [4, 8, 13, 16] have been described for the detection of RSV antigens in nasopharyngeal secretions. These methods have been found to be less sensitive than IFA to the mistreatment of the specimens during transportation, they allow easy testing of a large number of specimens, and they can be automated.

The sensitivity and the specificity of these assays depend largely on the quality of the antisera that are available. Monoclonal antibodies, since they are highly specific, well-characterized reagents and can be produced in great quantity with an uniform quality, may be preferred to animal antisera for use in antigen detection tests. In this report we describe a biotin avidin amplified ELISA employing exclusively monoclonal antibodies directed against the viral nucleocapsid.

1 Laboratoire de Virologie de la Faculté de Médecine et Groupe de Recherches U 74 de l'I.N.S.E.R.M., 3 rue Koeberlé, F-67000 Strasbourg

Results

Monoclonal Antibodies

Monoclonal antibodies must fulfil some conditions in order to be used in a sensitive solid phase ELISA suitable for diagnosis:

1. They must be directed against a viral component present in large amounts in the specimen. For RSV, monoclonal antibodies directed against the major component of the viral nucleocapsid, the 42,000 dalton polypeptide seem particularly appropriate since viral nucleocapsids are synthesized in excess during the replication cycle and accumulate to a high concentration in infected cells.
2. They must react in an ELISA using solubilized nucleocapsids as antigen and must have sufficient affinity in order to detect low concentrations of antigen.
3. Monoclonal antibodies directed against distinct and nonoverlapping antigenic sites on the viral nucleocapsid must be selected in order to develop an ELISA employing one monoclonal antibody for antigen capture and another monoclonal antibody, directed against a different epitope, for antigen detection.

Among twenty-five different monoclonal antibodies directed against RSV, we selected three hybridoma lines RSV4, RSV22, and RSV24 that secrete antibodies and fulfil these requirements.

These antibodies, IgG1, IgG1 and IgG2a respectively, reacted against the 42,000 dalton polypeptide as determined by radio immunoprecipitation using ^{14}C amino acid labeled RSV. Competitive antibody binding assays employing biotinylated antibodies showed that they are directed against three different epitopes on the viral nucleocapsid. On testing the reactivity of each of these monoclonal antibodies on 20 RSV strains isolated in tissue culture, no antigenic variation affecting these three epitopes has been observed. A mixture of the monoclonal antibodies has been found as efficient as a commercial bovine antiserum (Wellcome) for the detection of intracellular viral antigens in nasopharyngeal secretions by an indirect immunofluorescence assay.

These monoclonal antibodies have been purified from ascitic fluids by affinity chromatography on protein A sepharose according to the technique described by Ey et al. [5] and covalently labeled with biotin N hydroxy succinimide as described by Guesdon et al. [7].

Development of ELISA

For each monoclonal antibody, optimal coating of polyvinyl microtiter plates has been obtained with 0.05 *M* carbonate-bicarbonate buffer pH 9.6 and an incubation time of 2 h at 37 °C. Preliminary experiments have shown that only RSV4 and RSV22, when adsorbed to the solid phase, are able to bind large amounts of viral antigens. In contrast, the binding capacity of immobilised RSV24 is weak, and thus only RSV4 and RSV22 can be used for antigen capture.

The capability of each biotin conjugated antibody to detect antigen has been investigated by ELISA using polyvinyl microtiter plates coated with purified nucleocapsid. The binding to immobilized antigen of biotinylated RSV22 and

Table 1. Description of the Mc ELISA

Coating of polyvinyl microplates

- 100 µl per well of a mixture of RSV4 (1.25 µg/ml) and RSV22 (1.25 µg/ml) in carbonate buffer 0.05 *M* pH 9.6 during 2 h at 37 °C
- Washings with phosphate saline buffer (PBS) pH 7.2+0.05% of Tween 20
- Saturation of protein adsorption in situ with PBS+1% bovine serum albumin. 1 h at 37 °C. Washings

Incubation with specimens

- 100 µl per well of purified RSV nucleocapsid or sonicated nasopharyngeal secretions prepared as described by Sarkinen et al. [16] and diluted in PBS Tween. 16 h at +4 °C. Washings

Antigen detection

- 100 µl per well of biotinylated RSV 24 (1 µg/ml) diluted in PBS Tween. 2 h at 37 °C. Washings
- 100 µl of avidin-biotinylated peroxydase (Vector Laboratories) and prepared according to manufactor's instructions. Washings
- 150 µl per well of orthophenylenediamine substrate. 15 min in the dark
- 50 µl per well of SO_4H_2 2 N
- Lecture at 486 nm

RSV24 is significantly higher than that of biotinylated RSV4. Consequently only biotinilated RSV22 and RSV24 can be used as a second layer antibody.

Using RSV4 and RSV22 for antigen capture and biotinylated RSV22 and RSV24 for antigen detection, different ELISAs employing various combinations of the antibodies were developed in order to find the optimal combination for antigen detection. Serial two-fold dilutions of purified nucleocapsids (from 80 ng/ml to 0.625 ng/ml) were assayed to determine the minimum amount of antigen that could be detected. Maximum sensitivity was reached when a mixture of RSV4 and RSV22 was used for antigen capture and when biotinylated RSV24 was used for antigen detection. This ELISA is described in Table 1. The minimum amount of antigen that could be detected by this ELISA was between 1.25 and 2.5 ng/ml of protein. This assay is 2–4 times more sensitive than an ELISA employing polyclonal rabbit IgG directed against the viral nucleocapsid as the catching antibody and the same biotinylated immunoglobulin as the second layer antibody.

Testing of Clinical Specimens

The ELISA employing RSV4 and RSV22 for antigen capture and biotinilated RSV24 as the detector antibody (Mc ELISA) and the ELISA employing only rabbit anti-RSV nucleocapsid IgG (Pc ELISA) have been compared with an indirect immunofluorescence assay (IFA) on 54 nasopharyngeal secretions (Table 2).

The Mc ELISAs were as sensitive as the IFA and more sensitive that the PC ELISA. All 32 specimens scored as positive using IFA were also found positive with the Mc ELISA, and positive/negative (P/N) ratios higher than 3 have been observed in all these specimens. When tested with the PC ELISA, signifi-

Table 2. Comparison of an ELISA employing monoclonal antibodies (Mc ELISA) and an ELISA employing rabbit anti-RS nucleocapsid IgG (Pc ELISA) with IFA for the detection of RSV in clinical specimens

	IFA	
	Positive	Negative
Mc ELISA		
positive	32	0
negative	0	22
Pc ELISA		
positive	29	0
negative	3	22

cantly lower P/N ratios were found and three samples with P/N ratios lower than 2 must be considered as negative. Although all 22 specimens scored as negative by IFA were also negative by Mc ELISA with P/N ratios lower than 2, the specificity of the Mc ELISA has been confirmed by an inhibition experiment using rabbit anti-RSV nucleocapsid IgG and preimmune IgG.

Discussion

In order to develop a highly sensitive ELISA, the biotin avidin system has been chosen as an amplification mode. This system is based on the high affinity of avidin for biotin. One molecule of avidin can bind four molecules of biotin, and biotin can be conjugated to most proteins including antibodies and enzymes [17]. The biotin avidin amplification system has been employed in ELISAs using polyclonal antibodies [10, 18] and, more recently, monoclonal antibodies [1]. In our hands, ELISA employing biotin labeled antibodies and an avidin biotinylated peroxydase complex has proved to be 4–8 times more sensitive than a direct ELISA employing peroxydase or alkaline-phosphatase labeled antibodies.

Employing a mixture of RSV4 and RSV22 for antigen capture and biotinylated RSV24 for antigen detection, a highly sensitive and specific ELISA has been developed. This ELISA was able to detect as little as 1.25–2.5 ng of protein per milliliter and has been found to be as sensitive and specific as IFA for the detection of viral antigens in nasopharyngeal secretions.

Detection of RSV antigens by ELISA techniques has been reported by several groups. These assays employed either commercialy available antibodies [13] or antibodies produced by immunization with purified RSV [4], with antigens extracted from crossed immunoelectrophoresis gels [8] or with purified nucleocapsids [16]. When compared to IFA, these ELISAs showed different sensitivities. McIntosh et al. [13] and Hornleth et al. [8] found ELISA more sensitive than IFA. Sarkinnen et al. [16], employing antibodies directed against the viral nucleocapsid, developed an indirect RIA and an indirect ELISA with a sensitivity of 10–

30 ng/ml of viral nucleoprotein and found these immunoassays as specific and as sensitive as IFA. The ELISA employing rabbit immunoglobulins developed in this study, although slightly more sensitive than these immunoassay, was not able to detect viral antigens in all the IF positive specimens. Only the Mc ELISA, which is about 10 times more sensitive than the immunoassays described by Sarkinnen et al. was found to be as sensitive as IFA. Differences in the quality and in the preparation of clinical specimens can explain such discrepancies.

In conclusion, the monoclonal antibodies RSV4, RSV22, and RSV24 have proved to be at least as good as polyclonal antibodies for the diagnosis of RSV infections by IFA and by ELISA. IFA are particularly indicated when few specimens are to be tested, or when the specimens are collected late in the disease [6], but it cannot be performed when the cells in the nasopharyngeal secretions have been destroyed during specimen transportation. ELISA allows an easy testing of a large number of specimens, is less sensitive than IFA to the mistreatment of the specimen, but is frequently negative when the preparations are collected late in the disease [13].

References

1. Adler-Storthz K, Kendall C, Kennedy RC, Henkel RD, Dreesman GR (1983) Biotin-avidin-amplified enzyme immunoassay for detection of Herpes simplex virus antigen in clinical specimens. J Clin Microbiol 18:1329–1334
2. Bell DM, Walsh EE, Hruska JF, Schnabel KC, Hall CB (1983) Rapid detection of respiratory syncytial virus with a monoclonal antibody. J Clin Microbiol 17:1099–1101
3. Chanock RM, Kim HW, Brandt CD, Parott RH (1981) Respiratory syncytial virus. In: Evans AS (ed) Viral infections of humans, 2nd edn. Plenum, New York, pp 471–489
4. Chao RK, Fishau M, Schartzman JD, McIntosh K (1979) Detection of respiratory syncytial virus in nasal secretions from infants by enzyme-linked immunosorbent assay. J Infect Dis 139:483–486
5. Ey PL, Prowse SJ, Jenkin CR (1978) Isolation of pure IgG1, IgG2a and IgG2b immunoglobulins from mouse serum using protein A-sepharose. Immunochemistry 15:429–436
6. Gardner PS, McQuillin J, McGuckin R (1970) The late detection of respiratory syncytial cirus in cells of the respiratory tract by immunofluorescence. J Hyg 68:575–580
7. Guesdon J, Ternynck T, Avrameas S (1979) The use of the avidin-biotin interaction in immunoenzymatic techniques. J Histochem Cytochem 27:1137–1139
8. Hornsleth A, Friis B, Andersen P, Brenoe E (1982) Detection of respiratory syncytial virus in nasopharyngeal secretions by ELISA: comparison with fluorescent antibody technique. J Med Virol 10:273–281
9. Kaul A, Scott R, Gallager M, Scott M, Clement J, Ogra PL (1978) Respiratory syncytial virus infection: rapid diagnosis in children by use of indirect immunofluorescence. Am J Dis Child 132:1088–1090
10. Kendall A, Ionescu-Matiu I, Dreesman GR (1983) Utilisation of the biotin-avidin system to amplify the sensibility of the enzyme-linked immunosorbent assay (ELISA). J Immunol Methods 56:329–339
11. Kim HW, Wyatt RG, Fernie BF, Brandt CD, Arrobio JO, Jeffries BC, Parrott RH (1983) Respiratory syncytial virus detection by immunofluorescence in nasal sevretions with monoclonal antibodies against selected surface and internal proteins. J Clin Microbiol 18:1399–1404
12. Lauer BA (1982) Comparison of virus culturing and immunofluorescence for rapid detection of respiratory syncytial virus in nasopharyngeal secretions: sensitivity and specificity. J Clin Microbiol 16:411–412

13. McIntosh K, Hendry RM, Fahnestock ML, Pierik LT (1982) Enzyme-linked immunosorbent assay for detection of respiratory syncytial virus infection: application to clinical samples. J Clin Microbiol 16:329–333
14. McQuillin J, Gardner PS (1968) Rapid diagnosis of respiratory syncytial virus infection by immunofluorescent antibody techniques. Br Med J 1:602–605
15. Minnich LL, Ray CG (1982) Comparison of direct and indirect immunofluorescence staining of clinical specimens for detection of respiratory syncytial virus antigen. J Clin Microbiol 15:969–970
16. Sarkkinen HK, Halonen PE, Arstila PP, Salmi AA (1981) Detection of respiratory syncytial, parainfluenza type 2, and adenovirus antigens by radio immunoassay and enzyme immunoassay on nasopharyngeal specimens from children with acute respiratory disease. J Clin Microbiol 13:158–267
17. Wilchek M, Bayer EA (1984) The avidin-biotin complex in immunology. Immunol Today 5:39–43
18. Yolken RH, Leister FJ, Whitcomb LS, Santosham M (1983) Enzyme immunoassays for the detection of bacterial antigens utilising biotin labeled antibody and peroxydase biotin-avidin complex. J Immunol Methods 56:319–327

Identification and Serotyping of *Vibrio cholerae* 0:1 by Monoclonal Antibody Based Enzyme-Linked Immunosorbent Assays

B. GUSTAFSSON and T. HOLME[1]

Introduction

The disease cholera is caused only by *Vibrio cholerae* strains belonging to serogroup 0:1, characterized by the group-specific antigen A. This group is further divided into two major serotypes, Ogawa and Inaba, characterized by the type-specific antigens B and C respectively. These antigens are thermostable polysaccharides and part of the cell wall lipopolysaccharide (LPS) (Hisatsune and Kondo 1980; Mukerjee 1978). *V. cholerae* strains, other than serogroup 0:1, are designated as non-0:1 vibrios (Mukerjee 1978). The serological identification of *V. cholerae* 0:1 and its serotypes is based on the use of group-specific (anti-A) and type-specific (anti-B, anti-C) antisera. Today, polyvalent antisera are produced by immunizing rabbits with heat-killed vibrios of the two serotypes Ogawa and Inaba. Monospecific type sera are prepared by cross-absorption to remove antibodies directed against the group-specific antigen A (Mukerjee 1978). However, by this procedure there is also a concomitant loss of type-specific antibodies, making such antisera difficult to produce. Since as many as 60%–75% of all cholera cases are subclinical (Merson et al. 1978), it is important that sensitive and reproducible methods are available for the identification of *V. cholerae* 0:1. The aim of this study was to develop a highly specific monoclonal antibody based enzyme immunoassay for use in screening clinical specimens and for epidemiological purposes.

Production and Characterization of Monoclonal Antibodies

Antibody-secreting cell lines were produced by fusing SP2/O mouse myeloma cells with spleen cells from BALB/c mice previously immunized with *V. cholerae* 0:1 LPS (Gustafsson et al. 1982) or with whole bacteria inactivated with 1.5% formalin and heated to 65 °C for 60 min (Cryz et al. 1982). Cell culture supernatants were tested for specific antibodies in an LPS-enzyme-linked immunosorbent assay (Holmgren and Svennerholm 1973), modified for screening of hybridoma supernatants (Gustafsson et al. 1982) (Fig. 1). Hybrid cell lines of interest were recloned in 96-well microtiter trays by limiting dilution, using thymocytes as feeder layer. Antibodies were produced by growing cells in 250-ml tissue culture flasks in RPMI 1640 medium supplemented with 10% fetal calf serum and

1 Department of Bacteriology, Karolinska Institute, S-10401 Stockholm

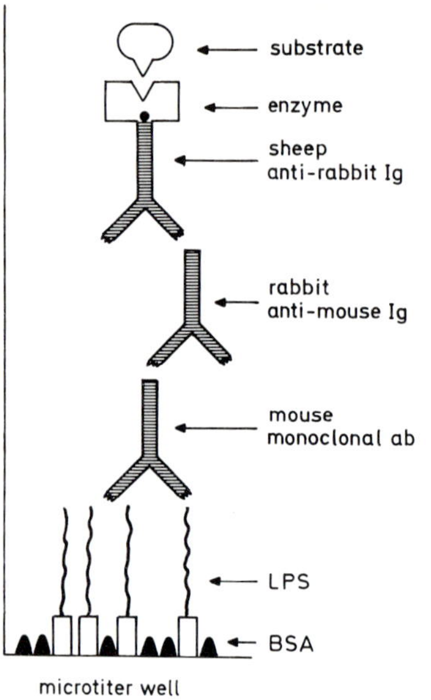

Fig. 1. LPS-ELISA principle. Microtiter wells were coated with Ogawa or Inaba LPS. Remaining binding sites were blocked by the addition of bovine serum albumin (BSA). The wells were then subsequently incubated with cell culture superntants, followed by incubations with rabbit anti-mouse immunoglobulin, sheep anti-rabbit immunoglobulin conjugated to peroxidase and substrate (1,2-phenylene-diamine-dihydrochloride)

Table 1. Characteristics of monoclonal antibodies and their titer in ELISA

Monoclonal antibody	Class or subclass	Light chain	Specificity	Titer[a]	
				In vitro	In vivo
110–1–5B	IgM	κ	A	64	32,000
112–3–2G	IgG3	κ	A	64	16,000
H 8	IgG3	λ	B	512	16,000
C6	IgG3	κ	C	128	8,000

[a] Endpoint titers of cell culture or ascites fluid in ELISA. Endpoint titers were defined as the highest dilution in a twofold serial dilution still giving an optical density at 450 nm of >0.2

L-glutamine (1 m*M*) or in pristane primed mice (Potter et al. 1972). Isotype and light chain produced by the hybridomas were determined by double-diffusion in agarose gel using rabbit antisera specific for IgM, IgG1, IgG2a, IgG2b, IgG3, and IgA, and for kappa and lambda light chains.

Characteristics of the monoclonal antibodies used in this study are summarized in Table 1. The antibodies H8 and C6 have been described in detail elsewhere (Gustafsson and Holme 1983).

Sandwich-ELISA

The monoclonal antibody 110-1-5B (capture antibody) was coated to the plastic of microtiter trays (Titertec PVC EIA 173; Flow Laboratories, Irwine, Scotland) by adding ascites antibodies diluted 1/1,000 in 50 m*M* carbonate buffer, pH 9.6

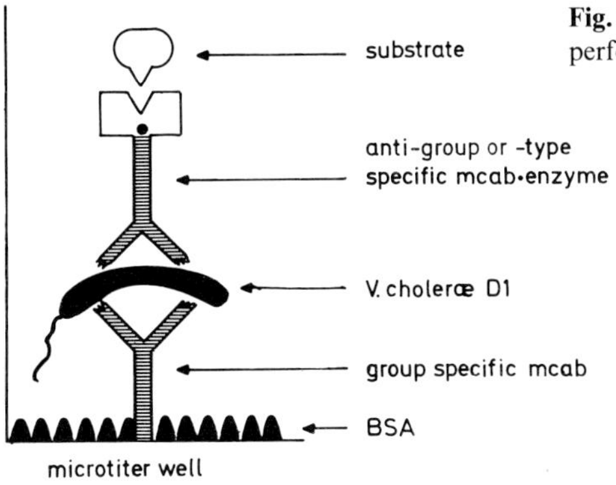

Fig. 2. Sandwich-ELISA principle. The assay was performed as described in the text

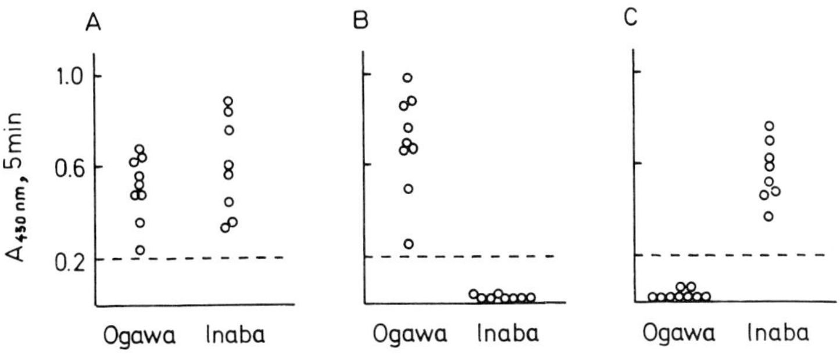

Fig. 3 A–C. Sandwich-ELISA absorbance values of *V. cholerae* 0:1 strains. The bacteria were cultivated overnight in nutrient broth (pH 7.2) at 37 °C. The assay was performed in microtiter wells coated with the group-specific antibody 110-1-5B using the peroxidase-conjugated antibodies 112-3-2G (**A**), H8 (**B**), and C6 (**C**) as second antibodies

to each well. Remaining binding sites were blocked by the addition of 5% BSA in 10 m*M* phosphate buffered saline (PBS) (pH 7.2)−0.05% Tween 20. Then 100 µl samples of bacteria suspended in PBS were added in duplicate to each well followed by incubation with 100 µl of the peroxidase-conjugated monoclonal antibodies 112-3-2G, H8 or C6 and 100 µl of 3 m*M* 1,2-phenylenediaminedihydrochloride in 40 m*M* Tris-hydrochloride buffer as substrate (Fig. 2). The optical density at 450 nm was measured in an Titertec Multiscan spectrophotometer. An optical density of >0.2 above background was considered a positive result.

Seventeen strains of *V. cholerae* representing 9 strains of serotype Ogawa and 8 strains of serotype Inaba were tested in the sandwich-ELISA. All strains, irrespective of serotype, showed positive reactions when 112-3-2G antibodies were used as second antibody-enzyme conjugate (Fig. 3 A). All Ogawa strains showed positive reactions when tested with H8 antibodies as second antibodies, while all Inaba strains were negative (Fig. 3 B). When C6 antibodies were used as second antibodies, all Ogawa strains were negative while all Inaba strains tested were positive (Fig. 3 C). By testing serial dilutions of bacteria it was found that a minimum of 3×10^6–4×10^7 CFU/ml was usually required to obtain a positive result in the sandwich-ELISA.

The sandwich-ELISA presented in this study is based on the use of monoclonal antibodies. Although the specificity of the assay will be superior to that of an assay utilizing polyclonal antisera, difficulties in coating affecting the efficiency of the capture antibody to the plastic of the microtiter wells may occur. A monoclonal antibody may have an excellent affinity to the antigen and still be useless as capture antibody if it is unable to stick to the plastic of the microtiter trays. In our experience, it is of great advantage to have several different monoclonal antibodies with specificity for the same antigen to choose from.

The second antibodies were conjugated to peroxidase. This minimizes the number of incubations needed to complete the assay compared to using a third anti-mouse IgG-enzyme conjugate together with unconjugated second antibodies. However, some monoclonal antibodies are difficult to conjugate to enzyme. This can be due to, for example, loss of specificity of the antibody during conjugation, formation of unstable conjugates or precipitation of the antibodies during purification of the antibodies from cell culture supernatant or ascites fluid. The introduction of a third anti-mouse IgG-enzyme conjugate will increase the number of incubations needed to complete the assay, but eliminate the need for different antibody-enzyme conjugates. A "general" anti-mouse IgG-enzyme conjugate can also be used in assays other than cholera serology.

LPS-ELISA Inhibition

Monoclonal antibodies can also be utilized in an LPS-ELISA inhibition test in which whole bacteria are used as antigen. In this assay, the monoclonal antibodies were preincubated with whole bacteria in glass tubes. Remaining antibodies, i.e. those not neutralized by the bacteria, were then measured in the LPS-ELISA (Fig. 1). The percentage inhibition was calculated from the ratio of the absorbance values obtained when measuring the amount of free antibodies after preincubation with bacteria and with PBS respectively. An inhibition of $>20\%$ was considered a positive result.

All 0:1 vibrios inhibited the monoclonal antibody 110-1-5B by $>20\%$ (Fig. 4A). All but two of the Ogawa strains and all but one of the Inaba strains tested inhibited the monoclonal antibody 112-3-2G (Fig. 4B). The type-specific antibody H8 was inhibited by all Ogawa strains while the Inaba strains showed no inhibition of this antibody (Fig. 4C). The type-specific antibody C6 was inhibited by Inaba strains only (Fig. 4D). The sensitivity of this assay was almost identical to the sandwich-ELISA when the antibodies 110-1-5B, H8 and C6 were used. The antibody 112-3-2G usually required bacterial concentrations of $>10^8$ CFU/ml to show a positive reaction.

Both the sandwich-ELISA and the ELISA-inhibition tests were highly specific for identification and serotyping of *V. cholerae* 0:1. Thirty-seven different strains of non-0:1 vibrios tested in the assays were all negative. Furthermore, cross-reactions of polyclonal cholera antisera with bacteria of the genus *Enterobacteriaceae* (Winkle et al. 1972) and with *Brucella abortus* (Eisele et al. 1948; Feely 1969) have been reported. The following strains have been tested in both assays and found negative: *E. coli* 0:5, 0:8, 0:55, 0:78, 0:85, 0:104, 0:105, 0:111, 0:150, *Salmo-*

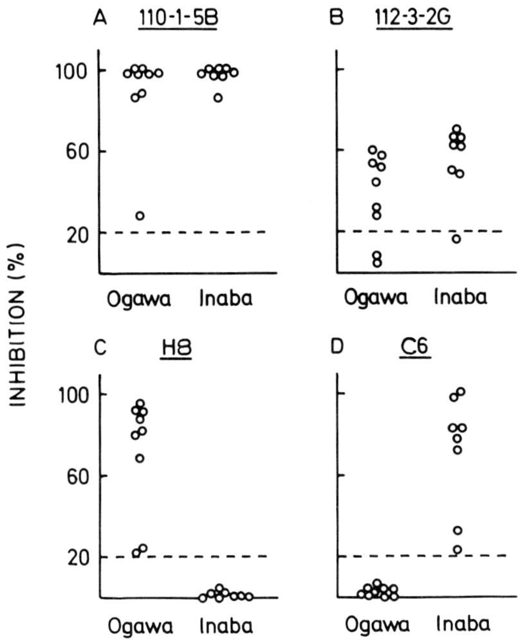

Fig. 4 A–D. Inhibition of the monoclonal antibodies 110-1-5B, 112-3-2G, H8, and C6 with strains of *V. cholerae* 0:1 in LPS-ELISA. The bacteria were cultivated overnight, in nutrient broth (pH 7.2) at 37 °C. The percentage inhibition was calculated as described in the text

nella bergen, S. urbana, S. enteritidis, S. paratyphi, S. godesberg, S. arizona, S. kaolack, S. kentucky, S. typhimurium, Shigella sonnei, Citrobacter freundii, and *Brucella abortus.*

Conclusions

The sandwich-ELISA and ELISA-inhibition tests represent highly selective methods for the identification and serotyping of *V. cholerae* 0:1. The microtiter application makes the methods suitable for screening samples from suspected cases as well as from the environment with a low consumption of reagents. The assays could probably also be useful in monitoring the somatic antigen content of genetically engineered vaccine strains.

Today, serological identification and serotyping of *V. cholerae* 0:1 is performed by slide agglutination, using polyclonal, absorbed rabbit antisera of varying quality (Donovan and Furniss 1982). The monoclonal antibodies used in this study offer a great advantage over rabbit antisera in that they can be used in slide agglutination without prior purification, either as cell culture supernatants or as ascites fluid (Gustafsson and Holme 1983; Holme and Gustafsson 1984).

Acknowledgements. We thank Ulla Björk for skillfull technical assistance. This study was supported by the Karolinska Institute, The Swedish Medical Research Council (project no. 656), and the Emil and Vera Cornell Foundation.

References

Cryz JR, Fürer E, Germanier R (1982) Effect of chemical and heat inactivation on the antigenicity and immunogenicity of *Vibrio cholerae*. Infect Immun 38:21–26

Donovan TJ, Furniss AL (1982) Quality of antisera used in the diagnosis of cholera. Lancet 2:866–868

Eisele CW, McCullough NB, Beal GA (1948) Brucella antibodies following cholera vaccination. Ann Intern Med 28:833–837

Feeley JC (1969) Somatic O antigen relationship of *Brucella* and *Vibrio cholerae*. J Bacteriol 99:645–649

Gustafsson B, Holme T (1983) Monoclonal antibodies against group- and type-specific lipopolysaccharide antigens of *Vibrio cholerae* 0:1. J Clin Microbiol 18:480–485

Gustafsson B, Rosen A, Holme T (1982) Monoclonal antibodies against *Vibrio cholerae* lipopolysaccharide. Infect Immun 38:449–454

Hitsatsune K, Kondo S (1980) Lipopolysaccharides of R mutants isolated from *Vibrio cholerae*. Biochem J 185:77–81

Holme T, Gustafsson B (1984) Monoclonal antibodies against group and type specific antigens of *Vibrio cholerae* 0:1. In: Macario AJL, Macario EC (eds) Monoclonal antibodies against bacteria. Academic, New York (to be published)

Holmgren J, Svennerholm A-M (1973) Enzyme-linked immunosorbent assays for cholera serology. Infect Immun 7:759–763

Merson MH, Black RE, Huq I (1978) Epidemiology of cholera and enterotoxigenic *Escherichia coli* diarrhea. In: Ouchterlony Ö, Holmgren J (eds) Cholera and related diearrheas. Karger, Basel, pp 34–45

Mukerjee S (1978) Principles and practice of typing *Vibrio cholerae*. In: Bergman T, Norris JR (eds) Methods in microbiology, vol 12. Academic, New York, pp 51–115

Potter M, Pumphrey JG, Walters JL (1972) Growth of primary plasmacytomas in the mineral oil-conditioned peritoneal cavity. JNCI 49:305–308

Winkle S, Refai M, Rohde R (1972) On the antigenic relationship of *Vibrio cholerae* to *Enterobacteriaceae*. Ann Inst Pasteur 123:775–781

The Role of Specific Antibodies in Different Immunoglobulin Classes in Rapid Diagnosis of Virus Infections

U. KRECH [1]

Introduction

The ultimate aim of the laboratory diagnosis is the demonstration of the aetiological agent. This, however, may prove to be difficult in certain conditions:

1. If the aetiological agent is no longer demonstrable in the acute stage of the disease (rubella-measles virus)
2. If the aetiological agent is difficult to isolate (VZV, papova virus)
3. If the aetiological agent persists in the host and may be excreted from time to time without being necessarily the aetiological agent for the disease present (CMV, EBV).

Aetiological diagnosis in the past depended on the demonstration of increases in antibody titer in two blood specimens taken at the appropriate intervals. With this technique, results can be expected at the earliest 10–14 days after the beginning of an acute infection, and therefore this technique is of little help in establishing a diagnosis in the early stage of the disease. After it was shown in the early 1960s that this technique makes it possible to demonstrate antibodies in the acute stage of infection in certain immunoglobulin classes and that these antibodies persist only during a limited period of time, it has been possible to make a rapid diagnosis in the acute stage of an infection testing for IgM antibodies in a single serum sample.

Using this technique to demonstrate IgM antibody, it has not only been possible to speed up viral diagnosis, but it has also been possible to differentiate the aetiology in clinical syndroms. For example, the occurrence of hepatitis B in children was literally unknown before the introduction of diagnostic procedures. Today it is recognized that the transmission of hepatitis B virus (HBV) occurs from the mother to the child with different frequency in different parts of the world during or shortly after delivery. This infection does not usually lead to clinically recognizable hepatitis in children; however, it is assumed that it may lead to chronic persistant hepatitis, and perhaps also to malignant degeneration later in life. With the introduction of early diagnosis in rubella by the IgM technique, it has been possible to make an aetiological diagnosis in the acute stage of the disease and to separate rubella from other diseases with exanthema, which is not always easy on clinical grounds alone. Furthermore, it has been possible to evaluate the risk of pre-natal infection and to take clinical action when desirable.

1 Institut für Hygiene und Mikrobiologie der Universität Bern, CH-3000 Bern

Technical Approaches

A great number of different procedures have been developed for the determination of IgM antibodies. Modifications and new approaches are continuously being evaluated. It is beyond the scope of this presentation to describe them all in detail and to evaluate their advantages and disadvantages. Efforts have been made to find suitable methods to isolate the IgM fractions from patient's serum and to develop indicator systems to make the antigen antibody reaction measurable. All these techniques have advantages and disadvantages. None of them is without pitfalls, and few are simple enough so that they can find a wide application in the non-specialized laboratory.

Diagnostic Interpretation

It may be seen in Table 1 that the persistance of antibodies in different immunoglobulins makes it possible to use them as a marker for immunity, for recent infection or for previous vaccination, reactivation or reinfection. In the absence of IgM antibodies, antibodies in the IgG fraction are considered a marker for past infection or vaccination, and are therefore interpreted as immunity. With the availability of monoclonals against IgG sub-classes, it has also become possible to study antibodies in the IgG sub-classes. From preliminary results, it would appear that antibodies against cytomegalovirus (CMV) [1] and hepatitis B [2] are predominantly in the IgG1 and IgG3 sub-classes, and that the minor sub-class IgG3 may contain more antibodies than the predominant sub-class IgG1. The antibody concentration in the IgG1 class tends to be even higher following reactivated infection than following primary infection. It is also interesting to note that IgG3 does not react with protein A and therefore is not removed from the patient's serum by protein A treatment.

Table 1. Properties of immunoglobulins related to diagnosis

Ig classes and sub-classes	Reaction		Placenta transfer	Marker
	with C3	with Protein A		
IgG 1	×	×	×	
IgG 2	×	×	×	Immunity
IgG 3	×		×	
IgG 4	×	×		
IgM 7–8 S			×	
IgM 1	×			Recent infection
IgM 2	×	×		Vaccination
				Reactivation
IgA 1				Cross-reaction
IgA 2		×		
IgAsec.				
IgE				Hypersensitivity symptoms
IgD				Chronic infection?

Antibodies in the IgM Fraction

Low molecular weight IgM antibody activity is found in the 7/8 S fraction. The low molecular weight was confirmed by gel filtration and the activity is more effectively blocked by anti-IgM than by purified anti-IgG. Monomeric IgM is not observed in sera of healthy new-born children and adults, but it is found in a high proportion of patients suffering from gammopathies. It has also been found in patients with (a) viral infection, such as hepatitis B carriers, (b) chronic active hepatitis [3, 4] and also (c) CMV infections. Low molecular weight IgM can be transferred via the placenta from the mother to the new-born and may confuse inter-uterine infection. Its possible role in the immune response still needs to be evaluated.

The presence of specific IgM (19 S) has been used to differentiate current or recent infection from remote infection. The presence of IgM in cord-blood is regarded as evidence of pre-natal infection. Specific IgM antibodies usually develop early in the course of infection, and after reaching a peak titer they drop under a detectable level within weeks. IgM antibody formation may not only follow primary infection, but it may also occur after recurrent infection and vaccination. The amount of IgM antibody produced differs from individual to individual, from disease to disease, and may also reflect clinical activity. It has been shown for instance that severely damaged new-born following intra-uterine CMV infection usually show high specific IgM antibody levels, while intra-uterine infection leading to clinically asymptomatic infection may show no demonstrable IgM levels. In cases of continued clinical activity following viral infection, IgM antibodies may persist over more than 8 months in patients with persistant hepatitis and for more than 2 years in patients with aggressive hepatitis [4]. The anti-IgM antibody with lymphocytic toxicity in serum of children with dengue fever is a complement-dependent cold reactive IgM antibody which interacts with peripheral blood cells of the patient and may modulate the humoral immune response of patients during the acute stage of infection. Blood groups, cold agglutimines, red cell antibodies, anti-IgG (rheumatoid factor) and cryo-globulines are IgM antibodies.

IgA Response

The IgA response follows very much the same pattern as the IgM response. Technically it is more difficult to obtain a pure IgA fraction than a pure IgM fraction. In some infections, IgA may appear slightly earlier than IgM [5]. In patients with herpes zoster, specific IgA is found in all patients, but some of them do not have a detectable level of specific IgM [6]. This may merely reflect the higher IgA concentration in the serum of most persons.

The demonstration of secretory IgA in naso-pharyngeal secretions (NPS) has been a convenient method in making a rapid diagnosis in patients with acute respiratory infection. The results correlate well with virus isolation. Secretory IgA antibodies appear to be localized and confined to the infected area. They are considered a marker for protection.

IgE Response

The search for specific IgE antibodies has been particularly rewarding in those diseases which show symptoms of hyper-sensitivity reactions, such as bronchiolitis and asthma. Specific IgE has been demonstrated in RSV infections associated with wheezing, bronchiolitis or asthma. Peak IgE titers and peak histamine concentrations correlated well with the degree of hypoxia [7]. In another study, IgE bound to naso-pharyngeal-epithelial cells was found in most patients with RSV infection during the acute stage. Continued presence of IgE was more common in RSV-induced bronchiolitis and asthma than in patients with mild upper respiratory infection. Persistence of IgE in the respiratory tract may also be responsable for recurrent episodes of wheezing that occur in many patients after RSV-induced bronchiolitis [8, 9].

High levels of IgE have also been recorded in the sera during the acute stage of patients with Japanese B-encephalitis. IgE declined sharply during early convalescence [10].

IgD Response

It is now well accepted that there are at least two sub-classes in the IgD immunoglobulin. However, its role in diagnostic work has not yet been extensively evaluated.

Problems in the Interpretation of IgM Responses

With the continued use of IgM response in clinical virology, it has become apparent that this technique has a number of pitfalls which are causing problems in the interpretation of IgM response. It is now obvious that IgM response may follow primary or secondary infection and may also follow vaccination. From the clinical point of view it would be very important to be able to differentiate at least between primary and secondary immune response.

IgM immune response is also observed among closely related viruses, such as among the members of the adeno group, the paramyxo group, and the herpes group of viruses. Homologus IgM antibody response is observed in both primary and secondary VZV infections. IgM response to HSV antigen has also been noted in a few patients with VZV who did not have IgG antibodies to HSV.

IgA antibodies to late antigen of VZV and HSV were demonstrable in latent and active infection with these viruses. As another example of cross-reaction among closely related viruses, IgM antibodies are found in most cases with acute adenovirus infections, but they can also be found in persons without fresh infection. It may be postulated that the IgM titers in these persons are due to the frequent stimulation of the immune system by the genus-specific determinant alpha of the hexon in repeated adenovirus infections.

However, cross-reactions have not only been observed among closely related viruses, but also among unrelated micro-organisms. Broad cross-reactions occur in a certain proportion of sera from all infections usually against more than one antigen. Over all frequency of cross-reaction was estimated to be in the range of 10%–20%, depending on the numbers of antigens tested. Cross-reactions may also occur if highly purified antigens are used, indicating that at lest some of the cross-reacting IgM antibodies were not directed against the antigen contaminants. IgM cross-reactivity was mainly found in patients with acute infections (T. Krech, unpublished data). It has been frequently observed in patients with immunological disorders and rarely in healthy blood donors.

From the data reported, it would seem that the immune response in different globulin classes offers a great opportunity to improve virus diagnosis. However, it must be realized that this technique is still complicated by a number of problems which have not yet been sufficiently investigated.

References

1. Linde GA, Hammarstrom L, Persson MA, Smith CI, Sundqvist VA, Wahren B (1983) Virus-specific antibody activity of different subclasses of immunoglobulins G and A in cytomegalovirus infections. Infect Immun 42(1):237–244
2. Morell A, Roth-Wicky B, Skvaril F (1983) Immunoglobulin G. Subclass restriction of antibodies against hepatitis B surface antigen. Infect Immun 39:565–568
3. Sjogren MH, Lemon SM (1983) Low-molecular-weight IgM antibody to hepatitis B core antigen in chronic infections with hepatitis B virus. J Infect Dis 148(3):445–451
4. Cappel R, Van Beers D, Maes F, Toppet M, Cadranel S (1981) Significance of persisting IgM anti-HBc antibodies in hepatitis B virus infection. J Med Virol 8(3):201–205
5. Sarov I, Haikin H (1983) Human cytomegolovirus specific IgA antibodies detected by immunoperoxidase assay in serum of patients with cytomegalovirus infections. J Virol Methods 6(3):161–169
6. Levy E, Sarov I (1981) Detection of specific IgA antibodies in serum of patients with varicella and zoster infections. Intervirology 15(2):103–110
7. Welliver RC, Wong DT, Sun M, Middleton E Jr, Vaughan RS, Ogra PL (1981) The development of respiratory syncytial virus-specific IgE and the release of histamine in nasopharyngeal secretions after infection. N Engl J Med 305(15):841–846
8. Welliver RC, Kaul TN, Ogra PL (1980) The appearance of cell-bound IgE in respiratory-tract epithelium after respiratory-syncytial-virus infection. N Engl J Med 303(21):1198–1202
9. Stempel DA, Clyde WA Jr, Henderson FW, Collier AM (1980) Serum IgE levels and the clinical expression of respiratory illnesses. J Pediatr 97(2):185–190
10. Pavri KM, Ramamoorthy CL, Thorje S (1980) Immunoglobulin E in patients with Japanese encephalitis. Infect Immun 28(1):290–291

Problems of Rubella Diagnosis by Various IgM Techniques and the Need for Test Combinations

G. ENDERS [1]

The role of specific IgM antibodies as a valuable marker was recognized for the diagnosis of acute postnatal and congenital rubella infections more than 15 years ago. Today IgM and/or IgA antibody detection plays a leading role in the diagnosis of a number of viral and other microbial infections. In no other infection, however, is the correctness of the IgM antibody test results of such importance as for the detection of acute rubella infection in pregnancy. An infant with congenital rubella syndrome is only one of the possible outcomes of infection in a pregnant woman. Therapeutic abortion is another serious consequence.

The most commonly used methods for IgM antibody detection today are summarized in Table 1. In the last 15 years the "gold standard" has been sucrose density gradient centrifugation (SDG) [29]. Pretreatment of patients' sera with heparin–manganese chloride ($MnCl_2$) to remove non-antibody inhibitors is recommended. Following the separation of antibodies from different immunoglobulin classes, the specificity of the IgM fractions has to be ascertained by additional techniques. This is usually done by radial immunodiffusion (RID) against anti-IgM and -IgG sera, by 2-mercaptoethanol reduction, and, of late, by testing the IgM fractions in the sensitive enzyme-linked immunosorbent assays for IgG detection [7]. The IgM and IgG concentrations of the fractions are routinely estimated by the short and long incubation hemagglutination inhibition test (HAI), by indirect immunofluorescence [2], and now also by the enzyme-linked immunosorbents assays [7]. Gel filtration [11,21] requires the preparation of the columns and their standardization with a rubella-specific IgM-positive serum and a known IgM-negative serum with a high HAI titer. In addition, pretreatment of patients' sera with $MnCl_2$ for removing the lipoproteins is necessary since they are eluted from columns of Sephadex G 200 and Sephacryl S-300 in the same fractions as IgM. As with SDG fractions, additional tests to ensure lack of contamination of IgM fractions with IgG, particularly in sera with low IgM and high IgG content, and assay of the fractions for IgM and IgG concentration (mostly done by HAI) must be performed. The most recent advances with the SDG and gel filtration techniques are the use of reorientation centrifugation [12, 18] with the vertical rotor, which appears to be as effective and much quicker than swing out centrifugation, and gel filtration with Sephacryl S-300 [21]. The latter offers considerable technical advantages over the use of S-200. Techniques such as immunoaffinity chromatography [12] and high pressure liquid chromatography [26] have also recently been applied for rubella IgM detection.

---

1 Virologisch-med. Diagnostisches Institut, Hölderlinplatz 10, D-7000 Stuttgart 1

Table 1. Main methods for rubella-specific IgM detection

Principles	Methods		Performance	
Serum fractionation	Sucrose density gradient ultracentrifugation Reorientation centrifugation	SDG	Subsequent test for verification of IgM specificity, e. g., RID, 2-ME, ELISA	Subsequent test for rubella antibody detection, e. g., HAI, IFT, ELISA
	Gel filtration with Sephacryl S-300	GF		
Solid-phase immunosorbent assays (indirect)	Enzyme-linked immunosorbent assay e. g., Enzygnost IgM, Rubazyme M	ELISA	Antigen and control antigen adsorbed to solid phase, e. g., plate or bead	Single or serial dilutions of patient's sera
	Radioimmunoassay, with I-125 labeled antibodies	RIA		
M-antibody-capture immunosorbent assay (direct)	Anti-μ-hemadsorption test	HIT	Hemadsorption	With red blood cells
	Anti-μ-HAI test	SPIT	Hemagglutination inhibition	
	M-antibody-capture enzyme immunoassay e. g., Rubenz M	MACEIA	Monoclonal antirubella IgG antibodies	
	M-antibody-capture radioimmunoassay	MACRIA		

These methods are now being replaced by a new generation of tests, the solid-phase immunosorbent assays. The may be divided into two categories. The first is the solid-phase antigen assay, in which semipurified rubella antigen [24], and in some tests such as ELISA Enzygnost and in certain radioimmunoassays [14, 19] also control antigen, is adsorbed to the solid phase (plastic plates or beads). Single dilutions or dilution series of the patient's sera are incubated. After washing, a specific anti-human IgM globulin which is labeled with an enzyme (alkaline phosphatase or peroxidase) or a radioactive marker (iodine 125) is added. The most favored tests of this kind are the commercially available ELISA Enzygnost and ELISA Rubazyme [7, 24] whereas the noncommercial RIA has no widespread application.

The second type of assay is the M-antibody capture immunosorbent assay. The solid phase (plastic plate or bead) is coated with antiserum to the Fc determinant of human IgM (anti-μ). The test is then performed in three stages, with washing between each stage. The anti-μ-coated solid phase is first incubated with a dilution of sera, then with crude rubella HA antigen, and finally with red cells (anti-μ hemagglutination inhibition test = SPIT [15], anti-μ-hemadsorption test = HIT [17]) or more recently with a monoclonal antirubella IgG antibody labeled with iodine 125 (MACRIA) [23] or the enzyme peroxidase (MACEIA) [3, 13, 16, 20, 28]. This latter technique is now commercially available as Rubenz M [20]. A two-stage anti-μ assay with enzyme-labeled rubella virus has thus far not been developed.

The validity and the main problems of the available methods for IgM detection are shown in Table 2. SDG and gel filtration are restricted to well-equipped centers. They are laborious, time consuming, and expensive. Furthermore, the various parameters of the main procedures and the additional tests are difficult to standardize, so that sensitivity and specificity vary between laboratories. In addition pretreatment of sera, particularly for gel filtration, and relatively large volumes of serum are needed, and only a few sera can be investigated simultaneously. SDG is generally more sensitive than gel filtration owing to the greater dilution of the serum sample in the latter. The results are reported as positive or negative and some laboratories report, for example, the HAI titers of the IgM fractions [7, 25].

The commercial ELISAs Enzygnost and Rubazyme M are relatively easy to perform and have shown good reproducibility and sensitivity [7]. The problem of false-positive IgM results due to pregnancy and infection-associated IgM rheumatoid factor (RF) in presence of specific IgG antibodies has been resolved by latex or lately by anti-IgG (RF) adsorption of the sera in the case of ELISA Enzygnost or incorporating aggregated IgG in the serum dilution in the case of the ELISA Rubazyme M [7, 24]. The problem of IgG antibody competition with IgM is possible but is seldom observed. A further advantage of these two ELISAs is that the same principle is applicable to IgG antibody detection using a specific anti-IgG conjugate. So IgM and IgG antibody detection can be done at the same time and in the case of the ELISA Enzygnost on the same plate. The results are read photometrically and are available within 5 h in printed form. They are quantitatively reported as optical density (OD) values or titers for the ELISA Enzygnost and as indices for the ELISA Rubazyme M. The indices in the Rubazyme

Table 2. Validity and main problems of rubella IgM detection methods

Technique	Validity	Major problems	Expression of results
SDG GF	Technique for reference laboratories Variable sensitivity and specificity depending on performance	Additional tests required, time consuming, expensive, large volumes of sera, small series only	Negative/positive (titers)
ELISA Enzygnost Rubazyme M	Technique for routine and rapid diagnosis Good reproducibility and sensitivity, more sensitive than SDG, stable reagents, large series possible	Rheumatoid factor, resolved by anti-IgG adsorption	Optical density values, titers, indices
HIT, SPIT	Additional test in experienced laboratories	Time consuming, less reproducible than ELISA	Negative/positive (titers)
MACRIA	Technique for individual centers Very sensitive with monoclonal antirubella IgG antibodies, continuous supply	Equivocal results can be false-positive, more time consuming than ELISA, short shelf life of iodine-125 conjugate, radioactive waste	Arbitrary units
MACEIA Rubenz M	Technique for routine diagnosis Reliable and specific with monoclonal antirubella IgG antibodies, stable reagents	Occasional false-negative results	Arbitrary units

are obtained by dividing the optical density of the patient's serum by the optical density of the kitinternal low positive control serum. For Rubazyme M a neutralization test with which positive IgM results can be confirmed has also been developed. The radioimmunoassays (RIA or IRMA) appear very sensitive but have only a restricted use due to the need for prestandardization of the various reagents and the difficulties and hazards faced, particularly in iodine-125-labeling of the antisera. The results are reported as specific binding index (SBI) calculated as the binding ratio (BR) against test antigen and divided by the BR against control antigen.

The anti-μ hemadsorption (HIT) and anti-μ hemagglutination inhibition (SPIT) tests are quite valuable and cheap additional Tests; however, they are laborious, require technical expertise and are less sensitive than the commercially available ELISAs. The reading of the results is difficult and subjective and they are reported as positive or negative or in titers [7, 17]. The M-antibody capture radioimmunoassay applied in a few centers in England (MACRIA) was initially used with a radioiodine-labeled antirubella hyperimmune rabbit serum [23]. The latter has been replaced by monoclonal antirubella IgG antibody prepared from ascitic fluid of mice, so offering significant advantages in ensuring a continued supply of a reliable reagent and reducing the amount of radioactivity used [27]. It appears to be the most sensitive test, though a 3%–5% false-positive rate occurs in the equivocal range. The amount of the rubella-specific IgM antibodies may be quantitated by comparing the results with a standard calibration curve obtained by assaying a dilution series of a standard positive control serum arbitrarily defined as 100 units. The current control series contains 40, 10, 3.3, and 1 arbitrary units (a.u.), with a cut-off point of 3.3 for definitely positive sera [20].

Since most laboratories are not equipped for RIA work, the same test principle using an enzyme label is preferred. The commercially produced MACEIA Rubenz M with enzyme-linked monoclonal antirubella IgG antibody has been shown to be reliable and specific [20]. False-positive results appear infrequent but false-negative results may be possible owing to the present excessively high cut-off level of 5 a.u. of the low-positive control serum. When red cells or radioactively labeled antibody are used, the assays appear unaffected by rheumatoid factor (RF). However, when the antibody is enzyme labeled, false-positive reactions may occur which are due to bound IgM-RF reacting with aggregated IgG in the antirubella enzyme conjugate [16]. This may be overcome by incorporating aggregated IgG in the serum diluent, using an enzyme-conjugated F $(ab)_2$ antirubella antibody [13, 28]. This problem has not been noted with the monoclonal antirubella IgG antibody [20].

All present tests for rubella serology to measure HAI, IgG, and IgM antibodies suffer from the fact that neither the biochemical nature of the rubella antigens nor the biological function of the different surface antigens for detecting early and permanent immunity has been definitely defined. So far two distinct structural glycopolypeptides designated E1 and E2 in the surface and one single structural nonglycolysated protein C associated with the viral genome of the rubella virus have been identified. By use of a series of monoclonal antibodies to map the functional activities of the virus to specific regions of the structural polypeptides it was found that E1 contains at least three domains, of which two were shown to possess hemagglutinating and hemolysing but no neutralization activity. The single antigenic domain defined to date on E2 is inaccessible to monoclonal antibodies when E2 glycoprotein assumes a confirmation characteristic of the mature virion [30, 31].

In the various tests mostly crude or semipurified rubella HA antigens grown in different tissue cultures with various strains and prepared in different fashions are employed. Also the antibodies used for labeling are polyvalent anti-human IgM or IgG hyperimmune sera from different species. In the one instance where a monoclonal antibody is available [27] it has been produced against crude rubella HA antigen. Solution of the rubella serology test problem can be expected if the biochemical and functionally defined rubella antigens are used in the next generation of rubella tests.

Until this time each new rubella diagnostic method or kit has to be thoroughly compared against the well-established techniques with which the presently known facts concerning the antibody response to rubella infection have been accumulated. These reference methods are still SDG for detection of IgM antibodies and the HAI test for detecting persisting IgG antibodies. The latter has been shown to measure, in general, the same antibody quality as the cumbersome neutralization test with which it is very difficult to obtain reproducible results.

Therefore in the last 3–4 years we have compared for IgM antibody detection the diagnostic value of SDG against the commercial, indirect ELISA Enzygnost IgM and Rubazyme M, the noncommercial anti-μ hemadsorption test (Hit) and, more recently, the commercial IgM antibody capture ELISA Rubenz M (MACEIA).

The interassay variation of the tests was assessed with laboratory-own and kit-internal reference sera as well as with repeatedly tested patients' sera (Table 3). We found good reproducibility of SDG IgM values with laboratory-own reference serum by testing the fractions in the short and long incubation HAI test. Furthermore, we found that testing the SDG fractions in the two ELISAs for IgM and IgG antibodies assured the specificity of the IgM fraction and was more sensitive than testing the fractions by HAI. Also ELISA Enzygnost and Rubazyme showed a fair reproducibility with the same reference serum in 205 and

Table 3. Reproducibility of results with various IgM tests using reference sera (interassay results)

Method		Reference serum	Mean value	1s	CV	*n*
Sucrose density gradient	short HAI	Patient no. 57419	Titer $2^{-3.5}$	$2^{\pm 0.6}$	17.1	58
	long HAI		Titer $2^{-5.1}$	$2^{\pm 0.7}$	13.7	58
ELISA Enzygnost	Behring	Control no. 4	Titer $2^{-8.0}$	$2^{\pm 0.5}$	6.3	205
			OD 0.452	0.171	37.8	205
ELISA Rubazyme M	Abbott	Control no. 4	Index 2.031	0.273	13.4	119
		Low positive[a]	OD 0.500	0.076	15.2	149
		High positive[a]	Index 2.910	0.580	19.9	149
Anti-μ-hemadsorption test		Control no. 4	Titer $2^{-9.2}$	$2^{\pm 1.7}$	18.5	12
		Control no. 7	Titer $2^{-12.3}$	$2^{\pm 0.6}$	4.7	13

[a] From Abbott Diagnostics; all other reference sera from our own laboratory

149 subsequent assays respectively. The same was observed with repeatedly tested sera. The reproducibility of the Hit with a co-efficient of variation of the mean titer of 19% was also acceptable. The reproducibility of the MACEIA Rubenz M could not be assessed since the formulation of this test is still being improved. In general, the reproducibility of results in all tests is much better with sera of high IgM concentration than with those of low concentration. Therefore both types of sera must be included for reference purposes [7].

From our experience with the various IgM tests, with regard not only to good reproducibility but also to performance, technical demand, and subjectivity of reading, the commercial ELISAs scored best. The reliability, limits, pitfalls, and advantages of these newer tests in routine rubella diagnosis can, however, be judged only after testing large numbers of sera from patients with different rubella problems and a known history, and comparing the results with those of the established methods [9].

The main diagnostic problems in respect of rubella concern pregnant women. They are listed in Table 4. In all of these instances IgM antibody testing is employed. For the interpretation of the positive or negative IgM antibody findings with the various IgM tests, the development and persistence of IgM following acute infection, vaccination, and congenital infection has to be known.

Table 4. Diagnostic problems with rubella during pregnancy, and the risk of rubella embryopathy

- Acute rubella at various stages of gestation
- Rubella immunity status
- Rubella reinfection following previous vaccination
- Accidental vaccination
- Congenital rubella

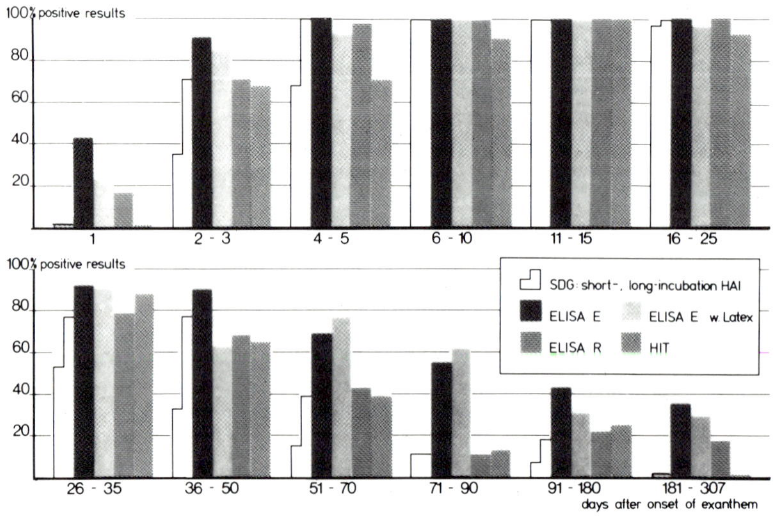

Fig. 1. Comparison of rubella IgM antibody detection with various methods in relation to onset of exanthem (649 sera)

The comparison of the sensitivity of rubella IgM antibody detection with the various IgM methods following acute infection is summarized in Fig. 1. The more sensitive test for the detection of IgM antibodies in the first 3 days and after the 51st day after onset of exanthem appeared to be the ELISA Enzygnost without latex or RF adsorption. After latex or RF adsorption, used in order to avoid false-positive results due to rheumatoid factor, ELISA Enzygnost and Rubazyme M showed similar sensitivity and were both more sensitive than SDG or Hit. In a more recent comparison including the Rubenz M ELISA this test proved to be more sensitive than all the other IgM tests investigated [9a].

Between the 6th and 25th day after onset of exanthem all tests, with the exception of one case with HIT, were able to demonstrate IgM antibodies. Between the 26th and 35th day after onset of exanthem there were a few IgM antibody-containing sera in which IgM was not detectable by SDG or Rubazyme M. Between the 36th and 70th day the detectability of IgM antibody by all tests declined. After that time IgM antibody detection was exceptional, and in the majority of cases was discovered only by ELISA Enzygnost without latex adsorption.

The detection of IgM antibody as measured by the various IgM methods following vaccination with the now mostly used RA 27/3 HDC vaccine is shown in Fig. 2 in comparison with IgM antibody development following acute rubella infection. The development of the HAI and IgG antibodies is also shown. As can be seen, IgM antibody production in vaccinees is less regularly detectable with any of the IgM tests than it is following natural infection. Also, the IgM titers or indices were at all times lower in vaccinees than in patients following acute rubella. Between 15 and 25 days following vaccination, IgM antibodies were detectable in approx. 60%–80% with the IgM tests. Thereafter the detection rate declined but in individual cases IgM antibodies were still detectable up to 307 days or even longer; this can pose a diagnostic problem if such low-positive IgM anti-

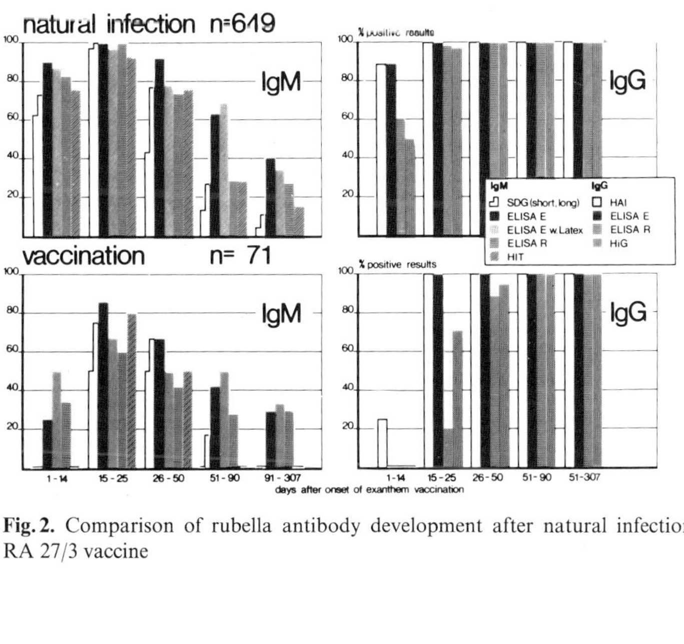

Fig. 2. Comparison of rubella antibody development after natural infection and vaccination with RA 27/3 vaccine

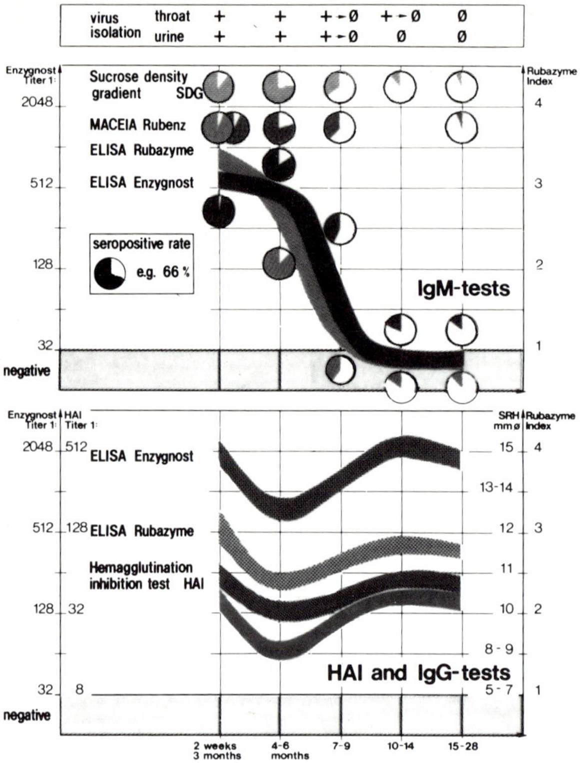

Fig. 3. Rubella IgM, HAI, and IgG antibody findings with various methods in 28 cases of rubella embryopathy (67 sera)

body levels are detected in a pregnant woman without using an IgM and IgG test combination and clinical information. For IgM antibody detection early and late after vaccination (as after natural rubella infection), SDG and Hit were found to be less sensitive than either of the ELISAs.

A comparison of the IgM tests and also the HAI and IgG tests for the detection of congenital rubella infection at different times after birth is shown in Fig. 3. Sixty-seven sera from 28 children with typical symptoms of rubella embryopathy taken at between 10 days and 28 weeks of age were investigated.

In cases with definite symptoms of rubella embryopathy, IgM antibodies in high concentration could be detected between the age of 2 weeks and 3 months after birth in 90%–97% by each of the IgM tests. Between 4 and 6 months the IgM detection rate declined to 78%–84%, between 7 and 9 months to 35%–43%, between 10 and 14 months to 11%–18%, and between 15 and 28 months to 4%–15%. As after acute infection and vaccination, the three ELISAs proved more sensitive than SDG and Hit (not shown in the figure) in detecting low IgM concentrations. However, minimum IgM concentration was sometimes detected by only one of the three ELISAs. The total correlation of IgM results by SDG and the two ELISAs Enzygnost and Rubazyme M for the 67 sera of the 28 children with rubella embryopathy is shown in Table 5. The correlation between SDG and ELISA Enzygnost and Rubazyme M, at 80.6% and 83.6% respectively,

Table 5. Comparison of SDG with two commercial ELISAs for IgM antibody detection in 28 infants with rubella embryopathy, 67 sera

	Concordant			Discordant			
SDG	∅	+	±	±	∅	∅	Correlation
ELISA	∅	+	±	+	(+)	+	
ELISA Enzygnost	11	43	0	4	7	2	54/67 80.6%
ELISA Rubazyme	12	43	1	3	3	5	56/67 83.6%

Table 6. Correlation of rubella SDG antibody results with results from other IgM methods ($n=445$)

	IgM tests versus	Sucrose density gradient			
		Positive (%)	Negative (%)	Borderline (%)	n
Sera from	ELISA Enzygnost	98	54	5	432
total	ELISA Enzygnost/Latex	94	58	17	266
rubella	ELISA Rubazyme	95	70	18	445
group	Anti-μ-hemadsorption test	91	85	–	230

appears unsatisfactory but can be explained by the different sensitivity of the various IgM tests for detecting low IgM concentrations, such as those present in older children with rubella embryopathy.

The correlation of IgM results by SDG and the other IgM tests was evaluated with 445 sera for the total rubella group. This included sera from cases of acute rubella, routine diagnosis, vaccinees, and congenital infection since this represents the material with which the laboratory has to deal in everyday diagnostics (Table 6). In the event of positive SDG results there was a 91%–98% correlation with the other tests. With negative SDG results, however, only 54%–85% of results with the other tests were negative, while when SDG IgM results were borderline, just 5%–18% of the results in the other IgM tests were borderline.

The excellent correlation of 96.5% between IgM results yielded by the two ELISAs Enzygnost and the improving Rubazyme M with a further 409 sera tested from cases of acute rubella and routine diagnosis is shown in Table 7. However, individual problems were observed with both tests. This finding is substantiated by employing an additional IgM test such as the newly introduced M-antibody capture MACEIA Rubenz M. The total correlation between the IgM results of Rubenz M with the limited number of 230 sera tested was 88% with ELISA Enzygnost and 87% with Rubazyme M. In the event of positive IgM results by Rubenz M, the correlation with the other two tests was 89%, and 83% respectively, whereas when Rubenz M results were negative 94% and 97% of the results of the other two tests were negative, indicating a higher sensitivity of Rubenz M in comparison with ELISA Enzygnost and Rubazyme M in their present formulation.

Table 7. Comparison of results with various IgM ELISAs in 409 sera from total rubella group

		ELISA Rubazyme		
		No. positive	No. negative	Correlation
ELISA Enzygnost	No. positive	163	5	97%
	No. negative	9	232	96%
	Correlation	93%	98%	97%
		MACEIA Rubenz M		
		No. positive	No. negative	Correlation
ELISA Enzygnost	No. positive	148	4	98%
	No. negative	23	55	70%
	Correlation	87%	94%	88%
ELISA Rubazyme	No. positive	142	2	99%
	No. negative	28	58	67%
	Correlation	83%	97%	87%

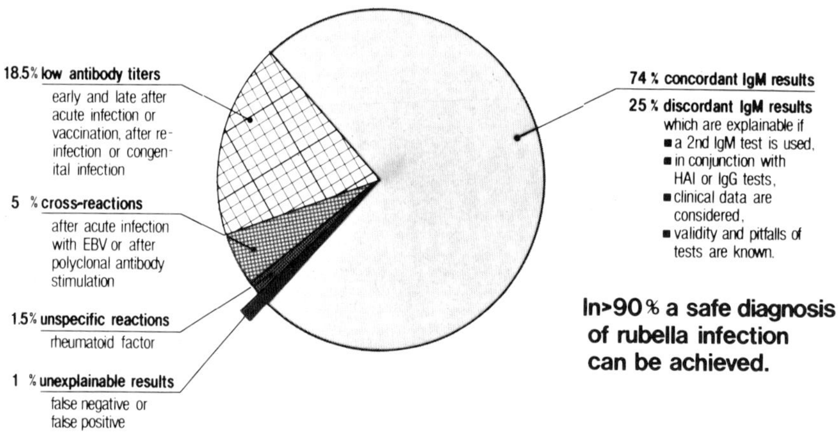

Fig. 4. Discordant results of four IgM methods in comparison with results from SDG (sera from total rubella group, $n = 445$)

The overall correlation between the results of SDG, considered the standard method, and the other IgM tests does not appear satisfactory (Fig. 4). Altogether, there were discrepant results in 26%. Analysing the reasons for the discrepant results (by considering the clinical history), it was found that 18.5% could be explained by the presence of low IgM antibodies early and late after acute infection, vaccination, reinfection following earlier vaccination and in infants with congenital rubella more than 6 months of age, and the higher sensitivity of the ELISAs. In 5%, specific low IgM antibodies were found by all or two of the three ELISAs in nine cases of acute Epstein-Barr infection (EBV) and in two cases of a presumably polyclonal antibody response in rubella immune patients. In the EBV cases there were also specific IgM antibodies detectable for cytomegalovirus and in the two latter cases there were specific IgM antibodies for three additional antigens: measles, mumps, and varicella zoster [10]. One mechanism for this cross-reactivity could be polyclonal stimulation of B-lymphocytes by EBV infection or another infection in rubella immune patients [1, 22]. Alternatively, IgM antibody against cellular protein, e.g., the antigen against which the heterophil antibody is directed, may be reacting with host antigens incorporated in the rubella virion. Particularly sera from patients with infectious mononucleosis may thus give positive reactions in many other tests for microbial specific IgM [1, 10, 22]. The occurrence of such cross-reactions shows that the detection of rubella IgM cannot be used as the sole indicator of recent rubella infection.

In 1.5% the discordant results from SDG were explainable by false-positive results with ELISA Enzygnost, including after latex adsorption, on account of high levels of rheumatoid factor. Only 1.3% of the discrepant results were not explainable.

This analysis shows that no IgM test is perfect and that for a reliable diagnosis of rubella infection in pregnancy quite frequently the use of a second IgM test is necessary, that IgM antibody detection should be done in conjunction with HAI and/or IgG tests, that the clinical data must be considered, and that the validity and the pitfalls of the tests employed must be known [1, 20]. In our experience, when these conditions are met a reliable diagnosis of rubella infection in pregnancy can be achieved in > 90% of cases.

Figure 5 shows the serological possibilities and limits for reliable diagnosis of acute rubella infection in pregnancy with the various IgM tests in conjunction with the HAI and the presently used IgG tests whose values have been comparatively investigated by us in a similar way to those for the most widely used IgM tests. A negative IgM antibody result with any of the listed IgM tests more than 4–6 weeks and with the Rubenz M more than 6–8 weeks [9a]. After a history of contact or rubella-like symptoms does not exclude a recent rubella infection. Also a borderline IgM result with any of the IgM tests in the absence of a history of recent rubella contact or symptoms may be false-positive. Therefore the need to perform IgM assays in conjunction with the HAI and/or IgG antibody tests is obvious.

It is easy to see that the HAI test and the ELISA Enzygnost IgG on the one hand and Rubazyme and the single radial hemolysis test (SRH) on the other hand are comparable in terms of detection of early rubella antibodies after onset of symptoms. For a rapid diagnosis of acute rubella in pregnancy in the 1st to 3rd week after onset of symptoms with one serum sample a safe test combination is the use of HAI plus SRH or Rubazyme G plus one of the IgM ELISAs. After the 2nd to 4th week, when HAI and IgG antibody values which are individually variable have reached their maximum levels and IgM antibodies, if detectable, may only be present in minimum concentration, two of the IgM tests in conjugation with the HAI or IgG tests and clinical data should be used in reaching a decision on whether to terminate pregnancy [4].

The test combinations for immunity screening in early pregnancy are shown in Fig. 6. Such antenatal screening programmes are carried out in several European countries. Usually no clinical information is provided; therefore with one

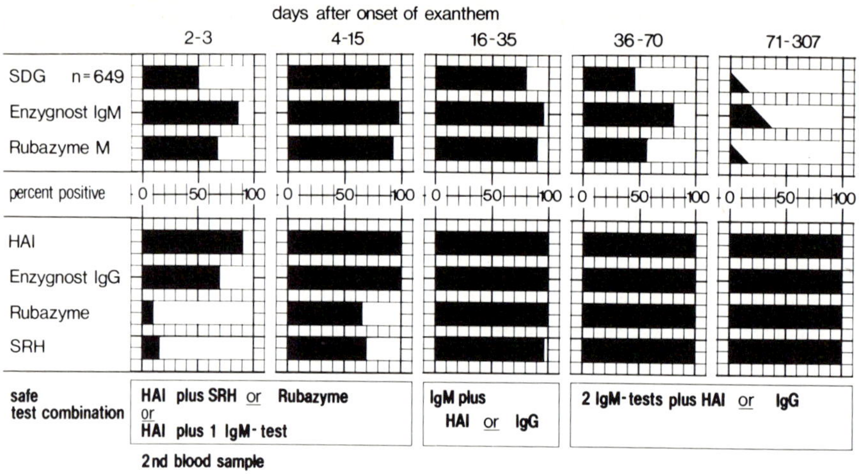

Fig. 5. Rubella IgM and IgG antibody development after acute infection (792 sera from 463 patients) and possible test combinations

test combination:		**HAI or IgG plus IgM**	- antibodies may be in a rising phase - excludes recent infection in the last 4-5 weeks
HAI titers SRH diam.	**results**	**IgM negative**	**IgM positive**
1:<8 <5mm	negative * < 10 U/ml	control up to 18th week, vaccinate after delivery.	obtain all possible clinical data for decision of: **termination of pregnancy**
1:8 5-7mm	equivocal 10-30 U/ml	use 2nd IgG test to support "immunty". vaccinate after delivery	
1:16 8-9mm	low positive 30-100 U/ml		
1:32-128 10-12 mm	positive 100-360 U/ml	immune	
1:≥256 ≥13mm	high positive ≥360 U/ml	immune	

Fig. 6. Reliable antenatal screening for rubella immunity (women in early pregnancy). * Units for HAI and SRH calibrated with the British Ref. Serum 69/6

test or one serum sample one cannot reliably establish whether the HAI titers or IgG antibody values are not in a rising phase or, when high antibody levels are found, whether there has not been a recent infection, particularly in times when rubella virus is circulating and the chance of rubella contact is high. An additional IgM test which proves negative would be a safer way of attesting immunity [4, 5]. If low-level IgM antibodies are detected, it has to be established whether they remain from a previous rubella infection or vaccination. With equivocal or low-level HAI titers of 1:8 or 1:16 a second IgG test must be done; when the results of these tests agree, this supports protection better than does the result of only

Table 8. Detection of rubella infection. In pregnancy (*top*) and in newborn and infants (*bottom*)

		Acute infection	Immunity	Reinfection	Accidental vaccination
Serology	HAI and/or IgG or 2 sera	Rise	Positive	Rise	Rise
	IgG 1 serum	Neg. – pos.	Positive	Pos. – high pos.	Low pos. – pos.
	IgM 1 serum	positive Fetoscopy: IgM in cord blood	Negative	Neg. – low pos.	Low pos.

		0–7 months	8 months – 2 years	3–10 years
Serology	HAI and/or IgG	Decrease and rise	Present	Decrease
	IgM	Positive	Low positive – negative	Negative
Virus isolation, antigen detection	Throat secr., CSF, urine, tissues,	Positive	Positive – negative	Negative
	B- or T-lymphocytes	Positive	Positive	?
CMI detection	Spec. immune compl.	Positive	Positive	Positive – negative

one test [8]. Even though we do not know how well, particularly low-level vaccine-acquired antibodies determined with the presently available tests will protect against reinfection and its possible consequences [6].

Most of the other rubella diagnostic problems mentioned above, like reinfection after previous vaccination and immunity status before accidental vaccination, can be solved with the test combinations described above (Table 8). The same is true for the detection of congenital rubella infection in symptomatic and asymptomatic infants at birth and later. However, for determining whether the infant is still infectious, virus isolation in tissue culture is used in addition to serology.

For the past 10–15 years IgM antibody detection with a great number of techniques has been performed for confirming acute rubella in pregnancy and congenital rubella. However, with the first generation techniques such as SDG and gel filtration and similar methods, IgM diagnosis was limited to a few well-equipped centers. With the introduction of second generation techniques such as radioimmunoassays and, particularly, commercial enzyme-linked immunoassays, many laboratories without previous experience in the problems of rubella diagnosis can carry out IgM antibody testing. Even though the commercial tests perform well and are generally more sensitive than SDG of gel filtration, there are limits and pitfalls with all of the IgM tests, including SDG and gel filtration. Therefore a reliable diagnosis of the stage of rubella infection, e.g., in pregnancy, can sometimes only be made using two of the newer IgM tests in conjunction with the determination of antibodies in the HAI and/or IgG tests and the use of clinical information.

References

1. Best JM, Palmer SJ, Morgan-Capner P, Hodgson J (1984) A comparison of Rubazyme-M and MACRIA for the detection of rubella-specific IgM. J Virol Methods 8:99–109
2. Cradock-Watson JE, Ridehalgh MKS, Pattison JR, Anderson MJ, Kangro HO (1979) Comparison of immunofluorescence and radioimmunoassay for detecting IgM antibody in infants with the congenital rubella syndrome. J Hyg (Lond) 83:413–425
3. Diment JA, Chantler SM (1981) Enzyme immunoassay for detection of rubella-specific IgM antibody. Lancet I:394–395
4. Enders G (1984) Rubella infections. Med Lab (Stuttg) [Suppl] I:58–71
5. Enders G (1985) Serological test combinations for safe detection of rubella infections. (Intern. Symposium on the Prevention of Congenital Rubella in Washington 13.–14. 3. 84). Rev Infect Dis 7 [Suppl 1] (to be published)
6. Enders G, Calm A, Schaub J (1984) Rubella embryopathy after previous maternal rubella vaccination. Infection 12:96–98
7. Enders G, Knotek F (1985) Comparison of the performance and reproducibility of various serological methods and diagnostic kits for the detection of rubella antibodies. J Virol Meth (to be published)
8. Enders G, Knotek F, Greeck A (1985) Comparison of negative, low-positive and positive hemagglutination inhibition rubella antibody titers by single radial hemolysis test, ELISA Enzygnost IgG and ELISA Rubazyme IgG. J Virol Meth (submitted for publication)
9. Enders G, Knotek F, Pacher U (1985) Comparison of various serological methods and diagnostic kits for the detection of acute, recent and previous rubella infection, vaccination and congenital infections. J Med Virol 12 (to be published)

9a. Enders G (1985) Detection of IgM antibodies against Rubella virus: Comparison of two indirect ELISAS and an anti-my-EIA. J Med Virol (to be published)

10. Enders G, Biber M (1985) Reactivity for rubella- and cytomegalo-specific IgM in sera from patients with infectious mononucleosis. J Clin Microbiol (submitted for publication)
11. Gupta JD, Peterson V, Stont M, Murphy AM (1971) Single sample diagnosis of recent rubella by fractionation of antibody on Sephadex G-200 column. J Clin Pathol 24:547–550
12. Huschka U, Maier J, Rautenberger EW, Doerr HW (1982) Rapid separation of immunoglobulin M by immunoaffinity chromatography for detection of specific antibodies to Rubella and *Treponema pallidum*. Eur J Clin Microbiol 1:118–121
13. Isaac M, Payne RA (1982) Antibody class capture assay (ACCA) for rubella-specific IgM antibody. J Med Virol 10:55–64
14. Kalimo KOK, Meurman OH, Halonen PE, Ziola BR, Viljanen MK, Granfors K, Toivanen P (1976) Solid-phase radioimmunoassay of Rubella virus immunoglobulin G and immunoglobulin M antibodies. J Clin Microbiol 4:117–123
15. Krech U, Wilhelm JA (1979) A solid-phase immunosorbent technique for the rapid detection of rubella IgM by haemagglutination inhibition. J Genet Virol 44:281
16. Kurtz JB, Malic A (1981) Rubella-specific IgM detected by an antibody capture assay/ELISA technique. J Clin Pathol 34:1392–1395
17. Logt JTH van der, Loon AM van, Veen J van der (1981) Hemadsorption immunosorbent technique for determination of Rubella immunoglobulin M antibody. J Clin Microbiol 13:410–415
18. Luton P, Ridgway GL (1979) Use of the ultracentrifuge vertical rotor in the detection of rubella specific IgM on a sucrose density gradient. J Clin Pathol 32:931
19. Meurman OH, Viljanen MK, Granfors K (1977) Solidphase radioimmunoassay of rubella virus immunoglobulin M antibodies: comparison with sucrose density gradient centrifugation test. J Clin Microbiol 5:257
20. Morgan-Capner P (1984) The detection of rubella-specific antibody. PHLS Microbiol Digest 1:6–11
21. Morgan-Capner P, Davies E, Pattison JR (1980) Rubella-specific IgM detection using Sephacryl S-300 gel filtration. J Clin Pathol 33:1082–1085
22. Morgan-Capner P, Tedder RS, Mace JE (1983) Rubella-specific IgM reactivity in sera from cases with infectious mononucleosis. J Hyg (Cambr) 90:407–413

23. Mortimer PP, Tedder RS, Hambling MH, Shafi MS, Burkhardt F, Schild U (1981) Antibody capture radioimmunoassay for anti-rubella IgM. J Hyg (Cambr) 86:139–153
24. National External Quality Assessment Scheme (NEQAS) (1982) Operated by the PHLS Division of Microbiological Reagents and Quality Control, Central Public Health Laboratory. Rubella serology IgM. Distribution No. 224
25. Pattison IR (1982) Laboratory investigation of rubella. In: Public Health Laboratory Service Monograph, Series No. 16
26. Sann G, Schneider G, Loeke S, Doerr HW (1983) Rapid fractionation of serum immunoglobulins by high pressure liquid gel permeation chromatography. Application to routine serologic procedures. J Immunol Methods 59:121–127
27. Tedder RS, Yao JL (1982) The production of monoclonal antibodies to rubella haemagglutinin and their use in antibody-capture assays for rubella-specific IgM. J Hyg (Camb) 88:335–350
28. Vejtorp M (1981) Solid phase anti-IgM ELISA for detection of rubella specific IgM antibodies. Acta Path Microbiol Scand B 89:123–128
29. Vesikari T, Vaheri A (1968) Rubella: A method for rapid diagnosis of recent infection by demonstration of the IgM antibodies. Br Med J 2:221–223
30. Waxham MN, Wolinsky JS (1983) Immunochemical identification of Rubella virus hemagglutinin. Virology 126:194–203
31. Waxham MN, Wolinsky JS (1985) A model of the structural organization of rubella virions. (Intern. Symposium on the Prevention of Congenital Rubella, Washington, 13.–14. 3. 84) Rev Infect Dis 7 [Suppl 1] (to be published)
32. Ziegelmaier R, Behrens F, Enders G (1980) ELISA: Demonstration of IgG and IgM antibodies in infections with Cytomegalo (CMV) and Rubella virus. Med Lab (Stuttg) 8:16–25

The Role of Specific Antibodies of Different Immunoglobulin Classes in the Rapid Diagnosis of Systemic Mycotic Infections

L. KAUFMAN [1]

Antibodies represent a major effector mechanism of the immunologic response. Many function to opsonize the tissue-form cells of pathogenic fungi for phagocytosis. IgA, IgE, IgG, and IgM class antibodies are known to react with fungal antigens. The most frequently detected specific antibodies are of the IgG class. These antibodies have both diagnostic and prognostic value. IgM-specific antibodies play a more restricted diagnostic role. They are particularly useful for the early detection of candidiasis, coccidioidomycosis, and histoplasmosis. Circulatory IgG and IgM antibodies are elicited by specific fungal antigens in individuals with such symptomatic systemic fungal infections as aspergillosis, blastomycosis, candidiasis, coccidioidomycosis, cryptococcosis, histoplasmosis, and paracoccidioidomycosis. These antibodies are detected by a variety of old and new serologic tests. They do not appear to play a major protective role in host immunity, in contrast to the T-lymphocytes, which are of paramount importance in protecting the host against systemic mycotic infections. I will appraise the value of these antibodies, describe their sequence of occurrence during the course of infection, and where possible, characterize the immunoglobulin classes associated with them.

Aspergillosis

Aspergillosis, caused by *Aspergillus fumigatus* and other species of *Aspergillus*, may manifest itself as an allergic bronchopulmonary disease (ABA), aspergilloma (fungus ball), or invasive disease. Antibody tests have been most successful in diagnosing ABA and aspergilloma. Approximately 75% of patients with ABA and over 90% of those with fungus ball produce diagnostic precipitins detectable by immunodiffusion (ID) tests. These tests are of less value in diagnosing invasive aspergillosis, which occurs mainly in the immunocompromised host. In ABA as well as in other allergic diseases, high titers of IgE antifungal antibodies occur in sera [23]. These antibodies, however, are not always specific for aspergillosis, as many patients with ABA demonstrate high serum titers of anti-*Candida* IgE as well [4].

1 U.S. Department of Health and Human Services, Public Health Service, Division of Mycotic Diseases, Center for Infectious Diseases, Centers for Disease Control, Atlanta, GA 30333, USA

IgG precipitins are most frequently demonstrated in ABA by using culture filtrate or mycelial extract antigens [5, 12, 28]. IgG *A. fumigatus* antibodies were also readily detected by enzyme-linked immunosorbent assay (ELISA) in 11 ID precipitin-positive serum samples from asthmatic patients. Depending on the antigens used, ELISA test results compared well to those obtained with ID tests [21].

In aspergillomas where *Aspergillus* spp. colonize a preformed pulmonary cavity, IgG class antibodies are also readily detectable [5, 12].

Antibodies are infrequently detected in patients with invasive aspergillosis, particularly those who are immunosuppressed. Using the ID test, we detected precipitins in 68% of 44 patients who had invasive aspergillosis with or without apparent altered immunologic status. In some situations, a three- to fivefold concentration of serum specimens will demonstrate precipitins which are otherwise undetectable. Antigens produced during infection may bind serum antibodies and thus not permit detection of the latter by conventional antibody tests. Early studies indicated that antibodies detected in patients with invasive aspergillosis are of the IgG class [12, 25]. This was recently confirmed by Wilson and Hearn [30], who, using an ELISA technique with cell wall antigens derived from *A. fumigatus*, demonstrated specific IgG antibodies in the sera of patients with aspergillosis.

Blastomycosis

Blastomycosis a systemic disease of North America, Africa, the Middle East, and India. It is caused by the fungus *Blastomyces dermatitidis*. Specific complement fixation (CF) and ID tests, using a purified culture filtrate antigen of *B. dermatitidis* (designated A), have proven valuable for diagnosing blastomycosis. CF tests demonstrating a sensitivity of about 70% and ID tests demonstrating a sensitivity of approximately 80% are totally specific [9]. Increased sensitivity has been achieved with ELISA; however, this benefit has been compromised by nonspecific reactions occurring with *Histoplasma capsulatum* antibodies in histoplasmosis case sera. The *B. dermatitidis* A antibody as well as the cross-reacting *H. capsulatum* antibody appear to be of the IgG class.

Candidiasis

Systemic candidiasis is an opportunistic infection caused primarily by *Candida albicans*. The infection usually arises from an endogenous source in a compromised host. The level of antibody detected depends on the test sensitivity and the patient's immunocompetence.

IgM antibodies may be detected hy hemagglutination with *C. albicans*-sensitized erythrocytes in the early stages of systemic candidiasis. Later, candidal IgG antibodies detectable by indirect fluorescent antibody (IFA) tests predominate [11, 16]. In studying sera from candidiasis patients in 1970, Lehner [13], also found IgG to be the principal immunoglobulin class of antibody reacting in the IFA test. Significant levels of IgG were found in 78% of sera from patients with candidiasis. IgA and IgM antibodies, however, were also shown to react. It was later reported [14] that *C. albicans* precipitins are of the IgG class. Agglutinins were found to be predominantly IgM, although IgG and IgA proteins were also noted. Weiner and Yount [29] found mannan antigenemia in immunosuppressed patients with early invasive disseminated candidiasis. This may result in immune complex formation, after which levels of candidal antimannan antibody of the IgG class increase. IgG antibodies against *C. albicans* cytoplasmic antigen detected by ELISA were found in patients with candidiasis. Serum samples positive for *C. albicans* precipitins by counter immunoelectrophoresis (CIE) were also positive by ELISA [22]. Specific IgA antibody to mannan has also been noted to increase early in the course of invasive candidiasis [6].

Coccidioidomycosis

Coccidioidomycosis, caused by the dimorphic fungus *Coccidioides immitis*, is prevalent in the arid and semiarid areas of the New World. Antibody responses in coccidioidomycosis have been thoroughly studied and are of proven diagnostic and prognostic value [8]. Radioimmunoelectrophoretic studies have revealed that, in coccidioidomycosis, antibody activity is associated with IgA, IgM, and IgG globulins. IgM activity parallels tube precipitin reactivitiy, while IgG parallels the CF titer. IgA globulin does not appear to associate with either CF or tube precipitin test reactivity [24]. Specific diagnostic IgM antibodies are detected within 1–2 weeks after the onset of symptomatic illness [27], with about 90% of symptomatic patients testing positive in the 2nd to 4th week. This IgM antibody, responding to a heat-stable coccidioidin antigen, is of diagnostic value in early acute coccidioidomycosis. It declines and disappears as the infection clears or progresses. IgM antibody is detected in serum by specific tube precipitin and ID tests and by the simple but less specific latex agglutination test. IgM is virtually absent in cerebrospinal fluid specimens (CSF) from patients with meningeal coccidioidomycosis, and the tests described are consequently of no value in diagnosing this form of the disease.

Within 1 to 2 months of clinical onset, IgG antibodies to *C. immitis* are produced in the infected patient. The IgG response can be quantitated by CF and ID tests with a heat-labile coccidioidin antigen in both serum and CSF specimens. Titers are not only diagnostic but may reflect the severity of infection.

A specific antibody to a second heat-stable antigen of *C. immitis* found infrequently in humans has been shown to be useful for identifying suspected *C. immitis* isolates by exoantigen studies [10]. This antibody has been shown to be of the IgM class (P. Cifuentes, unpublished data 1982).

Cryptococcosis

Cryptococcosis caused by *Cryptococcus neoformans* is a major cause of pulmonary and central nervous system disease in the debilitated and immunocompromised host. Maximal diagnosis is achieved through the use of antigen and antibody tests. The most widely used immunodiagnostic test for cryptococcosis ist the latex agglutination test for cryptococcal antigen. Antibody tests, however, are important in diagnosing early and localized cryptococcosis and with specimens from cryptococcosis patients who are antigen-negative. Positive antibody reactions are usually indicative of a favorable prognosis. Whole yeast cell antigens are used in the IFA and tube agglutination tests for *C. neoformans* antibody. The IFA test is widely used for detecting cryptococcal antibody. Using standard IFA and tube agglutination tests, Blumer and Kaufman [2] failed to detect early IgM *C. neoformans* antibodies. Immunoglobulin IgG *C. neoformans* antibody is detectable with the IFA test in the sera of about 54% of cryptococcosis patients. Cross-reacting antibodies from patients with blastomycosis and histoplasmosis are also of the IgG class [7].

IgG *C. neoformans* antibody is detected by tube agglutination tests in about 44% of cryptococcosis patients. Patients with cryptococcal IgG serum antibodies have a favorable prognosis and are likely to recover. These antibodies are relatively uncommon in CSF specimens.

Interest in detecting an early IgM antibody response in cryptococcosis prompted Reiss et al. [19] to evaluate an ELISA with *C. neoformans* galactoxylomannan antigen. IgM antibodies were found in 21.8% of 55 cryptococcosis patients examined. The IgM antibody was also present in two patients with cryptococcal antigenemia. IgM cross-reacting antibodies were also noted in patients with blastomycosis, candidiasis, and histoplasmosis. These results were simliar to those noted by Kaufman and Blumer [7] with IgG antibodies in the IFA test.

Histoplasmosis

Histoplasmosis, caused by the dimorphic fungus *Histoplasma capsulatum*, is a disease found worldwide. Serologic tests using yeast- and mycelial-form (histoplasmin) antigens are important in establishing a diagnosis. The CF test with yeast-form antigen has a sensitivity of about 94% with culture-proven case sera. ID test results usually correlate with histoplasmin CF test results and show a sensitivity of about 85%. In an early study, Chandler et al. [3] found complement fixing and -precipitating antibody to be primarily in the IgG class and latex-agglutinating antibodies in the IgM class. Although IgA levels increased in patients with histoplasmosis, no specific antibody was found in the IgA fraction. Studies with acute and chronic case sera indicated that CF activity against both mycelial- and yeast-form antigens occurs in the IgG fraction. CF antibodies were demonstrable 14 days after exposure. In one chronic case, CF antibodies against both CF antigens were found in the IgM. The ID reaction appeared to be of the IgG type and correlates with the histoplasmin CF reactions. CIE assays for *H. capsulatum* precipitins similarly appear to detect specific IgG antibodies [17].

The sequential immunoglobulin response in acute histoplasmosis appears to be similar to that of other infectious diseases in that there is an early IgM response, followed by an increase in IgA and IgG. Baum et al. [1] noted that during acute histoplasmosis, IgM antibodies may be detected by CF tests, but IgG antibodies predominate in most cases. Similarly, recent studies with histoplasmin antigens also indicate that significant elevations in IgM and IgG antibody responses occur. Studies by Wheat et al. [31] suggest that histoplasmosis may be diagnosed by the IgM response during the early stages of infection before IgG antibody appears. It has not been possible to confirm this, however, since most of the patients they studied were presented late in their illness and were already IgG positive.

Paracoccidioidomycosis

Paracoccidioidomycosis is an important systemic mycotic infection endemic in Central and South America. The disease, caused by the dimorphic fungus *Paracoccidioides brasiliensis*, may be accurately diagnosed serologically. CF and ID tests demonstrate a sensitivity of over 90% and 94% respectively with proven case sera. CF IgM antibodies are important in the recognition of early acute paracoccidioidomycosis [26]. Although IgM antibodies may occur in the early stages of infection, as in histoplasmosis, IgG antibodies appear to predominate at the time of the diagnosis. A close correlation exists between CF titers and IgG concentrations. This suggests that IgG is the functional immunoglobulin in the CF test for paracoccidioidomycosis.

Both IgM and IgG antibodies have been detected in paracoccidioidomycosis case sera by using an indirect immunoperoxidase technique and yeast-form cells of *P. brasiliensis* as an antigen [18]. IgG responses correlate with ID test results. IgM antibodies appear to be relevant to active disease and can be used to distinguish symptomatic patients from healthy subjects and those with inactive disease. In contrast to these findings, Restrepo et al. [20] reported that the stage and degree of *P. brasiliensis* infection appear to be associated with the relative concentration of serum IgG antibodies. IgM antibodies were found in 68% of *P. brasiliensis*-infected patients studied with immunofluorescence techniques. The presence of these antibodies, however, was not related to IgM levels, the clinical picture (i.e., disease activity or relapse), or to the presence of precipitating antibodies [15].

Zygomycosis

Zygomycosis is an acute fungal infection caused by opportunistic pathogens in the class Zygomycetes. CIE, ID, and IFA tests are used to serodiagnose this disease. The techniques, however, have not proven to be highly sensitive. Studies with sera from a fatal case of craniofacial zygomycosis indicated that the IFA test is more sensitive than the CIE test for detecting antibody to *Rhizopus* spp. High titers of serum IgA and IgM were noted in this acute disease. The role of IgA and IgM antibodies in such infections warrants investigation [32].

Conclusion

Substantial progress has been made over the last 20 years in elucidating the class of immunoglobulins responsible for specific antibody activitiy in different mycotic infections. Circulating antibodies, consisting of IgA, IgG, and IgM classes, occur in persons infected by fungal pathogens. The predominant class is dependent on the host response, the portal of entry for the etiologic fungus, and whether the affected site is invaded or colonized. Other important factors are the stage, type, duration, and severity of infection. IgM appears to be associated with early antibody responses in coccidioidomycosis, histoplasmosis, and paracoccidioidomycosis. Most efforts to develop procedures which consistently and reliably detect such early antibody responses and enable diagnosis of early acute disease have had limited success, with the exception of coccidioidomycosis. Specific IgG class fungal antibodies are diagnostically important and are readily detected serologically in sera from patients with aspergillosis, candidiasis, coccidioidomycosis, cryptococcosis, histoplasmosis, and paracoccidioidomycosis. The diagnostic value of IgA antibodies is still unclear. In some fungal antibody tests, such as the tube precipitin and latex agglutination tests, only IgM has been found to be reactive. The use of more sensitive enzyme immunoassays and other tests directed toward detection of defined, specific, fungal antibodies could permit the early identification of fungal infections in immunosuppressed patients and in those patients for whom results are negative by conventional serologic tests. Sensitive IgM-specific tests might also allow a diagnosis in immunocompetent patients before their IgG antibodies become apparent.

The appearance and persistence of specific immunoglobulin class antibodies may not only be of diagnostic importance but may also aid in determining the clinical form of the disease (i.e., acute, chronic, or disseminating), the severity of infection, and the need for antifungal therapy. As the immunoglobulin class of specific antibodies is determined and its role in the host response is clarified, it will become possible to develop sensitive tests to measure them, thus enhancing the value of mycoserologic tests.

References

1. Baum GL, Daniel TM, Rice E (1973) Immune globulin characterization of complement fixing antibodies in histoplasmic infection. Chest 64:16–20
2. Blumer SO, Kaufman L (1977) Characterization of immunoglobulin classes of human antibodies to *Cryptococcus neoformans*. Mycopathologia 61:55–60
3. Chandler JW Jr, Smith TK, Newberry W Jr. Chin TDK, Kirkpatrick CH (1969) Immunology of the mycoses. II. Characterization of the immunoglobulins and antibody responses in histoplasmosis. J Infect Dis 119:247–254
4. Edge G, Pepys J (1980) Antibodies in different immunoglobulin classes to allergic respiratory disease. Clin Allergy 10:45–48
5. Gordon MA, Lapa EW, Kane J (1977) Modified indirect fluorescent antibody test for aspergillosis. J Clin Microbiol 6:161–165
6. Holmberg K (1978) Serological diagnosis of systemic candidiasis and invasive aspergillosis. Scand J Infect Dis [Suppl] 16:26–32
7. Kaufman L, Blumer S (1978) Cryptococcosis: the awakening giant In: The black and white yeasts. Proceedings of the fourth international symposium on mycoses, scientific publ. no. 356. Pan American Health Organization, Washington DC, pp 176–182
8. Kaufman L (1980) Serodiagnosis of fungal diseases. In: Manual of clinical microbiology, 3rd ed. Am Soc Microbiol, Washington DC, pp 628–646

9. Kaufman L (1983) Mycoserology: its vital role in diagnosing systemic mycotic infections. Jpn J Med Mycol 24:1–8
10. Kaufman L, Standard P, Padhye A (1983) Exoantigen tests for immunoidentification of fungal cultures. Mycopathologia 82:3–12
11. Koniger G, Adam (1973) Serologischer Nachweis systemischer *Candida* Infektionen mit Hilfe von Antikörpertiterrelationen. Klin Wochenschr 51:437–444
12. Kurup VP, Fink JN (1978) Evaluation of methods to detect antibodies against *Aspergillus fumigatus*. Am J Clin Pathol 69:414–417
13. Lehner T (1970) Serum fluorescent antibody and immunoglobulin estimations in candidiasis. J Med Microbiol 3:475–481
14. Lehner T, Buckley HR, Murray IG (1972) The relationship between fluorescent, agglutinating and precipitating antibodies to *Candida albicans* and their immunoglobulin classes. J Clin Pathol 25:344–348
15. Mota FT, Franco MF (1979) Observoes sobre a pesquite de anticorpos IgM anti-*Paracoccidioides brasiliensis*, por immunofluorescecia no soro de pacientes com paracoccidioidomicose. Rev Inst Med Trop San Paulo 21:82–89
16. Muller HL (1974) IgM- and IgG-antibodies against *Candida* polysaccharide in the serodiagnosis of candidiasis. Bull Soc Franc Mycol Med 3:51–52
17. Picardi JL, Kauffman CA, Schwarz J, Phair JP (1976) Detection of precipitating antibodies to *Histoplasma capsulatum* by counterimmunoelectrophoresis. Am Rev Respir Dis 114:171–176
18. Pons I, Gimenez M, Guileron C, Szarfman A (1976) La tecnica de la immunoperoxidase en la deteccion de anticuerpos especificos en la infeccion humana por *Paracoccidioides brasiliensis*. Medicina 36:510–512
19. Reiss E, Cherniak R, Eby R, Kaufman L (1984) Enzyme immunoassay detection of IgM to galactoxylomannan of *Cryptococcus neoformans*. Diagn Immunol 2:109–115
20. Restrepo A, Restrepo M, Restrepo F De, Aristizabal LH, Moncada LH, Velez H (1978) Immune responses in paracoccidioidomycosis. A controlled study of 16 patients before and after treatment. Sabouroudia 16:151–163
21. Richardson MD, Stubbins JM, Warnock DW (1982) Rapid enzyme-linked immunosorbent assay (ELISA) for *Aspergillus fumigatus* antibodies. J Clin Pathol 35:1134–1137
22. Richardson MD, Smith M, Warnock DW (1983) Further evaluation of factors affecting a rapid ELISA procedure for detection of IgG antibodies to *Candida albicans*. Mycopathologia 82:167–173
23. Rosenberg M, Patterson R, Mintser R, Cooper BJ, Roberts M, Harris KE (1977) Clinical and immunologic criteria for the diagnosis of allergic bronchopulmonary aspergillosis. Ann Intern Med 86:405–414
24. Sawaki Y, Huppert M, Bailey JW, Yagi Y (1966) Patterns of human antibody reactions in coccidioidomycosis. J Bacteriol 91:422–427
25. Schaefer JC, Yu B, Armstrong D (1976) An aspergillus immunodiffusion test in the early diagnosis of aspergillosis in adult leukemia patients. Am Rev Respir Div 113:325–329
26. Singer LM, Fava Netto C (1972) Conglutinating complement-fixation reactions in paracoccidioidomycosis. In: Paracoccidioidomycosis. Proceedings of the first Pan American symposium, sci publ no. 254. Pan Am Health Organization, Washington DC, pp 227–238
27. Smith CE, Saito MT, Simons SA (1956) Pattern of 39,500 serologic tests in coccidioidomycosis. JAMA 160:546–552
28. Turner-Warwick M, Citron KM, Carrol KB, Heard BE, Mitchell DN, Pepys J, Scadding JG, Sontar CA (1975) Immunologic lung disease due to aspergillus. Chest 68:346–355
29. Weiner MH, Yount WH (1976) Mannan antigenemia in the diagnosis of invasive *Candida* infections. J Clin Invest 58:1045–1053
30. Wilson EV, Hearn VM (1982) A comparison of surface and cytoplasmic antigens of *Aspergillus fumigatus* in an enzyme-linked immunosorbent assay (ELISA). Mykosen 25:653–661
31. Wheat LJ, Kohler RB, French MLV, Garten M, Kleiman M, Zimmerman SE, Schlech W, Ho J, White A, Brahmi Z (1983) Immunoglobulin M and G histoplasmal antibody response in histoplasmosis. Am Rev Respir Dis 128:65–70
32. Yankey R, Abraham AA (1983) Serological studies of a case of fatal craniofacial mucormycosis. Mycopathologia 82:105–109

The Role of Specific Antibodies in Different Immunoglobulin Classes in the Rapid Diagnosis of Diseases Caused by Treponemes

F. MÜLLER [1]

Introduction

Treponemal infections like sexually transmitted syphilis, endemic syphilis, yaws, or pinta can be cured in all stages by specific treatment. However, these infections can also cure without any therapy, i.e., spontaneously. In the early stages of treponematoses, particularly in sexually transmitted syphilis, the demonstration of the causative agent in lesions is the diagnostic method of choice. But identification of *T. pallidum* is often unsuccessful because of several eventualities. In many individuals where there are clinical or epidemiological grounds for suspecting a treponemal infection, it is therefore necessary to provide the diagnosis from the humoral immune response of the infected organism.

From experimental observations Bienenstock and Bloch [4] and Franklin [5] have already postulated a direct correlation between persistence of the antigen in the host and synthesis of specific IgM class antibodies. The findings of Atwood and Miller [2] in treated and untreated human syphilis caused by a viable complex of antigens (*T. pallidum*) point in the same direction.

Today, immunological observations of syphilitic patients before and after penicillin treatment and compared with patients who remained untreated for years after infection lead to the conclusion that the demonstration of treponemal IgM antibodies is of particular importance in judging the need of treatment [7, 12, 27].

Demonstration of IgM and IgG Antibodies to Treponemes in the Different Stages of Infection

In all infectious diseases the synthesis of different classes of antigen-specific immunoglobulins depends on the stage of infection or their complete healing (lack of antigen stimulus). In syphilis, one of the common treponemal diseases, the immunological response of the host to antigen activity can be observed very early in the course of infection. The reason for this is the very early appearance of skin or mucous membrane lesions which lead patients to consult a medical doctor.

1 Abteilung für Infektionsimmunologie am Hygienischen Institut der Freien Hansestadt Hamburg, Gorch-Fock-Wall 15/17, D-2000 Hamburg 36

In primary syphilis almost exclusively treponemal IgM antibodies are found in the patient's sera. During the course of untreated infection (secondary syphilis) not only IgM but also treponemal IgG antibodies are synthesized and can be demonstrated in serum [27, 36]. In the following years, the untreated infection passes to a stage of clinical latency. In this stage different antibody constellations can be observed depending on the persistance of the treponemes in the host. Treponemal IgG antibodies can be demonstrated in all patients, but treponemal IgM antibodies can be shown in the serum of some of them additionally, probably as a result of antigenic activitiy. In patients with adequately treated or spontaneously cured infection, treponemal antigens are no longer present in the host and therefore no treponemal IgM is synthesized. Patients of these two groups cannot be differentiated on the basis of their humoral immune response.

Molecular Characterization of Treponemal Protein Antigens Inducing Antibody Synthesis

In most assays used for the demonstration of treponemal IgM or IgG antibodies, the antigens are combinations of several components of *T. pallidum* which till now have not been analyzed in detail. Consequently we tried to find out which of the protein antigens of *T. pallidum* will be precipitated by IgM and/or IgG antibodies of sera from syphilitic patients.

T. pallidum (Nichols strain) was labeled internally with [^{35}S] methionine. Radiolabeled treponemes were solubilized by sodium dodecyl sulfate (SDS) Triton X-100. SDS polyacrylamide-gel electrophoresis of treponemal materials followed by autoradiography revealed 32 proteins which could be distinguished by their molecular weights [18].

Using double antibody radioimmunoprecipitation we were able to show that treponemal IgM antibodies in the sera from patients with early primary syphilis precipitate at least five protein antigens and from patients with secondary syphilis about 23 protein antigens. Investigating sera from adequately treated syphilitic patients, no IgM precipitation of treponemal protein antigens could be demonstrated. In contrast, almost no reaction was observed when IgG antibodies in sera from patients with primary syphilis were used under the same experimental conditions. But in sera from patients with secondary syphilis a specific IgG reaction with a great number of protein antigens was observed, comparable to the IgM reaction of the same sera. Sera from patients with adequately treated syphilis show a weak IgG reaction with about 13 protein antigens (Table 1).

Conversely it can be concluded that a great number of protein antigens induce the synthesis of treponemal IgM and IgG antibodies in the infected host. Furthermore, the production of treponemal IgM antibodies is interrupted when the antigenic stimulus is eliminated either by sufficient treatment or spontaneous cure. Insofar these results are in accord with observations we had previously made using serological methods [27].

Table 1. Protein antigens of *T. pallidum* precipitated by IgM or IgG antibodies in pooled serum from syphilitic patients

Molecular weight	Total protein profile of [^{35}s]-methionine-labeled *T. pallidum*	Precipitated treponemal antigens					
		Untreated primary		Untreated secondary		Treated secondary infection	
		IgM	IgG	IgM	IgG	IgM	IgG
200,000	+						
140,000	+						
115,000	+++			+	+		
94,000	++			+	+		
79,000	+++			++	++		++
73,000	++			++	++		
69,000	+++	+		++	++		++
63,000	+++			++	++		
59,000	+++			++	++		
53,000	++						
49,500	+++			+++	+++		+
46,000	+++			+++	+++		+
43,500	+++			+++	+++		
40,000	+++			+++	+++		
38,500	+++	+		+++	+++		++
37,000	+++	+++	+	+++	+++		+++
35,000	+++	+++		+++	+++		+++
33,000	+++	+++	+	+++	+++	+	+++
30,500	+++	++	+	+++	+++		+++
29,000	+++			++	++		+
28,000	+++			++	++		+
26,000	+++			++	++		+
24,500	+						
24,000	+						
22,500	+++			++	++		
21,500	+						
19,500	+						
18,000	++			++	++		
16,500	+++			++	++		
15,500	+++			+++	+++		++
14,500	+						
13,500	+						

Cross-reacting Treponemal Antigens and Antibodies

It has been known for a long time that pathogenic treponemes contain antigens which evidently also occur in nonpathogenic treponemes. These "common" antigens have been analyzed on a molecular level [6, 16, 18]. In the infected organism they induce the synthesis of cross-reacting antibodies. Such antibodies produce many uncertainties in the serodiagnosis of treponematoses because they influence the specificity of serological tests.

This problem was first recognized by the observation that the fluorescent treponemal antibody (FTA) test, an indirect immunofluorescence assay, is distorted by a high rate of biologically false reactive results due to cross-reacting treponemal antibodies. It was the idea and the merit of Hunter et al. [8] to solve the problem by absorption of cross-reacting treponemal antibodies using an ultrasonicate of *T. phagedenis*, i.e., the introduction of the fluorescent treponemal antibody absorption (FTA-ABS) test in the serodiagnosis of treponematoses. By this modification the specificity of test results was increased considerably.

With respect to the Treponema pallidum haemagglutination assay (TPHA) we have made the hitherto unexplained observation that human sera, containing only antibodies to *T. phagedenis* antigens with high titer, do not agglutinate the antigen-coated erythrocytes. This phenomenon might possibly be explained by the particular composition of *T. pallidum* antigens on the surface of test erythrocytes and might be responsible for the high specificity of this assay of more than 99%.

For other serological tests, like the enzyme-linked immunosorbent assay (ELISA), no investigations have been performed in the past to elucidate whether a preabsorption of sera with cross-reacting antigens is necessary to achieve reliable test specificity. It therefore does not seem to have been sufficiently investigated whether the ELISA, using a flagella antigen of *T. phagedenis* [29], will result in an increased rate of biologically false results. The antigen of the assay belongs not to *T. pallidum*-specific but only to cross-reacting treponemal antigens.

Nonreactivity of Treponemal IgM Antibody Tests

In the past months a few reports have been published on patients with typical symptoms of untreated tertiary syphilis or neurosyphilis who had high serum titers of treponemal IgG antibodies but surprisingly not treponemal IgM antibodies [20, 24, 32]. In this connection the summarizing report of Uhr and Möller [27] on the regulatory effect of the antibody on the immune response should be called to mind. Taking the observations of Aurajo et al. [3] in newborn rabbits as a basis for further considerations, it can be suggested that extremely high IgG antibody levels of the same specificity might induce an in vivo suppression of IgM synthesis in human infections also.

With respect to human syphilis we observed in about 2% of the patients in the late stages of untreated infection an antibody constellation which was characterized by a TPHA-IgG titer of $\geq 1/20{,}000$ associated with no or only low treponemal IgM antibody level in the region of borderline results of test sensititivy. It seems to be possible that the lack of treponemal IgM antibody in serum in these patients might be explained by an in vivo suppression of the synthesis.

However, a dislodgement of IgM antibodies from the antigen in the presence of high IgG antibody concentrations of the same specificity can also be observed in vitro when the total serum is investigated. In the demonstration of specific IgM antibodies this prozone phenomenon requires quantification of the assay used. In tests which use the separated IgM fraction of the serum, this phenomenon does not occur.

Effect of Specific Treatment on the Humoral Immune Response to Treponemes

In no other infectious disease has the effect of specific treatment on the humoral immune response been investigated more thoroughly than in syphilis. As a result of sufficient therapy not only in the primary but also in the secondary stage of infection, antigen-specific IgM antibodies disappear from the patient's sera within 3 to a maximum of 12 months and are no longer detectable by the assays currently used (titers $<1/5$).

Treponemal IgG antibody titers may fall following successful therapy. In some patients with early treated primary syphilis no treponemal IgG antibodies could be detected by the TPHA or the FTA-ABS test 1–3 years after adequate treatment. On the other hand, many patients with late treated or spontaneously cured infections maintain high levels of treponemal IgG, produced by memory cell clones, for decades, perhaps for life. Persistence of this "serum scar" depends on the time interval between infection and first specific treatment, i.e., it is related to the intensitiy of the antigenic stimulus on the immune system of the host.

Assays for the Rapid Serological Diagnosis of Acute Treponemal Infection

The most commonly used standard tests in the serodiagnosis of acute treponemal infections like the TPHA or FTA-ABS tests are unsuitable for differentiation between immunoglobulin classes in total patient's sera; the antigen combinations of both tests react with treponemal IgM and IgG antibodies, respectively.

During the last 10 years several techniques for the demonstration of treponemal IgM antibodies have been recommended. Shortly after the first description of the IgM-FTA-ABS test, using total patient's serum, a μ-chain-specific FITC-labeled antihuman IgM serum and indirect immunofluorescence technique [1, 9, 33, 34, 35] this assay was vehemently criticized because of several theoretical shortcomings [10, 17, 28, 30]. These criticisms have been confirmed in the meantime [14, 19, 36]. For false reactive results, rheumatoid factors are a cause of great importance. False nonreactive results are induced by competition between high treponemal IgG and low IgM antibody titers on the antigen. If indirect immunofluorescence is to be used at all for detection of treponemal IgM antibodies it is necessary to separate the IgM fraction from the serum and to investigate the isolated IgM.

In further development of the method of Atwood and Miller [2], we have described a microtechnique of serum gel filtration which results in sufficiently pure IgM fractions [19, 23, 26]. This assay was called separated IgM-FTA-ABS or 19S(IgM)-FTA-ABS test. Since 1974 more than 20,000 sera from syphilitic patients have been investigated for treponemal IgM antibodies in our laboratory. We were able to show that by gel filtration and investigation of the separated IgM fractions false reactive as well as false nonreactive results could be almost completely eliminated. The test has a specificity and sensitivy of more than 97% in all stages of syphilis. We therefore consider this assay to be currently the method of choice [14, 20]. However, it is time-consuming, requiring highly qualified personnel and expensive equipment. It can therefore only be performed under special laboratory conditions. To find a method which is easier to perform, several other techniques for separation of IgM from the total serum were recommended in recent years – sucrose density gradient ultracentrifugation, immunoaffinity, ion exchange, or high-pressure liquid chromatography (HPLC). Of these the method to be recommended is HPLC [14].

The principles of a new technique, a solid-phase immunosorbent test (SPIT) were first described by Krech and Wilhelm [11]. Schmidt [31] has tried to adopt this technique for estimation of IgM antibodies in treponematoses. The solid phase hemadsorption assay (SPHA) has the adventage of easy performance; it can be automated. But both sensitivity and specificity of the assay are low when performed following the description of Schmidt [14, 22].

Recently our group has reported on experiments to demonstrate IgM antibodies to *T. pallidum* by the enzyme-linked immunosorbent assay (ELISA), using different antigen preparations and several techniques [13, 21, 24, 25]. We found that specificity and sensitivity of the TP-IgM-ELISA depend on the composition and purity of antigens and on the size of the antigen-coated surface which reacts with the patient's serum.

Our observations on the specificity, sensitivity, and antibody fixing capacity using different assays for identification of *T. pallidum*-specific IgM antibodies are summarized in Table 2. After the 19S(IgM)-FTA-ABS test, a TP-IgM-ELISA currently being developed and using relatively large antigen-coated beads yields high sensitivity and specificity in early treponemal infections (not more than 2 years' duration). This assay is furthermore distinguished by an antibody fixing capacity which is ten times higher that of the 19S(IgM)-FTA-ABS test.

Table 2. Specificity, sensitivity, and antibody fixing capacity of assays for the demonstration of *T. pallidum*-specific IgM antibodies

Assay	No. of investigated sera	Specificity (%)	Sensitivity (%)	Antibody fixing capacity
IgM-FTA-ABS test (in total serum)	1520	97	46	+
19S(IgM)-FTA-ABS test (in the fractions after gel filtration of sera)	1616	>97	>97	++
SPHA (IgM solid-phase hemabsorption assay)	763	92	68	++
TP-IgM-ELISA (microenzyme-linked immunosorbent assay in microtiter plates)	696	>97	>97	++
TP-IgM-EIA (enzyme immunoassay using antigen-coated beads)	1192	97	93–>97[a]	+++

[a] Depending on the stage of infection (lower sensitivity in patients with late latent treponemal infections and treponemal reinfections)

Conclusions

Differentiation of IgM and IgG class immunoglobulins in treponemal infections is of great diagnostical importance. In most patients the demonstration of treponemal IgM antibodies can be interpreted as a symptom for need of treatment. Follow-up studies have shown that this antibody class diminishes in the serum of treated patients to nondetectable values in direct connection with adequate therapy. The same decrease of IgM antibodies to *T. pallidum* occurs in the serum of spontaneously cured patients with treponematoses.

Pathogenic and nonpathogenic treponemes consists of "common" antigens. In serological tests for treponematoses it is therefore necessary to preabsorbe patient's sera with an ultrasonicate of *T. phagedenis*. In assays without preabsorption of sera it can be suspected that specificity might be decreased due to cross-reacting antibodies.

In patients with untreated late infections (more than 2 years' duration), the IgM immune response can be suppressed in vivo by a high IgG antibody level of the same specificity. This possibility should be kept in mind when decisions for need of treatment are considered.

The sensitivity and specificity of chromatographic methods for the demonstration of treponemal IgM antibodies have proved to be superior to the solid-phase techniques. However, assays using IgM serum fractions are time consuming and require expensive equipment as well as highly qualified personnel. Newer techniques like the ELISA are still under investigation and need further modification to become as sensitive and specific as those assays using separated IgM serum fractions.

References

1. Alford CA, Polt SS, Cassady GE, Straumfjord JV, Remington JS (1969) M-fluorescent treponemal antibody in the diagnosis of congenital syphilis. N Engl J Med 280:1086–1091
2. Atwood WG, Miller JL (1969) Fluorescent treponemal antibodies in fractionated syphilitic sera. The immunoglobulin class. Arch Dermatol 100:763–769
3. Aurajo FG, Remington JS (1975) IgG antibody suppression of the IgM antibody response to Toxoplasma gondii in newborn rabbits. J Immunol 115:335–338
4. Bienenstock J, Bloch KJ (1966) Some characteristics of human immunoconglutinin. J Immunol 96:637–645
5. Franklin EC (1968) Structure and function of immunoglobulins. NY J Med 68:411–422
6. Hanff PA, Miller JN, Lovett MA (1983) Molecular characterization of common treponemal antigens. Infect Immun 40:825–828
7. Herbst BR, Goerz G, Müller F (1979) Diagnostischer und therapeutischer Aussagewert des IgM-FTA-ABS und IgM-FTA-19S-Tests bei der Syphilis. Aktuelle Dermatol 5:175–183
8. Hunter EF, Deacon WE, Meyer PE (1964) An improved FTA test for syphilis, the absorption procedure (FTA-ABS). Public Health Rep 79:410–412
9. Johnston NA (1972) Neonatal congenital syphilis: diagnosis by the absorbed fluorescent treponemal antibody (IgM) test. Br J Vener Dis 48:465–469
10. Kaufman RE, Olansky DC, Wiesner PJ (1975) The FTA-ABS (IgM) test for neonatal congenital syphilis. A critical review. J Am Dis Assoc 1:79–84
11. Krech U, Wilhelm JA (1979) A solid-phase immunosorbent technique for the rapid detection of rubella IgM by haemagglutination inhibition. J Gen Virol 44:281–286

12. Leyh F, Müller F (1978) Bewertung der Syphilis-Therapie durch immunologische Verlaufskontrollen. Hautarzt 29:[Suppl II] 82–83
13. Lindenschmidt EG, Laufs R, Müller F (1983) Microenzyme-linked immunosorbent assay for the detection of specific IgM antibodies in human syphilis. Br J Vener Dis 59:151–156
14. Loeke S, Huschka U, Sann G, Schroeter R, Doerr HW (1984) Comparison of five methods for the detection of Treponema pallidum-specific IgM antibodies. Eur J Sex Trans Dis 1:191–194
15. Luger A, Schmidt B, Spendlingwimmer I (1977) Quantitative evaluation of the FTA-ABS-IgM and VDRL test in treated and untreated syphilis. Br J Vener Dis 53:287–291
16. Lukehart SA, Baker-Zander SA, Gubish ER (1982) Identification of Treponema pallidum antigens: comparison with nonpathogenic treponeme. J Immunol 129:833–838
17. Mohr FW, Schneweis KE (1976) Der FTA-ABS-IgM-Test im Vergleich zur quantitativen Cardiolipin-Reaktion und zum TPHA-Test. Ärztl Lab 22:305–318
18. Moskophidis M, Müller F (1984) Molecular analysis of immunoglobulin M and G immune response to protein antigens of Treponema pallidum in human syphilis. Infect Immun 43:127–132
19. Müller F (1982) Der 19S(IgM)-FTA-ABS-Test in der Serodiagnostik der Syphilis. Immun Infekt 10:23–34
20. Müller F (1983) Immunologische Grundlagen, Ergebnisse und Grenzen der Syphilis-Serodiagnostik. Laboratoriumsmed 7:12–16
21. Müller F (1983) Der Treponema pallidum IgM Enzyme-linked immunosorbent assay (TP-IgM-ELISA). Nachweis erregerspezifischer IgM-Antikörper im Serum und Liquor cerebrospinalis bei Syphilis-Patienten mit und ohne Beteiligung des ZNS an der Infektion. Z Hautkr 58:1689–1708
22. Müller F, Lindenschmidt EG (1982) Demonstration of specific 19S(IgM) antibodies in untreated and treated syphilis. Comparative studies of the 19S(IgM)-FTA test, the 19S(IgM)-TPHA test, and the solid phase haemabsorption assay. Br J Vener Dis 58:12–17
23. Müller F, Loa PL (1974) Neue Möglichkeiten in der immunologischen Diagnostik der Treponemen-Infektion (Syphilis). Infection 2:127–131
24. Müller F, Moskophidis M (1984) Evaluation of an enzyme immunoassay for immunoglobulin M antibodies to Treponema pallidum in human syphilis. Br J Vener Dis 60:288–292
25. Müller F, Moskophidis M, Prange HW (1984) Demonstration of locally synthesized immunoglobulin M antibodies to Treponema pallidum in the CNS of patients with untreated neurosyphilis. J Neuroimmunol 7:43–54
26. Müller F, Oelerich S (1978) Ein modifiziertes Verfahren des IgM-FTA-19S-Tests zum Nachweis kompetitiv gehemmter Antikörper bei der Syphilis. Ärztl Lab 24:386–391
27. Müller F, Oelerich S (1979) Korrelation immunologischer Parameter zu den Stadien der apparenten und der klinisch stummen Syphilis. Dermatol Monatsschr 165:385–395
28. O'Neill P, Nicol CS (1972) IgM class antitreponemal antibody in treated and untreated syphilis. Br J Vener Dis 48:460–466
29. Pedersen NS, Petersen CS, Axelsen NH (1982) Enzyme-linked immunosorbent assay for detection of immunoglobulin M antibodies against Reiter treponeme flagellum in syphilis. J Clin Microbiol 16:608–614
30. Reimer CB, Black CM, Phillips DJ, Logan LC, Hunter EF, Pender BJ, McGrew BE (1975) The specificity of fetal IgM: antibody or anti-antibody? Ann NY Acad Sci 254:77–93
31. Schmidt BL (1980) Solid-phase hemadsorption: a method for rapid detection of Treponema pallidum-specific IgM. Sex Transm Dis 7:53–58
32. Schroeter R, Wirth H, Petzold D (1984) Tertiär-Syphilis mit biologisch falsch negativem 19S(IgM)-FTA-ABS-Test. Hautarzt (to be published)
33. Scotti AT, Logan LC (1968) A specific IgM antibody test in neonatal congenital syphilis. J Pediatr 73:242–243
34. Scotti AT, Logan LC, Caldwell JG (1969) Fluorescent antibody test for neonatal congenital syphilis. A progress report. J Pediatr 75:1129–1134
35. Sepetjian M, Tissot-Guerraz F, Monier JC, Nivelon JL, Thivolet J (1970) Investigation of a specific IgM antibody test in neonatal congenital syphilis. Br J Vener Dis 46:18–20
36. Shannon R, Booth SD (1977) The patterns of immunological response at various stages of syphilis. Br J Vener Dis 53:281–286
37. Uhr JW, Möller G (1968) Regulatory effect of antibody on the immune response. Adv Immunochem 8:81–127

NMR in the Study of Cells and Tissues

H. RÜTERJANS and H. P. JURETSCHKE [1]

In the last years NMR spectroscopy has emerged as a non-destructive and non-invasive method for studying the metabolism of cells and tissues in vivo [1]. Today, the control of enzyme activity of isolated systems by the use of inactive precursor forms through covalent modification or allosteric effectors is quite well understood, sometimes even in great detail. Now, the biochemist, microbiologist or clinician is becoming increasingly interested in studying the organization and the control of integrated metabolic pathways in the intact organism (in vivo).

NMR spectroscopy with its diverse techniques renders the physiological state of a living system amenable to investigation in a non-destructive manner. It is even used as an imaging tool in clinics as an alternative to other methods. A combination of the spatial resolution capability of NMR with its capacity to depict the metabolic state of a specific organ or tissue in a non-invasive manner seems to have a highly promising outlook.

Some of the nuclei used as probes for in vivo biochemical studies are collected in Table 1, featuring some important NMR properties as well. Most of these nuclei have a spin quantum number 1/2, which is favourable because of the narrower line width of the resonances.

Table 1. Biochemically interesting nuclei and their NMR properties

Nucleus	Spin quantum number	Natural abundance	Relative sensitivity at constant field[a]
^{1}H	1/2	99.98	100
^{2}D	1	0.0156	1.5×10^{-4}
^{13}C	1/2	1.1	1.6×10^{-2}
^{15}N	1/2	0.36	3.7×10^{-4}
^{19}F	1/2	100	83
^{23}Na	3/2	100	9.3
^{31}P	1/2	100	6.6

[a] Relative to an equal number of protons, multiplied by the percentage natural abundance

NMR spectroscopy faces two basic problems: resolution and sensitivity. For good resolution fairly mobile molecules are required in addition to the spin quan-

1 Institut für Biophysikalische Chemie, Johann-Wolfgang-Goethe-Universität, Theodor-Stern-Kai 7, Geb. 75A, D-6000 Frankfurt 70

tum number 1/2. Fast molecular motion averages out dipole-dipole interactions and hence gives narrow lines. Thus, in ^{31}P NMR spectra only resonances from phosphate metabolites in the cytoplasma are observed, while the phosphorus atoms in the immobile DNA or in phospholipids produce lines too broad to be detected.

NMR spectroscopy is an inherently insensitive method. With regard to the signal-to-noise ratio, various factors have to be considered:
- Nucleus under study and its relative sensitivity
- Isotopic enrichment where possible
- Concentration of investigated metabolites
- Sample volume
- Magnetic field strength of the spectrometer
- Splitting and width of the spectral lines
- Relaxation times T_1 and T_2
- Total accumulation time.

The many factors affecting the S/N ratio do not allow giving exact estimates for the required sample concentration. In microbiological studies of cell cultures, usually suspensions with high cell densities (up to 10^{12} cells/ml) are prepared by low speed centrifugation. At these concentrations the suspensions generally are still fluid enough to stir and aerate, and hence metabolic processes requiring continuous oxygen supply, for example, are possible. In order to increase the time available for data accumulation, working at 20 °C may decrease the metabolic rates in a favourable manner [2].

A suitable nucleus is of paramount importance when considering sensitivity, but several factors limit freedom of choice. Thus, the proton having by far the highest sensitivity has two major drawbacks:

Its spectra are extremely complex due to the overlapping signals from the many compounds present; thus, the metabolites of interest are likely to be masked by the resonances of other cell components.

The large solvent signal (110 *M* water protons) poses high demands on dynamic range and requires sophisticated techniques if specific metabolites are to be monitored at the millimolar level.

^{31}P NMR is not affected by an overwhelming solvent signal or by low natural abundance. Biochemically interesting molecules, like ATP, phosphorcreatine or sugar phosphates, occur at concentrations high enough to be easily detected by ^{31}P NMR. Not surprisingly, most of the pioneer studies on in vivo biochemistry have used ^{31}P as a probe. Many phosphorus-containing compounds are involved in the energy delivering, storing, and transforming pathways of living systems, and hence ^{31}P NMR is a very apt way to monitor the energetic state of living cells, tissues or organs.

Additionally, the chemical shift value of the inorganic phosphate signal is dependent on pH, and thus the observation of this resonance allows for a non-invasive pH determination. The appearance of various phosphate signals sometimes indicates the presence of compartments with diverse pH values inside a cell [3].

^{13}C has a natural abundance of 1.1% and, hence, isotopic enrichment is essential in order to obtain spectra within reasonable accumulation times. Spectral resolution is generally better than for ^{1}H NMR; and in contrast to ^{31}P NMR, a wide range of compounds can be detected.

Another interesting nucleus is ^{15}N, especially as no suitable radioactive isotope of nitrogen for radiotracer experiments is available. The main problem of ^{15}N NMR is its very low sensitivity and natural abundance. Thus, only studies using highly enriched isotopic compounds are feasible.

One advantage common to ^{13}C and ^{15}N NMR is the possibility of selecting specific metabolic pathways by choosing appropriate compounds for enrichment.

Using in vivo ^{15}N NMR spectroscopy we have studied the nitrogen metabolism of *Saccharomyces cerevisiae* in a series of experiments trying to elucidate the flow of nitrogen along the various competing pathways, and the extent to which alternative pathways of nitrogen metabolism can occur in vivo under a variety of culture conditions. Commercially available baking yeast was incubated for different time periods in media containing 10 m*M* $^{15}NH_4Cl$ as sole source of nitrogen. For NMR measurements the cells were collected by low-speed centrifugation. The full viability of the yeast cells before and after the NMR experiment was checked according to established procedures (cell counting and plating).

Figure 1 shows the 50.65 MHz ^{15}N NMR spectrum of a *S. cerevisiae* cell suspension at 10 °C. Well-resolved resonances of various ^{15}N-labeled metabolites were observed. The assignment of the ^{15}N signals to the various intracellular intermediates and products was accomplished through their chemical shift values and coupling constants. The different chemical species vary in their spin-lattice

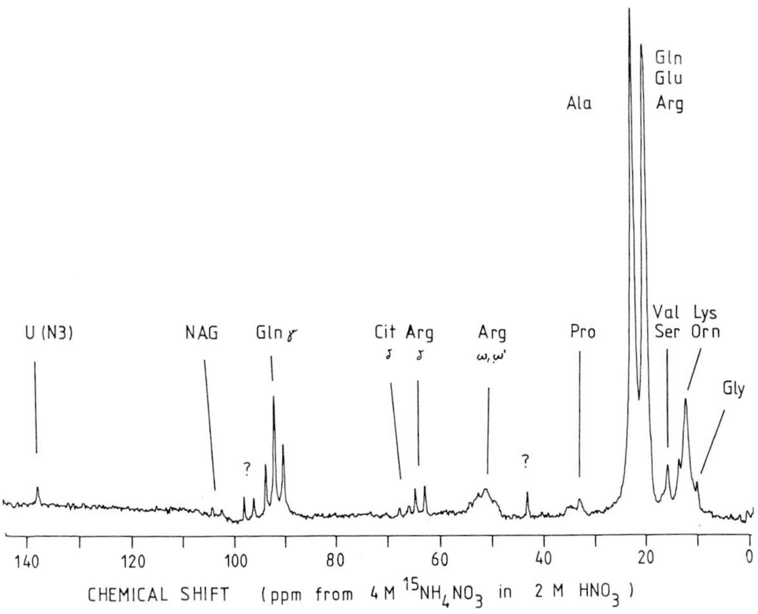

Fig. 1. Proton-coupled ^{15}N NMR spectrum of a suspension of intact *Saccharomyces cerevisiae* cells incubated for 90 min in a medium containing $^{15}NH_4Cl$. Spectral parameters: 10 mm o.d. sample tube, inverse gated decoupled spectrum at 50.65 MHz; T = 10 g“, 5,000 scans; total accumulation time, 1.2 h. *Cit,* citrulline; *NAG,* N-acetyl-D-glucosamine; *U(N3),* uridine N3

relaxation times and nuclear Overhauser enhancement factors and, hence, accurate correlation of the intracellular concentrations with peak intensities was not possible.

The α-amino nitrogen atoms of several amino acids were observed: proline, alanine, glutamate and glutamine, arginine, valine and/or serine, and glycine. Also the ε-amino group of lysine and/or ornithine and the amide group of glutamine (triplet at 92 ppm) are detected. Ammonium is incorporated into cells by glutamate dehydrogenase and by glutamine synthetase, which explains the strong intensity of the signals due to these two amino acids. Alanine results from the transamination of pyruvate. Because alanine can be reconverted to glutamic acid with release of pyruvic acid, a key intermediate in carbon metabolism, alanine may be a convenient reservoir of both the amino groups and pyruvic acid.

In addition to arginine, citrulline, an intermediate in arginine biosynthesis, can be detected.

The resonance at 137 ppm was assigned to the N3 of uridine and probably arises from uridine disphosphoacetylglucosamine. The tiny doublet labelled NAG in Fig. 1 can be assigned to *N*-acetylglucosamine, either as a free metabolite or as a component of the cell wall.

The influence of growth conditions on nitrogen metabolism was studied in another experiment. *S. cerevisiae* cells were allowed to grow for 6 h in a nitrogen-free medium in order to deplete their intracellular nitrogen pools. The cells were harvested and resuspended in media which contained 10 m*M* glutamine, 10 m*M* arginine or 10 m*M* histidine as sole nitrogen source. After a 90-min incubation period, cycloheximide was added. The different cultures were transfered to another medium containing 10 m*M* $^{15}NH_4Cl$ as sole nitrogen source. After 30-min incubation the cells were harvested and placed in 10-mm NMR sample tubes. The ^{15}N NMR spectra obtained are shown in Fig. 2.

Using glutamine as the nitrogen source, only a small amount of alanine is produced compared to the experiments with arginine and especially histidine. In the spectrum for alanine, the signal due to the ε-amino groups of lysine and ornithine (at 12 ppm) is hardly to be seen; no glycine (at 10 ppm) is detected at all. These amino acids are produced in higher amounts especially when histidine is the sole nitrogen source.

Another interesting feature is the synthesis of asparagine detected in the spectra for arginine and histidine. In these spectra the signal group centred at 92 ppm consists of two partially overlapping triplets differing by 0.3 ppm in their chemical shift values, as expected for the amide group signals of glutamine and asparagine [3].

The utilitiy of ^{15}N NMR spectroscopy is clear for monitoring the turnover of nitrogenous metabolites, such as glutamine and glutamate, or the way in which nitrogen from such sources is distributed among the various pathways.

Although the ^{15}N NMR spectra proved to be highly informative, the neccessary accumulation times are nevertheless quite long: about 40–80 min when measuring in the inverse gated decoupled mode or 10–20 min in the broad-band decoupled mode. We would like to report on a new technique for studying ^{13}C or ^{15}N enriched metabolites which uses the sensitivity of ^{1}H NMR and allows the detection of ^{1}H NMR spectra in less than 1 min.

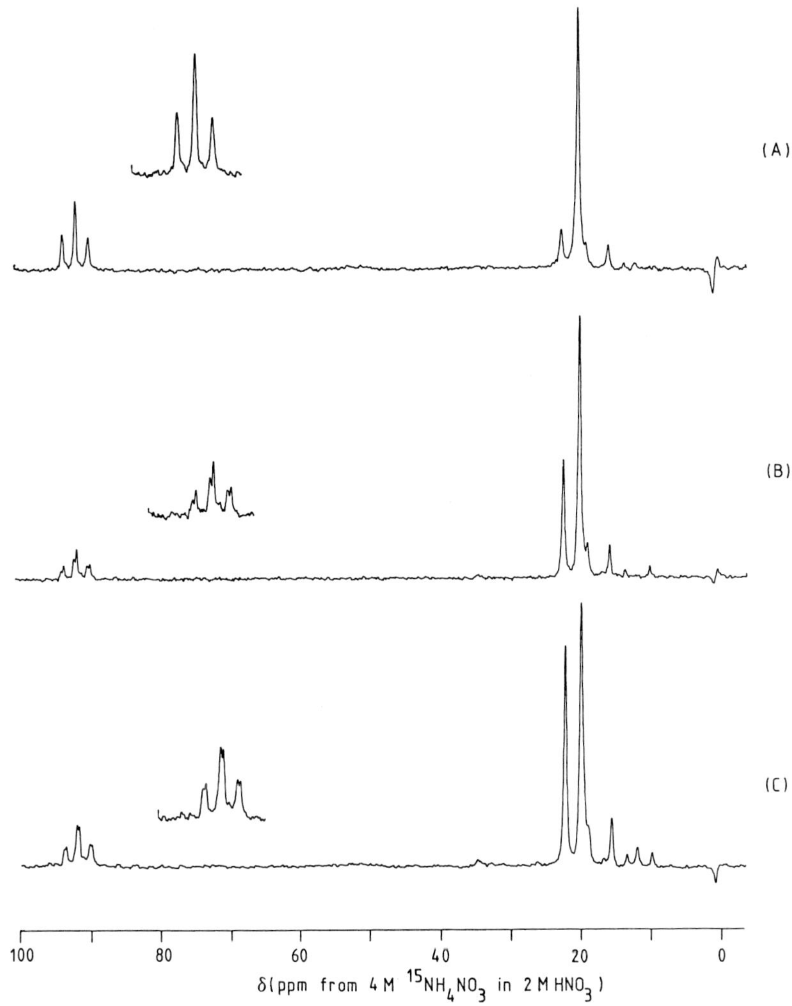

Fig. 2. Proton-coupled ^{15}N NMR spectra of suspensions of intact *S. cerevisiae* cells incubated for 1 h in medium containing $^{15}NH_4Cl$ after preincubation in media containing (*A*) 10 m*M* glutamine, (*B*) 10 m*M* arginine, and (*C*) 10 m*M* histidine as sole nitrogen source. After preincubation and addition of cycloheximide (20 mg/l), the cells were collected and resuspended in the medium containing ^{15}N. Spectral data, see Fig. 1

As already mentioned (vide supra), in order to use ^1H NMR for metabolic studies, a technique is necessary which, first, reduces the solvent signal intensity and, secondly, offers the possibility of following specific metabolites in a selective manner.

The problem posed by the water signal intensity may be solved by using the 2-1-4 Redfield sequence [4] which does not excite the water protons, or any of various recently published pulse sequences. By a suitable choice of pulse angles and precession delays, these sequences succeed in arranging the water proton magnetization along the longitudinal axis, while the magnetization of the interesting proton nuclei remains in the $x'y'$ plane of the rotating frame of reference. We routinely use the sequence of strong pulses suggested by Bleich and Wilde [5].

The problem of selectivity can be solved by using multipulse techniques based on multiple quantum NMR which allow editing the ^1H NMR spectrum in such a way so as to observe only those resonances arising from scalar coupling to another nucleus Y (^{13}C or ^{15}N, for example) [6].

The pulse sequence we use is the following:

$$90°(^1\mathrm{H,x}) - D1 - 90°(^1\mathrm{H,y}) - D2 - 90°(^{15}\mathrm{N}) - 90°(^{15}\mathrm{N},\pm) - D2 - AQ(\mathrm{BB}).$$

The first two 90° ^1H pulses, applied along the x or the y axis, and the intervening delay $D1$ describe the water signal suppression technique already discussed. The delay $D2$ equals $(2\,J_{N-H})^{-1}$ where J_{N-H} is the coupling constant between the nuclei ^1H and ^{15}N. Changing the value of this delay allows selecting compounds with the corresponding, specific coupling constant. Cycling the phase of the second 90° ^{15}N pulse in four steps will cancel the signals of protons not coupled to ^{15}N. During the acquisition time AQ, broad-band decoupling of the ^{15}N nuclei is possible and increases the sensitivity of the experiment.

In Fig. 3 the 500 MHz ^1H NMR spectrum of a *S. cerevisiae* cell suspension labeled with $^{15}NH_4Cl$ is shown. The spectrum was obtained using the multiple quantum pulse sequence. The delay D was set to select for a coupling constant of 90 Hz.

In addition to the water signal suppression scheme, the double quantum filter drastically diminished the intensitiy of the water resonance at 4.8 ppm. Two doublets at 6.9 and 7.6 ppm with a 90-Hz coupling constant were detected. These doublets are due to the two amide protons of glutamine. Due to the strong double-bond character of the C–N link, the free rotation of the $-NH_2$ group is

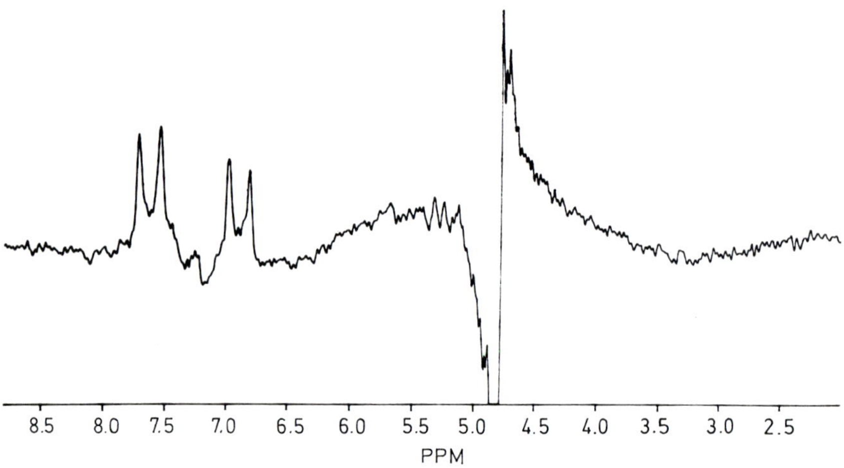

Fig. 3. 500 MHz ^1H NMR spectrum of a *S. cerevisiae* cell suspension pulse-labeled with $^{15}NH_4Cl$. The two doublets are due to the anisotropic amide protons of intracelluar glutamine. The spectrum was obtained with the pulse sequence discussed in the text. Spectral data: 256 scans, T = 20 °C, total accumulation time, 25.6 s; 5-mm o.d. NMR tube

severely hindered and, hence, the magnetic effect of the neighbouring carbonyl group renders both amide protons anisotropic. This was also shown by selective decoupling experiments (not shown).

As shown in Fig. 1 and 2, not all the ^{15}N label is incorporated into glutamine alone. The signals of the protons bound to other ^{15}N-labeled molecules are not observed as the coupling constants J_{N-H} of these species do not have the correct size of 90 Hz.

The spectrum shown in Fig. 3 is the sum of 256 scans acquired with a repetition time of 0.1 s. Hence, the total accumulation time was 25.6 s. The cell suspension (10^{12} cells/ml) was contained in a 5-mm o.d. NMR tube.

The high sensitivity and selectivitiy of this experiment seem to fulfill the requirements needed to use ^{1}H NMR spectroscopy for in vivo metabolic studies. Additionally, this pulse sequence can be used in reverse correlated two-dimensional spectroscopy, in which the indirect detection of the ^{15}N NMR spectrum via multiple quantum NMR is achieved. This facilitates the assignment of resonances to specific metabolites and, hence, the interpretation of the spectra.

References

1. Gadian DG (1982) NMR and its application to living systems. Clarendon, Oxford
2. Ugurbil K, Shulman RG, Brown TR (1979) In: Shulman RG (ed) Biological applications of magnetic resonance. Academic, New York
3. Martin GJ, Martin ML, Gouesnard JP (1981) ^{15}N NMR spectroscopy. Springer, Berlin Heidelberg New York
4. Redfield AG, Kunz SD, Ralph EK (1975) J Magn Reson 19:114–117
5. Bleich H, Wilde J (1984) J Magn Reson 56:154–155
6. Bax A, Griffey RH, Hawkins BL (1983) J Magn Reson 55:301–315

Rapid Microbial Identification by Circular Intensity Differential Scattering

C. T. GREGG, D. M. McGREGOR, W. K. GRACE, and G. C. SALZMAN [1]

Introduction

As one of the last platform speakers in this fascinating conference, I have the opportunity to summarize what we have heard during the past few days as a background for my own presentation. I would like to begin by asking a rhetorical question: "Who cares about rapid methods of microbial identification anyway?"

We have heard several answers to that question during this conference, but one of the best is that implied by the chairman of this session, Dr. Isenberg, when he said at the previous Conference on Rapid Methods: "It is quite proper to question the relevance of clinical microbiology when the clinician must rely on an educated guess ... to intercede effectively during the crucial early hours of disease manifestations" [1]. In a similar vein just last year, Drs. Tilton and Balows alluded "to the day when clinical microbiology will be a primary diagnostic service rather than a confirmatory one, and when clinical microbiologists and infectious disease specialists can point to significant reductions in morbidity and mortality" [2].

In addition to the medical and humanitarian interest in reducing morbidity and mortality, there is the very real economic impact of rapid microbial identification. In the context of diagnostically related groups (DRGs) in the US and similar cost containment measures elsewhere, this becomes important as well.

The economic impact does not arise from the modest part that microbial identification plays in the cost of illness resulting from infectious disease, since this is rarely more than 10% of the total cost. Rather, rapid identification has an economic impact because it reduces hospital stays and simultaneously reduces the quantities of intravenous fluids and antimicrobials required and the time spent in isolation rooms.

The best relevant data I know of comes from Dr. Marilyn Menegus and her colleagues at Strong Memorial Hospital in Rochester, New York, and deals with the readmission of neonates to the hospital in the first month of life ([3] and personal communication from Dr. Menegus, March, 1984). Considered on a year-round basis, about 45% of these readmissions are due to suspected sepsis; this figure rises to 70% during the summer months. These children are routinely treated as if they had bacterial disease, that is, they are given intravenous fluids and antimicrobials until bacterial cultures are shown to be negative.

1 Experimental Pathology Group, Life Sciences Division, Los Alamos National Laboratory, Los Alamos, NM 87545, USA

Table 1. Rapid methods of microbial identification

Test	Requirements	
	Time (h)	Organisms/ml
Coagglutination	1	10^6–10^7
Counterimmunoelectrophoresis	1	10^6–10^7
Fluorescent antibody staining	1	$>10^5$
ELISA (or ELISA A/B)	4–6	10^3–10^4
Radioimmunoassay	2–40	10^3–10^4
Circular intensity differential scattering (CIDS)	<0.2	$< 5 \times 10^5$

As you might suspect, about 70% of the summer admissions are due to nonpolio enteroviruses. In one group of 33 children, only one had a bacterial infection, 20 had isolatable viruses, and 12 had presumed viral infections. It is probable that many of the latter were infections caused by rotaviruses that cannot be cultivated. The important point is that those children in whom a virus was identified in 2 days or less had a shorter hospital stay than those children for whom isolation required 3 days or more (2.7 and 3.8 days respectively). This is a statistically significant difference resulting in corresponding reductions in intravenous fluids administered, isolation charges, and so on. Consequently, rapid microbial identification does matter for financial as well as medical and humanitarian reasons, as I am sure everyone would agree.

We might begin with a tabulation of what seems to me to be the best established, rapid methods of microbial identification. In Table 1, the first three methods, coagglutination, counterimmunoelectrophoresis, and fluorescent antibody staining, all require about an hour to perform and require more than 10^5 organisms per ml – how much more is a matter of some controversy.

ELISA, especially when coupled with the avidin/biotin probe, requires 4–6 h and as few as 10^3 organisms per ml. The avidin/biotin probe could in my opinion be applied with equal success to RIA, with similar sensitivities and times of 2–40 h, depending on how the test is carried out.

Circular Intensity Differential Scattering

The method described on the last line – circular intensity differential scattering, or CIDS for short – is the technique that I will describe. It is one of the few really new approaches to microbial identification to have come into existence in the past several decades, and while many questions concerning its use have yet to be answered, my colleague Dr. Salzman and I feel strongly that this method has the potential of revolutionizing clinical microbiology. It requires times on the order of minutes and, at the moment, something like 5×10^5 organisms per ml. We feel that further modifications of the instrument can improve the sensitivity by an order of magnitude or more.

Several generations of instruments are planned. The first of these is a static instrument designed to identify bacteria in pure culture or viruses in typical clinical specimens in a static cuvette system. This machine presently exists in prototype and is that on which most of the data I will present was taken. A second-generation machine employs the same principle but with the addition of a flow cytometer so that organisms can be examined one at a time. This machine is intended for bacterial identification. It will be particularly useful in identifying the components in a polymicrobic infection.

Unfortunately, the CIDS method involves concepts that are not the sort most microbiologists learn at their mother's knee. And the translation of these concepts into a working instrument involves some complex electronics. As if those were not obstacles enough, the results are expressed is in a form that is unfamiliar.

Thus, there are basic communications problems to overcome. A wonderful Peanuts cartoon depicts Charlie Brown saying, "There's no problem too big for me to run away from." To some extent, that is what I propose to do. I will concentrate on the results we have obtained and say comparatively little about the instrumentation or the theory. Fortunately, Dr. Salzman, who built the instrument and who is one of the few people who really understands the theory, is available to answer questions in these areas.

Since it is important to understand both the similarities and the differences between CIDS and circular dichroism (CD), the two properties may be usefully compared. CD is the differential *absorption* of left and right circularly polarized light and occurs only at an absorption band of the molecule in question, while CIDS is the differential *scattering* of left and right circularly polarized light and occurs throughout the spectrum, rather than only at absorption bands. CIDS, which is sensitive to long-range order in chiral biological macromolecules, is symbolically represented in Fig. 1.

My discussion will deal with the differential scattering of polarized light. This phenomenon, however, is not to be confused with the total light scattering that was investigated so extensively by Wyatt and others. Total light scattering depends only on the size, shape, and average refractive index of microorganisms and did not prove useful for microbial identification. CIDS involves light, alternatively right and left circularly polarized, impinging on a microbial sample. The sample differentially scatters left and right circularly polarized light with the re-

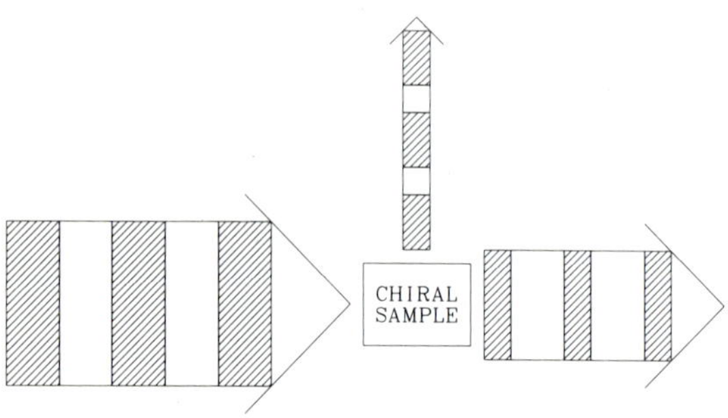

$$\mathrm{CIDS}(\theta) = (I_L(\theta) - I_R(\theta)) / (I_L(\theta) + I_R(\theta))$$

Fig. 1. The CIDS principle

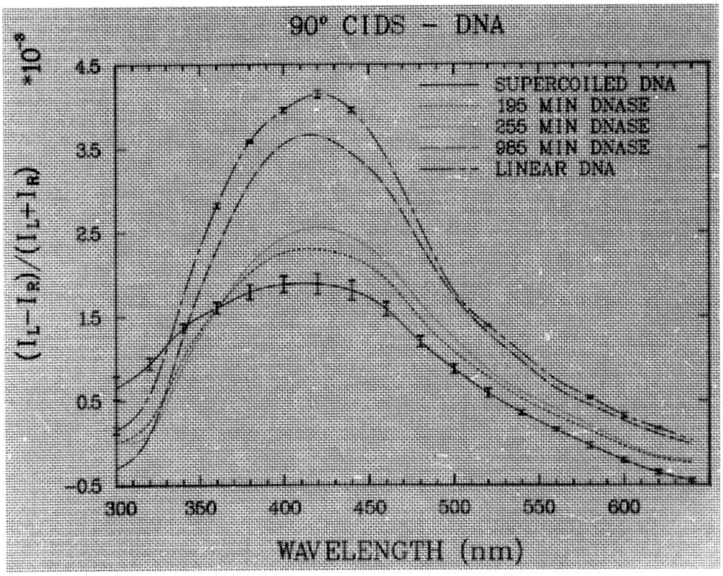

Fig. 2. Effect on the CIDS spectra as a function of wavelength of DNAse treatment of highly supercoiled plasmid DNA. The spectrum of calf thymus DNA (*linear, small fragments*) is shown for comparison

sult that the scattered light is enriched in one component or the other. It is this differential scattering that we are measuring.

What is the physical basis of these measurements in terms of microbial identification? We do not know all the answers to that question yet, but provisionally, we feel that we are measuring the three-dimensional "packaging" of helical molecules, largely that of the microbial genome. Theoretically, the CIDS spectrum is known to depend upon the pitch and radius of the scattering helix. Figure 2 shows some of our early work, in which the CIDS spectrum of highly purified and supercoiled DNA from plasmid pBr 322 was taken as a function of wavelength, the DNA then treated with DNAse, and the spectrum retaken at intervals. The latter was finally compared with the CIDS spectrum of commercial calf thymus DNA as a marker for short, linear fragments of DNA. It is clear that the CIDS spectrum changed progressively as the DNA uncoiled.

In the laboratory, the CIDS spectrum can be taken as a function of wavelength, as in the data just shown. It can also be taken as a function of scattering angle and as a function of matrix element, a point I will return to in a moment.

Figure 3 shows in block diagram form the geometry of the instrument we are currently using to make measurements as a function of scattering angle. It is this instrument that was used to obtain most of the data presented here. I will enumerate briefly the components and their function.

Light from a laser or any other convenient source passes through a polarizer, then through a photoelastic modulator that modulates the polarization at a selected frequency. This alternately left and right circularly polarized light then im-

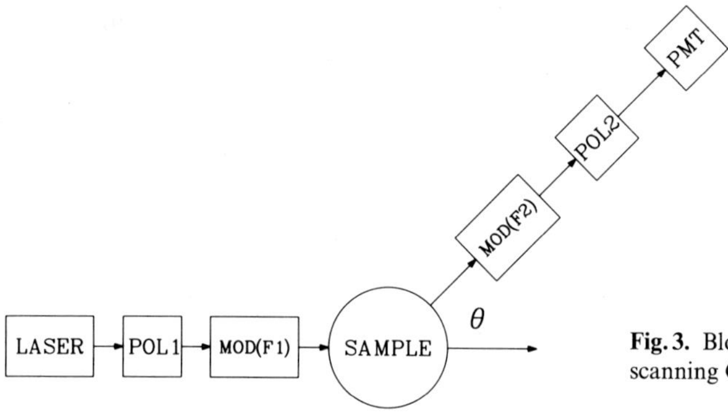

Fig. 3. Block diagram of the angular-scanning CIDS spectrometer

F_{11}(DC)	F_{12}(–)	F_{13}(100)	F_{14}(50)
F_{21}(–)	F_{22}(–)	F_{23}(–)	F_{24}(–)
F_{31}(94)	F_{32}(–)	F_{33}(6)	F_{34}(44)
F_{41}(47)	F_{42}(–)	F_{43}(53)	F_{44}(3)

Fig. 4. The Mueller matrix

pinges on a microbial sample in a cuvette, and the light is differentially scattered. The detection arm rotates under computer control, stopping at the specified angle and taking data for a specified length of time. The detector arm also contains a photoelastic modulator, a polarizer, and finally, a photomultiplier tube. This device measures the intensities of the scattered left and right circularly polarized light. The spectrometer that we use is a variant of one designed by Thompson et al. at the University of Texas [4]; other pioneering work in this area has been done by Hunt and Huffman [5] and by Bickel and his colleagues at the University of Arizona [6, 7] and especially by Bustamante, Tinoco, Maestre, and their co-workers at the University of California, Berkeley [8–14]. Some of our own work has been published as well [15, 16]. Additional theoretical work has been published by Zeitz et al. [17].

One additional complexity must be discussed before we turn to the data. To put it as simply as possible, the polarization properties of the light beam at each point in the instrument is described by a Mueller matrix (see Fig. 4). The Mueller matrix consists of 16 elements described as S_{11}, S_{12}, ect. Technically, only the S_{14} element of the Mueller matrix is the CIDS parameter, so we are being imprecise in describing the results of using other matrix elements as CIDS. Unfortunately, a better name for the technology is not available at the moment.

Each of these matrix elements can be determined by measuring different frequencies in the output signal. In the instrument that Dr. Salzman is now building as a prototype for a commercial instrument, it will be possible to measure eight of the 16 Mueller matrix elements, each of which provides some additional information facilitating microbial identification. This is an extraordinarily powerful tool the uses of which we have only begun to explore.

To summarize then, "CIDS" spectra – with CIDS in quotation marks – can be measured as a function of wavelength, scattering angle, and/or matrix element, and a number of matrix elements can be measured (virtually) simultaneously. This array of measurements potentially gives the method resolving power for microbial identification that we have only just begun to explore.

Experimental Data

I would now like to present some representative data that we have taken over the past couple of years. Because of the limitations of the laboratory we now occupy, we were unable to use live pathogens, so much of the data I will present is on vaccine preparations or other kinds of inactivated material. This is obviously a source of artifacts. We now have a better machine and better ways of expressing the data, and we will shortly begin to repeat some of these experiments in the facilities that we plan to move into in August. The data I present here must therefore be regarded as preliminary and are offered as "proof of principle" experiments rather than as definitive ones.

Figure 5 depicts CIDS spectra as a function of scattering angle for three type A influenza virus vaccine preparations in egg allantoic fluid given to us by our colleagues in the Center for Infectious Diseases of the Centers for Disease Control (CDC), Atlanta. Although these are certainly not pure virus preparations, the spectra of the three viral strains are clearly separable. At the time this work was done, it took about 20 min to run each of these spectra, and the presence of egg allanotic fluid did not prevent the three type A viruses from giving distinct CIDS spectra.

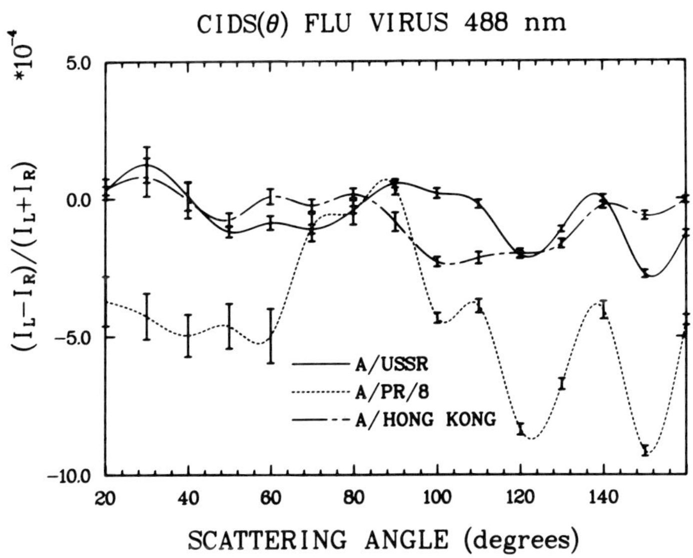

Fig. 5. CIDS spectra as a function of angle at 488 nm of three type A influenza virus vaccine preparations. Good discrimination is obtained among these three virus preparations around 60°, 110°, and 150°

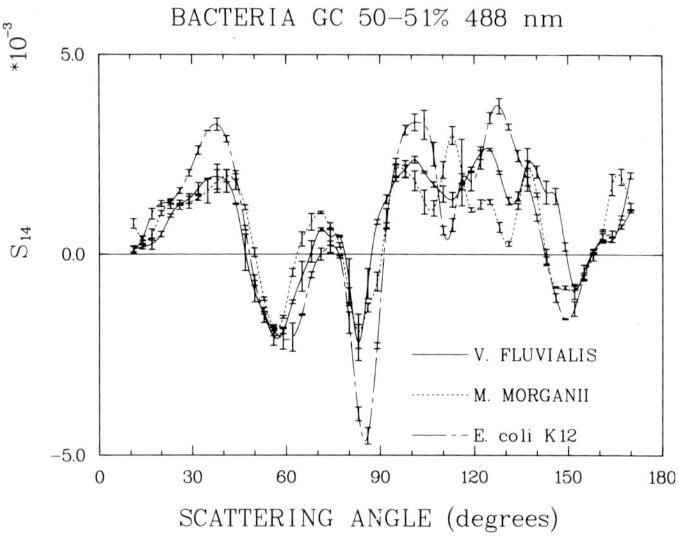

Fig. 6. The CIDS spectra at 488 nm as a function of angle for three bacteria of very nearly the same GC content. The spectra were run on separate suspensions of each bacterium in pure culture. Resolution among the three spectra is best at 120°–140°

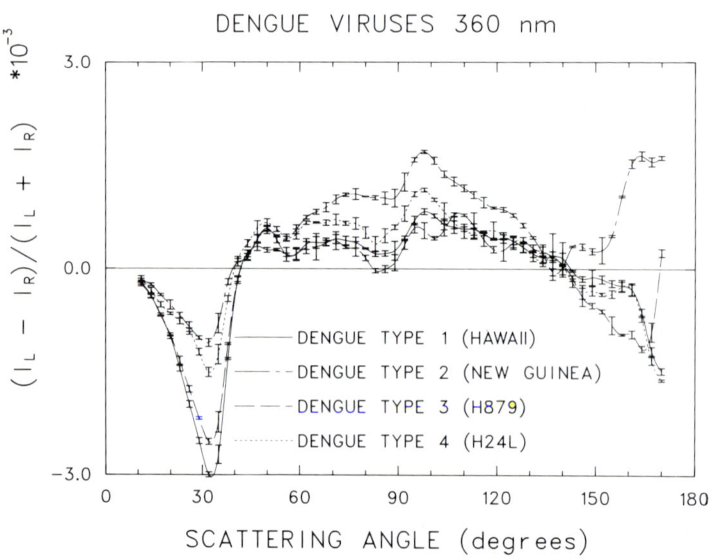

Fig. 7. The CIDS spectra at 360 nm of four types of dengue fever virus vaccines. The best resolution among the four viral vaccine types is obtained near 30°

Our colleagues at CDC wondered if perhaps the differences in the CIDS spectra were due to differences in guanine-cytosine (GC) content, as in the flow cytometric method of Van Dilla and his colleagues [12]. To determine this, they sent us samples of three very different bacteria that have essentially the same GC content. These are pure bacterial cultures run separately, with the CIDS spectra presented as a function of scattering angle. Figure 6 shows angular regions, particularly between 130° and 140°, in which the separation among the spectra is good. This finding is also highly reproducible.

Figure 7 shows the CIDS spectra as a function of wavelength for the four types of dengue fever virus as vaccine preparations, also supplied by CDC. Particularly in the region around 30°, the four viral types are readily distinguishable within the experimental variation indicated by the error bars. Note that all the dengue fever preparations have large negative signals in this region of the spectrum.

Similar data for four types of encephalitis virus vaccines from CDC are depicted in Fig. 8. Again, the region of maximum resolution is about 30°. The difficulty in distinguishing St. Louis encephalitis virus from dengue fever virus serologically is well-known. The CIDS spectra of the vaccine preparations are, however, very different.

We have also done some recent experiments with crude hepatitis B viral preparations kindly given us by Warner-Lambert Corporation and compared their spectra with those of dengue fever virus vaccine and eastern encephalitis vaccine. The findings are illustrated in Fig. 9. Here, we have employed the additional discrimination provided by measuring two matrix elements, S_{14}, the true CIDS element, and S_{34}. When both matrix elements are employed there is good separation

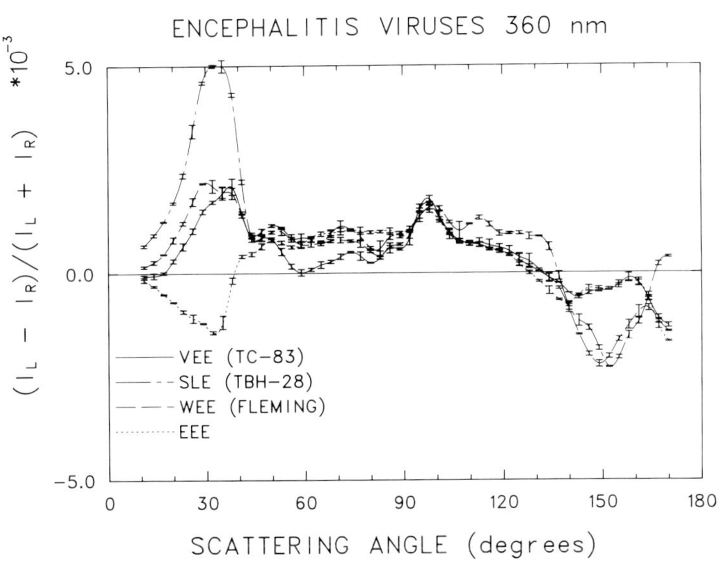

Fig. 8. The CIDS spectra at 360 nm of four types of encephalitis virus vaccines. Best resolution among the four viral vaccine types is obtained near 30°

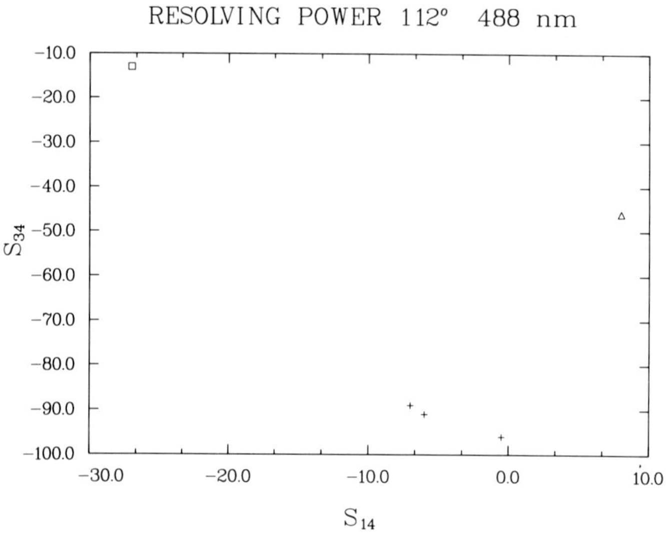

Fig. 9. The resolution obtained at 488 nm and 112° of a crude, formalin-treated hepatitis B virus preparation (+) (Dane particles, surface antigen, etc.), a dengue fever vaccine preparation (□), and a Venezuelan equine encephalitis vaccine preparation (△) when the "CIDS" signal from two Mueller matrix elements, F_{14} (true CIDS) and F_{34}, are employed

among these three types of preparation. The use of additional matrix elements should give increasingly enhanced resolution and, I remind you, we have eight matrix elements that we can measure simultaneously.

We have attempted to determine the sensitivity of the CIDS technique for determining hepatitis virus in clinical material. According to theory the CIDS signal – you will recall that it is a ratio of the sums and differences of intensities – should be independent of concentration. When we plotted the CIDS signal as a function of hepatitis DNA concentration we got usable CIDS spectra down to levels of 1 ng of hepatitis B viral DNA/ml, with no real effort made to maximize sensitivity.

Most people expect that a proper dilution curve should go through zero. With CIDS, however, a dilution curve is a straight line parallel to the x axis until the point at which the signal to noise ratio is so unfavorable that the signal can no longer be distinguished.

I repeat that this apparatus is far from being a system optimized for maximal sensitivity. Dr. Salzman feels that we can improve the sensitivity by at least an order of magnitude by some relatively simple modifications of the instrument.

Summary

We feel that this technique has immense potential although there are many obvious questions still be to be answered. You will be hearing from us at intervals as we extend our understanding of the strengths and limitations of this technique for rapid identification of microorganisms in clinical material.

Acknowledgements. The authors wish to thank the Life Sciences Division Office and the National Security Programs Office of the Los Alamos National Laboratory for financial assistance and encouragement. We are also deeply indebted to Dr. Walter Dowdle, Director of the Centers for Infectious Diseases, Centers for Disease Control, Atlanta and his colleagues for valuable advice and material. This work was performed under the auspices of the US Department of Energy and was supported in part by Grant GN26857 from the National Institute of General Medical Sciences, and the National Institutes of Health.

References

1. Isenberg H (1982) Clinical significance of automation in microbiology. In: Tilton RC (ed) Rapid methods and automation in microbiology. American Society for Microbiology, Washington DC, pp 96–97
2. Tilton RC, Balows A (1983) Introduction: body fluids and infectious disease – clinical and microbiological advances. Am J Med 75:1
3. Jenista JA, Prather SP, Powell KR, Menegus MA (1983) Virus cultures in the reduction of neonatal morbidity. Pediatr Res 17:224A (abstracts no 824)
4. Thompson RC, Bottiger JR, Fry ES (1980) Measurement of polarized light interactions via the Mueller matrix. Appl Opt 19:1323–1332
5. Hunt AJ, Huffman DR (1973) A new polarization-modulated light scattering instrument. Rev Sci Instrum 44:1753–1762
6. Bickel WS, Davidson JF, Huffman DR, Kilkson R (1976) Application of polarization effects in light scattering: a new biophysical tool. Proc Natl Acad Sci USA 73:496–490
7. Bickel WS, Stafford ME (1981) Polarized light scattering from biological systems: a technique for cell differentiation. J Biol Phys 9:53–66
8. Maestre MF, Gray DM, Cook RB (1971) Magnetic circular dichroism study in synthetic polynucleotides, bacteriophage structures and DNAs. Biopolymers 10:2537–2535
9. Bustamante C, Maestre MF, Tinoco I Jr (1980) Circular intensity differential scattering of light by helical structures. I. Theory. J Chem Phys 73:4273–4281
10. Bustamante C, Maestre MF, Tinoco I Jr (1980) Circular intensity differential scattering of light by helical structures. II. Applications. J Chem Phys 73:6046–6055
11. Bustamante C, Maestre MF, Tinoco I Jr (1981) Circular intensity differential scattering of light by helical structures. III. A general polarizability tensor and anomalous scattering. J Chem Phys 74:4839–4850
12. Maestre MF, Katz J (1982) A circular dichroism microspectrophotometer. Biopolymers 21:1899–1908
13. Tinoco I Jr, Bustamante C, Maestre MF (1982) Chiroptical methods and their applications to biomolecular structures. In: NATO Adv Study Inst Ser A 45:269–303
14. Van Dilla MA, Langlois RG, Pinkel D, Yajko D, Hadley WK (1983) Bacterial characterization by flow cytometry. Science 220:620–622
15. Salzman GC, Griffith JK, Gregg CT (1982) Rapid identification of microorganisms by circular intensity differential scattering. Appl Environ Microbiol 44:1081–1085
16. Salzman GC, Gregg CT (1984) Current and experimental methods of rapid microbial identification. Bio/Technology March
17. Zeitz S, Belmont A, Nicolini C (1983) Differential scattering of circularly polarized light as a unique probe of polynucleosome superstructures. A simulation by multiple scattering of dipoles. Cell Biophys 5:163–187

An Instrument for Virus Identification by Polarized Light Scattering: A Preliminary Report

G. C. Salzman, W. K. Grace, D. M. McGregor, and C. T. Gregg [1]

Biological macromolecules are asymmetric structures. As such, they interact differently with left and right circularly polarized light. Circular dichroism (CD), the differential absorption of left and right circularly polarized light, probes changes in the secondary and tertiary structure of molecules in solution. Circular intensity differential scattering (CIDS), which is the differential scattering of left and right circularly polarized light, probes the higher order structure of macromolecular aggregates [1, 2]. CIDS is given as the amount of light scattered when the incident beam is left circularly polarized minus that scattered when the incident beam is right circularly polarized, divided by the total amount of light scattered by the object, in our case a virus or bacterium.

CIDS is just one element of a 4×4 matrix called the Mueller matrix describing the scattering of light from an object at a particular angle and wavelength. All of the information in a beam of light at a given wavelength is contained in a four-element vector called the Stokes vector: I, Q, U, and V. I is the intensity of the light in a beam and is usually normalized to one, Q, U, and V describe the polarization properties of the light beam. Q is the tendency toward horizontal linear polarization. It is $+1$ for horizontal linearly polarized light and -1 for vertical linearly polarized light. U is the tendency toward linearly polarized light tilted at $+45°$ to the horizontal scattering plane. It has values ranging from $+1$ $(+45°)$ to -1 $(-45°)$. V is the tendency toward right circular polarization. For a right circularly polarized beam $V = +1$ and for a left circularly polarized beam $V = -1$. In our instrument the Stokes vectors of the incident and scattered light are manipulated by a series of optical elements represented by 4×4 Mueller matrices so that the intensity component of the stokes vector of the scattered light contains information about the scattering sample.

Figure 1 shows a schematic of the photopolarimeter. Figure 1 A shows the coordinate system. θ is the polar scattering angle in the horizontal scattering plane. γ is the azimuthal angle indicating the preferred axes for the optical elements. Figure 1 B shows a schematic of the instrument. The argon laser beam passes through a polarizer oriented at γ_1, then through a photoelastic modulator with its fast axis oriented at γ_2 and with sinusoidally varying retardance amplitude δ_2. It is driven at a frequency f_2. The Stokes vector of the incident light beam is now elliptically polarized. The light scatters from a homogeneous suspension of microorganisms in a cylindrical cuvette. The light scattered at an angle θ passes through

1 Experimental Pathology Group, Life Sciences Division, Los Alamos National Laboratory, Los Alamos, NM 87545, USA

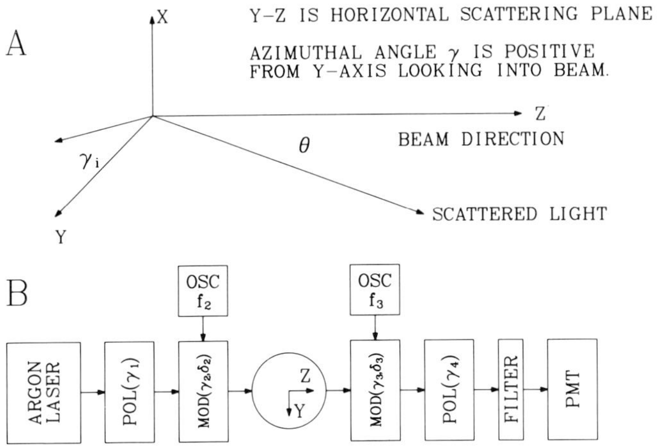

Fig. 1. A Coordinate system and **B** schematic of static Mueller matrix photopolarimeter

S_{11}(DC)	S_{12}(–)	S_{13}(100)	S_{14}(50)
S_{21}(–)	S_{22}(–)	S_{23}(–)	S_{24}(–)
S_{31}(94)	S_{32}(–)	S_{33}(6)	S_{34}(44)
S_{41}(47)	S_{42}(–)	S_{43}(53)	S_{44}(3)

Fig. 2. Mueller matrix for $\gamma_1 = \gamma_4 = +45°$ and $\gamma_2 = \gamma_3 = 90°$

S_{11}(DC)	S_{12}(100)	S_{13}(–)	S_{14}(50)
S_{21}(94)	S_{22}(6)	S_{23}(–)	S_{24}(44)
S_{31}(–)	S_{32}(–)	S_{33}(–)	S_{34}(–)
S_{41}(47)	S_{42}(53)	S_{43}(–)	S_{44}(3)

Fig. 3. Mueller matrix for $\gamma_1 = \gamma_4 = +90°$ and $\gamma_2 = \gamma_3 = +45°$

a second photoelastic modulator with its fast axis at azimuthal angle γ_3 and its retardance amplitude at δ_3 and then through a second polarizer with its passing axis at γ_4. The light finally passes through a laser line filter and impinges on the cathode of a photomultiplier tube. This instrument is conceptually similar to one developed by Thompson et al. [3] based on earlier work by Bickel et al. [4] and Hunt and Huffman [5]. Modulation frequency f_2 is 50 kHz and f_3 is 47 kHz.

The use of two frequencies produces intensity modulation in the output at a variety of frequencies consisting of linear combinations of the two reference frequencies. Each element of the Mueller matrix for the sample appears at a unique modulation frequency. With $\gamma_1 = \gamma_4 = +45°$ and $\gamma_2 = \gamma_3 = 90°$, the Mueller matrix for the sample is as shown in Fig. 2. S_{11} is the total scattering intensity and appears as a DC component at zero frequency. S_{14} is the CIDS and appears at 50 kHz. The other polarization sensitive element of particular interest is S_{34}, which appears at 44 kHz.

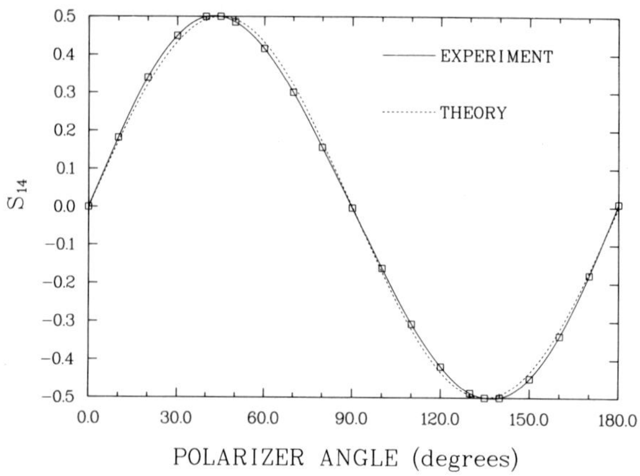

Fig. 4. S_{14} calibration

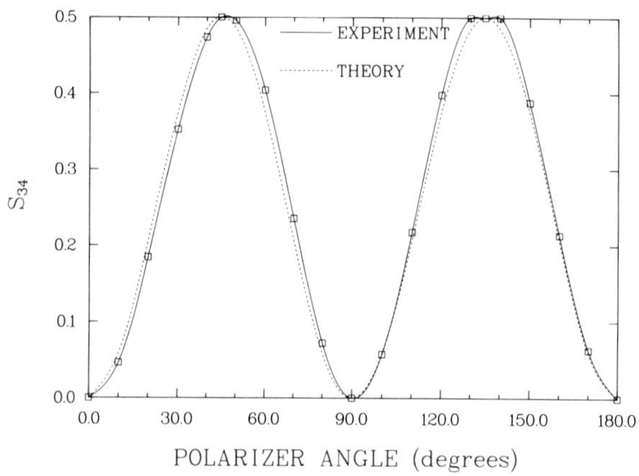

Fig. 5. S_{34} calibration

Figure 3 shows a sample Mueller matrix for the case $\gamma_1 = \gamma_4 = +90°$ and $\gamma_2 = \gamma_3 = +45°$. In this case, matrix elements in row three and column three are missing and S_{24} appears at 44 kHz instead of S_{34}.

Each matrix element can be calibrated by replacing the sample by optical elements with known Mueller matrices. S_{14} is calibrated by using a quarter wave plate with its fast axis at 0° followed by a polarizer rotated through 180°. The functional dependence should be $0.5 \sin 2\theta$. Figure 4 shows this calibration for S_{14}. Figure 5 shows a similar calibration for S_{34} where the functional dependence should be $0.5 \sin^2 2\theta$.

Experimental data obtained from this instrument is presented in the paper by Gregg et al. in this volume.

Acknowledgements. The authors would like to thank Mr. Richard D. Hiebert for his assistance. This work was performed under the auspices of the US Department of Energy and was suported in part by National Institutes of Health Grant GM 26857 and by the US Army Chemical Research and Development Center.

References

1. Bustamante CA, Tinoco I Jr, Maestre MF (1982) Circular intensity differential scattering of light. IV. Randomly oriented species. J Chem Phys 76:3440–3446
2. Zeitz S, Belmont A, Nicolini C (1983) Differential scattering of circularly polarized light as a unique probe of polynucleosome superstructure. A simulation by multiple scattering of dipoles. Cell Biophys 5:163–187
3. Thompson RC, Bottinger JR, Fry ES (1980) Measurement of polarized light interactions via the Mueller matrix. Appl Opt 19:1323–1332
4. Bickel WS, Davidson JF, Huffman DR, Kilkson R (1976) Application of polarization effects in light scattering: a new biophysical tool. Proc Natl Acad Sci USA 73:486–490
5. Hunt AJ, Huffman DR (1973) A new polarization-modulated light scattering instrument. Rev Sci Instrum 44:1753–1762

A New Method for the Rapid Identification and Differentiation of Pathogenic Microorganisms Using Fourier Transform Infrared Spectroscopy

P. GIESBRECHT, D. NAUMANN, H. LABISCHINSKI, and G. BARNICKEL [1]

Introduction

In spite of several attempts to introduce modern physical methods for the detection and identification of pathogenic bacteria, most routine identifications are presently based on microscopic, biochemical, and physiologic methods. Some emphasis has been put on the development for ready-to-use commercially available test kits, for example for the differentiation and identification of enterobacteria and for some other microorganisms, which result in identification within 24 h (and in certain cases also within 5–6 h) after obtaining isolated single colonies. The need for rationalization and speeding up of the identification procedure has led to the development of systems which allow easy and semiautomatic operation, but at relatively high cost. Nevertheless, there is an urgent need for new developments in this field which will have to fulfill the following requirements:

1. Rapid and secure identification of all phathogenic bacteria
2. Simple operation, using a uniform procedure
3. Identification down to the strain level, if desired
4. Simple standardization and full computerizability
5. Reasonable price – performance correlation.

In the following we will demonstrate that the modern technique of Fourier transform infrared spectroscopy (FT-IR) seems to be capable of fulfilling the bulk of these conditions and might provide a future alternative to existing procedures.

Materials and Methods

FT-IR spectra have been recorded on a 10MX-E spectrometer (Nicolet Instruments, Madison, USA). The bacteria stemmed either from the strain collection of the Robert Koch Institute of the Federal Health Office of Germany or were clinical isolates from the Rudolf Virchow Hospital in Berlin, Dept. of Microbiology. Prior to all infrared measurements, a standard experimental procedure concerning culture media, incubation time, harvest of bacteria, sample preparation, measuring time, and spectral resolution was applied. The details of the preparation procedure, of the techniques for recording the IR spectra, and of mathematical procedures will be published elsewhere [5].

1 Robert-Koch-Institut des Bundesgesundheitsamtes, Nordufer 20, D-1000 Berlin 65

Results and Discussion

The new method essentially takes advantage of the so-called Fourier-transform infrared spectra of microorganisms, which can at present be recorded within less than a minute with extremely high sensitivity and wavelength reproducibility [1], even when investigating only micrograms of the bacteria. Applying FT-IR microscopes, even nanograms of material may be studied with ample quality.

The infrared absorption spectra of most materials consist of a large number of absorption bands. It is well known that these bands are due to the vibrational and rotational motions of the molecules which are excited by the absorption of infrared radiation. As a matter of fact, each type of molecule, with the exception of optical enantiomers, possesses a characteristic IR absorption spectrum. It is for this reason that IR spectroscopy has become one of the powerful analytic tools, especially in the chemical and pharmaceutical industries. It was, therefore, quite

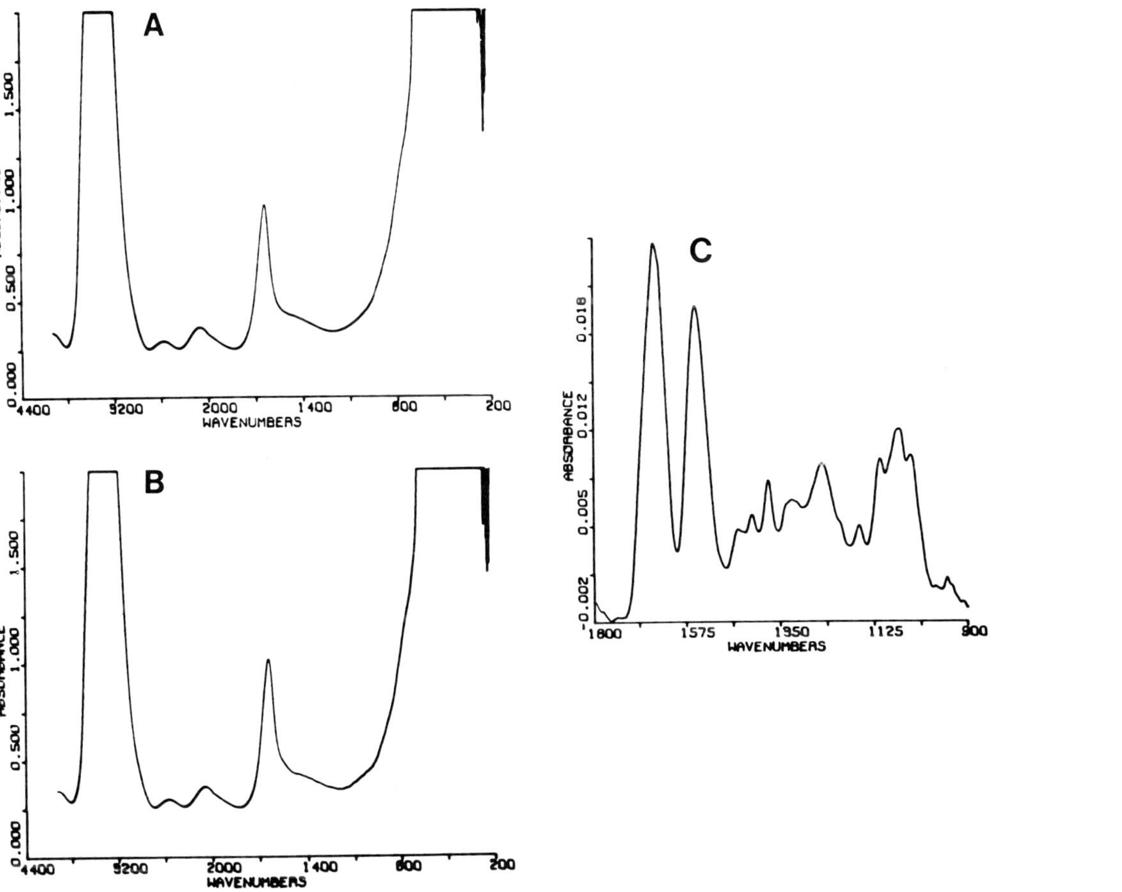

Fig. 1 A–C. Infrared spectra demonstrating sensitivity and solvent subtraction capabilities of FT-IR. **A** pure water; **B** water plus suspended cell wall peptidoglycan; **C** subtraction of spectrum **A** from spectrum **B**, resulting into the spectrum of cell wall peptidoglycan

natural to ask whether or not this method can also be applied to the differentiation and identification of pathogenic bacteria. However, although several attempts were undertaken as long ago as the late 1950s [6], no successful routine technique could be developed (because of the time-consuming preparation procedures, the very low scanning speed of conventional IR systems available at that time, and the difficulties in differentiating the complex and relatively similar spectra). Bearing in mind the outstanding progress of modern FT-IR in terms of sensitivity, wavelength accuracy, and recording speed, we have tried to apply this new technique to the identification of bacteria and bacterial components [2–4].

In order to provide an impression of what is now within reach of FT-IR, an example of our studies on bacterial cell walls in aqueous suspensions [2] is given (Fig. 1). Such a problem was not accessible before the introduction of modern FT-IR. Both spectra (Fig. 1 A = pure water, Fig. 1 B = water plus peptidoglycan) proved to be extremely similar and any trial to obtain significant information about this cell wall component from the solution spectrum alone failed. Owing to the absorbance subtraction capabilities of FT-IR, digital subtraction of water from the solution revealed the vibrational spectroscopic characteristics, i.e., the very IR absorption spectrum of peptidoglycan (Fig. 1 C).

Infrared spectra of intact bacteria revealed a highly complex absorption pattern (Fig. 2). Whatever all these absorption maxima may signify [5], they comprise the particular vibrational spectroscopic characteristics of all individual constituents of the bacteria (i.e., cell walls, proteins, membranes, nucleic acid, etc.), indicating that these spectra might, possibly, be specific enough to serve as a kind of "spectral fingerprint." If this were indeed the case [3, 4, 6], the comparison of experimentally measured unknown spectra with those stored on a reference data base should result in the identification of bacteria. Since spectral data storage is

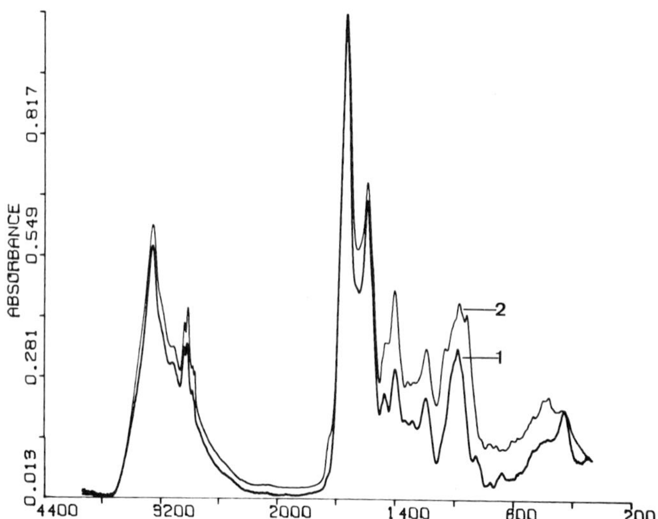

Fig. 2. Survey spectra of two different samples of bacteria. *1, Streptococcus pyogenes; 2, Klebsiella pneumoniae*

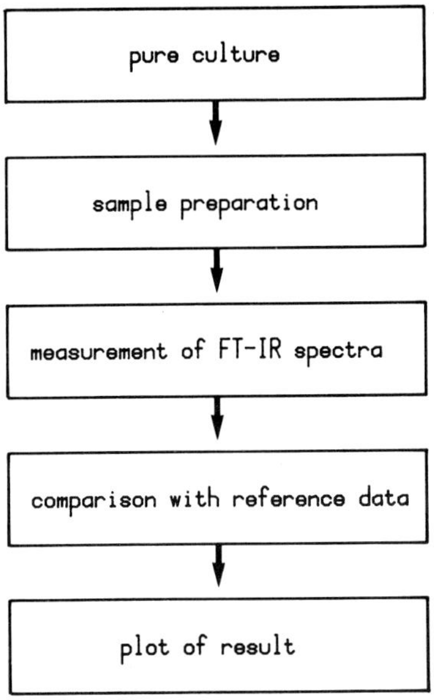

Fig. 3. Scheme of identification and differentiation procedure via FT-IR

now possible when applying FT-IR techniques, any searching procedure can be accomplished automatically, if desired.

In order to test these possibilities, we have developed a procedure, the principal aspects of which are schematically shown (Fig. 3). Similar to conventional identification methods, we start with single colonies from pure cultures of the microorganisms [5]. The procedure – sample preparation and data collection – is completed within minutes and the spectra obtained are then compared rapidly with those stored on a reference data base. The final result would be available under our conditions (which are not yet optimized) 10–15 min after obtaining the pure culture, independent of whatever kind of special microorganisms has to be identified [5].

To test the reproducibility of the method, spectra of *Staphylococcus aureus* (strain Pelzer) were measured from different colonies of different pure cultures (Fig. 4). We have selected the spectral range between 1,200 and 900 cm^{-1}, where the spectral features are dominated by the various bacterial polysaccharidal components [3]. The obvious similarity of the spectral contours of all spectra proves the outstanding reproducibility of the method.

To solve the question in which way spectral differences of the different bacteria can be quantified, we applied a fast-working cross-correlation algorithm resulting in a so-called "differentiation index" D, which – after appropriate scaling – can adopt values between 0 and 1,000 [5]. To avoid artifacts which may be introduced by baseline shifts due to different film thickness and diffuse scattering at the sample surface, we used derivatives of the spectra as a data base for spectral comparison [5]. Per definitionem, identical spectra must result in a value of zero; the comparison of similar spectra must yield low values for D; and dissimilar spectra must result in high values.

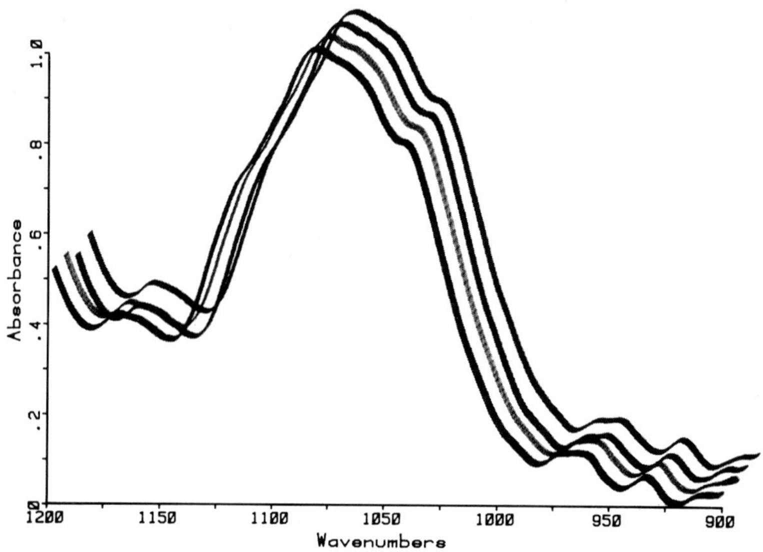

Fig. 4. Spectral contours of four independently measured samples of *Staphylococcus aureus*. In order to demonstrate the excellent reproducibility of the FT-IR technique, the individual spectra have been shifted against each other along both the absorbance and the wave number axis

Table 1. Reproducibility test[a]

Sample no.	1	2	3	4	5	6	7	8
1	0							
2	1	0						
3	1	1	0					
4	2	1	2	0				
5	2	3	2	5	0			
6	2	1	2	2	3	0		
7	2	2	2	1	4	1	0	
8	2	2	2	3	2	2	2	0

[a] Differentiation indices of independently harvested cultures of *Staphylococcus aureus* strain Pelzer grown on bacto peptone agar

All D-values of the cross-correlation chart (Table 1) of the differentiation indices of eight independently measured samples of *S. aureus* (strain Pelzer) ranged well below 10, indicating again the outstanding reproducibility of the procedure.

In a following step, the spectral properties of a variety of bacteria of different genera (*Staphylococcus aureus, Escherichia coli, Klebsiella pneumoniae, Salmonella typhimurium*) were analyzed (Fig. 5), again choosing the spectral range from 1,200 to 900 cm^{-1}. The spectral contours are rather different and hence (cf. Table 2) the computed D-values proved to be very high, each ranging above a value of 100.

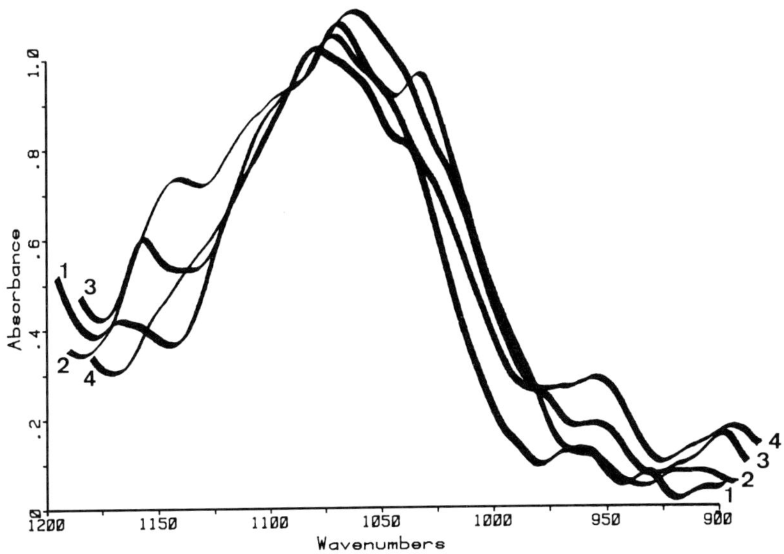

Fig. 5. Spectral contours of four selected bacterial samples of different genera (spectral range: 1,200–900 cm^{-1}, presentation analogue to Fig. 4). *1, Staphylococcus aureus; 2, Klebsiella pneumoniae; 3, Escherichia coli; 4, Salmonella typhimurium*

Table 2. Differentiation indices of four selected bacteria of different genera

Sample	*Staphylococcus aureus*	*Klebsiella pneumoniae*	*Escherichia coli*	*Salmonella typhimurium*
Staphylococcus aureus	0			
Klebsiella pneumoniae	307	0		
Escherichia coli	150	304	0	
Salmonella typhimurium	184	268	124	0

Stimulated by these results, we extended our efforts to the differentiation of species within a given genus. Figure 6 depicts the spectra of two different streptococci of different serologic groups A and B. Although the spectra appeared to be rather similar, characteristic deviations between the two types of streptococci were observed: major differences predominantly occurred between 1,200 and 900 cm^{-1}, indicating the well-known fact that different carbohydrates are present in the cell walls, of these two groups of streptococci.

In order to find out whether even different strains of a species can be differentiated by this new technique, the differentiation indices of nine different strains of *S. aureus* and – for comparison – of *S. epidermidis* were calculated (Table 3). All D-values ranged above 10 (cf. also Table 1); consequently all different strains could be differentiated. It should be noted that the correlation of all strains of *S. aureus* to *S. epidermidis* resulted in distinctly higher D-values than the cross-correlation of the *S. aureus* strains between each other.

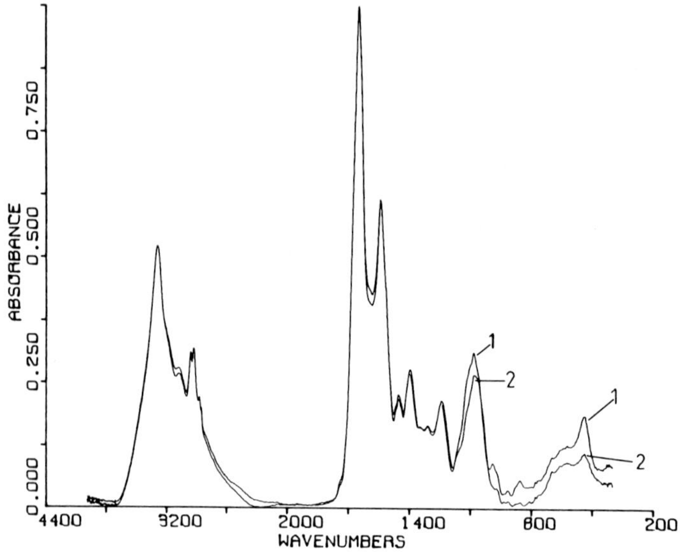

Fig. 6. Survey spectra of different species of streptococci. *1, Streptococcus pyogenes,* group A; *2, Streptococcus agalactiae,* group B. Major deviations between the two spectra occur between 1,200 and 900 cm^{-1}, where the characteristic absorptions resulting from polysaccharidal components dominate the spectra

Table 3. Differentiation indices of different strains of *Staphylococcus aureus*[a]

Sample	SG 511	52A5 G	Pelzer	Ox-ford	Clin. isol.	H, RKI	H, GB	H/28	H/4B	*S. epiderm.*
SG 511	0									
52A5 G	21	0								
Pelzer	17	11	0							
Oxford	18	16	16	0						
Clin. isol.	19	16	12	15	0					
H, RKI	18	31	15	21	17	0				
H, GB	29	20	23	7	24	36	0			
H/28	19	22	13	11	17	11	17	0		
H/4B	27	21	16	12	17	25	11	10	0	
S. epiderm.	53	72	55	47	50	55	54	56	50	0

[a] For comparison, the differentiation indices of a strain of *Staphylococcus epidermidis* are included in the correlation chart

Furthermore, our trials in the identification of different *Salmonella* strains were successful, too (Table 4). The D-values of the different serologic types of *Salmonella typhimurium* are given in comparison to another species (*Salmonella minnesota*); their D-values clearly indicate the possibility of differentiation and identification applying FT-IR techniques. We succeeded even in differentiating a *Salmonella* strain which differed only by carrying an additional O-antigen missing in the other strains. These data indicate that, at least in this case, one single O-antigen is sufficient to create differences detectable via this new technique.

Table 4. Differentiation indices of different serological types of *Salmonella typhimurium*[a]

Sample	*S. typhimurium* var. *Copenhagen*[b]	*S. typhimurium* var. *Copenhagen*[c]	*S. typhimurium*[d]	*S. minnesota*
S. typhimurium var. *Copenhagen*[b]	0			
S. typhimurium var. *Copenhagen*[c]	9	0		
S. typhimurium[d]	57	31	0	
S. minnesota	79	143	87	0

[a] For comparison, the differentiation indices of a strain of *Salmonella minnesota* are included in the correlation chart
[b] Devoid of O 5 antigen
[c] Devoid of O 5 antigen, but carrying a plasmid
[d] Containing the O 5 antigen and carrying the same plasmid as c

Table 5. Identification example[a]

Selected reference organisms from the data file	Differentiation index
Staphylococcus epidermidis	313
Staphylococcus staphylolyticus	279
Staphylococcus aureus	230
Streptococcus pyogenes	183
Bacillus subtilis	167
Streptococcus agalactiae	161
Klebsiella pneumoniae	159
Salmonella typhimurium	102
Salmonella minnesota	78
Escherichia coli	19

[a] The differentiation indices of an unknown clinical isolate in respect to some selected microorganisms from the reference data base are listed

Based on these first results, we tried to identify unknown bacterial species. After measuring the FT-IR spectrum of an unknown clinical isolate and comparing its spectral data with those stored on our reference data base, the differentiation indices were calculated (Table 5). The lowest value found was 19, obtained in respect to *E. coli*; as all the other D-values ranged far above this value, the unknown microorganism had to be identified as *E. coli*. This result was confirmed independently by conventional physiologic methods.

Conclusions and Summary

The data presented here demonstrate that the new Fourier transform infrared spectroscopic technique proved to be appropriate, at least within the experimental data so far available, for:

1. Very rapid identification of pathogenic bacteria within minutes after obtaining the pure culture
2. Application to all pathogenic bacteria so far tested, performing a standard procedure
3. Differentiation of bacteria down to the strain level
4. Computerizability
5. Simple operation procedure, suitable for routine application, with a high sample throughput which may be executed automatically.

We think that the scope of future application of the FT-IR technique might be:

1. Rapid diagnosis, because identification of microorganisms is possible within minutes after obtaining the pure culture
2. Epidemiologic studies, because differentiation down to the strain level is possible
3. Microbiologic research, because the technique is capable of detecting even minute variations during cell growth, cell metabolism, and chemical and antibiotic treatment [2, 4] of bacteria.

Now we have to improve the experimental and instrumental parameters. For instance, the search for a series of discriminating, group-specific regions within the IR spectra of bacteria could result in further speeding-up of the method. However, the most important task is to test this procedure over a long period of time, parallel to classical diagnostic methods, to rule out any weaknesses.

References

1. Ferraro IR, Basile IL (eds) (1979, 1982) Fourier transform infrared spectroscopy, application to chemical systems and techniques using FT-IR, vols 1–3. Academic, New York
2. Naumann D (1984) Some ultrastructural information on intact, living bacterial cells. Infrared Physics 24:233–238
3. Naumann D, Barnickel G, Bradaczek H, Labischinski H, Giesbrecht P (1982) Infrared spectroscopy, a tool for probing bacterial peptidoglycan. Eur J Biochem 125:505–515
4. Naumann D, Labischinski H, Barnickel G, Giesbrecht P (1983) A novel and rapid method for identifying important cell wall parameters: Fourier transform infrared spectroscopy. In: Hakenbeck R, Höltje JV, Labischinski H (eds) The target of Penicillin. de Gruyter, Berlin, pp 85–90
5. Naumann D, Labischinski H, Barnickel G, Giesbrecht P (1984) A new method for the rapid identification of microorganisms: Fourier transform infrared spectroscopy. (In preparation)
6. Norris KP (1959) Infrared spectroscopy and its application to microbiology. J Hyg 57:326–345

The Use of Microcalorimetry and Impedance in Clinical Microbiology

J. A. MORELLO [1]

Introduction

During the past 15 years, several investigators have examined novel methods and instrumentation for use in clinical microbiology that do not rely simply on automation of standard techniques. Two of the systems studied, microcalorimetry and impedance, are described here: their principles of action, clinical laboratory applications, advantages, and limitations.

Microcalorimetry

Principle

In calorimetry, heat quantities or heat effects are measured either by measuring temperature change on a thermometer or by recording thermoelectric effects generated in thermoelements by heat passage. Although calorimeters were used during the eighteenth and nineteenth centuries, the instruments available measured only large temperature changes. Modern microcalorimeters have increased sensitivity and are able to detect heat changes taking place in micromolar quantities of reactants with heat effects in the microwatt range [1]. In the instruments most commonly used for microbiological measurements, the heat produced during a reaction flows through thermoelectric couples, resulting in an electrical current that is directly proportional to the heat flow. The thermocouple is usually a plate made from semiconductive material and produces a weak potential (millivolts) that is amplified to a signal suitable for a recorder or other reading device.

During their growth and metabolism, microorganisms produce small amounts of heat, primarily as a result of enzymatic catabolic reactions [2]. If these heat changes are recorded and plotted against time, a heat profile, or thermogram, is obtained. The shape of the thermogram varies throughout the microbial growth cycle, beginning during log phase, rising as growth continues in the culture, then declining as metabolic activity decreases and finally stops.

1 Clinical Microbiology Laboratories and Department of Pathology, University of Chicago Medical Center, 5841 South Maryland Avenue, Chicago, Ill 60637, USA

Clinical Laboratory Applications

In clinical microbiology, microcalorimetry has been used for microbial identification, antibiotic susceptibility testing, and detection of significant bacteriuria. These applications take advantage of the characteristic heat profiles poduced by microbial species during metabolism, as well as the inhibition of microbial heat production that results from contact with inhibitory antimicrobial agents. Russell et al. [3] devised a multichannel microcalorimeter that could handle 50 specimens simultaneously. Each specimen was inoculated into a small volume of culture medium and inserted into a preassigned chamber in the instrument, which operated at 37 °C. Once the sample was placed in the microcalorimeter, all control was assumed by an integral computer system that handled data acquisition, processing, and display.

Russell's studies centered primarily on differentiating species of *Enterobacteriaceae* by their characteristic heat profiles when they were growing in brain heart infusion broth. Beginning with an inoculum of approximately 500 organisms, he found that between 10^5 and 10^6 organisms per ml were needed before it was possible to detect heat production. Most cultures generated 40–60 microcalories per s per ml, and the thermogram returned to the baseline after 5–7 h. The configuration of the heat profile varied among species. Depending on the amplitude and timing of the thermal events, it was possible to distinguish microorganisms from each other. For some species, a characteristic profile was seen only 3 h after heat production began, but for others, a period of as long as 14 h was needed for identification. Although an initial study of 200 cultures provided promising results, further work with a larger number of isolates indicated that very similar profiles could be generated by distinctly different organisms. In addition, the thermograms for strains of the same organism often exhibited great diversity. More profile uniformity was seen when brain heart infusion broth was replaced by a specially formulated culture medium.

In a smaller microcalorimetry study of 17 *Enterobacteriaceae* strains (12 species) growing in brain heart infusion broth, Herman et al. [4] also found that the best discriminators for organism identification were the maximum height (amplitude) and length of the thermogram, as well as the overall shape of the curve profile. Measurements of the total heat evolved were not discriminatory. Although these workers incubated their strains at 30 °C and used a different instrument for analysis, they obtained profiles similar to those of Russell et al. [3] for *Escherichia coli, Klebsiella pneumoniae,* and *Enterobacter cloacae*. Curves for *Serratia marcescens, Hafnia alvei, Citrobacter,* and *Enterobacter aerogenes,* however, differed under these conditions, perhaps reflecting different optimal growth temperatures for these organisms.

Beezer et al. [5] detected significant numbers of bacteria ($>10^5$) in urine specimens by a microcalorimetric method. They incubated an aliquot of the specimen in nutrient broth for 3.5 h before taking a reading in the instrument. The amplitude of the ensuing thermogram was related to the number of organisms originally present in the specimen. For 299 urine specimens, the false-negative rate was 1%, and the false-positive rate was 15%.

In a study applying microcalorimetry to antibiotic susceptibility testing, Binford et al. [6] tested eight antibiotics against 15 different bacteria isolated from urine. When the tests were performed under anaerobic conditions, the patterns of inhibition of metabolic heat production showed a correlation of 77% with results of the disk agar diffusion method. When certain discrepant tests were repeated under aerobic conditions, agreement with the disk method rose to 87%. In addition, differences between bactericidal and bacteriostatic drugs were seen. Others have used this technique to study the effects of various antibiotics on bacteria [7].

Advantages and Limitations

From the studies described, it is evident that microcalorimetry is adaptable to many clinical laboratory procedures, and results are usually available on the same day the test is set up. In general, microcalorimetry work has been conducted in only a few laboratories in Europe and the United States. In most instances, different methods and instrumentation have been used to perform the analyses; thus, neither standardized media or test procedures for performing these assays nor reference thermograms for interpretation are available. Except for the instrument used by Russell [3], automated, computerized microcalorimeters have not been described, and the procedures are time-consuming and labor intensive. Although the technique may be useful for research on the metabolic effects of microbial growth enhancers and inhibitors, its current application to the clinical microbiology laboratory is minimal.

Impedance

Principle

As with microcalorimetry, studies of electrical measurements for monitoring microbial growth date back at least to the last century [8]. When microorganisms grow and metabolize in a culture medium, the end products liberated change the chemical composition of the medium. The changes in medium composition are accompanied by changes in impedance, which is the resistance to the flow of an alternating current through a conducting material. In general, as microorganisms grow, the impedance of the medium decreases with time. The physical principles of impedance are described elsewhere [9].

A variety of physical and chemical factors that are not related to microbial growth may also affect impedance. These include temperature fluctations, evaporation, absorption of gases, and spontaneous changes in chemical composition that may occur as the medium ages [9, 10]. To compensate for these changes, a reference chamber containing uninoculated medium is usually monitored in parallel with the growth chamber. In this case, impedance is recorded as the ratio of the impedance in the reference chamber to the sum of the impedance in the reference chamber and the impedance in the growth chamber. Other factors affecting impedance include the nature of the electrodes and culture medium used and the growth characteristics of the microorganism under study [9]. Gold and stainless steel electrodes are most satisfactory because they produce stable base lines and detect small changes in impedance that result from microbial metabolism and growth. The sensitivity of impedance measurements can be increased by dialyzing culture media to reduce its ionic strength; however, this procedure may also delay microbial growth and subsequent impedance changes. Most commercially available culture media are satisfactory for monitoring microbial growth by impedance. Detection time may be delayed in cultures where microorganisms do not grow in proximity to the electrode, e.g., where they grow as discrete microcolonies or as a surface pellicle. In these instances, agitation of the culture vessel improves detection time.

Clinical Laboratory Applications

Using a variety of instruments for impedance measurements, workers have detected the presence of microorganisms in blood specimens, quantified organisms in urine specimens, identified microorganisms, and tested their antimicrobial susceptibility.

The Bactometer

In the United States, most reported clinical laboratory investigations of impedance were performed with the Bactometer Microbial Monitoring System (Bactometer, Inc.). Originally devised as an 8-sample system (Bactometer 8), newer models of the instrument monitor 32 (Bactometer 32) or 480 (Bactometer 120) samples simultaneously. The instrumentation has been described in detail by Gall and Curby [8]. Briefly, in the Bactometer 8 and 32, each sample chamber is paired with a reference chamber containing uninoculated medium. Changes in the impedance ratio between the two chambers are measured and recorded on a strip chart. In the instrument that monitors 480 specimens, a reference chamber is available for each 15-sample chamber. A microprocessor interprets the data and displays results on a video terminal or prints them on a hard-copy printer. Incubators are a component of the Bactometer 32 and 120 systems.

When tests are performed, each chamber in the Bactometer is monitored at approximately 96-s intervals. Initially, as the temperature of the broth equilibrates to that of the incubator, impedance changes related to temperature fluctuations are detected. The measurements then return to baseline readings until sufficient microbial growth has occurred to cause significant impedance change. In the Bactometer, the threshold reading signifying microbial growth is defined as an accelerating change in the impedance that is equal to or greater than 0.8% of the base line impedance when the measuring signal is at 2,000 Hz [10]. This corresponds to between 10^6 and 10^7 microorganisms per ml.

Blood Culture Studies. In 1977, Specter et al. [11] reported on a comparative study of the Bactometer 32 and the BACTEC 225 (Johnston Laboratories, Towson, Maryland) instruments for detecting microorganisms in blood cultures. Using sterile blood specimens inoculated with 10 different species of gram-positive and gram-negative bacteria, they detected growth within 6–18 h of incubation. The initial inoculum was calculated to be 0.8 to 1.5 bacteria per ml of culture broth. Gram-negative organisms (all members of the *Enterobacteriaceae*) were detected in an average of 6–9 h, whereas gram-positive organisms (*Staphylococcus aureus*, alpha- and beta-hemolytic streptococci, and pneumococci) required 12–18 h for growth detection. When the culture medium was gassed with CO_2, detection time of these gram-positive cocci was reduced to 7–14 h. In a study of anaerobic bacteria, a greater inoculum was used (10–50 organisms per ml of blood culture). *Clostridium perfringens* was detected in an average of 3 h and *Bacteroides fragilis* in 9 h. *Propionibacterium* and *Bacteroides melaninogenicus*, however, required 48 h for detection. In each instance, these detection times were comparable to those obtained with the BACTEC system.

In a more recent Bactometer study [12], Buckland et al. detected an inoculum of 10^2 colony-forming units per ml of bacteria in 6 h and 10^6 colony-forming units per ml in 2 h. The organisms used were *S. aureus*, *Haemophilus influenzae*, and *E. coli*. When comparing impedance and conventional broth methods for 189 blood cultures, of which 18 were positive, these authors detected growth at least 12 h earlier by the impedance method.

Detecting Microorganisms in Urine Specimens. Because impedance detection time is related to the number of microorganisms present in a sample, an approximate microbial quantification can be made; i.e., larger numbers of organisms are detected in shorter incubation periods. The time at which readings are made is critical, however, because insignificant numbers of organisms present in a specimen can multiply to significant numbers with prolonged incubation.

Zafari and Martin [13] inoculated a mixture of 0.5 ml of urine specimen and 0.5 ml of trypticase soy broth medium into Bactometer modules. Using a cutoff incubation time of 4 h, they detected 34 of 35 urine specimens that had more than 10^5 organisms per ml by standard culture methods. In a total sample of 156 urines, agreement with culture was 96.8%; there were four false-positive and one false-negative results. When a 2-h cutoff period was examined, agreement with culture results was 98.1%: only one false-positive result was obtained, but two positive specimens were missed (i.e., false-negatives). In another clinical study using similar conditions, Throm et al. [14] concluded that a 2–3 h incubation period was satisfactory for detecting clinically significant numbers of aerobic organisms in urine specimens.

Cady et al. [15] plotted impedance detection times for 1,133 urine samples against the initial concentration of organisms as determined by plate counts. When the data were examined using a 2.6-h cutoff point, agreement with standard culture methods was 95.8%; there were 22 false-positive and 26 false-negative results. With a 3.5-h cutoff point, agreement was 93.9% with 55 false-positive and 14 false-negative results.

Bacterial Identification. Preliminary data from Hadley [9] and Cady [10] have indicated that the shape of the impedance response may be characteristic for some microorganisms. The responses depend on the culture medium and electrodes used.

Antimicrobial Susceptibility Testing. In the presence of antibiotics, susceptible organisms have prolonged detection times or reduced impedance responses as compared with reactions in the absence of antibiotics. The lowest concentration of antibiotic that prevents either impedance detection or a defined impedance change is the minimal inhibitory concentration (MIC). A study by Colvin and Sherris [16] indicates that the Bactometer can be used for determining organism MIC. Chambers of the instrument were filled with varying dilutions of antibiotic in broth, inoculated, and monitored against organism-free reference chambers. When they compared results of overnight broth dilution tests with overnight impedance readings, agreement within one twofold dilution was 93%. In order to obtain a rapid, 6-h result, they increased the inoculum from 10^6 to 10^7 organisms. Agreement of this rapid test with the standard overnight MIC dilutions ($\pm$ one twofold dilution) was 74%. Most discrepancies occurred with tests of ampicillin, tetracycline, and polymyxin E.

Other Impedance Systems

Various workers have used impedance systems developed in their own laboratories for microbial studies. Ur and Brown [17] detected small inocula of several bacteria and a *Mycoplasma* species within 2 h, and demonstrated the inhibitory effect of antibiotics. Wheeler and Goldschmidt [18] quantified organisms from urine specimens by first filtering out leukocytes and debris in a coarse filter, washing the filtrate, and resuspending the microorganisms in distilled water. Using a nomograph, they related electrical measurements made in their instrument at an input frequency of 10 Hz to measurements made previously with known concentrations of microorganisms. Good correlation with standard laboratory methods was achieved, but the results were available in much shorter time periods. In another impedance study, after routine blood culture broth bottles were inoculated with patient's blood, Kagan et al. [19] removed 30 ml of the broth and subjected it to a lysis-filtration procedure. The filter pad was put into a culture bottle connected by stainless steel electrodes to an impedance measuring device. In a study of 264 blood culture specimens, the impedance method detected 36% more positive cultures than a conventional broth method (49/53 compared with 30/53). More fully automated refinements of this promising system that have been proposed have not yet been achieved.

Advantages and Limitations

As with microcalorimetry, impedance measurements have several applications for the clinical microbiology laboratory, and in most instances, the results are available within short time periods. The Bactometer 32 system has been commercially available; however, no new applications or further development for its clinical use are in progress. The Bactometer 8 and 32 are limited by their small test capacity. For these instruments, uninterrupted on-line monitoring of the culture vessel is an advantage, but the work-load volume is restricted. The Bactometer 120 can handle 480 specimens in four incubators; however, for continuous monitoring, dedicated incubators must be used with attachments that allow the elec-

Table 1. Summary of microcalorimetry and impedance

Characteristic	Microcalorimetry	Impedance
Principle	Measurement of heat generated during microbial growth and metabolism	Measurement of impedance decrease resulting from chemical changes in culture medium when organism metabolizes
Limit of detection (threshold)	10^5 and 10^6 organisms per ml	10^6 and 10^7 organisms per ml
Applications [reference]	Microbial identification [3, 4] Bacteriuria detection [6, 7] Antibiotic susceptibility tests [5]	Blood culture monitoring [11, 12, 19] Bacteriuria detection [13–15, 18] Bacterial identification [9, 10] Antimicrobial susceptibility tests [16]
Advantages	Several clinical laboratory applications Same-day result	Several clinical laboratory applications Same-day result Frequent, continuous monitoring
Limitations	No standardized media or instrumentation No reference thermograms available No automated system available Time-consuming, labor-intensive procedures	Characteristics of microbial growth may delay detection No automated system available Manual systems labor-intensive

trodes to be plugged into the control unit. The prototype incubators were large and required an inordinate amount of laboratory space. Other systems that have been described depend on equipment not usually available in the clinical microbiology laboratory and also require the time and interest of experienced individuals to compile and "debug" the units. Table 1 summarizes the features of both microcalorimetry and impedance.

Conclusions

Microcalorimetry and impedance systems were among the first automated instruments to be developed for the microbiology laboratory. Their instrumentation and applications were described in detail at the first of these international symposia in Stockholm in 1973. These innovative approaches to microbiologic diagnoses rely on principles of microbial growth that have been known for centuries but not previously exploited for routine use. Although initial studies appeared promising, practical refinements of these systems have not materialized. Instead, other instruments that were under development at the same time are now in widespread use throughout the United States. These include the BACTEC 460 system for rapid detection of microorganisms in blood and other sterile body fluids; the AutoMicrobic System (Vitek, Inc., Hazelwood, Missouri), Autobac IDX (General Diagnostics, Morris Plains, New Jersey), and Avantage (Abbott Diagnostics Division, Irving, Texas) for same-day identification and antimicrobial susceptibility tests of bacteria, and urine screening; and instruments such as the Lumac (3M, St. Paul, Minnesota) that detect significant numbers of organisms in urine by bioluminescence measurements.

Use of these instruments has increased the relevance of diagnostic microbiology because a majority of cultures can now be completed and the results reported within 24 h. The current trend is toward even more rapid assays, such as immunologic tests with monoclonal antibodies and DNA probes that detect specific microbial antigens or DNA directly in patient specimens [20]. At present, such areas appear to be more challenging and fruitful avenues for study than continued research with instruments that have failed to become clinically useful despite more than a decade of investigation.

References

1. Levin K (1977) Calorimetry, a time-honored technique with a potential in analytical work and cellular biology. Clin Chem 23:929–937
2. Forrest WW (1972) Microcalorimetry. In: Norris JR, Ribbons DW (eds) Methods in microbiology 6B. Academic, London, pp 286–318
3. Russell WJ, Farling SR, Blanchard GC, Boling EA (1975) Interim review of microbial identification by microcalorimetry. In: Schlessinger D (ed) Microbiology – 1975. American Society for Microbiology, Washington DC, pp 22–31
4. Herman JPM, Jakubczak E, Izard D, Leclerc H (1980) Apport de la microcalorimétrie dans la taxonomie de quelques groupes d'*Enterobacteriaceae*. Can J Microbiol 26:413–419
5. Beezer AE, Bettelheim KA, Al-Salihi S, Shaw EJ (1978) The enumeration of bacteria in culture media and clinical specimens of urine by microcalorimetry. Science Tools 25:6–9
6. Binford JS, Binford LS, Adler P (1973) A semiautomated microcalorimetric method of antibiotic sensitivity testing. Am J Clin Pathol 59:86–94
7. Arhammer M, Mårdh P-A, Ripa T, Andersson K-E (1978) Microcalorimetric study of the effects of cephalexin and cephaloridin on *Escherichia coli* and *Staphylococcus aureus*. Acta Pathol Microbiol Scand [B] 86:59–65
8. Gall LS, Curby WA (1980) Instrumented systems for microbiological analysis of body fluids. CRC, Boca Raton
9. Hadley WK, Senyk G (1975) Early detection of microbial metabolism and growth by measurement of electrical impedance. In: Schlessinger D (ed) Microbiology – 1975. American Society for Microbiology, Washington DC, pp 12–21
10. Cady P, Dufour SW, Shaw K, Kraeger SJ (1978) Electrical impedance measurements: rapid method for detecting and monitoring microorganisms. J Clin Microbiol 7:265–272
11. Specter S, Throm R, Strauss R, Friedman H (1977) Rapid detection of bacterial growth in blood samples by a continuous-monitoring electrical impedance apparatus. J Clin Microbiol 6:489–493
12. Buckland A, Kessock-Philip S, Bascomb S (1983) Early detection of bacterial growth in blood culture by impedance monitoring with a Bactometer model 32. J Clin Pathol 36:823–828
13. Zafari Y, Martin WJ (1977) Comparison of the BACTOMETER microbial monitoring system with conventional methods for detection of microorganisms in urine specimens. J Clin Microbiol 5:545–547
14. Throm R, Specter S, Strauss R, Friedman H (1977) Detection of bacteriuria by automated electrical impedance monitoring in a clinical microbiology laboratory. J Clin Microbiol 6:271–273
15. Cady P, Dufour SW, Lawless P, Nunke B, Kraeger SJ (1978) Impedimetric screening for bacteriuria. J Clin Microbiol 7:273–278
16. Colvin HJ, Sherris JC (1977) Electrical impedance measurements in the reading and monitoring of broth dilution susceptibility tests. Antimicrob Agents Chemother 12:61–66
17. Ur A, Brown DF (1975) Impedance monitoring of bacterial activity. J Med Microbiol 8:19–28
18. Wheeler TG, Goldschmidt MC (1975) Determination of bacterial cell concentrations by electrical measurements. J Clin Microbiol 1:25–29
19. Kagan RL, Schuette WH, Zierdt CH, MacLowry JD (1977) Rapid automated diagnosis of bacteremia by impedance detection. J Clin Microbiol 5:51–57
20. McCarthy LR (1983) The impact of new technology on clinical microbiology. Clin Microbiol Newsl 5:68–72

The Photometric Analysis of Microbial Growth for the Identification and Antimicrobial Susceptibility Testing of Bacterial Pathogens

S.C. EDBERG [1] and S.A. BERGER [2]

Photometric instruments have been used in the microbiology laboratory for over 50 years. Although the ability of light energy to traverse a colloidal suspension of microorganisms would seem to be directly transformable to the number of organisms per a given volume, the physical and mathematical complexity of photometric measurements applied to bacteriology has only recently been delineated. The basic challenge for photometry was to relate the properties of the microbial particle such as its shape, size, and refractive index to the angular distribution of scattered light. Transforming measurements of scattered light into usable information required knowledge of the light's wavelength, amplitude, phase, and polarization emanating in each direction from the scatterer.

The scattering of light occurs when an incident beam of either white or monochromatic light hits a particle. Based on the contour of the area it hits, the size of the particle, the refractive index of the particle, the quality of the surrounding medium, and the position of the collecting aperture, a certain amount of this light will leave the path from which it arose (see Fig. 1). The amount of light that does not reach the photoelectric cell, or is scattered, when the incident beam is at 0° is a measure of turbidity. Therefore, turbidity and light scattering are equivalent when the origin and detector are in a straight line. When the detector and the origin are not in a straight line one encounters not only the scattering of light due to the above-described conditions but exacerbation of some of these factors. Unfortunately, much of the literature does not make a distinction between turbidity and angular light scattering [6].

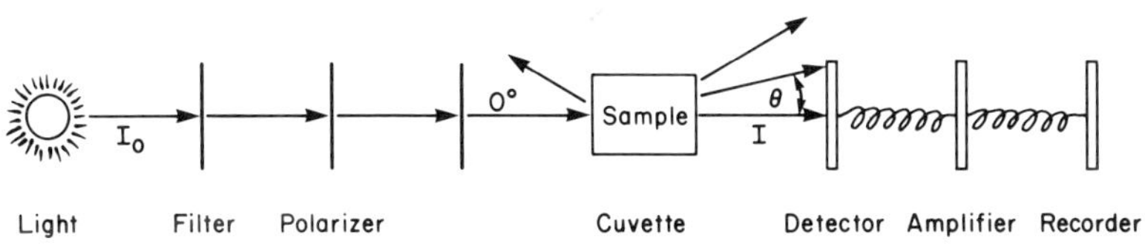

Fig. 1

All photometric instruments are composed of the same basic modules. Each has a light source generator, albeit a laser beam, white light, or light at a single wavelength. Each has a mechanism to columnate the light. Each has some form

1 Department of Laboratory Medicine, Yale University School of Medicine, New Haven, CT 06510, USA

2 Tel Aviv Medical Center, 2 Telalim, Tel Aviv, Israel

of chamber, or cuvette, in which the specimen is placed. Each has a detector. In all instruments the larger the number of particles there are in the light path the smaller the amount of light that reaches the detector. This relationship is geometric because the intensity of unscattered light decreases exponentially as the number of bacteria increase linearly. This is illustrated by considering a cuvette containing a bacterial suspension with a high enough concentration of particles or turbidity to reduce the emerging light by 0.9. If one were to place an identical tube in the path of the emerging light it would also be reduced by 0.9. The amount of light falling on the detector after two successive passages would, therefore, be $0.1 \times 0.1 = 0.01$.

Mathematically, the intensity of unscattered light, I, is equal to the intensity of incident light, Io, multiplied by $10^{(-N/No)}$, where N is the number of bacteria at the completion of a measurement time, t, and No is the number of bacteria at the beginning of the measurement time to. For the sake of simplification this relationship, governed by the parameters of the Beer-Lambert Law, is generally written

$$-\log I/Io = N/No \; .$$

Most instruments have a scale that reads directly $\log Io/I$, which is expressed as optical density (O.D.) or absorbance (A).

In the clinical laboratory we wish to take measurements in the range where a graph of optical density versus cell numbers yields a straight line. At high bacterial concentrations optical density is lower than predicted by the Beer-Lambert formula. This occurs because there is rescattering of light among the colloidal bacteria. Therefore, for every optical instrument a standard calibration curve must be established [1, 5].

Experimentally, bacterial growth may be measured by the increase in bacterial numbers, or mass, over time. During the time the cells are dividing with a constant doubling time the number of microbes increases exponentially. At the starting time, to, there are a certain number of bacteria, No. These cells are in every stage of the division cycle and the increase in numbers will be continuous. If t2 is the average amount of time between doubling, the number of bacteria at any given time is

$$N = (No)(10)^{to/t2} \; ,$$

where to/t2 is the number of doubling times that have occurred.

The number of bacteria at any given time may be calculated from the rearrangement of this equation to $\log N = \log No + to/t2$. This equation is a straight line when $\log N$ is plotted against t [1, 5, 6]. From the slope of this graph one can calculate the amount of time necessary to growth $\times$ fold.

From these parameters one can calculate the amount of time required to double the number of microbes, or the generation time. In order to make quantitative photometric measurements one must utilize an instrument that takes readings at short intervals over at least 6–8 h.

There are several theories governing the relationship between turbidity and the numbers of mass of colloids. For large spheres and modified spheres the Mie theory is most appropriate. For most clinical microbiological uses the Rayleigh-

Debye theory of light scattering is used. Two basic conditions must be met in order for the Rayleigh-Debye theory to be used. First, turbidity must be measured as a function of the ratio of particle size to wavelength with only the maximum turbidity employed. Second, the refractive indexes of the solvent and particle must not be significantly different. In liquid media during log phase growth both conditions are fulfilled. The Rayleigh-Debye theory has the further utility of being valid at both 0° and various angles [1–3].

Many factors must be considered in relating passaged light to numbers of particles in suspension. It is beyond the scope of this work to examine this relationship in detail but the following formula should be noted and used as a basis for evaluating systems:

$$R\theta = \frac{2nV^2\pi^2(m-1)^2P(\theta)}{(\lambda')^4} \times (1+\text{COS}^2\theta),$$

where $R\theta$ equals reduced intensity for unpolarized light scattered at an angle θ from the incident beam, n equals the number of particles per milliliter of suspension, V equals volume of a particle, m equals u'/u_0 where u' is the refractive index of the particle and u is the refractive index of the surrounding medium, lambda is the wavelength of light, and $P(\theta)$ is a function accounting for scattering. The function $P(\theta)$ can be utilized to convert $R(\theta)$ at 0° to an $R(\theta)$ at any given angle [1].

The Rayleigh-Debye theory defines several pertinent physical conditions that must be considered when designing an instrument:

1. Angle of Scattering to the Photocell

After the light has passed through the sample it is scattered in a cone to the photocell (see Fig. 1). The size of a cone may have a significant impact on the efficiency of the photometer and its ability to detect changes in the suspension [2].

2. Size of the Microbe

It is often assumed that the size and refractive index of the population of a given species does not appreciably affect the outcome of the direct relationship between optical density and cell numbers. Within the confines of the Rayleigh-Debye theory this relationship is correct at 0° light scattering, but deviation of linearity may exert profound defects. The larger the angle beta (see Fig. 2), the greater will be the effect of cell size on this relationship [2, 3].

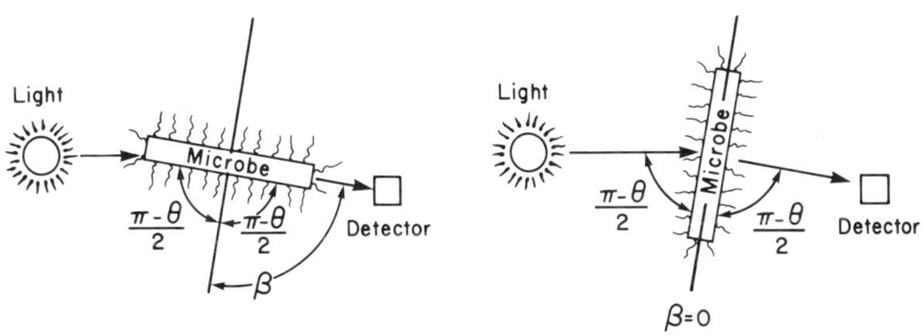

Fig. 2

It is known that young cultures have a larger size distribution of cells than older cultures. Furthermore, the maximum size of cells in cultures is not predictable since each species may exhibit considerable heterogeneity. There are clone to clone differences also. The majority of rapidly growing bacteria reach their maximum size in 4–6 h in log phase growth. It has been noted, for example, that *Staphylococcus aureus* at 4–6 h is twice the size of cells at 24 h and that the bacilli of *Escherichia coli* in a 2-h culture were three times longer than those in a 24-h culture.

The refractive index may also change over time. Younger cells are more opaque than are older cells. Perhaps this reflects a change in the density of ribosomes or some function of the cell membrane.

A small degree of polydispersity does not, however, appreciably alter the intensity of light scattering. One definition of absorbency is independent of particle size, shape, or orientation:

$$A = \frac{32}{18} \times \pi^3 \times \frac{(du/dc)^2}{No} \times \frac{q^2 v}{(\lambda')^4},$$

where du/dc describes change in the refractive index of the particle, q^2 is a measurement of the mass of the particle, v is particles per unit volume, uo is the refractive index at to, and λ is wavelength.

Large particles, such as fungi or bacterial aggregates, require a modification of this formula known as the Jobst approximation which states:

$$A = \frac{9\pi}{12} \times \frac{(du/dc)^2}{\mu o} \times \frac{q^2 v}{r^{2/3}(\lambda')^2},$$

where all symbols are the same as the previously described relationship and r equals the radius of the particle.

In evaluating an instrument one should utilize these relationships with the prime consideration that they apply best at 0° light scattering. As the angle between incident and recorded light increases, the changes in bacterial size and shape become more significant. One should also consider that turbidity is more a function of bacterial mass than bacterial numbers. Since the ratio of mass to numbers is fairly constant during logarithmic growth, for the purposes of microbiological applications this distinction is not necessarily important [3].

3. *Orientation of the Particles*

Bacteria, during their growth cycle, are randomly oriented. Even cocci, when they grow in clusters or chains, may be considered random. If the orientation of the colloidal suspension of microbes becomes polarized a birefringent (for example, under the influence of an electric current or heat differences), the $P(\theta)$ will increase in importance [2, 3].

Utility of Turbidimetric Measurements

Turbidity can, with an accurary at least equal to any of the other commonly utilized methods, determine the numbers of particles in suspension. Why, then, have bacteriologists been reluctant to employ this technique? In many laboratories it is still the custom, even though more cumbersome and not more accurate, to perform pour plates to count bacteria. The major reason lies in the difference between static and kinetic measurements. Until the mid-1970s microbiologists generally utilized single point static measurements to determine microbial numbers. Beginning in the 1970s the introduction of the widely available instruments Autobac (General Diagnostics, Morris Plains, N.J.) and Abbott MS-2 (Abbott Instruments, Dallas, Texas) made semikinetic and kinetic measurements available on a widespread basis. Kinetic measurements provide continuous measurements of microbial turbidity and are able to apply the above-described formulas directly once the microbes enter log phase growth.

Static Measurements

Inaccuracies yielded by static measurements are invariably due to faulty technique or lack of proper calibration of the instrument. One of the most common errors is the vigorous shaking of the cuvette before insertion into the photometer. Small air bubbles are formed in the broth suspension and produce false readings. Small air bubbles rise in the suspension with the large ones, leaving the optical pathway within 10–15 s. Thus, the usual practice of measuring turbidity immediately after shaking contributes a substantial and variable error. Automated instruments have eliminated this problem by determining turbidity over a very short time period and by constantly shaking the contents of the cuvette.

Another problem particular to turbidimetric measurements is the error introduced by flow birefringence. The turbidity of suspensions of rod-shaped bacteria is more difficult to measure accurately than is the turbidity of cocci because of the flow birefringence resulting from the movement of the rods. Small temperature gradients are established that cause the bacteria to flow. Again, automated instruments have obviated this problem by maintaining the solution at a constant temperature.

The instrument utilized can exert significant variations in static photometric measurements. The major variables inherent in photometer design include the size of the cuvette, the shape of the cuvette, the slit width (the width of the slit that columnates the light directed towards the suspension), the nature of light employed (e.g., monochromatic, white, laser), the energy of the light beam directed towards the bacterial suspension, the distance the detector is from the bacterial suspension, the angle of the cone of scattered light from the bacterial suspension to the detector, and the sensitivity of the detector. In most instruments and most static measurement situations deviation from the Beer-Lambert Law generally occurs at an optical density reading of 0.3 or greater and when the wavelength of light is equal to or less than 500 nm. Failure to account for this factor is common [1–5].

The major use of static photometric measurements is in the reading of microtiter trays. American Microscan was the first widely available microtiter tray able to be read automatically utilizing a combination of turbidity plus spectrophotometric color changes. The microtiter tray consisted of a clear plastic sheet containing 96 wells. Each well contained a particular concentration of a given antibiotic or a particular biochemical test. The microbial isolate was inoculated into all wells and the system incubated for 18–24 h. After incubation the microtiter tray was placed on the reader and turbidimetric measurements obtained. A decrease in turbidity in a given well compared with a control well with no antibiotic indicated sensitivity of the organism; no significant change in turbidity compared with a control indicated resistance. The Microtiter System is in widespread use today, with automated readers available from a variety of companies.

Kinetic Measurements

Kinetic measurements are more suited to rapid processing in the Clinical Microbiology Laboratory. By measuring continuously, kinetic measurements eliminate such vagaries as bubble formation, change in temperature, flow birefringence, uncleanliness of cuvettes, and cumbersome manipulation of data. Kinetic instruments can directly electronically analyze and compare the growth in a control chamber versus growth in another chamber under exactly the same conditions in the presence of an agent that may have an effect on the growth of the organism. Most commonly kinetic instruments have been developed for the Clinical Microbiology Laboratory for the determination of antimicrobial susceptibility testing.

Although not in widespread commercial use, the Autoturb (Eli Lilly), designed and developed for the study of antibiotic-producing organisms, first put into practice many of the theoretical and applied principles later adopted by other instruments. This instrument was able to assess quantitatively the antibiotic-producing properties of a particular isolate against a large number of test bacterial pathogens. The machine determined when the test bacterial pathogens were in log phase growth and over a period of 4–6 h sampling measured antibiotic activity.

In the mid-1970s the first commercial instrument for the determination of antimicrobial susceptibility testing for the routine clinical microbiology laboratory was introduced. The Autobac (developed by Pfizer, Groton, Ct. and now part of General Diagnostics, Morris Plains, N.J.) utilized white light with measurements taken at 30° light scattering.

A long plastic cylinder divided into compartments composed the incubation-cuvette system. Each compartment, in effect, was its own cuvette. All compartments except one contained an antibiotic; the last compartment was a growth control to which the turbidity in all other compartments was compared. Turbidity was standardized utilizing a single measurement photometer and the inoculum distributed amongst the compartments in the cylinder. The cylinder was placed on a shaker to insure proper mixing and aeration and incubation was allowed to proceed for 4–6 h. The technologist manually removed the cylinders and inserted them into a reading chamber. The instrument automatically read turbidity and compared it with the growth control. If a sufficient decrease in turbidity occurred, the organism was considered to be sensitive to the particular antibiotic; if there was no major change in turbidity compared with the control, the organism was considered resistant to the antibiotic.

In the late 1970s Abbott Laboratories (Irving, Texas) introduced the first automated kinetic measurement device for the determination of antimicrobial susceptibilities. This machine, developed by Akro-Medic, utilized, like the Autobac, a long cylinder made of plastic with divided sections to act as cuvettes. Unlike the Autobac, there was an upper chamber attached to the bottom cuvettes that served as an initial growth phase to insure that the microbes were in log phase growth before the inoculum came into contact with the antibiotic. The cuvette was inoculated by the technologist and inserted into a combination incubator, shaker, and photometer. The instrument monitored the growth in the upper chamber until log phase was achieved; through a series of vaccuum and pressure changes the inoculum was brought into contact with each antibiotic-containing cuvette. Every few minutes a light-emitting diode (red light) was pulsed through each cuvette and a reading taken. Utilizing these kinetic measurements the instrument could, based on the Rayleigh-Debye Theory, calculate the quantitative sensitivity of the microbe to the given antibiotic. It compared the change in the rate of growth in the antibiotic-containing cuvette with the change in the rate of growth in the control. The MS-2 has continued to evolve and now Abbott is marketing its Quantum system. This system, based on spectrophotometric measurements, can identify Enterobacteriaceae, *Pseudomonas,* and yeasts within 4–6 h of incubation.

Other instruments such as the Vitek (McDonell-Douglas) utilize photometric measurements for the identification, antibiotic sensitivity testing, and counting of microbial pathogens. It is expected that the newer microprocessor will soon eliminate the need for large, central instruments and permit the clinical laboratory greater flexibility for photometric interpretation on a decentralized basis.

References

1. Beattie WH, Jung HC (1968) Determination of concentrations by angular light scattering for the turbidimetric titration of polymers. J Colloid Interface Sci 27:581–590
2. Koch AL (1968) Theory of the angular dependence of light scattered by bacteria and similar-sized biological objects. J Theor Biol 18:133–156
3. Koch AL (1961) Some calculations of the turbidity of mitochondria and bacteria. Biochim Biophys Acta 51:429–441
4. Beattie WH (1965) Turbidimetric titration method for determining solubility distributions of polymers. J Polymer Sci 3:527–549
5. Koch AL, Ehrenfeld E (1968) The size and shape of bacteria by light scattering measurements. Biochim Biophys Acta 165:262–273
6. Kavenagh F (ed) (1972) Analytical microbiology. Acedemic, New York, pp 43–121

The Application of Impediometry to Rapid Antibiotic Susceptibility Testing and Early Detection of Bacterial Growth

M. BAKHTIAR, S. SELWYN, and A. UR [1]

Introduction

Over the last few years there has been considerable interest in the application of impedance measurements for monitoring microbial activity (Ur and Brown 1974). A number of publications have appeared which report that the potential sensitivity of such systems may surpass any of the more classical bacteriological techniques (Bakhtiar and Selwyn 1982). Yet despite these encouraging results the technique until now has not entered into general microbiological use but has instead remained in the domain of scientific curiosity. It is therefore of interest to examine whether this delay has been caused by inherent deficiency in the method or is due to the normal process of transformation of a technical innovation into a commercially available system.

We have recently examined the modern computerised version of the Bactobridge, which is the latest model of equipment developed to apply impediometry to clinical and industrial microbiology. The results and ease of operation tempted us very quickly to extend the tests further in both medical and non-medical microbiology. We report here examples of these applications.

Materials and Methods

Equipment

The Bactobridge model U3 incorporates a Hewlett-Packard computer with integrated printer and screen (Fig. 1) and has an expandable number of channels up to a maximum of 128. In the present series of tests we limited the number of channels to ten. The conductivity cells were tubular, having stoppers at either end, and had a sample volume of 2.2 ml (Fig. 2).

1 Department of Medical Microbiology, Charing Cross and Westminster Medical School, Horseferry Road, London SW1P 2AR, Great Britain

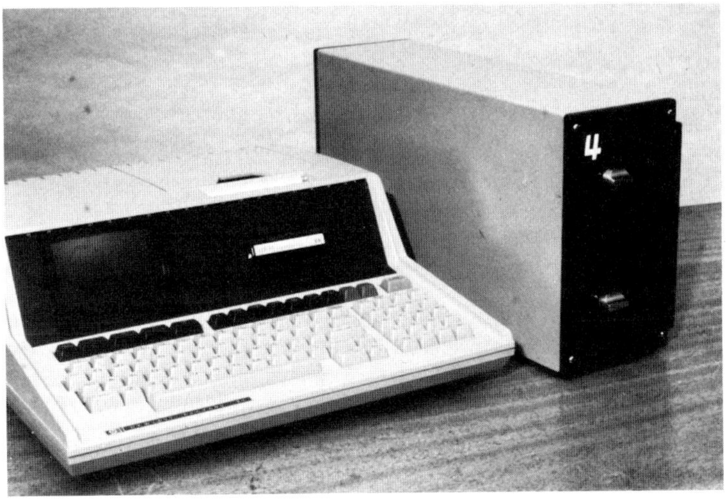

Fig. 1. The U3 Bactobridge, incorporating the Hewlett-Packard computer

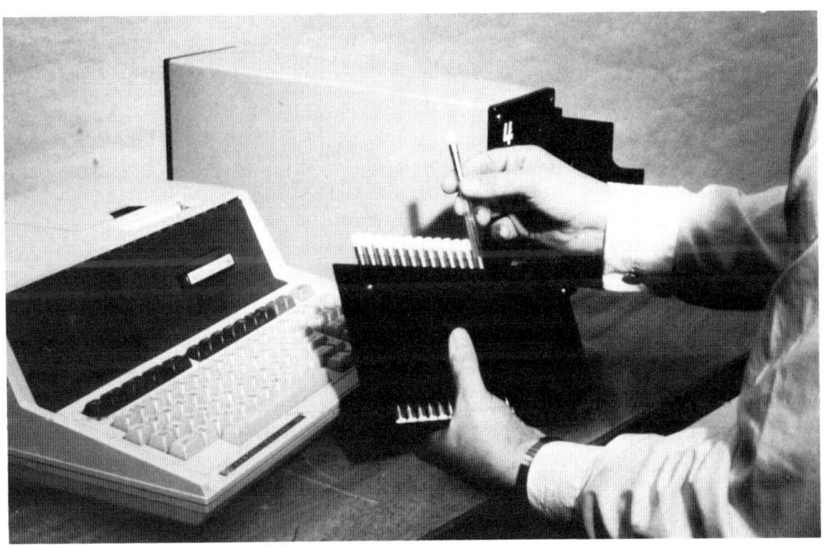

Fig. 2. The Bactobridge reaction and control cells

Micro-Organisms and Media

The standard medium used for the tests was PPLO broth (Oxoid). Other culture media tested were Iso-sensitest broth (Oxoid), and also locally produced nutrient broth, peptone water and glucose broth (Cruickshank et al. 1975). All the bacterial strains used in this study were isolated from clinical specimens received in the Department of Microbiology, Westminster Medical School. The standard strains were obtained from the bacterial collection of this Department.

Test Procedures

Early Detection of Micro-Organisms

Overnight cultures of bacterial growth were diluted in PPLO broth to give a series of inocula from 10^2 to 10^7 colony-forming units per millilitre (CFU/ml). The final preparations were transferred to each of the conductivity cells and the tests were set for 8 h. The growth in each cell was monitored during this period.

Study of Bacterial Growth in Different Liquid Media

Overnight cultures of *Escherichia coli* B86, *Klebsiella aerogenes* K37, *Staphylococcus aureus* ST3, and *Pseudomonas aeruginosa* PS39 were diluted in PPLO broth, nutrient broth, Iso-sensitest broth, glucose broth and peptone water to give an inoculum of 10^5 CFU/ml in each case. These were then transferred to the conductivity cells and the activity related to impedance changes was studied at the different sensitivity levels of the Bactobridge during a period of 8 h.

Detection of Bacteria in Urine

Several urine samples which were received for microbiological investigation in this Department were transferred in 2.2 ml volumes to the conductivity cells. The presence or absence of bacterial growth in each urine sample was studied over a period of 6 h. Aliquots of each sample were diluted 1 : 10 in PPLO broth, and then transferred to the cells. The results were compared with those of undiluted urine.

Antibiotic Susceptibility Tests

Determination of minimum inhibitory concentration (MIC) was carried out using the Bactobridge. A series of dilutions of each antibacterial agent was prepared in PPLO broth, and inocula of the test organisms were added from overnight cultures to give a final population density of 10^5 CFU/ml. These preparations were transferred to the cells in 2.2 ml volumes, and their activities were studied in comparison with the control cell. The MIC was the lowest concentration of the antibiotic associated with no significant impedance change.

Results

Detection of Viable Micro-Organisms

The general sensitivity of the system is shown in the initial studies on the growth of *E. coli* in PPLO broth using six channels (Fig. 3). The curves from each channel are represented with an ordinate of 1% of the impedance change. Channel 1, containing sterile PPLO medium in both the control and the reaction cells, shows stability of $<0.2\%$ of the ordinate over the entire 8 h. Channel 2, having 100 CFU/ml showed an exponential increase of the curve which was clearly evident by 5 h, thus allowing early detection of growth activity in the culture. Channels 3, 4, 5,

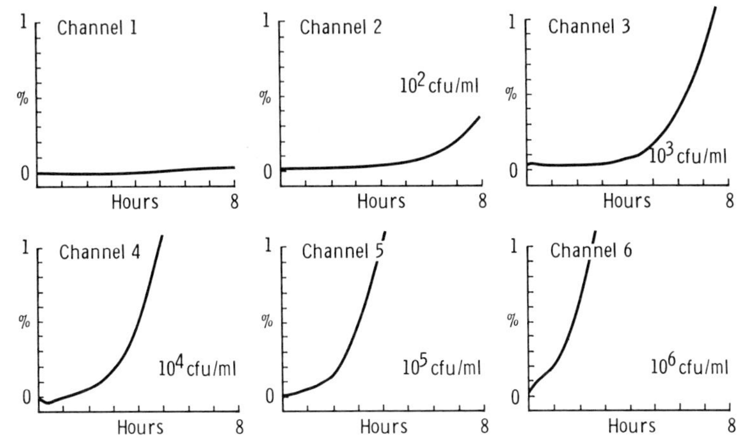

Fig. 3. Activity of different inoculum sizes of *E. coli* in PPLO broth

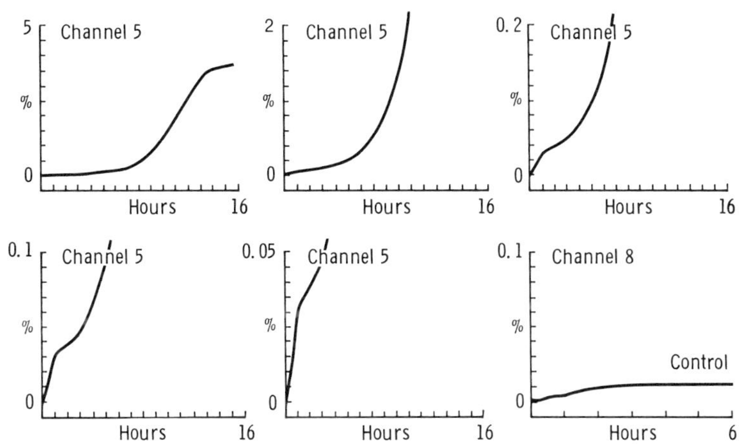

Fig. 4. The activity of 10^5 CFU/ml of *E. coli* in PPLO broth displayed at different amplifications of impedance change (channel 5). The uninoculated control is represented in channel 8 at a sensitivity of 0.1%

and 6 showed similar exponential curves but these were detectable even earlier, at, respectively, 3 h, 1.5 h, 1 h, and 10 min after the test was started. During the first 15 min these curves showed occasional waves of up to 5% of the scale. These did not appear to be artefacts but to reflect metabolic events.

Figure 4 demonstrates a particularly valuable feature of the equipment used – the ease with which the sensitivity can be selected at will. Depending on the choice of ordinates for the display, the curves can show the entire amplitude of the impedance changes, or, when high sensitivity is required, the early part only of the activity is shown in an amplified form. The selection of increasing sensitivity over a 500-fold or even greater scale is eventually limited only by residual "noise." A 100-fold range is represented in Fig. 4, where with an ordinate repre-

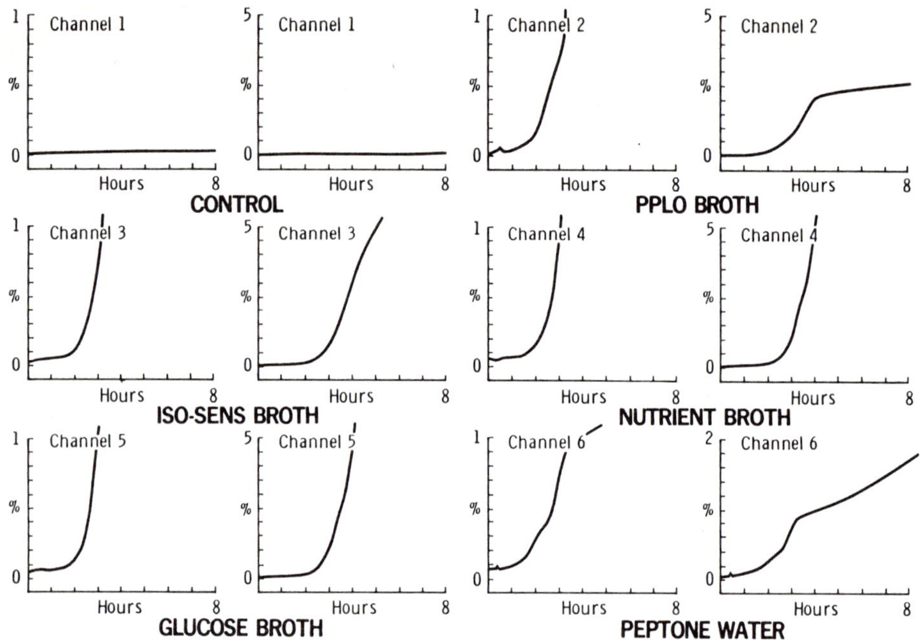

Fig. 5. Activity of 10^5 CFU/ml of *E. coli* in different culture media

senting 5% of impedance change the exponential increase of the curve is not detectable until about 5–6 h; this exponential curve becomes evident progressively earlier, within the first hour with an increase of sensitivity to 0.2%, and almost immediately at 0.1% and 0.05%. The virtual absence of noise is shown in Fig. 4, channel 8 when the activity was displayed at 0.1% impedance change in the control cells.

The results of tests using different culture media are presented in Fig. 5, where the same organism, *E. coli*, was used in a final inoculum in the different media of 10^5 CFU/ml. Channel 1 was used as a general control monitoring the Bactobridge output where the cells contained PPLO medium in the reaction sample against the same medium in the control, and shows no detectable exponential curve at either 1% or 5% sensitivity. Similar growth curves were obtained in each of the media used, although there were individual variations in characteristics such as amplitude and the speed of rise. For example, growth in PPLO broth (channel 2) gave a curve which at 4 h had already changed to a very shallow increase, whilst with nutrient broth the curve was still increasing very steeply at 4 h. However, the first part of these two curves, even when examined at 1%, shows very similar results and detection time. On the other hand comparison of the early part of the curves obtained with peptone water and glucose broth shows that, at 1% sensitivity, the detection of growth in peptone water is noticeable 30 min before it is observed in glucose broth.

The results obtained in tests on four specimens of urine are shown in Fig. 6. The specimen of urine represented in channel 7 produced an exponential curve

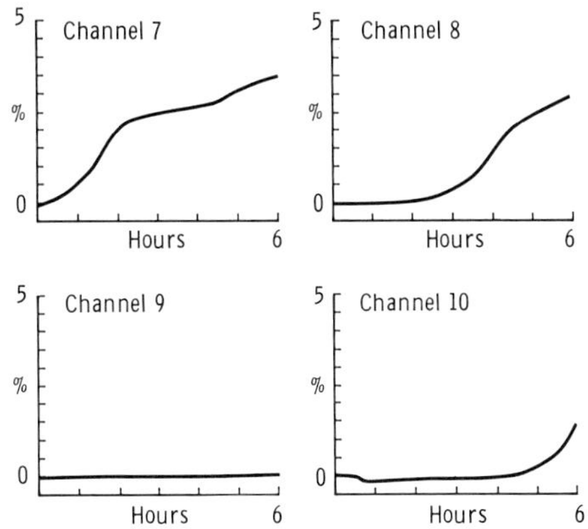

Fig. 6. Detection of significant bacteriuria

detectable as early as 30 min even at 5% display scale. The same urine was subsequently shown by conventional microbiological examination to contain "more than 10^5" CFU/ml *E. coli*. In channel 8 another sample of urine gave an exponential curve detectable only at 2 h. This urine, when checked by the routine microbiological laboratory, was found to contain 10^5 CFU/ml of *E. coli*. The urine sample presented in channel 9 did not produce any exponential curve even after 6 h; after routine examination this urine was reported as "clear." The curve in channel 10 shows an exponential growth curve apparent at 4.5 h, and this urine specimen was subsequently found to have less than 10^3 CFU/ml of mixed organisms on routine examination.

Antibiotic Susceptibility Tests

Representative results of these tests are shown in Fig. 7, where, for ease of comparison, a scale of 3% impedance change has been used. Channel 1 contained the general control with sterile PPLO broth. Channel 2 was the specific control for the antibiotic test where the culture did not contain any antibiotic and the result therefore represents the unhindered growth of the organism. Channels 3, 4, 5, and 6 represent the growth curves of the same inoculum as in channel 2, but in the presence of an increasing concentration of the new quinolone, enoxacin, from 0.01 mg/litre to 0.1 mg/litre. Although the tracing in channel 3 at first glance appears to be a normal growth curve, it does, in fact, show a difference in the growth rate when compared in detail with channel 2. The curve in channel 4, with a drug concentration of 0.02 mg/litre is very similar to the one in channel 3 except for the period after the 4th hour, when the curve in channel 4 climbs less steeply than that in channel 3.

When 0.05 mg/litre of the drug was used (channel 5), considerable flattening of the curve was observed, and after 6 h it had hardly reached the amplitude attained by the culture in channel 2 within 2 h. This inhibitory effect was more pronounced at a drug concentration of 0.1 mg/litre (channel 6). The tracing showed only minimal activity which appeared as a straight line with a slight gradient. As

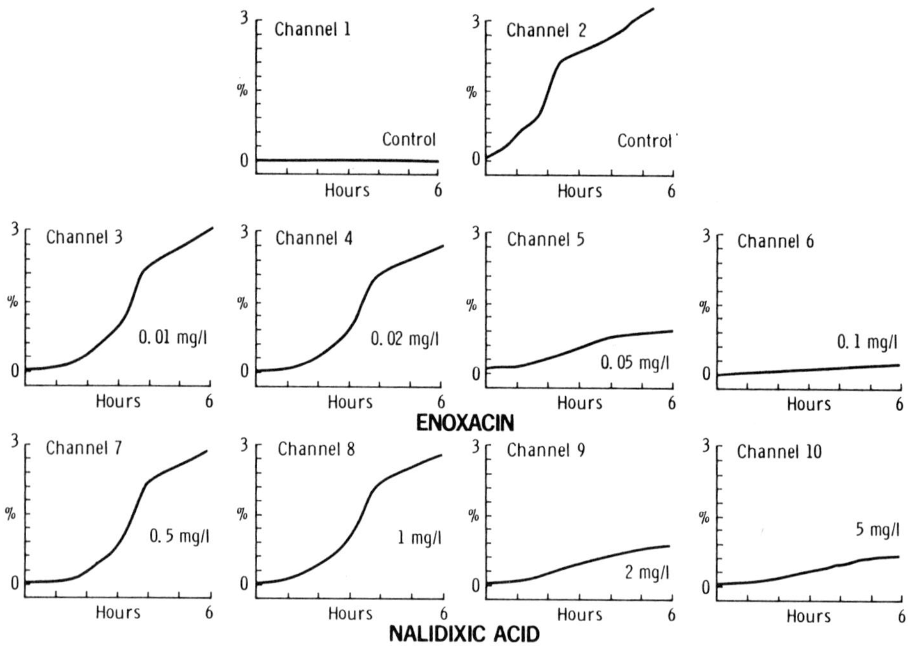

Fig. 7. Determination of MICs of two quinolone derivatives against *E. coli*

there is no growth of organism at this concentration, it is taken to be the MIC for this drug. The curves in channels 7–10 represent a similar test but with increasing concentration of the original quinolone, nalidixic acid, from 0.5 mg/litre in channel 7 up to 5 mg/litre in channel 10. The curves in channels 7 and 8 are very similar to those in channels 3 and 4, showing only a limited effect of the drug on the culture as compared with its uninhibited growth in channel 2. However, in channels 9 and 10 there is a marked flattening of the curve, and indeed in channel 10 the exponential component is hardly evident. The concentration in channel 10 (5 mg/litre) was therefore considered to be the MIC for this drug.

Discussion

Any attempt to replace the established conventional microbiological procedures with new automated or semi-automated methods must be viewed with extreme caution if the simplicity and reliability of these systems are to be preserved. The new highly mechanised and automated techniques should be economical in labour and time, easily monitored, and yet give results which are at least as reliable as the conventional techniques.

Ur and Brown (1974), using a prototype of the Bactobridge, originally described the application of impedance changes to several microbiological investigations. We have extended the possibilities and usefulness of their approach with the aid of a more sophisticated version of this equipment.

The results in Fig. 3 show the reliability of this system in detecting the activity of a small number of organisms in an artificially contaminated sample. The high stability of the system and the suppression of noise enable a detailed study of bacterial growth to be made at different levels of sensitivity over a 100-fold or greater range (Fig. 4). When, however, the sample to be examined has a high level of contamination, exceeding 10^4 CFU/ml, it is usually possible to see evidence of activity even in the first few minutes of the test at 1% impedance change scale.

Although earlier work tended to use PPLO broth as a standard growth medium for this technique, we have explored the effectiveness of other media. The results, in Fig. 5, show that a satisfactory response was obtained with each of the five media tried. There were, however, interesting variations noted in the character of the curves. Some of these differences related to the speed with which the exponential curves were initially detected, but others related to the amplitude of the curves. While all these media gave satisfactory results, it may be useful for specific purposes to exploit the differences observed between media. Thus where the purpose of a test is the early detection of a very small number of organisms it may be useful to select peptone water; conversely, when working with gas-producing organisms, particularly at higher population levels where speed of detection is not crucial, it may be better to avoid this medium. Obviously, the selection of optimal media for specific tests depends not only on the purpose of the test but also on the nature of the organism.

The results in this study often relate to cultures of *E. coli* as this was our standard model. However, the tests were extended to *Klebsiella* species, *Staph. aureus,* and *Pseudomonas* species which gave very similar results.

As a major part of the workload in a clinical microbiological laboratory involves the routine examination of urine, we felt this should be one of the first fields in which to test the Bactobridge. The results shown in Fig. 6 agreed with those obtained by conventional bacteriological tests involving quantitative culture procedure. However, our results were obtained much earlier and would have been available to the clinician within a few hours of collecting the specimens. We believe that by using a 5% sensitivity of impedance change, significant bacteriuria can be detected in a specimen within a few hours – and, indeed, at higher population levels within 30 min. This must be contrasted with conventional microbiological methods for testing urine, the results of which are not usually available for at least 18 h.

The advantages are not limited to the early detection of significant bacteriuria. The method is readily adapted to the performance of direct antibiotic sensitivity tests on specimens of infected urine, and in this way can provide the clinician with comprehensive information for diagnosis and therapy within a very short time.

The rapid determination of bacterial susceptibility to antimicrobial agents should, as in the case of urine examination, greatly ease the burden of the clinical microbiological laboratory, by saving much of the time and labour which is at present expended on conventional tests. In addition to obtaining MICs the equipment can readily be used for break-point determinations. Moreover, the technique can be applied to the problem of antibiotic assays, if a suitable series of controls is used.

Outside the medical field many further applications have already been investigated by us, for example in relation to bacterial contamination of milk and other foods. Clearly, the range of microbiological applications of the Bactobridge is extremely wide, and the equipment seems ideally suited for use in the routine laboratory.

References

Bakhtiar M, Selwyn S (1982) Comparative studies on the bactericidal activities of tetracyclines, chloramphenicol and other "bacteriostatic" antibiotics. In: Periti P, Grassi GG (eds) Current chemotherapy and immunotherapy. Proceedings of 12th International Congress of Chemotherapy, vol 1. The American Society for Microbiology, Washington DC, p 76

Cruickshank R, Duguid JP, Marmion BP, Swain RHA (1975) Medical microbiology: the practice of medical microbiology, 12th edn. Churchill Livingstone, Edinburgh, vol 2

Ur A, Brown DFJ (1974) Rapid detection of bacterial activity using impedance measurement. J Biomed Eng 9:18–20

Chromatographic Analysis for Identification of Microorganisms and Diagnosis of Infection: An Introduction

C. W. Moss [1]

Since its introduction in 1905, chromatography has played an important role in microbiology. Over the years, microbiologists have used chromatographic procedures for isolation, purification, and separation of bacterial components, byproducts, and products from metabolism. Column and paper chromatography were the principal techniques used until the introduction of gas-liquid chromatography (GLC) and thin-layer chromatography (TLC) in the 1950s. The accuracy, speed, sensitivity, and simplicity of GLC and TLC resulted in a major expansion of chromatography to essentially all areas of microbiology.

The concept of using gas-liquid chromatography as an analytical tool for identification of microorganisms was advanced in the early and middle 1960s. Results from early studies of organic acids and other microbial metabolites, cellular fatty acids, and products resulting from the pyrolysis of whole bacterial cells demonstrated the feasibility of this approach and provided a basis for subsequent detailed investigations of a large number of microorganisms. The reliability, precision, and accuracy of recent studies have been improved significantly by continued advances in GLC instrumentation and columns, by introduction of accurate and reliable columns and instruments for high performance liquid chromatography (HPLC), and by a variety of analytical techniques employed in the identification of components separated by chromatographic processes, including mass spectrometry and nuclear magnetic resonance spectroscopy. These combined techniques have also been used for direct analysis of body fluids in order to discover specific chemical markers associated with etiologic agents of infection.

Within the short period of this seminar, only five major areas of chromatographic applications in microbiology will be presented. Four colleagues who have made significant contributions to this field will share their particular experience and findings in the following reports.

1 Centers for Disease Control, Center for Infectious Diseases, Division of Bacterial Diseases, Biochemistry Laboratory, Atlanta, GA 30333, USA

Chromatographic Analysis of Bacterial Cellular Components

C. W. Moss [1]

Introduction

A major application of chromatography in microbiology has been the study of the chemical composition of microorganisms. Data from these studies have demonstrated the value of chemical information for differentiation of closely related groups or species of microorganisms. The chemical classes studied in greatest detail are cellular fatty acids (and other lipids) and respiratory quinones. Methodology and chromatographic techniques for determination of these compounds and examples of their usefulness in taxonomy are discussed in this report.

Cellular Fatty Acids

The lipids of bacteria are found in the cell wall/cell membrane fraction where the fatty acids are bonded to other cellular components. For determination of total fatty acids, the acids must be liberated or cleaved from other cellular components by hydrolysis. Thus, early studies which utilized extraction with lipid solvents did not remove all acids, particularly tightly bound hydroxy fatty acids. The particular hydrolysis procedure used to liberate the fatty acid markedly influences the final results. In a detailed comparison of various hydrolysis procedures, our laboratory found extensive destruction of cyclopropane acids with acid but not with base hydrolysis [13]. The data in Table 1 clearly show the destructive effects of acid hydrolysis on the 19-carbon cyclopropane acid (19 cyc) in *Brucella suis*. With basic hydrolysis, the relative concentration of 19 cyc was approximately 40%, but this was reduced to 0% with 3*N* HCl in methanol and 8% with 3*N* aqueous HCl. In some cases, not all of the hydroxy acids are liberated by basic hydrolysis, but a useful compromise for recovery of most of the fatty acids of bacteria is hydrolysis with 15% NaOH [13]. After hydrolysis, the fatty acids are converted to methyl esters and analyzed by gas liquid chromatography (GLC). GLC is the preferred chromatographic method because of its accuracy, speed, sensitivity, and the excellent separation efficiency which is now obtained with fused-silica capillary columns [19]. As a result of the excellent resolution with capillary columns, identification is accomplished with a high degree of accuracy by comparing retention times of separated methyl esters to those of known standards (Supelco, Inc., Bellefonte, Pa.).

1 Centers for Disease Control, Center for Infectious Diseases, Division of Bacterial Diseases, Biochemistry Laboratory, Atlanta, GA 30333, USA

Table 1. Comparison of hydrolysis procedures on the cellular fatty acid composition of *Brucella suis*

Hydrolysis procedure[a]	Fatty acids as % of total					Total artifacts
	16:0[b]	17:0	18:1	18:0	19 cyc	
5% NaOH	6	3	38	7	42	4
15% NaOH	6	3	37	6	45	3
3*N* HCl*	6	3	27	6	–	57
3*N* HCl	6	3	37	7	8	39

[a] For details of hydrolysis procedures see [13]
[b] See definitions in Table 2

Data from cellular fatty acid analyses have provided valuable information and classification of a number of microorganisms. Our studies with *Legionella* have shown that all members of this genus can be distinguished from other gram-negative bacteria by their cellular fatty acid content [19, 20, 23]. The fatty acid composition of each of the ten named *Legionella* species are shown in Table 2. The data show that the species are qualitatively similar, characteristic features being the presence of branched-chain acids (a-15:0, i-16:0, a-17:0) together with relatively large amounts of unsaturated and monounsaturated 16-carbon acids (16:0, 16:1) and the absence or occurrence of only trace amounts of hydroxy acids. Quantitative differences permit broad grouping of the species. The presence of i-16:0 as the major acid distinguishes *L. pneumophila* and *L. oakridgensis* from other species; these two species can be differentiated by the presence of larger amounts of 18:0 and smaller amounts of i-14:0 and a-15:0 in *L. oakridgensis*. *L. bozemanii*, *L. dumoffii*, *L. jordanis*, *L. gormanii*, and *L. wadsworthii* form another large group and are distinguished from *L. pneumophila* and *L. oakridgensis* by the presence of a-15:0 as the major component and generally larger amounts of a-17:0. *L. micdadei* is quantitatively most similar to the *L. bozemanii* group but differs from all other species by the presence of small amounts (3%) of a-17:1 acid. *L. longbeachae* contains approximately equal amounts of a-15:0, i-16:0, and 16:0, whereas *L. feeleii* contains about equal amounts of a-15:0, i-16:0, 16:1, and 16:0. These data for *Legionella* have proved to be invaluable in the rapid screening of suspect isolates, in defining new serogroups, and in discovering new species. Similarly, useful information has been obtained for other microorganisms by a number of workers [2, 11, 14, 16, 30, 31]. Presently, many laboratories routinely test for cellular fatty acids, and these data are included in most taxonomic studies.

In addition to cellular fatty acids, other lipids which are also easily detected and resolved with chromatographic procedures have proved to be of value in taxonomy. For example, the discovery of high concentrations of sphingophospholipids in the cellular lipids of a group of glucose-nonfermenting, gram-negative rods resulted in designation of *Sphingobacterium* as a new genus [30, 31]. Chromatographic analysis of characteristic lipid components of mycobacteria provide useful information for rapid identification and differentiation of these medically important bacteria [14, 17, 26].

Table 2. Cellular fatty acid composition of named *Legionella* species

Species	Fatty acid composition (%)																	
	i14:0	14:0	a15:0	15:1	15:0	i16:1	i16:0	16:1	16:0	a17:1	a17:0	17cyc	17:0	i18:0	18:1	18:0	19:0	20:0
L. pneumophila	8	T[a]	14	T	1	2	32	13	10	T	11	3	1	T	–[a]	2	1	2
L. oakridgensis	T	T	2	2	1	2	26	18	11	1	5	9	3	2	1	13	2	2
L. bozemanii	4	T	31	T	T	T	17	11	12	T	23	2	T	T	T	T	T	T
L. dumoffii	2	T	26	T	1	T	14	16	9	T	22	5	3	T	1	T	1	T
L. gormanii	5	1	25	2	1	T	20	15	10	T	13	1	3	T	T	2	1	1
L. jordanis	2	T	44	1	T	1	18	6	3	1	20	1	1	–	–	1	T	1
L. adsworthii	1	T	35	1	T	T	10	12	6	–	23	2	13	T	T	2	T	T
L. micdadei	T	T	40	T	1	T	11	10	10	3	24	T	1	–	T	T	T	T
L. longbeachae	3	T	13	2	1	T	19	31	15	T	9	2	2	1	T	3	1	2
L. feeleii	3	1	18	3	3	T	17	17	21	T	6	–	1	T	–	2	1	2

Numbers to the left of the colon refer to the number of carbon atoms; numbers to the right refer to the number of double bonds; i, indicates a methyl branch at the iso carbon atom; a, indicates a methyl branch at the anteiso carbon atom; and cyc refers to a cyclopropane ring.

[a] T, <1%; –, not detected

Isoprenoid Quinones

Isoprenoid quinones are widely distributed in nature and have been shown to be useful for identifying and classifying some bacteria [7, 29]. They are divided into two major structural groups: naphthaquinones, which include menaquinones and demethylmenaquinones, and benzaquinones, which include ubiquinones. Menaquinones (vitamin K_2) and ubiquinones (coenzyme Q) occur most frequently in bacteria and are located in the bacterial plasma membrane, where they function in electron transport. Menaquinones have a 1,4-naphthoquinone ring nucleus, whereas ubiquinones have a 1,4-benzoquinone ring. Both menaquinones (MK) and ubiquinones (QN) have polyprenyl side chains which vary both in the number of isoprene units and in the degree of unsaturation. This structural variation in the polyprenyl side chain is the basis for bacterial classification.

Isoprenoid quinones are soluble in lipid solvents and are extracted directly from microorganisms with diethyl ether, acetone, chloroform, ethanol, hexane, or a combination of these solvents. The isoprenoid quinones are separated from other components in the extraction mixture by thin-layer chromatography (TLC) on silica gel, and the individual quinones are further separated by reverse-phase thin-layer chromatography (RPTLC). Recently, our laboratory along with others have used reverse-phase high-performance liquid chromatography (RPHPLC) to resolve both ubiquinones and menaquinones [5, 6, 21, 27]. The RPHPLC technique has the advantage of speed, sensitivity, resolution, and accurate quantitative measurements. In addition, individual isoprenoid quinones can be collected during the separation with RPHPLC and analyzed by other techniques such as mass spectrometry (MS) and nuclear magnetic resonance spectrometry (NMR).

Most microorganisms contain isoprenoid quinones with fewer than ten isoprene units in the polyprenyl side chain [7, 29]. *Legionella,* however, is a major exception, in that all known species contain major amounts of ubiquinone (QN) with ten or more isoprene units in the side chain [12, 20]. On the basis of qualitative and large quantitative differences in ubiquinone content, the ten named *Legionella* species can be placed into five distinct groups (Table 3). Group 1 contains only *L. pneumophila,* which is characterized by major amounts (designated by 4) of Q12 with smaller amounts (designated by 1) of Q11 and Q13. The four species in group 2 generally contain about equal amounts of Q9, Q10, Q11, and Q12, with only small amounts of Q13. The two species in group 3 contain Q10 as the major component, moderate amounts (designated by 2) of Q9 and Q10, and only small amounts of Q12. *L. micdadei,* and *L. jordanis,* the two species in group 4, contain about equal amounts of Q12 and Q13, small amounts of Q11, but none (or only traces) of Q9, Q10, or Q14. *L. feeleii* (group 5) differs from all other *Legionella* by the presence of Q14, with Q13 as the major component, and the absence (or only traces) of Q9, Q10, Q11, and Q12. Thus, both ubiquinones and cellular fatty acids provide useful chemical markers for rapid identification of *Legionella.*

The usefulness of isoprenoid quinone data in taxonomy will increase with advances in separating techniques and procedures. For example, Carlone and Anet [5] using RPHPLC recently discovered a methyl-substituted menaquinone-6 (2[5 or 8]-dimethyl-3-farnesyl-farnesyl-1,4-naphthoquinone) in *Campylobacter*, sub-

Table 3. Grouping of named *Legionella* by ubiquinone (QN) content

Group	Ubiquinone						
	Q9	Q10	Q11	Q12	Q13	Q14	Species
1	–	–	1	4	1	–	*L. pneumophila*
2	4	4	4	4	1	–	*L. bozemanii* *L. longbeachae* *L. dumoffii* *L. gormanii*
3	2	4	2	1	–	–	*L. oakridgensis* *L. wadsworthii*
4	–	–	1	3	4	–	*L. micdadei* *L. jordanis*
5	–	–	–	–	4	2	*L. feeleii*

Numbers refer to relative amounts of ubiquinones with the major component designated by 4. 1, 2, and 3 refer to peaks with 25%, 50%, and 75% of the area of a peak designated 4, respectively

sequent studies in this laboratory have shown that this component is present in all *Campylobacter* species and that it appears to be a useful marker of the genus [22]. Recently, we have used supercritical fluid chromatography to resolve each component of a mixture containing menaquinones MK3–MK10 and ubiquinones Q6–Q10 in a single chromatographic run (D. Gere and C. W. Moss, 1984, unpublished). Using this procedure and other advancements in chromatography, it is likely that other unique marker quinones will be discovered.

Carbohydrates

Compared to studies on cellular fatty acids and quinones, there have been relatively few studies to evaluate the value of carbohydrate composition for taxonomy of microorganisms. Reasons for these limited studies include one or more of the following technical considerations:

(a) difficulties in developing rapid and simple hydrolysis procedures with little or no destruction or conversion of sugars during the hydrolysis procedure;
(b) lengthy and detailed derivatization procedures prior to analysis by GLC;
(c) the formation of multiple derivatives from a single sugar during derivatization, leading to multiple peaks upon subsequent GLC analysis; and
(d) Difficulty in identification and quantitation of individual sugar components due to multiple peaks and peaks which often overlap with other sugars in a complex biological mixture.

Despite these difficulties, data from several studies have demonstrated the value of carbohydrate analysis in distinguishing various streptococci [1, 24], nocardiae [15], mycobacteria [2], and other microorganisms [28]. However, extensive application of carbohydrate data for taxonomic purposes will require the development of much simpler sample preparation and derivatization techniques as well as improved chromatographic procedures.

Proteins, Peptides, and Amino Acids

Chromatographic processes have been used extensively for isolation, purification, and identification of proteins, peptides, and amino acids of microorganisms. The major application for these compounds in chemotaxonomy has been in comparing the amino acid composition of isolated cell walls [3, 4]. Recently, however, there has been considerable interest in evaluating the chemical relatedness of microorganisms by comparing their soluble cellular protein patterns after electrophoresis on polyacrylamide gels. These electrophoretic protein profiles have provided additional valuable information in a number of taxonomic studies [8, 10, 18].

The continued use of chromatography in microbiology is assured by developments in the field of biotechnology. The central component in molecular biology is the individual microorganism through which recombinant DNA techniques are applied. The ultimate product of these gene manipulations is a polypeptide or protein which must be isolated and characterized by procedures such as ultracentrifugation, electrophoresis, and gravity-fed column chromatography. These important techniques, which continue to undergo refinement, are now supplemented and often supplanted by HPLC. HPLC encompasses a wide variety of separation modes (i.e., reversed-phase, size-exclusion, ion-exchange) and detection methods (i.e., absorbance, fluorescence, electrochemical, post-column reaction) which provide speed, simplicity, and versatility in solving various separation problems [25]. As the field of biotechnology continues to advance, so will the application of HPLC and other techniques in separation science.

References

1. Aluyi HS, Drucker DB (1983) Trimethylsilyl-sugar profiles of *Streptococcus milleri* and *Streptococcus mitis*. J Appl Bacteriol 54:391–397
2. Alvin C, Larsson L, Magnusson M, Mardh PA, Odham G, Westerdahl G (1983) Determination of fatty acids and carbohydrate monomers in micro-organisms by glass capillary gas chromatography: analysis of *Mycobacterium gordonae* and *Mycobacterium scrofulaceum*. J Gen Microbiol 129:401–405
3. Becker B, Lechevalier MP, Gordon RE, Lechevalier HA (1964) Rapid differentiation between *Nocardia* and *Streptomyces* by paper chromatography of whole-cell hydrolysates. Appl Microbiol 12:421–423
4. Berd D (1973) Laboratory identification of clinically important aerobic actinomycetes. Appl Microbiol 25:665–681
5. Carlone GM, Anet FAL (1983) Detection of menaquinone-6 and a novel methyl-substituted menaquinone-6 in *Campylobacter jejuni* and *Campylobacter fetus* subsp *fetus*. J Gen Microbiol 129:3385–3393
6. Collins MD (1982) A note on the separation of natural mixtures of bacterial menaquinones using reverse-phase high-performance liquid chromatography. J Appl Bacteriol 52:457–460
7. Collins MD, Jones D (1981) Distribution of isoprenoid quinone structural types in bacteria and their taxonomic implications. Microbiol Rev 45:316–354
8. Ferragut C, Izard D, Gavini F, Kersters K, Delay J, Leclerc H (1983) *Klebsiella trevisanii*: a new species from water and soil. Int J Syst Bacteriol 33:133–142
9. Fox A, Morgan SL, Hudson JR, Zhu ZT, Lau PY (1983) Capillary gas chromatographic analysis of alditol acetates of neutral and amino sugars in bacterial cell walls. J Chromatogr 256:429–438

10. Hanna J, Neill SD, O'Brien JJ, Ellis WA (1983) Comparison of aerotolerant and reference strains of *Campylobacter* species by polyacrylamide gel electrophoresis. Int J Syst Bacteriol 33:143–146
11. Jantzen E, Berdal BP, Omland T (1979) Cellular fatty acid composition of *Francisella tularensis*. J Clin Microbiol 10:928–930
12. Karr DE, Bibb WF, Moss CW (1982) Isoprenoid quinones of the genus *Legionella*. J Clin Microbiol 15:1044–1048
13. Lambert MA, Moss CW (1983) Comparison of the effects of acid and base hydrolyses on hydroxy and cyclopropane fatty acids in bacteria. J Clin Microbiol 18:1370–1377
14. Lechevalier MP (1982) Lipids in bacterial taxonomy. In: Laskin AI, Lechevalier HA (eds) CRC handbook of microbiology, 2nd edn. CRC, Boca Raton, pp 436–508
15. Lechevalier MP, Lechevalier H, Horan AC (1973) Chemical characteristics and classification of nocardiae. Can J Microbiol 19:965–972
16. Mayberry WR, Lambe DW, Ferguson KP (1982) Identification of *Bacteroides* species by cellular fatty acid profiles. Int J Syst Bacteriol 32:21–27
17. Minnikin DE, Goodfellow M (1980) Lipid composition in the classification and identification of acid-fast bacteria. In: Goodfellow M, Board RG (eds) Microbiological classification and identification. Academic, London, p 189
18. Moore WEC, Hash DE, Holdeman LV, Cato EP (1980) Polyacrylamide slab gel electrophoresis of soluble proteins for studies of bacterial floras. Appl Environ Microbiol 39:900–907
19. Moss CW (1981) Gas-liquid chromatography as an analytical tool in microbiology. J Chromatogr 203:337–347
20. Moss CW, Bibb WF, Karr DE, Guerrant GO, Lambert MA (1983) Cellular fatty acid composition and ubiquinone content of *Legionella feeleii* sp. nov. J Clin Microbiol 18:917–919
21. Moss CW, Guerrant GO (1983) Separation of bacterial ubiquinones by reverse-phase high performance liquid chromatography. J Clin Microbiol 18:15–17
22. Moss CW, Kai A, Lambert MA, Patton C (1984) Isoprenoid quinone content and cellular fatty acid composition of *Campylobacter*. J Clin Microbiol 19:772–776
23. Moss CW, Weaver RE, Dees SB, Cherry WB (1977) Cellular fatty acid composition of isolates from Legionnaires' disease. J Clin Microbiol 6:140–143
24. Prichard DG, Coligan JE, Speed SE, Gray BM (1981) Carbohydrate fingerprints of streptococcal cells. J Clin Microbiol 13:89–92
25. Regnier FE (1983) HPLC of proteins, peptides, and polynucleotides. Anal Chem 55:1299A–1306A
26. Shaw N (1974) Lipid composition as a guide to the classification of bacteria. Adv Appl Microbiol 17:63–108
27. Tamoaka J, Katayoma-Fujimura Y, Kuraishi H (1983) Analysis of bacterial menaquinone mixtures by high performance liquid chromatography. J Appl Bacteriol 54:31–36
28. Wilkinson SG, Carby KA (1971) Amino sugars in the cell walls of *Pseudomonas* species. J Gen Microbiol 66:221–227
29. Yamada Y, Inouye G, Tahara Y, Kondo K (1976) The menaquinone system in the classification of coryneform and nocardioform bacteria and related organisms. J Gen Appl Microbiol 22:203–214
30. Yabuuchi E, Moss CW (1982) Cellular fatty acid composition of strains of three species of *Sphingobacterium* gen. nov. and *Cytophaga johnsonae*. FEMS Microbiol Lett 13:87–91
31. Yabuuchi E, Kaneko T, Yano I, Moss CW, Miyoshi M (1983) *Sphingobacterium* gen. nov., *Sphingobacterium spiritivorum* comb. nov., *Sphingobacterium multivorum* comb. nov., *Sphingobacterium mizutae* sp. nov., and *Flavobacterium indologenes* sp. nov.: glucose-nonfermenting gram-negative rods in CDC groups IIk-2 and IIb. Int J Syst Bacteriol 33:580–598

Analysis of Bacterial Metabolites and Components by Computerized GC/MS System – From Shorter Chain Acids to Very Long-Chain Compounds up to C_{80}

I. YANO [1]

Introduction

Recent advances in analytical chemistry have enabled us to obtain a good deal of information on the structure of organic compounds synthesized by microorganisms which may contribute to the taxonomy and physiology. Mass spectrometry is a very useful technique and, when coupled with the resolution power of gas chromatography, provides the most precise information on the structure of small quantities of bacterial metabolites or cellular constituents. We have mainly developed the analytical method for shorter chain compounds, such as fermentation products – short-chain acids, alcohols, and amines – and for longer ones, such as long-chain fatty acids, long-chain alcohols, long-chain bases, ceramides, and diglycerides derived from membrane lipids. Furthermore, we have tried to establish the analytical method for "myocolic acid," which is the most characteristic, long-chain fatty acid in acid-fast bacteria and related taxa.

We introduce a new technique for the classification or identification of bacteria, mainly using gas chromatography–mass spectrometry (GC/MS) computer systems.

Sphingolipid Analysis in *Sphingobacterium*

Sphingolipids are known to be one of the important components of membrane lipids in higher animals and plants. In contrast, they are extremely rare in prokaryotic cells and are reported to occur in only limited genera of bacteria. Previously, we found that the cellular lipids of *Flavobacterium devorans* ATCC 10829, which belongs to the Pseudomonas like group IIK-type 1, contains a unique sphingoglycolipid, ceramide glucuronic acid [9]. Group IIK-type 2 strains, very much like group IIK-type 1, have also been characterized by the presence of sphingophospholipids [10]. On the basis of the presence of this unique compound and numerical taxonomical studies of 92 phenotypic characters, Yabuuchi et al. [8] placed the group IIK-type 2 organisms into a new genus and proposed the name *Sphingobacterium* gen. nov. Since *S. spiritivorum* KM2138 was designated as the nomenclatural type for the species and three biovars are regarded as composing three separate species [biovar 1, *S. spiritivorum* (including

1 Niigata University School of Medicine, Niigata 951, Japan

Table 1. Long-chain base composition of the total lipids from 16 strains of *Sphingobacterium* species. Reprinted from [11]

Long-chain base	*S. spiritivorum*							*S. multivorum*					*S. mizutae*			
	2134	2138	2148	3101	3102	3103	3104	926	2087	2532	2812	2813	1203	2055	2274	2289
d-16:1 (?)	tr	–	–	tr	tr	tr	tr	tr	tr	tr	tr	tr	tr	tr	tr	tr
d-16:0	4.0	4.0	3.6	6.5	5.7	5.5	5.2	5.5	3.8	5.7	5.2	14.6	2.1	2.3	2.9	1.9
d-17:Obr	84.0	72.8	85.1	81.5	83.3	83.9	86.1	67.3	74.5	49.4	66.7	69.3	86.3	87.4	73.0	76.4
d-17:0	6.3	18.4	1.8	5.0	3.3	5.5	1.5	9.1	8.3	20.6	7.5	2.9	5.3	6.9	5.0	17.8
d-18:1 (?)	tr	tr	tr	tr	tr	tr	tr	tr	tr	tr	tr	2.9	tr	tr	tr	tr
d-18:0	5.7	4.8	9.5	7.0	7.7	5.1	7.2	14.5	13.4	15.8	20.7	10.2	4.1	3.4	16.8	3.8
d-19:Obr	tr	–	tr	tr	tr	tr	tr	tr	tr	6.5	tr	tr	tr	tr	tr	tr
d-20:0	tr	tr	tr	tr	tr	tr	tr	3.6	tr	1.0	tr	tr	2.1	tr	2.2	tr

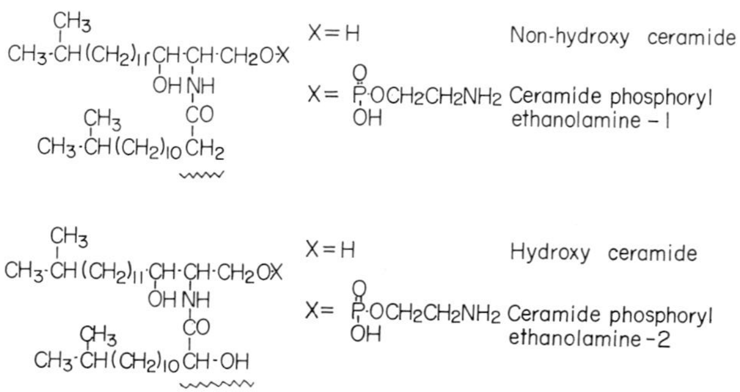

Fig. 1. Structure of ceramides and ceramide phosphorylethanolamines in *Sphingobacterium* species

KM2138); biovar 2, *S. multivorum* (including KM926), and biovar 3, *S. mizutae* (including KM1203)], we examined the basic structure of sphingophospholipids in these bacteria. As a result, it was revealed that the long-chain base composition of cellular lipids of 16 representative strains, including *S. spiritivorum, S. multivorum,* and *S. mizutae,* were very similar to each other, and the most abundant long-chain base was branched-chain C_{17} sphinganine (most probably 15-methylhaxadecasphinganine), usually with lesser amounts of C_{16} and C_{18} homologs, (Table 1, Fig. 1). Furthermore, from the GC/MS analysis of intact ceramides, the major molecular species of the ceramides were identified as LCB-d-iso 17:0-20H iso 15:0FA, LCB-d-iso 17:0-iso 15:0FA, and LCB-d-n 16:0-iso 15:0FA [11, 12].

Analysis of α-Oxidation Products

α-oxidation of long-chain fatty acids has been established in higher plants, mammalian tissues, and yeasts. However, very little has been determined about the operation of the α-oxidation system in bacterial cells. Previously, we found that 2-hydroxy fatty acids were distributed in certain strains belonging to the Actinomycetales [13, 14]. We reported that palmitic acid was actively converted to 2-hydroxy palmitic acid, which was then converted to pentadecanoic acid by the resting cell suspension and by the cell-free extracts of *Arthrobacter simplex* [15–17]. Recently, many reports have appeared on the occurrence of 2-hydroxy fatty acids in gram-negative bacterial lipids. Another unique aspect of *Sphingobacterium,* which is a gram-negative, glucose nonfermentative (absolutely aerobic) rod, is that it possesses 2-hydroxy 13-methyltetradecanoic (isopentadecanoic) acid as the major hydroxy fatty acid. Since the concentration of 2-hydroxy 13-methyltetradecanoic acid in cellular lipid is extremely high (up to 30% of the total fatty acids), we examined the operation of the α-oxidation system of long-chain fatty acids using ^{14}C- or stable isotope-labeled precursors.

It was demonstrated that 1-^{14}C-13-methyltetradecanoic acid (iso 15:0) was converted rapidly and specifically to 2-hydroxy 13-methyltetradecanoic acid (2-OH iso 15:0) in both the growing and the resting cells under the aerobic conditions, which is evidence of 2-hydroxylation of long-chain fatty acids. On the other hand, 1-^{14}C-acetate was widely incorporated into major fatty acids of this bacterium: palmitic, hexadecenoic, 13-methyltetradecanoic, 2-hydroxy 13-methyltetradecanoic, 3-hydroxy palmitic, and 3-hydroxy 15-methylhexadecanoic acids. Direct conversion of exogenous nonhydroxy fatty acid to 2-hydroxy fatty acid was also demonstrated by the incubation of nonlabeled or 14d_3-myristic acid, and the resultant 2-hydroxy or 14d_3-2-hydroxy myristic acid was identified by GC/MS. The enrichment of deuterium atoms was also estimated by mass fragmentography. To obtain direct evidence for aerobic 2-hydroxylation of fatty acids, the bacterial suspension was incubated in an $^{18}O_2$ atmosphere instead of $^{16}O_2$ for 1–4 h, and the resultant fatty acids were analyzed by GC/MS and mass fragmentography.

The results showed that the ^{18}O atom was incorporated specifically as the oxygen atom of the hydroxyl group at the 2 position of 2-hydroxy fatty acid(s)

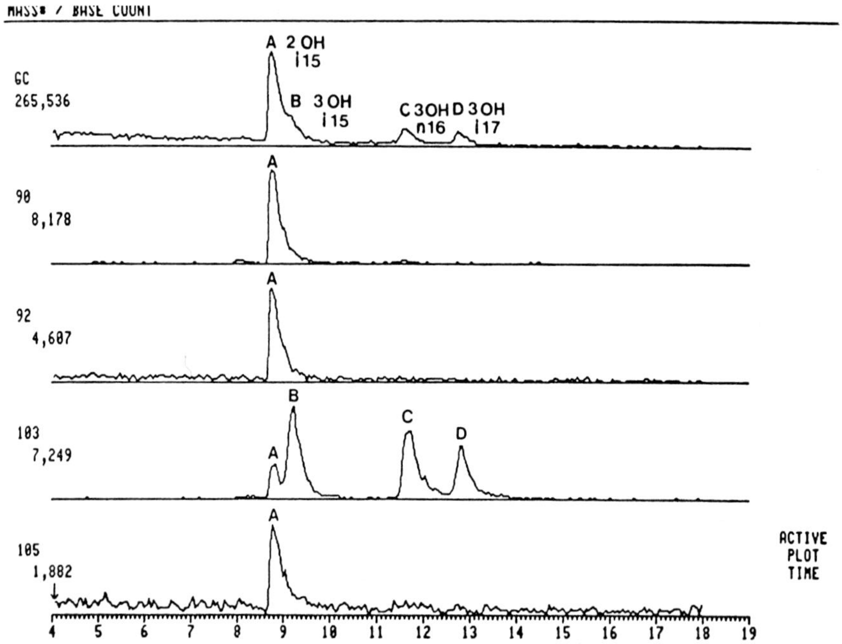

Fig. 2. Mass fragmentograms of ^{18}O-labeled hydroxy fatty acid methyl esters in *Sphingobacterium* after incubation in $^{18}O_2$

Table 2. ^{18}O-Incorporation into 2- or 3-hydroxy fatty acids

	Peak		$(M)^+$	$(M+1)^+$	(M-59)	M-(H_2O)	103	90
A	(2-OH-i15)	EI	272		213	254	103	90
			274		215	(no 256)	105	92
		CI		273	213	255		
				275	215	(no 257)		
B	(3-OH-i15)	EI	272				103	None
			(no 274)				(no 105)	
		CI		273		255		
				(no 275)		(no 257)		
C	(3-OH-n16)	EI	286				103	None
			(no 288)				(no 105)	
		CI		287		269		
				(no 289)		(no 271)		
D	(3-OH-i17)	EI	300				103	None
			(no 302)				(no 105)	
		CI		301		283		
				(no 303)		(no 285)		

and not into 3-hydroxy or nonhydroxy fatty acids (Fig. 2, Table 2). These results indicate that the α-oxidation system in this bacterium proceeds exclusively by an oxygenase-type reaction.

Determination of Molecular Species Composition of Mycolic Acids by GC/MS Computer System

Mycolic acids are a series of high molecular weight 3-hydroxy fatty acids with a long-chain alkyl branch at the 2 position. Although mycolic acid was first isolated from *Mycobacterium tuberculosis* in 1938, these types of fatty acid are now known to be widely distributed in *Mycobacterium, Nocardia, Corynebacteria,* and related taxa. Since their structure varies widely depending on the bacterial genus and species or environmental conditions, the development of appropriate analytical method was considered essential. However, until recently, attempts at the gas chromatographic analysis of mycolic acids have met with only limited success because of the instability of these fatty acids at high temperatures. We have tried and established the gas chromatographic separation and mass spectrometric identification of mycolic acids ranging from C_{20} to C_{80} in *Mycobacteria, Nocardia, Rhodococcus, Corynebacteria, Bacterionema* and related groups [18–20, 6, 7, 3, 4]. Using this method, the molecular species of mycolic acids were clearly separated as the trimethylsilyl ether derivatives of the methyl esters, according to their total carbon numbers. In most rapidly growing *Mycobacteria,* such as *M. smegmatis, M. phlei, M. chelonei, M. fortuitum, M. pelegrinum,* the major species of α-mycolic acids (M_1) were found to be C_{72}, C_{74}, C_{76}, C_{77}, C_{78}, and C_{80} dienoic (or dicyclopropanoic) monocarboxylic acids, centering at C_{76} or C_{78}, while in *M. vaccae, M. diernhoferi, M. parafortuitum, M. aurum, M. rhodesiae,* and *M. gilvum,* they were found to be C_{68}, C_{70}, C_{72}, C_{74}, and C_{76}, centering at C_{72} or C_{74}, both of which possessed a saturated C_{20} or C_{22} alkyl branch at the 2 position (Figs. 3–5). Furthermore, α'-mycolic acid (M'_1) composition also differed distinctively according to the species of rapidly growing *Mycobacteria: M. smegmatis* and *M. chelonei* possessed $C_{62:1}$ and $C_{64:1}$ most abundantly, *M. vaccae* and *M. parafortuitum* $C_{58:1}$ and $C_{60:1}$. It was noted that α'-mycolic acids of *M. fortuitum* consisted of $C_{66:2}$ and $C_{68:2}$.

Conversely, the slowly growing species, such as *M. tuberculosis, M. bovis,* and the *M. avium-intracellulare* (MAIS) complex, possessed C_{78}, C_{80}, C_{82}, and C_{84} dienoic α-mycolic acids, with a considerable amount of odd-numbered carbon mycolic acids, centering at C_{80}. The dicarboxy mycolic acid composition, which is derived from ester mycolic acid and is shorter than the corresponding α-mycolic acids, was also characteristic for each species of *Mycobacteria:* The MAIS complex possessed the longest homologs such as C_{65}, C_{67}, and C_{69} dicarboxylic acids, while *M. phlei* had C_{58}, C_{60}, and C_{62} as the major components. Other subclasses of mycolic acids obtained from *Mycobacteria,* such as dihydroxy, epoxy, methoxy, and monoketo monohydroxy acids, were also separated gas chromatographically and identified mass spectrometrically.

Based on the molecular species composition, the average carbon numbers of the mycolic acids from 16 strains of *Nocardia* and related taxa (*Rhodococcus* and

A
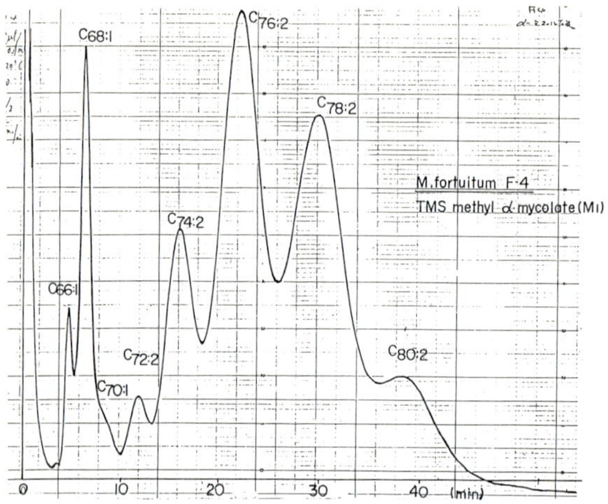

Fig. 3. Gas chromatograms of TMS methyl mycolates from *M. fortuitum* **A** and *M. avium-intracellulare* complex **B**

B
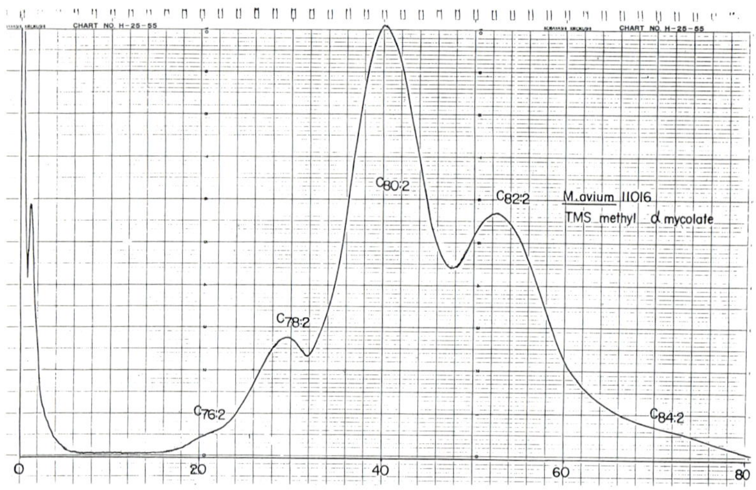

the *Gordona* complex) were determined. Each species of bacterium was demonstrated to possess a characteristic profile of mycolic acid composition, and the species were shown to be classified approximately into at least five groups: (1) C_{28-48} (average carbon numbers C_{34-37}), mostly saturated for *Rhodococcus erythropolis* and several *Rhodococcus* species; (2) C_{34-50} (average C_{43-45}) monoenoic and dienoic for *R. rubrum*, *R. luteum*, and related species; (3) C_{44-58} (average C_{51-52}) monoenoic to tetraenoic for *N. asteroides* and "true" nocardiae; (4) C_{52-60} (average C_{57-58}) monoenoic to pentaenoic for *R. terrae*; (5) C_{56-68} (average C_{62-64}) monoenoic to hexaenoic for *R. bronchialis*, *R. sputi*, and *N. polychromogenes* [13]. The most unusual mycolic acids were those from *Gordona (R) aurantiaca*, a psychrophilic acid-fast bacterium isolated initially from a patient with a pulmonary infection [2, 1]. The mycolic acids from *G. aurantiaca* were extremely complex and consisted of 30 or more molecular species differing in numbers of carbon atoms and double bonds [5]. The total carbon atoms of my-

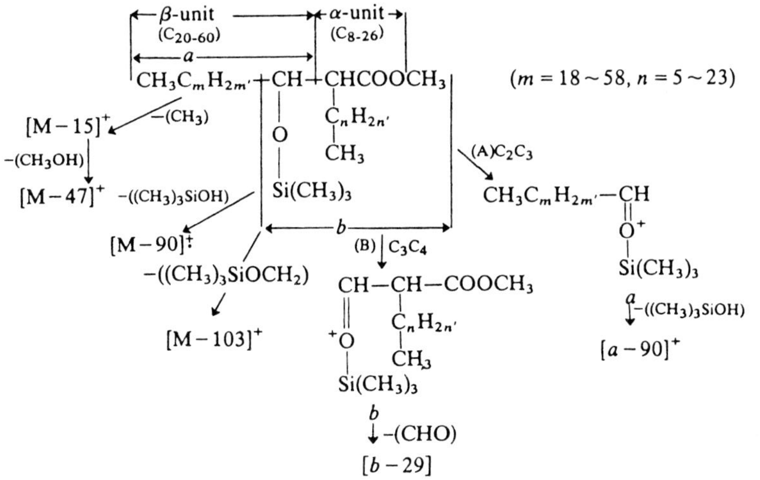

Fig. 4. Mass fragmentation of TMS methyl mycolate in EI mass spectrometry

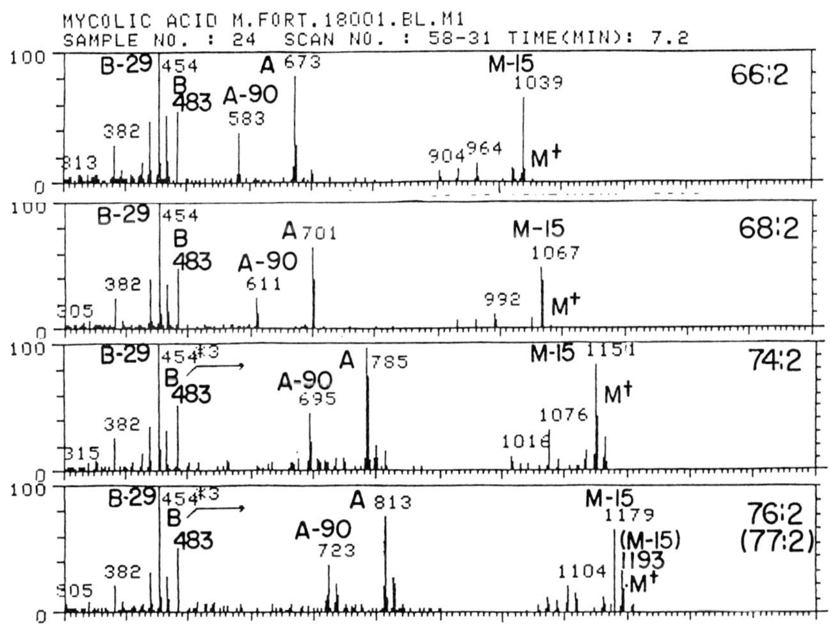

Fig. 5. Mass spectra of TMS-methyl mycolate from *M. fortuitum*

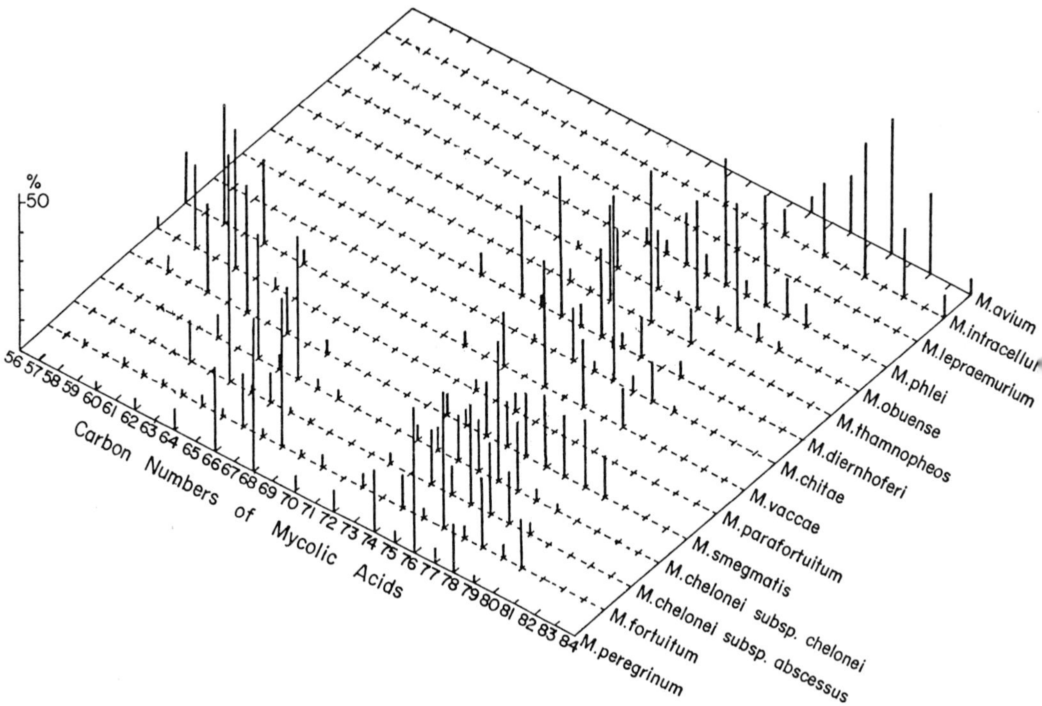

Fig. 6. Molecular species composition of α-mycolic acids (M + M1′) in rapidly or slowly growing *Mycobacteria*

colic acids ranged C_{62-78}, and the number of double bonds from one to seven. According to the number of carbon atoms in the mycolic acids, *G. aurantiaca* is close to *Mycobacteria*; from the number of double bonds, it is close to *Nocardia* and *Rhodococci*. From the structure of the α-branch, *G. aurantiaca* appeared to be situated at an intermediate position essentially between *Mycobacteria* and *Nocardia*. The molecular species composition of α-mycolic acids in *Mycobacteria* is demonstrated in Fig. 6.

These results indicate that structural determination of mycolic acids by GC/MS can be one of the most useful and precise criteria for the chemotaxonomy or identification of *Mycobacteria, Nocardia,* and their related taxa.

References

1. Goodfellow M, Orlean PAB, Collins MD, Alshamaony L, Minnikin DE (1978) Chemical and numerical taxonomy of strains received as *Gordona aurantiaca*. J Gen Microbiol 109:57–68
2. Tsukamura M, Mizuno S (1971) A new species *Gordona aurantiaca* occurring in sputa of patients with pulmonary disease. Kekkaku 46:95–98
3. Tomiyasu I, Toriyama S, Yano I, Masui M (1981) Changes in molecular species composition on nocardomycolic acids in *Nocardia rubra* by the growth temperature. Chem Phys Lipids 28:41–54
4. Tomiyasu I (1982) Mycolic acid composition and thermally adaptative changes in *Nocardia asteroides*. J Bacteriol 151:828–837

5. Tomiyasu I, Yano I (1984) Separation and analysis of novel polyunsaturated mycolic acids from a psychrophilic, acid-fast bacterium, *Gordona aurantiaca*. Eur J Biochem 139:173–180
6. Toriyama S, Yano I, Masui M, Kusunose M, Kusunose E (1978) Separation of C_{50-60} and C_{70-80} mycolic acid molecular species and their changes by growth temperatures in *Mycobacterium phlei*. FEBS Lett 95:111–115
7. Toriyama S, Yano I, Masui M, Kusunose E, Kusunose M, Akimori N (1980) Regulation of cell wall mycolic acid biosynthesis in acid-fast bacteria. J Biochem 88:211–221
8. Yabuuchi E, Yano I, Kaneko T, Ohyama A (1981) Classification of group IIK-2 and related bacteria. In: Reichenbach H, Weeks OB (eds) The Flavobacterium-Cytophaga Group. Verlag Chemie, Weinheim (GBF Monograph Series No. 5)
9. Yamamoto A, Yano I, Masui M, Yabuuchi E (1978) Isolation of a novel sphingoglycolipid containing glucuronic acid and 2-hydroxy fatty acid from *Flavobacterium devorans* ATCC 10829. J Biochem 83:1213–1216
10. Yano I, Yamamoto A, Ohno Y, Masui M, Yabuuchi E (1977) Occurrence of sphingophospholipids in Gram-negative, nonfermentative (aerobic) rods. Proc Japan Conf Lipid Biochem 19:19–22
11. Yano I, Tomiyasu I, Yabuuchi E (1982) Long chain base composition of strains of three species of *Sphingobacterium* gen. nov. FEMS Microbiol Lett 15:303–307
12. Yano I, Imaizumi S, Tomiyasu I, Yabuuchi E (1983) Separation and analysis of free ceramides containing 2-hydroxy fatty acids in *Sphingobacterium* species. FEMS Microbiol Lett 20:449–453
13. Yano I, Furukawa Y, Kusunose M (1969) Occurrence of α-hydroxy fatty acids in Actinomycetales. FEBS Lett 4:96–98
14. Yano I, Furukawa Y, Kusunose M (1970) α-Hydroxy fatty acid-containing phospholipids of *Nocardia leishmanii*. Biochim Biophys Acta 202:189–191
15. Yano I, Furukawa Y, Kusunose M (1969) Conversion of palmitic acid to α-hydroxypalmitic acid in *Corynebacterium simplex*. Biochim Biophys Acta 187:166–168
16. Yano I, Furukawa Y, Kusunose M (1971) α-Oxidation of long-chain fatty acids in cell-free extracts of *Arthrobacter simplex*. Biochim Biophys Acta 239:513–516
17. Yano I, Furukawa Y, Kusunose M (1971) Conversion of α-hydroxypalmitic acid to pentadecanoic acid by resting cells of *Arthrobacter simplex*. J Gen Appl Microbiol 17:429–432
18. Yano I, Saito K, Furukawa Y, Kusunose M (1972) Structural analysis of molecular species of nocardomycolic acids from *Nocardia erythropolis* by the combined system of gas chromatography and mass spectrometry. FEBS Lett 21:215–219
19. Yano I, Saito K (1972) Gas chromatographic and mass spectrometric analysis of molecular species of corynomycolic acids from *Corynebacterium ulcerans*. FEBS Lett 23:352–356
20. Yano I, Kageyama K, Ohno Y, Masui M, Kusunose E, Kusunose M, Akimori N (1978) Separation and analysis of molecular species of mycolic acids in Norcardia and related taxa by gas chromatography mass spectrometry. Biomed Mass Spectrom 5:14–24

Gas Chromatographic Analysis of Clinical Samples for Rapid Diagnosis of Infection

L. LARSSON [1]

Introduction and Instrumental Requirements

Introduction

As an aid in the rapid diagnosis of infectious diseases, gas chromatography (GC) can be used to analyse clinical samples, viz. body fluids and tissue homogenates, to detect chemical compounds that may serve as markers of infection. These chemical markers may be either the constituents or the products of the microbes responsible for the infection, or substances originating from a specific body response to the infection. Unlike many other analytical methods, GC is ineffective in the analysis of high molecular weight compounds, and thus its use is restricted to the study of low or medium molecular weight markers.

Since this communication is intended as a brief review of methods for direct GC analysis of clinical specimens in the diagnosis of infection, no studies on short-term incubated cultures by GC as a rapid diagnostic alternative to "traditional" methods will be discussed here.

Technical Aspects of the Gas Chromatographic System

The GC system used must be capable of detecting specific chemical marker(s) in clinical specimens, often of very complex chemical composition. The required degree of sensitivity and specificity of the analytical system varies according to the diagnostic purpose. Because of their great separation efficiency, capillary GC columns are generally to be preferred to packed columns when a high degree of specificity is required. In the analysis of comparable samples, capillary columns also promote sensitivity better than packed columns, as the peaks are narrower and more prominent, and thus easier to discern. For high sensitivity, samples should be applied to capillary columns by splitless or on-column injection, rather than by split injection – a technique which results in only around 1%–5% of the total material injected being actually introduced into the column for separation and identification. The splitless injection technique can be used in conjunction with an automatic injection unit, and is therefore to be preferred in routine, high sensitivity capillary GC analysis [14].

---

1 Department of Medical Microbiology, University of Lund, Sölvegatan 23, S-223 62 Lund

The choice of GC detector is of utmost importance for satisfactory analytic (and diagnostic) results. *Flame ionization* (FI) detectors are very common in the gas chromatographs used in clinical microbiology laboratories. This type of detector responds to all organic molecules, with a typical lowest detectable amount of injected sample of around 1 ng (depending upon the molecular structure). The *electron capture* (EC) detector responds with extremely high sensitivity – in the low-picogram range – to compounds possessing high electron affinity. For nonelectrophilic compounds, the sensitivity is very poor, however, and substances used as chemical markers of infection are usually derivatized to form halogen-containing derivatives prior to analysis, thereby rendering them susceptible to sensitive EC detection. Finally, detection using *mass spectrometry* (MS) is possible by adapting the GC/MS to selected ion monitoring (SIM). In this technique the MS, instead of scanning continuously over a given mass range, is focused at only one or a limited number of selected masses which are typical for the compound(s) under study [22]. The resulting detector will possess an extremely high selectivity and sensitivity. In studies using electron impact (EI) or chemical ionization (CI) MS SIM, the sensitivity is often found to be roughly equal to that of EC detection with halogen-containing derivatives. The sensitivity can be further increased to a considerable extent with more sophisticated MS techniques such as negative ion (NI) MS SIM, though this method has hitherto only rarely been used in studies of bacterial constituents and products. Among the range of techniques of GC detection, MS adapted to SIM – owing to its extreme sensitivity and selectivity – is preferred when trace amounts of specific chemical markers are to be analyzed in complex clinical samples.

Applications: Flame Ionization Detection

Diagnosis of Anaerobic Infection

GC analysis of short-chain fatty acids in clinical specimens such as pus, pleural and joint aspirates, sputum, etc. can be used for determining whether a given infection is caused by aerobic or by anaerobic bacteria. The analyses are made either after acidification and extraction of the specimens [11, 30–32], by direct injection [24], or by using head-space GC analysis [19]. In general, either the absence of short-chain fatty acids or the presence of acetic acid only is a reliable indicator of the presence of aerobic organisms, whereas a mixture of such acids – volatile as well as nonvolatile – indicates the presence of anaerobic bacteria, either alone or in combination with aerobic organisms. Identification of the causative anaerobic organisms' genus or species cannot as a rule be made in direct analysis, however [31]. Although both false negative results (often for *Bacteroides fragilis*) and false positive results have ben reported, the diagnostic value of this simple GC method is indisputable.

Differentiation Between Bacterial and Nonbacterial Inflammation

Increased concentrations of lactic acid are often found in clinical specimens from patients with, for example, meningitis, peritonitis, septic arthritis, and pleural infection. Thus measurements of lactic acid concentrations in certain body fluids may constitute an important adjunct to other methods for rapidly detecting bacterial infection. However, false negative results may occur, for example, in cases of gonococcal infection and in early stages of bacterial meningitis [5, 8]. Simple GC methods for measuring lactate are available, although many laboratories use alternative (commercially available) lactate test kits based upon specific biochemical reactions [5].

Diagnosis of Disseminated *Candida* Infection

Several fungal species produce D-arabinitol [3]. Increased concentrations of arabinitol and mannose have been reported in sera from rats and rabbits experimentally infected with *C. albicans* and from patients with disseminated *Candida* infection [21, 27]. These compounds are therefore useful chemical markers in the rapid diagnosis of candidiasis, although relatively high concentrations of arabinitol were also found recently in sera from patients with sarcoidosis [12]. Hitherto studies have included both D and L forms of arabinitol and mannose in the measurements. However, use of an optically active stationary phase in the GC column would make possible specific analysis of the D forms only. Such a methodological improvement should significantly increase the precision of diagnostic results.

Other Examples

Reports have been published describing the potential value of GC/FI analysis, e.g., in detecting *Clostridium difficile* by demonstrating isocaproic acid in stools [25], and in the diagnosis of meningitis [1, 4, 13]. It is unfortunate that in many cases the diagnostic uses of these methods have not so far been confirmed by studies in other laboratories.

Applications: Electron Capture and Mass Spectrometric Detection

3-Hydroxytetradecanoic Acid

This fatty acid (beta-hydroxymyristic acid) is a constituent of the endotoxin in gram-negative bacteria. In the direct analysis of serum, this compound has obvious potential as a chemical marker for septicaemia caused by gram-negative bacteria. Using EI SIM, as little as 200 fmol of this hydroxy acid could be detected, and the recovery for lipid A and endotoxin, when added to rabbit serum, was 80%–90% [18]. An increase of the sensitivity would be achieved by using CI or NI SIM as the detection system, and by removal of interfering serum lipids prior to analysis. Such lipids would be easy to remove by extraction because β-hydroxymyristic acid is linked to a polysaccharide structure in lipid A and thus not extractable by common organic solvents.

Muramic Acid

Muramic acid – a component of peptidoglycan – is found in nature exclusively in the cell walls of bacteria and blue-green algae. Demonstration of this compound in clinical samples from patients with arthritis and other conditions would be extremely valuable in evaluating the etiological agent(s) of the diseases. Using SIM detection, muramic acid was demonstrated in tissues and joints of rats in which arthritis had been induced by the injection of group A streptococcal cell wall fragments, but was not found in normal control rats [10], or in synovial tissues from patients with rheumatoid arthritis [26]. Again, the use of more sensitive MS detection techniques (NI SIM) is to be recommended in future studies. Alternative constituents of the peptidoglycan such as D-amino acids and (for gram-negative bacteria) diaminopimelic acid would also be interesting to study with sensitive GC/MS analysis.

3-Hydroxydodecanoic Acid

Together with β-hydroxymyristic acid, this fatty acid is present in *Neisseria* in relatively large amounts. A GC assay was constructed utilizing EC detection of the heptafluorobutyryl ether butyl ester derivative, and was capable of detecting 10 pg of 3-hydroxydodecanoic acid [29]. Clinical studies showed that this fatty acid could be detected in vaginal fluid specimens from female contacts of men with gonorrhoea. Further studies will be necessary to evaluate this method of rapid gonorrhoea diagnosis.

Arabinitol

Confirmatory results on the value of arabinitol level measurements for diagnosis of *Candida* infection (see above) have been achieved using SIM. The advantages of SIM over FI detection in determining small increments of the arabinitol concentration in serum were demonstrated [28].

Tuberculostearic and Mycocerosic Acids

10-Methyloctadecanoic acid (tuberculostearic acid) is a branched-chain fatty acid that is present in microorganisms belonging to *Actinomycetales*, e.g., in all species of *Mycobacterium*. In a series of studies [2, 15–17, 20, 23], we demonstrated the value of this fatty acid as marker for mycobacterial infection in direct SIM analyses of sputum from patients with pulmonary tuberculosis and cerebrospinal fluid (CSF) from cases of tuberculous meningitis. This technique enables the time required for the diagnosis of mycobacterial infections to be reduced from several weeks to a few hours. In summary, we have found that

Mycobacteria grown in vivo produce tuberculostearic acid

Using SIM, this fatty acid can be demonstrated in sputum and CSF from patients with mycobacterial infection even in cases where no acid-fast rods can be detected in direct microscopy studies

Initiation of drug therapy before samples are taken for analysis does not jeopardize the possibility of subsequent tuberculostearic acid detection

Tuberculostearic acid is not detected in specimens from patients with non-tuberculous meningitis or pneumonia, or from healthy persons that are vaccinated against tuberculosis

The method involves autoclaving of the specimens, lipid extraction, acidic methanolysis, and extraction of the methyl ester fraction. For sputum specimens, a cleaning-up step using thin-layer chromatography is recommended. The final extract used for analysis must often be concentrated to 5–10 µl in order to detect tuberculostearic acid, after which 1 µl is injected splitless into a capillary column. The MS is normally focused at mass numbers 312 (EI), 313 (CI, isobutane), and 330 (CI, ammonia).

Our present work in this research area aims to investigate the possibilities of using EC detection for the rapid screening of specimens and NI SIM for definite diagnosis, utilizing analysis of pentafluorobensyl or trichloroethyl esters of tuberculostearic acid in both cases.

We have also used C32 mycocerosic acid, which is exclusively present as a diagnostic marker in a limited number of mycobacterial species (including *Mycobacterium tuberculosis*). It was possible to detect this acid in 5-day incubated cultures of sputum, i.e., weeks before mycobacterial colonies could be observed on the slants of the Löwenstein-Jensen medium [17].

Compounds Originating from Body Response to Infection

Methods have been developed for GC separation and EC detection of free amines, hydroxy acids, and fatty acids present in various clinical specimens such as urine, CSF, pleural fluid, and serum. The potential of this approach for rapid diagnosis of infectious diseases has been reported (for an extensive review see [9]). Most of the chromatographic peaks of diagnostic value probably represent substances that originate from a specific body response to the infection as they can not be identified as cellular constituents or be found in spent culture medium.

The chemical structures of the markers used have so far remained largely unidentified. However, presence in CSF of arachidonic acid and, at least for patients from certain geographical areas, 3-(2′-ketohexyl)-indoline has been reported as a typical marker for tuberculous meningitis [6]. Promising results for diagnosis of viral and cryptococcal meningitis have also been reported, as well as the method's potential value in differentiating infectious diseases by analysis of serum. Thus, chromatograms where the major peaks were assumed to represent hydroxy acids in sera from patients with Rocky Mountain spotted fever, chickenpox, and

N. meningitidis infection, were readily distinguishable from each other [7]. Although this technique not only seems to be easy to use as MS is not involved, but has also given convincing results, it will probably not become widely employed until confirmatory results have been reported from other laboratories in independent studies.

Acknowledgements. Grants from the Swedish National Association against Heart and Chest Diseases and from the Swedish Medical Research Council (grant no. 5965) are gratefully acknowledged.

References

1. Amundson S, Braude AI, Davis CE (1974) Rapid diagnosis of infection by gas-liquid chromatography: analysis of sugars in normal and infected cerebrospinal fluid. Appl Microbiol 28:298–302
2. Bergman R, Larsson L, Odham G, Westerdahl G (1983) Selected ion monitoring for rapid differentiation of mycobacteria isolated from domestic animals. J Microbiol Methods 1:19–22
3. Bernard EM, Christiansen KJ, Tsang S-F, Kiehn TE, Armstrong D (1981) Rate of arabinitol production by pathogenic yeast species. J Clin Microbiol 14:189–194
4. Brook I (1979) Abnormalities in spinal fluid detected by gas liquid chromatography in meningitis patients. Chromatographia 12:583–586
5. Brook I (1981) The importance of lactic acid levels in body fluids in the detection of bacterial infections. Rev Infect Dis 3:470–478
6. Brooks JB, Edman DC, Alley CC, Craven RB, Girgis NI (1980) Frequency-pulsed electron capture gas-liquid chromatography and the tryptophan color test for rapid diagnosis of tuberculous and other forms of lymphocytic meningitis. J Clin Microbiol 12:208–215
7. Brooks JB, McDade JE, Alley CC (1981) Rapid differentiation of Rocky Mountain spotted fever from chicken pox, measles and enterovirus infections and bacterial meningitis by frequency-pulsed electron capture gas-liquid chromatographic analysis of sera. J Clin Microbiol 14:165–172
8. Curtis GDW, Newman RJ, Slack MPE (1983) Synovial fluid lactate and the diagnosis of septic arthritis. J Infect 6:239–246
9. Edman DC, Brooks JB (1983) Gas-liquid chromatography-frequency pulse-modulated electron-capture detection in the diagnosis of infectious diseases. J Chromatogr 274:1–25
10. Fox A, Schwab JH, Cochran T (1980) Muramic acid detection in mammalian tissues by gas-liquid chromatography-mass spectrometry. Infect Immun 29:526–531
11. Gray J, Perks WH, Birch J (1980) Direct gas-liquid chromatography of clinical material as an aid in the diagnosis of anaerobic lung infections. Curr Chemother 1:540–543
12. Karam GH, Elliott AM, Polt S, Cobbs CG (1984) Elevated serum D-arabinitol levels in patients with sarcoidosis. J Clin Microbiol 19:26–29
13. LaForce FM, Brice JL, Tornabene TG (1979) Diagnosis of bacterial meningitis by gas-liquid chromatography. II. Analysis of spinal fluid. J Infect Dis 140:453–464
14. Larsson L, Odham G (1984) Gas chromatography. In: Odham G, Larsson L, Mårdh P-A (eds) Gas chromatography/mass spectrometry applications in microbiology. Plenum, New York, pp 7–26
15. Larsson L, Mårdh P-A, Odham G (1979) Detection of tuberculostearic acid in mycobacteria and nocardiae by gas chromatography and mass spectrometry using selected ion monitoring. J Chromatogr 163:221–223
16. Larsson L, Mårdh P-A, Odham G, Westerdahl G (1980) Detection of tuberculostearic acid in biological specimens by means of glass capillary gas chromatography/electron and chemical ionization mass spectrometry, utilizing selected ion monitoring. J Chromatogr 182:402–408
17. Larsson L, Mårdh P-A, Odham G, Westerdahl G (1981) Use of selected ion monitoring for detection of tuberculostearic and C32 mycocerosic acid in mycobacteria and in five-day-old cultures of sputum specimens from patients with pulmonary tuberculosis. Acta Pathol Microbiol Scand [B] 89:245–251
18. Maitra SK, Schotz MC, Yoshikawa TT, Guze LB (1978) Determination of lipid A and endotoxin in serum by mass spectroscopy. Proc Natl Acad Sci USA 75:3993–3997

19. Mårdh P-A, Larsson L, Odham G (1981) Head-space gas chromatography as a tool in the identification of anaerobic bacteria and diagnosis of anaerobic infections. Scand J Infect Dis [Suppl] 26:14–18
20. Mårdh P-A, Larsson L, Hoiby N, Engbaek HC, Odham G (1983) Tuberculostearic acid as a diagnostic marker in tuberculous meningitis. Lancet 8320:367
21. Marier RL, Milligan E, Fan Y-D (1982) Elevated mannose levels detected by gas-liquid chromatography in hydrolysates of serum from rats and humans with candidiasis. J Clin Microbiol 16:123–128
22. Odham G, Larsson L, Mårdh P-A (1984) Quantitative mass spectrometry and its application in microbiology. In: Odham G, Larsson L, Mårdh P-A (eds) Gas chromatography/mass spectrometry applications in microbiology. Plenum, New York, pp 303–333
23. Odham G, Larsson L, Mårdh P-A (1979) Demonstration of tuberculostearic acid in sputum from patients with pulmonary tuberculosis by selected ion monitoring. J Clin Invest 63:813–819
24. Phillips I, Taylor E, Eykyn S (1980) The rapid laboratory diagnosis of anaerobic infection. Infection 8:S155–S158
25. Potvliege C, Labbé M, Glupczynski Y, Yourassowsky E, Grokaert F (1983) Detection of *Clostridium difficile* in stools by agar culture, enrichment broth culture, and gas-liquid chromatography. 1st European congress of clinical microbiology, 17–21 October, Bologna
26. Pritchard DG, Settine RL, Bennett JC (1980) Sensitive mass spectrometric procedure for the detection of bacterial cell wall components in rheumatoid joints. Arthritic Rheum 23:608–610
27. de Repentigny L, Kuykendall RJ, Reiss E (1983) Simultaneous determination of arabinitol and mannose by gas-liquid chromatography in experimental candidiasis. J Clin Microbiol 17:1166–1169
28. Roboz J, Suzuki R, Holland JF (1980) Quantification of arabinitol in serum by selected ion monitoring as a diagnostic technique in invasive candidiasis. J Clin Microbiol 12:594–602
29. Sud IJ, Feingold DS (1979) Detection of 3-hydroxy fatty acids at picogram levels in biological specimens. A chemical method for the detection of *Neisseria gonorrhoeae*? J Invest Dermatol 73:521–525
30. Thadepalli H, Gangopadhyay PK (1980) Rapid diagnosis of anaerobic empyema by direct gas-liquid chromatography of pleural fluid. Chest 77:507–513
31. Watt B, Geddes PA, Greenan OA, Napier SK, Mitchell A (1982) Can direct gas-liquid chromatography of clinical specimens detect specific organisms? J Clin Pathol 35:706–708
32. Watt B, Geddes PA, Greenan OA, Napier SK, Mitchell A (1982) Gas-liquid chromatography in the diagnosis of anaerobic infections: a three year experience. J Clin Pathol 35:709–714

Pyrolysis Gas-Liquid Chromatography for Identification of Microorganisms

N. J. Stern[1]

Abstract

Pyrolysis gas-liquid chromatography (PGLC) of bacteria is a technique which produces pyrolyzed bacterial components with specifically defined patterns. In theory, these patterns when analyzed statistically produce a "fingerprint" for each organism tested. The technique lends itself to automated bacterial identification since systems are available to automatically prepare samples and pyrolyze and chromatograph the resulting volatiles. The data are analyzed and objective bacterial classification is a theoretical possibility. Because of the inherent variability in staining reactions, erroneous interpretations of fermentation patterns, morphology variability, etc., microbiologists have hoped that PGLC might provide a simple and impartial means of bacterial identification. Through the use of this technique, researchers have reported success in differentiating bacterial genera, species, and serotypes. However, reported differentiations have been based upon the use of limited numbers of strains analyzed. With a greater number of organisms tested, increased difficulty could be anticipated in correctly categorizing bacteria through PGLC analysis. Data also exist suggesting that when diversely related bacteria are analyzed by PGLC, the similarities between bacteria may be greater than the a priori taxonomic discriminations, resulting in an inaccurate PGLC classification for a given isolate. This chapter points out both the strong and weak points for this method of bacterial analysis.

Introduction

Traditional criteria used for the identification of bacteria employ subjective evaluations of morphological appearances, staining reactions, fermentation patterns, etc. Variability in the interpretation of these criteria may be due to decomposition and inconsistencies of staining materials or bacteriological media. Therefore, inaccurate assessment of these taxonomic tests may result. The substantial uncertainties involved with bacterial identification are best understood after gaining familiarity with the degree of subjectivity involved in interpreting many of these tests.

1 Meat Science Research Laboratory, Agricultural Research Service, US Department of Agriculture, Beltsville Agricultural Research Center, Beltsville, MD 20705, USA

Over the past 20 years, studies using pyrolysis gas-liquid chromatography (PGLC) of microorganisms have attempted to automate and objectify bacterial classification. Very important progress has been made in this area of research, and the vast majority of published studies report success in discriminating genera, species, serotypes, and strains of specified bacteria. However, there are still serious unanswered questions that remain to be addressed before PGLC can be embraced as a reasonable alternative to current, accepted classification approaches. The purposes of this chapter are: (1) to review the methods employed, (2) to examine the variety of appropriate statistical analyses, and (3) to consider the published data and resulting interpretations.

Materials and Methods

Sample Preparation

Various studies have pointed out important differences in pyrochromatogram patterns occurring whenever growth parameters are manipulated. For example, Drucker and Gibson (1982) determined that major changes in PGLC analyses of *Streptococcus faecalis* resulted when the organism was grown at 22 °C vs. 37 °C. In that study, additional but less important changes in PGLC patterns were ascribed to modifying atmospheric conditions (aerobic vs. anaerobic), length of culture incubation (24 h vs. 70 h), growth medium (MacConkey's vs. blood agar plates), and less severe temperature alterations (45 °C vs. 37 °C). The authors used data obtained from PGLC and compared mean analyses of each growth condition by Pearson coefficient of linear correlation. As a comparison, standard growth conditions (37 °C, aerobic atmosphere, blood agar, 24 h) yielded a coefficient of 1.0, growth at 22 °C produced a coefficient of 0.21, at 45 °C a coefficient of 0.81, 70-h incubation a coefficient of 0.78, MacConkey's agar a coefficient of 0.70, and anaerobiosis a coefficient of 0.56. Data obtained from the PGLC of *Staphylococcus aureus* yielded a coefficient of 0.55. These data indicate that the lack of correlation found with taxonomic differences is similar to the lack of correlation brought about by changing the temperature for growth.

Oxborrow and colleagues (1977 a) state that species of *Bacillus* sampled at different incubation times had different pyrochromatograms and could be compared only with chromatograms of cultures of equal age. In another report (Oxborrow et al. 1977 b), they found little or no differences between the chromatograms of individual *Bacillus* species grown on six lot numbers of the same medium. When chromatograms of three different media were compared, differences were observed in the areas most characteristic of individual species.

Gutteridge and Norris (1980) reported that changing the bacterial culture medium significantly altered the quantitative aspects of the pyrochromatograms for the three organisms they were testing. They also stated that the influence of culture time and incubation temperature were less significant than the influence of the culture medium. In this study, they observed that growth on different media produced no unique pyrolysis products, but rather changed the existing peak heights and ratios of the peak heights. Therefore, the statistical analysis used to discriminate among the test organisms was similarly influenced.

The above studies help to explain why poor intra- and interlaboratory reproducibility exists and why it is to be expected for PGLC analysis whenever all sampling parameters are not carefully monitored. It should be recognized that there is no one universal growth condition that is capable of cultivating all bacteria, and therefore unbiased preparation of samples for PGLC analysis seems impossible. However, PGLC comparisons could still be plausible among organisms capable of growth under similar conditions.

My laboratory has made comparisons of selected organisms incubated under constant growth conditions for each set of studies. Preparation of each bacterial sample for analysis entailed inoculating a fixed amount of stock culture into identical aliquots of medium, all from the same batch, and growing the static culture at 23 °C for 48 h (Stern et al. 1979). Subsequently, the cells were washed three times using 100-ml volumes of 0.05% formaldehyde and harvested by centrifugation. The pellet was resuspended in 3 ml of the same diluent and was lyophilized and stored under dark conditions. We repeated the entire procedure for each strain to be analyzed by PGLC to ensure reproducibility of results. I should like to emphasize this point, as some investigators have made replicate PGLC analyses from the same sample, which could only test the reproducibility of the thermal degradation of the bacteria. The reproducibility of the entire procedure must be evaluated by taking into account the possible variations that can be attributed to variations in cell growth as well as the preparatory steps. Differences in PGLC data between bacteria must be greater than differences resulting from experimental variation.

Pyrolysis Unit

The pyrolysis unit used in my laboratory consists of a Chemical Data Systems model 100 Pyroprobe (Oxford, Pennsylvania) and a Perkin-Elmer platinum coil probe (Norwalk, Connecticut). Each freeze-dried bacterial sample is packed into a clean quartz tube and inserted into the coil probe with clean forceps. The coil probe interfaced with the pyrolysis unit fits directly to the injection port of a Perkin-Elmer Sigma 3 gas-liquid chromatograph (Norwalk, Connecticut). Pyrolysis is carried out at 900 °C for 20 s in a continuous flow of helium gas. The injector temperature is set to 250 °C, allowing the volatile pyrolysate compounds to reach the column.

Pyrolysis is a different process as compared with combustion. Combustion occurs when a product is degraded through the application of both heat and oxygen, and the end product typically is comprised of small, comparatively nondescript compounds such as carbon dioxide, water, and ammonia (Gutteridge and Norris 1979). On the other hand, pyrolysis occurs as a result of heat breaking down relatively weaker bonds of complex materials. This results in a series of more definite compounds with comparatively greater molecular weights. Compounds such as acetonitrile, toluene, ethyl benzene, pyridine, pyrrole, furfuryl alcohol, phenol, and cresol are major pyrolysates (Abbey et al. 1981). These compounds should give evidence of the original composition of the microorganism, and hence the term "fingerprinting" has been proposed (Reiner 1967).

Several factors involved in the pyrolysis step of PGLC influence the data produced. Cone and Lechowich (1970) point out that the consistency of packing the sample into the probe appears to influence the pyrogram, since the pyrolysis products of samples packed too firmly were not swept instantaneously into the column after pyrolysis, and those packed too loosely were often blown from the probe by the carrier gas. Use of the pyroprobe coil and the resulting concomitant carbonization of the coil is known to influence the pyrogram patterns (Gutteridge et al. 1979). With increased carbonization, a longer period of time is needed to reach the same desired pyrolysis temperature. Varying the period of temperature increase could produce a different quantity of pyrolysate products and comparison of data is confused by probes of varied age. It is also clear that the pyrolysis temperature must be the same to make comparisons between bacterial PGLC patterns.

Both filament and Curie-point pyrolysis have been employed in PGLC analyses. The filament pyrolysis unit has been used in my laboratory and positions the sample within a quartz tube, within the coiled filament. Curie-point pyrolysis places the sample directly onto the ferromagnetic wire, which is then placed in the work coil of a high-frequency oscillator (Gutteridge and Norris 1979). Major concerns with both methods include the carbonization issue previously discussed, as well as the rapid removal of pyrolyzates from the heating chamber onto the chromatographic column. If there is considerable "dead volume," the pyrolyzates may become involved in secondary reactions and yield different chromatographic patterns (Needleman and Stuchberg 1977).

Gas-Liquid Chromatography

Most PGLC studies of microorganisms have employed polar columns coated or packed with Carbowax 20M. However, Needleman and Stuchberg (1977) determined that similar numbers of nonpolar pyrolyzate products could be generated and detected on an Apiezon L packed column as could be detected using a Carbowax 20M column of the same dimensions. One of the major advantages in using Carbowax 20M is the tolerance of high (up to 200 °C) temperatures, which is useful in eluting high molecular weight compounds from the column. However, even with relatively high temperature tolerance, Carbowax 20M does begin to bleed from a column at temperatures as low as 160 °C. With this "bleed off," the resolving capacity of any column is diminished and therefore the retention times of the pyrolyzates will necessarily vary with the column age. Therefore, it seems valuable to evaluate other GLC stationary phase material that can better withstand the high temperatures needed to elute the high molecular weight pyrolyzates.

Through the pyrolysis process, high molecular weight components are swept onto the chromatographic column. Such compounds need high temperatures to maintain volatility and percolate through the stationary phase. Because such temperatures cannot be employed without destroying the resins, the high molecular weight compounds collect at the beginning of the column. These compounds then interfere with the "clean" resolution of subsequent pyrolysis analyses (Needleman and Stuchberg 1977). In my laboratory, we have attempted to remedy this situation by breaking off the first inch or two of our capillary column (50 m in length) after each of 12–24 analyses. In this manner, we can remove the tarry build-up without greatly sacrificing the resolving power of our column. This approach seems to work well for five to ten "breaks," but the resolving power of the column does eventually deteriorate. Quinn (1974) used a disposable 25-ft precolumn after every 50–100 analyses to prolong the life of the analytical column. Stack et al. (1977) regularly repacked the first 10 cm of their column and frequently cleaned all connections to extend the life of their packed column.

Both packed and capillary columns have been used in PGLC studies. In some of the original work on the application of PGLC for bacterial identification, Reiner (1967) employed a packed column which resolved relatively few pyrolyzate components but enabled discrimination among those species tested. Many workers continued with this approach, while others believed the resolving power of a capillary column would enhance distinguishing the fine components of bacterial samples. It was thought that a more discriminating pyrochromatogram fingerprint would lead to better identification. Oyama and Carle (1967) were the first to employ a Carbowax 20M capillary column in PGLC studies on bacterial discriminations. Their approach was then refined by Quinn (1974), who optimized the resolving capacity of her chromatograph to elute approximately 200 peaks from the microorganisms sampled. The trend has been toward using capillary columns, although both expense and short column life are considerations.

A variety of oven temperature programs have been used, ranging from 40 °C to 200 °C, with various lengths of isothermal holding and rates of temperature increase. Because such a diversity of compounds are produced during pyrolysis of bacteria, any temperature program should attempt to account for separation of both the low molecular weight and high molecular weight compounds. By and large, the low-weight compounds are eluted unseparated in the first portion of the chromatograph, while separation of the heavier compounds is accomplished through temperature programming (Fig. 1). It is thought that the heavier compounds are more representative of unique structures that might aid in the identification of the bacteria.

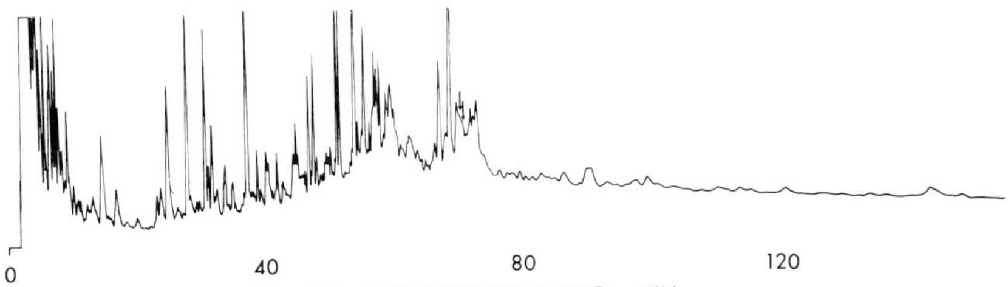

Fig. 1. Representative tracing of PGLC elution patterns of *Salmonella typhi*. (Reprinted with permission from Appl Environ Microbiol 38:1098)

Data Analysis

Raw Data

Use of a constant amount of material for pyrolysis was considered necessary to allow comparisons between pyrochromatograms (Cone and Lechowich 1970). This is due to the predominance of quantitative rather than qualitative differences that occur in bacterial pyrochromatograms (Quinn 1974). The process of weighing microgram quantities of lyophilized organisms can be quite tedious and necessitates the use of uncommon weighing devices (Emswiler and Kotula 1978). Alternatively, areas or peak heights can be normalized to eliminate the effect of sample size (Gutteridge et al. 1979).

Next, bacterial samples are analyzed by comparing either the areas under given peaks or the time slices of selected intervals. Some of the earliest PGLC workers used subjective discrimination of peak patterns to discriminate among the samples analyzed. Subsequently, more objective statistical analyses were employed. Emswiler and Kotula (1978) and others (Stern et al. 1980) determined the area under the curve for each 1-min interval of the pyrochromatograms. These data were subjected to stepwise discriminant analysis to determine which variables could best discriminate between the samples.

Multivariate Statistical Approaches

With the multitude of peaks produced in PGLC and large numbers of samples processed, sophisticated statistical analyses are required. In the case of either time-slice or peak area characterization of an organism, each sample may be described as a point in a multidimensional space (MacFie et al. 1978). The number of dimensions in this space is at least as large as the maximum number of sample components or time slices measured. The principal components analysis obtains successive axes to illustrate maximal variation among the samples. The majority of the variations among the samples may be represented by only three or four of these axes. Because the first two axes (the first and second principal components) are selected by accounting for maximal variation among the sample plots, the resulting two-dimensional graph often reveals both clustering of like samples and distances between dissimilar samples.

In canonical variates analysis, a mathematical distance, Mahalanobis D^2 (Mahalanobis 1936), is calculated to describe each grouping of samples in the above mentioned multidimensional space relative to one another. This analysis attempts to maximize the variation between the groups being compared in relation to the variation seen within each group being analyzed. In this manner, a new set of axes is generated to describe the mathematical distance between the groups in question. This new multidimensional space is expressed as a set of canonical variate axes which transposes multidimensional space into two dimensions, manifesting maximal Mahalanobis D^2 in two dimensions.

Stepwise discriminant multiple regression analysis (Dixon 1975) attempts to determine which criteria (time slice or peak area) best differentiate the a priori groups tested. The mean values and standard deviations for the data points of each group are calculated and F-statistics determined for each criteria. Computer analysis then selects the peak (or time slice) that enables intergroup discrimination on the basis of maximal inter- to intragroup variance. In a stepwise procedure, the computer selects the most discriminating peak and sequentially continues this process of selection from the most significant remaining peaks. The procedure continues until none of the remaining peaks (or time slices) account for further discrimination between the groups being compared.

Data Interpretation

Successful Differentiations

At least three studies exist that report useful discrimination among species of *Clostridium* with PGLC analysis. Cone and Lechowich (1970) reported that subjective visual interpretation of differences between pyrochromatogram patterns enables discrimination among *C. botulinum* types A, B, and E. They could not, however, detect toxin production by the organism through the use of PGLC methodology. Next, Reiner and Bayer (1978) reported that PGLC enabled differentiation among various toxin-producing types of *C. botulinum* and *C. sporogenes*. To accomplish this differentiation, peaks occurring at the high boiling end of the pyrochromatogram were used to distinguish subjectively between the proteolytic and nonproteolytic strains of *C. botulinum*. In the most detailed and objective study on PGLC of *C. botulinum* and related organisms, Gutteridge and colleagues (1980) were able to substantiate the capability of the methodology. They used 65 strains of clostridia, including *C. botulinum* types A–F, *C. sporogenes* and other putrefractive anaerobes. The PGLC analyses were carried out in duplicate with a packed column. Twenty-six consistent peaks from replicate pyrochromatograms were averaged and the data were analyzed using canonical variates and stepwise discriminant analysis. Using this approach, the groups were discriminated into the following categories: (a) *C. botulinum* type A, proteolytic types B and F, and *C. sporogenes*; (b) *C. botulinum* types C and D; and (c) *C. botulinum* type E and nonproteolytic types B and F.

Similarly, the genus *Bacillus* has been studied by several investigators to determine whether PGLC could discriminate among species (Chaolin et al. 1981; O'Donnell et al. 1980a, b; Stern 1981). In these studies, as in the case with clostridia, the species of bacilli were discriminated either by subjective evaluation of peak heights, or through more sophisticated statistical analysis. Other papers describe the use of PGLC to identify *Vibrio cholerae* (Haddadin et al. 1973), differentiate and characterize species of *Salmonella* (Emswiler and Kotula 1978), and differentiate selected *Enterobacteriaceae* (Stern et al. 1979).

Limitations

As is the case with many other interesting concepts, PGLC of bacteria has limited applications. The enthusiasm of the original workers for this area of study was based on the ability to use the method to discriminate between select samples analyzed. It must be recognized that the successes reported were based on limited numbers of bacterial samples, and therefore the appropriateness of the technique would have to be proven using a great number of unknown bacterial isolates.

Several publications hint at the boundaries of PGLC in bacterial taxonomic classification. MacFie and co-workers (1978) analyzed five genera of relatively common food spoilage bacteria by PGLC. Their data were initially evaluated using principal component analysis with the result that members of the same genus appeared scattered in three or even four of the quadrants defined by the first two principal component axes. Discrimination among the five genera was improved when the data were subjected to canonical variates analysis. Considerable overlapping among three of the five species was illustrated.

In my laboratory, we reported (Stern et al. 1980) the ability of PGLC to discriminate between invasive and noninvasive *Yersinia enterocolitica* in only 20 of 29 analyses. Similarly, other researchers have stated concern over the usefulness of the method. Wasserfallen and Rinderknecht (1978) state "the species of bacteria is difficult to differentiate by pyrolysis gas chromatography because of the limited resolution of gas chromatography, and as a consequence there are difficulties with the treatment of pyrograms by integration systems." Garbary and Mortimer (1978), studying the application of PGLC to algal taxonomy, suggest that the technique does not satisfactorily elucidate systematic relationships. Britz and Steyn (1979) stated that their results "showed that the pyrolysis profiles were of no apparent taxonomic value for differentiation of the species or strains of the proprioni-bacteria." When 18 foodborne bacterial strains were analyzed by PGLC and statistical procedures, only 63% were correctly classified to the genus level when compared with one another (Stern 1982). Finally, a telling indicator of the limitations of this method was described by Oyama and Carle (1967), who found pyrochromatograms of bovine insulin and albumin to be remarkably similar to the patterns produced by PGLC analyses of a variety of bacteria.

The following concerns remain as obstacles to the realistic application of PGLC as an alternative taxonomic system:

1) Consistency of bacterial culture age
2) Consistency of growth medium
3) Consistency of growth temperature
4) Consistency of lyophilized sample packing
5) Column age and deterioration
6) Carbonization of pyroprobe
7) Secondary reactions within "dead-space"
8) Resin bleed-off at high temperatures
9) Elution of high molecular weight pyrolyzates
10) Expense of the required systems.

References

Abbey LE, Highsmith AK, Moran TF, Reiner EJ (1981) Differentiation and characterization of *Klebsiella pneumoniae* strains by pyrolysis-gas-liquid chromatography-mass spectrometry. J Clin Microbiol 13:313–319

Britz TJ, Steyn PL (1979) Identification of proprioni bacteria by pyrolysis gas-liquid chromatography. S Afr J Dairy Technol 11:119–122

Chaolin L, Liansheng J, Lianyun D, Yuchen O (1981) A preliminary report on the identification of *Bacillus thurengiensis* subspecies by pyrolysis gas-liquid chromatography. Acta Microbiol Sinica 21:457–463

Cone RD, Lechowich RV (1970) Differentiation of *Clostridium botulinum* types A, B, and E by pyrolysis-gas-liquid chromatography. Appl Microbiol 19:138–145

Dixon WJ (1975) Stepwise discriminant analysis. In: BMDP biomedical computer programs. University of California Press, Berkeley, pp 411–452

Drucker DB, Gibson LF (1982) Pyrolysis gas chromatography of *Streptococcus faecalis*: effect of cultural conditions on pyrochromatograms. Microbios 33:93–100

Emswiler BS, Kotula AW (1978) Differentiation of *Salmonella* serotypes by pyrolysis-gas-liquid chromatography of cell fragments. Appl Environ Microbiol 35:97–104

Garbary D, Mortimer M (1978) Use and analysis of pyrolysis-gas-liquid chromatography in algal taxonomy. Phycologia 17:105–106

Gutteridge CS, Norris JR (1979) The application of pyrolysis techniques to the identification of microorganisms. J Appl Bacteriol 47:5–43

Gutteridge CS, Norris JR (1980) Effect of different growth conditions on the discrimination of three bacteria by pyrolysis gas-liquid chromatography. Appl Environ Microbiol 40:462–465

Gutteridge CS, MacFie HJH, Norris JR (1979) Use of principal components analysis for displaying variation between pyrograms of micro-organisms. J Anal Appl Pyrol 1:67–76

Gutteridge CS, Mackey BM, Norris JR (1980) A pyrolysis gas-liquid chromatography study of *Clostridium botulinum* and related organisms. J Appl Bacteriol 49:165–174

Haddadin JM, Stirland RM, Preston NW, Collard P (1973) Identification of *Vibrio cholerae* by pyrolysis gas-liquid chromatography. Appl Microbiol 25:40–43

MacFie HJH, Gutteridge CS, Norris JR (1978) Use of canonical variate analysis in differentiation of bacteria by pyrolysis gas-liquid chromatography. J Gen Microbiol 104:67–74

Mahalanobis PC (1936) On the generalized distance in statistics. Proc Natl Inst Sci India 2:49–55

Needleman M, Stuchberg P (1977) The identification of microorganisms by pyrolysis gas-liquid-chromatography. In: Jones CER, Cramers CA (eds) Analytical pyrolysis. Elsevier Scientific, Amsterdam, pp 77–78

O'Donnell AG, Norris JR, Berkeley RCW, Claus D, Kaneko T, Logan NA, Nozaki R (1980a) Characterization of *Bacillus subtilis, Bacillus pumilus, Bacillus licheniformis,* and *Bacillus amgloliquifaciens* by pyrolysis gas-liquid chromatography, deoxyribonucleic tests, and API systems. Int J Syst Bacteriol 30:448–459

O'Donnell AG, MacFie HJH, Norris JR (1980b) An investigation of the relationship between *Bacillus cereus, Bacillus thurengiensis,* and *Bacillus mycoides* using pyrolysis gas-liquid chromatography. J Gen Microbiol 119:189–194

Oxborrow GW, Fields ND, Puleo JR (1977a) Pyrolysis gas liquid chromatography of the genus *Bacillus*: effect of growth time on pyrochromatogram reproducibility. In: Jones CER, Cramers CA (eds) Analytical Pyrolysis: proceedings of the third international symposium on analytical pyrolysis, 7–9 September 1976. Elsevier Scientific, Amsterdam, pp 69–76

Oxborrow GS, Fields ND, Puleo JR (1977b) Pyrolysis gas liquid chromatography of the genus *Bacillus*: effect of growth media on pyrochromatogram reproducibility. Appl Environ Microbiol 33:865–870

Oyama VI, Carle GC (1967) Pyrolysis gas chromatography application to life detection and chemotzonomy. J Gas Chromatogr 5:151–154

Quinn PA (1974) Development of high resolution pyrolysis gas chromatography for the identification of micro-organisms. J Chromatogr Sci 12:796–806

Reiner EJ (1967) Studies on differentiation of micro-organisms by pyrolysis gas-liquid chromatography. J Gas Chromatogr 5:65–67

Reiner E, Bayer FL (1978) Botulism: a pyrolysis-gas-liquid chromatographic study. J Chromatogr Sci 16:623–629

Stack MV, Donoghue HD, Tyler JE, Marshall M (1977) Comparison of oral streptococci by pyrolysis gas-liquid chromatography. In: Jones CER, Cramers CA (eds) Analytical Pyrolysis. Elsevier Scientific, Amsterdam, pp 57–68

Stern NJ (1981) Discrimination of *Bacillus cereus* from selected *Bacillus* spp. by pyrolysis gas-liquid chromatography. J Food Sci 46:1427–1429

Stern NJ (1982) The inability of pyrolysis gas-liquid chromatography to differentiate selected food-borne bacteria. J Food Prot 45:229–233

Stern NJ, Kotula AW, Pierson MD (1979) Differentiation of selected *Enterobacteriaceae* by pyrolysis gas-liquid chromatography. Appl Environ Microbiol 38:1098–1102

Stern NJ, Kotula AW, Pierson MD (1980) Virulence prediction of *Yersinia enterocolitica* by pyrolysis gas-liquid chromatography. Appl Environ Microbiol 40:646–651

Wasserfallen K, Rinderknecht F (1978) Möglichkeiten der Unterscheidung von Bakterien mit Hilfe der Currie-Punkt-Pyrolyse, der Gas-Chromatographie und der Computerauswertung der Pyrogramme. Chromatographia 11:128–136

Taxonomic Calculations Based on Fatty Acid Spectra of Bacteria. Requirements for Chromatographic Analysis of Fatty Acids

L. V. ANDREEV [1]

A hierarchic scale of significance of bacterial properties and, therefore, a specific view on the direction in which the diagnostic of bacteria should primarily be developed, exists in all major fields of biotechnology and medicine. Such an approach is most useful for rapidly obtaining optimal results, but on the other hand leads to a distorted overall perception of the whole variety of bacteria and to an overestimation of the role of individual links of the natural system of bacteria, based on their momentary importance for human society. It is consequently apparent that further increases in the rapidity, reliability, and sensitivity of identification methods should rest upon the theory of diversity of bacteria, i.e. their systematics.

In this respect, great importance is to be ascribed to bacterial lipid composition, which is a polymodal criterion, i.e. it includes the characters which reflect the taxonomic status of bacteria of different ranks in the taxonomic hierarchy. Bacterial lipid composition makes it possible not only to attribute bacteria to certain genera or families, but also to identify strains. For example, the chromatogram of fatty acids (FA) presented in Fig. 1 suggests several conclusions: (1) the ratio between [c18 : 1ω9] and {[c19 : 1ω7] + [Δ19 : 0]} indicates that the bacterium is gram-negative; (2) the presence of 3-hydroxy myristic and myristic acids [3-(OH)14 : 0 and 14 : 0] points to Enterobacteriaceae [5]; and (3) the relationship between [c16 : 1ω7], [Δ17 : 0], and [c18 : 1ω7] allows species and infraspecies differentiation [3].

The techniques of for determining the lipid composition are readily accessible, rapid and quite sensitive. For instance, the chromatogram presented in Fig. 1 was obtained by an exceptionally simple method involving the procedure of lipid transmethylation by tetramethyl ammonium hydroxide (TMAH) in a gas chromatograph sample injector [10, 14]. To dissolve bacterial cells in 20% aqueous solution of TMAH (1 min, 105 °C), to inject the solution into a sample injector heated to 360°–380 °C and to separate FAs one needs as little as 10 min. The above chromatogram was obtained using the biomass of bacteria grown on nutrient agar in the form of colonies 1 mm in diameter.

Determination of the phospholipid composition of bacteria is also simple and may be performed by TLC. Living cells are applied directly onto a sorbent layer and extracted with organic solvent [13]. High performance thin-film chromatog-

1 Institute of Biochemistry and Physiology of Microorganisms, USSR Academy of Science, Pushchino, Moscow region, 142292, USSR

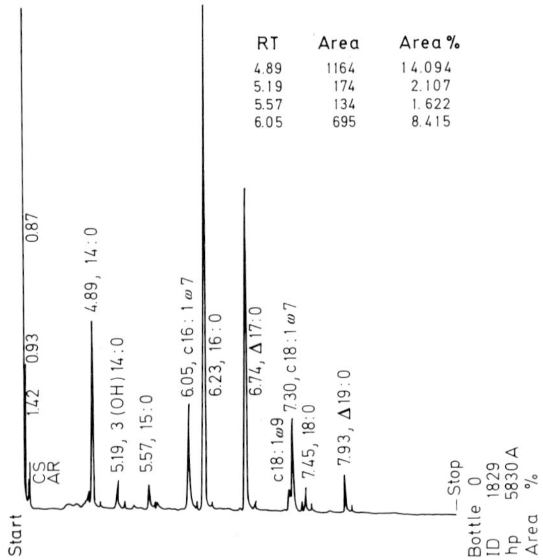

Fig. 1. Chromatogram of *E. coli* VKM B-115 fatty acids obtained by transmethylation by TMAH [10, 14] on a glass capillary column 10 m × 0.2 mm, coated with OV-101

raphy [4, 6] enables one to carry out the analysis within several minutes, with the sensitivity being greatly increased.

As a taxonomic character lipid composition has, along with its many advantages, some essential drawbacks which limit its application in the identification and classification of bacteria. These disadvantages are obviously of a temporary nature since they are mainly the result of an absence of scientifically grounded recommendations concerning the interpretation of bacterial lipid composition; this may vary markedly even under strictly standardized growth conditions and depends on various environmental factors. Noteworthy variations in FA contents may be neglected or are considered differently by different researchers. There is no unanimity even as far as the accuracy of FA analysis is concerned. In some laboratories FA are assayed to an accuracy of 1%, in others to three decimal places.

The investigations carried out by us over the last 15 years have demonstrated the possibility of adopting a strictly quantitative approach to the interpretation of lipid spectra of different systematic groups of bacteria, based on uniform theoretical principles, and their application in the identification and classification of eubacteria. The conceptual framework of this approach is as follows. In eubacteria, whose membrane system excludes the possibility of anatomic compartmentation of membrane enzymic processes, lipids serve as mediators which are responsible for the physiologically expedient coupling of the activity of membrane enzymes and the state of the pool of coenzymes and some important low-molecular metabolites. This coupling is performed by a special mechanism, previously described as a principle of quasi-equilibrium [3, 7, 8], as well as by a special organization of phospholipids in native membranes – polylipids [2, 3].

This theory has been discussed previously in more detail [2, 3, 7, 8]. In the present report we shall consider some practical consequences which have proved useful in the elaboration of methods of rapid taxonomic differentiation of bacteria by their FA spectra. Based on this theory, biosynthesis of FA is believed to be determined by partial quasi-equilibrium processes, which can be divided to six basic groups. Such a division seems to be a rough abstraction, because in fact each process includes several stages involving various enzymes. However, as was shown when studying a large number of eubacteria, such an abstraction appears very useful in the description of the qualitative and quantitative composition of FA and the regularities of its variations at a cell population level. The above-mentioned six processes are:

1. F_m – chain initiation
2. A_{mn} – elongation of FA chain by successive addition of C_2-fragments of malonyl-CoA or (in some cases) acetyl-CoA
3. W_n – dehydration of hydroxy acids to monounsaturated *cis*- and *trans*-acids
4. H_n – hydrogenation of *trans*-monounsaturated acids to saturated acids
5. D_{mn} – desaturation of saturated acids
6. M_n – methylenation of double-bonds of monounsaturated FA

Index "m" is to characterize the nature of the primer acid which gives rise to a fatty acid chain. In even straight chain acids "m" is expressed as "e"; in odd normal acids as "o"; in odd-iso acids as "l"; in even-iso acids as "v"; and in odd-anteiso acids as "i". Index "n" stands for the number of C atoms of the primer acid which undergoes transformations.

According to the principle of quasi-equilibrium, the gross biosynthesis of lipids follows the mass action law, or, in other words, the FA concentration is proportional to the product of concentrations of the components involved in the synthesis of a given acid, raised to the powers which are equal to their stoichiometric coefficients. For instance, the concentration of i14:0 is proportional to the isobutyrate concentration multiplied by the malonate concentration raised to 5, because synthesis of i14:0 starts with the interaction between isobutyryl-CoA and acyl-carrier protein, and stops after the addition of the 5th C_2-fragment [16].

However, the principle of quasi-equilibrium is not applicable in the above form, because of incomparable dimensions of concentrations of the FA and components involved in the quasi-equilibrium processes of their synthesis. This problem may be overcome by using relative coefficients. For example, the ratio between [i14:0] and [i15:0], expressed by F_{vl_5}, is proportional to the ratio between concentrations of iso-butyryl-CoA and iso-valeryl-CoA in bacterial cell juice, multiplied by the quasi-equilibrium constant K_{vl_5}.

Similarly, the coefficient A_{mn} of the chain extension process is determined as the ratio between the concentrations of the fatty acids with n+2 and n C atoms of the same family F_m. Theoretically, this coefficient must be proportional to the ratio between the concentrations of malonyl-CoA (or acetyl-CoA) and free coenzyme A.

The coefficient W_{16} is calculated as the ratio between the concentrations of palmitelaidic and palmitoleic acids; the coefficient H_{16}=[16:0]/[tr16:1ω7];

$D_{e_{18}}$=[c18:1ω9]/[18:0]; M_{16} and M_{18} in gram-negative bacteria =[Δ17:0]/[c16:1ω7] and [Δ19:0]/[c18:1ω7]respectively; $A_{euc_{16}}$=[c18:1ω7]/[c16:1ω7], etc.

As we have revealed [7], the coefficients F, A, W, H, D, and M change strictly regularly during the growth and development of eubacteria, and in dependence upon variations in environmental parameters. For instance, in gram-negative bacteria, the product of the coefficients

$$I_T = A_{e_{14}} \cdot A_{euc_{16}} \cdot W_{16} \cdot H_{16} \cdot D_{e_{18}}^{-1} \cdot M_{16} \tag{1}$$

changes strictly regularly during batch culturing according to the equation:

$$I_T = I_{T\infty}^{\frac{t-\tau}{t}}, \tag{2}$$

where t is the time which has passed since inoculation, $I_{T\infty}$ is the constant which depends on strain peculiarities, and τ is the constant for bacteria of the same family, which has the dimensions of time. The τ value does not depend on culturing medium and increases with the decrease of growth temperature. For example, for Enterobacteriaceae τ=3.4 h at 37 °C and 5.8 h at 29 °C. As was previously shown [3, 7], the product of coefficients A, W, H, and M displays strictly regular changes with the alterations in cell concentration in batchwise growing gram-negative culture.

Studies of the bacteria of different systematic groups have demonstrated the physiologic–biochemical and taxonomic specificity of the above-mentioned types of partial quasi-equilibrium processes which may be described strictly quantitatively. It is noteworthy from the practical viewpoint that the above regularities most often hold true with bacteria grown on solid media. This, in combination with the methods which allow for the precise quantitative analysis of lipid composition of single bacterial colonies, provides a broad range of prospects for the elaboration of rapid methods of identification.

To demonstrate the quasi-equilibrium principle in bacterial populations grown on solid media, we shall consider the FA composition of three *Arthrobacter* strains grown in test tubes with slant agar at 19 °C or 29 °C over 63 h (Table 1). Among FA synthesized by these bacteria, there are four pairs of FA, each of which belongs to one of four families: F_e, F_v, F_l, and F_i. Each acid in a pair differs from the other by 2 C units.

For each strain grown at a certain temperature one may calculate four coefficients A, e.g. $A_{e_{14}}$=[16:0]/[14:0], $A_{v_{14}}$=[i16:0]/[i14:0], etc. It appears that the A values for each family of acids, determined at 19 °C ($A_{(19)}$) and 29 °C ($A_{(29)}$) for these three *Arthrobacter* strains, are interrelated as follows:

$$A_{(19)} = 0.63 \cdot A_{(29)}^{1.10}. \tag{3}$$

This equation accurately describes the experimental results, though the two extremes of $A_{(19)}$ (minimum and maximum) differ from each other by nearly 150 times.

Determination of invariant characteristics of FA spectra affords the scientifically grounded taxonomic calculations. Usually such calculations are most useful for identification at the intrageneric level. To allocate bacteria to genera, the qual-

Table 1. Fatty acid composition of *Arthrobacter aurescens* VKM 653 (ATCC 13344), *A. oxydans* VKM 663 (ATCC 14358) and *A. nicotianae* VKM (ATCC 15236)

Strain	Growth temperature	Relative percentage of fatty acids							
		e		v		i		l	
	(°C)	14:0	16:0	i14:0	i16:0	a15:0	a17:0	i15:0	i17:0
VKM 653	19	2.00	2.49	0.54	3.11	77.37	11.22	3.01	0.19
	29	1.42	2.57	0.69	5.09	68.82	18.03	2.57	0.28
VKM 663	19	3.54	3.82	0.97	5.08	72.57	9.95	3.63	0.14
	29	2.12	3.51	1.12	8.07	63.71	16.43	4.58	0.45
VKM 803	19	5.25	7.51	2.96	9.22	57.43	7.68	9.31	0.63
	29	3.28	6.98	1.82	7.84	57.57	13.69	7.69	1.14

itative peculiarities of FA spectra may be used. In some cases, minimal additional information is sufficient, e.g. the presence of endospores is important evidence in the differentiation between *Bacillus* and *Kurthia,* whose FA spectra are practically identical.

According to FA spectra, all obligately and facultatively aerobic eubacteria may be divided to six groups: three basic (V, B, O) and three mixed groups, the so-called "centaurs" (VB, BO, VO). The groups V and O differ from each other by virtue of the ratio of *cis*-octadecenoic acid concentrations. All bacteria of group V contain minor quantities of oleic (c18 : 1ω9) acid (usually from 0.2% to 3%) and major amounts of *cis*-vaccenic (c18 : 1ω7) acid (10%–90%). In the bacteria of group O, 18 : 1 is presented only by oleic acid, whereas the bacteria of group B do not have octadecenoic acids and possess branched acids. When distinguishing between V and O aerobic eubacteria, one should bear in mind that up on ageing of the population, the concentration of 18 : 1 decreases, followed by the production of methylenated or methylated derivatives with 19 carbon atoms. Concentrations of these derivatives must be summed up with the concentrations of the octadecenoic acids they have been derived from.

Bacteria of the mixed groups contain combinations of the differential FAs. The fatty acid spectra of the OV type include, along with 18 : 1ω7, increased amounts of 18 : 1ω9. These bacteria differ from groups V and O by virtue of the presence of dihydrosterculinic (*cis*-9,10-methylene octadecanoic) acid in aged cultures. The bacteria of VB and BO groups have FA of the V and O type spectra plus branched iso- and antheiso-acids. Type VB is the least stable: under certain culturing conditions, e.g. when carbohydrates are used as growth substrates, it may turn in to the V type.

Apart from these 6 types, about 20 subtypes may be distinguished. The number of subtypes may increase markedly if, on typing, hydroxy acid composition is taken into consideration [19]. Below we mention some of the peculiarities taken into consideration when dividing bacteria into subgroups. (a) Unlike Enterobacteriaceae, the bacteria *Pseudomonas* and of some genera of Vibrionaceae synthesize, along with c16 : 1ω7 and c18 : 1ω7, substantial amounts of palmitelaidinic and *trans*-vaccenic acids. (b) Bacteria *Caulobacter, Asticcacaulis,* and

Prosthemicrobium are characterized by the presence of *cis*-11-methyloctadecenic acid [1], which is absent from some other differentiable gram-negative bacteria. (c) *Hyphomicrobium*, for instance, contains (to 10% rel.) one isomer of *cis, cis*-octadecadienic acid [9] on growth both on pepton-containing media and methanol. Some species of obligate methanotrophic bacteria, which assimilate methane carbon via serine pathway, synthesize several 18:2 isomers [18].

After grouping and subgrouping it is possible to allocate bacteria to genera. Further differentiation may be made based on quantitative calculations according to previously constructed diagnostic schemes which are, as a rule, specific for each group and subgroup. The principal but not the only strategy is to find the invariants of FA spectra in the conditions which provide maximal variability in the contents of FAs. Such an approach was applied by us to the species differentiation of *Thermus* [12].

Another method of calculation was used for the differentiation of *Kurthia* (the experiments were performed jointly with Dr. N. V. Cherevach). We studied the FA composition of 32 *Kurthia* strains. Three strains were obtained from NCIB, ATCC and CCM; others were isolated and described by Belikova et al. [15]. The bacteria were grown in test tubes with slant plain agar for 1–5 days at temperatures within the range 15° to 37 °C (for some strains to 44 °C). In these bacteria, which belong to type B, the contents of some branched fatty acids vary quite widely. However, by plotting the values of $\ln F_{vl_5} = \ln \frac{[i15:0]}{[i14:0]}$ against the values of ln $A_{v_{14}} \cdot A_{l_{15}} \cdot A_{i_{15}}$, we may distinguish between three groups of strains, separated by hiatuses (Fig. 2). This method of grouping wholly agrees with the results of DNA–DNA hybridization of *Kurthia* strains recently reported by Cherevach et al. [17]. The group numbering corresponds to that used in the work mentioned: I = pigmentless strains; II and III = yellow-pigmented strains capable (II) and incapable (III) of growth at 44 °C.

Coefficients A and F and their combinations are basic means for species and generic differentiation of bacteria of group B. For example, an exponential rela-

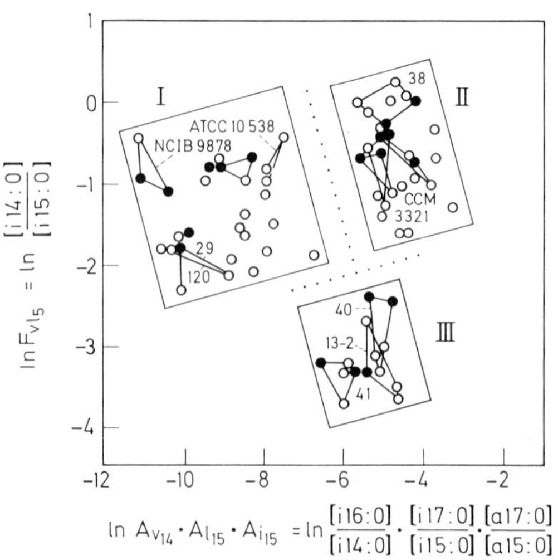

Fig. 2. Grouping of *Kurthia* strains based on their FA spectra. Bacteria were grown on slant plain agar in test tubes for 1 day at temperatures from 15 °C to 37 °C (or 44 °C). *Dark circles* correspond to the results obtained at 25 °C and during the growth period from 2 to 5 days. Circles which refer to one and the same strain are connected by *straight lines*. Some strains are designated by figures

tionship between F_{vi_5} and $A_{i_{15}}$ allows for differentiation of *Arthrobacter* from *Curtobacterium, Microbacterium,* and phytopathogenic "*Corynebacterium.*" In this case neither the growth temperature nor the culture age are taken into consideration. Coefficient D is rarely used for differentiation of bacteria belonging to group B, because of the intraspecies variability of these coefficients, as is the case, for example, with *Promicromonospora citrea* [11]. They decrease with the increase of growth temperature and depend on the stage of development of the culture.

The advantage of taxonomic calculations based on the quasi-equilibrium principle is that they provide a possibility to regulate the interpretation of FA spectra of eubacteria. This, however, does not imply any restriction in the choice of possible techniques. On the contrary, there are, for example, at least half a dozen of techniques for the differentiation of *Kurthia* which lead to the same result. For instance, the degree of homology of *Kurthia* fatty acid spectra may be determined by the equation:

$$h_{FA_{xr}} = 100 \cdot \exp\left(-\frac{1}{n} \cdot \Sigma \left|\ln \frac{I_x}{I_r}\right|\right), \tag{4}$$

where $h_{FA_{xr}}$ is the percent of homology of FA spectra of a strain "x" and a reference strain "r"; I_x and I_r are their quasi-equilibrium coefficients, respectively; and n is the number of coefficients used in the calculation.

Figure 3 illustrates the distribution of 30 *Kurthia* strains according to the degree of relatedness to two reference strains: a type strain NCIB 9878 and strain 38 isolated and described by Belikova et al. [15]. The degree of homology was determined based on three elongation coefficients $A_{v_{14}}$, $A_{l_{15}}$, and $A_{i_{15}}$. Three curves which may be described by the equations

$$h_{9878} \cdot h_{38} = 0.152\,, \tag{5}$$

$$h_{9878} = 0.152 \cdot h_{38}\,, \tag{6}$$

$$h_{9878} = 0.245 \cdot h_{38} \tag{7}$$

correspond to three groups of strains: equation (5) to group I, (6) to group II, and (7) to group III.

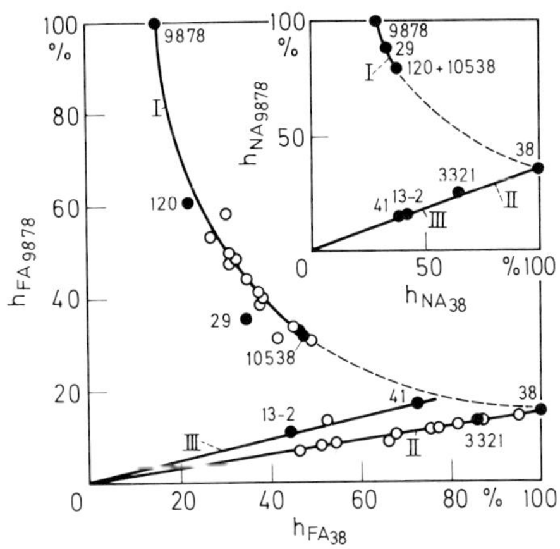

Fig. 3. Homology among *Kurthia* strains by the ir FA spectra (h_{FA}) and DNA–DNA hybridization (h_{NA}). The $h_{NA_{9878}}$ and $h_{NA_{38}}$ values are taken from [17]. The h_{FA} values were calculated based on the fatty acid spectra of bacteria grown on slant plain agar at 25 °C for 1 day. *Dark circles* correspond to the strains for which both h_{FA} and h_{NA} were determined. Strain numbers are designated by figures

Figure 3 also indicates that the degree of homology of fatty acid spectra of *Kurthia* correlates perfectly with the results of DNA–DNA homology studies.

Variability of the bacterial lipid composition, which usually complicates the identification, may be used to advantage, increasing the reliability of identification schemes. This is achievable with the help of the approaches based on the quasi-equilibrium principle, which make it possible to use data obtained in various laboratories even if there are appreciable variations in the contents of the components of FA spectra.

However, the above advantages may not be realized unless some requirements in respect of representation of the results of FA analyses and, consequently, the instrumentation employed for such analyses, are met: (a) the total FA composition of bacterial population cells, including both extractable and firmly bound lipids, should be considered; (b) complete analysis of the isomer composition of both unsaturated and branched FA must be performed; (c) the analysis accuracy must be no lower than to 0.1% rel.; and (d) the representation of the results as the average of parallel measurements is inadmissible. Unfortunately, these requirements are rarely met.

The examples considered above are only a few cases which demonstrate the possibility of taxonomic calculations based on FA spectra of eubacteria. Such calculations made after the grouping and subgrouping of eubacteria according to the qualitative composition of their fatty acids offer the possibility of developing rapid automated methods of identification based on the use of a single character. The accessibility of this apporach may be demonstrated by the example of the species identification of obligate methanotrophic bacteria [18]. The identification potential of their lipid composition appeared to be no lower than that of antigenic characteristics, DNA–DNA homology, protein electrophoretic patterns, etc. As for the simplicity and rapidity of the technique under consideration, it is beyond compare.

References

1. Akimov VN, Andreev LV (1983) Influence of growth conditions on fatty acid composition of stalked bacteria of *Caulobacter* genus. Proceedings of FEMS int symp on environmental regulation of microbial metabolism, Centre of Biological Research, Pushchino, p 61
2. Andreev LV (1978) Possible mechanism of energy accumulation by protobionts and its evolution to contemporary forms of life (in Russian). Abstr All-Union conference on evolutionary biochemistry and origin of life, EGU, Erevan, pp 43–44
3. Andreev LV (1982) General regularities and specific peculiarities of lipid biosynthesis and metabolism in prokaryotes (in Russian). In: Imshenetsky AA (ed) Biosynthesis and metabolism of microbial lipids. Vaskhnil, Moscow
4. Andreev LV (1982) High performance thin-film chromatography in biological studies. J Liq Chrom 5:1573–1582
5. Andreev LV (1983) Express diagnostic of enterobacteria by gas chromatography (in Russian). Abstr I All-Union conference on chromatography in biology and medicine. USSR Ministry of Health, Moscow
6. Andreev LV (1983) Method of separation and detection of substances on silica gel thin films in-pregnated with silver nitrate and phosphatotungstic acid. J High Res Chrom CC 6:575–576
7. Andreev LV (1984) Quasi-equilibrium as a general principle of regulation of lipid composition of eubacteria. In: Dawes EA, Tempest DW, Kulaev IS (eds) Environmental regulation of microbial metabolism. Academic Press, London to be published

8. Andreev LV, Afinogenova AV, Romay Penabad Z, Lambina VA (1983) Peculiarities of fatty acid composition of *Bdellovibrio*. Folia Microbiol 28:28–35
9. Andreev LV, Akimov VN, Galchenko VF (1977) Complex use of chromatographic techniques to detect octadecadienic acids in bacteria (in Russian). Abstr All-Union conference on chromatography in biology and medicine. USSR Ministry of Health, Moscow, p 39
10. Andreev LV, Belyakovich TG (1982) High performance thin layer chromatography coupled with glass capillary chromatography. J High Res Chrom CC 5:100–101
11. Andreev LV, Evtushenko LN, Agre NC (1983) Fatty acid composition of *Promicromonospora citrea* (in Russian). Mikrobiogiya 52:58–63
12. Andreev LV, Egorova LA, Loginova LG (1981) Fatty acid composition of obligately and extremely thermophilic bacteria of genus *Thermus* (in Russian). Prikl Biokhim Mikrobiol 17(3):368–374
13. Andreev LV, Galchenko VF (1983) Phospholipid composition and differentiation of methanotrophic bacteria. J Liq Chrom 6(14):2699–2707
14. Andreev LV, Sklifas AN (1977) Taxonomic aspects of bacterial lipid metabolism (in Russian). In: Blokhina IN (ed) Biofizika i Biokhimiya Mikroorganizmov, ser biol GGU, Gorky, pp 95–99
15. Belikova VL, Cherevach NV, Baryshnikova LM, Kalakoutskii LV (1980) Morphological and physiologo-biochemical characteristics of *Kurthia zopfii*. Mikrobiologiya USSR 49:55–58
16. Kaneda T (1977) Fatty acids of the genus *Bacillus*: an example of branched-chain preference. Bacteriol Rev 41(2):391–418
17. Cherevach NN, Tourova TP, Belikova VL (1983) DNA–DNA homology studies among strains of *Kurthia zopfii*. FEMS Microbiol Lett 19:243–245
18. Galchenko VF, Andreev LV (1984) Taxonomy of obligate methanotrophs. In: Crawford RL, Hanson RS (eds) Microbiol growth on C_1-compounds. ASM, Washington, DC, pp 269–275
19. Moss CW (1978) New methodology for identification of non-fermenters: gas-liquid chromatographic chemotaxonomy. In: Gilardi GL (ed) Glucose nonfermentative gram negative bacteria in clinical microbiology. CRC Press, Boca Raton, pp 171–201

Extraction and Chromatographic Analysis of Characteristic Mycobacterial Lipids

D. E. MINNIKIN, G. DOBSON, and J. H. PARLETT [1]

Introduction

Members of the genus *Mycobacterium,* which includes many important pathogens, are often difficult to classify and identify because of the great variety in their growth rates and biochemical activity [12]. A unifying feature of mycobacteria, however, is the possession of both an internal plasma membrane and an outer membrane based on a matrix of covalently bound, high molecular weight, long-chain, 2-alkyl-branched 3-hydroxy acids, termed mycolic acids [20]. This unusual type of outer membrane is restricted to mycobacteria and related mycolic acid-containing actinomycetes such as corynebacteria, nocardiae, and rhodococci [23]. The mycobacteria, in particular, synthesize a range of free lipids, associated with the outer-membrane organelle, whose unusual structures indicate enormous chemosystematic potential. Chemical characters, such as lipids, are of particular value in taxonomic studies of inert slow-growing organisms, including examples such as *Mycobacterium leprae* which have not been routinely cultivated on laboratory media [9, 30].

Various types of chromatography are the methods of choice for the rapid small-scale analysis of mycobacterial lipids. Efficient use of lipid analysis for identification requires that the minimum amount of biomass be used and that the chemical and chromatographic procedures be so streamlined that multiple samples can be processed quickly without complex manipulations. Various levels of chemotaxonomic information can be obtained by lipid analysis; the degree of detail which is necessary depends on the particular problem. At the lowest level, the presence or absence of a particular lipid may be sufficient to identify an isolate. Characteristic lipid profiles or patterns, which can be subjectively recognized, form the next level of information, and the majority of lipid studies yield such results. In certain cases, however, the profiles are so complex that objective numerical analysis is necessary, using suitable statistical procedures [11].

For many bacterial taxa, three coherent sets of lipid classes are routinely analyzed, comprising long-chain fatty acids, nonpolar isoprenoid quinones, and polar lipids [21, 22, 24]. Mycobacterial free lipids, however, span the whole range of lipid polarities, and the long-chain fatty acids include complex mixtures of mycolic acids and other unusual acids. Early, large-scale analyses of mycobacterial lipids involved differential solvent extraction leading to wax and fat fractions

1 Department of Organic Chemistry, The University, Newcastle upon Tyne, NE1 7RU, Great Britain

which were used as a source of material for the structural analysis of the main lipid types [2]. Such methods proved inappropriate for rapid, small-scale analyses. This paper outlines a recently developed lipid extraction scheme which allows the reproducible preparation of nonpolar and polar lipid fractions and includes special thin-layer chromatographic systems designed to span the whole range of mycobacterial lipid polarities [8].

Lipid Extraction

The efficient extraction of free lipids from biomass requires the use of simple, small-scale procedures that can be applied reproducibly to all samples regardless of their physical condition. An ideal method would also allow all the potentially characteristic lipids to be extracted from a single small sample of biomass, and it would cover the whole range of lipid polarities in order to allow the detection of previously unknown lipid molecules. The well-established lipid extraction procedure of Bligh and Dyer [3] is particularly attractive for the purpose, as it can handle wet or dry organisms and has an integral aqueous wash of the final organic phase to remove nonlipid contaminants. The procedure that has been developed [8, 29], involving two sequential extraction steps, is summarized in Fig. 1. In the first step, a biphasic mixture of petroleum ether and aqueous methanol is used to give a very rapid, yet efficient extraction of relatively nonpolar lipids in the upper petroleum ether phase. The residue, composed of partially extracted cells and the lower aqueous methanolic phase, is repeatedly extracted with a monophasic mixture of chloroform, methanol, and water. The combined extracts are then diluted with chloroform and water to give two layers, the lower containing the polar lipids.

The above procedure is designed to use very simple laboratory apparatus. The recommended extraction vessels are 13 × 100-mm disposable culture tubes, and satisfactory mixing is obtained with tube rotators which can be constructed from unsophisticated stirrer motors. Precise accuracy in measuring solvents is unimportant, and very simple laboratory centrifuges are adequate for assisting the separation of phases. The method was developed specifically for the rapid extraction

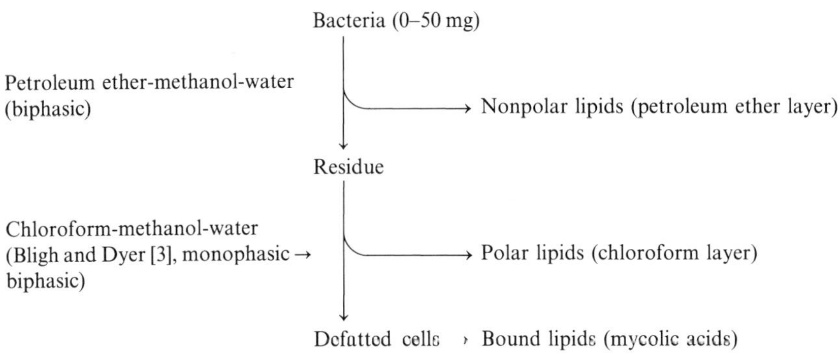

Fig. 1. Scheme for extraction of free lipids

of isoprenoid quinones [29] which did not totally survive the complete Bligh and Dyer procedure. It had been previously shown [16] that nonpolar mycobacterial lipids such as dimycocerosates of phthiocerol and glycosyl phenolphthiocerol could be extracted with petroleum ether, so the extension of the present method to mycobacterial lipids [8] was clearly indicated. The advantage of obtaining the total lipids in two separate extracts is that less lipid is wasted using thin-layer chromatographic systems of different polarities.

Free Lipid Analysis

Thin-layer chromatography (TLC) has been developed for recording patterns of the whole range of mycobacterial lipid polarities. TLC is rapid, economical, and in the two-dimensional mode combined with the use of spray reagents, specific for particular functional groups. It also has high discriminating power. The scheme developed for the resolution of the total polarity range of mycobacterial lipids [8] is outlined in Table 1. A selection of the lipid types analyzed in the various TLC systems is given in Table 1, and detailed examples of typical patterns obtained are available elsewhere [8]. An important aspect of the analysis scheme is that it produces semiquantitative data which is of value not only in assessing

Table 1. Scheme for thin-layer chromatographic analysis of mycobacterial free lipids

Lipids (from 50 mg dry cells)	TLC plates (20 μl each) (*n*)	TLC system[f]	Lipid types
Nonpolar (in 250 μl $CHCl_3$)	1[a]	A	Menaquinones Phthiocerol dimycocerosates Triacylglycerols
	2[a, b]	B	Trehalose mycolipenates
	2[a, b]	C	Glycosyl phenolphthiocerols Free fatty acids
	2[a, b]	D_1	Trehalose dimycolates Sulfolipids
Polar (in 250 μl $CHCl_3$–CH_3OH, 2:1)	3[a, b, c]	D_2	Glycopeptidolipids
	4[a, b, d, e]	E	Phospholipids Glycolipids

[a] Molybdophosphoric acid universal spray [8]
[b] α-Naphthol-sulfuric acid for sugars [8]
[c] Peptidolipid spray [8]
[d] Ninhydrin for amino groups followed by lipid phosphate spray [8]
[e] Periodate-Schiff for α-glycols [8]
[f] TLC two-dimensional systems: A, Petroleum ether (b.p. 60°–80°) – ethyl acetate (98:2 × 3) and petroleum ether-acetone (98:2); B, petroleum ether-acetone (92:8 × 3) and toluene-acetone (95:5); C, chloroform-methanol (96:4) and toluene-acetone (80:20); D_1 and D_2, chloroform-methanol-water (100:14:0.8) and chloroform-acetone-methanol-water (50:60:2.5:3); E, chloroform-methanol-water (60:30:6) and chloroform-acetic acid-methanol-water (40:25:3:6)

Table 2. Phthiocerol dimycocerosates of mycobacteria [10, 19, 26, 33]

Mycocerosate[a]		Phthiocerol A[b]	Phthiodiolone A[b]
C_{27}–C_{32}	*M. bovis*	C_{34}, C_{36}	C_{33}, C_{35}[c]
C_{27}–C_{32}	*M. microti*	C_{34}, C_{36}	C_{33}, C_{35}[c]
C_{27}–C_{32}	*M. tuberculosis*	C_{34}, C_{36}	C_{33}, C_{35}[c]
C_{27}–C_{32}	*M. kansasii*		C_{25}, C_{27}
C_{30}–C_{34}	*M. leprae*	C_{30}, C_{32}	ND
C_{27}–C_{30}	*M. marinum*	C_{28}, C_{30}	C_{27}, C_{29}

ND, not determined

[a] Mycocerosate $CH_3 \cdot (CH_2)_n \cdot (CH_2 \cdot \overset{CH_3}{\overset{|}{C}H} \cdot)_x \cdot CO-$
n, 16–20, x, 3,4

[b] Phthiocerol A $CH_3 \cdot CH_2 \cdot \overset{OCH_3}{\overset{|}{C}H} \cdot$ } $(CH_2)_4 \cdot \overset{OH}{\overset{|}{C}H} \cdot CH_2 \cdot \overset{OH}{\overset{|}{C}H} \cdot (CH_2)_{12-22} \cdot CH_3$

Phthiodiolone A $CH_3 \cdot CH_2 \cdot \overset{O}{\overset{||}{C}} \cdot$ }

[c] Structure only determined in *M. tuberculosis*

which lipids are consistently of value as chemotaxonomic characters but also in giving an appreciation of the interrelation and possible role of such cellular components.

The least polar TLC system (Table 1) is used for the identification of menaquinones, triacylglycerols, and particularly, dimycocerosates of the phthiocerol family [8]. The latter waxes can be isolated, degraded, and their long-chain components analyzed by gas chromatography in combination with mass spectrometry [10, 19, 26]. A summary of the distribution of these waxes is shown in Table 2. Menaquinones can be isolated and identified by TLC [6], high-performance liquid chromatography [17], or mass spectrometry [5]. Mycobacteria usually contain a dihydrogenated menaquinone with nine isoprene units [20, 21].

The second least polar TLC system (Table 1) is used only for the identification in *M. tuberculosis* of acylated trehaloses whose acyl constituents are mycolipenic and mycolipanolic acids [8, 20, 31]. The third least polar TLC system (Table 1) provides for the separation of glycosyl phenolphthiocerol dimycocerosates [8], whose limited distribution is summarized in Table 3. The glycolipids from *M. bovis* and *M. marinum* cochromatograph, but they give distinctly different staining reactions with α-naphthol-sulfuric acid spray for sugars.

The next most polar TLC solvent comination is used to analyze both the nonpolar and polar lipid extracts (Table 1). In the nonpolar extract, the so-called "cord factors," which are dimycolates of trehalose [20], are usually detected, but their quantitative variability makes them unreliable as chemotaxonomic characters. Nonpolar extracts of *M. tuberculosis* may also contain acylated trehalose sulfates [8, 20] and trehaloses acylated with mycolipenic, mycolipanolic, and mycosanoic acids, which are related, methyl-branched, fatty acids [8, 20, 31]. The lat-

Table 3. Glycosyl phenolphthiocerols of mycobacteria [8, 20, 23]

Taxon	Sugar unit	Phenolphthiocerol dimycocerosate
M. bovis	2-*O*-Me-D-rhamnose	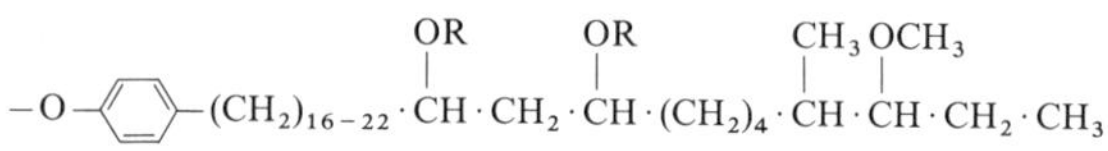
M. marinum	3-*O*-Me-L-rhamnose	
M. kansasii	2-*O*-Me-fucose, 2-*O*-Me-rhamnose, 2,4-*O*-Me_2-rhamnose	
M. leprae	3,6-*O*-Me_2-glucose, 2,3-*O*-Me_2-rhamnose, 3-*O*-Me-rhamnose	

R, mycocerosate (see Table 2)

Fatty acyl – Phe(D) – aThr(D) – Ala(D) – alaninol(L) – 3,4-di-*O*-methyl rhamnose

(aThr) | [6-Deoxytalose – rhamnose – $(sugar)_x$]-*O*-Acetyl

Fig. 2. Generic structure of mycobacterial glycopeptidolipids [4]

ter two classes of lipids are also detected in polar lipid extracts of *M. tubercolosis,* this being one of the few examples where the extraction procedure does not give a clear division of the constituent lipids. The TLC system under consideration (Table 1) is used for the detection in polar lipid extracts of the characteristic glycopeptidolipids found in the *M. avium, M. intracellulare, M. scrofulaceum* complex [4, 8]. The patterns obtained for the native lipids are very complex, and mild alkaline deacylation removes acetyl groups from these lipids and produces simplified patterns of both nonpolar and polar glycopeptidolipid cores [4, 8]. The relatively nonpolar core lipids are common to most strains, but the polar versions differ in their terminal sugar constituents, enabling the various serotypes to be distinguished by TLC [4]. An example of such glycopeptidolipids is given in Fig. 2.

The most polar TLC system (Table 1) allows the identification of the common mycobacterial phospholipids, such as diphosphatidylglycerol, phosphatidylethanolamine, phosphatidylinositol, and mono- and diacylated forms of phosphatidylinositol di- and pentamannosides [8, 20]. In addition, the TLC system shows the presence of characteristic polar glycolipids, such as polar trehalose derivatives characteristic of *M. kansasii* [8, 13]. TLC analysis of similar unidentified, characteristic, polar glycolipids has been developed as a reliable means of mycobacterial identification [4, 14, 35].

Bound Lipid Analysis

A variety of methods have been used to release covalently bound mycolic acids from whole organisms or defatted cells remaining after lipid extraction, as outlined in Fig. 1. Acid methanolysis of dry biomass [25] was developed as a rapid, routine method for preparation of mycolic acid methyl esters from mycobacteria and related nocardiae, rhodococci, and corynebacteria. This method has some disadvantages for mycobacterial mycolates containing cyclopropane rings, as these mycolates may be partially degraded under acid conditions [27]. A procedure involving alkaline degradation with methanolic tetramethylammonium hydroxide and treatment of the resulting salts with iodomethane in dimethylformamide gives reliable methanolysates of mycobacteria [27, 28]. Alternatively, digestion with aqueous methanolic potassium hydroxide and acidification gives free mycolic acids which can be converted to methyl [8] or 4-nitrobenzyl esters [32] by phase-transfer–catalyzed reaction with iodomethane or 4-nitrobenzyl bromide. Hydrolysis with potassium hydroxide in methyl cellosolve is also an efficient method for releasing mycolic acids [7].

Characteristic patterns of mycobacterial methyl esters can be recorded by two-dimensional TLC [20, 21, 25, 28] and also by several one-dimensional systems [7, 25]. Quantitative profiles of mycolic acids have been recorded by high-performance liquid chromatography of 4-bromophenacyl esters [34, 36], pentafluorobenzoates of methyl esters [9], and tert-butyldimethylsilyl ethers of 4-nitrobenzyl esters [32]. Gas chromatography-mass spectrometry of trimethylsilyl ethers of methyl mycolates has also been developed for the analysis of mycolic acid profiles [38]. The distribution of the various mycolic acid types has been given in detail elsewhere [7, 8, 28].

Conclusions

The methods outlined in this paper allow the efficient analysis of all the characteristic mycobacterial free and bound lipids. By simple qualitative TLC analysis

Table 4. Representative mycobacterial lipid chemotypes [8]

Mycolic acids[a]	Taxon	Free lipids	
α	*M. fallax*		
α, α′	*M. chelonae*		
α, keto	*M. leprae* *M. bovis* BCG[b]	Glycosyl phenol-phthiocerols (*M. leprae*, *M. bovis* BCG, *M. bovis*, *M. kansasii*, *M. marinum*)	Phthiocerol dimycocerosates (*M. leprae* to *M. tuberculosis*)
α, keto, methoxy	*M. bovis* *M. kansasii* *M. marinum* *M. tuberculosis*	Sulfolipids, trehalose mycolipenates (*M. tuberculosis*)	
α, α′, epoxy	*M. fortuitum*[c] *M. smegmatis*		
α, keto, wax ester	*M. avium* *M. intracellulare* *M. scrofulaceum*	Glycopeptidolipids	

[a] α-Mycolates lack oxygen functions in addition to 3-hydroxy acid unit; α′-mycolates have lower molecular weight. Detailed mycolate structures are given elsewhere [20, 23, 28]

[b] Certain BCG substrains have *M. bovis* mycolate pattern [8]

[c] α′-Mycolates not positively characterized

of patterns of these lipids [8] it is possible to distinguish mycobacterial species as shown by the examples in Table 4. Having established the essential lipid profile of a particular strain, it is then possible to increase the power of lipid analysis by obtaining quantitative profiles of, for example, the mycocerosates and phthiocerols [10, 19, 33] and mycolic acids [9, 32, 38]. Gas chromatographic analysis of long-chain fatty acids from whole organisms has also been developed [1, 7, 15, 18, 37] as a method for identifying mycobacteria. This type of analysis includes the use of selected ion-monitoring gas chromatography-mass spectrometry for the detection of specific components, such as tuberculostearic and mycocerosic acids [18].

Acknowledgements. Support is acknowledged from the British Leprosy Relief Association, Medical Research Council (G974/522/S and G8216538 SB), and Science and Engineering Research Council (GRA 88651). Aspects of the work were carried out in collaboration with M. Goodfellow and S.M. Minnikin.

References

1. Anderson O, Jantzen E, Closs O, Harboe M, Saxegaard F, Fodstad F (1982) Fatty acid and polar lipid analysis as tools in the identification of *Mycobacterium leprae* and some related slow-growing mycobacterial species. Ann Microbiol (Paris) 133B:29–37
2. Asselineau J (1966) The bacterial lipids. Hermann, Paris
3. Bligh EG, Dyer WJ (1959) A rapid method of total lipid extraction and purification. Can J Biochem Physiol 37:911–917
4. Brennan PJ, Heifets M, Ullom BP (1982) Thin-layer chromatography of lipid antigens as a means of identifying non-tuberculous mycobacteria. J Clin Microbiol 15:447–455
5. Collins MD, Pirouz T, Goodfellow M, Minnikin DE (1977) Distribution of menaquinones in actinomycetes and corynebacteria. J Gen Microbiol 100:221–230
6. Collins MD, Shah HN, Minnikin DE (1980) A note on the separation of natural mixtures of bacterial menaquinones using reverse-phase thin-layer chromatography. J Appl Bacteriol 48:277–282
7. Daffé M, Lanéelle MA, Asselineau C, Lévy-Frébault V, David H (1983) Intérêt taxonomique des acides gras des mycobactéries: proposition d'une méthode d'analyse. Ann Microbiol (Paris) 134B:241–256
8. Dobson G, Minnikin DE, Minnikin SM, Parlett JH, Goodfellow M, Ridell M, Magnusson M (1984) Systematic analysis of complex mycobacterial lipids. In: Goodfellow M, Minnikin DE (eds) Chemical methods in bacterial systematics. Academic, London, pp 237–265
9. Draper P, Dobson G, Minnikin DE, Minnikin SM (1982) The mycolic acids of *Mycobacterium leprae* harvested from experimentally infected nine-banded armadillos. Ann Microbiol (Paris) 133B:39–47
10. Draper P, Payne SN, Dobson G, Minnikin DE (1983) Isolation of a characteristic phthiocerol dimycocerosate from *Mycobacterium leprae*. J Gen Microbiol 129:859–863
11. Drucker DB (1981) Microbiological applications of gas chromatography. Cambridge University Press, Cambridge
12. Goodfellow M, Wayne LG (1982) Taxonomy and nomenclature. In: Ratledge C, Stanford JL (eds) The biology of the mycobacteria, vol 1. Academic, London, pp 471–521
13. Hunter SW, Murphy RC, Clay K, Goren MB, Brennan PJ (1983) Trehalose – containing lipoolgosaccharides. A new class of species – specific antigens from *Mycobacterium*. J Biol Chem 258:10481–10487
14. Jenkins PA (1981) Lipid analysis for the identification of mycobacteria: an appraisal. Rev Infect Dis 3:862–865

15. Julák J, Tureček F, Mikova Z (1980) Identification of characteristic branched – chain fatty acids of *Mycobacterium kansasii* and *gordonae* by gas chromatography-mass spectrometry. J Chromatogr 190:183–187
16. Koul AK, Gastambide-Odier M (1977) Microanalyse rapide de dimycocérosate de phthiocérol, de mycosides et de glycérides dans les extraits à l'éther de pétrole de *Mycobacterium kansasii* et du BCG, souche Pasteur. Biochimie 59:535–538
17. Kroppenstedt RM (1982) Separation of bacterial menaquinones by HPLC using reverse-phase (RP 18) and a silver-loaded ion exchanger as stationary phases. J Liq Chromatogr 5:2359–2367
18. Larsson L, Mårdh PA, Odham G, Westerdahl G (1981) Use of selected ion monitoring for detection of C_{32} mycocerosic acid in mycobacteria and in five-day-old cultures of sputum specimens from patients with pulmonary tuberculosis. Acta Pathol Microbiol Scand [B] 89:245–251
19. Mallet AI, Minnikin DE, Dobson G (1984) Gas chromatography-mass spectrometry of *tert*-butyldimethylsilyl ethers of phthiocerols and mycocerosic alcohols from *Mycobacterium tuberculosis*. Biomed Mass Spectrom 11:79–86
20. Minnikin DE (1982) Lipids: complex lipids, their chemistry, biosynthesis and roles. In: Ratledge C, Stanford JL (eds) The biology of the mycobacteria, vol 1. Academic, London, pp 95–184
21. Minnikin DE, Goodfellow M (1980) Lipid composition in the classification and identification of acid-fast bacteria. In: Goodfellow M, Board RG (eds) Microbiological classification and identification. Academic, London, pp 189–256
22. Minnikin DE, Goodfellow M (1981) Lipids in the classification of *Bacillus* and related taxa. In: Berkeley RCW, Goodfellow M (eds) The aerobic endospore-forming bacteria. Academic, London, pp 59–90
23. Minnikin DE, O'Donnell AG (1984) Actinomycete envelope lipid and peptidoglycan composition. In: Goodfellow M, Mordarski M, Williams ST (eds) The biology of the actinomycetes. Academic, London, pp 337–388
24. Minnikin DE, Goodfellow M, Collins MD (1978) Lipid composition in the classification and identification of coryneform and related taxa. In: Bousfield IJ, Callely AG (eds) Coryneform bacteria. Academic, London, pp 85–160
25. Minnikin DE, Hutchinson IG, Caldicott AB, Goodfellow M (1980) Thin-layer chromatography of methanolysates of mycolic acid-containing bacteria. J Chromatogr 188:221–233
26. Minnikin DE, Dobson G, Hutchinson IG (1983) Characterization of phthiocerol dimycocerosates from *Mycobacterium tuberculosis*. Biochim Biophys Acta 753:445–449
27. Minnikin DE, Minnikin SM, O'Donnell AG, Goodfellow M (1984) Extraction of mycobacterial mycolic acids and other long-chain compounds by an alkaline methanolysis procedure. J Microbiol Methods 2:243–249
28. Minnikin DE, Minnikin SM, Parlett JH, Goodfellow M, Magnusson M (1984) Mycolic acid patterns of some species of *Mycobacterium*. Arch Microbiol 139:225–231
29. Minnikin DE, O'Donnell AG, Goodfellow M, Alderson G, Athalye M, Schaal A, Parlett JH (1984) An integrated procedure for the extraction of bacterial isoprenoid quinones and polar lipids. J Microbiol Methods 2:233–241
30. Minnikin DE, Dobson G, Draper P (1985) Characterization of *Mycobacterium leprae* by lipid analysis. Acta Leprologica 2:113–120
31. Minnikin DE, Dobson G, Sesardic D, Ridell M (1985) Mycolipenates and mycolipanolates of trehalose from *Mycobacterium tuberculosis*. J Gen Microbiol 131 (to be published)
32. Minnikin DE, Dobson G, Goodfellow M, Draper P, Magnusson M (1985) Quantitative comparison of the mycolic and fatty acid composition of *Mycobacterium leprae* and *Mycobacterium gordonae*. J Gen Microbiol 131 (to be published)
33. Minnikin DE, Dobson G, Goodfellow M, Magnusson M, Ridell M (1985) Distribution of some mycobacterial waxes based on the phthiocerol family. J Gen Microbiol 131 (to be published)
34. Qureshi N, Takayama K, Jordi HC, Schnoes HK (1978) Characterization of the purified components of a new homologous series of α-mycolic acids from *Mycobacterium tuberculosis* H37Ra. J Biol Chem 253:5411–5417
35. Schrt I, Küppler W, Lange A (1982) Thin-layer chromatography of mycobacterial lipids and its importance for the differentiation and identification of mycobacteria. Z Erk Atmungsorgane 158:233–242

36. Steck PA, Schwartz BA, Rosendahl MS, Gray GR (1978) Mycolic acids: a reinvestigation. J Biol Chem 253:5625–5629
37. Tisdall PA, Roberts GD, Anhalt JP (1979) Identification of clinical isolates of mycobacteria with gas liquid chromatography alone. J Clin Microbiol 10:506–514
38. Toriyama S, Yano I, Masui M, Kusunose E, Kusunose M, Akimori N (1980) Regulation of cell wall mycolic acid biosynthesis in acid-fast bacteria. I. Temperature-induced changes in mycolic acid molecular species and related compounds in *Mycobacterium phlei*. J Biochem 88:211–221

Progress in Methods for More Rapid Isolation of Agents Responsible for Bacteremia

J. M. MATSEN [1]

Introduction

The processing of blood cultures is undergoing rapid change, as is apparent from a steady proliferation of information regarding blood volumes, media, media volumes, blood to media ratios, media additives, subculturing philosophy, collection systems, and media container permutation, as well as innovative total system approaches.

Broth Media Formulation and Additives

Progress in broth media formulation can be considered as both system and nonsystem related. One pertinent advance in "nonsystem" media regards supplemented peptone broth. Information now available indicates that this broth has an advantage over other media in the isolation of streptococci and anaerobic organisms (Reimer et al. 1983). Other media advances are primarily system related.

Though the emphasis of this paper is on the progress in methods for more rapid isolation of microorganisms in blood cultures, it should be noted that media additives have been a major focus of investigational efforts during the recent past. A recent study of hypertonic media suggests that hypertonicity should be reserved for unusual circumstances or where strongly suspected organism presence is not borne out by routine culture methodology (Reimer et al. 1983; Welch and Kelly 1979).

Antimicrobial Inhibitors. Additives for the inactivation of antimicrobials or the use of Thiol Broth as a specific antimicrobial inhibitor have received considerable scrutiny. These antimicrobial-inactivating methods not only hold promise for the increased isolation of organisms which might otherwise be negatively affected by the presence of the antimicrobial. Ideally such systems might also allow a more rapid growth of cultures, which could conceivably expedite the isolation of pathogens.

Thiol Broth. Difco Laboratories manufactures Bacto-Blood Culture Bottle Thiol Broth, which uses sodium polyanetholsulfonate to process patient specimens

1 Departments of Pathology and Pediatrics, University of Utah School of Medicine, Salt lake City, UT 84132, USA

when the patient from whom the culture is taken is receiving antimicrobial therapy. Murray and Niles (1982) report that the Thiol formulation indeed inactivates penicillin and aminoglycoside antimicrobial groups but has no effect on other antimicrobials. Further, these authors note that it was not possible to reproduce this effect when a variety of other media were tested.

Antimicrobial Removal Device. The antimicrobial removal device (ARD) is a method based on the premise that resins will attract and bind antimicrobial substances so as to inactivate up to 100 μg/ml of these agents in blood. The ARD, first described by Wallis et al. (1980), employs a vial containing cationic and polymeric adsorbent resins in saline with sodium polyanetholsulfonate. After inoculation into the ARD vial a 10-ml aliquot of blood is then rotated for 15 min to insure optimal adsorbtion. The blood is then withdrawn for processing by whichever routine culture system the user may prefer.

The literature regarding the ARD is somewhat controversial. Washington (1982) reports that the "customary microscopic indicators of growth, i.e., turbidity, hemolysis, colonies, become impossible to detect in broth cultures of processed blood, and routine ('blind') aerobic and anaerobic subcultures must be performed at frequent intervals to detect growth."

Wright et al. (1982) report that when the use of the ARD was compared with cultures of unprocessed blood there was no statistically significant change in the isolation rate for any organism group, regardless of whether or not the patient was on antimicrobial therapy. Furthermore, these investigators found significantly fewer ($p<0.05$) patients for whom the only positive cultures were made with ARD-processed blood as compared to those whose only positive culture was with unprocessed blood.

In a review of the literature relating to resin-containing methods, Hansen et al. (1983) conclude that such methods can improve detection of microorganisms in patients with clinically significant bacteremia who are receiving antimicrobials. They also stress that such methods should always be used in conjunction with conventional culture means and that the more rapid detection times reported using resin-containing systems to inactivate antibiotics may be artificially skewed. Because of the factors described above by Washington (1982), the studies which have demonstrated more rapid isolation have used an enhanced subculturing frequency, that would not be routinely used in most laboratories.

BACTEC 16B Medium. Recent studies by a number of investigators (Hansen et al. 1983; Smith and Eng 1983; Appelbaum et al. 1983; Doern and Gantz 1983) have shown the efficacy of the 16B medium, a resin-containing media modification, when compared with the routinely used BACTEC 6B medium. Also recently developed is the new 17D medium, a resin-containing anaerobe version medium.

Sodium Polyanetholsulfonate. It should be mentioned that, among other properties, sodium polyanetholsulfonate also manifests the capacity to inactivate certain antimicrobials, such as the aminoglycoside and polymyxin groups. However, the concentration usually used ($<0.03\%$) is less than optimal for complete inactivation.

Systems-Related Methods of Microbial Isolation

The BACTEC System

The BACTEC system, a radiometric detection system based on the release of $^{14}CO_2$ during bacterial metabolism of radioactively labeled nutrients, has seen a number of recent changes and advances. Not the least of these have been the development of resin-containing blood culture bottles (16B and 17D) and a new anaerobic blood culture bottle (7D), which has shown impressive results and represents a considerable improvement over earlier BACTEC anaerobic culture media. The BACTEC system is widely used in the United States and Canada, with most users indicating 50%–65% of organisms detected within 24 h of system inoculation. In this respect, the BACTEC system does as well as any currently available system for the consistent, rapid detection of organisms in bacteremia. In a recent study by McLaughlin et al. (1983), the BACTEC system was compared with the lysis-centrifugation technique (Isolator). Of 91 blood specimens obtained that yielded clinically significant organisms, 52 were positive by both systems, 27 were positive by the Isolator system, and 12 were positive with the BACTEC system only. Of the 57 organisms detected by both systems, 28 were detected simultaneously, 13 were detected earlier by the Isolator system, and 16 were detected earlier by the BACTEC system. Isolated colonies were obtained earlier by the Isolator system in 40 cases and by the BACTEC system in five cases. The additional benefits of standardization and automation in the processing of blood cultures are very attractive and compelling features of the BACTEC system for many users.

BACTEC Anaerobe Media (7D and 17D). We have previously mentioned the development of the 16B and 17D BACTEC media. The 17D medium is so new that there is no currently available literature on it. The only comparison of the 7D and 17D media I was able to find is a report given at the last meeting of the American Society for Microbiology, held in March of this year. In the study they reported on, Reynolds et al. (1984) had detected 77 total isolates, with four detected in the 7D medium only, 17 detected in the 17D medium only, and 48 detected in both. They report that an additional 10% of bacteremic episodes were detected through using the 17D media.

BACTEC Hypertonic 8B Culture Medium. LaScolea et al. (1983) have reported on the advantages of the BACTEC hypertonic 8B culture in the detection of *Hemophilus influenzae* bacteremia in children. They report that within the first 12 h. *Hemophilus influenzae* was detected in the hypertonic medium in 48.5% of the positive cultures, compared with 35% for the aerobic medium. Importantly, their study indicates that, after the first 12 h, the hypertonic medium yielded positive results sooner than the aerobic medium, the difference being statistically significant ($p < 0.01$). Their conclusion is that the BACTEC 8B medium offers a distinct advantage in the early detection of *Hemophilus influenzae* bacteremia, and they feel it can be recommended as a routine test for those age groups in which this pathogen plays a predominant role.

BACTEC Computerization. Another recent innovation with regard to the BACTEC system is the use of microcomputers to control the BACTEC 460 instrument, to receive and analyze data from it, and to prepare summary reports and reviews. The addition of the microcomputer virtually eliminates all manual data recording and precludes transcription errors.

The Isolator System

Description of the Isolator System. The Isolator system is a lysis-centrifugation system in which the lysing agent is present in the tube into which the blood sample from the patient is collected. The centrifugation step separates the organism fraction (the concentrate) from the serum supernatant. The concentrate is removed and inoculated onto primary isolation plate media, which are then incubated for organism detection and isolation. Several studies indicating the advantages of the Isolator system in the detection of organisms have been published. A recent report by Kelly et al. (1983) reports a four-institution collaborative study. Of 154 clinically significant isolates, 89% were detected by the Isolator system, and 73% were detected by a two-bottle broth culture method. A total of 15 polymicrobial cultures were encountered, with the lysis-centrifugation technique detecting 93% of the organisms in these cultures, while the broth culture method detected only 20%. Isolated colonies of clinically important organisms were available 30 h earlier with the lysis-centrifugation technique. In another study from the Mayo Clinic, Henry et al. (1983) report on a controlled evaluation of 6,010 blood cultures, with the yield of clinically significant microorganisms being greater from the lysis-centrifugation system than from either a nonvented vacuum bottle containing trypticase soy broth or a vented bottle containing biphasic brain heart infusion medium. They noted a significantly increased frequency in the isolation of *Staphylococcus aureus* and *Candida* species with the Isolator system and a significantly decreased time required for the detection of *Staphylococcus aureus, Pseudomonas aeruginosa,* and *Candida species.*

Isolator Contamination. One problem with the Isolator system that has been reported is that of plate contamination. Data from our own laboratory indicate that during the collaborative study in wich we participated in 1981, we had 7+% contamination. During all of 1982 and 1983, during which time about 35% of our blood cultures were performed with the Isolator system, we had a contamination rate of 1.3%. The significant decrease in contamination in our own institution as well as other institutions where similar decreases have occurred is undoubtedly due to such factors as technologist familiarity with the system, increased awareness that plates are held 5 days, and initial processing in a laminar flow hood.

Isolator Special Organism Considerations. The literature relating to the Isolator system does not report consistently on the potentially decreased recovery of *Streptococcus pneumoniae* and aerobic bacteria, which some investigators consider a problem. The study by Henry et al. (1983) referred to previously indicates that the Isolator should be used together with a bottle-containing broth suitable for the recovery of these organisms.

An additional advantage of the Isolator system is the enhanced recovery of yeasts and fungi. Recent reports by Bille et al. (1983), Kiehn et al. (1983), and Bille et al. (1984) indicate that the Isolator system is an excellent recovery system for these organisms and might even be the best system currently available. Pierce et al. (1983) have also shown an impressive yield of mycobacteria from mycobacteremia situations.

Isolator Initial Processing. Concern has arisen over the laborintensive initial processing of the Isolator system, which has in the past required the use of a needle syringe procedure that is relatively cumbersome and tedious. The DuPont Company has very recently developed a processing system for the Isolator that facilitates the initial handling of extractions from the microbial tubes. The system, called the Isostat, consists of a manually operated press device, a single-use access and venting cap, supernatant pipettes, and transfer pipettes. All processing steps are carried out via top entry into the tubes. The system provides a 40%–50% reduction in the time required to process the Isolator tube and in addition, provides a more standardized result in the removal and inoculation of the concentrate to plate media, requiring less technologist manipulation of the tubes. An additional, potentially important advantage of this new system is the increased safety achieved by the elimination of all needles and syringes. The simplicity of the system may also in fact make it possible to utilize the Isolator microbial tubes more extensively, since it is now more convenient for night and evening laboratory personnel to process blood specimens received outside normal laboratory hours.

Finally, the manufacturer of the Isolator system has now made available a 10-ml tube as well as a 7.5-ml tube. One concern I have with regard to the Isolator system stems from the excellent studies now being reported on the volume of blood which is optimally necessary for the recovery of organisms in bacteremia. The rate of incidence in more than 50% of adult bacteremia cases is less than one organism per ml, which is cause for concern. One study not yet performed with the Isolator system is to compare the recovery achieved by using two Isolator tubes with the recovery resulting from the use of only a single tube. Such a study needs to be done.

The Septi-chek System

Septi-chek System Description and Comparisons. The Septi-chek system utilizes an agar paddle positioned in a screw-on device above a trypticase soy broth blood culture bottle. The principle is similar to that of the biphasic blood culture bottle. The agar surfaces are inoculated by tilting the bottles, thus eliminating subculture. This system has been shown to result in an enhanced recovery for anaerobic organisms. In a study reported by Henry et al. (1984) in which the Septi-chek system was compared with a vented biphasic brain heart infusion blood culture bottle, Septi-chek had a significantly higher isolation rate for gram-negative organisms in general and the *Enterobacteriaceae* in particular when compared to a more conventional bottle blood culture system. Septi-chek also had a significantly increased rate of isolation for streptococci.

Pfaller et al. (1982) conducted a study in which they compared the Septi-chek with a trypticase soy two-bottle broth blood culture system. They demonstrated that 299 of 413 organisms were detected in the Septi-chek, whereas only 253 organisms were detected in the two-bottle system.

Because of its pertinence to those of us evaluating blood culture systems, a direct comparison in a large sample study of the Septi-chek and the Isolator systems should be carried out.

Subculture and Staining Methods

Subcultures. In discussing the rapid isolation of blood culture organisms, we should focus on the important role played by early subculture in cases where a conventional two-bottle system is utilized. Todd and Roe (1975) first directed attention to the principle of the early subsulture. Subsequent studies by Silva and Washington (1980) and Hawkins et al. (1983) have provided additional perspectives of relevance.

Acridine Orange Staining. Kronvall and Myhre (1977) pointed out the advantages of using acridine orange staining with body fluid cultures. Their report was followed by McCarthy and Senne (1980), who demonstrated that acridine orange stain is as satisfactory as a 48-h subculture for detecting organisms present in blood cultures. Mirrett et al. (1982) further demonstrated the advantages of acridine orange over methylene blue and Gram stains. Cashman et al. (1983) demonstrated the value of the acridine orange stain when used at the time the Isolator tubes are initially processed. The tubes in their study had been held overnight prior to processing.

Tierney et al. (1983) stained 2,205 microscopically negative blood cultures with acridine orange after 6–17 h of inoculation and incubation and found that this is as sensitive as an early subculture in detecting a positive blood culture. Of the 179 positive blood cultures in their sample, 30 (16.8%) were detected by acridine orange alone, 19 (10.6%) were detected by early subculture alone, 84 (46.9%) were detected by both techniques, and 46 (25.7%) were not detected by either method. For selected high-risk patients, acridine orange may also be helpful in processing blood much as the initial buffy coat smear has been used. This should be done only in those patients for whom there is a high expectation of large organism numbers in blood.

Other Developmental Systems

Potential exists for direct organism detection through the measurement of specific organism components of by-products. Antigen-antibody detection systems, chromatographic and mass spectrometry approaches, and nucleic acid assay methods have limited current applicability and will require considerable development before they become widely available. A recent report by Molin et al. (1983) reported encouraging results utilizing a bioluminescent assay of bacterial ATP. As early as 1975, Ur and Brown reported success utilizing electrical impedance. This was followed by Specter et al. (1977). More recently an article by Buckland et al. (1983) is optimistic concerning the detection of organisms by impedance methodology.

Summary

New modifications in media formulations appear to have enhanced the capacity for organism detection with the BACTEC system. There is also sufficient evidence to indicate that, in comparison with conventional systems, both the Septi-chek and the Isolator systems have increased detection rates. However, it may be that a dual or triple approach is required for optimum recovery of organisms. A multiple approach of this type must attempt to cover all ranges of the microbial pathogen spectrum, including anaerobic organisms, other fastidious organisms, and those organisms whose enhanced recovery seems linked to the newly developed systems. Such organisms include staphylococci, yeasts, and fungi. Recent data indicate that by covering all ranges of the spectrum with the help of optimal culture methodology, one can also significantly increase the overall yield and rapidity of organism detection.

References

Appelbaum PC, Beckwith DG, Dipersio JR, Dyke JW, Salventi JF, Stone LL (1983) Enhanced detection of bacteremia with a new bactec resin blood culture medium. J Clin Microbiol 17:48–51

Bille J, Stockman L, Roberts GD, Horstmeier CD, Ilstrup DM (1983) Evaluation of a lysis – centrifugation system for recovery of yeasts and filamentous fungi from blood. J Clin Microbiol 18:469–471

Bille J, Edson RS, Roberts GO (1984) Clinical evaluation of the lysis-centrifugation blood culture system for the detection of fungemia and comparison with a conventional biphasic broth blood culture system. J Clin Microbiol 19:126–128

Buckland A, Kessock-Philip S, Bascomb S (1983) Early detection of bacterial growth in blood culture by impedance monitoring with a bactometer model 32. J Clin Pathol 36:823–828

Cashman JS, Boshard R, Matsen JM (1983) Viability of organisms held in the isolator blood culture system for 15 h and their rapid detection of acridine orange staining. J Clin Microbiol 18:709–712

Doern GV, Gantz NM (1983) Detection of bacteremia in patients receiving antimicrobial therapy: an evaluation of the antimicrobial removal device and 16B medium. J Clin Microbiol 18:43–48

Hansen SL, Hetmanski J, Stewart BJ (1983) Resin-process methods for improved isolation of organisms from blood and other body fluids. Am J Med July 28:31–36 (special issue)

Hawkins BL, Peterson EM, de la Maza LM (1983) Rapid detection of positive blood cultures. J Clin Microbiol 18:716–718

Henry NK, McLimans CA, Wright AJ, Thompson RL, Wilson WR, Washington JA II (1983) Microbiological and clinical evaluation of the isolator lysis-centrifugation blood culture tube. J Clin Microbiol 17:864–869

Henry NK, Grewell CM, McLemans CA, Washington JA II (1984) Comparison of the Roche septi-chek blood culture bottle with a brain heart infusion biphasic medium bottle and with a tryptic soy broth bottle. J Clin Microbiol 19:315–317

Kelly MT, Fojtasek MF, Abbott TM, Hale DC, Dizikes JR, Boshard R, Buck GE, Martin WJ, Matsen JM (1983) Clinical evaluation of a lysis – centrifugation technique for the detection of septicemia. JAMA 250:2185–2188

Kiehn TE, Wong B, Edwards FF, Armstrong D (1983) Comparative recovery of bacteria and yeasts from lysis – centrifugation and a conventional blood culture system. J Clin Microbiol 18:300–304

Kronvall G, Myhre E (1977) Differential staining of bacteria in clinical specimens using acridine orange buffered at a low pH. Acta Pathol Microbiol Scand [B] 85:249–254

LaScolea LJ Jr, Sullivan TD, Dryja D, Neter E (1983) Advantages of BACTEC hypertonic culture medium for detection of *Haemophilus influenzae* bacteremia in children. J Clin Microbiol 17:1177–1179

McCarthy LR, Senne JE (1980) Evaluation of acridine orange stain for detection of microorganisms in blood cultures. J Clin Microbiol 11:281–285

McLaughlin JC, Hamilton P, Scholes JV, Bartlett RC (1983) Clinical laboratory comparison of lysis-centrifugation and BACTEC radiometric blood culture techniques. J Clin Microbiol 18:1027–1031

Mirrett S, Lauer BA, Miller GA, Reller LB (1982) Comparison of acridine orange, methylene blue, and Gram stains for blood culture. J Clin Microbiol 15:562–566

Murray PR, Niles AC (1982) Inactivation of penicillins by thiol broth. J Clin Microbiol 16:982–984

Pfaller MA, Sibley TK, Westfall LM, Hoppe-Bauer JE, Keating MA, Murray PR (1982) Clinical laboratory comparison of a slide blood culture system with a conventional broth system. J Clin Microbiol 16:525–530

Pierce PF, DeYoung DR, Roberts GD (1983) Mycobacteremia and the new blood culture systems. Ann Intern Med 99:786–789

Reimer LG, Reller LB, Mirrett S, Wang WL, Cox RL (1983) Controlled evaluation of hypertonic sucrose medium at a 1:5 ratio of blood to broth for detection of bacteremia and fungemia in supplemented peptone broth. J Clin Microbiol 17:1045–1049

Reynolds JK, Siders JA, Van Dierendonck MB, Allen SD (1984) Evaluation of the BACTEC 17D anaerobic resin medium for the detection of bacteremia. Abstracts of the annual meeting of the American Society for Microbiology, St. Louis, 4–9 March. Abstract no. C-12, p 238

Silva HA, Washington JA II (1980) Optimal time for routine early subculture of blood cultures. J Clin Microbiol 12:445–446

Smith SM, Eng RHK (1983) In vitro evaluation of the BACTEC resin-containing blood culture bottle. J Clin Microbiol 17:1120–1126

Specter S, Throm R, Strauss R, Friedman H (1977) Rapid detection of bacterial growth in blood samples by a continuous-monitoring electrical impedance apparatus. J Clin Microbiol 6:489–493

Tierney BM, Henry KN, Washington JA II (1983) Early detection of positive blood cultures by the acridine orange staining technique. J Clin Microbiol 18:830–833

Todd JK, Roe MH (1975) Rapid detection of bacteremia by an early subculture technique. Am J Clin Pathol 64:694–699

Ur A, Brown DFJ (1974) Impedance monitoring of bacterial activity. J Med Microbiol 8:19–28

Wallis C, Melnick JL, Wende RD, Riely PE (1980) Rapid isolation of bacteria from septicemic patients by use of an antimicrobial agent removal device. J Clin Microbiol 11:462–464

Washington JA II (1982) The role of the microbiology laboratory in the diagnosis and antimicrobial treatment of infective endocarditis. Mayo Clin Proc 57:22–32

Welch DF, Kelly MT (1979) False-positive macroscopic appearance of blood cultures in sorbitol-containing hypertonic medium. J Clin Microbiol 9:635–636

Wright AJ, Thompson RL, McLimas CA, Wilson WR, Washington JA II (1982) The antimicrobial removal device: a microbiological and clinical evaluation. Am J Clin Pathol 78:173–177

Mechanised Plate Spreading – From Concept to Realisation

N. B. Coupe [1] and P. J. L. Sequeira [2]

Introduction

Primary culturing to produce isolates for identification and enumeration has long been recognised as a time-consuming process, with the classical loop and Petri plate techniques still occupying up to a third of the total technician time available in the busy clinical bacteriology department. It is still true to say that bacteriology remains very much a handcraft, having lagged far behind other branches of pathology in the development of mechanised and automated techniques. Although automatic methods based on growth rates in selective media are becoming more common, there is still room for many devices to improve efficiency in the microbiology laboratory.

The development of our mechanised system for spreading Petri plates resulted from the need to release more skilled technician time from routine tasks in a laboratory with increasing work loads and where space and staff were at a premium and budgets were constantly under pressure: plating out became an obvious target for mechanisation.

The resulting development programme was aimed at devising a compact system capable of coping with the wide range of specimen types found in the typical clinical laboratory and which, by speeding up routine spreading, would reduce the need for more staff.

The plate spreader which evolved from the development programme at Withington Hospital made savings of up to 30 s per plate and gave consistent and reproducible separations even when operated by unskilled staff.

The machine was considered to be suitable for commercial exploitation and further development work at Denley resulted in the device being marketed under the name "Autospreader."

The Concept

The overall philosophy of the development programme was to keep the system as simple as possible (in what is essentially a complex manual procedure) by using the dexterity of the operator at those stages which would otherwise become difficult and therefore expensive to mechanise: the system had to operate with the

1 Denley Instruments Limited, Natts Lane, Billingshurst, Sussex RH14 9EY, Great Britain
2 Withington Hospital, Manchester, Great Britain

traditional cost-effective, readily available and well proven poured Petri plate in order to interface readily with the accepted routine of the laboratory.

The wide range of specimen types, their consistencies and the variety of containers in which they were sent to the laboratory determined that inoculation of the plates should be by hand, using sterile disposable swabs or loops and using the specimen as received; to have decided otherwise would have involved a time-consuming aseptic pretreatment stage to transform each specimen, whatever its consistency, into an inoculum which could be sampled and transferred automatically. It was also decided that, for simplicity in design, media selection and plate labelling should also be manual procedures.

These decisions meant that any real time saving had to come from the spreading and loop sterilisation stages only; this was estimated to mean a saving of between 20 and 30 s per plate compared with hand spreading with a single loop. Such a gain was considered to be worthwhile provided that the mechanised spreader was able to operate automatically and spread one plate unattended while the operator prepared the next.

The Specification

The following specification for a plate spreader to be used in the clinical laboratory was outlined:

1. It should be small and preferably portable – so that it could be located at the work station where space was invariably at a premium.

2. The operation should allow plates to be spread in sequence and as soon as possible after inoculation – to reduce absorption of the inoculum into the agar.

3. The spreading sequence should be entirely automatic and initiated by loading the plate into the instrument – so that the operator would be free to prepare the next plate during the operating cycle.

4. The machine should be capable of "queuing" a limited number of plates – to absorb any short-term build-up when, for example, more than one plate was being inoculated from a single specimen.

5. The spread plates should be automatically stacked – so that the operator would be able to easily remove groups of four or five plates by hand.

6. The loop system should be capable of representatively spreading low count inocula and also give a very high ultimate dilution to produce discrete colonies from inocula containing up to 10^8 or 10^9 organisms per millilitre.

7. The loop system should be capable of being sterilised after each plate in a controlled manner to minimise spatter and eliminate cross contamination.

8. The system should be capable of adjustment to allow for the variations in size of "standard" 100-mm Petri plates from different manufacturers.

9. A throughput of between 300 and 400 plates per hour should be possible – being the maximum achieved by a team of three skilled operators with one selecting the media and labelling, the next applying the inocula and the last spreading the plates.

The Initial Prototype

The Plate Drive

Spiral plating was the method chosen as the basis of operation for the system because of the potentially simple design that this would allow. The possibility of producing plate rotation without using a turntable was investigated and a practical solution evolved using two continuous, flexible belts between which the dish was held, with the weight of the plate keeping it in contact with the tray chute on which the belt drives were mounted (Fig. 1).

With this mechanism the plate could first be moved into the spreading zone by driving one belt in the forward direction, then rotated by driving the second belt at the same speed but in the opposite direction. Slow movement back out of the zone could then be achieved by driving the second belt at a slightly faster speed; fast movement out was achieved by stopping the first belt.

By inclining the chute at about 30° it was possible to slide plates in at the top and gravity feed them into the drive system. As the drive engaged the Petri plate and revolved it into the spreading zone, the lid was removed and retained in the stacking area by two side rails which diverged slightly upwards from the inclined chute.

A simple trigger initiated a timer which controlled the sequencing of the motors and thus the plate movements; the trigger also acted to stop the plate when it was returned from the spreading zone so that it could be lifted out of the drive belts mechanically to join up with the waiting lid and be stacked into an open-fronted hopper.

The feed and drive system proved extremely effective and remained substantially unchanged until the second production series.

The Loop System

The plate was rotated about eight times by the belt drive while being returned slowly out of the spreading zone, and using a single loop it was possible to produce good spiral tracks from the perimeter of the plate to the centre during this

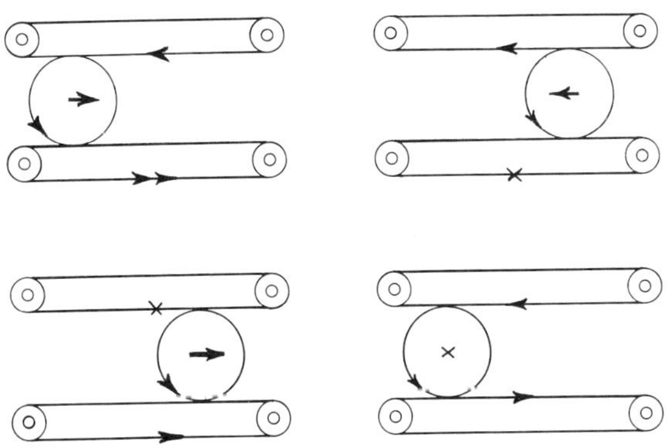

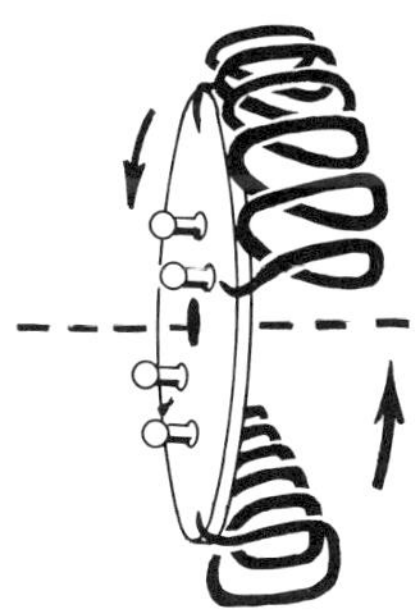

Fig. 2. The loop system

◂ **Fig. 1.** The belt drive system

movement. With small quantities of inoculum placed about 1 cm from the plate perimeter, good pick-up of inoculum onto the loop was obtained but poor colony separation resulted, particularly when spreading high count inocula. The need for a multiple loop system was thus established and, after many experiments, a novel spreading head made from a flattened coil of nichrome wire formed into an arc was developed (Fig. 2). Each "loop" on the coil could be used in turn by moving through the arc and the coil could be sterilised by heating electrically.

Two of these loop systems were mounted on the perimeter of a disc of insulating material attached to one end of a shaft which could be rotated by means of a drive motor. This motorised assembly was mounted over the spreading zone, so that it could be lowered to rotate one of the flattened coils in contact with the agar surface during the period when a plate was being revolved out of the spreading zone. The action was thus to bring each "loop" on the coil sequentially into contact with the surface. The flattened coil was made wide enough to ensure that the tracks produced by the "loops" overlapped and thus gave very effective dilution of the inoculum from the plate perimeter to the centre in a series of spiral bands.

When the plate was safely out of the way and being stacked, the contaminated coil was sterilised and the second coil rotated into position ready for use while the first coil was cooling. The alternate use of two coils meant that a plate could be spread about every 10 s.

The motorised spreader assembly was carefully counterbalanced, rather like the arm of a record player, and spring loaded to ensure constant pressure on the agar when the loop was lowered onto the surface for spreading. The floating action of the head allowed plates with differing levels of agar to be used and compensated for any undulations in the agar surfaces due to uneven pouring.

Springs were used to adjust the contact pressure instead of weights because they gave increased inertia to the system and removed the tendency for the loops to skip when plates were poured out of level.

The Preliminary Trials

After proving that the principles involved actually worked in practice, a second prototype was produced for comparative trials with conventional hand plate spreading in the hospital bacteriology department. On this model, the drive system was altered so that the plate was revolved twice in a fixed position at the start of spreading to ensure that the inoculum was spread as a peripheral band on the agar surface before movement of the plate out of the zone produced a spiral track. This improved the distribution of colonies for low count inocula and gave better separation with high count inocula.

In the laboratory trials satisfactory spreading was achieved with all but the occasional swab from very heavily infected patients in the burns unit. The machine appeared to cope equally well with samples ranging from faeces to sputum and produced consistent results with urine. With samples having counts in excess of 10^8 organism per millilitre, only small areas of discrete colonies were found when the area of inoculation extended more than 10 mm from the edge of the plate. The importance of placing the inoculum at the periphery of the plate was stressed and results improved immediately.

To assess the value of the machine in routine use it was installed in the Gonococcus Laboratory, which provided a large throughout of uniform material. It was entirely satisfactory for this work and the technicians quickly adjusted to examining the plate from the centre outwards rather than by the zones they were used to seeing on hand-spread plates. The machine was also used in this laboratory for spreading secondary cultures.

On the whole the mechanical performance of the machine during these latter trials was good, considering that it was very much a prototype. However, it was apparent that several improvements were needed to increase the reliability of the instrument.

Variations in the diameters of plates routinely purchased were frequent and improvements were made to the method of adjusting the chute to reduce jamming problems. Such jams caused loop damage and the spreader head was redesigned so that it could be readily replaced. The stacking device also needed to be redesigned to incorporate a slipping clutch which prevented the occasional jammed plate from becoming shattered and thus potentially hazardous. The clutch was also adjusted so that it would slip when the stack of plates reached the top of the hopper. This slipping action, whether caused by a plate jam or a full hopper, automatically stopped the machine. Plates showed a tendency to slip or jump out of contact with the belts and so adjustment points and side rails were incorporated in the design to correct these faults.

The only major disaster resulted from an undetected break in a spreader loop resulting in failure to sterilise and thus in cross contamination between plates. A circuit to detect loop failure was included to stop the machine and signal that a failure had occurred. In order to start using the machine again it was necessary to switch it off and then on again to clear the control system and set into operation a cycle to ensure that both loops were sterile. Until this sterilising cycle was completed the instrument would not accept plates into the spreading area.

With these improvements the machine was incorporated into the day to day routine of the laboratory and proved to be such a valuable aid that it was thought probable that many more laboratories would benefit from having one.

Denley were enthusiastic partners in re-engineering the device and, for the first production series, adopted the operating systems and controls very much as they were from the final prototype which had proved so successful in routine use.

The product was well received in the British market but many problems occurred because setting up the instrument for plates from different manufacturers proved to be too difficult for untrained operators; control over the spring pressure applied to the head was too coarse and occasionally caused the agar in the plate to be "ploughed" very effectively by the loop system. Despite such disappointments the Autospreader built up a very loyal group of supporters and the overall concept was felt to have been proven as acceptable.

There followed a redevelopment program based on the field experience gained with the first series of machines. The number of controls was reduced and adjustment of the chute and side rails simplified. The spring tensioning was improved so that adjustment was easier and fine control of the pressure between the head and the agar surface could be achieved. This improved spreading performance considerably and enabled spreading to be achieved with agar concentrations as low as 1.2% w/v.

The drive system was also simplified by using only one belt, driven by a reversible motor. Plate movement was controlled by a combination of the belt drive and a roller operated sequentially by a motorised cam. This system gave much better adjustment for setting the relative positions of plate and loop during spreading and better control over the plate movement. This also virtually eliminated the earlier problem of the plate slipping out of the belt drive.

Above all the sequence of the plate and head movements and of the stacking procedure was monitored by opto-sensors instead of by microswitches. The logic was made completely electronic, and this in turn made control of the loop heating cycle much easier and improved the sensitivity in detecting loop failure by breakage or short circuiting. The electrical and electronic layouts were much improved and the final product received certification by the Canadian Standards Association after testing at the British Standards Test Laboratories in England. The product met every detail of the original specification and in its fully developed state is a reliable and cost-effective laboratory aid. The final product which evolved is shown in Fig. 3.

The sole remaining problem to be solved is that of enabling the system to be able to spread urine samples evenly over the agar surface rather than as a continuously diluted spiral track, as it has not been fully accepted that the larger the area which appears free from colonies the lower the original concentration.

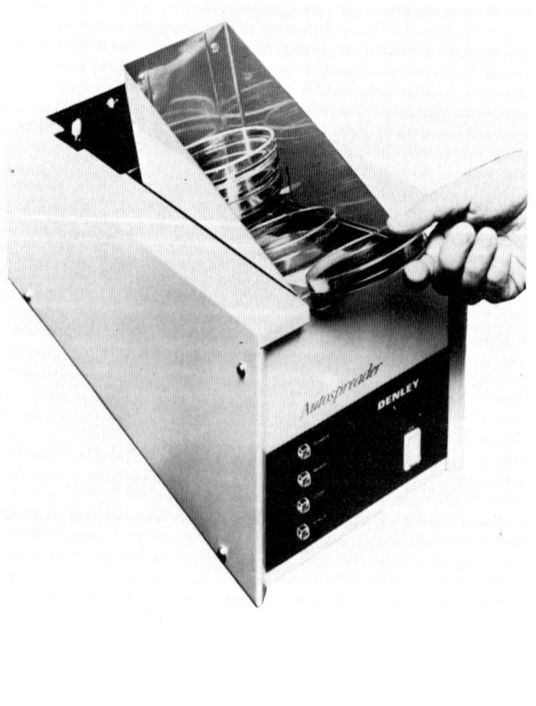

Fig. 3. The Denley Autospreader

COBAS Bact: The Centrifugal Approach to Clinical Bacteriology

A. BALOWS [1]

Introduction

In the past few years, interest in developing and applying automation for use in the clinical microbiology laboratory has risen considerably. Such advances are necessary if laboratories are to provide physicians with timely, standardized, and reproducible laboratory test results. With the attainment of this goal, it should be possible for clinical microbiology to become more relevant and useful and for the microbiologist to provide the required support for rapid diagnosis and therapy. This applies particularly to in vitro results of antibiotic susceptibility testing and identification of the etiological agent. Additionally, because of cost constraints, it has become important to relieve laboratory personnel of many routine tasks and thus make it possible to increase productivity without increasing staff.

COBAS Bact (Figs. 1–4) was designed primarily to help the laboratory meet these demands. Currently it can perform 5-h susceptibility tests on all fast-growing, nonfastidious bacteria and identify members of the family Enterobacteriaceae. The instrument also can be used to analyze growth curves according to the needs of the user; other applications are in various stages of development. Thus, the basic instrument, as presently designed and with modifications as required by the specific needs of the user, will be able to perform a variety of tests.

The instrument with the incubator and the transport system has a capacity of 50 test specimens (Fig. 5). During the testing process, the specimens are shaken and maintained at a constant temperature to ensure optimum conditions for bacterial growth. Following the testing, thermal disinfection of the transport system at 80 °C is accomplished on command.

The specimen for any type of analysis is introduced into the center of a designated rotor, which is then closed with an adhesive label containing patient identification and program to be performed (Fig. 6). The rotor is then placed into the instrument (see Fig. 4) and taken up by the transport system into a cassette. During actual operation of the system, the cassette is closed (Fig. 7), and the rotor stays within its cassette during the entire analysis. The rotor circulates through the incubator in such a way as to reach the input/output and measuring position every 20 min (Fig. 8).

1 Centers for Disease Control, Center for Infectious Diseases, 1600 Clifton Road N.E., Atlanta, GA 30333, USA

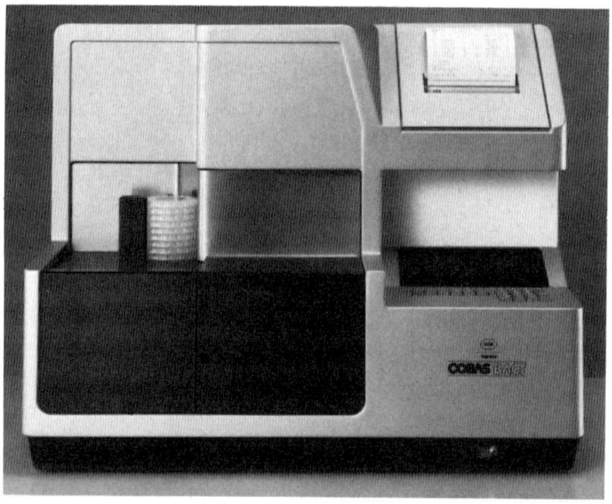

Fig. 1. COBAS Bact – entire instrument

Fig. 2. Close-up of keyboard for program selection

Fig. 3. Printer

Fig. 4. Input stack for introduction of specimens

Fig. 5. Contained within the instrument is the ▶ shaking incubator with a capacity of 50 specimens and the transport mechanism to the measuring position

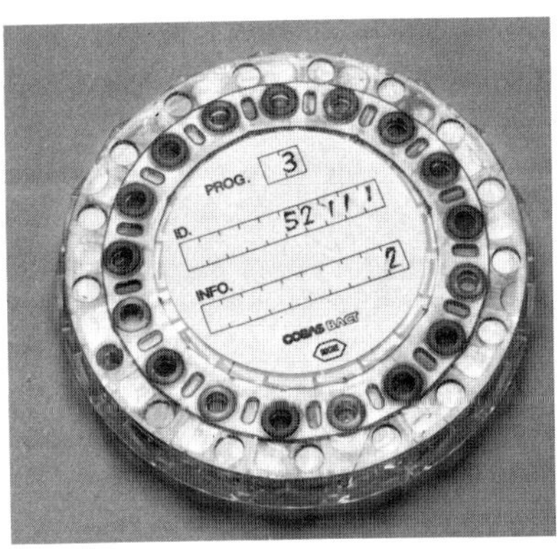

Fig. 6

Fig. 7 ▶

Fig. 6. A rotor to which the bacterial inoculum has been added and an adhesive label with patient of specimen identification and desired program

Fig. 7. The rotor is held in place by a closed cassette (shown in the open position here) that holds the rotor in place during incubation and transport to the measuring position

Fig. 8. One rotor is in place in the centrifuge, a second rotor is ready to be returned to its cassette, and the third rotor (after final reading) is being discarded

After the initial 20-min incubation, the rotor reaches the measuring station for the first time, the specimen previously added to the center of the rotor is distributed into the individual compartments of the rotor by centrifugal force, and the initial measurements are made. Each medium-containing compartment of the rotor is filled with exactly 300 µl of the bacterial suspension to be tested. Excess fluid is driven into a specially developed overflow compartment.

Two rotors are available: one for susceptibility testing designed to accept the desired antibiotic impregnated discs and the other for identification of the Enterobacteriaceae (Figs. 9 and 10). A third rotor with special media is being developed for nonfermenters, since these organisms usually require a heavier inoculum and a 6-h incubation time.

Specimens may be introduced into the system in random order and at any time. The system is ideally suited for either same-day (walk away) or overnight operation; it operates fully automatically once the test specimens have been introduced. After completion of the 5-h cycle, the rotor is ejected automatically into the waste compartment (Fig. 11), and the results are printed out (Fig. 4). This printout format is used to establish the identification data base. It is anticipated that the individual substrate data will not appear in the routine printout.

The "free program" is used for research purposes; this allows the investigator to program and perform measurements at four different wavelengths for a preselected length of time (up to 254 20-min cycles), using one or more of the available centrifugation modes.

Direct connection to a central data-processing unit is possible with a bidirectional interface. The system stores all data even in case of a power failure and informs the user of the duration of the failure and the lowest temperature of the incubator during the power interrruption. The user can then decide whether the results are valid or need to be repeated.

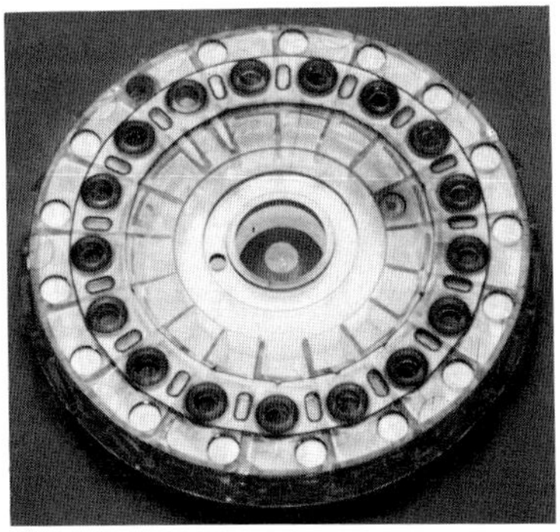

Fig. 9. The rotor for identification of Enterobacteriaceae in which sterile, air-dried substrate with suitable indicator is contained in each compartment

Fig. 10. A close-up view of a segment of the identification rotor before substrates have been added

Fig. 11. Waste compartment with discarded rotors ready for autoclaving

Identification of Bacterial Isolates

The following substrates are incorporated in the disposable identification rotor: 11 carbohydrates (glucose, adonitol, arabinose, cellobiose, inositol, lactose, malonate, rhamnose, sorbitol, mannitol, citrate), three amino acids (arginine, ornithine, lysine), and two other metabolic end-product tests (urea and H_2S). In the production of the identification rotor, all substrates are loaded in liquid form, the rotor is then air-dried with sterile air and packaged for storage until ready to be used. Shelf-live studies are under way but a 6-month or greater shelf-life is anticipated.

Test Performance

A well-isolated colony representing the organism to be identified is selected from a primary isolation plate, and a spot indole test is performed manually using *p*-dimethylaminozimtaldehyde in hydrochloric acid. With this reagent, indole-positive organisms produce a blue to blue-green color, and indole-negative organisms produce a faint yellow or red color. The result of the indole test is keyed into the COBAS Bact at the time the rotor is placed in the instrument.

Using the same or morphologically identical colonies, a bacterial suspension corresponding to a 1.0 McFarland standard (about 10^8 CFU/ml) is prepared in 5 ml nonbuffered physiological saline solution. The suspension is pipetted into the rotor (Fig. 12), and the rotor is then sealed with an adhesive label containing relevant patient information.

After the first 20 min of incubation, the rotor reaches the measuring station and it is centrifuged to distribute exactly 300 μl into each of the 16 compartments. The growth extinction in the citrate compartment is measured at this time, since the result is based on the measurement of the turbidity in this compartment, while all the other reactions are based on indicator changes following changes in pH resulting from bacterial substrate utilization. In this way, the citrate compartment also serves as inoculum control.

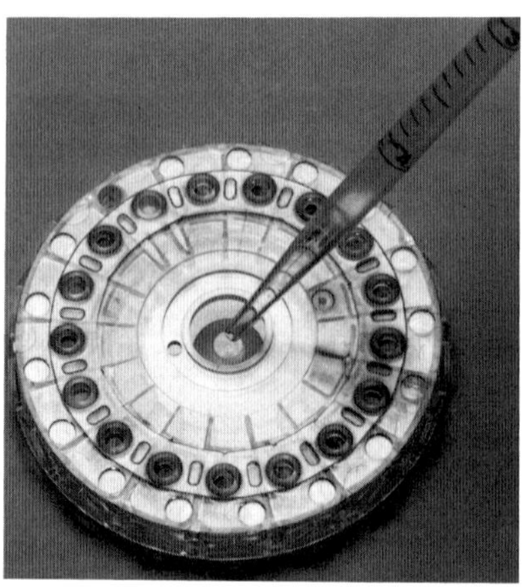

Fig. 12. The identification rotor showing addition of bacterial suspension

After 4 h of incubation all compartments are measured at four different wavelengths (605 nm, 564 nm, 546 nm, and 430 nm). At each wavelength, five flash readings with a xenon lamp are made and averaged by the instrument's computer. The readings permit the analyses of all possible color changes occurring in the identification rotor.

Establishment of the Data Base

Results for the development of the identification data base have been collected in four European and two American microbiology laboratories to allow for strain variation from various geographical regions. The number of strains of each species of the Enterobacteriaceae used for the initial establishment of the data base corresponds approximately to their frequency of isolation in a routine clinical laboratory. The following 13 genera and 36 species can be identified: *Escherichia coli; Shigella dysenteriae, S. flexneri, S. boydii, S. sonnei; Edwardsiella tarda; Salmonella typhi, S. paratyphi A, Salmonella* spp; *Arizona* sp; *Citrobacter diversus, C. freundii, C. amalonaticus; Proteus vulgaris, P. mirabilis; Morganella morganii; Providencia rettgeri, P. Alcalifaciens, P. stuartii; Enterobacter cloacae, E. aerogenes, E. sakazakii, E. intermedium, E. agglomerans, E. gergoviae; Hafnia alvei; Serratia marcescens, S. odorifera, S. rubidaea, S. liquefaciens; Klebsiella pneumoniae, K. ozaenae, K. oxytoca, K. rhinoschleromatis; Yersinia enterocolitica, Y. pseudotuberculosis.*

Reproducibility

The reproducibility of the individual substrate compartments of the COBAS Bact identification rotor has been tested with six ATCC strains (*Escherichia coli, Enterobacter cloacae, Shigella sonnei, Salmonella typhimurium, Klebsiella pneumoniae, Proteus vulgaris*) on different days in two laboratories.

The results are as follows.

Intralaboratory Reproducibility (Table 1). There appears to be good to excellent reproducibility except for rhamnose, citrate, and mannitol.

Interlaboratory Reproducibility (Table 1). Here there is good to excellent reproducibility except for urea, citrate, H_2S, and mannitol. Rhamnose shows acceptable reproducibility but low reactivity.

Analysis of the lower levels of reproducibility has revealed the following. Urea discrepant results are primarily due to klebsiellae that give a weak urease reaction in the 5-h incubation. Additional strains are being tested to allow for recalibration of the spectrophotometric readings. The discrepant results obtained with rhamnose, citrate, H_2S, and mannitol are not due to variation in the ability of the organism to utilize the substrates but rather to a need for recalibration of the sensors and revision of the interpretive criteria. Both of these categories of lower interlaboratory reproducibility results have been corrected in the instrument, and data from repeat reproducibility tests are being generated.

Table 1. Reproducibility of COBAS Bact identification system

		Intralaboratory reproducibility (%)[a]	Interlaboratory reproducibility %[b]	
			Lab 1	Lab 2
1	Glucose	98	96	100
2	Adonitol	100	99	100
3	Argenine	97	95	99
4	Arabinose	99	98	99
5	Lysine	97	97	98
6	Cellobiose	99	98	100
7	Inositol	100	100	100
8	Ornithine	98	97	100
9	Lactose	99	100	98
10	Urea	95	90	100
11	Malonate	100	100	100
12	Rhamnose	82	81	83
13	Citrate	82	87	77
14	Sorbitol	98	100	97
15	H_2S	93	100	85
16	Mannitol	88	93	83

[a] Forty rotors used for each of six ATCC bacterial strains ($n = 240$)

[b] Twenty rotors used for each of six ATCC bacterial strains in two different laboratories ($n = 120$)

The generally acceptable reproducibility of these preliminary data supports a prediction of a reliably functioning, truly automated system that can be programmed to provide rapid identification, antibiotic susceptibility data, and other analyses. The currently developed and programmed international data base will be updated periodically as needed or as variant strains appear; this updating can be done very simply by exchanging one of the memory modules of the system. Clinical data are now being collected in several institutions in order to test the data base and to optimize the system before it is put on the market, which is planned for the fall of 1984 in Europe and early 1985 in the USA.

Conclusion

This instrumental approach and concept merit consideration and attention, since the system will be the first fully automated identification and susceptibility system that requires no further manual manipulation after the inoculation of the rotor. The flexibility of the system allows the COBAS Bact to be adapted to the individual needs of each laboratory rather than having the laboratory adjust its work schedule to meet the continuous monitoring requirements of a semiautomated instrument.

Short-Term Methods for Isolation and Enumeration of Bacteria in the Respiratory Tract

A. TURANO and F. PIRALI [1]

Introduction

Interest in quantitative respiratory bacteriology derives from the opinion that, as in other clinical specimens, e.g., urine, quantitative data may have a pathogenetic value in the complete and quick diagnosis of certain bacterial bronchopneumopathies. The concentration of a single bacterial species isolated from sputum, tracheoaspirates, bronchoaspirates, or other bronchial material, at least in some cases, could allow some real difficulties in respiratory bacteriology, such as the contamination by oral microflora, to be surmounted. In fact, with this method it is easier to detect the species which has a pathogenetic role in a bronchopneumopathy, because this species is more numerous than "colonizing" bacteria of the respiratory tract. Furthermore, the same quantitative examination allows the ratios of the different bacterial species to be studied, thus providing an exact map of the complex bacterial ecology of the respiratory tract.

Numerous authors [1–9] over the past 30 years have studied the quantification of bronchial microflora often with contrasting results in relation to the different pathologies and methods used. There is a strong need for a quantification method allowing large number of clinical specimens to be quickly and easily praessed. In our institute, we have developed a simple method for counting the microorganisms in the respiratory tract. Here we present the most important technical aspects and some results of its application in the diagnosis of bronchopneumopathies. The method of quantifying respiratory tract microflora may be applied to bronchial secretions such as sputum, tracheoaspirates, bronchoaspirates, or material derived by laryngeal puncture.

Material and Methods

The clinical specimens collected in a perspex sterile container (LP 202038) and quickly sent to the laboratory. It is then put inside a sterile plastic bag and fluidized with 0.1% dithiotreitole solution (Sputasol Oxoid, code SR 89/54560) at a dilution of 1 : 2 or 1 : 10 and homogenized mechanically for 2 min in a Stomacher (Stomacher Lab. Blender). After the preparation, the material is seeded by a 10-ml calibrated loop, and parallel streaks are made on sheep blood agar plates (Sclavo code 50856), salt mannitol agar (BBL code 21173), McConkey agar (BBL

1 Institute of Microbiology, Brescia University, Piazza Spedali Civili 1, I-25100 Brescia

Table 1

Bacterium	Patients studied	Broncho-pneumonia	Chronic bronchitis	Pulmonary emphysema	Negative	Pleural metastasis	Pleural empyema	TBC	Infection in chronic bronchitis
E. coli	13	7, 8, 8, 8, 7, 7[a]	6, 5, 5	5, 5	4, 6				
Klebsiella	10	8, 8, 8, 8, 7	4		4, 5			5	7
Pseudomonas	25	8, 8, 8, 5, 8, 7, 8	6, 4, 5	5			5		7, 7
		5, 6, 4, 5, 7, 6, 7, 5	5, 5, 5						
Enterobacter	7	8, 8	6, 5		5	6			7
Acynetobacter	6	8, 8, 8, 8, 4, 5							
Serratia	3	8, 8			8				
Candida	16	8, 8, 6	5, 3, 3, 6, 6	6	6			6	
			6, 4, 5, 5, 7						
Staphylococcus aureus	5	8, 7	8		4, 8				
Streptococcus pneumoniae	8	8, 8, 8, 8					5		8, 6, 6
Streptococcus viridans	11		7, 8, 8, 5	7					6
Streptococcus-β haemoliticus	1		5, 8, 7, 7, 6						
Proteus mirabilis	1		6	6					
Staphylococcus albus	1							8	
Total	107								

[a] Correlation between log bacterial concentration per milliliter of sputum or tracheoaspirate and clinical status: 3, 10^3, 4 10^4, 5, 10^5, 6, 10^6, 7, 10^7, 8, 10^8

code 21172), Sabouraud agar (BBL code 10048), and chocolate agar with Isovitalex (BBL code 10005) plus factor V (Difco Laboratories code 1622-35). The streaks are made by loading the clinical specimen into the loop. One specimen is used on each agar plate. The loop then progressively discharges its contents onto the agar plate, and the bacterial growth is seen as a number of streaks proportional to the initial bacterial concentration.

We have made numerous determinations in comparison with the agar dilution method and we have fixed the numerical value corresponding to the growth on each streak. If the bronchial secretion has been diluted at 1:10 the values are roughly 100,000 on the first streak, 1,000,000 on the third streak, 10,000,000 on the sixth streak, 100,000,000 on the 13th streak, and intermediate values on the other streaks. This method, performed by a technician with microbiological experience, is sufficiently sensitive and reproducible to be routinely used in our institute for all bronchial material submitted for microbiological examination. With this method, we have verified the clinical utility of the quantitative data in respiratory microbiology, correlating the bacterial concentration in sputum and tracheoaspirates with the clinical status of 107 patients in the Brescia Hospital Medical School from 1983 to 1984. The results are shown in Table 1.

Discussion and Conclusion

The method described above and routinely used in our institute allows the quantification of a mixed bacterial flora. For its standardization and use in respiratory microbiology, it is necessary that fluidization and homogenization of the bronchial material be thorough. In comparison with other quantification methods, it presents the advantage of being easy and quick. The simultaneous use of selective and nonselective media on the same plate allows comparisons of the total and partial bacterial concentrations for each species to be made. By means of the streaks, the progressive dilution of the bronchial material generally allows isolated bacterial colonies to form on the plates in 12–18 h in biochemical assays.

Our data seem to demonstrate that bacteria may be responsible for a lung infection only if they are present in sputum or bronchoaspirates within certain concentrations. However, there are some exceptions, such as in patients with immunodeficiency, where lower microorganism concentrations are sufficient to be considered a sign of potential pathogenicity.

Special remarks must be made concerning patients affected by chronic bronchopneumopathy or patients undergoing antibiotic therapy, where a strict correlation between quantitative microbiological data and clinical status is not valid. Naturally, judgement is influenced by the available clinical information and previous microbiological reports. Furthermore, by comparing the data from different times, it is possible to detect quantitative modifications in the bacterial species present and a shift toward the normal complement of flora. This gives the possibility of following up the clinical evaluation of a bronchopneumopathy.

The levels of growth are specific for the isolated gram-positive and gram-negative species and show that simple identification of a bacterium supposed to be a pathogen in sputum is not always sufficient proof of its etiological role.

Our results are not complete; they only represent a stage in solving the problems of quantitative respiratory bacteriology. However, in these results there do appear to be definite pathogenetic implications. Research in this area is difficult and the problems are diverse, but the work presented here does seem to represent an important step in respiratory bacteriology.

References

1. May R (1952) The bacteriology of chronic bronchitis. Lancet 11:1206
2. Louria DB, Kaminsky T (1962) The effects of four antimicrobial drug regimens on sputum superinfection in hospitalized patients. Am Rev Respir Dis 85:649
3. Wilson MJB (1973) Quantitative sputum culture as a mean of excluding false positive reports in the routine microbiology laboratory. Am Rev Respir Dis 108:171
4. Crofton J (1969) The defences of the respiratory tract in respiratory diseases. Blackwell, Oxford
5. Borderon E, Leprince A, Gueveler C, Borderon JC (1981) Valeur des examens bacteriologiques des secretions tracheales. Rev Fr Mal Respir 9:229–239
6. Gevaudan MJ (1976) Analyse quantitative de la flore bacterienne isolee de l'expectoration des bronchiteux chroniques. Rev Fr Mal Respir 4:861–870
7. Corda M (1975) Interesse pratico della batteriologia quantitativa in pneumologia. Annali Sclavo 17:3
8. Le Lezzin M (1976) Analyse quantitative de la flore bacterienne des crachat en routine hospitaliere. Rev Fr Mal Respir 4:379–388
9. Chaste A (1980) La numeration bacterique des expectoration. Rev Fr Mal Respir 8:393–402

Some New Media for Isolation and Rapid Identification of Bacteria

G. Satta [1], G. Grazi [2], O. Soro [2], R. Pompei [1], and P. E. Varaldo [2]

Summary

We have developed new media for routine clinical microbiology, some of which allow reliable identification of some clinically important species and some of which facilitate identification of bacterial species. A medium was developed for the separation of *Staphylococcus aureus* from other staphylococcal species. This medium is rendered highly selective by addition of 7.5% sodium chloride and is prepared in such a way that it allows direct detection of several different biochemical properties of the colonies. The medium was used for isolation and rapid identification of *S. aureus* from 4,000 clinical specimens. Identification of this microorganism by simple observation of the colonies grown on the selective medium resulted very reliable. Another medium was developed for rapid separation of staphylococci from enterococci isolated from human urine. This medium separated the mentioned species by testing bacteriolytic activity of the colonies under particular conditions. A third medium was prepared to facilitate identification of *Haemophilus* colonies. This medium exploits the fact that *Haemophilus* strains are intensely phosphatase positive while most of the strains generally present in the upper respiratory tract either do not show or show lower phosphatase activity.

Introduction

The identification of bacterial strains is time consuming, costly, and lengthy when compared to other analyses of medical interest [6]. There is therefore widespread interest in the development of methods of analysis which overcome these problems; their lengthiness is definitely the most important problem since the faster clinical analyses allow diagnosis to be made and therapy to be set up the more useful they are [4–6].

For the rapid identification of bacteria, an ideal condition would be that of identifying bacterial strains simply by observing the colonies grown on the plates onto which specimens have been inoculated. This is of course a very difficult task, which might however be achieved more easily for micro-organisms for which highly selective media can be prepared. In addition, the preparation of media

1 Institute of Microbiology and Virology, University of Cagliari, Via G. T. Porcell 4, I-09100 Cagliari
2 Institute of Microbiology, University of Genoa, I-Genoa

which, although they may not allow direct identification, could facilitate it by enabling a direct detection of one or more useful properties of the colonies would be rather desirable [10, 15].

In this work we describe the use of three different media, two of which permit direct identification of the colonies grown and one of which facilitates identification of strains of some species.

Materials and Methods

Bacterial Strains. All strains used were isolated from specimens of human origin obtained by the clinical microbiology laboratories of our Institute. Strain identification was performed by standard methods [7]. Staphylococci were also identified as described by Varaldo et al. [16].

Medium Composition and Preparation. The two media for rapid identification of *Staphylococcus aureus* strains (NCB1 and NCE) were prepared and used as previously described [10]. The medium for rapid identification of enterococci (ES) was prepared by adding heat killed *Micrococcus luteus* cells at a concentration of 1 optical unit per millilitre to a standard agar medium which adequately supported growth of both staphylococci and enterococci. The medium was then made selective by addition of 6.5% sodium chloride. The preparation of heat-killed *M. luteus* cells and evaluation of bacteriolytic activity were as described by Satta et al. [12]. The medium for easier identification of both *Haemophilus influenzae* and β-hemolytic streptococci (PE) was a modification of the MGP medium for phosphatase detection [11].

Results

Medium for Rapid Identification of *S. aureus* Strains

As mentioned in the materials and methods section, this medium has been described elsewhere. In this study we have further evaluated the validity of two different versions of it (NCB1 and NCE). The NC media are highly selective for the presence of a high salt concentration and reveal the properties of bacterial colonies reported in Table 1. The possibility of reliably identifying *S. aureus* colonies by this medium is based on several facts: as previously shown by us and by others [10], low doses of lysostaphin inhibit *S. aureus* and *S. simulans*, but do not inhibit strains of other species; *S. simulans* does not ferment maltose which is fermented by *S. aureus* strains; in the special test only *S. aureus* gave a positive phosphatase reaction; and lecithinase is produced by *S. aureus*, but not by other staphylococci. Identification schemes using the NC media are given in Table 2 together with an evaluation of reliability of identification with the various possible results obtained.

When first described the NC media were evaluated using a relatively small number of human specimens [10]. By extending their use to a very large number

Table 1. Properties of the NC media

Characters tested by the medium	Reaction	Evidentiation of the character tested in medium	
		NCB 1	NCE
Maltose fermentation	Positive	Colonies and surface plate zones around them turn from blue to ochre yellow	Colonies and surface plate zones around them turn from pale orange to intense red
	Negative	Colony and plate color is blue	Colony and plate color is pale orange
Lysostaphyn sensitivity	Positive	No growth around the paper strip containing lysostaphyn	No growth around the paper strip containing lysostaphyn
	Negative	Colonies grow around the strips	Colonies grow
Specific phosphatase	Positive	Colonies and surface plate zones around them are greenish-yellow	Colonies pale red and surrounded by a yellow halo
	Negative	Colonies and surface zones are ochre-yellow	Colonies are intense red and surrounded by an intense red halo
Lecithinase activity	Positive	Not tested	Colonies surrounded by opaque halos
	Negative	Not tested	No halos around colonies

Table 2. Scheme for identification of *S. aureus* colonies on the NCE media

Medium	Reactions				Identification	Comments
	Malt. ferm.	Phosph.	Lysost.	Lecith.		
NCB1	+	+	+	NT	*S. aureus*	Very reliable
	+	−	+	NT	*S. aureus*	Unsatisfactory
	+	+	−	NT	Non-aureus	Unsatisfactory
	+	−	−	NT	Non-aureus	Very reliable
	−	−	+	NT	*S. simulans*	Reliable
NCE	+	+	+	+	*S. aureus*	Very reliable
	+	−	+	+	*S. aureus*	Reliable
	+	+	+	−	*S. aureus*	Reliable
	+	+	−	+	*S. aureus*	Reliable
	+	−	−	−	Non-aureus	Very reliable
	+	−	−	+	Non-aureus	Reliable
	−	−	+	−	*S. simulans*	Reliable

of specimens we have now been able to confirm their great reliability in *S. aureus* identification. Table 3 shows in fact that of a total of 4,215 specimens containing staphylococcal colonies, separation into *S. aureus* and non-*S. aureus* on the NCB1 medium required a coagulase test in only 47 cases, which is slightly more than 1% of the total. Table 4 shows that the NCE medium allowed a further inprovement in reliability and identified *S. aureus* colonies as reliably as the most reliable conventional methods, in no case requiring an additional test to be performed.

Table 3. Identification of *S. aureus* strains in 4215 specimens of human origin by the NCB1 medium

No. of strains	Reactions observ. in NCB1 medium			Other tests performed	Ident. on NCB1	Indent. by conv. meth.
	Malt.	Phosph.	Lysost.			
2149	+	+	+	None	*S. aureus*	*S. aureus*
14	+	–	+	Coag.[a]	*S. aureus*	*S. aureus*
11	+	–	+	Coag.	Non-aureus	Non-aureus
10	+	+	–	Coag.	Non-aureus	Non-aureus
12	+	+	–	Coag.	*S. aureus*	*S. aureus*
21	–	–	+	None	*S. simulans*	*S. simulans*
1998	+	–	–	None	Non-aureus	Non-aureus

[a] Coag. = coagulase

Table 4. Identification of *S. aureus* strains in 1047 specimens of human origin by the NCE medium

No. of strains	Reactions observ. in NCBR medium				Other tests performed	Ident. on NCBR	Ident. by conv. meth.
	Malt.	Phosph.	Lysost.	Lecith.			
543	+	+	+	+	None	*S. aureus*	*S. aureus*
18	+	+	+	–	None	*S. aureus*	*S. aureus*
4	+	+	–	+	None	*S. aureus*	*S. aureus*
6	+	–	+	+	None	*S. aureus*	*S. aureus*
3	+	–	–	+	None	Non-aureus	Non-aureus
5	+	+	–	–	None	Non-aureus	Non-aureus
7	–	–	+	–	None	*S. simulans*	*S. simulans*
461	+	–	–	–	None	Non-aureus	Non-aureus

Medium for Rapid Identification of Enterococci

We showed previously that all staphylococci produce bacteriolytic enzymes which are different for the various species and can be identified by simple methods [12, 14, 16]. This finding prompted us to study possible production of similar enzymes in other gram-positive and gram-negative strains. By testing the bacteriolytic activity of several streptococcal strains of various groups we found that the colonies of the group A strains formed opaque zones of lysis, most group D strains caused very clear zones of lysis, while no strain of the other groups showed lytic activity (Table 5). It is also interesting to note (Table 6) that the bacteriolytic group D strains all belonged to species of the enterococcal group (*S. faecalis, S. faecium,* and *S. durans*) while the nonbacteriolytic strains belonged to non-*Enterococcus* species (*S. bovis* and *S. equinus*). We then developed a medium (called ES) where both staphyloccoci and Enterococci grow optimally, but all of the latter strains and none of the former show bacteriolytic activity. This medium was made selective by addition of sodium cloride and was used for rapid identification of enterococci in human urine. Table 7 shows that all enterococci isolated from 3,000 urine samples could be correctly identified by simply observing properties

Table 5. Bacteriolytic activity of streptococci of various groups

Streptococci	No. of strains	Lytic activity
Group A	35	+[a]
Group B	21	–
Group C	7	–
Group D	51	+++
Group D	6	–
Group H	15	–
Group K	9	–

[a] Opaque zone of lysis

Table 6. Species identification of the bacteriolytic and nonbacteriolytic group D Streptococci

No. of strains	Lytic activity	Species identification
38	++	*Streptococcus faecalis*
12	++	*Streptococcus faecium*
1	+	*Streptococcus durans*
5	–	*Streptococcus bovis*
1	–	*Streptococcus equinus*

Table 7. Identification of enterococci in human urine by the ES medium

No. of strains	Lytic properties in the ES medium	Identification on Es medium	Species identification by conv. methods
191	++	Enterococci	*S. faecalis*
83	++	Enterococci	*S. faecium*
6	–	Staphylococci	Group B *Streptococcus*
116	–	Staphylococci	*Staphylococcus* (various species)

of the colonies grown after inoculation of the specimens on ES plates. Assignment to *Staphyloccocus* genus was less reliable since some of the nonbacteriolytic strains were identified as streptococci, not belonging to the *Enterococcus* group. However, these strains amounted to only 1.5% of the total.

Medium for Facilitated Identification of Both *Haemophilus* and *β*-Hemolytic Streptococci

We have previously described a new method (MGP) for testing phosphatase activity in bacteria [11]. In this method the phosphatase positive strains form intense green colonies while the other strains form unstained colonies. The MGP method has now been adapted to be performed on blood agar plates and used to evaluate phosphatase activity of some important pathogens of the upper respiratory tract.

Table 8. Colony staining and hemolytic activity of *H. influenzae* and β-hemolytic streptococci on PE blood agar plates

Strains	No. of strains	Colony staining	β-hemolysis
H. influenzae	20	Deep green	–
Group A strep.	17	Deep green	++
Group B strep.	24	Deep green	+/–
Group C strep.	18	Very light green	++
Group D strep.	38	Very light green	++
Group G strep.	9	Very light green	++
Group G strep.	8	Deep green	++
Group G strep.	4	Deep green	+/–

Table 9. Identification of group A streptococci by testing bacitracin sensitivity on PE blood agar plates

Strains	No. of		Colony stain (of sensit. strains)	β-hemolysis (of sensit. strains)
	bacit. res.	bacit. sens.		
Group A	–	37	Deep green	Clearly evident
Group B	29	6	Deep green	Virtually absent
Group C	33	5	Very light green	Clearly evident
Group D	26	2	Very light green	Clearly evident
Group G	16 (2 strains)	4	Very light green	Clearly evident
	(2 strains)		Deep green	Virtually absent

Table 8 shows that all strains of both *H. influenzae* and β-hemolytic streptococci of groups A, B, and G formed intense green colonies while β-hemolytic streptococci of groups C and D formed unstained colonies. It is also interesting to note that in this medium the group B strains, as opposed to all others, showed very light hemolytic activity.

Isolation of *Haemophilus* strains is based on the use of paper discs containing the X and V factors and Gram staining of colonies grown around the discs [7]. The PE blood agar medium could facilitate identification of such strains since among colonies grown around the paper disc only the intense green ones can be of *H. influenzae*. To evaluate this possibility we inoculated 623 human sputum samples both on traditional and on PE blood agar plates. A total of 22 *H. influenzae* strains were isolated on the former plates as opposed to 30 isolated on the latter.

The PE blood agar plates were finally used to test bacitracin sensitivity of a total of approximately 200 β-hemolytic streptococci belonging to 5 different groups. The apparent sensitivity of all strains turned out to be virtually identical on PE blood agar plates and on standard plates and this medium proved of great utility in recognizing bacitracin-sensitive non-group-A streptococci (Table 9). From a total of 17 non-group-A β-hemolytic strains that were sensitive to bacitracin all could be easily identified (as non-group-A) either because of loss of hemolytic activity or due to formation of unstained colonies.

Discussion

All the different media evaluated in this work proved to be very useful. The NC media have in particular proved capable of permitting a quite reliable identification of *S. aureus* strains by simply observing properties of the colonies grown after inoculation of the clinical specimens. Of the two NC media, the NCE proved to be particularly reliable since it permitted correct identification of all *S. aureus* strains without performing any additional tests. However, the NCB medium also showed a high reliability since it permitted correct identification of almost 99% *S. aureus* strains. It is probably worth mentioning that in absolute terms the NC media are not more expensive than other selective media for *S. aureus*. However, their preparation may be slightly more complex so that they would probably be of greater utility if made commercially available.

The practical utility of SE medium was demonstrated by the fact that it permitted correct identification of all enterococci isolated from as many as 3,000 human urine samples. Enterococci and staphylococci, which both grow on the SE medium, are the most important gram-positive pathogens of the human urinary tract and account for a significant percentage of urinary tract infections, particularly in hospitalized patients [3]. Considering the cost of the SE medium, which is not higher than that of any standard bacteriology medium, a most profitable use of the SE medium would be to inoculate it with urine that appears to be contaminated by gram-positive micro-organisms after observation of a gram-stained smear under a light microscope [2]. After incubation of the plates one could very easily identify enterococci if colonies surrounded by a zone of bacteriolysis are observed and establish the presence of staphylococci if the growth of colonies which are not lysing the *micrococcus* is observed. We have already used the SE medium according to this scheme and have proved its utility [9]. In addition to this, the SE medium could also be of great utility in environmental microbiology where enterococci are sought as indicators of contamination by fecal material [1].

The PE blood agar did not permit direct identification of bacterial species but facilitated identification of both *H. influenzae* and group A β-hemolytic streptococci. This medium allows growth of all strains that thrive on standard blood agar plates and is not significantly more expensive than the latter. It can therefore replace the standard blood agar plates with the advantages of facilitating identification of group A streptococci very reliably when used for testing bacitracin sensitivity. Apart from this, the medium also permits a fairly reliable identification of β-hemolytic strains of group B, whose importance as human pathogens appears to be ever increasing [8]. Other simple methods that permit reliable identification of the various streptococcal groups have been developed and made commercially available recently [13]. However, they are much more expensive and time consuming than blood agar plates.

Acknowledgement. This work was supported by grant No. 83.00694.52 from *Consiglio Nazionale delle Ricerche*.

References

1. American Public Health Association (1976) Standard methods for the examination of water and wastewater, 14th edn. American Public Health Association, Washington DC
2. Appelbaun PC, Ohnstead CC (1982) Evaluation gram-stain screen and micro-ID methods for direct identification of *Enterobacteriaceae* from urines. Med Microbiol Immunol (Berl) 170:173–184
3. Asscher AW (1980) The challange of urinary tract infections. Academic, London
4. Eng J, Brunn JN (1982) A clinical evaluation of the vacutainer 20 ML tube for aerobic cultures. Acta Pathol Microbiol Scand [B] 90:261–265
5. Finegold SM, White ML, Zinnet I, Winn WR (1976) Rapid diagnosis of bacteriuria. Appl Microbiol 18:458–463
6. Gall LS, Curby WA (1980) Instrumented systems for microbiological analysis of body fluids. CRC, Boca Raton
7. Lennette EH, Balows A, Hanseler W, Truant J (eds) (1980) Manual of clinical microbiology. American Society for Microbiology, Washington DC
8. Patterson MJ, Hafelz A (1976) Group B Streptococci in human disease. Bacteriol Rev 40:774–796
9. Satta G, Foddis G, Marcialis MA, Pompei R (1983) Rapid bacteriological diagnosis of urinary tract infections by the application of some novel concepts. 13th International Congress of Chemiotherapy, Vienna, August 1983
10. Satta G, Grazi G, Varaldo PE, Soro O, Romanzi CA (1978) A new selective medium for the isolation and rapid identification of coagulase-positive and coagulase-negative *Staphylococcus aureus* strains. International meeting on food microbiology and technology Tabiano B (Parma) April, 1978
11. Satta G, Grazi G, Varaldo PE, Fontana R (1979) Detection of bacterial phosphatase activity by means of an original and simple test. J Clin Pathol 32:391–395
12. Satta G, Varaldo PE, Grazi G, Fontana R (1977) Bacteriolytic activity in staphylococci. Infect Immun 16:32–42
13. Soro O, Grazi G, Varaldo PE (1981) Comparison of three commercially available kits for routine grouping of streptococci on the basis of biochemical tests, coagglutination and latex agglutination. Microbios Lett 18:117–124
14. Valisena S, Varaldo PE, Satta G (1982) Purification and characterization of three bacteriolytic enzymes excreted by *Staphylococcus aureus*, *Staphylococcus simulans* and *Staphylococcus saprophyticus*. J Bacteriol 151:636–647
15. Varaldo PE, Grazi G, Cisani G, Satta G (1979) Routine separation of staphylococci from micrococci based on bacteriolytic activity production. J Clin Microbiol 9:147–148
16. Varaldo PE, Grazi G, Soro O, Cisani G, Satta G (1980) Simplified lyogroup system a new method for routine identification of staphylococci. Description and comparison with three other methods. J Clin Microbiol 12:63–68

Current Trends and Practical Application in Rapid Viral Diagnosis

M. GRANDIEN [1]

Modern medicine now offers antiviral drugs for treatment of certain severe viral infections and specific immune globulins or vaccines for effective prophylactic measures to the exposed. This increases the demand for a precise etiological diagnosis obtainable early in the course of a disease and within a few hours of a patient's admission to the hospital, thus allowing intervention in the treatment of the patient or contacts. Also, in diseases where such treatment is not available, a rapid viral diagnosis gives the physician confidence that he has done what is necessary for successful management of his patient.

Methods

The conventional virus diagnosis is performed by virus isolation in tissue culture or by demonstration of an antibody titre rise in the patient's serum. At best, virus isolation gives a diagnosis after 1–3 days (influenza, herpes simplex), while it usually takes 1–3 weeks. The demonstration of an antibody titre rise has to wait for the convalescent serum taken after about 2 weeks.

The rapid virus diagnosis works either with techniques which can detect virus, viral antigens or genomes directly in the human specimen or with methods which can determine virus-specific IgM in a single serum sample. The diagnosis may be obtained in a few hours. Most of the methods are simple, reliable and relatively inexpensive, which also make them suitable for epidemiological screenings.

The Immunofluorescence Technique

The cells in the specimen are spread over several small test areas and fixed in acetone, which makes them permeable to reagents. Virus-specific antibodies diffusing into the cells bind fluorescein to the viral proteins, giving a green fluorescence in blue or ultraviolet light. The specimen can be investigated for several agents at the same time and the investigation takes 1–3 h. The method requires high-quality reagents and experienced immunofluorescence (IF) microscopists.

1 Department of Virology, National Bacteriological Laboratory, S-105 21 Stockholm

Immunoassays

The immunoassays (IA) ELISA, RIA, and TR-FIA are sensitive, specific, and suitable for large-scale investigation; ELISA is most frequently used. A common test system employs antibodies on a solid phase capturing viral antigens from the specimen. These antigens are then indicated by the addition of enzyme-labelled antibodies. Subsequently, a substrate is added causing a colour to develop due to the action of the enzyme present in the test well. The colour can be seen by the naked eye; a spectrophotometer increases the accuracy and sensitivity of the method, which can be performed in microtitre plates. The results are obtained in 5–18 h. In the time-resolved fluoroimmunoassay (TR-FIA) the construction of the assay is in principle the same as described for ELISA, though fluorescence is used for the indication of results. A fluorescent substance, conjugated to antibodies used in the second step, is excited by a short light pulse, and after a short delay the specific fluorescence is measured in the fluid phase.

Electron Microscopy

The virus particles are identified in electron microscopy (EM) by their size and structure. Viruses of the same family can be differentiated by the addition of specific antibodies which aggregate the virus particles, immune EM (IEM), or which exclusively bind them to the EM grid, solid phase EM (SPIEM).

Other Methods

During the last few years several other techniques have been described for the detection of virus directly in the clinical specimen. Latex agglutination, passive haemagglutination and immunodiffusion methods are used in some laboratories. A few newly developed techniques are of special interest.

Nucleic Acid Hybridization for Detection of Viral DNA

Modern molecular biology can be used to produce large amounts of viral gene probes in bacteria. These probes react with the original DNA fragment used for their production and after isotope labelling they can be used for the identification of virus genomes in a clinical specimen [13]. The results are usualy obtained after 1–2 days. The method has been used for identification of adeno-, cytomegalo-, and parvovirus. Its use will increase with the ability to produce probes to other viruses.

Detection of Specific Viral Enzyme Activity

The method has been used for rapid diagnosis and differentiation of herpes simplex virus types 1 and 2 and varicella-zoster virus. The isotope-labelled thymidine analogue iododeoxyuridine-125 is used as a substrate in a system which selectively measures viral enzyme activity. Subsequently, antibodies to the specific enzymes are used to differentiate the virus infection [6].

Demonstration of Virus-Specific IgM in Serum

Because of the localization or pathogenesis of some viral diseases no suitable specimen can be obtained for diagnosis. In others, no method may be available for the detection of virus. In these cases the demonstration of virus-specific IgM antibodies in a single serum specimen may rapidly give an accurate and early diagnosis.

The result of an IgM test can be influenced by the presence of the rheumatoid factor or large amounts of specific antibodies if IgG, causing false positive and false negative results, respectively. To avoid this, the IgM capture (μ-capture) method has been developed: antibodies to human IgM capture IgM from serum. The viral specificity of the captured IgM is shown by the addition either of enzyme-labelled viral antigens for ELISA [2, 8] or of virus and red blood cells [7] for the solid-phase or the haemadsorption immunosorbent techniques (SPIT or HIT).

While the domenstration of virus-specific IgM in a single serum specimen may give a reliable diagnosis, the finding of specific IgG does so only exceptionally. Whether characterization of the IgG subclass of an antibody response is useful for making a diagnosis remains to be shown.

Application in Viral Disease

Acute Respiratory Infections

Respiratory viral infections cannot be reliably differentiated by their symptoms, and a precise diagnosis from the virus laboratory is essential for treatment and epidemiological measures.

Viruses

Respiratory syncytial virus (RS virus) appears in epidemics during the cold or rainy season. It is the main cause of severe respiratory infections in small children. Occasionally, severe infections and death are seen in older patients. Ribavarin has successfully been used for treatment of the disease.

Influenza A. During an epidemic a "flu" can be clinically diagnosed. The accuracy of such a diagnosis is only about 80%. Small children with influenza may initially get febrile convulsions, others pseudo croup and a few severe bronchitis or pneumonia. An antiviral drug, amantadine hydrochloride, has been shown to be effective in prophylaxis and treatment of influenza A.

Influenza B usually occurs endemically and can likewise give severe disease. Abdominal pain is not unusual and occurs especially in children.

Adenovirus gives a multitude of different symptoms from the respiratory tract. To some extent this depends on age of the patient and serotype of the virus. The infection may sometimes present as pharyngitis with exanthema and can easily be clinically incorrectly diagnosed as a bacterial infection.

Parainfluenza Virus. There are four types of parainfluenza virus. They mainly infect children. Types 1 and 2 are common causes of pseudo croup in children, while type 3 may produce severe disease in small children similar to what is seen in RS virus infections.

Measles. An atypical measles sometimes needs to be verified. The diagnosis may lead to prophylactical measures such as providing immunoglobulin to the exposed or isolating children in risk groups.

Diagnostic Use of the Different Techniques

In acute viral respiratory infections infected epithelial cells are found in respiratory secretions. The best specimen for viral diagnosis is the nasopharyngeal secretion (NPS) though sputum and lung tissue may also be used. The cells in the specimen are used for IF diagnosis, while the secretion as such is diagnosed by immunoassay (ELISA, RIA, TR-FIA).

IF. The IF technique has been shown to be reliable, reproducible, and simple; it is also suitable for diagnosis in areas distant from a virus laboratory. Important pioneer work was done by Gardner and McQuillin, who established the technique for collection and preparation of specimens and defined the quality of reagents to be used [3]. The European Group for Rapid Viral Diagnosis has recommended this method to WHO and to their own members and arranged a supply of quality-controlled reagents for diagnosis of respiratory viral infections. An international study was done with this method on the epidemiology of certain respiratory infections: influenza A, RS, and parainfluenza virus types 1 and 3. One of the four viruses was found in 25% of the almost 8,000 children with respiratory infections investigated [10]. Table 1 shows diagnostic results from the author's laboratory. Only about 8% of the specimens are unsuitable for investigation because of too few cells in the secretion. About one-third of the tested specimens contained a virus which could be diagnosed by this method. The specimens were investigated for viruses relevant for the age of the patient and the time of the year.

The interpretation of the staining results is facilitated by the characteristic localization and appearance of the fluorescence in the different viral infections. Figure 1 shows infected cells in NPS: influenza B, RS virus, and adenovirus.

Table 1. Diagnosis of respiratory infections by immunofluorescence in nasopharyngeal cells

Specimens	4038
Poor quality	– 309 (7.5%)
RS virus	655
Influenza A	225
Influenza B	123
Adenovirus	51
Parainfluenza 1	38
Parainfluenza 3	79
Others	20
Total of positives	1191 (29%)

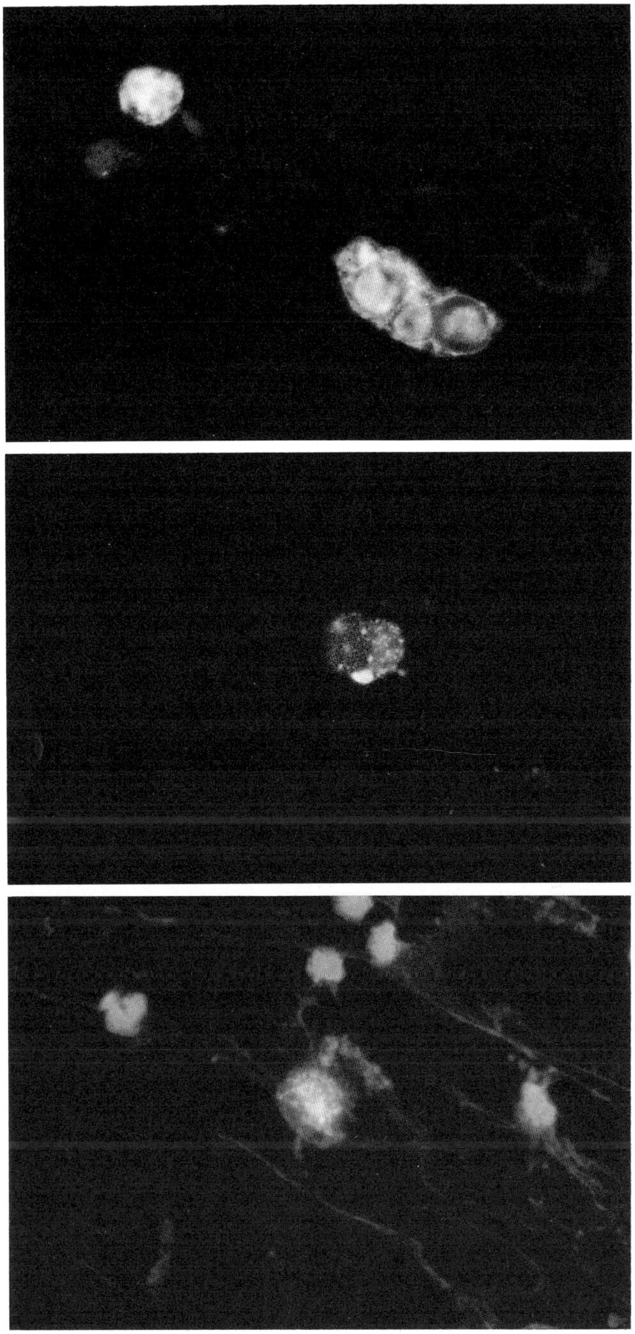

a

b

c

Fig. 1 a–c. Immunofluorescence staining of nasopharyngeal cells for diagnosis of respiratory infections: **a** influenza, **b** RS virus, and **c** adenovirus

Immunoassays. ELISA and RIA have both been described with regard to the diagnosis of respiratory viruses in clinical specimens. They have been shown to be sensitive and reproducible methods of diagnosis [11]. In my own experience ELISA has been a little less sensitive than IF for the diagnosis of RS virus, while influenza A and parainfluenza 3 seem to be detected with the same sensitivity. The main advantage of the method is the objective recording of results by spectrophotometers and the availability of automated systems for performing the test.

In the TR-FIA described with regard to the diagnosis of respiratory viruses by Halonen and co-workers [5], the sensitivity of the test seemed to be somewhat higher than that of ELISA; however, this did not significantly influence the number of positives found.

Diarrhoeal Disease

Viruses are the most comon cause of diarrhoeal disease in children. They may also cause gastroenteritis and diarrhoea in adults. Nosocomial infections are difficult to avoid. The cause can be diagnosed in 70% of children with diarrhoea.

Viruses Causing Gastroenteritis

Several viruses have been found in faecal specimens. While rota-, enteric adeno-, and Norwalk-like viruses have been shown to cause diarrhoea, other viruses found in faeces have not with certainty been associated with human disease.

Rotavirus is the main cause of gastroenteritis in children. The virus occurs mainly during the winter season. Children under the age of 2 years have the highest incidence of infections.

Enteric Adenovirus. Only a few of the more than 40 serotypes of adenovirus cause gastroenteritis. Recently, new serotypes have been described which are difficult or impossible to grow in tissue culture, but cause gastroenteritis in children [14]; these viruses are called enteric adenovirus (serotypes 40 and 41).

Norwalk or Norwalk-like Viruses. Several newly described viruses belong to this group. They may cause outbreaks of gastroenteritis in institutions, schools, or generally and mainly affect students and adults but also children.

Other Viruses. Astro-, calici-, corona-, and parvoviruses which are sometimes found in faecal specimens have not been associated with human gastroenteritis. Among the enteroviruses, coxsackie- and polioviruses do not usually cause diarrhoea. Certain ECHO viruses have been found in outbreaks, though it has not been possible to show that they were the cause of the disease.

Use of Methods

The majority of viruses causing diarrhoea cannot be isolated in tissue culture. Many of them were first found by EM. Sensitive IA and EM are used for diagnosis directly in the faecal specimen.

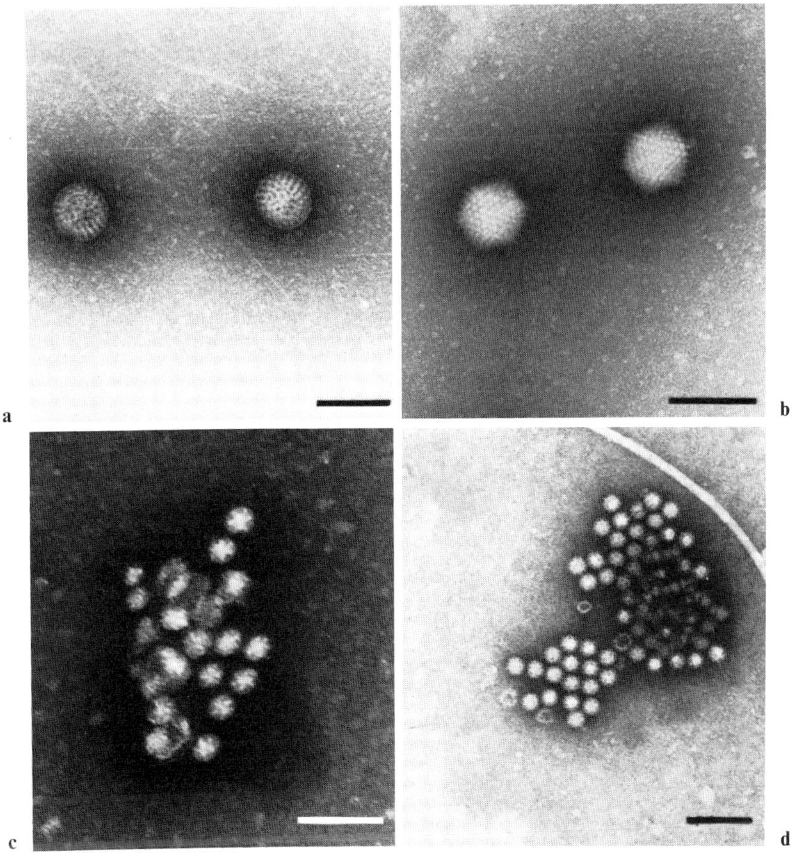

Fig. 2 a–d. Virus in faecal specimens found by EM: **a** rotavirus, **b** adenovirus, **c** astrovirus, and **d** small round viruses, structured (Norwalk-like viruses) (Photo: L. Svensson)

EM is used without specific reagents when a small number of specimens are to be investigated. Some of the viruses are easily recognized after contrast staining (Fig. 2). Rotavirus has a diameter of 70 nm and a characteristic wheel-like pattern. Adenovirus has a diameter of 70 nm and gives a hexagonal impression; the projection (fibre and knob) are usually not seen. Small round viruses with a diameter of 20–25 nm sometimes appear with a visible structure (e.g. Norwalk virus) while in other specimens they lack characteristic structures and are difficult to recognize.

IA such as ELISA, RIA, and TR-FIA are more suitable than EM for large-scale diagnosis. The specific reagents necessary for performance of the assay may be difficult to produce as the viruses which are to be diagnosed, are often difficult to grow in tissue culture.

Most laboratories have ELISA for the diagnosis of rotavirus, either as a commercial kit or with reagents of their own. ELISA is also used for the diagnosis of adenovirus by use of group-specific reagents. Differentiation between the two serotypes of enteric adenovirus (types 40 and 41) can also be done with ELISA. An IA for the diagnosis of Norwalk agent has been described but is not in comon use due to difficulties in producing specific reagents.

Other Methods. A handy latex agglutination test is now commercially available for rotavirus diagnosis. The accuracy may not be as high as for the described tests. This is partly compensated by the simplicity of the test and its immediate result.

Viral Infections in Skin and Mucous Membranes

A variety of viral infections are associated with exanthema. These have to be differentiated from exanthemas caused by other microorganisms such as fungi and bacteria as well as from exanthemas caused by drugs.

Viruses Causing Exanthema

The exanthema can appear either with or without vesicles.

Herpes simplex Virus (HSV) may cause severe infections in immunosuppressed patients as well as in infants during the neonatal period. The rapid confirmation of a suspected infection is a necessity for antiviral therapy.

Varicella. A rapid diagnosis of a varicella or zoster infection and an immediate result of a varicella immunity test may be of the utmost importance for the patient or his contacts.

Measles is usually easy to diagnose clinically. There are, however, atypical infections which have to be diagnosed before taking prophylactic measures, such as the administration of specific immunoglobulins to contacts and the isolation and vaccination of certain risk groups.

Rubella in the pregnant mother may constitute a real risk to the fetus in early pregnancy. The rapidly obtained diagnosis or immunity test result is of critical importance to the management of the patient.

Other viruses such as Epstein-Barr virus (EBV) and cytomegalovirus (CMV) may occasionally cause exanthemas. Human parvovirus causing erythema infectiosum (fifth disease) appears endemically or in outbreaks. It is important to diagnose all of these viruses in order to exclude other etiological agents such as rubella.

Use of Methods

The choice of method is correlated to the type of lesion in the skin.

Vesicular Exanthemas. Viral antigens are present in the vesicles caused by HSV or varicella-zoster virus. Diagnosis can be obtained by using the IF technique to examine cells scraped from the base of the vesicle [3] or biopsies of the skin [9] or immunoassays in the vesicle fluid.

In the author's laboratory IF has been used since 1977 for this type of diagnosis. The investigation has become increasingly popular because of its rapid an-

Table 2. The increasing number of skin scrapings received for rapid diagnosis of vesicular exanthema by IF (herpes simplex and varicella-zoster virus)

Period	No. of specimens	Positives
1977/78	33	16 (48%)
1978/79	83	29 (34%)
1979/80	146	58 (33%)
1980/81	332	120 (28%)
1981/82	594	249 (40%)
1982/83	769	405 (52%)
1983/84	1170	597 (51%)

Table 3. Comparison of results obtained by IF for herpes simplex virus and virus isolation for 247 specimens. Agreement is 90%

	Virus isolation	
	+	−
IF+	75	17
IF−	9	146

swer (3 h) and its reliability. Table 2 shows the trend during the last seasons. Positive results are seen in 50% of the specimens. Only a very small portion of test samples cannot be investigated because of too few intact cells. A comparison with results from the routine diagnostic department shows the good correlation between results from IF and virus isolation for HSV: 85%–90% agreement (Table 3). The results of ELISA used to detect HSV antigen also seem to correlate those from virus isolation, though the sensitivity is generally only 70% compared to virus isolation.

Virus-specific antibodies of IgG and IgM are detected by ELISA both in HSV infections and varicella. An IgM antibody response is not always seen in reactivations.

Exanthema Without Vesicular Lesions. Measles can be diagnosed by the detection of viral antigens in nasopharyngeal cells during the first 3–4 days as described in the previous section. For diagnosis later in the disease the demonstration of the development of specific IgM can be used. Rubella cannot yet be reliably diagnosed by methods for antigen detection. Specific IgM appears early and its detection in single serum specimens is used for early rapid diagnosis.

Erythema infectiosum is diagnosed by detecting IgM antibodies to the human parvovirus, which has been shown to cause the disease [1]. For the majority of other diseases with exanthema, e.g. the enteroviruses, no reliable techniques are available for rapid diagnosis of the virus.

Infections in the Central Nervous System

Efficient antiviral agents are now available for treatment of the herpes simplex encephalitis. Tick-borne spirochetes sensitive to antibiotics have been found to cause central nervous system (CNS) infections [12]. All this has also increased the demands for early and rapid diagnosis for other forms of viral encephalitis.

Viruses

Acute CNS infection may be the sole site of an infection, as in many arboviral infections and in the herpes simplex encephalitis. It may also occur as a manifestation of a systemic disease which is seen in e.g. mumps. While infectious virus is present in the cerebrospinal fluid (CSF) in meningitis, this is not usually the case in encephalitis caused by HSV or rabies.

Acute herpes encephalitis (also called acute necrotizing encephalitis) is usually caused by HSV type 1. In the neonatal period severe general HSV infection may sometimes occur with spread of the virus to the CNS. In about 85% of these cases HSV type 2 is the cause of infection.

In its classical form with temporal or fronto-temporal localization, the encephalitis may be easily diagnosed. The accuracy of the clinical diagnosis is, however, only about 50%–60%, which increases the need for a precise viral diagnosis from the laboratory.

Arbovirus. Many viruses in the large group of arboviruses cause encephalitis, e.g. the tick-borne encephalitis (TBE) and in the far east the Japanese encephalitis (JE). In the USA the La Crosse, St. Louis and equine encephalitis viruses may cause severe infections. The differentiation of these diseases from HSV encephalitis may sometimes be difficult.

Rabies. The early diagnosis of rabies in a human may sometimes be missed because of uncharacteristic symptoms. In cases of suspected human exposure no effort should be avoided to diagnose the biting animal, as efficient hyperimmunoglobulins and vaccines have been shown to prevent the disease.

Other CNS Infections. Measles, influenza, and varicella may sometimes be complicated by an encephalitis. Enterovirus infections and mumps both cause meningoencephalitis, mumps as a frequent and enterovirus as a less frequent complication to the infection.

Use of Methods

The possibility of rapid diagnosis varies with the type of infection [4]. Detection of viral antigens can be done either in brain material or in other material from the site of infection.

Detection of Viral Antigens in Biopsies. Brain biopsy is performed in some countries for the diagnosis of herpes encephalitis. The brain material is used for the detection of HSV by IF, EM, and ELISA. Attempts at virus isolation are also made. A diagnosis is obtained in about 70% of the HSV infections by examination of the biopsy.

In rabies the central vital parts of the brain are first infected and biopsies are therefore less suitable. Virus spread occurs early from the infected brain by the cranial nerves, and the virus is found in skin and mucous membranes of the head, in cornea and in salivary glands. From skin biopsy (from the neck behind the ear) suitable material is obtained for the investigation with IF. The result is ready after 2–4 h.

Intrathecal Antibody Production. A reliable diagnosis of an acute infection in the CNS can also be obtained by the detection of intrathecal antibody production. This is best done by sensitive methods like ELISA, although other serological techniques can be used, e.g. IF or even complement-fixation tests. Serum and cerebrospinal fluid are sampled on the same day and virus-specific IgG (and/or IgM) antibody titres are determined. Normally, the antibodies occur in a certain proportion in the serum and CSF. This proportion is disturbed by the local production of antibodies. Also, in encephalitis caused by other agents this principle can be used for diagnosis (measles, varicella).

Detection of Virus Specific IgM. In primary herpes simplex infections IgG and IgM both appear at the end of the first week. In reactivation IgM as a rule appears in severe infections. The high sensitivity of ELISA makes it a suitable test to detect these infections.

In TBE no brain material can be sampled for a reliable diagnosis. Most patients develop specific IgM at, or soon after, the infection. As there is an early dominance of specific IgM over IgG in this disease, treatment of serum for selective destruction of IgM with 2-mercaptoethanol and subsequent antibody detection is a suitable method. ELISA and other IAs have been described.

In mumps no methods are available for an early rapid detection of virus or viral antigen in CSF. The specific IgM appears after 3–5 days of illness and can be detected by ELISA or SPIT.

References

1. Anderson MJ, Davis LR, Jones SE, Pattison JR, Serjeant GR (1982) The development and use of an antibody capture radioimmunoassay for specific IgM to a human parvovirus-like agent. J Hyg Camb 88:309–324
2. Duermeyer W, Wielaard F, Veen J van der (1979) A new principle for the detection of specific IgM antibodies applied in an ELISA for hepatitis A. J Med Virol 4:25–32
3. Gardner PS, McQuillin J (1980) Rapid virus diagnosis – application of immunofluorescence, 2nd ed. Butterworth, London
4. Grandien M, Olding-Stenkvist E (1984) Rapid diagnosis of viral infections in the central nervous system. Scand J Infect Dis 16:1–8
5. Halonen P, Meurman O, Lövgren T, Hemmilä I, Soini E (1983) Detection of viral antigens by time-resolved fluoroimmunoassay. In: Bachman PA (ed) New developments in diagnostic virology. Springer, Berlin Heidelberg New York, p 133 (Current topics in microbiology and immunology, vol 104)
6. Källander C, Gronowitz S, Olding-Stenkvist E (1983) Rapid diagnosis of varicella-zoster virus infection by detection of viral deoxythymidinekinase in serum and vesicle fluid. J Clin Microbiol 17:280–287
7. Krech U, Wilhelm JA (1979) A solid-phase immunosorbent technique for the rapid detection of rubella IgM by hemagglutination inhibition. J Gen Virol 44:281–286
8. Loon M vaan, Logt JTM van der, Veen J van der (1981) Direct enzyme-linked immunosorbent assay that uses peroxidase-labelled antigen for determination of immunoglobulin M antibody to cytomegalovirus. J Clin Microbiol 13:416–422
9. Olding-Stenkvist E, Grandien M (1976) Early diagnosis of virus-caused vesicular rashes by immunofluorescence on skin biopsis. 1. Varicella, zoster and herpes simplex. Scand J Infect Dis 4:27–35
10. Ørstavik I, Grandien M, Halonen P, et al. (1980) Rapid immunofluorescence diagnosis of respiratory syncytial virus infections among children in European countries. Lancet II:32

11. Sarkkinen HK, Halonen PE, Arstila PP, Salmi AA (1981) Detection of respiratory syncytial, parainfluenza type 2, and adenovirus antigens by radioimmunoassay on nasopharyngeal specimens from children with acute respirtory disease. J Clin Microbiol 13:258–265
12. Sköldenberg B, Gårde A, Carlström A, Stiernstedt G, Kolmodin G, Nord CE (1983) Chronic meningitis caused by a penicillin-sensitive micro-organism? Lancet II:75–78
13. Stålhandske P, Petterson U (1982) Identification of DNA viruses by membrane filter hybridization. J Clin Microbiol 15:744–747
14. Uhnoo I, Wadele G, Svensson L, Johansson M (1984) Importance of enteric adenoviruses 40 and 41 in acute gastroenteritis in infants and young children. J Clin Microbiol 20:365–372

Respiratory Viral Antigen Detection by Solid-Phase Immunoassays: Comparison with Other Diagnostic Tests and Application for Routine Diagnosis

H. Sarkkinen [1]

Introduction

In many viral diseases, infectious virus or viral antigens are shed into body fluids during the acute phase of illness and may be detected by various methods. Indeed, electron microscopy, immunofluorescence (IF), counterimmunoelectrophoresis, complement fixation, latex agglutination, immune adherence hemagglutination, and many other methodologies have been suggested as rapid techniques for viral diagnosis. Most of these techniques have not been more widely used in diagnostic virology because their relative insensitivity, complicated nature, and labor intensity for routine work. Methods based on labeled antibodies, including radioimmunoassay (RIA), enzyme-immunoassay (EIA), and related techniques have, however, proven to be practical for routine diagnosis of hepatitis [1, 2] and viral gastroenteritis [3], and their advantages over the traditionally used virus isolation and viral antibody measurements are evident. Therefore, applications of these techniques to other diagnostic problems have now emerged. The present report reviews some of the data obtained in our laboratory concerning the use of indirect solid-phase immunoassay for the detection of respiratory virus antigens in nasopharyngeal aspirates. The 3½-year experience of the use of these techniques in day-to-day viral diagnosis of respiratory infections is discussed.

Methods

Indirect Immunoassays for Respiratory Virus Antigens

Nasopharyngeal specimens (NPS) were collected by suction with a disposable mucus extractor (Vycon, Ecouen, France), mainly from hospitalized children with acute respiratory infections. After collection the NPS were transported without cooling to Department of Virology within 2–72 h depending on the time and place of specimen collection. At the laboratory, all NPS were diluted 1 : 5 to 1 : 10 with phosphate buffered saline (PBS), pH 7.2, containing 20% inactivated fetal calf serum, 2% Tween 20, and 10^{-4} *M* merthiolate. Specimens were then sonicated (Branson Instruments Co., Stanford, Conn., USA) for 30 s to 3 min to solubilize the mucus and disrupt the cells.

1 Department of Virology, University of Turku, Kiinamyllynkatu 10, SF-20520 Turku 52. Present address: Department of Clinical Microbiology, University of Kuopio, PL 6, SF–70221 Kuopio

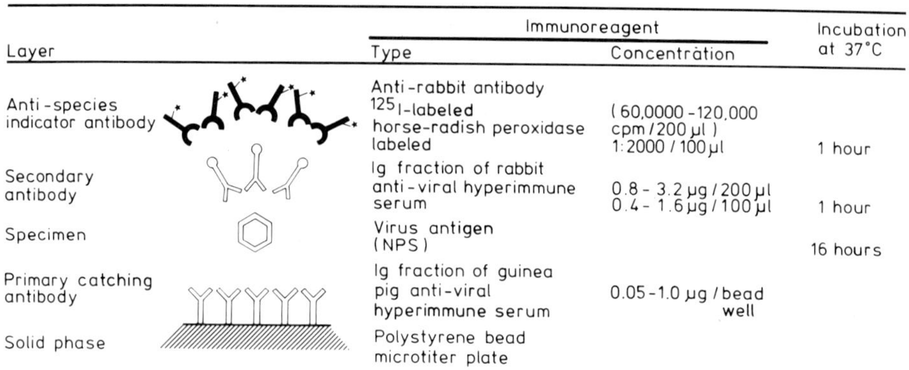

Fig. 1. Principle of indirect solid-phase immunoassay for detection of respiratory virus antigens in nasopharyngeal aspirates

Table 1. Variance in radioimmunoassays (RIA) and enzyme-immunoassays (EIA) for viral antigen detection measured as coefficient of variations. Adenovirus (AD), influenza A and B virus (INFA, INFB), parainfluenza type 1, 2, and 3 (P1, 2, 3), and respiratory syncytial virus (RSV)

Test for	Specimen	Pos/Neg	Intraassay ($n=10$)		Interassay ($n=10$)	
			RIA	EIA	RIA	EIA
AD	Nasopharyngeal secretions	+	5.1	9.6	21.7	25.5
		–	8.4	6.0	31.0	46.1
INFA	Nasopharyngeal secretions	+	9.1	17.0	15.5	16.4
		–	10.8	13.0	31.7	31.8
INFB	Nasopharyngeal secretions	+	8.4	5.9	41.9	13.9
		–	7.8	8.0	26.9	43.5
P1	Nasopharyngeal secretions	+	14.0	8.0	23.0	20.7
		–	9.0	10.0	35.7	55.4
P2	Nasopharyngeal secretions	+	14.9	16.0	12.0	16.4
		–	15.8	8.0	39.9	31.0
P3	Nasopharyngeal secretions	+	6.2	12.0	31.4	22.2
		–	10.8	11.0	26.6	35.3
RSV	Nasopharyngeal secretions	+	5.0	6.0	19.6	22.9
		–	7.1	8.0	32.3	32.8

The method for the detection of adenovirus, influenza A and B, parainfluenza types 1, 2, and 3, and respiratory syncytial virus (RSV) antigens was a standard indirect solid-phase RIA or EIA [4–6] (Fig. 1). Viral antigens in sonicated NPS were captured on a solid-phase precoated with antiviral guinea pig total immunoglobulin fraction (Ig). This was followed by sequential additions of antiviral rabbit Ig and either ^{125}I-labeled (RIA) or horse-radish peroxidase labeled (EIA, Orion Diagnostica, Helsinki, Finland) antirabbit antibodies. In EIA an additional incubation with a substrate solution of orthophenylene diamine was required. The test was considered positive if the counts per minute or the absorbance value of the specimen was two or more times higher than that of the buffer blanks. In addition, borderline positives (binding values 2–5 times that of the buffer blanks) were retested in a separate blocking test [5]. The guinea pig and rabbit antiviral antibodies used as the coating and secondary antibodies were produced by intradermal immunizations with purified virions or structural proteins [4–6].

The sensitivities of the RIA/EIA assays varied between 1 and 30 ng/ml of viral proteins as determined by control titrations of purified viruses. The specificities of the assays were proved by cross-testing purified viral antigens and clinical specimens containing high concentrations of viral antigens. Further, the specificities of the bound radio or enzyme activity of specimens with low positive reactions were confirmed by testing in a blocking test as indicated above. The reproducibilities of the assays were determined by measuring the intra-assay and interassay variances for each test in 10 parallel and 10 successive tests, respectively (Table 1).

Other Tests

The IF for adenovirus, influenza A and B, parainfluenza types 1, 2, and 3, and RSV were done according to Gardner and McQuillin [7]. The reagents for IF were obtained from Wellcome Laboratories or from the members of the European Group for Rapid Viral Diagnosis [4–6].

The EIA antibodies to RSV and adenovirus were measured as described by Meurman et al. [8, 9].

Results

Comparison of Indirect Immunoassays with Other Diagnostic Tests

Indirect IF has been used as a rapid diagnostic test for respiratory virus infections in our laboratory since 1979. The RIAs and EIAs were initially compared with IF using known virus positive and virus negative NPS [4–6]. These primary studies, which included about 300 patients, indicated a relatively good correlation among the three test systems. The studies also suggested that the quality of the nasopharyngeal specimen was a critical factor influencing the end result. It was established and later confirmed by others [10] that nasopharyngeal aspirates with high antigen concentration were superior to throat swabs. Furthermore, sonication of the NPS proved to be more efficient than chemicals such as *N*-acetylcysteine [11] and deoxycholate for homogenization of the mucus in the aspirates. On the basis of these preliminary results, a further clinical study was done to evaluate the diagnostic efficacies of IF and RIA in day-to-day work. The final results shown in Table 2 indicated that RIA was superior to IF for detection of adenovirus and influenza A and B virus antigens, but that a considerable number

Table 2. Nasopharyngeal specimens tested simultaneously by immunofluorescence (IF) and radioimmunoassay (RIA), November 1980 – October 1981

Total no. tested	Virus	Positive			
		RIA/IF	RIA only	IF only	Total[a]
1053	Adenovirus	23	7	0	30
	Influenza A	11	2	0	13
	Influenza B	9	6	0	15
	Parainfluenza 1	1	2	2	5
	Parainfluenza 2	0	0	1	1
	Parainfluenza 3	13	4	3	20
	Respiratory syncytial	65	6	6	77

[a] Positive either in RIA or IF or both

of discrepancies were found with parainfluenza viruses and RSV. Most of the discrepancies were explained by the fact that many of the latter specimens gave borderline positive values in either assay. However, two specimens were strongly positive for RSV in IF by both IF and RIA immunoreagents but repeatedly negative in RIA. Some of the NPS with adequate numbers of desquamated epithelial cells reacted strongly in adenovirus RIA, but were repeatedly negative in IF. Thus, although differencies in reagents and in quality of the specimens explained in part the contradictionary results, these results indicated that the assays were suboptimal. These results are in agreement with those of other groups [10, 12] who have also found similar results with RSV IF and EIA. It may be that different types of viral infections favor one type of assay over the other. Thus, in a lytic type of infection, such as adenovirus, in which large amounts of extracellular antigens are produced, RIA/EIA may be a better assay since it detects both soluble and cell-associated antigens.

The diagnostic efficacy of antigen-RIA was also compared to that of immunoglobulin class-specific antibody-EIA using serum pairs from 52 (adenovirus) and 33 (RSV) infected children [8, 9]. The highest rate for adenovirus diagnosis (85% of the cases detected) was obtained by antigen-RIA followed by IgG, IgM, and IgA antibody-EIA tests that detected infection in 77%, 48%, and 37%, respectively, of the same patients. In the case of RSV the IgG antibody-EIA detected 91% of the cases followed by antigen-RIA, then IgM and IgA antibody-EIA which gave 79%, 70%, and 48% detection rates. The diagnostic efficacy of antigen RIA was greatly dependent on the time of specimen collection. From specimens collected during the first 5 days 84% were positive in adenovirus RIA and 88% and RSV RIA, whereas only 50% of the specimens collected at 6–10 days after onset of disease were positive. These results clearly pointed out that the viral excretion is greatest during the acute phase of the disease, and that if NPS are collected early in disease the diagnostic efficacy of antigen detection exceeds that of antibody measurements.

Application of Indirect Immunoassays for Day-to-Day Diagnosis of Respiratory Infections

NPS from patients with acute respiratory virus infections were sent to our diagnostic unit mainly from children in pediatric wards in southwestern Finland (population of about one million). A routine screening of NPS for respiratory viruses by indirect RIA was started in October 1980. However, during the first year, NPS were partly tested also in IF (see the previous Section). In January 1984 EIA replaced RIA as the routine test.

For convenience tests were run in a single dilution and each NPS was tested for the seven viruses. To offer service to clinics as rapidly as possible the NPS were tested on a daily basis as follows: all NPS that came to our laboratory before 1 p.m. were tested the same day from Monday to Friday. Results were reported by telephone the next day (Tuesday to Saturday) between 1 and 5 p.m. and subsequently by mail. To secure maximum rapidity in reporting, the results that had to be repeated were telephoned as uncertain positives or negatives without delay, and the final, confirmed results were reported later. The total time for performing the tests and reporting the results was thus about 28 h (transportation time excluded).

The number of specimens tested each month varied from approximately 100–200 NPS in nonepidemic periods and up to 400–500 NPS in periods of respiratory epidemics. During peak RSV epidemic periods about 30 NPS were tested daily. In addition, approximately 8% of the NPS had to be retested (by blocking test), either to confirm the specificity of a weak positive reaction or to rule out nonspecific reactions detected as a simultaneous binding in all assays. Thus, the practical capacity of the system, about 20–25 NPS/day/one technician (maximum 35–40 NPS/day/one technician), corresponding to 400–500 NPS/month/one technician, was usually quite sufficient.

Table 3. Respiratory virus infections, diagnosed by radioimmunoassay antigen detection in the Department of Virology, University of Turku. Adenovirus (AD), respiratory syncytial virus (RSV), influenza A virus (INFA), influenza B virus (INFB), parainfluenza 1, 2, and 3 virus (P1, 2, 3)

		No. of specimens	Total no. of positives (%)	Positive for						
				AD	RSV	INFA	INFB	P1	P2	P3
1980	Nov	95	6 (6.3)	1					1	4
	Dec	101	4 (4.0)	2						2
1981	Jan	145	12 (8.3)	3		8				1
	Feb	121	10 (8.3)	1		6				3
	Mar	104	9 (8.7)	4	1					4
	Apr	85	8 (9.4)		2					6
	May	71	9 (12.7)	2	6					1
	Jun	92	6 (6.5)	3					2	1
	Jul	68	3 (4.4)	1	2					
	Aug	70	7 (10.0)	4	2				1	
		952	74 (7.8)	21	13	14			4	22

Table 4. Respiratory virus infections, diagnosed by radioimmunoassay antigen detection in the Department of Virology, University of Turku. Adenovirus (AD), respiratory syncytial virus (RSV), influenza A virus (INFA), influenza B virus (INFB), parainfluenza 1, 2, and 3 virus (P1, 2, 3)

		No. of specimens	Total no. of positives (%)	Positive for						
				AD	RSV	INFA	INFB	P1	P2	P3
1981	Sep	149	22 (14.8)	5	16			1		
	Oct	357	148 (41.5)	9	136			3		
	Nov	493	216 (43.8)	17	193			6		
	Dec	295	73 (24.7)	4	62			7		
1982	Jan	309	78 (25.2)	31	36			10		1
	Feb	277	66 (23.8)	14	9			27	9	7
	Mar	240	61 (25.4)	12	3			16	16	14
	Apr	183	33 (18.0)	8		2	1	4	8	10
	May	217	30 (13.8)	12			4	1	6	7
	Jun	142	10 (7.0)	3			6	1		
	Jul	103	12 (11.7)	6			2		2	2
	Aug	130	7 (5.4)	6						1
		2895	756 (26.1)	127	455	2	13	76	41	42

Table 5. Respiratory virus infections, diagnosed by radioimmunoassay antigen detection in the Department of Virology, University of Turku. Adenovirus (AD), respiratory syncytial virus (RSV), influenza A virus (INFA), influenza B virus (INFB), parainfluenza 1, 2, and 3 virus (P1, 2, 3)

		No. of specimens	Total no. of positives (%)	Positive for						
				AD	RSV	INFA	INFB	P1	P2	P3
1982	Sep	98	6 (6.1)	5						1
	Oct	138	13 (9.4)	13						
	Nov	211	15 (7.1)	13						2
	Dec	183	18 (9.8)	16					1	1
1983	Jan	239	21 (8.8)	12	1	4				4
	Feb	306	58 (19.0)	3	7	38	1			9
	Mar	338	106 (31.4)	7	18	25	22			34
	Apr	248	79 (31.9)	7	20	2	25			25
	May	119	28 (23.5)	12	8		1			7
	Jun	166	23 (13.9)	4	16					3
	Jul	118	17 (14.4)	5	11					1
	Aug	122	11 (9.0)	7	4					
		2286	395 (17.3)	104	85	69	49		1	87

Until May 1984, 8,298 NPS had been studied (Table 3–7). On the average, about 20% contained virus, but during peak epidemic seasons more than 40% of the NPS were positive. About 97% of the virus positive NPS were from children. The few diagnosed adult cases were usually either influenza or adenovirus infections from army trainees or elderly patients. As expected, RSV was the most commonly found virus, with typical fall epidemics in 1981 and 1983, missing epi-

Table 6. Respiratory virus infections, diagnosed by radioimmunoassay or enzyme-immunoassay antigen detection in the Department of Virology, University of Turku. Adenovirus (AD), respiratory syncytial virus (RSV), influenza A virus (INFA), influenza B virus (INFB), parainfluenza 1, 2, and 3 virus (P1, 2, 3)

		No. of specimens	Total no. of positives (%)	Positive for						
				AD	RSV	INFA	INFB	P1	P2	P3
1983	Sep	179	10 (5.6)	3	7					
	Oct	216	45 (20.8)	8	37					
	Nov	366	136 (37.2)	11	123			1	1	
	Dec	425	186 (43.8)	3	183					
1984	Jan	309	88 (28.5)	13	75					
	Feb	167	28 (16.8)	11	14	1		1	1	
	Mar	268	46 (17.2)	17	8	2	6	4	1	8
	Apr	235	49 (20.9)	14	3		10	7	2	13
		2165	588 (27.2)	80	450	3	16	13	5	21

Table 7. Respiratory virus infections, diagnosed by radioimmunoassay or enzyme-immunoassay antigen detection in the Department of Virology, University of Turku. November 1980 – April 1984. Adenovirus (AD), respiratory syncytial virus (RSV), influenza A virus (INFA), influenza B virus (INFB), parainfluenza 1, 2, 3 virus (P1, 2, 3)

No. of specimens	Total no. of positives (%)	Positive for						
		AD	RSV	INFA	INFB	P1	P2	P3
8298	1813 (21.8)	332	1003	88	78	90	52	172

demics in 1980 and 1982, and a uncommon spring epidemic in 1983. Adenovirus infections were constantly observed, but major influenza epidemics were not seen during the study period. Parainfluenza type 3 virus infections were also detected each year, whereas parainfluenza types 1 and 2 were less common findings.

Concluding Remarks

Our aim has been to provide a rapid diagnostic service of respiratory virus infections for the clinics on a daily basis. Earlier, from 1979 until 1981, we used indirect IF along the lines of Gardner and McQuillin [7] for this work and achieved excellent results. Our diagnostic burden increased during this period and pointed out the disadvantages of IF; the specimen handling and the time consuming IF microscopy. Therefore, indirect RIA/EIA assays suitable for screening large numbers of clinical specimens were developed.

The comparisons of RIA/EIA with IF and viral serology by us [4–6, 8, 9] and IF, EIA and related methods with virus isolation by others [10, 11, 13–16] proved that all these methods are bound to miss some of the positive cases and that one type of test found suitable for diagnosis of a specific respiratory virus infection may not necessarily be the best choice for some other type of infection. Thus, the choice between these tests is not always easy given the availability of laboratory equipment and personal. Keeping in mind the limitations of other diagnostic techniques, it is my opinion that many practical aspects favor the use of RIA/EIA-type of tests in day-to-day diagnostic service. First, the specimen handling is simple for RIA/EIA since the viral antigens in NPS are stable. Our experience and that of others [10] indicate that NPS can be stored at ambient temperatures at least for a few days without impairing the test results. This is an advantage at night and during weekends, and for hospitals remote from a virus laboratory. Second, using these assays a considerably expanded work load can be accommodated. Finally, interpretation of the results takes only a few minutes by experienced, but not necessarily academic personnel or through the use of computerized systems. The requirement for overnight incubation, which seriously delays reporting the results, may be overcome in future by using more sensitive indicator systems such as those based on fluorescence probes [13, 17].

In conclusion, the extensive use of RIA/EIA has assured us that these types of assays are practical and efficient for diagnosis of respiratory infections. Also, our experience, particularly with adenovirus infections [18] has shown that the information obtained by these techniques is useful in clinical work. Therefore, assays for additional respiratory viruses should be added to this diagnostic repertoire. Indeed, indirect immunoassays have been developed recently for two coronaviruses and parainfluenza type 4 viruses in our laboratory. Whether or not rhinoviruses could be detected by these or other techniques, such as recombinant DNA technology [19], remains to answered. In any case, the clinical significance of rapid viral diagnosis is evident particularly in the field of pediatric infectious diseases.

Acknowledgements. The excellent technical assistance of Ms. Kaija Johansson and Ms. Pia Saranpää is gratefully acknowledged. These studies were supported by grants from the Sigrid Juselius Foundation and Emil Aaltonen Foundation.

References

1. Ling CM, Overby LR (1972) Prevalence of hepatitis B virus antigen as revealed by direct radioimmune assay with ^{125}I-antibody. J Immunol 109:834–841
2. Mathiesen LR, Feinstone SM, Wong DC, Skinhoej P, Purcell RH (1978) Enzyme-linked immunosorbent assay for detection of hepatitis A antigen in stool and antibody to hepatitis A antigen in sera: comparison with solid-phase radioimmunoassay, immune electron microscopy and immune adherence hemagglutination assay. J Clin Microbiol 7:184–193
3. Sarkkinen HK, Tuokko H, Halonen PE (1980) Comparison of enzyme immunoassay and radioimmunoassay for detection of human rotaviruses and adenoviruses from stool specimens. J Virol Methods 1:331–341
4. Sarkkinen HK, Halonen PE, Arstila PP, Salmi AA (1981) Detection of respiratory syncytial, parainfluenza type 2, and adenovirus antigens by radioimmunoassay and enzyme immunoassay on nasopharyngeal specimens from children with acute respiratory disease. J Clin Microbiol 13:258–265
5. Sarkkinen HK, Halonen PE, Salmi AA (1981) Detection of influenza A virus by radioimmunoassay and enzyme-immunoassay from nasopharyngeal specimens. J Med Virol 7:213–220

6. Sarkkinen HK, Halonen PE, Salmi AA (1981) Type-specific detection of parainfluenza viruses by enzyme-immunoassay and radioimmunoassay in nasopharyngeal specimens of patients with acute respiratory disease. J Gen Virol 56:49–57
7. Gardner PS, McQuillin J (1980) Rapid virus diagnosis. Application of immunofluorescence. Butterworths, London
8. Meurman O, Ruuskanen O, Sarkkinen H (1983) Immunoassay diagnosis of adenovirus infections in children. J Clin Microbiol 18:1190–1195
9. Meurman O, Sarkkinen H, Ruuskanen O, Hänninen P, Halonen P (1984) Diagnosis of respiratory syncytial virus infection in children: comparison of viral antigen detection and serology. J Med Virol 14:61–65
10. Mckintosh K, Hendry RM, Fahnestock ML, Pierik LT (1982) Enzyme-linked immunosorbent assay for detection of respiratory syncytial virus infection: application to clinical samples. J Clin Microbiol 16:329–333
11. Chao RK, Fishaut M, Schwartzman JD, McKintosh K (1979) Detection of respiratory syncytial virus in nasal secretions from infants by enzyme-linked immunosorbent assay. J Infect Dis 139:483–486
12. Hornsleth A, Friis B, Andersen P, Brenoe E (1982) Detection of respiratory syncytial virus in nasopharyngeal secretions by ELISA: comparison with fluorescent antibody technique. J Med Virol 10:273–281
13. Berg RA, Rennard SI, Murphy BR, Yolken RH, Doli R, Straus SE (1980) New enzyme immunoassays for measurement of influenza A/Victoria/3/75 virus in nasal washes. Lancet 1:851–853
14. Yolken RH, Torsch VM, Berg R, Murphy BR, Lee YC (1980) Fluorometric assay for measurement of viral neuraminidase-application to the rapid detection of influenza virus in nasal wash specimens. J Infect Dis 142:516–523
15. Harmon MW, Pawlik KM (1982) Enzyme immunoassay for direct detection of influenza type A and adenovirus antigens in clinical specimens. J Clin Microbiol 15:5–11
16. Harmon MW, Russo LL, Wilson SZ (1983) Sensitive enzyme immunoassay with B-D-galactosidase-Fab conjugate for detection of type A influenza virus antigen in clinical specimens. J Clin Microbiol 17:305–311
17. Halonen P, Meurman O, Lövgren T, Hemmilä I, Soini E (1983) Detection of viral antigens by time-resolved fluoroimmunoassay. Springer, Berlin Heidelberg New York Tokyo, pp 133–145 (Current topics in microbiology and immunology, vol 104)
18. Ruuskanen O, Sarkkinen H, Meurman O, Hurme P, Rossi T, Halonen P, Hänninen P (1984) Rapid diagnosis of adenoviral tonsillitis: a prospective clinical study. J Pediatrics 104:725–728
19. Hyypiä T, Stålhandske P, Vainionpää R, Petterson V (1984) Detection of enteroviruses by spot hybridization. J Clin Microbiol 19:436–438

Coagglutination Procedures and Affinity Chromatography in Antigen and Antibody Detection

M. C. Goldschmidt [1]

Introduction

Protein A

One of the virulence characteristics of *Staphylococcus aureus* is the elaboration of a protein, "protein A" (PA), which has the capability of binding nonspecifically to the Fc portion of several classes of immunoglobulins, especially to IgG from humans and many animals. Thus, the "destructive" portion (Fab) of the IgG is kept extended outward away from the cells, while the bacteria, once surrounded by heterogeneous immunoglobulin molecules, appear more like "self" than "foreign" to host surveillance systems. It is astonishing and fascinating to observe how this immunoglobulin combining ability of PA has been applied to many diverse characterization and rapid detection procedures, several of which will be discussed. Only a few of many references can be cited in this paper. Journals that usually have articles concerned with protein A include: J Clin Microbiol, Acta Pathol Microbiol Immunol Scand Sect B, J Immunol Methods, Infect Immun, Eur J Clin Immunol, and Clin Exp Immunol.

PA (42,000 MW) is usually covalently bound to the staphylococcal cell wall. However, it can be partially or totally released by some strains (including their protoplasts) during growth. The Cowan I strain produces a large amount of PA [about 1.7% (dry weight) of total protein] and has been used in the majority of studies involving whole staphylococci. Although PA reacts with several classes of immunoglobulin, such as IgM, this paper will deal only with PA-IgG complexes. An excellent review by Langone [1] covers production, physicochemical characteristics, reactivity with immunoglobulins from various animal species, etc. Other publications by Facklam et al. [2], Kronvall [3], Goldschmidt [4], and Forsgren and Sjöquist [5] provide additional information.

Coagglutination

PA-containing staphylococci bound to IgG are called "sensitized staphylococci." This is a "pseudoimmune" reaction in that the actual binding of PA with the Fc portion of the molecule is not dependent on the antigenic specificity of the anti-

1 Dental Science Institute, Dental Branch and Graduate School of Biomedical Sciences, The University of Texas Health Science Center at Houston, P.O. Box 20068, Houston, TX 77225, USA

body or whether it is free or already coupled to its homologous antigen. Immune complex formation appears to enhance the binding between PA and several poorly reactive IgG types. The term coagglutination (COAG) has also been applied to reactions involving only purified PA as well as the whole staphylococci.

As with any immune reaction, the specificity of the antigen–antibody reaction determines in large part the resultant sensitivity of the assay. One great advantage of COAG is the capability of identifying the "target" organism or antigen in a mixed culture, on the primary isolation plate, or in a clinical specimen. This is, of course, due to the specificity and extreme sensitivity of the antigen–antibody reaction. Similarly, the "target" antibody can be pulled out of a mixture, detected, and even purified.

Immunoamplification

When various markers such as fluorescent compounds, radionuclides, or enzymes have been complexed directly with antibodies, loss of antibody reactivity often occurs. However, when the marker is combined with PA or staphylococcal cells instead, antibody reactivity is not affected after the two are coupled. The desired staphylococcal or PA-marker can be used with *any* desired antibody and is therefore a much more flexible reagent. The antibody can even be eluted and the PA-marker reused. This is especially true of PA-Sepharose columns. Since the PA-antibody complex forms practically instantaneously, lengthy and complicated procedures are not necessary. Some of these marker labels combined with PA or staphylococci include fluorescent compounds (fluorescein isothiocyanate), radionuclides (such as ^{125}I), enzymes (horseradish peroxidase), Sepharose CL-4B, and sheep erythrocytes for rosette formation [6–9]. Applications involving many of these immunoamplification techniques will be discussed throughout this article. Most of these PA-markers can be obtained commercially or made in the laboratory.

Rapid Identification of Microorganisms and Other Types of Antigen

Identification from Pure Cultures of Microorganisms

One of the first and still most common uses of COAG involves the rapid identification of pure cultures of microorganisms. A partial list is presented in Table 1. The number of organisms identified by this method constantly increases. The β-hemolytic streptococci [10], *Neisseria gonorrhoeae* [11], and *Haemophilus influenzae* type b [12] are now routinely identified by COAG methods in many clinical laboratories. In addition to being an easy and rapid procedure, 1 ml of antiserum, enough for 20–50 conventional tests, will yield enough COAG reagent for 400–1,000 tests. Several commercial kits are available for these and other organisms.

Table 1. Microorganisms detected by coagglutination

Bacteria	*Viruses*
Actinomyces israelii	Avian reticuloendotheliosis
A. viscosus	Bovine rhinotracheitis
Brucella canis	C-type
Chalamydia trachomatis	Denque
Escherichia coli (04, enterotoxin)	Herpes simplex types 1 and 2
Haemophilus influenzae	Influenza
Klebsiella pneumoniae	Rotavirus
Mycobacteria	Rubella
Mycoplasma	Varicella zoster
Neisseria gonorrhoeae	
N. meningitidis	*Fungi*
Pasteurella multocida	*Candida albicans*
Peptostreptococcus species	*Histoplasma capsulatum*
Salmonella species	*Cryptococcus neoformans*
Serratia marcescens	
Shigella flexineri	*Protozoa*
S. sonnei	*Acanthamoeba culbertsoni*
Streptococcus pyogenes	*Entamoeba histolytica*
S. mutans	*Naegleria fowleri*
S. pneumoniae	*Leishmania*
Vibrio cholerae	

Table 2. Identification upon primary isolation

Name	Immunological group detected	Method
Actinomyces viscosus	Surface antigens	PA–FITC[a]
Candida albicans	Mannans, other surface antigens	COAG[b] PA–FITC
Escherichia coli	04, heat-labile, enterotoxin	COAG
Haemophilus influenzae	B antigens, (c, a, d, e, f, pool)	COAG[c]
Neisseria gonorrhoeae	W, J, M antigens	COAG[c]
N. meningitidis	A, B, C antigens	COAG
Salmonella species	02, 04, 09, Vi, D, d; polyvalent antisera to 1–11, 13–16, 19, 22, 23; synthetic disaccharide-bovine serum albumin conjugate; lipopolysaccharide	COAG
Shigella flexneri	B antigen	COAG
S. sonnei	D antigen	COAG
Streptococcus pyogenes (β-hemolytic)	A, B, C, F, G antigens	COAG[c]
S. pneumoniae	Surface antigens	COAG[c]
Herpes simplex virus	Type 1 antigens	COAG
Adenovirus	Structural protein	COAG

[a] PA–FITC: purified protein A coupled to fluorescein isothiocyanate
[b] COAG: coagglutination with sensitized whole cells of *Staphylococcus aureus*, Cowan I strain
[c] Commercial kits available

Identification upon Primary Isolation

In addition to the identification of purified bacterial cultures, killed sensitized staphylococci have been added to the surface of primary isolation plates. Agglutination occurred around colonies of homologous organisms such as salmonellae and shigellae [13]. Many of the microorganisms detected in this way are listed in Table 2.

Similarly, layering of sensitized staphylococci over tissue cultures (hopefully inoculated with viral containing material), results in an agglutination reaction over the cells containing viruses, such as herpes simplex [14]. Most authors report that COAG appears to be a simpler procedure to perform than immunofluorescence or immunoelectrophoresis (etc.) and is equal in sensitivity. In addition, very little specialized equipment is usually required for most COAG reactions.

Identification from Clinical Specimens

As can be seen in Tables 3 and 4, many organisms can now be directly identified from clinical specimens by COAG. These include, for example, *Haemophilus influenzae* type b from cerebrospinal fluid [15] and group B β-hemolytic streptococci from urogenital and gastric swabs [16]. Rapid identification and detection of herpes simplex virus, varicella zoster virus, and rotavirus directly from clinical specimens within a few hours of their receipt has been reported by several authors [17, 18]. The first two viruses have been detected in concentrations of less than

Table 3. Direct isolation from clinical specimens containing bacteria

Organism or antigens detected	Clinical specimen	Method
Actinomyces viscosus	Plaque	PA–FITC[a]
Haemophilus influenzae (a, b, c, d, e, f)	Cerebrospinal fluid, urine, blood	COAG[b,c]
H. pleuropneumoniae	Lung biopsy	COAG
Neisseria meningitidis (A, B, C, Y, W135)	Cerebrospinal fluid	COAG[c]
Pasteurella multocida	Blood, liver extract	COAG
Salmonella species	Food, urine	COAG
Streptococcus pyogenes (β-hemolytic A, B, C, F, G)	Cerebrospinal fluid, urogenital and gastric swabs, blood, placenta, pharyngeal aspirates	COAG[c]
S. mutans	Plaque	PA–FITC
S. pneumoniae	Cerebrospinal fluid, sputum, blood	COAG

[a] PA–FITC: purified protein A coupled to fluorescein isothiocyanate
[b] COAG: coagglutination with sensitized whole cells of *Staphylococcus aureus* Cowan I strain
[c] Commercial kits available

Table 4. Direct isolation from clinical specimens

Organism or antigens detected	Clinical specimen	Method
Bovine rhinotracheitis virus	Serum	PA–HPO[a]
Herpes simplex virus types 1 and 2	Genital, oral, finger	COAG[b]
Rotavirus	Fecal extracts	PA[c]
Rubella virus	Serum	PA–HPO
Varicella zoster virus	Genital, oral	COAG
Antigranulocyte antibodies	Serum	^{125}I–PA[d]
Human IgG	Urine, cerebrospinal fluid	COAG
Major histocompatibility complex antigens, guinea pig class I and class II (also with peritoneal blood leukocytes or spleen cells as target cells)	Guinea pig alloimmune serum	PA–SBRC[e]
Monoclonal antibodies reacting with PA	Ascites fluid, tissue culture fluids	PA-Sepharose CL–4B[f] PA–FITC
Staphylococcal enterotoxins A, B, C, D, E	Foods	COAG
Tumor antigen-immune complexes	Plasma	PA

[a] PA–HPO: purified protein A coupled to horseradish peroxidase
[b] COAG: coagglutination with sensitized whole cells of *Staphylococcus aureus,* Cowan I strain
[c] PA: purified protein A
[d] ^{125}I–PA: ^{125}I coupled to purified protein A
[e] PA–SRBC: protein A coupled to sheep erythrocytes
[f] PA-Sepharose CL–4B: protein A coupled to Sepharose CL–4B columns

10 cells/ml in glutaraldehyde-fixed clinical specimens. After incubation with immune serum, they were then mixed with formalinized staphylococci and vacuum filtered through a 5-μm filter. The filter was clarified and then examined for numbers of sensitized staphylococci attached to the cell surfaces [17]. Concentrations of rotavirus as low as 10 ng/ml were detected by serological trapping on electron microscope grids precoated with PA-antibody [18]. *Histoplasma capsulatum* and *Candida* were also identified.

Identification of Other Types of Antigen

As indicated in Table 4, many other types of antigen can be easily detected by COAG reactions. The immunologically active excreted factor (EF) of *Leishmania* has been monitored in this way [19]. Two adenovirus structural proteins and a herpes simplex glycoprotein have been purified by sensitized staphylococci bound to plastic tissue culture plates [20]. Sensitized staphylococci have been used as a reagent in a rapid radioimmunoassay for human leukocyte interferon [21].

Protein A coupled to periodate-oxidized ferromagnetic iron-dextran particles and immunoglobulins have been used as reagents to label antigen sites on human erythrocytes for visualization by scanning or transmission electron microscopy [22]. As can be seen in Fig. 1, we have found that PA-FITC plus rabbit IgG antibody to cell surface antigens of *Bifidobacterium bifidum* can also be used to visualize the antigen–antibody reaction by transmission electron microscopy.

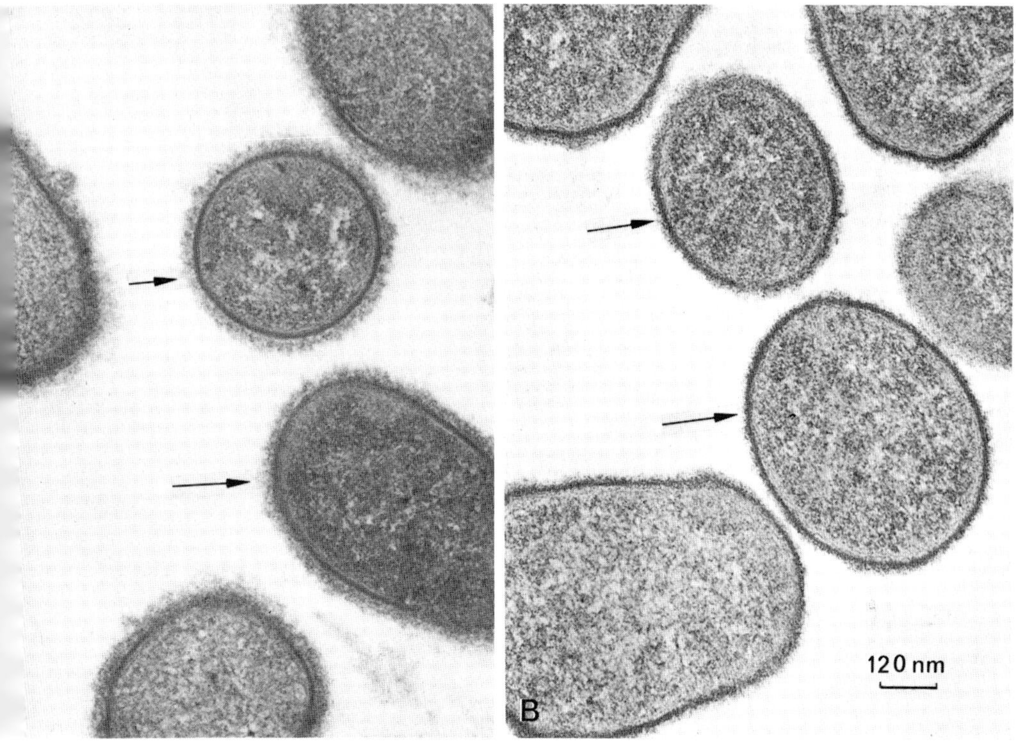

Fig. 1 A, B. Electron micrograph of *Bifidobacterium bifidum* cells. **A** cells incubated with rabbit homologous antiserum + PA-FITC; **B** control cells

We have used flow cell cytometry as a rapid method for characterizing and quantitating antigen–antibody reaction using an *Actinomyces viscosus*-polyclonal homologous rabbit antibody PA-FITC complex. A Coulter Counter EPICS-5 fluorescence-activated flow cell cytometer with an argon laser was used. The bacteria were observed for "green" fluorescence at 488 nm gated on 90° lightscatter signals according to the method of Warner et al. [23]. As can be seen in the fluorescence histogram in Fig. 2, *Actinomyces viscosus* cells fluoresced when complexed to homologous antiserum-PA-FITC (curve A) while normal rabbit serum (curve B) combined with PA-FITC but did not complex with *Actinomyces viscosus* and no fluorescence was observed. The curves in Fig. 3 indicated that quantitative data can be obtained from a comparison of different antibody dilutions. Quantitation of this same antibody–antigen can also be obtained using a modified tube enzyme-linked immunosorbent assay (PA-HPO). In brief, *Actinomyces viscosus* cells were fixed to plastic tubes. Bovine serum albumin was added as a blocking agent. PA-HPO-*A. viscosus* antibody was then added. After a suitable incubation, the substrate, 2,2′-azino-di-{3-ethyl benzthiazoline-6} sulphonic acid (or "ABTS"), and H_2O_2 were added [24]. A green-blue color developed which was read at 419 nm.

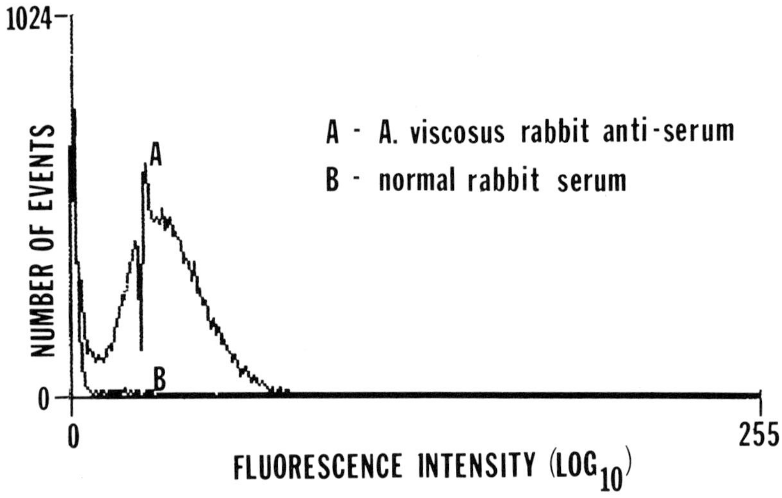

Fig. 2. Fluorescence histograms obtained with a Coulter Counter EPICS 5 fluorescence-activated flow cytometer. *A*, *Actinomyces viscosus* cells incubated with rabbit homologous antiserum (1 : 5,000) + PA-FITC; *B*, *A. viscosus* cells incubated with normal rabbit serum + PA-FITC

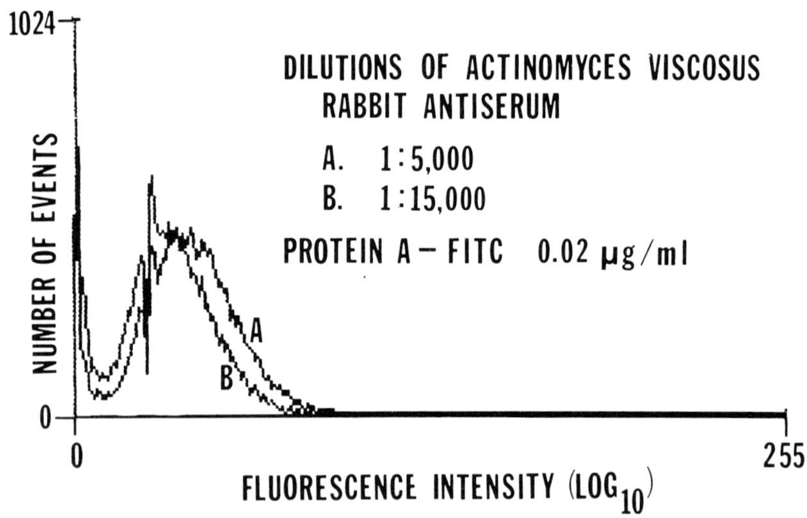

Fig. 3. Fluorescence histograms obtained with a Coulter Counter EPICS 5 fluorescence-activated flow cytometer. *A*, *Actinomyces viscosus* cells incubated with rabbit homologous antiserum (1 : 5,000) + PA-FITC; *B*, *A. viscosus* cells incubated with rabbit homologous antiserum (1 : 15,000) + Pa-FITC

Detection of Antibodies

Detection of Antibodies in Serum

Human IgG in urine and cerebrospinal fluid has been detected using COAG and observing changes in optical density [25]. Antibodies to infectious bovine rhinotracheitis virus were detected using PA-HPO in a "macro" and "micro" ELISA assay [9].

Anaphylactic (IgG_1) and complement-fixing (IgG_2) antibodies against class I major histocompatibility complex guinea pig antigens have been recognized using sheep erythrocytes coated with PA [26]. Sera from patients with *Candida albicans* infections reacted positively using PA-FITC and the homologous *Candida* antigens (Goldschmidt and Hopfer, unpublished data). Horseradish peroxidase conjugated to PA was used in an ELISA assay to detect measles virus antibodies [27].

Staphylococcus aureus Cown I strain was used to absorb specific immune complexes from antiserum to poliovirus to which ^{35}S-labeled N or H antigens had been added [28].

Detection of Monoclonal Antibodies

There are some classes of mouse immunoglobulins (e.g., IgG_{2a}, IgG_{2b}) which will couple with PA. Thus, these classes of monoclonal antibodies can be detected by procedures involving staphylococcal cells or PA-markers. Table 5 lists some of these antibodies.

Dr. Dorothy Lewis (Department of Immunology, Baylor College of Medicine, Houston, Texas) and I prepared monoclonal antibodies to *Actino-*

Table 5. Detection of monoclonal antibodies

Specimen	Immunological group detected	Method
Shigella flexneri	O-antigenic polysaccharides II: 3, 4, 7, 8; 7, 8	COAG[a]
Neisseria gonorrhoeae	Protein I, outer membrane proteins	COAG
Actinomyces viscosus	Surface antigens	PA–FITC[b]
Mouse hybridoma antibodies	IgG	PA–HPO[c]
Human immunoglobulin secreting cells	IgG	PA–SRBC[d]
Aggregated and soluble immune complexes	IgG	COAG
Mouse hybridoma-produced monoclonal antibodies from tissue-culture fluid or ascites fluid	IgG reacting with PA	PA-Sepharose[e] CL–4B

[a] COAG: coagglutination with sensitized whole cells of *Staphylococcus aureus,* Cowan I strain
[b] PA–FITC: purified protein A coupled to fluorescein isothiocyanate
[c] PA–HPO: purified protein A coupled to horseradish peroxidase
[d] PA–SRBC: purified protein A coupled to sheep erythrocytes
[e] PA-sepharose CL–4B: purified protein A coupled to Sepharose CL–4B

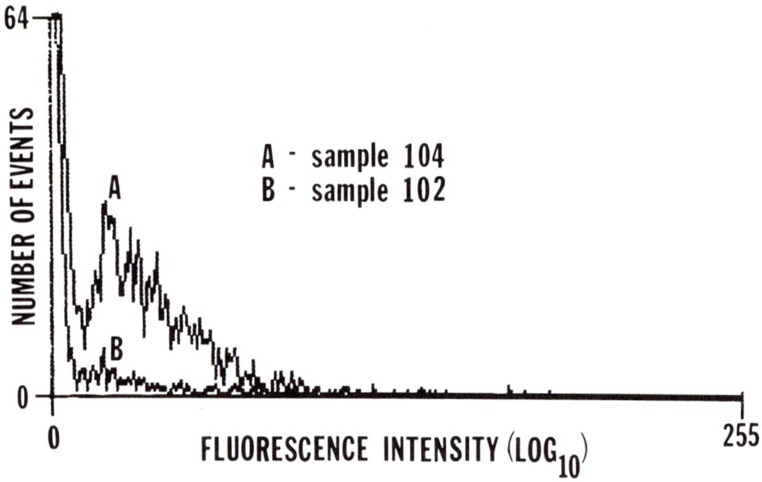

Fig. 4. Fluorescence histograms obtained with a Coulter Counter EPICS 5 fluorescence-activated flow cytometer. Hybridomas were made against *Actinomyces viscosus*. *A*, *A. viscosus* + hybridoma supernatant fluid from well 104 + PA-FITC; *B*, *A. viscosus* + hybridoma supernatant fluid from well 102 + PA-FITC

myces viscosus following the hybridoma procedure of Goding [29] in her laboratory using a P3-X63-Ag 8 myeloma cell line. Supernatants were tested by the PA-FITC flow cell cytometry methods previously described [23]. Figure 4 shows the fluorescence histograms from two of the resultant hybridoma supernatants. Thus, PA-reactive monoclonal antibodies could easily and rapidly be detected and assayed by this method. Increasingly, monoclonal antibodies are being detected and characterized by various COAG methods. For example, monoclonal antibodies against a chlamydial antigen [30] and against *Neisseria gonorrhoeae* [31] were screened using ^{125}I-PA. COAG using staphylococcal cells was also used in the latter study [31]. Ascites fluid containing monoclonal antibodies raised against *Shigella flexneri* O-antigenic polysaccharides were used to sensitize whole cells of *Staphylococcus aureus* Cowan I in COAG reactions against various strains of *S. flexneri* [32].

Affinity Chromatography

In the last several years, the production of monoclonal antibodies by mouse hybridoma formation has become a fairly common procedure. The resultant hybridomas can be grown either in tissue culture or as mouse ascites tumors. A recently developed method of purifying these immunoglobulins in a rapid single step procedure involves the use of PA-Sepharose column chromatography. Mouse IgG_{2a}, IgG_{2b}, IgG_3, and some IgG_1 will bind to PA as well as some IgM. After the PA-Sepharose column is charged with the sample, the antibodies can be eluted when a series of buffers of increasing acidity are added [33]. Figure 5, kindly provided by Dr. T. Douglas, depicts the use of a PA-Sepharose Cl-4B column to elute IgG_{2a}

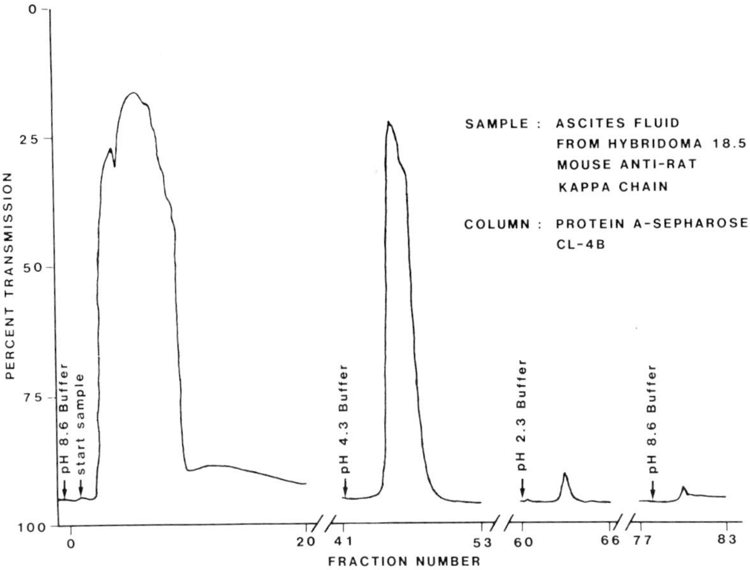

Fig. 5. Elution of IgG_{2a} from ascites fluid from a mouse hybridoma line 18.5 prepared against anti-rat κ-chains using a protein A-Sepharose CL-4B column

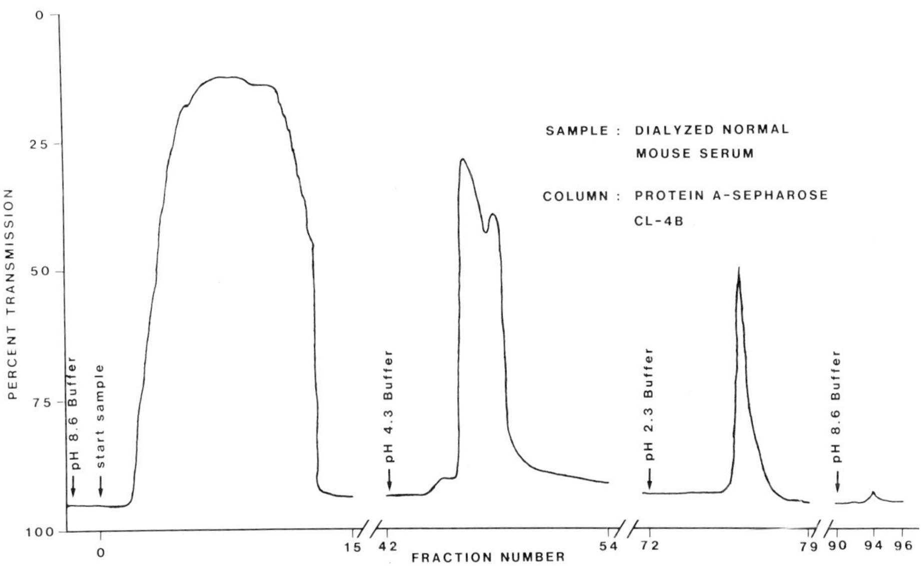

Fig. 6. Elution of IgG from dialyzed mouse serum using a protein A-Sepharose CL-4B column

using ascites fluid from mouse hybridoma line 18.5 prepared against anti-rat κ-chains. Figure 6 shows dialyzed normal mouse serum similarly treated. Several workers have used this method to remove immunoglobulins from hybridoma culture fluids [32, 34]. PA-Sepharose has also been used to purify antigen–antibody complexes consisting of monoclonal antibodies to *Pseudomonas aeruginosa* exotoxin A [35], anti-Ia murine antibody MKD6, and Fab fragments of this antibody [36]. Using both PA-HPO and an anti-PA-HPO antibody in a modified ELISA assay considerably enhanced detection of the latter immunoglobulins.

Immunoblotting and Immunoprecipitation Reactions

Immunoblotting Procedures

Recently immunoblotting procedures on nitrocellulose paper have used PA-immunoamplification techniques to detect antigens or antibodies. Hybridoma supernatants spotted on nitrocellulose have been detected using rabbit anti-mouse IgG coupled to PA-HPO [37]. Viral proteins have been transferred to nitrocellulose sheets and detected using PA-HPO and ^{125}I-PA [38]. *Mycoplasma pulmonis* antigens were also characterized in this manner using iodinated PA [39].

Immunoprecipitation Procedures

When *Legionella pneumophila* antigens were cloned in *Escherichia coli,* radioimmunoprecipitation was performed with adsorbed rabbit antisera and PA-Sepharose [40]. Hyperimmune serum to purified type 1 fimbriae of *E. coli* was mixed and immunoprecipitated with formalin-fixed *Staphylococcus aureus* [41]. The use of COAG in these types of procedures has also become routine in many laboratories. In fact, many scientists no longer bother to reference the method.

Immunoabsorption as a Treatment for Cancer

Circulating immune complexes of tumor antigen–antibody have been detected in human gastrointestinal [42], breast [43], and melanoma cancers [44]. Their presence is generally correlated with a poor prognosis. In addition, antibodies sometimes made against the tumor antibodies themselves worsen the situation. Both staphylococcal cells and PA have been immobilized on solid phase supports and successfully used to trap circulating immune (tumor-antibody) complexes from the plasma of these patients [45]. Decreases in tumor burden and tumor necrosis have resulted. Thus, there appears to be a very important role for PA in the immunotherapy of cancers.

Future Applications

The fact that many COAG reactions can be performed relatively inexpensively and without extensive instrumentation makes this procedure an important candidate for teaching microbiology on the high school and college levels, for aiding diagnoses in space, and for planning microbiological projects and field trips to remote sites. The areas of marine biology, environmental testing, and forensic medicine, as well as the food and pharmaceutical industries, will surely increase their use of this technique. Cell biologists should take advantage of the exquisite sensitivity of the antigen-PA (±markers)–monoclonal antibody complexes to probe molecular reactions and more easily identify cellular components.

Summary

We have only partially documented the development of rapid detection and characterization methods which take advantage of the ability of protein A and/or most strains of *Staphylococcus aureus* to combine with several types of immunoglobulins, including some classes of monoclonal antibodies. This trait has been used in many creative ways, including antigen or antibody detection, immunoamplification procedures, affinity chromatography for IgG purification, assay of monoclonal antibodies, and as an immunosorbent in cancer therapy. The simplicity of the method lends itself to basic and applied laboratories the world over. The fact that many COAG reactions can be performed without expensive and extensive instrumentation makes it an even more valuable technique. Thus, the future of protein A as an immunoreagent appears bright indeed!

Acknowledgements. The authors wishes to acknowledge the help of Mr. M. Campbell with the electron micrography and photography. Dr. Dorothy Lewis of Baylor College of Medicine kindly opened her laboratory to us and bore the brunt of the work involved in the flow cell cytometry and hybridoma projects. Dr. Tommy Douglas of the Graduate School of Biomedical Sciences, University of Texas Health Science Center at Houston, provided the data for the PA-Sepharose CL-4B figures.

References

1. Langone JL (1982) Protein A of *Staphylococcus aureus* and related immunoglobulin receptors produced by streptococci and pneumococci. Adv Immunol 32:157–252
2. Facklam R, Laurell G, Lind I (1980) Recent developments in laboratory identification techniques. Excerpta Medica Internat Cong Ser 519:1–209
3. Kronvall G (1973) A rapid slide agglutination method for typing pneumococci by means of a specific antibody adsorbed to protein A-containing staphylococci. J Med Microbiol 6:187–190
4. Goldschmidt MC (1982) Use of coagglutination techniques in the rapid identification of microorganisms. In: Tilton RC (ed) Rapid methods and automation in microbiology. Am Soc Microbiol, Washington, pp 202–206
5. Forsgren A, Sjöquist J (1966) Protein A from *Staphylococcus aureus*. I. pseudoimmune reaction with human gamma globulin. J Immunol 97:822–827
6. Lazar GS, Gaidulis L, Henke M, Blume K (1984) A sensitive screening method of detecting antigranulocyte antibodies employing radiolabeled staphylococcal protein A. J Immunol Methods 68:1–9
7. Trienekens PH, Faulkner J, Bird AG (1984) Addition of polyethylene glycol 6,000 improves the sensitivity of the protein A plaque assay for the detection of human immunoglobulin secreting cells. J Immunol Methods 66:341–347
8. Goldschmidt MC, O'Neill PA (1981) A rapid coagglutination method to detect oral *Actinomyces*. J Dent Res 60:487
9. Beccaria E, Ferrari A, Nachtmann C, Boniolo A, Bovis M, Zannino M, Petruzzelli E (1982) Rapid detection of antibodies to infectious bovine rhinotracheitis by "macro" and "micro" ELISA. Dev Biol Stand 52:141–146

10. Damask LJ, Montoya O, Axelrod JL (1979) Rapid slide agglutination test for Lancefield grouping of streptococci. Arch Pathol Lab Med 103:456–458
11. Tompkins DS, Nehaul BB, Smith CA, Cooke EM (1981) Evaluation of the Phadebact gonococcus test in the identification of *Neisseria gonorrhoeae* in a routine diagnostic laboratory. J Clin Pathol 34:1106–1109
12. Quadri SM, Talbot D (1983) Evaluation of a coagglutination test for the rapid serotyping of *Haemophilus influenzae*. Med Microbiol Immunol (Berl) 172:161–164
13. Edwards EA, Hilderbrand RL (1976) Method of identifying *Salmonella* and *Shigella* directly from the primary isolation plate by coagglutination of protein A-containing staphylococci sensitized with specific antibody. J Clin Microbiol 3:339–343
14. Lancz GJ, Specter SC (1982) A simple and rapid test for the identification of clinical herpes simplex virus isolates. J Med Virol 10:11–15
15. Burdash NM, Smith KA, Welborn AL (1982) Rapid detection of *Haemophilus influenzae* type b in cerebrospinal fluid by commercial coagglutination and latex agglutination kits. Eur J Clin Microbiol 1:131–133
16. Slifkin M, Freedel D, Gil GM (1982) Direct serogrouping of group B streptococci from urogenital and gastric swabs with nitrous acid extraction and the Phadebact streptococcus test. Am J Clin Pathol 78:850–853
17. Dishon T, Morgensen SC (1983) Rapid and direct detection of herpes simplex virus and varicella-zoster virus antigens in clinical specimens by staphylococcal reagent and membrane filtrations. Eur J Clin Microbiol 2:581–587
18. Nicolaieff A, Obert G, Regenmortel MH van (1980) Detection of rotavirus by serological trapping on antibody coated electron microscope grids. J Clin Microbiol 12:101–104
19. Dishon T, Slutzky GM, Elon J, Greenblatt CL (1981) Coagglutination and indirect hemagglutination in the detection of an excreted immunologically active substance from *Leishmania*. Israel J Med Sci 17:245–248
20. Randall RE (1983) Preparation and uses of immunoabsorbent monolayers in the purification of virus proteins and separation of cells on the basis of their cell surface antigens. J Immunol Methods 60:147–165
21. Daubas P, Mogensen E (1982) A radioimmunoassay of human leukocyte interferon using protein A-containing *Staphylococcus aureus*. J Immunol Methods 48:1–12
22. Molday RS, Mackenzie D (1982) Immunospecific ferromagnetic iron-dextran reagents for the labeling and magnetic separation of cells. J Immunol Methods 52:353–367
23. Warner NL, Daley MJ, Richey J, Spellman C (1979) Flow cytometry analyses of murine B cell lymphoma differentiation. Immunol Rev 48:197–243
24. Groome NP (1980) Superiority of ABTS over Trinder reagent as chromogen in highly sensitive peroxidase assays for enzyme linked immunoadsorbent assays. Clin Chem Clin Biochem 18:345–349
25. Singh AK, Makker SP (1983) A rapid tube ELISA for human IgG in body fluids using *Staphylococcus aureus*. Clin Chem Acta 129:271–277
26. Neveu T, Lefroit-Joliy M, Voisin GA (1983) A rapid and reliable in vivo method for anti-GPLA (class I and class II antigens) antibody titration and GPLA typing. J Immunol Methods 59:205–215
27. Rice P, Rice A, Casali P, Oldstone MBA (1983) A new solid-phase enzyme-linked immunosorbent assay for specific antibodies to measles virus. J Infect Dis 147:1055–1059
28. Vrijsen R, Rombaut B, Boeye A (1983) A simple protein A micro-immunoprecipitation method; assay of antibodies to the N and H antigens of poliovirus. J Immunol Methods 59:217–220
29. Goding JW (1980) Antibody production by hybridomas. J Immunol Methods 39:285–308
30. Caldwell HD, Hitchcock PJ (1984) Monoclonal antibody against a genus-specific antigen of *Chlamydia* species: location of the epitope on chlamydial lipopolysaccharide. Infect Immun 44:306–314
31. Tam MR, Buchanan TM, Sandstrom EG, Holmes KK, Knapp JS, Siadak AW, Nowinski RC (1982) Serologic classification of *Neisseria gonorrhoeae* with monoclonal antibodies. Infect Immun 36:1042–1053
32. Carlin NIA, Lindberg AA (1983) Monoclonal antibodies specific for O-antigenic polysaccharides of *Shigella flexneri:* clones binding to II, II:3,4, and 7,8, epitopes. J Clin Microbiol 18:1183–1189

33. Oi VT, Hertzenberg LA (1980) Immunoglobulin-producing hybrid cell lines. In: Michell BB, Shigi SM (eds) Selected methods in cellular immunology. Freeman, San Francisco, pp 351–372
34. Underwood PA, Kelly JF, Harman DF, MacMillan HM (1983) Use of protein A to remove immunoglobulins from serum in hybridoma culture media. J Immunol Methods 60:33–45
35. Galloway DR, Hedstrom RC, Pavlovskis OR (1984) Production and characterization of monoclonal antibodies to exotoxin A from *Pseudomonas aeruginosa*. Infect Immun 44:262–267
36. Holbeck SL, Nepom GT (1983) Enhanced detection of immunoglobulin binding by a modified ELISA. J Immunol Methods 60:47–52
37. Beyer CF, Beyer V (1984) A "dot-immunobinding assay" on nitrocellulose membranes for the determination of the immunoglobulin class of mouse monoclonal antibodies. J Immunol Methods 67:79–87
38. Ramirez P, Bonilla JA, Moreno E, Leon P (1983) Electrophoretic transfer of viral proteins to nitrocellulose sheets and detection with peroxidase-bound lectins and protein A. J Immunol Methods 62:15–22
39. Nichols EJ, Kenny GE (1984) Immunochemical characterization of a heat-stable surface antigen of *Mycoplasma pulmonis* expressing both species-specific and strain-specific determinants. Infect Immun 44:355–363
40. Engleberg NC, Drutz DJ, Eisenstein BI (1984) Cloning and expression of *Legionella pneumophila* antigens in *Escherichia coli*. Infect Immun 44:222–227
41. Dodd DC, Eisenstein BI (1982) Antigenic quantitation of type 1 fimbriae on the surface of *Escherichia coli* cells by an enzyme-linked immunosorbent inhibition assay. Infect Immun 38:764–773
42. Ray PK, Idiculla A, Mark R, Rhoads JE, Thomas H, Bassett JG, Cooper DR (1982) Extracorporeal immunosorption of plasma from a metastatic colon carcinoma patient by protein A-containing nonviable *Staphylococcus aureus*. Clinical evaluation of the patient's response. Cancer 49:1800–1809
43. Terman DS, Shearer WT, Ayus JC, Lehane D, Young JB (1981) Necrotizing tumoricidal response after plasma perfusion over immobilized protein A: initial experience in human breast adenocarcinoma. Clin Res 29:441A
44. Gupta RK, Leitch AM, Morton DL (1983) Detection of tumor associated antigen in eluates from protein A columns used for *ex vivo* immunoadsorption of plasma from melanoma patients by radioimmunoassay. Clin Exp Immunol 53:589–599
45. Barnes BC (1981) Plasma immunoabsorption: alteration of humoral immune components as a treatment for cancer. Cancer Bull 33:278–281

Instrumental Methods for Antigen Detection

R. C. Tilton [1], O. Steingrimsson [2], and R. W. Ryan [1]

Introduction

The detection of microbial antigens by immunological methods has proven to be a clinically relevant and reliable method for the rapid diagnosis of infectious disease. Traditionally, the laboratory has isolated the microorganism from clinical specimens, and, if necessary, determined its antibiotic susceptibility profile. If cultural microbiology was either unsuccessful or beyond the competence of the routine laboratory, then patients were tested for specific antibody response. While important, this process is time-consuming, requiring specimens submitted 2–3 weeks apart unless an increase in IgM antibody can be observed.

A variety of immunological techniques are now available to detect microbial antigens directly in body fluids. They include counterimmunoelectrophoresis (CIE), latex agglutination (LA), coagglutination (CA), enzyme immunoassay (EIA), fluorescent immunoassay (FIA), and radioimmunoassay (RIA). CIE, LA, and CA are manual techniques which, as yet, have not been automated. EIA, FIA, and RIA, however, are more aptly suited to semiautomation and several instruments are commercially available for detection of microbial antigen. Not all of these instruments automate the entire process of EIA, FIA, and RIA; rather, they are sophisticated detection instruments which utilize digital circuitry for process and analytical control. Table 1 lists those instruments commercially available for antigen detection.

One of the instruments, the Quantum II (Abbott) has been evaluated in our laboratory for the detection of gonococcal and chlamydial antigens directly from clinical specimens. It should be noted at the outset that the *Chlamydia* reagents provided by Abbott Laboratories were for experimental use only and may not be the ones that will eventually become commercially available. The Gonozyme reagents, on the other hand, were approved products purchased by the investigators.

The study was a collaborative one with Dr. Olafur Steingrimsson at the University of Iceland. Iceland is a technically, socially, and culturally sophisticated country. The laboratory at the University of Iceland has been culturing for *C. trachomatis,* a reportable infection, since the end of 1981. In 1982, there were 2,500 cultures with 500 positives (20%) and 4,000 cultures in 1983 with 700 positives

1 Department of Laboratory Medicine, University of Connecticut Health Center, Farmington, CT 06032, USA
2 University of Iceland, Reykjavik, Iceland

Table 1. Instruments commercially available for antigen detection

Instruments	Technology	Antigen(s) detected
Quantum II (Abbott Laboratories)	EIA (heterogenous)	Group A streptococci, *Neisseria gonorrheae, Chlamydia trachomatis,* hepatitis A and B components, rotavirus, respiratory syncytial virus, herpes virus
TDX (Abbott Laboratories)	Flurorescence polarization immunoassay (FPIA)	Gentamicin, tobramycin, netilmycin, streptomycin, kanamycin, amikacin, vancomycin
EMIT (Syva Co.)	EIA (homogenous)	Gentamicin, tobramycin, amikacin, netilmycin
IDX (International Diagnostic Technologies)	Solid-phase fluoroimmunoassay (FIAX)	CMV, rubella, herpes, toxoplasma, rubella IgM antibodies (antigens to become available)
Fluorostat, TDA series (Ames Co.)	Quenching fluoroimmunoassay	Amikacin, dibekacin, gentamicin, kanamycin, netilmycin, sisomycin, vancomycin, tobramycin
Valek	Time-resolved immunofluorescence	Influenza virus (A, B), parainfluenza virus 1, 2, 3, adenovirus, respiratory syncytial virus, rotavirus
Ortho Micro ELISA	ELISA (Dynatech reader)	Herpes, RSV, *Chlamydia* (to be available)

(18%). If the number of specimens received in January and February 1984 are compared with the number received the previous year, a 50% increase is seen. If this trend continues, over 6,000 specimens will be received in 1984. In the first year, about half of the specimens were from symptomatic patients and the rest were collected from screening of high-risk populations. Last year, the screening was stepped up and now all pregnant women attending the maternity clinic of the University Hospital are screened for *Chlamydia trachomatis.*

Of interest also is the fact that the population of Iceland was just over 233,000 in 1983. If one assumes that sexually transmitted disease is most prevalent in certain age-groups and that the majority of the specimens are taken from those age-groups, it appears likely that the University of Iceland is examining a significant proportion of the population likely to be carrying the disease. For example, the number of individuals aged 15–24 years was approximately 40,000 in 1983. Because of the effort involved in doing cultures for *C. trachomatis,* it seems unlikely that any other population in the world is going to be examined to the same extent (of course this may change with the new methodologies). Whether this extent of culturing is going to result in a lower incidence of *Chlamydia* or not seems to be of some importance in trying to determine exactly to what extent the population is being screened and what effect it has on the reservoir of the disease.

For these reasons, this evaluative study was conceived. Few patients should be lost to follow-up because of the size and relative transparency of the population.

Materials and Methods

Swabs provided by Abbott Laboratories (STD-Pen, males; STD-EZE, females) and Culturettes (Marion Laboratories) were used to collect duplicate specimens for *Chlamydia* detection from the female cervix and the male urethra. A third swab (Culturette) was collected for culture of *Neisseria gonorrheae*. This swab was always the last one taken, except where pus was obvious on the cervix or in the urethra. In this event, the pus was removed with a swab prior to *Chlamydia* culture. Specimens for *N. gonorrheae* were immediately brought to the laboratory, gram-stained, and cultured on chocolate agar plates and modified Thayer-Martin agar plates. After incubation in 5% CO_2, the presumptive positive isolates were tested for acid production from glucose, maltose, sucrose, and lactose.

The second swab (Culturette) was placed in a sucrose-phosphate buffer on ice and delivered to the laboratory for culture. If the specimen was not immediately cultured, it was frozen at -70 °C. *Chlamydia* were cultured as follows. The specimen was agitated with glass beads. To a monolayer of McCoy cells, 0.5 ml of the specimen was added. The cells were centrifuged at $3{,}000 \times g$ at 35 °C for 1 h. The supernatant was aspirated and replaced with maintenance media containing cyclohexamide. The tubes were incubated at 35 °C for 48–72 h and stained with iodine. The specimens negative at 72 h were passed once.

The first swab taken (STD-PEN, STD-EZE) was stored at 4 °C for no longer than 3 days. The collected swabs were flown to the University of Connecticut Health Center for Chlamydiazyme and Gonozyme tests. Swabs placed on the airplane on Thursday afternoon in Iceland were received in the laboratory in Connecticut on Friday morning and tested that day. Controls, in which duplicate specimens were cultured in Iceland and Connecticut, showed no significant deterioration of the specimen. The test protocol (Chlamydiazyme, Gonozyme) was that provided by the manufacturer except that both tests were performed on a single swab. Results can be seen in Tables 2–4.

Results

Table 2 shows a comparison of a rapid EIA with culture for *C. trachomatis*. While the specificity of the EIA is similar for males and females, there is a marked difference in sensitivity – 100% for women with cervicitis and 66% for men with urethritis. Table 3 compares these data with those previously presented in three independent studies. There were no false-negative EIA results in females, but 13 occurred in males. This may be related to the specimen collection procedure. The data suggests that a threshold concentration of antigen is relatively easy to collect in females from vigorous swabbing or scraping of the cervix. Similar rigorous procedures designed to collect urethral mucosal epithelium in the male are painful and are not well tolerated. Predictive values (PV) from a positive result range from 92.8% to 94.1%, and for a negative result from 80.6% to 100% for females. These indices answer the question: if the patient has a positive test, how likely is he/she to have the disease? If the patient has a negative test, how likely is it that he/she will be free of disease? While predictive values are important statistical information, it must be recognized that predictive values change as a function of

Table 2. Comparison of Chlamydiazyme and culture for *Chlamydia trachomatis*

	Female	Male	Total
Total number of patients	39	80	119
Overall agreement	38/37 (97.4%)	65/80 (81.3%)	86.6%
Sensitivity	100%	66%	76.3%
Specificity	95.8%	96.4%	96.2%
Predictive value			
Positive	94.1%	92.8%	93.3%
Negative	100%	80.6%	85.5%
Prevalence	40.0%	41.0%	40.7%

Table 3. Comparison of Chlamydiazyme results

Reference	Sensitivity (%)	Specificity (%)	Patients
Abstr. Ann. Mtng. ASM (1983) p. 253 Leman C et al. Abbott Laboratories	Equivalent to cell culture	97.0	845 females (multicenter study)
Abstr. Ann. Mtng. ASM (1983) p. 254 Jones MF et al. Mayo Clinic	83.9	92.1	154 males 142 females (18.9% incidence)
Abstr. Ann. Mtng. ASM (1983) p. 254 Baselski V et al. University of Tennessee	74.7	96.9	512 females (28.5% incidence)
Tilton RC et al. (present study) University of Connecticut School of Medicine	100 (females) 66 (males) 76.3 (overall)	95.8 (females) 96.4 (males)	39 females 80 males Prevalence (40.7%)

Table 4. Comparison of Gonozyme and culture for *Neisseria gonorrheae*

	Female	Male	Total
Total number of patients	28	51	79
Overall agreement	71.4%	92.2%	84.8%
Sensitivity	100%	100%	100%
Specificity	74.1%	91.6%	84.8%
Predictive value			
Positive	38.4%	63.6%	50%
Negative	100%	100%	100%
Prevalence	13.8%	12.7%	13.2%

the prevalence of the disease in the population. In the present study, the overall PV for a positive is 93.3%, based on a disease prevalence of 40.7%. The Icelandic population sampled was obviously one of high risk. In a population at low risk in which disease prevalence was 4%, not 40%, PV positive would be lowered to 51.6%. Neither sensitivity nor specificity of the test are affected by disease prevalence. Ferraro and Kunz (1982) suggest that the PV positive of a test may be improved by: (1) increasing the test specificity, or (2) increasing the apparent disease prevalence by preselection of test subjects using another test or by only evaluating the high-risk group. In terms of this study, if Chlamydiazyme were used to screen the normal, low-risk population, a negative test would accurately predict the absence of disease, particularly for females. A positive test, however, may not be a reliable predictor of disease presence. It should also be noted, however, that the one false-positive EIA in this study was a female who had been previously treated for chlamydial and gonococcal infection. This suggests that there was residual antigen and that the culture was falsely negative. Undoubtedly, such cases will skew the results in favor of the noncultural test.

There is an advantage in performing two tests for sexually transmitted diseases rapidly on the same swab. Table 4 shows the results of the EIA for *N. gonorrheae* (Gonozyme) performed on the same swab as for Chlamydiazyme. The test is better for males than females. While Gonozyme is a sensitive test (100%), it lacks specificity for females (74.1%), but is quite adequate for males (91.6%). The significance of this observation is that while a negative Gonozyme will rule out gonorrheae in both females and males, a positive test on a female is not reliable and should be followed up by culture. This is not unlike the results of a Gram stain of cervical secretions in the female patient. The reports of Papasian et al. (1984) and Danielsson et al. (1983) are at odds. Papasian et al. (1984) reported sensitivities and specificities of 97.3% and 95.8% for males and 79.2% and 87.2% for females. This is in contrast to Danielsson et al. (1983) who reported sensitivities and specificities of 90.9% and 100% for females and 83.3% and 94.3% for males, respectively, and Aardoom et al. (1982) who reported sensitivities and specificities for males and females of 100%, 100% and 86%–91.7% and 89.7%–96,7%. Stamm et al. (1984), in a similar study, reported Gonozyme results on 1,171 men and 723 women seen at a sexually transmitted disease clinic. Gonozyme was essentially equivalent to Gram stain in men (sensitivity, 94%; specificity, 98%). In females, sensitivity was 78% and specificity 98% compared with cervical culture. The sensitivity of Gram stain in females was only 48%. Our data is distinct in that the EIA sensitivity for both males and females agrees with cultural results but diminished specificity is observed when testing females by solid-phase EIA. With the exception of one patient, no clinical data are available to suggest that the females showing false-positive EIAs were either currently on antigonococcal therapy or were "test of cure" specimens.

This study suggests that with the provisos already discussed, both the Chlamydiazyme and Gonozyme EIA tests are effective substitutes for cultural methods and can be performed on the same swab.

References

Aardoom HA, Hoop DD, Iserief COA, et al. (1982) Detection of *Neisseria gonorrheae* antigen by a solid phase immunoassay. Br J Vener Dis 58:359–362

Danielsson D, Mai H, Forslin L (1983) Diagnosis of urogenital gonorrheae by detecting gonococcal antigen with a solid phase immunoassay (Gonozyme™). J Clin Pathol 36:674–677

Ferraro MJB, Kunz LJ (1982) Predictive value of microbiologic diagnostic tests. In: Lorian VJ (ed) Significance of medical microbiology in the care of patients. Williams and Wilkins, New York

Papasian CJ, Bartholomew WR, Amsterdam D (1984) Validity of an enzyme immunoassay for detection of *Neisseria gonorrheae* antigens. J Clin Microbiol 19:347–350

Stamm WE, Cole B, Fennel C, et al. (1984) Antigen detection for the diagnosis of gonorrheae. J Clin Microbiol 19:399–403

Detection of Group A Streptococcal Antigen by Coagglutination and Latex Agglutination

M. A. Gerber [1]

The original method for the serological grouping of streptococci was developed by Lancefield [21] and involved hot-acid extraction of group-specific antigen followed by precipitin testing with group-specific antisera in capillary tubes. Over the years, this method has been modified by the use of several different extraction procedures, including hot-formamide extraction [16], autoclaving [28], nitrous acid extraction [12], *Streptomyces albus* lytic enzyme extraction [24], and pronase B extraction [9]. In addition to the classical method of serogrouping by precipitation, fluorescent-antibody techniques have also been developed [26]. While accurate and reliable, all of these methods are relatively laborious, time consuming, and require well-trained laboratory personnel.

In 1973, Christensen et al. [7] described a simple coagglutination procedure for the rapid serogrouping of streptococci. The agglutinating reagent consisted of formaldehyde and heat-treated Cowan I staphylococci which had been coated with group-specific rabbit antisera by means of the Fc-protein A reaction with γ-globulin. Overnight Todd-Hewitt broth cultures of streptococci were treated with trypsin and then allowed to react with this coagglutination reagent on a glass slide while observing for agglutination with the naked eye. Serogrouping by coagglutination was performed on 179 stains of streptococci and showed complete agreement with serogrouping by the Lancefield precipitin test. The coagglution method required the presence of 10^8–10^9 organisms/ml and was accurate, rapid, and simple to perform, although some cross-reactivity between strains of *Streptococcus pneumoniae* and the group C reagent was noted. Carlson and McCarthy [4] developed a similar coagglutination procedure for serogrouping streptococci, but used a *Streptomyces albus* – lysozyme enzyme rather than a trypsin extraction. When tested on overnight Todd-Hewitt broth cultures, this method produced results that were comparable to the results of Lancefield precipitin testing. Edwards and Larson [10] described a modification of the coagglutination method in which the sensitized protein A-containing staphylococci were applied directly to a colony of streptococci on a blood agar plate. The plate was then rocked while observing for agglutination. However, other investigators [4, 20] attempting to use this modification found that the results were more difficult to interpret than when the test was performed on a glass slide.

Arvilommi [1] in 1976 used the first commercial coagglutination reagent – Phadebact Streptococcus Test (Pharmacia Diagnsotics, Uppsala, Sweden) – to

1 Department of Pediatrics, University of Connecticut School of Medicine, Farmington, CT 06032, USA

serogroup overnight Todd-Hewitt broth cultures of 126 clinical isolates of β-hemolytic streptococci and found a close correlation between the coagglutination results and the results of the Lancefield precipitin test. Although almost all of the group B strains could be serogrouped by coagglutination without pretreatment, about 60% of the other serogroups required pretreatment with trypsin in order to eliminate nonspecific agglutination. Hahn and Nyberg [18] performed a similar investigation using the Phadebact test and overnight Todd-Hewitt broth cultures of streptococci and found a 98.7% agreement between the results of the Phadebact and Lancefield precipitin tests. However, they did not encounter any auto or nonspecific agglutination and found no need to pretreat the specimens with trypsin. Similarly, Stoner [34] found a 100% correlation between Phadebact grouping of overnight Todd-Hewitt broth cultures and Lancefield precipitin grouping with only 5% of the strains requiring pretreatment with trypsin.

In 1978, Slifkin et al. [29] described a modification of the Phadebact test which allowed the serogrouping of β-hemolytic streptococci directly from the isolation plate using a minimum of five isolated colonies. The serogrouping results with this direct method coagglutination were comparable to the results of the coagglutination procedure performed on a 4-h or 24-h Todd-Hewitt broth culture, as well as to the results of the Lancefield precipitin test. Slifkin and Interval [30] later demonstrated that the organisms in a single, isolated, β-hemolytic streptococcal colony or in mixed growth on a blood agar plate could be accurately serogrouped with the Phadebact test if pretreated with a micronitrous acid extraction.

The Streptosec test (Organon Teknika, Belgium) is another commercial coagglutination kit; it was introduced in 1981 and consists of a plastic slide with dried sensitized protein A-containing staphylococci. This kit has a shelf life of 1.5 years at 4 °C compared with a shelf life of 3 months for most coagglutination reagents. The results of the Streptosec test compared favorably with the results of the Lancefield precipitin test in identifying groups A, B, C, and G streptococci when performed on overnight Todd-Hewitt broth cultures or suspensions of colonies taken directly from blood agar plates [13].

In 1978, Lue et al. [22] described a simple latex agglutination test for the rapid serogrouping of streptococci. This procedure employed a trypsinized suspension of one to five colonies of streptococci taken off of blood agar plates which was then mixed with latex particles sensitized with rabbit antisera to groups A, B, C, and G streptococci. Despite the presence of some cross-reactions, this test compared favorably with the fluorescent-antibody and Lancefield precipitin tests. The following year, Facklam et al. [15] used the first commercial latex agglutination reagent – Streptex (Wellcome Reagents, England) – for the rapid serogrouping of streptococci. This procedure involved an initial pronase extraction, could identify groups A, B, C, D, G, and F, and was comparable to the Lancefield precipitin test for the serogrouping of β-hemolytic streptococci from blood agar plates as well as from overnight Todd-Hewitt broth cultures. However, the Streptex kit was not as accurate in serogrouping the non-β-hemolytic streptococci, particularly *S. bovis*. Others have noted similar problems when using the Streptex kit to serogroup β-hemolytic nonenterococcal group D streptococci [35]. Castle et al. [5] performed a comparison of the Streptex kit and the Lancefield precipitin test using β-hemolytic streptococci from primary plates, purity plates, and overnight broth cultures. They found that the Streptex serogrouping was accurate for groups A, B, C, F, and G, but not for group D, that the Streptex could be performed on as few as five colonies, and that the Streptex did not require a pure culture.

The SeroSTAT kit (Scott Laboratories, Fiskeville, RI) is another commercial latex agglutination test, which uses an initial trypsinization step only if cross-reactions or equivocal reactions are noted without it. In 1982, Keville and Doern [19] compared the SeroSTAT and Streptex kits to the Lancefield precipitin test using 96 strains of β-hemolytic streptococci (groups A, B, C, F, and G) isolated from overnight cultures on blood agar plates. Approximately half of the strains showed cross-reactions when the SeroSTAT was performed without trypsinization. The investigators therefore elected to trypsinize all strains before performing the SeroSTAT test. Both the Streptex and the SeroSTAT kits produced results which were comparable to the results of the Lancefield precipitin test except in the identification of group F strains, for which the Streptex kit was not as accurate (SeroSTAT had no group F reagent at this time).

Over the past few years, a number of comparison studies have been performed of the various coagglutination and latex agglutination procedures available for the rapid serogrouping of streptococci. Slifkin and Pouchet-Melvin [31] performed a comprehensive evaluation of the Phadebact, SeroSTAT, and Streptex kits and compared them to the Lancefield precipitin test. The Phadebact was used to group β-hemolytic streptococci in pure culture directly from the isolation plate (direct method) or from 4-h or 24-h Todd-Hewitt broth cultures. The SeroSTAT test was used only for pure cultures by the direct method and the Streptex test was used for both pure cultures and mixed cultures by the direct method and for overnight Todd-Hewitt broth cultures. Although the Streptex was associated with more false-negative responses than the other two kits, the results of all three rapid procedures compared favorably with the results of the Lancefield precipitin test. While the Streptex kit offered a direct method on both isolated and mixed cultures, the Streptex direct method on mixed cultures was associated with more cross-reactions than the direct methods for the other two kits, which employed only pure cultures. The Phadebact and SeroSTAT reagents appeared to be more sensitive and to produce a faster agglutination than the Streptex reagent. The Streptex kit required a 1-h pronase extraction, while the other two kits required no extraction. The Streptex kit provided serogrouping of group D and group F β-hemolytic streptococci, while the Phadebact and SeroSTAT kits offered neither at this time. The authors concluded that the Phadebact kit was the most versatile by offering not only a direct and a 24-h grouping method, but also a 4-h method (the Streptex and SeroSTAT kits have subsequently introduced a 4- and 24-h method).

Burdash et al. [3] performed a similar evaluation of the Phadebact, SeroSTAT, and Streptex kits and compared them with the Lancefield precipitin test in the grouping of 150 β-hemolytic streptococci from pure cultures. Only the 4-h Phadebact, the Streptex direct method, and the SeroSTAT direct method were used. All three procedures showed good correlation with the Lancefield precipitin test but the SeroSTAT kit was less accurate than the other two kits, particularly in the identification of group G streptococci. Although initially performed without an extraction procedure, the SeroSTAT test had to be repeated for 15% of the isolates after trypsinization because of cross-reactions or equivocal reactions. The Streptex and the new Phadebact group D reagents both correctly identified all 55 isolates of group D streptococci. The SeroSTAT kit did not contain a group D reagent. The authors felt that the agglutination reactions were easier to read with the Phadebact kit than with the other two kits. Recently, Evins and co-workers [14] compared the performance of the Phadebact kit and the Streptex kit (now employing a *Streptomyces griseus* enzyme rather than a pronase extraction) with the Lancefield precipitin test in the serogrouping of 33 blind-coded reference strains of streptococci. In addition, the Streptex kit was used to group 14 recent clinical isolates of group D streptococci. The Phadebact kit correctly identified all of the group A, B, C, and G strains using the direct, 4-h, and 24-h methods. The Streptex kit correctly identified all of the group A, B, C, G, and F strains using the direct method on pure cultures; however, it failed to identify many *S. bovis* and *S. faecalis* strains and cross-reacted with some viridans streptococci.

In summary (Table 1), the Phadebact, Streptex, and SeroSTAT kits all produce results that are comparable to the results of Lancefield precipitin testing, but possess several advantages over the traditional procedure, including the use of overnight blood agar cultures, no requirement for special equipment or supplies, and more rapid availability of results. Although at one time the Phadebact test was the only one that offered a direct plate, 4-h broth, and 24-h broth procedure, all three tests now offer all of these options. The streptex is the only one of the three tests that routinely requires an antigen extraction procedure. This procedure increases the performance time and complexity of the Streptex when compared with the other two tests. The SeroSTAT, however, does occasionally require an initial trypsinization. The Phadebact and SeroSTAT direct plate methods must be performed on pure culture, while the Streptex direct plate

Table 1. Comparison of three commercial kits for the rapid serogrouping of streptococci

Kit	Identifiable groups	Antigen extraction	Procedures available	Direct plate specimen	Accuracy
Phadebact	A, B, C, D, G	None	Direct from plate 4-h broth 24-h broth	Pure culture	Comparable to LPT[a] Group D reagent – new, few data available
Streptex	A, B, C, D, F, G	1 h *S. griseus* lytic enzyme	Direct from plate 4-h broth 24-h broth	Pure *or* mixed culture	Comparable to LPT *except* Group D reagent – less accurate Mixed culture specimen-less accurate
SeroSTAT	A, B, C, F, G	1 h trypsinization for ~ 15% of strains	Direct from plate 4-h broth 24-h broth	Pure culture	Comparable to LPT Group F reagent – new, few data available

[a] Lancefield precipitin test

method can be performed on either pure or mixed cultures. When performed on mixed cultures, however, this procedure has produced less accurate results. The SeroSTAT is the only one of the three tests that does not contain a group D reagent. However, the Phadebact group D reagent was only recently introduced and little data regarding its accuracy are available, while investigations of the Streptex group D reagent suggest that it is less accurate than the other reagents in the Streptex kit. The Phadebact is the only one of the three tests that does not contain a group F reagent, although the SeroSTAT group F reagent was only recently introduced and little data are available regarding its accuracy.

Throat cultures on blood agar plates have been used to confirm the diagnosis of group A β-hemolytic streptococcal (GABHS) pharyngitis for over three decades [2]. However, the 24- to 48-h delay inherent in this culturing procedure has led many physicians to initiate antibiotic therapy prior to knowing the results of the throat culture. Since the great majority of patients with pharyngitis do not have GABHS pharyngitis [36], this approach results in a large number of patients receiving an unnecessary course of antibiotics. Those physicians who obtain throat cultures and wait the 24–48 h for the results before initiating antibiotic therapy may be faced with the problem of poor compliance in a patient who may already be improving. A simple, rapid, inexpensive test for diagnosing GABHS pharyngitis that could be completed while the patient waited would, therefore, be tremendously helpful to practicing physicians. Patients who had GABHS pharyngitis could then be placed on antibiotics immediately, while patients without GABHS pharyngitis could be reassured and unnecessary antibiotics avoided.

Recently, several coagglutination methods for the rapid identification of the GABHS antigen extracted directly from throat swabs have been described (Table 2). Slifkin and Gil [32] combined a nitrous acid extraction and a coagglutination method (using Phadebact Strap A reagent) to produce the micronitrous acid extraction–coagglutination test (MCT) for the rapid identification of GABHS directly from throat swabs. In their diagnostic microbiology laboratory, the MCT took approximately 20 min to perform, was positive in 48 (77%) of 62 patients with positive throat cultures, and produced no false-positive results in 311 patients with negative throat cultures. In all 14 patients with false-negative MCT results, the corresponding blood agar culture had less than 30 colonies of GABHS per plate. We examined the use of the MCT in a busy, pediatric, ambulatory clinic and compared it with blood agar cultures in 414 children with pharyngitis [17]. The MCT was positive in 78 (78%) of the 100 children with positive throat cultures and negative in 307 (98%) of the 314 patients with negative throat cultures. When throat cultures with less than 50 colonies of streptococci per plate were not considered positive, the MCT was positive in 68 (94%) of the 72 patients with positive throat cultures. Although rapid and inexpensive ($0.25 per test), the MCT is relatively complex and requires the use of special equipment (e.g., centrifuge, micropipettor) which would probably preclude its use by most practicing physicians. Otero et al. [27] recently described another coagglutination procedure for the rapid identification of GABHS directly from the throat swabs involving an extraction with an enzyme derived from *Streptomyces griseus* followed by a reaction with a laboratory-produced coagglutination reagent. This test was positive in 49 (89%) of the 55 patients with positive throat cultures and negative in

Table 2. Comparison of methods for the rapid identification of GABHS directly from throat swabs

Method	Extraction procedure	Agglutination procedure	Performance time	Sensitivity	Specificity	Use in diagnostic microbiology lab	Use in physician's office
MCT	Micronitrous acid	Coagglutination	20 min	77–78% (94–100%)[b]	98–100%	Practical	Impractical – complexity – equipment
Otero et al.	*S. griseus* Lytic enzyme	Coagglutination	60 min	89 % (100%)[c]	99%	Practical	Impractical – complexity – time – equipment
Edwards et al.[a]	Trypsin	Latex	60 min	84%	87%	Practical	Impractical – gargled specimen – complexity – time – equipment
Directigen	Enzyme	Latex	75 min	84–90% (95–97%)[c]	91–99%	Practical	Impractical – complexity – time – equipment
Culturette brand	Micronitrous acid	Latex	15 min	81–95% (95%)[c]	97–100%	Practical	Practical

[a] Uses gargled saline rather than a throat swab
[b] When only blood agar cultures with > 50 colonies of GABHS per plate are considered positive
[c] When only blood agar cultures with > 10 colonies of GABHS per plate are considered positive

480 (99%) of the 483 patients with negative throat cultures. In all six cases in which the rapid test gave a false-negative result, the corresponding blood agar culture had less than ten colonies of GABHS per plate. The complexity of this procedure, the requirement for special equipment, and the fact that it takes approximately 1 h to perform make it unlikely that practicing physicians would be able to employ it in their offices.

Several latex agglutination procedures for the rapid identification of the GABHS antigen extracted directly from throat swabs have also been described recently. In 1982, Edwards et al. [11] evaluated a rapid test which involved having patients gargle with approximately 10 ml of saline, followed by trypsinization of the gargled material, and then reaction of this material with sensitized latex particles (Streptex group A reagent). This test identified as positive 26 (84%) of the 31 patients who had positive throat cultures, while identifying as negative 19 (87%) of the 22 patients with negative throat cultures. The complexity of this procedure, the fact that it takes approximately 1 h to perform, the requirement for special equipment, and the fact that it uses gargled material rather than a throat swab make it impractical for use by physicians, especially those who are seeing children.

The Directigen Group A Strep Test Kit (Hynson, Westcott, and Dunning, Baltimore, MD) is a new commercial latex agglutination test for the rapid identification of GABHS directly from throat swabs. This kit consists of an enzymatic extraction of the organisms on the throat swab followed by a latex agglutination reaction. There have been two preliminary investigations of the Directigen test, both performed in large, diagnostic microbiology laboratories. In the first [8], the Directigen had a sensitivity of 90% and a specificity of 91% when compared with blood agar cultures of 429 throat swabs, 78 of which were positive for GABHS. In the other [23], the Directigen had a sensitivity of 90% and a specificity of 96% when compared with blood agar cultures of 222 throat swabs, 67 of which were positive for GABHS. The sensitivity of this test increased to 97% when cultures with less than ten colonies of GABHS per plate were not considered positive. The accuracy of the Directigen test was not affected by whether the throat swab was transported dry or in a transport media, or by whether or not the throat swab had been streaked onto a blood agar plate prior to the performance of the Directigen test. We evaluated the Directigen test in a busy pediatric office and found that it had a sensitivity of 84% and a specificity of 99% when compared with blood agar cultures of 263 throat swabs, 85 of which were positive for GABHS. When cultures with less than ten colonies of GABHS per plate were not considered positive, the sensitivity of the Directigen test was 95%. While easily adapted to a diagnostic microbiology laboratory, the complexity of this procedure, the requirement for special equipment (e.g., pipettes, test tubes, micropipettor, mechanical rotator), and the fact that it takes approximately 75 min to perform make it unlikely that a practicing physician would be able to employ the Directigen test routinely in their office. The time requirement alone would preclude the performance of this test while the patient waited for the results.

Another new commercial latex agglutination test for the rapid identification of GABHS directly from throat swabs is the Culturette Brand 10-min Group A Strep ID Kit (Marion Scientific, Kansas City, MO). This kit consists of a micronitrous acid extraction of the organisms on a throat swab followed by a latex agglutination reaction. Slifkin and Gil [33] evaluated this test in a large, diagnostic microbiology laboratory and found it to have a sensitivity of 95% and a specificity of 100% when compared with blood agar cultures of 557 throat swabs, 82 of which were positive for GABHS. The accuracy of this test was unaffected by storage of the throat swabs at room temperature for up to 3 weeks or by prior streaking of the throat swab on a blood agar plate. In addition, the micronitrous acid extraction destroyed the viability of potential bacterial pathogens, including *Mycobacterium tuberculosis,* so that used reagents could be discarded without concern about potential biohazards. Two similar investigations [25, 6] in other diagnostic microbiology laboratories found that the Culturette Brand test had a sensitivity of 81% and 93% respectively, and a specificity of 99% and 97% respectively. We evaluated the Culturette Brand test in a busy pediatric office and found that it had a sensitivity of 83% and a specificity of 99% when compared with blood agar cultures of 339 throat swabs, 108 of which were positive for GABHS. When cultures with less than ten colonies of GABHS per plate were not considered positive, the sensitivity of the Culturette Brand test was 95%. The Culturette Brand test was simple to perform, could be completed in approximately 12 min, was entirely self-contained with no additional equipment or supplies required, and could be performed in a physician's office while the patient waited for the results.

While both the Directigen and the Culturette Brand tests could be employed in a diagnostic microbiology laboratory, our experience would suggest that only the Culturette Brand test is practical for routine use in a physician's office. The results of preliminary investigations of the accuracy of both of these rapid tests are encouraging. However, before either is accepted as a substitute for the throat culture, further studies, particularly of the patients with negative rapid tests and minimally positive throat cultures (less than ten colonies of GABHS per plate), need to be performed.

References

1. Arvilommi H (1976) Grouping of beta-haemolytic streptococci by using coagglutination, precipitation or bacitracin sensitivity. Acta Path Microbiol Scand 84:79–84
2. Breese BB, Disney FA (1954) The accuracy of diagnosis of streptococcal infections on clinical grounds. J Pediatr 44:670–673
3. Burdash NM, West ME, Newell RT, et al. (1981) Group identification of streptococci. Evaluation of three rapid agglutination methods. Am J Clin Pathol 76:819–822
4. Carlson JR, McCarthy LR (1979) Modified coagglutination procedure for the serological grouping of streptococci. J Clin Microbiol 9:329–332
5. Castle D, Kessock-Philip S, Easmon CSF (1982) Evaluation of an improved Streptex kit for the grouping of beta-haemolytic streptococci by agglutination. J Clin Pathol 35:719–722
6. Chang MJ, Mohla C (1984) An evaluation of the Meritec-Strep™ Test Kit for the rapid detection of group A streptococcus antigen from throat swabs. (Abstract). Program and abstracts of the 84th annual meeting of the American Society for Microbiology, 4–9 March, p 255
7. Christensen P, Kahlmeter G, Jonsson S, et al. (1973) New method for the serological grouping of streptococci with specific antibodies adsorbed to protein A-containing staphylococci. Infect Immun 7:881–885

8. Contronì G, Cohen GJ, Mohla C, et al. (1983) An evaluation of a rapid latex agglutination test for the detection of group A streptococcal antigen in throat swabs. (Abstract). Program and Abstracts of the 23rd Interscience Conference on Antimicrobial Agents and Chemotherapy, 24–26 October, p 96
9. Ederer GM, Hermann MM, Bruce R, et al. (1972) Rapid extraction method with pronase B for grouping beta-hemolytic streptococci. Appl Microbiol 23:285–288
10. Edwards EA, Larson GL (1974) New method of grouping beta-hemolytic streptococci directly on sheep blood agar plates by coagglutination of specifically sensitized protein A-containing staphylococci. Appl Microbiol 28:972–976
11. Edwards EA, Phillips IA, Suiter WC (1982) Diagnosis of group A streptococcal infections directly from throat secretions. J Clin Microbiol 15:481–483
12. El Kholy A, Wannamaker LW, Krause RM (1974) Simplified extraction procedure for serological grouping of beta-hemolytic streptococci. Appl Microbiol 28:836–839
13. Engel HWB, Silfhout A van (1981) Simplified coagglutination test for serological grouping of beta-hemolytic streptococci. J Clin Microbiol 14:252–255
14. Evins GM, Pigott NE, Facklam RR, et al. (1983) The development by the Centers for Disease Control of a specification for streptococcal serogrouping kits and its application to Streptex and to the Phadebact Streptococcus Test. J Biol Standard 11:333–339
15. Facklam RR, Cooksey RC, Wortham EC (1979) Evaluation of commercial latex agglutination reagents for grouping streptococci. J Clin Microbiol 10:641–646
16. Fuller AT (1938) The formamide method for the extraction of polysaccharide from beta-hemolytic streptococci. Br J Exp Pathol 19:131–139
17. Gerber MA (1983) Micronitrous acid extraction-coagglutination test for rapid diagnosis of streptococcal pharyngitis. J Clin Microbiol 17:170–171
18. Hahn G, Nyberg I (1976) Identification of streptococcal groups A, B, C, and G by slide co-agglutination of antibody-sensitized protein A-containing staphylococci. J Clin Microbiol 4:99–101
19. Keville MW, Doern GV (1982) Comparison of the API 20S *Streptococcus* identification system with an immunorheophoresis procedure and two commercial latex agglutination tests for identifying beta-hemolytic streptococci. J Clin Microbiol 16:92–95
20. Kirkegaard MK, Field CR (1977) Rapid slide coagglutination test for identifying and typing group B streptococci. J Clin Microbiol 6:266–270
21. Lancefield RC (1933) A serological differentiation of human and other groups of hemolytic streptococci. J Exp Med 57:571–595
22. Lue YA, Howit IP, Ellner PD (1978) Rapid grouping of beta-hemolytic streptococci by latex agglutination. J Clin Microbiol 8:326–328
23. Maters AW, Charache P (1983) Latex agglutination for rapid detection of group A beta strep from direct culture swabs. (Abstract). Program and Abstracts of the 23rd Interscience Conference on Antimicrobial Agents and Chemotherapy, 24–26 October, p 96
24. Maxted WR (1948) Preparation of streptococcal extracts for Lancefield grouping. Lancet 2:255–256
25. Miceika B, Vitous A, Thompson KD (1984) A ten-minute latex agglutination test for the detection of group A streptococcal antigen from throat swabs. (Abstract). Program and abstracts of the 84th annual meeting of the American Society for Microbiology, 4–9 March, p 254
26. Moody MD, Ellis EC, Updyke EL (1958) Staining bacterial smears with fluorescent antibody. IV. Grouping streptococci with fluorescent antibody. J Bacteriol 75:553–560
27. Otero JR, Reyes S, Noriega AR (1983) Rapid diagnosis of group A streptococcal antigen extracted directly from swabs by an enzymatic procedure and used to detect pharyngitis. J Clin Microbiol 18:318–320
28. Rantz RA, Randall E (1955) Use of autoclaved extracts of hemolytic streptococci for serological grouping. Stanford Med Bull 13:290–291
29. Slifkin M, Engwall C, Pouchet GR (1978) Direct-plate serological grouping of beta-hemolytic streptococci from primary isolation plates with the Phadebact Streptococcus Test. J Clin Microbiol 7:356–360
30. Slifkin M, Interval G (1980) Serogrouping single colonies of beta-hemolytic streptococci from primary throat culture plates with nitrous acid extraction and Phadebact streptococcal reagents. J Clin Microbiol 12:541–545

31. Slifkin M, Pouchet-Melvin GR (1980) Evaluation of three commercially available test products for serogrouping beta-hemolytic streptococci. J Clin Microbiol 11:249–255
32. Slifkin M, Gil GM (1982) Serogrouping of beta-hemolytic streptococci from throat swabs with nitrous acid extraction and the Phadebact Streptococcus Test. J Clin Microbiol 15:187–189
33. Slifkin M, Gil GM (1984) Evaluation of the culturette brand ten minute group A Strep I.D. technique. J Clin Microbiol 20:12–14
34. Stoner RA (1978) Bacitracin and coagglutination for grouping of beta-hemolytic streptococci. J Clin Microbiol 7:463–466
35. Waitkins SA, Ratcliffe JG, Anderson RD, et al. (1979) Grouping of streptococci by Streptex. J Clin Pathol 32:1234–1236
36. Wannamaker LW (1972) Perplexity and precision in the diagnosis of streptococcal pharyngitis. Am J Dis Child 124:352–358

Growth-Independent Rapid Automated Enzyme Identification System

S. Bascomb [1]

Review of Current ID Systems

Bacterial identification is practised daily in the hospital diagnostic laboratory, and the need for identification of pathogens will probably continue as long as patients suffer from bacterial infections. Moreover, in the majority of cases the initial decision of the clinician regarding the treatment of a patient suspected of having a bacterial infection, is not based on identification performed in the diagnostic laboratory, because these results take too long to arrive. Laboratory identification is used for confirmation of the clinician's intuitive decision, and if the treatment prescribed was unsuccessful the identification results help in choosing a different treatment. With the exception of a few categories of specimens that can be characterised using immunological procedures applied directly to the specimen, results of tests are rarely available on the day the specimen arrives in the laboratory. Preliminary results are available after 24 h and final reporting is not possible until 2 or 3 days after receipt of the specimen. By comparison, results from the clinical chemistry laboratory are commonly available on the same day as the specimen is received and, if necessary, can be completed within a few hours. The question must be asked, "Why does clinical microbiology lag behind clinical chemistry in providing rapid service to the patient?" Or, "Why is completion time so much longer in the microbiology laboratory?"

Identification of bacterial isolates can be performed by one of the following methods: (1) conventional tests, (2) identification kits, (3) systems based on multiwell plates, and (4) instrumental identification systems.

Completion time, namely the time from receipt of the sample to reporting, has been discussed by Bascomb [11], who concluded that even methods which can perform actual testing in <24 h take at least 48 h for completion, additional time being required for preliminary testing or for preparation of sufficient inoculum.

Serological and nucleic acid hybridisation techniques which are faster may be used for the detection of certain pathogens. Unfortunately, at present the specificity and sensitivity of such methods are insufficient to justify abandonment of cultural methods. Moreover, the faster methods can only answer the question "Is organism A present in the specimen?". If the answer is negative, investigation must start again from the clinical specimen. With the increasing occurrence of infections caused by commensal organisms, identification methods which will an-

1 Department of Medical Microbiology, Wright Fleming Institute, St. Mary's Hospital Medical School, Paddington, London W2 1PG, Great Britain

swer the different question, namely "What is the identity of organisms present in the specimen?" are required.

A meaningful decrease in completion time will be obtained only with methods that are able to perform all tests with the bacteria from a single colony in <3 h. This colony must be taken from the primary isolation plate, and should contain sufficient organisms to perform all necessary testing. These requirements preclude growth-dependent tests, as such tests cannot be guaranteed to be complete within 3 h, because not all bacteria taken from an overnight culture show significant growth within this period. The above considerations led me to choose growth-independent enzyme tests for the development of a rapid bacterial identification system.

The advantages of enzyme tests over conventional ones have been discussed in detail [34, 5]. The advantages are related to the rationale of testing for individual enzyme activities, as opposed to testing for products of complete pathways (the basis of many conventional tests) or the detection of cell constituents (the basis of many chromatographic and serological techniques), and the increased sensitivity obtained by judicious choice of test conditions and use of sensitive equipment for measurement [11].

Application of Enzyme Tests to Characterisation of Bacteria

Types of Enzyme Test

Application of enzyme tests to classification or identification of bacteria can be divided into three categories:

1. Qualitative tests in which the presence of an enzyme is detected subjectively by observation of a change in the appearance (e.g. gas bubbles as an indication of catalase activity), a change in the colour [25, 41] or a change in the fluorescence [43] of bacteria–substrate mixture.
2. Study of iso-enzyme using electrophoretic techniques, in which similar enzymes performing the same catalytic activity are compared with regard to their electrophoretic mobility, as well as substrate, co-factor and inhibitor specificities [51, 26, 59].
3. Quantitative tests in which the activity of the enzyme is measured and expressed in terms of the quantity or rate of reaction-product formation [40, 48, 12, 13, 24].

Use of Synthetic Substrates

Detection of enzyme activity has been made easier by the introduction of a variety of synthetic chromogenic and fluorogenic substrates. A chromogenic molecule does not show significant absorbance in the conjugated form, but the unconjugated moiety, e.g. *p*-nitroaniline, released by enzyme activity, has a high absorption co-efficient; increase in absorbance of the reaction mixture can therefore be used as a measure of enzyme activity. In cases where the chromogenic moiety, e.g. nitrophenol or phenolphthalein, is pH sensitive it may be necessary to adjust pH before measuring absorbance. Other substrates include indoxyl, phenyl, naphthyl or naphthylamine groups in their molecules; the liberation of such substances is demonstrated either by spontaneous oxidation (indoxyl) or by addition of specific reagents (often diazo dyes) which produce colour [45]. Diazotisation in situ is also used to increase the sensitivity of measuring *p*-nitroaniline [33].

LeMinor and Ben Hamida [41] recommended the use of ortho-nitrophenol-β-D-galactopyranoside (ONPG) for the rapid detection of β-galactosidase in bacterial cultures. Kilian and Bülow [35] extended the range of nitrophenol derivatives used for enzyme assay. Phenolphthalein phosphate has been used to demonstrate phosphatase activity of staphylococci colonies [2]. A phenolphthalein derivative was used for detection of β-glucuronidase activity [15]. Bürger [15, 16] and Kersters and De Ley [34] described various methods for using β-naphthyl, indoxyl, and nitrophenyl derivatives for the detection of a variety of hydrolases and other catabolic enzymes in suspensions of resting bacteria as well as in cell-free extracts. They also described a few methods for the detection of peptidase activity using β-naphthylamine and *p*-nitroaniline derivatives. Muftić [45] reported synthesis of a large number of α- and β-naphthylamine derivatives and investigated the best conditions for diazotisation of naphthylamine for the detection of amino-peptidase activity. Buissière et al. [14] used naphthyl and naphthylamine derivatives for the study of hydrolases, aryl peptidases and proteases of bacterial suspensions. Most of these are now available in the API-ZYM kit. Lazdunski et al. [38] used *p*-nitroaniline, as well as β-naphthylamine derivatives, to study an *E. coli* amino-peptidase.

Fluorogenic substrates include fluorescent moieties such as fluorescein [39, 18], naphthyl, naphthylamine [58], 4-methylumbelliferone [43, 27], and 7-methyl-coumarin amide [60, 24]. In others fluorescence occurs on change from the oxidised to the reduced form, e.g. nicotinamideadenine dinucleotides (NAD, NADP). This change in fluorescence has been used for estimation of enzymes which utilise NAD or NADP as co-factors [42]. The same authors described a large number of enzyme assays which have been developed by coupling fluorogenic enzyme reactions to enzymic processes which have no fluorescent component. More details of fluorogenic assays can be found in Guilbault [30] and Leaback [39].

Activities of those enzymes which cause significant changes in the pH of the reaction mixture can be measured using chromatic [40, 54, 13] or fluorescent [21, 24] pH indicators.

Until now most enzymes whose activities have been used in identification have been of the catabolic type. Hydrolases like esterases, glycosidases, phosphatases, peptidases, and proteases have been most commonly studied. In a large number of the synthetic substrates available, the chromogenic moiety is attached to a simple moiety, e.g. amino acid, monosaccharide, fatty acid, phosphate or sulphate ion. These are suitable for the assay of dipeptidases, diglycosidases or esterases and phosphatases. Synthetic molecules are also available for the detection of endopeptidases [17] and a few are available for the assay of polysaccharide depolymerases [53].

A distinction may exist between conditions which are optimal for enzyme activity and those optimal for detection of the product. When these coincide it is possible to follow the release of the product during the reaction and to perform kinetic assay. When these conditions differ, enzyme activity is stopped, after a definite incubation period, by the addition of reagents which will reveal the presence of the reaction product. Measurements of activity are only possible by application of end point methods.

Differentiation of Bacteria by Qualitative Methods

Watson [57] reviewed the use of aminopeptidase testing in identification of microorganisms. The commercially available qualitative 4-h system API-ZYM (API Product Laboratory), which includes 14 hydrolases, 3 aryl peptidases, and 2 proteases, has been applied to the study of non-haemolytic streptococci [56], a variety of bacteria of clinical importance [32], gram-negative anaerobic bacteria [52], bacteroides [37], and oral and non-oral gram-negative bacteria [50]. Various other combinations of synthetic enzyme substrates were used for differentiation of species of *Neisseria* [19] and *Mycobacterium* [29,45].

These qualitative enzyme studies were useful in showing differences in the enzymatic capability of different organisms but none offered a workable identification system based solely on the detection of enzyme activities.

Differentiation of Bacteria by Quantitative Methods

Quantitative measurements of enzyme activities offer increased sensitivity but require some degree of automation if large numbers of determinations are needed. Five types of automated or partially automated equipment systems are available for enzyme assays, namely: continuous flow analysers, discrete analysers, fast centrifugal analysers, specialised analysers, and systems using multiwell plates and plate readers. The first four were developed for use in the clinical chemistry laboratory, the last one primarily for immunological assays.

Continuous Flow Methods

The use of continuous flow methods in bacteriology started in the field of microbiological assay [23]. The study of bacterial growth by such methods was reviewed by Ferrari and Marten [22]. Methods for measuring the enzyme activity of bacteria were developed for nitrophenol derivatives by Dealy and Umbreit [20], Bascomb [3,4], and Bascomb and Spencer [13]; for glutamate decarboxylase by Leclerc [40], Trinel and Leclerc [54, 55], and Bascomb and Spencer [13]; for ammonia-releasing enzymes by Bascomb and Grantham [12] and Bascomb and Spencer [13]; for dehydrogenases by Avanzini et al. [1], Roodyn and Maroudas [48]. Bascomb and Spencer [13] also measured the activity of oxidase and butanediol-producing enzymes. Bascomb [7] measured the activity of 17 enzymes fluorimetrically using a semi-automated continuous flow technique. Continuous flow methods are very versatile, but sample throughput is too slow to cope with the envisaged workload of the diagnostic laboratory [10].

Discrete Analysis

Manual Tests. Discrete analysis was used for measurement of enzyme activity in both manual and automated modes. Manual methods for measurement of nitrophenol-releasing enzymes were used by Kilian and Bülow [35]. Grange and Clark [28] used 4-methylumbelliferyl (4-MU) derivatives for the detection of hydrolases by measuring absorbance at 360 nm.

Grange [27] also used 4-MU derivatives but determined its release by the more sensitive measurement of its fluorescence. Fluorescence of β-naphthylamine was measured to determine aminopeptidase profiles of *Bacillus* sp. [58], *Erwinia* [44], enterobacteria [47], and *Pseudomonas* [31].

Use of Kem-O-Mat. Bascomb [6, 7] reported the use of the Kem-O-Mat Discrete Analyser (Coulter Electronics) for the assay of bacterial enzymes. Eight tests could be performed satisfactorily with this system, each using the equivalent of 1/20th of the cells present in an overnight colony (of aerobic or facultative anaerobic gram-negative bacteria). Incubation periods varied between 1 and 2 h. However, the number of enzyme tests that could be performed successfully with this system was limited. The Kem-O-Mat could therefore not be used on its own for identification, but when used in combination with automated and semi-automated continuous flow methods [7, 8] for identification, of 276 isolates falling into 16 genera, including fermentative and oxidative gram-negative organisms as well as staphylococci and streptococci, it provided 80% overall agreement with conventional identification. Taxa like *Escherichia, Providencia,* and *Pseudomonas* showed >90% agreement.

Use of Aminco Fluorocolorimeter. Godsey et al. [24] used an Aminco Fluorocolorimeter to monitor the fluorescence of 19 bacteria–substrate mixtures. A total of 539 clinical isolates belonging to ten species of Enterobacteriaceae were identified by this method, showing an 87% overall agreement with conventional identification. This method could in theory provide identification in 30 min, but it requires a heavy inoculum, as bacteria from one colony enable performance of four tests only. Additional time is required for inoculum preparation resulting in >48 h for completion.

Use of Multiwell Plate. Multiwell plates were designed for serological techniques, and a number of mechanical/automated devices are now available for rapid distribution of reagents in such plates and for reading absorbance or fluorescence of the contents of each well.

Bascomb described identifications of bacteria based on measurement of absorbance [10] and fluorescence [11] of bacterial–substrate mixtures in multiwell plates after incubation for 2 h.

In the fluorimetric system described [11], pre-prepared plates containing dried substrates were kept at 5°–8 °C. The distribution of bacterial suspension made from one colony and of reagents was still manual. The inoculated plates were incubated for 2 h, after which time a number of reagents were added and after an additional incubation of 15 min the fluorescence of each well was measured. Data acquisition was automated using a Commodore PET connected to the measuring instrument using an interactive program. The list of strains tested is entered after the reading of the final tray. Identification of each strain could be completed 2.5 h after receipt of the primary isolation plate.

Automated identification of bacteria requires a mathematical model suitable for the type of data available. The Bayes probability model developed by Lapage et al. [36] is used in most current identification systems which are based on qualitative conventional tests. However, such a model is not particularly suited to quantitative data. Mathematical models available for quantitative data have been described [49] and discussed [9, 11].

We used the Discriminant Function Analysis [46] model for identification based on fluorimetric measurement of enzyme activities. All the necessary coefficients were calculated using the SPSS Package [46], on a main frame computer.

An identification program for the Commodore PET Microcomputer (Model 4032) using the transformed values for each test and the above Discriminant Function coefficients [11] required 6 min for identification of each isolate.

Performance of Enzyme Identification System

A comparison between identifications based on rapid fluorescent enzyme tests and conventional methods, when applied to 1,543 cultures belonging to 24 genera, is given in Table 1, showing an overall agreement of 88%. Detailed examples of performance for selected taxa are given in Table 2.

Table 1. Percentage agreement[a] between rapid enzyme and conventional identifications of 1543 cultures

Genus	No. of cultures	Agreement (%)	Genus	No. of cultures	Agreement (%)
Proteus	230	93	*Aeromonas*	44	86
Pseudomonas	212	87	*Serratia*	39	90
Staphylococcus	146	96	*Shigella*	30	70
Escherichia	112	95	*Flavobacterium*	24	92
Streptococcus	107	98	*Pasteurella*	21	86
Enterobacter	103	81	*Achromobacter*	18	78
Klebsiella	90	93	*Hafnia*	17	100
Acinetobacter	74	84	*Yersinia*	15	93
Salmonella	64	67	*Chromobacterium*	12	100
Citrobacter	56	71	*Moraxella*	10	50
Providencia	53	87	*Plesiomonas*	8	87
Alcaligenes	51	80	*Edwardsiella*	7	86

[a] Overall agreement with conventional ID = 88.0%

Table 2. Agreement between rapid enzyme and conventional identifications of 193 *Klebsiella* and *Enterobacter* isolates

Enzyme ID \ Conventional ID	*K. oxyt.*	*K. pneum.*	*K. rhino.*	*E. aerog.*	*E. agglom.*	*E. aming.*	*E. cloac.*	*E. gergo.*	*E. inter.*	*E. sakaz.*
1. *K. oxytocum*	14				2					
2. *K. pneumoniae*	1	57		3	1		1			1
3. *K. rhinoscleromatis*		4	7							
4. *E. aerogenes*				18						
5. *E. agglomerans*		1			2					
6. *E. amingenus*					1	5				1
7. *E. cloacae*							20			1
8. *E. gergoviae*		1						11		
9. *E. intermedium*									5	
10. *E. sakazakii*							3			13
Others	1	4	–	–	3	–	8	2	–	2
No. tested	16	67	7	21	9	5	32	13	5	18
% correct species	88	85	100	86	22	100	62	85	100	72
% correct genus	94	91	100	86	33	100	72	85	100	83

Conclusions

Enzyme tests provide a suitable system for rapid identification of clinical isolates because they can be performed with bacteria from the primary isolation plate and take only 2.5 h. Agreement obtained with identification based on conventional tests is good enough in most taxa. However, some work is still required to improve the performance of the Enzyme Identification System for the few remaining species.

Acknowledgements. I thank Miss G. Parker and Mr. R. Dale for technical assistance, Mrs. K. Foster for computing assistance and Mrs. A. Chibah for preparation of the manuscript.

References

1. Avanzini F, Magnanelli D, Boffi V (1968) The automated determination of bacterial respiratory activity using tetrazolium red. In: Automation in analytical chemistry, Technicon Symposia 1967, vol II. Mediad, New York, pp 285–290
2. Barber M, Kuper SWA (1951) Identification of *Staphylococcus pyogenes* by the phosphatase reaction. J Pathol 63:65–68
3. Bascomb S (1976) Enzymatic activities of bacteria and their survival during treatments affecting cell integrity. Proc Soc Gen Microbiol 3:87
4. Bascomb S (1977) Rapid identification of bacteria from clinical specimens by continuous flow analysis. In: Johnston HH, Newson SWB (eds) Proceedings of the 2nd International Symposium on rapid methods and automation in microbiology. Learned Information (Europe), Oxford, p 53
5. Bascomb S (1980a) Identification of bacteria by measurements of enzyme activities and its relevance to the clinical diagnostic laboratory. In: Goodfellow M, Board RG (eds) Microbiological classification and identification. Academic, New York, pp 359–373

6. Bascomb S (1980 b) Rapid identification of bacteria of clinical importance by automated enzyme assay. In: International colloquium of the Société Francaise de microbiologie on méthodes rapides et automatiques en microbiologie alimentaire. Institute Pasteur, Lille, p 23
7. Bascomb S (1980 c) The identification of bacteria. UK Patent application No. 2,048,302A
8. Bascomb S (1981) Comparison of continuous flow and discrete analysis measurements of enyzme activities for rapid identification of bacteria. In: Abstracts of 3rd International Symposium on Rapid Methods and Automation in Microbiology. Washington DC, No. 102
9. Bascomb S (1983 a) Comparison of mathematical models for identification of bacteria using quantitative characters. In: Les Colloques de l'INSERM, Les bacilles a gram negatif d'entéret medical et en santé publique: taxonomie-identification-applications. INSERM, vol 114, pp 423–430
10. Bascomb S (1983 b) Application of automated enzyme methods to the identification of clinical isolates. In: Les Colloques de l'INSERM, Les bacilles a gram negatif d'intéret medical et en santé publique: taxonomie-identification-applications. INSERM, vol 114, pp 471–476
11. Bascomb S (1984) Rapid identification of bacteria by an automated multi-enzyme assay. In: Sanna A, Morace G (eds) New horizons in microbiology. Elsevier, Amsterdam, pp 241–255
12. Bascomb S, Grantham CA (1975) Application of automated assay of asparaginase and other ammonia-releasing enzymes to the identification of bacteria. In: Board RG, Lovelock DW (eds) Some methods for microbiological assay. Society for Applied Bacteriology Technical Series No. 8, Academic, London, pp 20–54
13. Bascomb S, Spencer RC (1980) Automated methods for identification of bacteria from clinical specimens. J Clin Pathol 33:36–46
14. Buissière J, Fourcard A, Colobert L (1967) Usage de substrates synthétiques pur l'étude de l'équipment enzymatique de microorganismes. CR Séances Acad Sci 264:415–417
15. Bürger H (1967 a) Biochemische Leistungen nichtproliferierender Mikroorganismen. II. Nachweis von Glycosid-Hydrolasen, Phosphatasen, Esterasen und Lipasen. Zentralbl Bakteriol [Naturwiss] 202:97–109
16. Bürger H (1967 b) Biochemische Leistungen nichtproliferierender Mikroorganismen. III. Nachweis von Arylsulfatasen und Peptidasen. Zentralbl Bakteriol [Naturwiss] 202:395–401
17. Claeson G, Aurell L, Karlsson G, Friberger P (1977) Substrate structure and activity relationship. In: Witt I (ed) New methods for the analysis of coagulation using chromogenic substrates. de Gruyter, Berlin, pp 37–54
18. Cundell AM, Pisani AM, Findl E (1979) A rapid method for detection of coliform bacteria. Dev Ind Microbiol 20:571–577
19. D'Amato RF, Eriquez LA, Tomfohrde KA, Singerman E (1978) Rapid identification of *Neisseria gonorrhoeae* and *Neisseria meningitidis* by using enzymatic profiles. J Clin Microbiol 7:77–81
20. Dealy JD, Umbreit WW (1965) The application of automated procedures for studying enzyme synthesis in *Escherichia coli*. Ann NY Acad Sci 130:745–750
21. Dyer DL (1971) Microbiological detection and identification system. US Patent No. 3,551,295
22. Ferrari A, Marten J (1972) Automated microbiological assay. In: Norris JR, Ribbons DW (eds) Methods in microbiology vol 6B. Academic, London, pp 331–342
23. Gerke JR, Haney TA, Pagano JR, Ferrari A (1960) Automation of the microbiological assay of antibiotics with an autoanalyser instrument system. Ann NY Acad Sci 87:782–791
24. Godsey JH, Matteo MR, Shen D, Tolman G, Gohlke JR (1981) Rapid identification of Enterobacteriaceae with microbial enzyme activity profiles. J Clin Microbiol 13:483–490
25. Gordon J, McLeod JW (1928) The practical application of the direct oxidase reaction in bacteriology. J Pathol 31:185–190
26. Goullet PH (1978) Characterisation of *Serratia marcescens, S. liquifaciens, S. plymuthica* and *S. marinorubra* by the electrophoretic patterns of their esterases. J Gen Microbiol 108:275–281
27. Grange JM (1977) A fluorigenic substrate for the rapid differentiation of *Mycobacterium fortuitum* from *Mycobacterium chelonei* on the basis of heat stable esterase activity. Tubercle 58:147–150
28. Grange JM, Clark K (1977) Use of umbelliferone derivatives in the study of enzyme activities of mycobacteria. J Clin Pathol 30:151–153
29. Grange JM, McIntyre G (1979) Fluorigenic glycosidase substrates: their use in the identification of some slow growing mycobacteria. J Appl Bacteriol 47:285–288
30. Guilbault GG (1973) Practical fluorescence, theory, methods and techniques. Dekker, New York

31. Hsu EJ, Godsey JH, Chang EK, Landuyrt SL (1981) Differentiation of pseudomonads by amplification of metabolic profiles. Int J Syst Bacteriol 31:43–55
32. Humble MW, King A, Phillips I (1977) API-ZYM: a simple rapid system for the detection of bacterial enzymes. J Clin Pathol 30:275–277
33. Ison C, Glynn AA, Bascomb S (1982) Acquisition of new genes by oral *Neisseria*. J Clin Pathol 35:1153–1157
34. Kersters K, De Ley J (1971) Enzymic tests with resting cells and cell-free extracts. In: Norris JR, Ribbons DW (eds) Methods in microbiology vol 6A. Academic, London, pp 33–52
35. Kilian M, Bülow P (1976) Rapid diagnosis of Enterobacteriaceae. I. Detection of bacterial glycosidases. Acta Pathol Microbiol Scand [B] 84:245–251
36. Lapage SP, Bascomb S, Willcox WR, Curtis MA (1970) Computer identification of bacteria. In: Baillie A, Gilbert RJ (eds) Automation, mechanisation and data handling in microbiology. Society for Applied Bacteriology Technical Series No. 4. Academic, London, pp 1–22
37. Laughon BE, Syed SA, Loesche WJ (1982) API-ZYM system for identification of *Bacteroides* spp., *Capnocytophaga* spp. and Spirochetes of oral origin. J Clin Microbiol 15:97–102
38. Lazdunski V, Busuttil J, Lazdunski A (1975) Purification and properties of periplasmic aminoendopeptidases from *Escherichia coli*. Eur J Biochem 60:363–369
39. Leaback DH (1975) An introduction to the fluorimetric estimation of enzyme activities, 2nd edn. Koch Light Laboratories Ltd
40. Leclerc H (1967) Mise en évidence de la décarboxylase de l'acide glutamique chez les bactéries à l'aide d'une technique automatique. Ann Microbiol (Paris) 112:713–731
41. LeMinor L, Ben Hamida F (1962) Advantage de la recherche de la β-galactosidase sur celle de la fermentation du lactose en milieu complex dans le diagnostic bactériologique, en particulier des Enterobacteriaceae. Ann Microbiol (Paris) 102:267–277
42. Lowry OH, Passioneau JV (1972) A flexible system of enzymatic analysis. Academic, New York
43. Maddocks JL, Greenan MJ (1975) A rapid method for identifying bacterial enzymes. J Clin Pathol 28:686–687
44. McIntyre JL, Huber D, Kuc J, Williams EB (1975) Aminopeptidase profiles of virulent and avirulent *Erwinia amylovora* and *Erwinia herbicola*. Phytopathology 65:1206–1212
45. Muftić M (1967) Application of chromogenic substrates to the determination of peptidases in mycobacteria. Folia Microbiol (Praha) 12:500–507
46. Nie NH, Hull CH, Jenkins JG, Steinbrenner K, Brent DH (1975) SPSS: Statistical package for social sciences. McGraw-Hill, New York
47. Peterson EH, Hsu EJ (1978) Rapid detection of selected Gram-negative bacteria by aminopeptidase profiles. J Food Sci 43:1853–1856
48. Roodyn DB, Maroudas NG (1968) Multiple enzyme analysis with computer processing of data. Anal Biochem 24:496–505
49. Sjostrom M, Kowalski BR (1979) A comparison of five pattern recognition methods based on the classification results from six real data bases. Anal Chimica Acta 112:11–30
50. Slots J (1981) Enzymatic characterisation of some oral and nonoral Gram-negative bacteria with the API-ZYM system. J Clin Microbiol 14:288–294
51. Stewart DB, Stewart DJ (1971) Esterases and phosphatases of *Escherichia coli* serotypes isolated from the pig. J Gen Microbiol 65:175–184
52. Tharagonnet D, Sisson PR, Roxby CM, Ingram HR, Selkon JB (1977) The API-ZYM system in the identification of Gram-negative anaerobes. J Clin Pathol 30:505–509
53. Trepeta RW, Edberg SC (1984) Measurement of microbial alpha-amylases with *p*-nitrophenyl glycosides as the substrate complex. J Clin Microbiol 19:60–62
54. Trinel PA, Leclerc H (1972) Automation de l'analyse bactériologique de l'eau. Étude d'un nouveau test spécifique de contamination fécale et de conditions optimales de mise en évidence. Water Res 6:1445–1458
55. Trinel PA, Leclerc H (1977) Automatisation de l'analyse bactériologique de l'eau: description d'une nouvelle méthod de colimetrie. Ann Microbiol (Paris) 128A:419–432
56. Waitkins SA, Ball LC, Fraser CAM (1980) Use of the API-ZYM system in rapid identification of α and non-haemolytic streptococci. J Clin Pathol 33:53–57
57. Watson RR (1976) Substrate specificities of amino-peptidases: a specific method for microbial differentiation. In: Norris JR, Ribbons DW (eds) Methods in microbiology vol 9. Academic, London, pp 1–14

58. Westley JR, Anderson PJ, Close VA, Halpern B, Lederberg EM (1967) Aminopeptidases profiles of various bacteria. Appl Microbiol 15:822–825
59. Williams RAD, Shah HN (1980) Enzyme patterns in bacterial classification and identification. In: Goodfellow M, Board RG (eds) Microbiological classification and identification. Academic, London, pp 299–318
60. Zimmerman M, Yurewicz E, Pate G (1976) A new fluorogenic substrate for chymotrypsin. Anal Biochem 78:258–262

An Automatic Micromethod for the Identification of Gram-Negative Bacilli by Carbon Substrate Assimilation Tests

J. Freney [1], P. Laban [2], M. Desmonceaux [2], H. Alexandre [1], B. Poggi [1], J. P. Gayral [2], and J. Fleurette [1]

Introduction

The biochemical properties of the non-fermentative Gram-negative bacilli differ markedly from those of the Enterobacteriaceae. Consequently, the conventional tests used for identifying Enterobacteriaceae were found unsuitable for studying the non-fermentative bacilli [35, 39, 40]. In 1966, therefore, Stanier and his colleagues [41] published a classification of the genus *Pseudomonas,* largely based upon the principle of carbon substrate assimilation. This landmark study has since been followed by many other similar studies of the pseudomonads [2, 34, 37, 38] and other non-fermentative Gram-negative bacilli [3–5, 18, 21, 26, 35, 39]. The use of carbon substrate assimilation tests was then extended to the study of Enterobacteriaceae and Vibrionaceae by Veron and Le Minor in 1975 [42–44] and by many other workers since [13–15, 20, 27, 29, 33, 36]. Veron and Le Minor showed that a classification of these organisms based on carbon substrate assimilation differed little from that obtained using classical cultural and biochemical characteristics [23]. Furthermore, assimilation tests tended to be more reproducible than conventional ones [44]. It was also noted that assimilation tests were suitable for incorporation into semi-automatic systems capable of detecting the degree of bacterial growth.

Whilst assimilation tests are highly suitable for the taxonomic study of Gram-negative bacilli, they have not yet found a place in the routine medical microbiology laboratory for the following reasons: they consume large amounts of medium; a basic medium suitable for both auxotrophs and prototrophs has been difficult to define; long incubation periods are sometimes required; and results may be difficult to interpret [22, 39].

In 1980, API introduced a micromethod for testing carbon substrate assimilation. Comprising 147 substrates and a synthetic medium, it gave results which generally agreed with those using conventional tests [16] and was successfully used in taxonomic studies [6, 7, 9, 21, 24, 25]. With the aim of developing a system for identifying Gram-negative bacilli of clinical origin, 60 substrates were then incorporated into a prototype, automatically read gallery [11, 12]. The 32 most discriminant substrates were then selected for inclusion in the ATB 32 GN gallery. The present study was designed to evaluate this gallery.

1 Laboratoire de Bactériologie, Faculté de Médecine Alexis Carrel, Rue Guillaume Paradin, F-69372 Lyon Cédex 8

2 Laboratoire de Recherche API, La Balme Les Grottes, F-38390 Montalieu-Vercieu

Materials and Methods

Strains

1,505 strains representing 85 species of Gram-negative bacilli were tested (Tables 1, 2). The strains had been obtained either from clinical samples (Edouard Herriot and Louis Pradel Hospitals, Lyon) or from the API research laboratory collection. Eighty-six reference strains were also tested (Table 3). Most of the clinical strains had first been identified using the API 20 E and API 20 NE galleries [32]. The other strains had been identified as previously described [19, 28, 30].

Assimilation Tests

The galleries were made of polyvinyl chloride and consisted of 32 cupules each holding 200 µl. The 32 dehydrated substrates (14 carbohydrates, 14 organic acids, and 4 amino acids) and their respective concentrations in the growth medium are shown in Table 4.

The growth medium was a chemically defined synthetic medium of known viscosity. It contained salts, vitamins, growth factors and amino acids and did not permit bacterial growth without the addition of a carbon substrate. Bacterial growth was detected automatically by the ATB 1 500 reader (API system). Results were read as positive and negative as follows. For each gallery, 32 values, varying between 1 and 124 and corresponding to the optical density of each cupule, were obtained. The scale was arbitrarily chosen but directly related to the optical density detected. The micro-computer linked to the reader then calculated a breakpoint value separating the 32 results in each gallery into two groups of minimal variance. Values above the breakpoint were then taken as positive and those below it as negative. By calculating a breakpoint for each gallery it was possible to test both strains which produced a light growth, e.g. *Shigella, Moraxella,* and *Flavobacterium,* and those which grew profusely, e.g. *Klebsiella* and *Enterobacter*.

Methods

Strains were first cultured on bromocresol purple lactose agar, or sheep blood agar for fastidious strains. Sufficient colonies of each strain were then suspended in 2 ml of physiological saline to produce an opacification of 0.5 on the MacFarland scale. Four drops of this suspension were then mixed with 6 ml of the basic culture medium. Finally, 135 µl of the inoculated medium were added to each cupule using either a semi-automatic pipette or a repeating dispenser (API system). The gallery was incubated at 32 °C and read automatically after 1 day or after 2 days for slowly growing bacteria.

Table 1. Comparison of identification results obtained using ATB 32 GN versus conventional methods with *Enterobacteriaceae* and *Vibrionaceae*

Species	No. of strains	Correctly identified			Incor-rectly identified	Not identified
		Total	24 h	48 h		
Aeromonas hydrophila	13	92[a]	84	8	8	
Citrobacter freundii	33	100	97	3		
Enterobacter aerogenes	22	100	82	18		
Enterobacter agglomerans	17	71	71			29
Enterobacter cloacae	41	90	90		5	5
Enterobacter gergoviae	14	100	71	29		
Enterobacter sakazakii	11	100	100			
Escherichia adecarboxylata	7	100	100			
Escherichia coli	27	92	88	4	4	4
Escherichia coli AD	13	100	100			
Escherichia vulneris	3	100	100			
Hafnia alvei	26	96	72	24		4
Klebsiella oxytoca	76	95	87	8	1	4
Klebsiella ozaenae	14	93	71	22		7
Klebsiella pneumoniae	83	88	81	7	6	6
Klebsiella rhinoscleromatis	10	100	20	80		
Klebsiella trevisanii	15	100	93	7		
Kluyvera ascorbata	5	100	100			
Kluyvera cryocrescens	7	86	86			14
Levinea amalonatica	11	100	91	9		
Levinea malonatica	19	100	95	5		
Plesiomonas shigelloides	10	100	100			
Proteus mirabilis	29	93	86	7	7	
Proteus morganii	26	96	96			4
Proteus rettgeri	18	100	89	11		
Proteus vulgaris	25	100	84	16		
Providencia alcalifaciens	18	100	100			
Providencia stuartii	25	100	56	44		
Rhanella aquatilis	6	100	100			
Salmonella arizonae	16	94	63	31	6	
Salmonella paratyphi A	15	100	93	7		
Salmonella spp	41	98	62	36	2	
Salmonella typhi	14	100	86	14		
Serratia liquefaciens	17	100	100			
Serratia marcescens	32	97	91	6		3
Serratia odorifera	10	90	20	70	10	
Serratia plymuthica	11	100	36	64		
Serratia rubideae	14	100	100			
Shigella boydii	12	75	(33+42[b])			25
Shigella dysenteriae	13	100	92	8		
Shigella flexneri	29	97[b]	76	21		3
Shigella sonnei	23	100	100			
Vibrio alginolyticus	10	100	70	30		
Vibrio cholerae	9	89	78	11		11
Vibrio parahaemolyticus	10	90	60	30	10	
Vibrio vulnificus	12	100	92	8		
Yersinia enterocolitica	21	91	86	5		9
Yersinia pseudotuberculosis	19	100	100			
Yersinia ruckeri	10	100	70	30		
Total (%)	962	95.4	82.9	12.5	1.7	2.9

[a] Percentage of positive strains [b] Identification to genus level

Table 2. Comparison of identification results obtained using ATB 32 GN versus conventional methods with Gram-negative bacilli other than *Enterobacteriaceae* and *Vibrionaceae*

Species	No. of strains	Correctly identified			Incorrectly identified	Not identified
		Total	24 h	48 h		
Achromobacter xylosoxidans	35	97[a]	71	26		3
Acinetobacter alcaligenes	8	100	87	13		
Acinetobacter anitratus	26	92	92			8
Acinetobacter haemolyticus	8	100	87	13		
Acinetobacter lwoffi	16	88	88		6	6
Alcaligenes denitrificans	19	100	42	58		
Alcaligenes faecalis/A. odorans	17	100	92	8		
Bordetella bronchiseptica	18	100	61	39		
Chromobacterium typhiflavum	8	100	100			
Chromobacterium violaceum	9	100	67	33		
CDC VE$_2$	11	100	100			
CDC Vd	11	91	9	82		9
CDC II B	20	95	90	5		5
EF$_4$ group	6	100	83	17		
Flavobacterium meningosepticum	11	100	(64+9[d])	(18+9[d])		
Flavobacterium multivorum	11	91	82	9		9
Flavobacterium odoratum	5	100	(80+20[d])			
Moraxella liquefaciens	5	80	80			20
Moraxella osloensis	10	100	100			
Pasteurella aerogenes	8	100	100			
Pseudomonas acidovorans	28	93	(75+18[b])			7
Pseudomonas aeruginosa	26	96	96			4
Pseudomonas alcaligenes	16	94	(0+94[c])			6
Pseudomonas cepacia	15	100	87	13		
Pseudomonas diminuta	17	94	76	18		6
Pseudomonas fluorescens	24	96	42	54		4
Pseudomonas maltophilia	22	91	14	77		9
Pseudomonas mendocina	7	100	100			
Pseudomonas paucimobilis	8	100	75	25		
Pseudomonas pickettii	17	100	70	30		
Pseudomonas pseudoalcaligenes	12	100	(75+25[c])			
Pseudomonas putida	22	100	82	18		
Pseudomonas putrefaciens	15	100	47	53		
Pseudomonas stutzeri	22	100	95	5		
Pseudomonas testosteroni	24	100	(79+21[b])			
Pseudomonas vesicularis	6	83	50	33		17
Total (%)	543	96.8	77.6	19.2	0.1	3.1

[a] Percentage of positive strains
[b] Identification to *Pseudomonas* spp
[c] Identification to alcaligenes group
[d] Identification to genus level

Table 3. Reference strains tested

a) Enterobacteriaceae	
Citrobacter freundii	ATCC 10787, ATCC 11102, ATCC 14235
Enterobacter aerogenes	ATCC 13048, ATCC 15038
Enterobacter cloacae	ATCC 13047, ATCC 222, ATCC 23355
Enterobacter sakazakii	NCTC 11467
Escherichia coli	ATCC 25922, ATCC 4157, ATCC 25190, ATCC 27165
Hafnia alvei	ATCC 25927
Levinea malonatica	ATCC 27157
Klebsiella ozaenae	ATCC 11297
Klebsiella pneumoniae	ATCC 27736, ATCC 8044, ATCC 8045, ATCC 9997, ATCC 13883
Klebsiella rhinoscleromatis	ATCC 6908, NCTC 5046
Proteus mirabilis	ATCC 4675, ATCC 21100
Proteus morganii	ATCC 25830, ATCC 25829
Proteus rettgeri	ATCC 14505, ATCC 9250
Proteus vulgaris	ATCC 6380, ATCC 13315, ATCC 6059
Providencia alcalifaciens	ATCC 9886, ATCC 12013
Providencia stuartii	ATCC 25825, ATCC 25826, ATCC 25827
Salmonella arizonae	ATCC 12323, ATCC 12325, ATCC 13314
Salmonella spp	ATCC 11997, ATCC 8388, ATCC 12002
Salmonella typhimurium	ATCC 14028, ATCC 13311
Serratia liquefaciens	ATCC 14460
Serratia marcescens	ATCC 264, ATCC 8100
Shigella boydii	ATCC 25930, ATCC 9207
Shigella dysenteriae	ATCC 11456, ATCC 13313
Shigella flexneri	ATCC 25929, ATCC 12661
Shigella sonnei	ATCC 25931, ATCC 11060
Yersinia enterocolitica	ATCC 23715
Yersinia pseudotuberculosis	ATCC 23207
b) Non-fermenting Gram-negative bacilli and Vibrionaceae	
Alcaligenes denitrificans	ATCC 15173
Alcaligenes faecalis/odorans	ATCC 8750, ATCC 19209, ATCC 15554
Bordetella bronchiseptica	NCTC 10539, NCTC 8751, NCTC 8344, NCTC 452
Chromobacterium violaceum	NCTC 9757
CDC II B group	NCTC 10795, NCTC 10796
Flavobacterium multivorum	NCTC 11033, NCTC 11034, NCTC 11343
Flavobacterium odoratum	NCTC 11180
Pseudomonas aeruginosa	ATCC 10145, ATCC 27853
Pseudomonas mendocina	CNCM 7519, CNCM 7250, CNCM 7521, CNCM 7522
Moraxella liquefaciens	CNCM 5413, CNCM 7041, CNCM 7042
Moraxella osloensis	CNCM 6815, CNCM 6835
Vibrio parahaemolyticus	CDC 1146–78, CDC 1106–78

ATCC, American Type Culture Collection (USA); *CDC,* Center for Disease Control (USA); *CNCM,* Collection Nationale de Culture de Microorganismes (France); *NCTC,* National Collection of Type Culture (England)

Table 4. The 32 substrates tested and their concentration (g carbon/litre) in the basic medium

Carbohydrates			Organic acids			Aminoacids		
Mannitol	MAN	2 g/litre	Itaconate	ITAT	1 g/litre	Serine	SER	2 g/litre
L-Rhamnose	RHA	2	Propionate	PROP	1	Histidine	HIS	2
D-Glucose	GLU	2	Suberate	SUBE	1	L-Alanine	ALA	2
N-Acetyl-glucosamine	NAG	2	Caprate	CAPR	0.5	Proline	PROL	2
Salicine	SAL	2	Malonate	MNT	3			
D-Ribose	RIB	2	Valerate	VALT	1			
D-Melibiose	MEL	2	Acetate	ACET	1			
Inositol	INO	2	Citrate	CIT	1			
L-Fucose	FUC	2	DL-Lactate	LACT	1			
D-Sucrose	SAC	2	2-Ketogluconate	2KG	1			
Sorbitol	SOR	2	5-Ketogluconate	5KG	1			
Maltose	MAL	2	DL-3-Hydroxy-butyrate	3OBU	1			
L-Arabinose	ARA	2	Parahydroxy-benzoate	pOBE	1			
Glycogen	GLYG	2	Metahydroxy-benzoate	mOBE	1			

Identification of Species

An experimental data base was formed from results obtained from both the present (Table 5) and previous studies [11, 12]. This consisted of 97 taxons grouped as follows:

- 70 taxons consisting of species, CDC groups, e.g. *P. aeruginosa,* CDC group Vd
- 14 taxons consisting of species biotypes, e.g. *P. fluorescens* 1, 2, 3
- 4 taxons, each composed largely of strains of the same species, but including some strains from other species in the same genus, e.g. *P. testosteroni*/sp., *S. flexneri*/sp.
- 5 taxons each grouping two species of the same genus, e.g. *A. faecalis/odorans; S. liquefaciens/plymuthica*
- 4 taxons each grouping several species of the same genus, e.g. *Pseudomonas* spp., *Flavobacterium* spp.

The data base was then processed using an identification programme based on the Bayes procedure.

The profile for each test strain was then analysed and the resultant identification compared with the reference identification. A strain could thus either be:

1. *Correctly Identified*
   a) After 24 h, if the profile corresponded with the reference identification or with a taxon as defined above
   b) After 48 h, if the profile corresponded with the reference identification or with a taxon as defined above, or if it fell between the reference identification and one or two other species
2. *Incorrectly Identified*
   a) After 24-h incubation, if the profile corresponded with that of a species other than the reference identification

b) After 48-h incubation, if the profile corresponded with that of a species other than the reference identification or if it fell between two or three species not including the reference identification

3. *Not Identified*

a) Where the profile included at least one result clearly at variance with an otherwise acceptable identification

b) Where the profile included several results which did not correspond precisely with any profile in the data base.

Results

1. Of the 1,505 strains tested, 95.8% were correctly identified (80.9% in 24 h and 14.9% in 48 h), 1.2% were incorrectly identified and 3% were not identified.

2. Of the 962 strains representing 49 species of Enterobacteriaceae and Vibrionaceae, 95.4% were correctly identified (82.9% in 24 h and 12.5% in 48 h), 1.7% were incorrectly identified and 2.9% were not identified (Table 1).

All strains of *Shigella dysenteriae* and *S. sonnei* were correctly identified to the species level. However, five of the nine strains of *S. boydii* and 28 of the 29 strains of *S. flexneri* were only identified to the genus level.

3. Of the 543 strains representing 36 species of Gram-negative bacilli other than Enterobacteriaceae and Vibrionaceae, 96.8% were correctly identified (77.6% in 24 h and 19.2% in 48 h), 0.1% were incorrectly identified and 3.1% were not identified (Table 2).

Seventy-five percent of *Pseudomonas acidovorans* strains and 79% of *P. testosteroni* strains were identified to the species level, whilst 18% and 21% respectively were only identified as members of the *Pseudomonas* spp. taxon. Ninety-four percent of strains of *P. alcaligenes* and 25% of strains of *P. pseudoalcaligenes* were identified only as members of the alcaligenes group (*P. pseudoalcaligenes/alcaligenes*). The remaining strains of *P. pseudoalcaligenes* were correctly identified to the species level.

Eighteen percent of *Flavobacterium meningosepticum* strains and 20% of *F. odoratum* strains were identified only to the genus level whilst the remainder were fully identified.

Seventeen of the 1,505 strains tested were given an identification different from the references identification (Table 6).

Discussion

The system correctly identified most of the species tested. The only species of which fewer than 90% of strains were identified were: *E. agglomerans* (71%), *K. pneumoniae* (88%), *K. cryocrescens* (86%), *V. cholerae* (89%), *A. lwoffi* (88%), *M. liquefaciens* (80%), and *P. vesicularis* (83%).

The failure to identify 29% of strains of *E. agglomerans* was probably due to the considerable heterogeneity of the species [8, 17].

Table 5. Percentages of positive reactions used in the experimental identification datab

	MAN	RHA	GLU	NAG	SAL	RIB	MEL	INO	FUC	SAC	SOR	MAL	ARA	ITAT	PR
Ps aeruginosa	94	1	100	80	1	5	0	1	1	0	3	5	1	100	10
Ps fluorescens 1	100	0	100	86	0	99	0	0	0	0	0	0	3	0	10
Ps fluorescens 2	100	50	100	100	0	99	0	50	0	0	0	0	0	75	10
Ps fluorescens 3	80	0	100	80	0	99	0	80	0	40	80	0	100	99	6
Ps putida 1	100	0	100	0	0	25	0	0	0	0	0	0	1	100	9
Ps putida 2	25	0	100	0	0	50	0	0	0	0	0	0	3	3	7
Ps putida 3	1	12	100	1	0	87	0	0	0	0	0	0	100	2	9
Ps cepacia	93	0	93	90	60	76	16	86	86	53	86	10	60	20	9
Ps acidovorans	94	0	1	0	0	0	0	11	0	0	0	0	1	100	10
Pseudomonas spp	1	0	0	0	0	0	0	0	0	0	0	0	0	91	9
Ps pseudoalca./alca.	0	0	14	7	0	7	0	0	0	0	0	0	0	50	6
Ps testoteroni/sp	1	0	1	0	0	0	0	0	1	1	0	1	1	32	8
Ps stutzeri	80	0	97	0	0	0	0	2	0	0	0	80	2	99	3
Ps mendocina	0	0	100	0	0	0	0	0	0	0	0	0	0	100	10
Ps maltophilia	4	2	95	100	70	3	78	0	0	79	0	98	1	1	9
Ps diminuta	3	3	7	3	3	3	3	0	0	15	0	3	7	0	3
Ps vesicularis	0	0	84	23	0	0	0	0	0	0	0	76	0	0	1
Ps pickettii	11	0	94	22	5	11	5	0	0	11	0	5	94	11	9
Ps paucimobilis	0	11	100	88	88	0	88	0	33	100	0	100	88	0	
Ps putrefaciens	0	0	0	100	0	0	0	0	0	0	0	0	13	0	9
Chrom violaceum	28	0	100	99	0	100	0	28	0	28	28	0	0	0	8
CDC group V-E1	87	87	100	12	100	99	0	100	100	0	0	75	100	0	5
CDC group VE-2	81	45	90	0	0	81	0	81	72	9	81	90	81	54	5
Bord bronchiseptica	0	0	0	0	0	0	0	0	0	0	0	0	0	85	8
CDC group IV-C2	16	0	0	0	16	0	16	0	0	16	0	0	0	83	5
CDC group IV-E	0	0	0	0	0	0	0	0	0	0	0	0	0	100	
Alc faecalis/odorans	0	0	1	0	0	0	0	0	0	0	0	0	0	7	9
Alc denitrificans	0	0	0	0	0	0	0	0	0	0	0	0	0	96	9
Achro xylosoxidans	1	3	98	5	3	6	1	0	1	3	1	6	1	96	9
DC group Vd	89	97	100	100	0	92	0	99	100	97	36	97	100	0	[illegible]
Fl.meningosepticum	100	0	100	98	2	0	0	0	5	0	0	50	2	0	
Fl.odoratum	0	0	0	0	0	0	0	0	0	0	0	0	0	0	
Fl.multivorum	1	5	100	99	100	1	100	1	0	100	0	100	99	0	
Fl.breve/sp	0	0	87	0	0	0	0	0	0	12	0	100	10	0	[illegible]
CDC group II-B	0	0	100	0	67	0	0	0	0	16	0	100	100	0	
Flavobacterium spp	20	10	99	0	10	0	0	0	0	10	0	99	0	0	
A.calc.var.anitratus	10	2	79	2	0	60	1	2	0	2	1	0	72	0	[illegible]
A.calc.var.lwoffi 1	1	0	1	1	1	1	0	1	13	1	1	1	1	1	
A.calc.var.lwoffi 2	1	0	1	1	1	1	0	1	13	1	1	1	1	1	
Ac.haemol./alcalig.	0	0	0	0	0	0	0	0	0	0	0	0	0	0	[illegible]
Moraxella osloensis	0	0	0	0	0	0	0	0	0	0	0	0	0	0	[illegible]
Mo liquefaciens	0	0	0	0	0	0	0	0	0	0	0	0	0	0	
Mo urethralis	0	0	0	0	0	0	0	0	0	0	0	0	0	100	1[illegible]
Past aerogenes	0	0	94	75	0	68	26	68	5	95	0	94	50	0	
Pasteurella EF4	0	0	100	0	0	28	0	0	0	0	0	0	0	0	
Aeromonas hydrophila	96	20	100	100	63	100	0	6	0	86	26	100	76	3	
Ples shigelloides	0	0	94	94	0	68	26	68	5	0	0	94	0	0	
Vib alginolyticus	93	0	100	98	0	31	6	0	6	96	6	87	6	0	
Vib paraheamolyticus	100	0	100	99	0	50	0	0	0	16	0	99	100	0	
Vibrio cholerae	100	0	100	100	0	100	0	0	0	100	0	100	0	0	
Vibrio vulnificus	58	0	100	100	8	100	0	0	16	0	0	100	8	0	
Cit amalonaticus	100	100	100	100	99	100	0	0	100	1	100	100	100	0	
Cit diversus	100	100	100	100	60	100	5	99	95	33	100	100	100	0	
Citrobacter freundii	100	99	100	100	0	100	75	66	96	74	99	92	100	0	
Ent aerogenes	100	100	97	97	100	100	94	97	100	100	100	100	100	0	
Ent agglomerans	100	90	99	99	99	100	1	99	1	99	1	99	99	0	
Ent cloacae	100	99	100	100	43	100	77	2	95	100	96	100	100	1	
Ent gergoviae 1	100	100	100	100	100	100	100	99	5	100	25	100	100	0	
Ent gergoviae 2	100	100	100	100	100	100	100	99	99	100	25	100	100	0	
Ent sak/cloacae	100	90	100	100	100	100	100	99	5	100	20	100	100	0	
Esc.adecarboxilata	100	100	100	100	100	100	100	0	0	38	0	100	100	0	
Escherichia coli 1	99	85	100	99	23	100	95	6	97	31	97	100	99	8	
Escherichia coli 2	93	44	100	99	13	100	5	6	93	10	75	100	100	0	
Esc.vulneris	100	100	100	100	100	100	100	0	0	38	0	100	100	0	
Hafnia alvei	100	99	100	100	18	100	0	0	87	0	0	100	96	3	
Klebsiella oxytoca	100	95	100	100	100	100	100	100	100	100	100	100	100	3	
Kl ozaenae	87	25	99	99	90	100	99	85	90	1	75	99	87	0	
Kl pneumoniae	100	96	100	100	100	100	100	100	96	100	100	100	99	0	
Kl rhinoscleromatis	100	100	100	99	100	100	96	99	66	100	100	100	99	1	
Kl trevisanii	100	96	100	100	100	100	100	100	96	100	100	100	100	0	
Kluyvera ascorbata	100	100	100	100	100	100	100	0	0	100	40	100	100	0	
Kluy.cryocrescens	100	100	100	100	100	100	100	0	50	55	50	100	100	0	
Morganella morganii	0	0	99	99	5	99	0	0	0	0	2	0	0	0	
Prot mirabilis	3	3	100	100	3	100	0	6	0	4	0	6	3	0	
Prot vulgaris	5	7	100	100	58	100	2	7	2	94	0	92	2	0	
Prov alcalifaciens	0	0	100	100	10	100	0	30	0	10	0	10	10	0	
Prov rettgeri	85	70	99	99	80	99	0	85	0	65	0	10	5	0	
Providencia stuartii	4	0	100	100	4	100	0	100	0	40	0	0	0	0	
Rahnella aquatilis	100	100	100	100	100	90	100	0	100	100	100	100	100	0	
Salmonella arizonae	100	100	100	100	0	100	71	28	100	0	100	100	100	28	
Salm gallinarum	100	0	100	95	0	50	0	0	50	0	0	75	20	0	
Salmonella para-A	100	100	100	100	0	100	100	0	100	5	95	100	100	0	
Salm choleraesuis/sp	100	99	100	100	0	100	99	0	100	0	100	100	0	0	
Salmonella typhi	100	1	100	100	0	100	99	1	1	0	99	100	0	0	
Salm typhimurium/sp	100	99	99	100	1	100	90	65	99	0	99	99	99	1	
Ser.liquef./plym.	100	3	100	100	100	100	95	96	92	100	94	100	95	0	
Ser.marcescens	100	5	100	100	98	100	5	91	98	98	99	100	5	0	
Ser.odorifera	100	99	100	100	90	100	100	100	99	60	99	90	100	0	
Serratia rubidaea	100	0	100	100	100	99	100	100	67	100	0	100	100	0	
Shigella boydii/sp	60	1	100	100	0	80	1	0	1	0	70	1	70	0	
Shigella dysenteriae	0	73	93	93	0	13	13	0	0	0	13	0	0	0	
Shigella flexneri/sp	79	0	100	92	0	1	8	0	0	0	8	16	10	0	
Shigella sonnei 1	100	1	100	100	0	100	0	0	16	0	0	100	100	1	
Shigella sonnei 2	100	99	100	100	0	100	0	0	99	0	0	99	100	1	
Y enterocolitica	100	0	100	100	16	99	0	33	1	90	100	99	98	0	
Y pseudotuberculosis	95	79	95	100	24	60	82	1	0	0	1	97	79	0	
Yersinia ruckeri	100	0	100	100	0	25	0	0	0	0	0	100	0	0	

able 4 for abbreviations)

APR	MNT	VALT	ACET	CIT	LACT	HIS	ALA	2KG	5KG	3OBU	GLYG	pOBE	mOBE	PROL	SER
.00	100	95	98	100	100	92	100	98	0	98	0	99	3	100	1
99	3	100	99	100	100	99	100	100	0	99	0	100	0	100	100
99	75	75	99	100	100	100	100	100	0	75	0	50	0	100	100
99	80	60	99	100	100	100	100	100	0	99	0	80	0	100	100
.00	100	99	100	100	100	25	100	99	0	99	0	75	0	100	100
.00	2	100	100	100	100	100	100	98	0	100	0	80	0	100	100
.00	100	100	100	100	100	100	100	100	0	100	0	100	0	100	100
86	93	96	100	93	100	96	100	90	50	96	3	96	76	100	83
55	83	99	99	64	100	100	92	0	0	100	0	94	99	100	0
41	0	100	100	50	100	100	36	0	0	100	0	100	100	100	0
96	0	50	79	43	93	64	96	0	0	43	4	1	1	100	86
39	0	100	100	1	100	5	0	5	0	99	0	100	1	99	1
88	91	44	100	94	100	5	100	0	0	99	77	19	1	100	5
.00	100	100	100	100	100	100	100	0	0	100	0	0	0	100	100
4	68	80	97	97	97	31	94	1	20	0	1	0	2	96	32
1	0	38	65	3	3	76	19	0	0	69	0	0	0	80	15
0	0	30	46	0	10	0	15	0	0	85	0	0	0	30	0
11	95	27	94	94	100	83	83	88	0	94	0	83	0	94	22
0	0	0	0	22	66	0	22	0	0	44	0	77	0	11	0
33	0	86	93	0	100	3	80	0	0	73	0	0	0	6	73
28	0	100	99	57	100	100	100	0	0	14	28	0	0	99	100
87	75	25	75	100	100	0	100	100	5	87	0	100	0	100	100
45	72	18	100	81	100	72	90	81	0	90	0	81	0	81	81
4	0	66	95	97	89	2	52	0	0	99	0	0	0	77	33
66	83	16	83	83	83	83	16	0	0	83	0	16	16	100	0
0	0	50	0	100	50	0	50	0	0	100	0	100	0	100	0
80	87	98	98	93	96	72	94	1	0	94	0	16	3	94	6
45	3	100	100	100	100	67	87	3	0	100	0	3	22	96	80
85	1	97	98	96	100	50	96	1	0	96	5	1	18	96	60
5	0	34	100	89	97	97	94	100	40	50	2	73	0	100	94
0	0	0	0	20	0	0	0	0	0	0	0	0	0	1	0
0	0	0	0	0	0	0	0	0	0	0	0	0	0	99	0
0	0	0	0	0	0	0	0	0	0	0	80	0	0	7	0
0	0	70	99	0	16	0	0	0	0	0	87	0	0	66	16
0	0	0	0	67	0	0	0	0	0	0	100	0	0	10	0
0	0	0	15	25	10	0	0	0	0	0	100	0	0	75	0
97	70	95	99	91	87	91	100	4	0	97	4	85	4	100	33
30	1	75	99	2	76	1	95	1	0	1	1	1	10	45	1
30	1	75	99	80	76	1	1	1	0	1	1	99	10	45	1
100	5	98	99	87	40	50	100	0	0	5	0	75	0	100	25
40	0	84	57	0	100	0	57	1	0	94	0	0	0	0	42
0	0	0	60	40	40	99	0	0	0	0	0	0	0	20	20
0	0	100	100	50	0	0	0	0	0	0	0	0	0	100	0
0	0	0	0	0	20	0	0	0	0	0	0	0	0	10	21
0	0	0	71	0	85	0	57	0	0	0	0	0	0	0	14
60	0	56	86	30	70	96	83	1	0	6	96	0	0	96	96
0	0	0	0	0	0	0	0	0	0	0	0	0	0	10	21
0	0	0	18	18	100	93	43	18	0	12	81	0	0	81	87
0	0	0	50	30	100	99	83	1	0	0	100	0	0	100	83
0	0	0	88	100	88	88	0	0	0	0	77	0	0	88	88
0	0	0	12	83	100	38	83	0	4	0	100	0	4	96	75
1	0	1	100	30	100	1	100	100	100	1	1	87	100	75	100
1	20	4	95	99	100	50	99	100	100	4	4	1	95	20	100
1	0	11	100	96	100	33	92	99	100	92	3	0	85	92	100
2	96	5	96	97	100	100	100	100	1	92	1	81	94	97	90
1	1	0	1	30	57	57	71	99	0	1	1	1	1	86	57
1	10	0	84	100	96	96	96	100	0	1	1	0	1	99	100
50	60	1	99	99	100	100	99	100	100	20	0	60	99	100	100
80	60	1	99	99	100	80	99	100	100	99	0	99	1	100	100
0	10	0	94	92	92	1	94	99	0	1	0	0	0	100	98
0	50	0	99	0	100	100	100	100	0	0	0	0	0	99	100
1	1	0	100	9	100	12	97	40	20	4	23	4	27	91	100
1	6	0	96	0	100	0	75	17	1	0	0	3	58	65	96
0	50	0	99	0	100	0	100	100	0	0	0	0	0	99	100
0	1	0	90	1	99	96	81	100	0	0	1	0	1	100	99
5	60	3	98	100	99	100	98	100	100	17	30	84	80	100	99
0	1	0	1	10	85	65	65	99	0	0	1	27	1	50	87
10	60	9	98	100	100	93	99	100	0	94	20	90	5	96	100
0	0	0	0	1	100	99	99	99	0	0	1	0	0	99	99
10	60	9	99	100	100	93	100	100	100	90	1	99	1	99	100
10	15	0	90	100	100	0	100	50	100	0	0	90	100	5	100
0	0	0	90	33	100	0	100	100	100	0	0	83	0	10	100
0	0	0	2	32	97	0	8	5	0	0	2	0	0	58	94
0	3	0	80	90	100	15	93	5	0	0	0	0	0	95	100
5	5	0	94	30	94	33	94	5	0	0	0	0	1	97	100
1	0	0	99	90	8	90	40	1	0	0	10	0	10	100	99
40	0	5	95	90	1	85	90	80	0	0	0	0	0	90	95
38	0	0	100	100	0	80	80	2	0	0	0	0	0	100	98
0	0	0	0	90	0	0	40	100	100	0	0	0	0	100	40
0	30	0	100	100	100	57	98	0	20	28	2	0	14	99	100
0	0	0	20	40	20	0	0	0	5	0	0	0	5	20	20
0	0	0	2	30	100	0	1	0	20	0	1	0	100	95	100
0	0	0	50	20	100	0	30	0	10	0	0	0	90	80	90
0	0	1	1	90	100	0	30	1	0	0	0	0	66	62	100
1	1	0	99	100	100	20	98	30	87	0	5	0	99	98	100
80	3	7	96	100	99	100	99	100	99	23	19	23	7	100	99
97	1	25	98	100	99	98	99	100	100	87	9	29	24	99	99
50	0	0	50	100	99	100	100	100	0	2	0	0	0	100	100
60	0	0	80	100	100	100	100	100	0	0	0	0	0	99	99
0	0	0	0	1	100	0	1	0	0	0	1	0	1	40	40
0	0	6	26	0	90	0	66	1	0	0	0	0	0	0	57
0	0	0	4	0	92	0	1	3	0	0	0	0	0	1	14
0	0	0	0	0	99	1	0	1	1	0	1	0	1	33	90
0	0	0	0	0	75	1	0	1	12	0	1	0	1	63	99
0	0	0	2	1	1	10	20	10	1	1	0	0	0	15	50
0	0	0	39	1	47	3	5	1	1	0	0	0	0	58	53
0	0	0	50	0	60	0	10	0	0	0	0	0	0	15	100

Table 6. Study of 17 strains for which ATB 32 GN and conventional methods gave different identifications

Conventional method	Number	ATB 32 GN	Discrepant tests
Aeromonas hydrophila	1	*Vibrio cholerae*	Salicine (−), Arabinose(−), Alanine (−), Caprate (−), Valerate (−)
Acinetobacter lwoffi	1	*Alcaligenes faecalis/ A. odorans*	DL-3-Hydroxybutyrate (+)
Enterobacter cloacae	2	*Klebsiella pneumoniae*	Salicine (+), DL-3-Hydroxy-butyrate (+)
Escherichia coli (ATCC 25290)	1	*Salmonella* spp	Arabinose (−), Proline (−)
Klebsiella oxytoca	1	*Klebsiella trevisanii/ Enterobacter gergoviae*	Metahydroxybenzoate (−)
Klebsiella pneumoniae	4	*Klebsiella trevisanii*	5-Ketogluconate (+)
Klebsiella pneumoniae (ATCC 8044)	1	*Klebsiella rhinoscleromatis*	Fucose (−), Acetate (−), Citrate (−), DL-3-Hydroxybutyrate (−), Parahydroxybenzoate (−)
Proteus mirabilis	2	*Proteus morganii*	Acetate (−), Alanine (−)
Salmonella arizonae	1	*Salmonella* spp	Inositol (+), Propionate (+), Metahydroxybenzoate (+)
Salmonella spp	1	*Salmonella arizonae*	Metahydroxybenzoate (−)
Serratia odorifera	1	*Klebsiella rhinoscleromatis*	Fucose (−), Sucrose (+), Citrate (−)
Vibrio parahaemolyticus	1	*Shigella boydii*	Maltose (−), Histidine (−), Alanine (−)

(+) assimilated; (−) not assimilated

One strain of *E. coli* ATCC 25290 was incorrectly identified after 24 h as a salmonella but correctly identified after 48 h. A correct identification was later achieved within 24 h by either increasing the concentration of the inoculum four-fold or by adding 1% Fildes extract to the growth medium.

Two strains of *Proteus mirabilis* were incorrectly identified as *P. morganii* after 24 h but correctly identified after 48 h. This was because the tests for acetate and alanine became positive late. As with the previously mentioned strain of *E. coli*, these strains were correctly identified in 24 h if the growth medium was heavily inoculated or supplemented with Fildes extract.

The four strains of *K. pneumoniae* which were incorrectly identified as *K. trevisanii* [10] appear to belong to a homogeneous group. They all grew at 44.5 °C, assimilated 5-ketogluconate and failed to acidify adonitol in conventional tests, the latter being found by Bagley et al. [1] in only 13.6% of strains and by Gavini et al. [15] in only 12.3%. The strain of *K. pneumoniae* ATCC 8044 identified as *K. rhinoscleromatis* seemed very unusual as it did not grow at 44.5 °C.

The data base for the salmonella provided for two taxons of *Salmonella* spp. The five strains of *S. choleraesuis* all belonged to the *S. choleraesuis*/sp. taxon whereas the 11 strains of *S. typhimurium* all belonged to the *S. typhimurium*/sp. taxon. Two strains of Salmonella were correctly identified to the genus level but incorrectly speciated.

Although some species were represented by only a small number of strains, nevertheless the ATB 32 GN gallery was clearly able to identify various species that have only recently been described or are rarely encountered in the routine laboratory [10, 31].

Overall, the results very largely agreed with those obtained using conventional methods. Those species which were sometimes incorrectly identified in 24 h will be the subject of a complementary study with the aim of modifying the growth medium to permit more rapid assimilation of certain substrates including the organic acids.

Conclusions

The ATB 32 GN gallery offers the following features:
1. A standard methodology
2. Automatic reading of growth density giving objective results
3. Analysis of the resulting profile by a micro-computer
4. Good agreement with reference methods
5. Capability of identifying many rarely encountered or difficult to identify species
6. Suitability for epidemiological studies due to the large number of substrates tested.

Carbon substrate assimilation tests are being used increasingly by taxonomists to define new species. The ATB 32 GN gallery is the first micromethod to use this approach. Practical and versatile, it promises to become one of the methods of choice for the routine identification of most Gram-negative bacilli.

Acknowledgements. This work has been supported by grants DGRST/EPR 1979/930-428 and ANVAR no. A80 11140 V 110 O. We thank Dr. L. D. Gruer for translating the paper.

References

1. Bagley ST, Seidler RJ, Brenner DJ (1981) *Klebsiella planticola* sp. nov.: a new species of Enterobacteriaceae found primarily in non-clinical environments. Curr Microbiol 6:105–109
2. Ballard RW, Palleroni NJ, Doudoroff M, Stanier RY, Mandel M (1970) Taxonomy of the aerobic *Pseudomonads/Pseudomonas cepacia, P. marginata, P. alliicola* and *P. caryophylli*. J Gen Microbiol 60:199–214
3. Baumann P, Doudoroff M, Stanier RY (1968 a) A study of the *Moraxella* group I. Genus *Moraxella* and the *Neisseria catarrhalis* group. J Bacteriol 95:58–73
4. Baumann P, Doudoroff M, Stanier RY (1968 b) A study of the *Moraxella* group II. Oxidase negative species (genus *Acinetobacter*). J Bacteriol 95:1520–1541
5. Bruun B (1983) Studies on a collection of strains of the genus *Flavobacterium*. 2. Nutritional studies. Acta Pathol Microbiol Immunol Scand [B] 91:35–41
6. De Ley J, Mannheim W, Segers P, Kersters K, Hinz KH, Lievens A (1983) Taxonomy of the genus *Bordetella*. In: Leclerc H (ed) Les bacilles à Gram négatif d'intérêt médical et en Santé Publique: Taxonomie. Identification. Applications. Les Editions INSERM, Paris, vol 114, pp 245–251
7. Descamps P, Veron M, Le Minor S, Buissiere J (1982) Phénotypes et marqueurs épidémiologiques de *Salmonella typhimurium*. Rev Epidemiol Santé Publique 30:423–435
8. Ewing WH, Fife MA (1972) *Enterobacter agglomerans* (Beijerinck) comb. nov. (the *Herbicola lathyri* bacteria). Int J Syst Bacteriol 22:4–11

9. Feltham RKA, Stevens M (1983) A taxonomic study of the genus *Serratia* with particular attention to the position of *Serratia fonticola*. In: Leclerc H (ed) Les bacilles à Gram négatif d'intérêt médical et en Santé Publique: Taxonomie. Identification. Applications. Les Editions INSERM, Paris, vol 114, pp 143–155
10. Ferragut C, Izard D, Gavini F, Kersters K, De Ley J, Leclerc H (1983) *Klebsiella trevisanii:* a new species from water and soil. Int J Syst Bacteriol 33:133–142
11. Freney J, Fleurette J, Laban P, Gayral JP, Desmonceaux M (1983) Système automatique d'identification des bacilles à Gram négatif. In: Leclerc H (ed) Les bacilles à Gram négatif d'intérêt médical en Santé Publique: Taxonomie. Identification. Applications. Les Editions INSERM, vol 114, pp 455–466
12. Freney J, Laban P, Desmonceaux M, Fleurette J (1983) Micromethod for carbon substrate assimilation by *Pseudomonas maltophilia*. Zentralbl Bacteriol Parasitenkd Infectionskr Hyg IA 255:479–488
13. Gavini F, Lefebvre B, Leclerc H (1976a) Positions taxonomiques d'entérobacties H_2S par rapport au genre *Citrobacter*. Ann Microbiol (Inst Past) 127A:275–295
14. Gavini F, Lefebvre B, Leclerc H (1976b) Etude taxonomique d'entérobactéries appartenant ou apparentées au genre *Enterobacter*. Ann Microbiol (Inst Past) 127B:317–335
15. Gavini F, Leclerc H, Lefebvre B, Ferragut C, Izard D (1977) Etude taxonomique d'entérobactéries appartenant ou apparentées au genre *Klebsiella*. Ann Microbiol (Inst Past) 128B:45–49
16. Gavini F, Izard D, Leclerc H, Desmonceaux M, Gayral JP (1980) Carbon sources assimilation tests: comparison between a conventional method and a microtechnique (API) in study of Enterobacteriaceae. Zentralbl Bacteriol Parasitenkd Infektionskr Hyg I C_I:182–187
17. Gavini F, Lefebvre B, Leclerc H (1983) Etude taxonomique des souches appartenant ou apparentées au genre *Erwinia*, groupe *Herbicola* et à l'espèce *Enterobacter agglomerans*. Syst Appl Microbiol 4:218–235
18. Gilardi GL (1971) Characterization of non fermentative non fastidious Gram negative bacteria encountered in medical bacteriology. J Appl Bacteriol 34:623–644
19. Gilardi GL (ed) (1978) Glucose non fermenting Gram negative bacteria in clinical microbiology. CRC Press, West Palm Beach
20. Grimont PAD, Grimont F, Dulong de Rosnay HLC, Sneath PHA (1977) Taxonomy of the genus *Serratia*. J Gen Microbiol 98:39–66
21. Holmes B, Ahmed MS (1981) Group EF_4: a pasteurella like organism. In: Kilian M, Frederiksen W, Biberstein EL (eds) *Haemophilus*, *Pasteurella* and *Actinobacillus*. Academic Press, London, pp 161–174
22. Hugh R (1978) Classical methods for isolation and identification of glucose non fermenting Gram negative rods. In: Gilardi GL (ed) Glucose non fermenting Gram negative bacteria in clinical microbiology. CRC Press, West Palm Beach, pp 1–13
23. Johnson R, Colwell RR, Sakazaki R, Tamura K (1975) Numerical study of the Enterobacteriaceae. Int J Syst Bacteriol 25:12–37
24. Joly-Guillou ML, Bergogne-Berezin E, Laban P, Gayral JP (1984) Etude bioenzymatique et lysotypique de deux cents souches d'*Acinetobacter calcoaceticus*. Pathol Biol (Paris) 32:177–181
25. Kersters K, Segers P, Gayral JP, De Ley J (1981) The taxonomy of the genus *Alcaligenes* studied by auxanography and DNA/r RNA hybridizations. IV. International conference on culture collections, 20–24 July 1981, Brno, Czechoslovakia
26. Kiredjian M, Popoff M, Coynault C, Lefevre M, Lemelin M (1981) Taxonomie du genre *Alcaligenes*. Ann Microbiol (Inst Past) 132B:337–374
27. Le Minor L, Veron M, Popoff M (1982) Taxonomie des *Salmonella*. Ann Microbiol (Inst Past) 133B:223–243
28. Mc Faddin JF (1980) Biochemical tests for identification of medical bacteria, 2nd edn. Williams Wilkins, Baltimore
29. Mc Kell J, Jones D (1976) A numerical taxonomic study of *Proteus–Providencia* bacteria. J Appl Bacteriol 41:143–161
30. Martin WJ, Washington JA II (1980) Enterobacteriaceae. In: Lennette EH, Spaulding EH, Truant JP (eds) Manual of clinical microbiology, 3rd edn. American Society for Microbiology, Washington DC, pp 195–219
31. Miller JM (1983) New genera and species of Enterobacteriaceae. Clin Microbiol Newsl 5:149–151

32. Monget D, Boeufgras JM, Desmonceaux M, Ponchon S, Guicherd M (1983) A new system for the identification of clinical isolates of non enteric Gram negative rods: API 20 NE. First european congress of clinical microbiology, Oct 1983, Bologna, Italy
33. Naemura LG, Bagley ST, Seidler RJ, Kaper JB, Colwell RR (1979) Numerical taxonomy of *Klebsiella pneumoniae* strains isolated from clinical and non clinical sources. Curr Microbiol 2:175–180
34. Palleroni NJ, Doudoroff M, Stanier RY, Solanes RE, Mandel M (1970) Taxonomy of the aerobic *Pseudomonads:* the properties of the *Pseudomonas stutzeri* group. J Gen Microbiol 60:215–231
35. Pichinoty F, Veron M, Mandel M, Durand M, Job C, Garcia JL (1978) Etude physiologique et taxonomique du genre *Alcaligenes: A. denitrificans, A. odorans* et *A. faecalis*. Can J Microbiol 24:743–753
36. Popoff M, Stoleru GH (1980) Position taxonomique des variants biochimiques de *Citrobacter freundii*. Ann Microbiol (Inst Past) 131A:189–196
37. Ralston E, Palleroni NJ, Doudoroff M (1973) *Pseudomonas pickettii,* a new species of clinical origin related to *Pseudomonas solanecearum*. Int J Syst Bacteriol 23:15–19
38. Ralston-Barrett E, Palleroni NJ, Doudoroff M (1976) Phenotypic characterization and deoxyribonucleic acid homologies of the *Pseudomonas alcaligenes* group. Int J Syst Bacteriol 26:421–426
39. Rarick HR, Riley PS, Martin R (1978) Carbon substrate utilization studies of some cultures of *Alcaligenes denitrificans, Alcaligenes faecalis,* and *Alcaligenes odorans* isolated from clinical specimens. J Clin Microbiol 8:313–319
40. Stanier RY (1976) Réflexions sur la taxonomie des *Pseudomonas*. Bull Inst Past 74:255–270
41. Stanier RY, Palleroni NJ, Doudoroff M (1966) The aerobic pseudomonads: a taxonomic study. J Gen Microbiol 43:159–271
42. Veron M (1975) Nutrition et taxonomie des Enterobacteriaceae et bactéries voisines I: méthode d'étude des auxanogrammes. Ann Microbiol (Inst Past) 126A:267–274
43. Veron M, Le Minor L (1975) Nutrition et taxonomie des Enterobacteriaceae et bactéries voisines II: résultats d'ensemble et classification. Ann Microbiol (Inst Past) 126B:111–124
44. Veron M, Le Minor L (1975) Nutrition et taxonomie des Enterobacteriaceae et bactéries voisines III: caractères nutritionnels et différenciation des groupes taxonomiques. Ann Microbiol (Inst Past) 126B:125–147

Rapid Typing of Pili of Pathogenic *Escherichia coli* by Dispersive Immunoelectron Microscopy

H. GELDERBLOM, L. BEUTIN, D. HADJIYIANNIS, and H. REUPKE [1]

Introduction

The enterobacterium *Escherichia coli* constitutes part of the ubiquitous normal human gut flora. Certain serotypes of *E. coli* are intestinal pathogens and were classified in enteropathogenic, enteroinvasive, and enterotoxigenic strains. These represent severe health hazards for humans and animals, especially in developing countries. In addition some strains of *E. coli* are known to cause disease of the human urinary tract, i.e. cystitis, pyelitis, and ultimately pyelonephritis.

Colonization of the intestine and the bladder is due to the ability of certain strains of *E. coli* to produce surface structures which serve as adhesins. Among these, several immunologically different types of filamentous fimbriae or pili have been classified recently [5]. These fimbriae enable the specific binding of bacteria to the surface of epithelial target cells. Some types of pili, CFA/I, CFA/II, and E8775 are encoded by high molecular weight plasmids and are often associated with the production of heat labile and/or heat-stable enterotoxin in human infections [2–4, 10]. The combination of colonization factor antigen (CFA) with enterotoxin enables *E. coli* to provoke a "cholera-like" diarrhoea in humans. Detection of fimbrial antigens might help to differentiate between pathogenic and nonpathogenic *E. coli*. Specific antisera reactive against different fimbrial antigens are commonly used in agglutination tests for identification of pathogenic *E. coli* isolates. In the course of setting up several of these agglutination tests, immunoelectron microscopy (IEM) was employed to control specificity of the typing reactions. Two basic modes of dispersive IEM were evaluated:

1. Bacteria were pre-adsorbed onto the EM grid and incubated with antisera and the appropriate immunoreagents followed by negative staining [1].
2. Suspensions of bacteria in small centrifuge tubes were immunolabelled and washed free from unbound antibodies by centrifugation followed by adsorption to EM grids and negative staining.

In the study presented, direct IEM using unlabelled rabbit antibody and indirect IEM using protein A-gold, anti-rabbit-IgG-gold, or anti-rabbit-IgG-ferritin conjugates as markers were used in parallel. Paying particular emphasis to practicability, reliability, and sensitivity, direct and indirect IEM were compared.

1 Robert-Koch-Institut des Bundesgesundheitsamtes, Nordufer 20, D-1000 Berlin 65

Materials and Methods

Bacterial Strains

All bacteria described here are isolates of *Escherichia coli*. The strains 539/83, 540/83, and 536/83 (all 0-serogroup 114) are stool isolates from hospitalized children in Berlin. E8775A and E87750/C (both 025:H42) were kindly provided by Dr. B. Rowe, London, U.K. E87750/C is a derivative of the original isolate E8775A, which had lost the fimbrial antigen described for the original strain [10]. The carrying strains H10407 (078:H12, CFA/I) and 20586 (06, CFA/II) have been described elsewhere [2, 3]. TPE210 and TPE214 are derivatives of H10407 and 20586 which have lost spontaneously their CFA-type fimbriae.

Culture Media

Bacteria were usually propagated in Tryptic Soy Broth (Difco Lab., Detroit, Michigan). For detection of pili the strains were grown as single colonies on CFA-agar [4] for 18 h at 37 °C.

Preparation of Rabbit Antisera

Rabbits were immunized several times with increasing amounts of viable bacteria which were grown on CFA-agar and diluted in phosphate-buffered saline (PBS) pH 7.2.

Blood was collected 7 days after the last booster injection. Serum was tested for 0-agglutination as described by Ørskov et al. [6].

Preparation of Fimbriae-Specific Antisera

Rabbit antiserum prepared against *E. coli* 540/83, which is fully fimbriated, was absorbed with 536/83, which is non-fimbriated. Absorptions were carried out by mixing 1 ml immune serum with 9 ml of a dense live culture of bacteria in PBS for at least 2 h. Absorption of the immune sera was repeated several times until the serum was monospecific for the pili of the vaccine strain. Antisera prepared against E8775A, H10407, and 20586 were absorbed in the same way with their corresponding derivatives lacking the fimbrial antigen. The presence of fimbriae on the bacterial surface was demonstrated by slide agglutination with specific antisera and by electron microscopy.

Electron Microscopy

Bacteria picked from single colonies were suspended in PBS containing 0.1% gelatine with the help of a thin platinum loop. This conditioned PBS was used throughout all preparation steps as diluent and washing buffer. Copper grids (G 400, 400 mesh, Science Services, Munich) covered by Pioloform (Pioloform F, Wacker Chemie, Munich; Stockem [9]) and reinforced with carbon were used after glow discharge in air (0.5 Torr for 1 min; Edwards 306 high vacuum and coating plant, Sussex, U.K.).

For the pre-absorption mode of IEM all preparation steps were performed on parafilm. EM grids were placed for 2 min on droplets (50 μl) of the bacterial suspension and then transferred for immune incubation onto droplets (20 μl) of diluted rabbit anti-pili serum (1:25, 1:250, 1:500, 1:1,000, 1:2,000, 1:4,000, 1:8,000 in PBS) for 20 min at room temperature. Non-bound antibodies were removed by washing the grid for a total of 5 min on five successive droplets of PBS. These specimens were then either negatively stained for direct evaluation or processed further for indirect labelling on droplets of one of the following marker reagents: anti-rabbit-IgG-ferritin (Cappel Laboratories, Cochranville, Pa., USA), anti-rabbit-IgG-gold 10 nm or protein-A-gold 10 nm (Janssen Life Sciences, Beerse, Belgium). These were applied in dilutions of 1:10 to 1:160 for 20 min. Non-bound reagent was removed by successive washing on five droplets of PBS and three droplets of double-distilled water before negative staining.

Immunolabelling of bacteria in suspensions was performed in conical plastic centrifuge tubes (Eppendorf microvial, 1.5 ml). Fifty microlitres of specific serum was mixed 1+1 with the bacterial suspension and left for 20 min at room temperature followed by two washings in PBS of the bacteria (8,000 *g* for 20 s in an Eppendorf 3200 desk centrifuge, Netheler and Hinz, Gerätebau, Hamburg). The second step for indirect labelling was performed correspondingly. After two additional washings in double-distilled water the bacteria were resuspended in water, adsorbed onto EM grids and contrasted. Negative staining was done with 1% phosphotungstic acid, pH 7.0, or preferentially with 1% uranyl acetate in water. The specimens were evaluated using a Siemens 101 or a Zeiss EM 10 A electron microscope operated at 80 kV.

Results

The presence of pili was easily detected by direct negative staining in all strains giving a positive slide agglutination. Usually more than 90% of the fimbriated bacteria showed these surface elements. Morphologically, there were no apparent differences between the CFA pili and those of strains 539/83, 540/83, and E8775, respectively. They radiate in high numbers from the bacterial body and appear to be more rigid and much shorter than F pili or flagella of *E. coli*. Pili measure 0.6–0.9 μm in length with a diameter of 8–9 nm and show a central stain-filled channel of 2–3 nm in width (Fig. 1). Strains E87750/C, TPE 210, TPE 214, and 536/83, which are seronegative, never showed fimbriae.

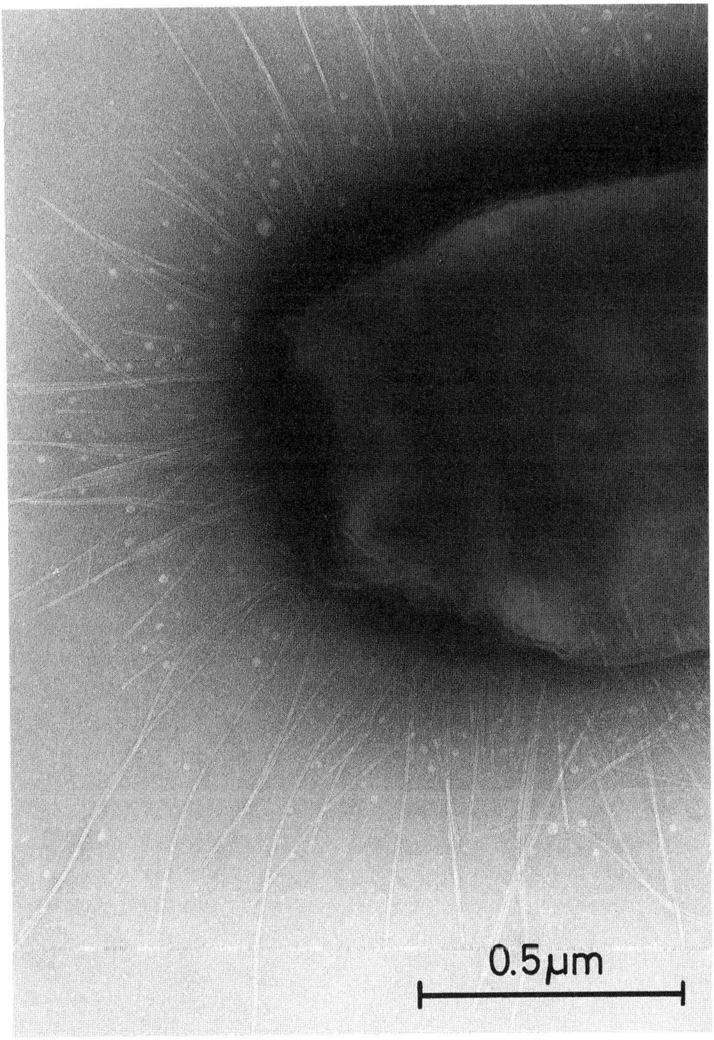

Fig. 1. Part of an *E. coli* cell, strain 540/83, showing an abundance of pili. Negative staining was performed with 1% uranyl acetate (also for Figs. 2–6)

As long as negative staining is well performed, the direct mode of immunolabelling led to unambiguous results (Fig. 2). The reaction of pili and antibodies in the direct method, however, becomes difficult to evaluate if the contrast stain is not spread evenly.

The faint antibodies attached to the pili become either obscured if covered by too much stain or remain undetectable if embedded too thinly. Under these conditions the indirect technique using electron-dense markers proved much more reliable (Fig. 3).

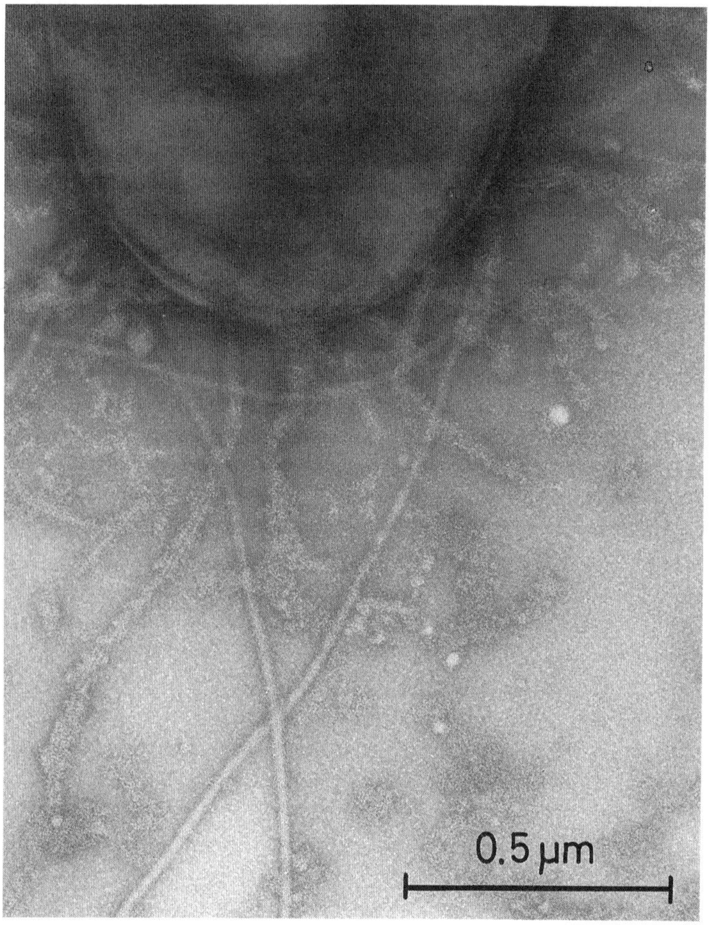

Fig. 2. Direct IEM: *E. coli* strain 20586 pre-adsorbed on to the EM grid and incubated with anti-CFA/II serum (1 : 25). While flagella remain unlabelled, the pili are heavily tagged by antibodies

Provided that typing sera of known specificity were used in the first incubation, the results of indirect IEM could be assessed even from badly stained specimens, just from the presence and distribution of the gold or ferritin label (Fig. 4).

When typing sera and immunoreagents were end-point titrated and evaluated with the direct and the indirect technique, decoration and cross-linking of pili by the unlabelled antibodies were detectable down to dilutions of 1 : 2,000. If diluted further, the reaction pattern became less distinct or negative. Specimens labelled indirectly with markers still revealed positive typing for at least two additional dilution steps of the first antibody. There were no obvious differences in sensitivity with either pre-adsorption or in-suspension labelling. Nevertheless, the latter technique has some advantages over the pre-adsorption method: (1) The non-spe-

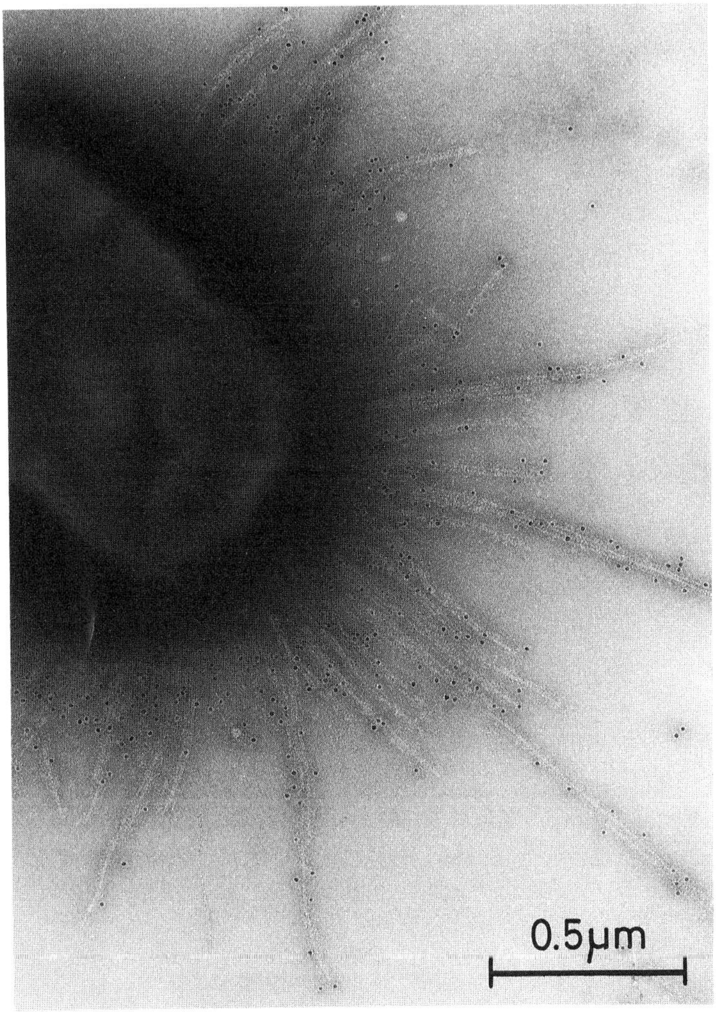

Fig. 3. Indirect IEM: *E. coli* strain 540/83 after pre-adsorption onto the EM grid and successive incubation with anti-540/83 serum (1:25) and protein A-gold 10 (1:10). The maker is closely associated with antibody-coated pili

cific background can be kept lower than with the pre-adsorption mode. Two cycles of washing of the incubated bacteria performed in less than 2 min rendered the specimens as free as washing on five successive droplets of the pre-adsorption specimens. (2) Positive sediments of indirectly labelled bacteria are clearly distinguishable by their slightly reddish colour from the white sediments observed with unlabelled specimens. On the other hand, the centrifugation method uses for convenience higher volumes of the reagents and in the EM pili sometimes look crumbled and attached to the bacterial body (Fig. 5).

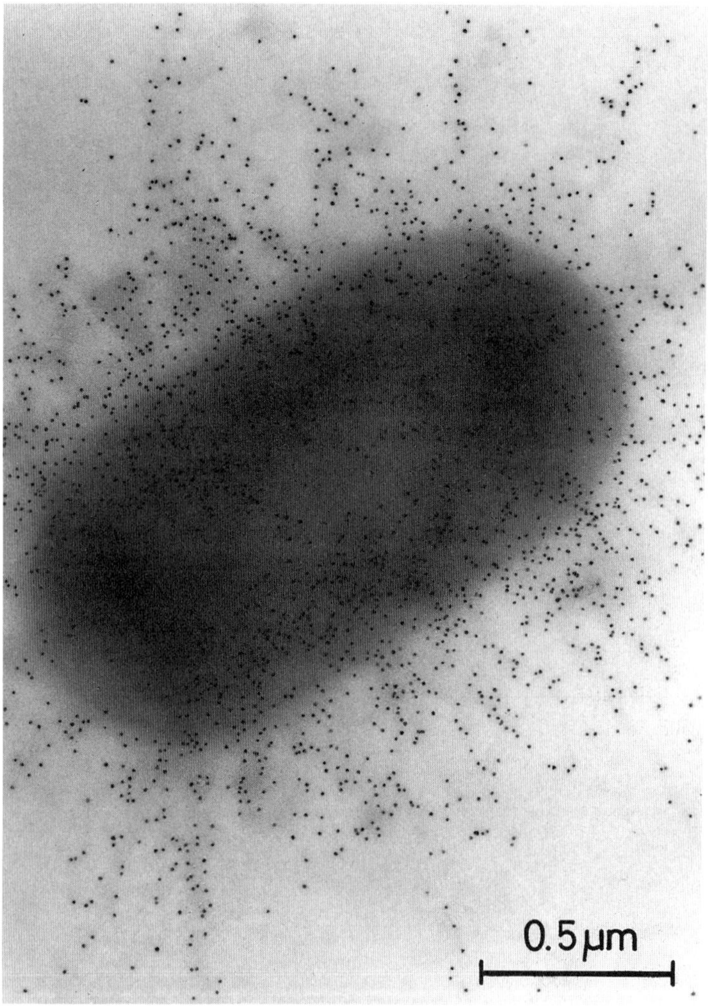

Fig. 4. Indirect IEM: *E. coli* strain 20586 pre-adsorbed onto the EM grid and incubated successively with anti-CFA/II serum (1 : 25) and protein A-gold 10 (1 : 10). Despite the unfavourable staining a positive typing is evident from the distribution of the gold marker

Two main requirements for indirect IEM are low non-specific binding and ready detectability of the marker systems employed. IgG-ferritin conjugates, which have been widely used [11], consistently showed a high background labelling (Fig. 6) when compared with IgG-gold or protein A-gold conjugates. Furthermore, the iron-containing core of the ferritin molecule is nowhere near as electron dense as the colloidal gold label introduced recently into IEM [1, 7]. IgG-gold conjugates exhibited a considerably lower background than the ferritin marker, while protein A-gold labelling led to satisfying results. In general, with antisera

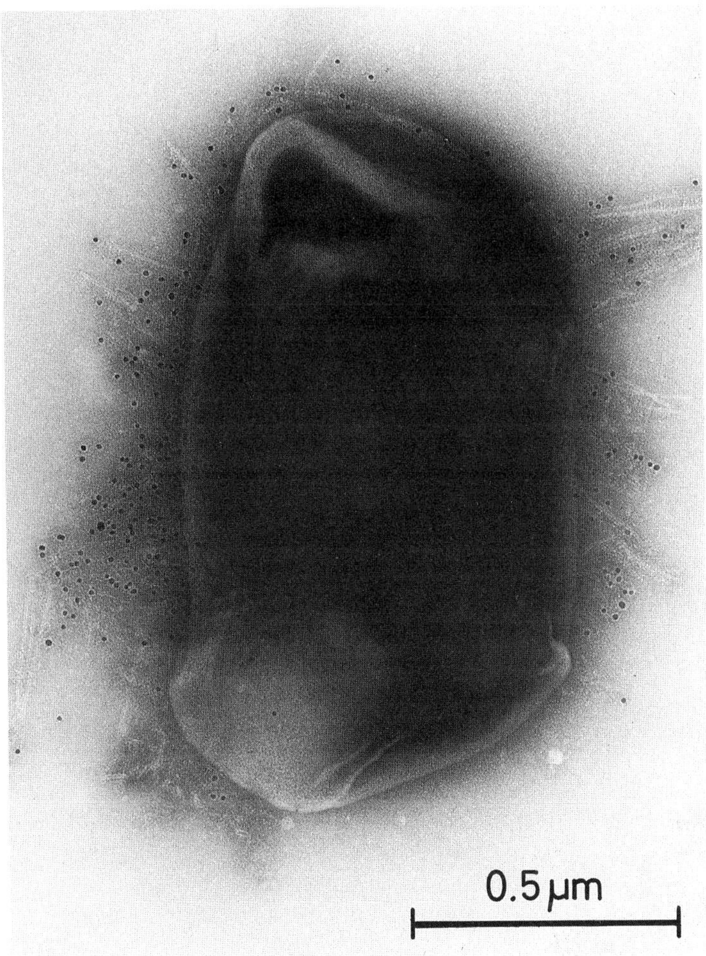

Fig. 5. Indirect IEM: *E. coli* strain 539/83 labelled in suspension with anti-540/83 serum (1 : 10) and protein A-gold 10 (1 : 10). Although the sedimentation steps of this in-suspension labelling procedure have distorted the pili heavily, a positive labelling is obvious

and all three marker systems washings with conditioned PBS were mandatory to reduce non-specific binding of antibodies and electron dense labels.

For both the pre-adsorption and the in-suspension labelling mode, the time needed for preparation was about 30 and 60 min for the direct and the indirect method, respectively.

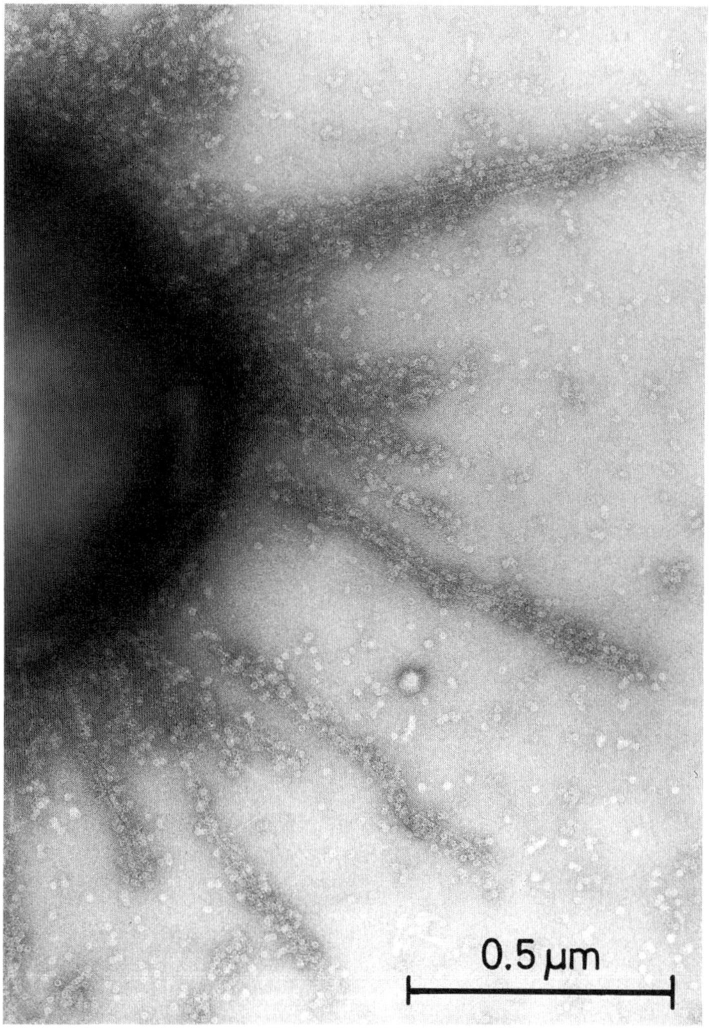

Fig. 6. Indirect IEM: *E. coli* strain 539/83 after pre-adsorption onto the EM grid incubated with anti-540/83 serum (1:10) and anti-rabbit-IgG-ferritin (1:20). The ferritin label tightly packed around the pili is less electron dense than colloidal gold. In addition the marker is also found in considerable amounts in the background

Discussion

Allowing at the same time direct observation of molecular structure as well as mapping of antigenic sites, IEM is a valuable tool for the determination of the relation between structure and function. The antigenic make-up of bacteria, cellular components or viruses can be studied successfully after incubation of the structures of interest in disperse solution with appropriate antisera followed by

EM screening. In this study eight pathogenic strains of *E. coli* were screened for the expression and specificity of fimbrial antigens using different antisera. Unwanted, non-specific "side-reactions" – against lipopolysaccharide antigens of the cell wall and due to unsufficient pre-absorption of the typing anti-pili sera – were reliably recognized by both direct and indirect IEM.

The use of unlabelled antibodies applied in the direct IEM technique allowed a rapid and clear-cut diagnosis of the pilus serotype as long as a high-quality negative staining was performed. Another interesting advantage of direct IEM is the high topological accuracy of the method provided that purified IgG or Fab is used [8]. On the other hand, indirect IEM, more time and material consuming, is mainly independent from the quality of staining. In accordance with Beesley [1] we therefore prefer this method as long as non-fractionated antisera are used for routine IEM screening. Comparing the marker systems employed, IgG-ferritin conjugates by far failed to meet the requirements for a dense probe and a low non-specific background. The background is only to some extent lower when the dense IgG-gold complexes are used. Protein A-gold with its low non-specific binding therefore appears to be the marker system of choice, having the additional advantages than it can be prepared in widely differing size classes and can be applied to detect IgGs from a large series of animal species [7].

Two modes of dispersive IEM were compared with each other:

1. The pre-adsorption technique of the bacteria onto the EM grid followed by incubation with the antisera.
2. Immunolabelling in suspension followed by washings which can be done rapidly using a small desk centrifuge. This procedure leads to a remarkably low non-specific background labelling; however, it needs a higher amount of reagents. In addition, the labile pili sometimes appear distorted, fragmented and attached to the cell wall of the bacteria.

Acknowledgements. The authors thank Mrs. G. Esche for typing the manuscript and Mrs. B. Jungnickl for photographic assistance.

References

1. Beesley JE, Betts MP (1984) Application of the colloidal gold cytochemical technique to antigens in suspension. Proc R Micr Soc 19:36–37
2. Evans DG, Evans DJ Jr (1978) New surface-associated heat-labile colonization factor antigen (CFA/II) produced by enterotoxigenic *Escherichia coli* of serogroups 06 and 08. Infect Immun 21:638–647
3. Evans DG, Silver RP, Evans DJ Jr, Chase DG, Gorbach SL (1975) Plasmid-controlled colonization factor associated with virulence in *Escherichia coli* enterotoxigenic for humans. Infect Immun 12:656–667
4. Evans DG, Evans DJ Jr, Tjoa W (1977) Hemagglutination of human group A erythrocytes by enterotoxigenic *Escherichia coli* isolated from adults with diarrhea: correlation with colonization factor. Infect Immun 18:330–337
5. Ørskov I, Ørskov F (1983) Serology of *Escherichia coli* fimbriae. Prog Allergy 33:80–105
6. Ørskov I, Ørskov F, Jann B, Jann K (1977) Serology, chemistry and genetics of O and K antigens of *Escherichia coli*. Bacteriol Rev 41:667–710

7. Roth J (1982) The protein-A gold (pAg) technique – a qualitative and quantitative approach for antigen localisation on thin sections. In: Bullock CR, Petrusz P (eds) Techniques in immunocytochemistry, vol 1. Academic, London, pp 107–133
8. Schwarz H, Riede I, Sonntag I, Henning U (1983) Degrees of relatedness of T-even type *E. coli* phages using different or the same receptors and topology of serologically cross-reacting cites. EMBO J 2:375–380
9. Stockem W (1970) Die Eignung von Pioloform F für die Herstellung elektronenmikroskopischer Trägerfilme. Mikroskopie 26:185
10. Thomas LV, Cravioto A, Scotland SM, Rowe B (1982) New fimbrial antigenic type (E 8775) that may represent a colonization factor in enterotoxigenic *Escherichia coli* in humans. Infect Immun 35:1119–1124
11. Walker PD, Short JS, Thomson RO, Roberts DS (1973) The fine structure of *Fusiformis nodosus* with special reference to the location of antigens associated with immunogenicity. J Gen Microbiol 77:351–361

Enzyme Immunoassays Using Fluorescent Substrates

R. H. YOLKEN [1]

Quantitative solid-phase immunoassays are attaining widespread usage for the direct detection of microbial agents in body fluids. While there are a number of formats in which solid-phase assays can be performed, most solid-phase assays utilized for the detection of microbial antigens are performed in a format in which an antigen is bound to a solid-phase surface coated with specific antibody. The bound antigen is then quantitated by subsequent reactions with labeled antibodies. The antigen is thus quantitated by the measurement of the label bound to the solid phase by the antigen-antibody reactions. The sensitivity of solid-phase immunoassays depends on a number of factors, the most important of which are the concentration and affinity of the immunoreactants utilized in the assay system [10, 16]. However, the sensitivity of the solid-phase immunoassays also depends to a great extent on the detection limit of the label which is utilized as the immunoglobulin marker. There have thus been a number of approaches which have been utilized to devise sensitive labeling systems for solid-phase immunoassay systems. One involves the use of high-energy markers which can be detected in low concentrations. The most widely used markers of this type are isotopes which emit measurable radioactivity [2, 11]. In addition, markers which generate fluorescent energy have also been utilized since, while they generate lower amounts of energy, they have longer half-lives and less potential for biohazard than radioactive isotopes [4]. The high-energy labels have the advantage that they do not require secondary reactions to be detected, and this can be utilized for rapid highly quantitative assays. However, the ultimate sensitivity of immunoassays which utilize these markers is limited by the detection level of the markers [8]. We have found that this level of sensitivity is often not adequate for the detection of small amounts of antigens in body fluids, especially in cases in which standard fluorescent labels are utilized.

An alternative approach to the development of sensitive immunoassay systems for the detection of microbial antigens in body fluids involves the use of molecules with enzymatic activity as immunoglobulin markers. Since small amounts of enzyme can catalyze the conversion of large amounts of substrate into measurable products, assays which utilize enzymatic markers offer the potential for a high degree of assay sensitivity [1, 13, 14]. In addition, enzymatically labeled immunoreactants have the advantage that they are stable for extended periods of time. However, the sensitivity of enzyme immunoassays which utilize substrates

1 Department of Pediatrics, Division of Infectious Diseases, The John Hopkins University School of Medicine, Baltimore, MD 21205, USA

that generate visibly colored molecules upon enzymatic activation is limited by the detection limit of these relatively low-energy markers. For example, while enzymes in common use have turnover rates of approximately 100,000, the fact that some commonly used substrates require concentrations of 10^{-6} mol/l to be detectable by colorimetric instrumentations limits the sensitivity of standard enzyme immunoassays to 10^{-11} mol/l or approximately 10^{-9} gm/ml of an analyte with a molecular weight of 100,000 [8]. This level of sensitivity is approximate equivalent to that of the nonenzymatic assay systems described above.

One possible method of markedly improving the sensitivity of immunoassay systems would be to combine the two types of assay described above. Thus, an assay system could be designed utilizing enzymatic labels which react with substrates that generate high-energy products upon enzymatic reaction. This form of assay system would have the advantages of enzymatic activation as well as the enhanced sensitivity inherent in the use of high-energy markers. While assay systems can be constructed utilizing radioactive substrates [7], difficulties inherent in the use of radioactivity and the need to separate enzymatic product from unreacted substrate limit the utility of this approach. However, the recent availability of large number of substrates which generate fluorescent molecules upon enzymatic interaction offers the possibility of the development of practical enzyme assay systems which utilize the fluorescent measurement of reaction endpoint [3, 5]. Such systems offer the advantages both of the stability and quantitative nature of fluorescent measurement and the catalytic magnification inherent in enzyme-substrate interaction. Some of the enzymes which can be utilized for this purpose are depicted in Table 1. The most widely used substrates have been those which generate the highly fluorescent ketone umbelliferone upon enzymatic cleavage, since this molecule could be measured at concentrations of 10^{-15} mol/l by standard laboratory fluorescence measurement instrumentation. The recent availability of instruments for the measurement of fluorescence in this range in microtiter plates has markedly expanded the utility of such systems by allowing for the accurate measurement of multiple reactions in short periods of time (Table 2) and for the performance of kinetic measurements of enzymatic activity. Umbelliferyl derivatives can be utilized as substrates for a wide range of glycosidic enzymes. However, the most widely utilized substrates are 4-methyl-umbelliferyl phosphate, which is a fluorescent substrate for alkaline phosphatase, and 4-methyl-umbelliferyl-galactoside, which is a substrate for β-galactosidase. Both β-galactosidase and alkaline phosphatase are enzymes which can be linked to immunoreactants by a number of chemical methods with a good degree of retention of enzymatic and immunological activity [16]. Thus, assay systems which utilize these enzymes and their fluorescent substrates can be formulated for the efficient detection of immunologically active antigens in a solid-phase immunoassay format. Studies performed in our laboratory indicate that commercially available umbelliferyl-galactoside substrates are more stable in solution and exhibit less background fluorescence than available umbelliferyl-phosphate substrates. However, the turnover rate of alkaline phosphatase-labeled immunoreactants is somewhat higher than that of equivalent immunoreactants labeled with β-galactosidase, rendering the two assays essentially equivalent in sensitivity. Fluorescent substrates have also been developed for other enzymes such as glucose oxidase [6] and

Table 1. Comparison of enzymes available for EIA

Enzyme	pH optimum	Specific activity (units/mg)	Molecular weight	Practical substrates available		
				VisuaL	Fluorescent	Radioaktive
Alkaline phosphatase	8–10	400	100,000	NP-PO_4	MU-PO_4 FM-PO_4	^{3}H-AMP ^{3}H-NP-PO ^{14}C-NP-PO
Peroxidase	5– 7	900	40,000	H_2O_2+5AS H_2O_2+OPD H_2O_2+ABTS	NADH HPA	N.A.
β-galactosidase	6– 8	400	540,000	NP-Gal	MU-Gal	^{3}H-GalP
Glucose oxidase	4– 7	200	160,000	Glu+5AS Glu+NBT Glu+MTT	NADH HPA	^{3}H-Glu
Catalase	6– 8	40,000	250,000	H_2O_2 [a]	N.A.	N.A.

Abbreviations: ABTS, 2,2-azino-di-(3 ethyl benzothiazolin sulfone-6) diammonium salt; NP-PO, nitrophenyl phosphate; 5AS, 4_5 Aminosalicylic acid; OPD, orthophenylene diamine; NP-Gal, nitrophenyl galactose; MU-OP, methyl umbelliferyl phosphate; FM-PO_4, fluorescein methyl phosphate; H_2O_2, hydrogen peroxide; HPA, β-hydroxyphenylacetic acid; NBT, P-nitro blue tetrazolium chloride; MTT, thiazolyl blue; ^{3}H-GalP, tritium-labeled β-galactose phosphate; ^{3}H-Glu, tritium-labeled glucose.

[a] Measured spectrophotometrically at $OD_{240\,nm}$

Table 2. Measurement of umbelliferone by microplate colorimeter

Umbelliferone (pg/wells)	Mean [a]	SD	CV (%)
96	1,587.3	71.1	4.4
24	390.1	11.6	2.9
6	100	0.9	0.9
1.5	25.3	1.56	6.1
Negative	0.75	0.66	–

[a] Mean of eight determinations

horseradish peroxidase [19], which are widely used in enzyme immunoassay (EIA) systems (Table 1).

EIAs utilizing fluorescent substrates are performed in a manner similar to that of EIAs which utilize standard substrates. However, the fluorescent substrates is substituted for the colorimetric one and the reaction is quantitated by means of a fluorescence measurement device. EIAs utilizing fluorescent substrates have thus been devised for the measurement of viral, bacterial, and fungal antigens. The performance characteristics of these immunoassays compared favorably with those of equivalent assay systems utilizing colorigenic substrates [9, 12, 17]. It should be noted, however, that the sensitivity of EIAs is determined by the reaction characteristics of the antigens and antibodies utilized as well as the efficiency of the labeling procedure. For this reason, the use of high-energy substrates does not necessarily result in the development of an assay system with increased sensitivity or without utilizing lower energy markers. In order to attain maximal sensitivity, the use of the high-energy substrates should be combined with additional measures directed at optimalizing the antigen-antibody kinetics of the assay system [8, 12]. This can be accomplished by the use of high-affinity antibodies to maximize antigen binding and by the use of immunoglobulin purification or monoclonal antibody techniques to increase the concentration of immunoglobulins directed at the antigen to be measured. Similarly, since the use of high-energy substrates will also increase the signal generated by nonspecific reactions, it is crucial that formats be utilized which minimize the amount of nonspecific reactions. This can best be accomplished by the use of the Fab portion of the immunoglobulin molecule to reduce the occurrence of nonspecific interactions with the Fc portion of the immunoglobulin molecule as well as by the incorporation of sufficient concentrations of detergents and carrier proteins to minimize nonspecific binding to the solid-phase surface. The use of such techniques increases the likelihood of achieving enhanced sensitivity by the use of the higher energy substrates. Even in situations in which enhanced sensitivity cannot be achieved, the use of the high-energy substrates allows maximal antigen-antibody interaction to be attained faster than can be accomplished utilizing the colorimetric substrates. Thi principle is demonstrated in Fig. 1. In an assay system for the detection of adenovirus antigen, the ultimate sensitivity of the EIA system is similar whether fluorescent or colorimetric substrates are used. However, in the case of fluorescent substrate, this level of sensitivity is attained following 5 min of incubation, while this level is not attained until after 4 h utilizing the colorimetric substrate. The use of fluorescent substrate can thus markedly decrease the amount of time required for the detection of the microbial antigens in body fluids. We have found this decreased reaction time to be particularly important in some reaction systems involving monoclonal antibodies, since detection reaction with low-affinity monoclonal antibodies can take extended periods of time to generate measurable reaction products. In addition, the use of high-energy substrate allows for the detec-

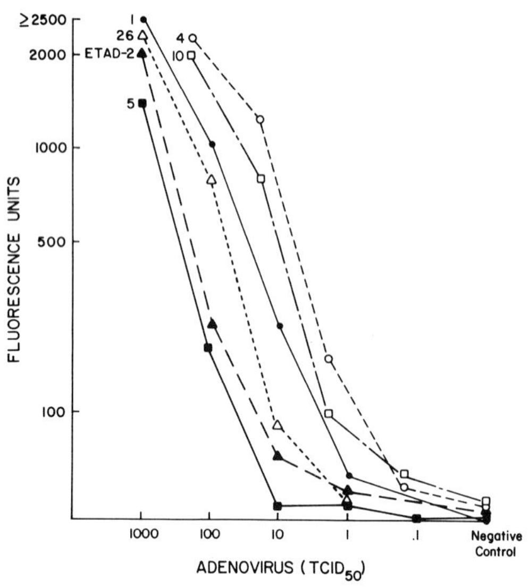

Fig. 1. Fluorescent enzyme immunoassay for adenoviruses using monoclonal antibody

tion of smaller quantities of substrate. This allows the EIAs to be performed with smaller quantities of immunoreactants than equivalent assays which utilize colorimetric substrates. This can be advantageous in situations in which the quantities of clinical specimens of immunoreactants are limited.

Fluorescent substrates can also be used for the direct detection of microbial enzymes in body fluids. The direct detection of microbial enzymatic activity in body fluids offers a number of advantages over standard solid-phase EIAs for the rapid diagnosis of infectious diseases [15]. For example, unlike immunoassays, direct enzymatic detection is not limited by the kinetic parameters of antigen-antibody reactions. In addition, the direct enzymatic detection formats do not require separation or washing steps to remove unbound immunoreactants. The principal limitation of direct enzymatic detection formats is the need to find enzymatic activities which are specific for the identification of the infecting microorganism. One enzyme which can be utilized as a marker of viral infections is neuraminidase, an enzyme which is expressed by pathogenic influenza viruses during the course of respiratory infection. Neuraminidase is also produced by host cells and some bacteria. However, the viral neuraminidase can be distinguished from other

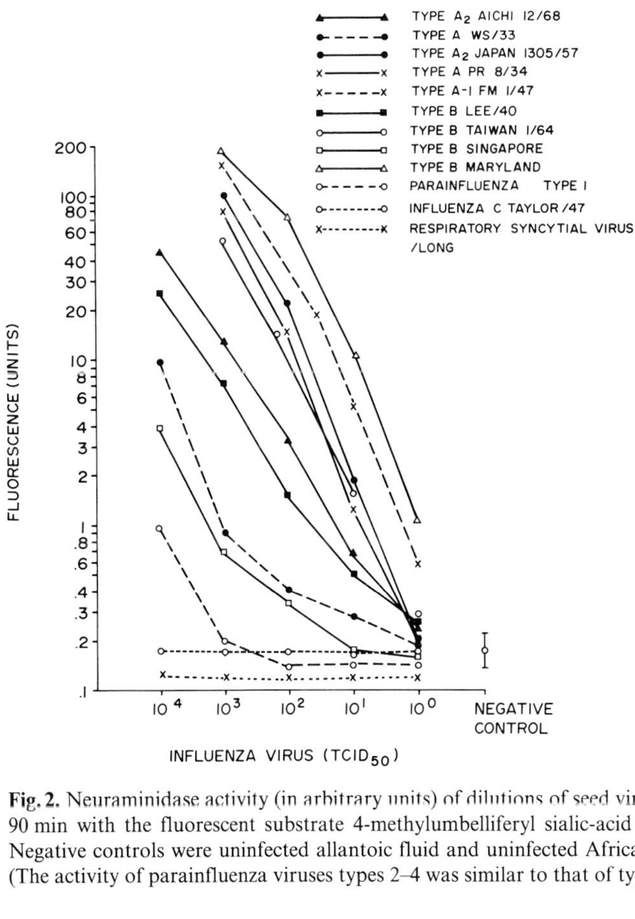

Fig. 2. Neuraminidase activity (in arbitrary units) of dilutions of seed virus preparations incubated for 90 min with the fluorescent substrate 4-methylumbelliferyl sialic-acid of *N*-acetyl-neuraminic acid. Negative controls were uninfected allantoic fluid and uninfected African green monkey kidney cells. (The activity of parainfluenza viruses types 2–4 was similar to that of type 1)

sources of neuraminidase activity by means of differences in pH optima and sensitivity to divalent cations. While there are numerous ways of measuring neuraminidase, one of the most sensitive and convenient involves the use of a 4-methyl-umbelliferylsialic acid substrate which is converted to the highly fluorescent 4-methylumbelliferyl substrate following reaction with the neuraminidase (Fig. 2). This system allows for the detection of picomoles of enzyme in body fluids by the simple mixing of fluid and substrate and measurement of the subsequent generation of fluorescent substrate in a fluorometer. The antigenic nature of the infecting influenza virus can then be determined by subsequent reaction with specific antibodies. This system has proved to be useful both in the monitoring of influenza infection and in the screening of tissue culture cells where the growth of influenza viruses [18]. There are additional enzymes, such as β-lactamase, which are produced by microbial organisms but are generally not present in noninfected body fluids. In addition, there are nucleic acid-processing enzymes, which are produced by herpes virus and other viral agents and can be distinguished from host and bacterial enzymes. The measurement of such enzymes by fluorescent substrates might thus allow for the widespread use of this system for the diagnosis of microbial infections directly in body fluids.

The techniques of both EIA and fluorescent immunoassay have proven to be versatile ones for the detection of microbial antigens directly in body fluids and in in vitro cultivation systems. Our studies indicate that certain aspects of each system can be combined to allow for the development of an assay system which retains the advantages inherent in both fluorescence and enzymatic markers. We believe that the development of more efficient antibodies and more convenient methods for measuring fluorescence in multiple specimens will allow for the continued application of this technique for the immunodetection of miocrobial antigens. In addition, the same principles can be applied to the improvement of other detection systems, such as nucleic acid hybridization using enzyme-linked probes. The availability of a wide range of sensitive immunoassay systems for the direct detection of microbial antigens in body fluids might allow for the more accurate diagnosis of infectious disease and the improved medical management of infected patients.

Acknowledgements. This work was supported by the National Institutes of Health, Bethesda, Maryland, Contract No 1 AI-22680. We thank Rhonda Reese for her expert secretarial assistance.

References

1. Engvall E, Perlmann P (1972) Enzyme linked immunosorbent assay. ELISA. III. Quantitation of specific antibodies by enzyme-linked anti-immunoglobulin in antigen coated tubes. J Immunol 109:129–135
2. Felber JP (1978) Radioimmunoassay in the clinical chemistry laboratory. Adv Clin Chem 20:129–179
3. Fernley HN, Walker PG (1965) Kinetic behavior of calf-intestinal alkaline phosphatase with 4-methyl-umbelliferyl phosphate. J Biochem 97:95–99
4. Gardner PS, McQuillin J (1968) Application of immunofluorescent antibody technique in rapid diagnosis of respiratory syncytial virus infection. Br Med J 3:340–343

5. Guilbault GG, Brignac P Jr, Juneau M (1968) New substrates for the fluorometric determination of oxidative enzymes. Anal Chem 40:1256–1263
6. Guilbault G (1968) Use of enzymes in analytical chemistry. Anal Chem 40:R459–R464
7. Harris CC, Yolken RH, Krokan H, Hsu IC (1979) Ultrasensitive enzymatic radioimmunoassays: Application for the detection of Cholera Toxin and Rotavirus. Proc Natl Acad Sci USA 76:5336–5339
8. Ishikawa E, Imagawa M, Nashida S (1983) Ultrasensitive enzyme immunoassays using fluorogenic, luminogenic, radioactive, and related substrates and factors to limit the sensitivity. In: Avrameas S, et al. (eds) Immunoenzymatic Techniques. Elsevier, pp 219–232
9. Konijn A, Levy R, Link G, Hershko C (1982) A rapid and sensitive ELISA for serum ferritin employing a fluorogenic substrate. J Immun Methods 54:297–307
10. Pesce AJ, Ford DJ, Gaizutis MA (1978) Qualitative and quantitative aspects of immunoassays. Scand J Immunol 8:1–6
11. Rosenthal JD, Hayashi K, Notkins AL (1973) Comparison of direct and indirect solid-phase microradioimmunoassay for the detection of viral antigens and antiviral antibody. Appl Microbiol 25:567–573
12. Shalev A, Greenberg A, McAlpine P (1980) Detection of attograms of antigen by a high-sensitivity enzyme-linked immunosorbent assay using a fluorogenic substrate. J Immun Methods 38:125–139
13. Voller A, Bartlett A, Bidwell DE (1976) Enzyme immunoassays for parasitic diseases. Trans R Soc Trop Med Hyg 70:98–106
14. Yolken RH (1980) Enzyme-linked immunosorbent assay (ELISA): A practical tool for rapid diagnosis of viruses and other infectious agents. Yale J Biol Med 53:85–92
15. Yolken RH (1981) Enzymatic analysis for rapid detection of microbial infection in human body fluids: an overview. Clin Chem 27(9):1490–1498
16. Yolken RH (1982) Enzyme immunoassays for the detection of infectious antigens in body fluids: current limitations and future prospects. Rev Infect Dis 4:35–62
17. Yolken RH, Stopa PH (1979) Enzyme-linked fluorescence assay: ultrasensitive solid-phase assay for detection of human rotavirus. J Clin Microbiol 10:317–321
18. Yolken RH, Torsch V, Berg R, Murphy BR, Lee YC (1980) Fluorometric assay for measurement of viral neuraminidase: Application to the rapid detection of influenzae virus in nasal wash specimens. J Infect Dis 142:516–523
19. Zaitsu K, Nakashima C, Akiyama S, Ohkura Y (1982) Sensitive fluorogenic substrate for peroxidase and its application to enzyme immunoassays. J Pharmacobiodyn 5:5–20

Identification of Bacteria by Flow Immunofluorescence

A. P. PHILLIPS and K. L. MARTIN [1]

Introduction

Flow immunofluorescence (FIF) analysis using commercially available flow cytometers is well established for the analysis of surface antigens on eukaryotic cells. However, the relatively small number of published reports of flow cytometry of bacteria have concentrated on DNA, RNA, and protein analysis, in the context of cell cycle studies, and there have been few FIF applications. The FIF technique was used by Steen et al. (1982) for *Mycobacterium leprae,* and by Ingram et al. (1982) for *Legionella pneumophilia.* A previous communication from this laboratory described the FIF analysis of *Bacillus* spores and vegetative cells (Phillips and Martin 1983). At that time, the 90° scatter signals of intact bacteria and cellular debris could not be distinguished, and attempts to detect narrow forward angle (NFA) scatter were unsuccessful. However, conditions have now been established for the acquisition of NFA scatter signals generated by bacteria, and this scatter can be used to trigger the fluorescence measurement of stained bacteria in immunofluorescence (IF) reaction mixtures analysed directly, i.e. without first washing to remove excess unreacted conjugate.

FIF Analysis Procedure

The flow cytometer used was an Ortho Cytofluorograf 50H; the 2 W argon-ion laser was operated at 488 mm with an output of 150 mW. The multichannel distribution analyser was used in the "multiplex" mode, so that data from the x axis photomultiplier (NAF scatter) and from the y axis photomultiplier (red and green fluorescence) were acquired in separate halves of the analyser memory; thus, histograms were available for both the x and y parameters. Histograms were plotted on an x/y plotter. The Cytofluorograf signals were also outputted to a Dec PDP 11/34 computer and stored on floppy discs for off-line calculation of statistical values.

IF reaction mixtures contained formalin-inactivated bacteria at a concentration of 5×10^6 organisms per millilitre and fluorescein conjugated anti-bacterial IgG at a concentration of 10 μg ml^{-1}. After incubation for 1 h, reaction mixtures were diluted 1:5 in PBS and 20 μl were analysed in the Cytofluorograf. Sample flow rate was monitored by timing the movement of the meniscus in a graduated

1 Chemical Defence Establishment, Porton Down, Salisbury, Wilts. SP4 OJQ, Great Britain

pipette fitted to the sample line; data were only accepted when the 20-μl sample was introduced into the cytometer over 30–40 s.

Sorting was performed on samples flowed in bulk at flow rates broadly similar to those used during analysis. The sorter was used without anti-coincidence, to give maximum recovery of particles. Sorted samples were collected on microscope slides, and the stained bacteria were enumerated by means of a Leitz fluorescence microscope.

Significance of Fluorescence Signals from Bacterial IF Mixtures

Figure 1 shows a fluorescence histogram for spores of *B. anthracis* Sterne stained with specific fluorescein conjugate. Autofluorescence reaction controls and fluorescein conjugate alone gave fluorescence signals limited to the lowest chan-

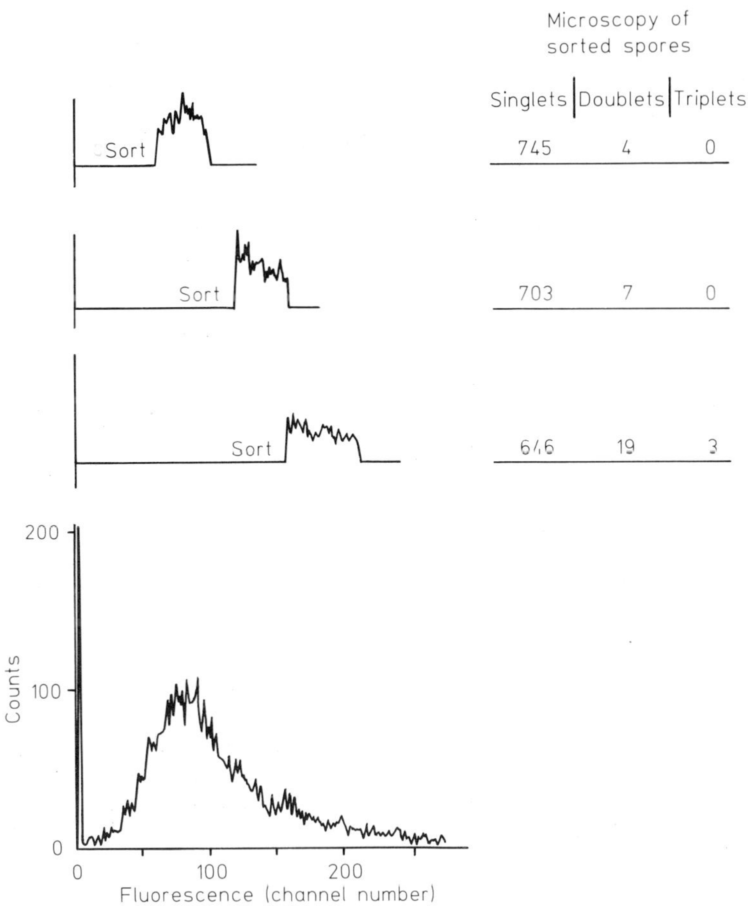

Fig. 1. Fluorescence histogram for spores of *B. anthracis* Sterne stained with specific fluorescein conjugate

nel of the histogram. The use of NFA scatter to trigger such fluorescence histograms constitutes a considerable advance over the previous method in which two samples (the reaction mixture and a centrifugation supernatant) had to be analysed in succession and the data subtracted in order to reduce the interference from stained cellular debris and from free conjugate (Phillips and Martin 1983). However, careful control of sample flow rate was still important with the new method; doubling the flow rate resulted in a massive signal in the lower fluorescence channels that tended to overlap with the left-hand edge of the fluorescence peak.

Two lines of evidence support the premise that this fluorescence peak was indeed due to the presence of stained bacteria. First, when IF reaction mixtures were centrifuged sufficiently to remove most of the bacteria, before analysis, the peak of the histogram was eliminated. The second type of evidence was derived by setting "windows" on selected regions of the fluorescence histogram, sorting onto microscope slides, and enumerating the stained bacteria by microscopy. The numbers of single and aggregated spares recovered from 10^3 fluorescent events sorted in three regions of the fluorescence histogram are shown in Fig. 1. The high percentage recovery of stained spores indicates that the fluorescent events recorded by the cytometer were indeed due to stained spores and were not caused by the stained cellular debris known to be present in such reaction mixtures. The paucity of spore aggregates in the sorted samples strongly suggests that the pronounced right-hand tail of the fluorescence histogram of the particular sample of spores was not the result of aggregation of spores but reflects a skew distribution of the total antibody bound to individual spores.

Suitability of FIF Techniques for Routine Analysis

With practice, it proved possible to analyse IF reaction mixtures at 5-min intervals. Cytometer signals were acquired by the PDP computer and stored for off-line statistical treatment if required. After analysing each sample there was time to output both the x and y histograms onto the x/y plotter to provide an immediate record; a few key integral values were also calculated using the Cytofluorograf distribution analyser. Fluorescence distributions could often be compared satisfactorily by subjective assessment of modes in terms of the displacement from the origin, without the need for subsequent (and rather time consuming) interrogation of the computer.

The data of Fig. 2 illustrate the sensitivity of the Cytofluorograf in detecting stained spores. As few as 500 *B. anthracis* Vollum spores stained with specific conjugate were analysed in the presence of 5,000 *B. subtilis* var *niger* and 5,000 *E. coli* MRE 162 organisms, two bacterial strains not stained by the anti-*B. anthracis* conjugate.

The resulting fluorescence histogram for the 20-μl sample is clearly distinct from the histogram for the control reaction mixture that lacks *B. anthracis* spores. Similar results were obtained for more dilute suspensions when 100 μl were analysed.

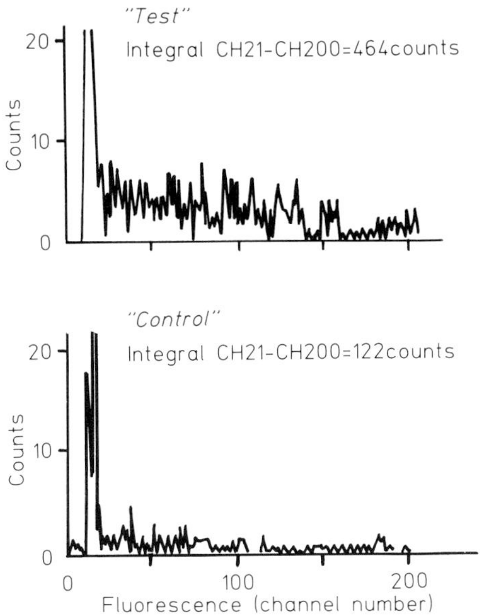

Fig. 2. Detection of stained spores using Cytofluorograf

The present FIF technique was highly reproducible. Typically, five replicate *B. anthracis* reaction mixtures in each of which at least 3,000 spore fluorescence events were recorded gave mean, median and mode values with coefficient of variation values within each set of 2.5%, 2.7%, and 5.5% respectively. Part of this small degree of variation must be attributed to variation in the volume of conjugate added to each reaction mixture.

Differentiation of Spores and Gram-Negative Bacteria

The NFA scatter histograms for *B. anthracis* Vollum spores and for *E. coli* organisms shown in Fig. 3 may be seen to overlap, but *E. coli* predominantly scatter light at the lower voltages. When NFA scatter was windowed so as to select particles with either low or high scatter values, the differential effect on fluorescence data for these two bacterial strains (see Table 1) was such that it would be possible

Table 1. Differentiation of fluorescence signals of stained *B. anthracis* spores and stained *E. coli* by windowing on NFA scatter

Stained bacteria	Fluorescence Counts, Ch 21-300		
	NFA ungated	NFA gated Ch 2-77	NFA gated Ch 89-512
B. anthracis spores	12,610	1,350	8,340
E. coli	5,810	3,880	630

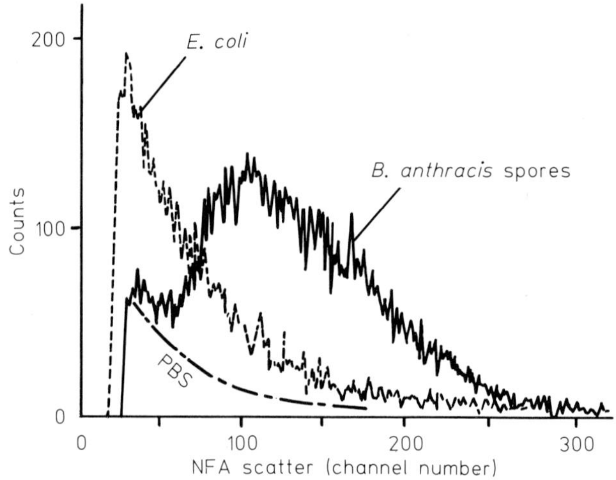

Fig. 3. NFA scatter histograms for *B. anthracis* Vollum spores and *E. coli* organisms

to distinguish stained spores and stained *E. coli* in a mixed population by the use of simultaneous equations. In the absence of a computer this would necessitate 2 separate analyses of the mixed sample under the different window conditions; for a Cytofluorograf linked to a computer it would be possible to perform this calculation on data from a single analysis.

Conclusion

Flow cytometry of bacterial samples provides a rapid method for quantitative measurement of the uptake of fluorescein conjugates on individual organisms in the population. Sample preparation is simple in that IF reaction mixtures can be analysed directly without the need to wash bacteria free of excess conjugate. As few as 500 stained bacteria can be detected in the presence of an excess of unstained bacteria. Samples can be analysed at the rate of about 12 per hour. Even at this rate of sample turnover, data can be acquired from 10^4 bacteria in each sample; this allows measurement of antigen expression among individual bacteria in heterogeneous populations to levels of accuracy and sensitivity not possible with photometric microscopy techniques (Phillips and Martin 1982; Phillips et al. 1983).

The obvious disadvantages of flow cytometry remain the high capital cost of the multi-purpose cytometers that are available, even if purchased without a sorter facility, and the high level of operator training needed.

References

Ingram M, Cleary TJ, Price BJ, Castro A (1982) Rapid detection of *Legionella pneumophilia* by flow cytometry. Cytometry 3:134–137

Phillips AP, Martin KL (1982) Assessment of immunofluorescence measurements of individual bacteria in direct and indirect assays for *Bacillus anthracis* and *Bacillus cereus* spores. J Appl Bacteriol 53:223–231

Phillips AP, Martin KL (1983) Immunofluorescence analysis of *Bacillus* spores and vegetative cells by flow cytometry. Cytometry 4:123–131

Phillips AP, Martin KL, Broster MG (1983) Differentiation between spores of *Bacillus anthracis* and *Bacillus cereus* by a quantitative immunofluorescence technique. J Clin Microbiol 17:41–47

Steen HB, Boye E, Skarstad K, Bloom B, Godal T, Mustafa S (1982) Applications of flow cytometry on bacteria: cell cycle kinetics, drug effects, and quantitation of antibody binding. Cytometry 2:249–257

The Principle of Time-Resolved Fluorometry

E. SOINI [1]

Lanthanides as Fluorescent Labels

Background interference is the most important problem in any fluorometric determination of material of biological origin. This is due to the fact that the analyte concentrations to be measured are often very low and the sample contains other fluorescent material besides the label itself, thus background interference often overlaps the analyte-specific signal. The situation is illustrated in Fig. 1, which shows that the fluorescence emission of blood serum covers the whole visible wavelength range of the electromagnetic spectrum, including excitation and emission peaks of fluorescein isothiocyanate (FITC) – the most common and traditional fluorescent probe.

Chelates of europium and terbium are a potential alternative to radioisotopic and established nonisotopic labels in immunoassays and other specific ligand assays. Europium and terbium form highly fluorescent chelates with many organic ligands. The main advantages of using lanthanide chelates as fluorescent probes in biological material are their high quantum yield, exceptionally large Stokes shift, narrow emission peaks, and optimal emission and excitation wavelengths. The emission peak of europium is above the end point of serum interference. Because of the extraordinarily large Stokes shift, the sharp emission peak of europium (614 nm) can easily be separated from scattering caused by excitation (340 nm) and from the interfering fluorescence of serum (400–600 nm). The excitation energy is absorbed by the ligand and then transferred to the chelated metal ion, which emits the energy at its characteristic wavelength. A good fluorescent lanthanide chelate is free from fluorescence emission of the ligand itself, and the energy is transferred to the lanthanide without any radiative transition.

The difference between europium and terbium is mainly based on their different excitation and emission wavelengths and their fluorescent properties in solution. The strongest fluorescent emission peaks of terbium chelates are at 490 and 545 nm. Terbium forms more stable chelates in water and, unlike europium, it requires no protection against quenching by water. The use of terbium as a label presents problems because the excitation maximum of terbium chelates is below 300 nm and plastics and optical glass reduce the excitation below 320 nm. Europium chelates are particularly suitable as probes as far as their fluorescence spectrum is concerned. Their absorbance at 340 nm is on the same level as that of the best organic fluorochromes, although their quantum yield is often lower. Their

1 Wallac Biochemical Laboratory, P.O. Box 10, SF-20101 Turku 10

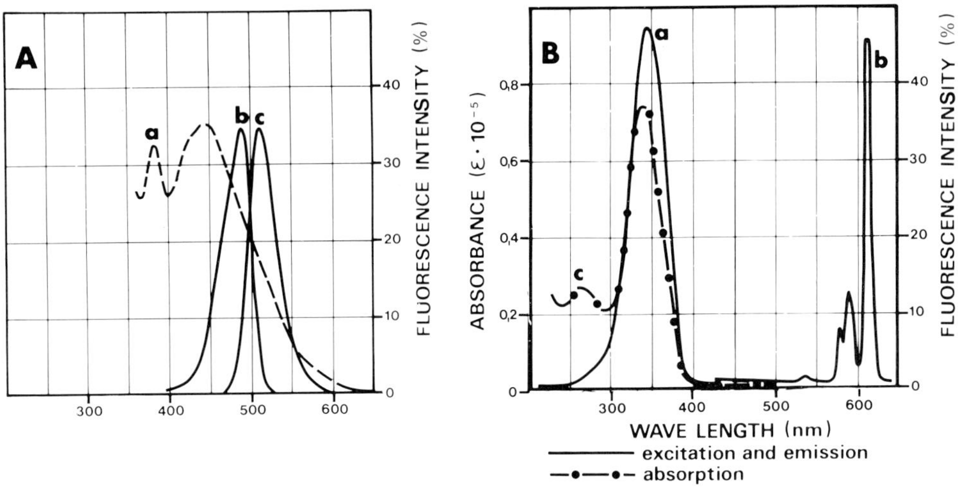

Fig. 1. A *a*, fluorescence emission of serum; *b*, excitation of FITC; *c*, emission of FITC. **B** *a*, excitation of europium; *b*, emission of europium; *c*, absorption of europium

excitation band is relatively broad and dependent solely on the ligand, therefore it is possible to use a broad excitation bandwith and consequently a lower excitation light source intensity. As already mentioned, the advantage lies in the very large Stokes shift (270 nm). In addition, europium chelates have a very narrow light emission peak, which is why background in the region of the fluorescence emission is very low and good separation of the excitation and emission peaks can be achieved with interference filters.

Time-Resolved Fluorometry of Lanthanide Chelates

Owing to natural fluorescence from various compounds in biological samples such as serum proteins, conventional fluorescent probes suffer from serious limitations of sensitivity. With lanthanide chelates, however, it is possible to reduce considerably the background level by selective detection of long-decay fluorescence. The fluorescence decay time of lanthanide chelates is often in the order of

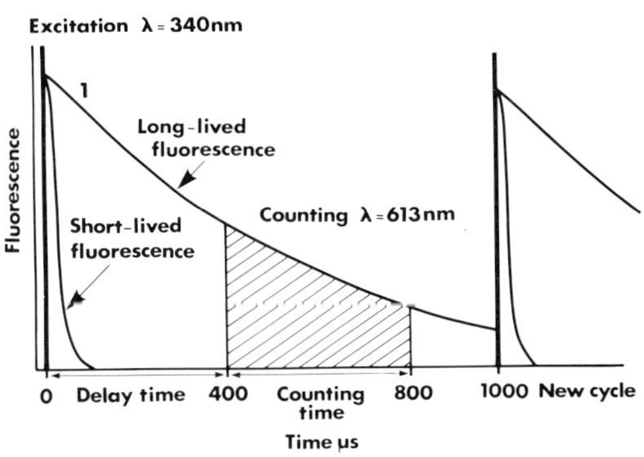

Fig. 2. Selected counting time (400–800 ms) of Eu chelate fluorescence emission (*1*) following an excitation light pulse. Background fluorescence ceases after a few nanoseconds

10–1,000 μs, whereas the decay time of natural fluorescence in a typical biological sample is in the order of 1–20 ns. For this reason, the pulsed light source, time-resolved fluorometers used with lanthanide chelates is potentially several orders of magnitude more sensitive than conventional fluorometers [1].

A pulsed light is used to produce an excitation pulse of short duration compared with the decay time of the lanthanide chelate. Certain europium chelates have a decay time of 500 μs and the interfering rapid decay of natural fluorescence can be discriminated against by not activating the detector until a delay of, e.g., 400 μs has elapsed after excitation of the lanthanide chelate (Fig. 2). As a result, the interfering background fluorescence is reduced by two to three orders of magnitude [2] and a time-resolved fluorometer covers a concentration range of lanthanide chelates which is not measurable using a conventional filter fluorometer or spectrofluorometer.

Time-Resolved Fluorometer

An automatic time-resolved fluorometer has been developed in our laboratory. This instrument, named the 1 230 Arcus Fluorometer, is equipped with a sample changer for sample tube racks and a built-in microcomputer. Two different types of sample holder can be used in the 1 230 Arcus Fluorometer: (1) a cuvette rack for ten sample tubes (outer diameter 12 mm, length 80 mm), and (2) a rack which can hold one 12-well microtitration strip. There is space for 30 racks on the con-

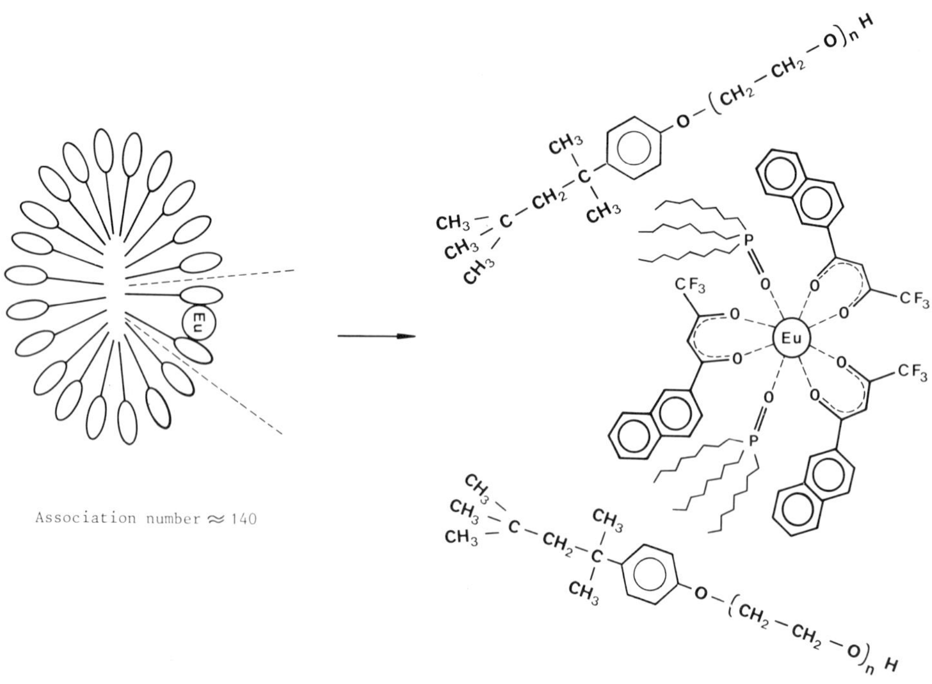

Fig. 3. Hypothetical structure of a fluorescent europium chelate. Europium is chelated with a β-diketone and trioctylphosphine oxide within a micellar structure of Triton X-100

veyor table, and thus the maximum sample capacity is 300 cuvettes or 360 microtitration wells. The two rack types can be mixed together. A xenon flashlamp is used as the pulsed excitation light source of the Arcus Fluorometer. The flash duration of the lamp is of the order of a few microseconds and a repetition rate of about 1,000 flashes/s is achieved (see Fig. 2).

The optimal excitation wavelength for europium chelates is 340 nm but other values are possible with different interference filters. The built-in manually operated filter changer holds two standard 1-in. filter discs.

As excitation of the sample takes place through the side of the tube, the generated fluorescence light is measured through the tube bottom. The right angle between the excitation source and the detector improves the distinction between the source and fluorescent light. A side window-type photomultiplier tube is operated with a red sensitive multialkali cathode in the single-photon mode. The output pulses of the photomultiplier tube are amplified by a fast preamplifier and then fed into the 50 MHz prescaler which divides them by 100 to lower the pulse frequency for the main scaler. After the counting period is completed, the microprocessor reads the contents of the scalers and stores the accumulated counts in the memory.

Time-Resolved Fluoroimmunoassay

Time-resolved fluorometry as applied to lanthanide chelates offers a new highly sensitive method for immunoassays called time-resolved fluoroimmunoassays (TR-FIA). If a europium ion is used as the label in immunoassay, it has to be strongly bound to an immunoreactive component. Strong europium binding can be achieved with different polycarboxylic acids such as EDTA, EGTA, HEDTA, and DTPA. An additional requirement for europium labeling is the introduction of a functional group to molecules such as EDTA, which allows conjugation to the immunoreactive component. Two useful chelate derivatives, aminophenyl-Eu-chelate and isothiocyanatophenyl-Eu-chelate, have been synthesized. The former is conjugated mainly to tyrosine residues of the protein (antibody) after diazotization, and the latter is bound to free amino groups of the protein (antibody) [3]. However, these labeled antibodies are essentially nonfluorescent and are virtually labeled solely by a metal label. In the nonfluorescent form, the metal is strongly bound to the antibody and it thereby provides the good stability needed in the immunoreaction.

It is evidently difficult to combine strong chelating capacity with good absorption and energy transfer properties. That is why the europium labeling and measurement procedures are separate: An antibody labeled with a nonfluorescent europium chelate is used in the immunoreaction, and later the europium is released and coupled to another chelating agent, which results in intense fluorescence. The measurement of labeled antigen-antibody conjugates is thus carried out after the separation step by adding another chelating agent – called the enhancement solution – which dissociates the europium ion from the antibody chelate complex and forms a β-diketone chelate of high fluorescence [3]. As europium quantitation in immunoassays occurs in aqueous solution, it was necessary to design an enhancement solution in which the europium is measured in order to dissolve the β-diketone required for energy absorption and transfer, as well as simultaneously to protect the ion from quenching by water molecules.

The enhancement solution contains a nonionic detergent, Triton X-100, which dissolves the sparingly soluble organic component in the micellar phase and excludes the quenching water from the chelated europium ion. Insulation from the aqueous solvent is further optimized by adding a synergistic agent, trioctylphosphine oxide, to the europium chelate solution. Figure 3 presents a schematic outline of the hypothetical micellar structure into which the fluorescent europium chelate, consisting of the europium ion, β-diketone, and the trioctylphosphine oxide molecules, is solubilized. The high sensitivity and long lifetime of europium fluorescence indicate that the quenching effect of water has been almost completely eliminated. When the optimized solution is used for the enhancement of europium fluorescence and a time-resolved fluorometer is used for measurement, europium can be quantitated over a concentration range of 5×10^{-14} to 10^{-7} mol. Thus, at least the same order of sensitivity is obtained as is commonly given for ^{125}I when measured by radioimmunoassay [3]. This unique procedure is called the dissociation-enhanced lanthanide fluoroimmunoassay (DELFIA™). (DELFIA is a trade mark of LKB-Wallac.)

By combining the use of europium chelates as probes with time-resolved fluorescence measurement, a very high detection sensitivity is obtained for the following reasons: (1) The specific activity of the tracer can be increased by using stronger excitation; (2) background fluorescence can be avoided by using time-resolved fluorescence detection; (3) lanthanides are not normally present in biological samples; (4) lanthanides do not interact with the sample.

Applications of DELFIA

A number of DELFIA methods have been developed, including hCG [4, 5], HBsAg [6], hAFP [7], TSH [8], phospholipase A_2 [9], ferritin [10], cortisol, estradiol. Viral antigens and antibodies have been determined using "direct" or "indirect" immunometric assays [11, 12].

The DELFIA procedure for peptides and proteins, hormones, as well as for viral and bacterial antigens and antibodies is a solid phase, two-side fluoroimmunoassay based on the sandwich technique. Two different antibodies, the first attached by adsorption to the wells of microtitration strips and the second labeled with europium chelate, are directed against two separate antigenic determinants of the analyte molecule. Undiluted serum, 25–50 µl, is sample to microtitration strips and the immunoreactions are carried out in one or two steps with 20 min–4 h incubation periods. In the final step, europium is released from the chelate by the enhancement solution and a new europium chelate is formed. The fluorescent signal is measured in the LKB-Wallac 1,230 ARCUS time-resolved fluorometer using a 1-s measuring period. The fluorescence signal is directly proportional to the concentration of the analyte. As an example, the standard curve of the HBsAg assay is shown in Fig. 4.

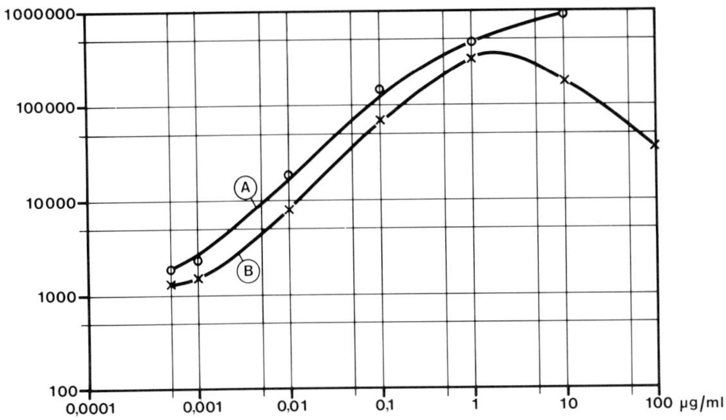

Fig. 4. Standard curve of DELFIA assay for HBsAg. *Curve A* is the normal two-step assay with 2 h + 2 h incubation periods, and *curve B* is a 50-min single-step assay

Discussion

Certain important performance parameters of time-resolved fluoroimmunoassays are summarized below and compared with other methods including enzyme immunoassay. The common important performance-limiting factors are the non-specific binding effects and the sample preparation error, which are not discussed in this context.

Time-resolved fluorometry offers an extraordinarily wide *dynamic range* ($1:10^4$) and linearity, which cover the whole concentration range of clinical importance without extra dilution of the sample. However, the high degree of linearity can be exploited only in the immunometric assays because the competitive binding assay principle works only within a very limited dynamic range.

A unique advantage of time-resolved fluorometry is the extremely high *specific activity* of the label in terms of emitted photons per second. This is due to the fact that the excited states of the fluorescent probes are created in the measuring position of the sample. The photon emission rate is a function of excitation intensity and decay time, and excitation itself can be augmented by increasing the intensity of the exciting radiation. The optimal decay time of fluorescent probes for time-resolved fluorometry is 10–1,000 µs. A typical specific activity of fluorescent probes is $< 10^7$ photons/s/pmol label compound. The specific activity of radioisotopic labels is normally in the order of one disintegration/s/pmol isotopic label. Consequently, the specific activity of fluorescent probes is more than a million times higher than that of radioisotopes and the lowest level of concentration can be measured in 1 s with satisfactory statistical precision.

Time-resolved fluoroimmunoassay has an extremely high *sensitivity* and provides a very high count rate with low instrumental background; concentrations as low as 10^{-17} mol/cuvette can be detected. However, it is important to realize that the limiting factor for the lowest level of detection in a practical time-resolved fluoroimmunoassay is not the instrumental background or the specific activity of the label, i.e., the signal to noise ratio, but the signal from the blank sample, which is determined in practice only by the nonspecific binding effects. An important prerequisite for the low blank signal is also the purity of the enhancement solution.

The extra step in the DELFIA *procedure,* addition of the enhancement reagent, is a disadvantage. The enhancement reaction is, however, very rapid and troublefree and is undoubtedly a simple step, unlike the activation of the enzyme reaction in photometric methods or the luminescence reaction in luminometric assays.

Coated-tube technology is well established in practical immunoassays and the europium-labeled antibody of DELFIA reacts with immobilized antibody-antigen on the solid phase. The *microtitration strips* have proved to be very practical and reliable, and they make sample processing simple and fast. Commonly available microtitration plate compatible sample preparation accessories, dispensors, and wash/aspirate devices can be used. Washing off the unreacted fraction in the microtitration wells is a step which requires particular care in the DELFIA method. Extremely high specific activity and a large excess of the reactant, which often contains more than 10^7 counts/s in one test tube, is used for achieving high sensitivity, large dynamic range, and rapid immunoreaction. As a consequence, the washing must be very efficient in order to reach a stable and low blank level. We normally recommend washing six times, using an appropriate washing/aspirating device.

The linear standard curve and the stable label provide simple *standardization* of the assay, and in this respect time-resolved fluorometry is superior to any photometric, luminometric, or radioisotopic method.

Good *intraassay precision* can be achieved with many different immunoassay methods. However, in photometric and especially nephelometric methods, the quality of cuvettes should be very carefully controlled. Conversely, quality errors which increase the scattered light emission are not important in time-resolved fluorometry. The fluorescent chelate is formed in solution after dissociation of the europium ion. This involves a significant advantage from the fluorometric point of view. The detection of the fluorescence signal in solution is much more stable and reproducible than in the solid phase on coated plastics, consequently the signal is not dependent on the homogeneity of the coating or on the position of the tube in the fluorometer detector. The photon collection geometry is also better and the tube gives a minor contribution to the background.

The temperature coefficient of the fluorescence signal when the enhancement solution is used is $-1\%/^{\circ}C$. For this reason, it is important that the samples are not exposed to a strong temperature change when loaded into the fluorometer. The inside temperature of the Arcus fluorometer is at most 2 °C above the room temperature.

The present DELFIA methodology is simple and straightforward for immunoassays which require high performance. It is highly insensitive to *interferences* from the sample because the methodology involves three points which may be sufficient to eliminate the interference problems: (1) Washing which cleans the cuvette of interfering constituents in the sample; (2) dissociation of the biochemically inert europium ion from the solid phase antibody-antigen complex, formation of a new chelate provides a stable and reproducible fluorescence; (3) the time-resolved fluorescence detection eliminates autofluorescence of the materials necessary in the system as well as possible interfering residues from the sample.

Acknowledgements. This paper has reviewed and referred to the results by a number of my colleagues at the LKB-Wallac Research Laboratories as well as to results of collaborative efforts. In the field of microbiological applications, the Department of Virology, University of Turku, Finland and the Finnish Red Cross Blood Transfusion Centre, Helsinki, Finland, have made a considerable contribution. I gratefully acknowledge the contributions of the Middlesex Hospital Medical School, Dept. of Molecular Endocrinology, London, UK; University of Turku, Dept. of Clinical Chemistry and Dept. of Pathology, Turku, Finland; University of Helsinki, Dept. of Obstetrics and Gynaecology, Helsinki, Finland; Medix Laboratories Ltd., Kauniainen, Finland; and Farmos Group, Diagnostics Factory, Oulunsalo, Finland (formerly Nordiclab).

References

1. Soini E, Hemmilä I (1979) Fluoroimmunoassay: Present status and key problems. Clin Chem 25:353–361
2. Soini E, Kojola H (1983) Time-resolved fluorometer for lanthanide chelates – a new generation of nonisotopic immunoassays. Clin Chem 29:65–68
3. Hemmilä I, Dakubu S, Mukkala V-M, Siitari H, Lövgren T (1984) Europium as a label in time-resolved immunofluorometric assays. Anal Biochem 137:335–343
4. Pettersson K, Siitari H, Hemmilä I, Soini E, Lövgren T, Hänninen V, Tanner P, Stenman U-H (1983) Time-resolved fluoroimmunoassay of human choriogonadotropin. Clin Chem 29:60–64
5. Stenman U-H, Myllynen L, Alftan H, Seppälä M (1983) Ultrarapid and highly sensitive time-resolved fluoroimmunometric assay for chorionic gonadotropin. Lancet 17:647–649
6. Siitari H, Hemmilä I, Soini E, Lövgren T, Koistinen V (1983) Detection of hepatitis B surface antigen using time-resolved fluoroimmunoassay. Nature 301:258–260
7. Suonpää M, Lavi J, Hemmilä I, Lövgren T (1985) A new sensitive assay of AFP using time-resolved fluorescence and monoclonal antibodies. Clin Chem Acta 145:341–348
8. Pettersson K, Söderlund J, Lövgren T (1984) A highly sensitive one incubation assay of hTSH using time-resolved fluorescence and monoclonal antibodies. Clin Chem (submitted for publication)
9. Eskola JU, Nevalainen TJ, Lövgren TN-E (1983) Clin Chem 29:1777
10. Mäentausta O, Lövgren T (1984) Time-resolved fluoroimmunoassay of ferritin. Clin Chem Acta (submitted for publication)
11. Meurman OH, Hemmilä IA, Lövgren TN-E, Halonen PE (1982) Time-resolved fluoroimmunoassay: a new test for rubella antibodies. J Clin Microbiol 16:920–925
12. Halonen P, Meurman O, Lövgren T, Hemmilä I, Soini E (1983) Detection of viral antigens by time-resolved fluoroimmunoassay. In: Bachman PA (ed) New developments in diagnostic virology. Springer, Berlin Heidelberg New York, pp 133–146
13. Hemmilä I, Viljanen M, Lövgren T (1984) Time-resolved fluoroimmunoassay (TR-FIA) of tetanus antibodies. Anal Chem 317:737–738

Microbiologic Applications of Fluoroimmunoassay

J. W. CHIDLOW [1] and D. S. SMITH [2]

The term immunofluorescence has been applied to the use of fluorescent-labelled antibodies for the specific detection and localisation of antigens in cells and tissues, as first described by Coons et al. (1942). The technique was placed on a sound practical basis by Coons and Kaplan (1950) with the introduction of fluorescein as label, and from that date immunofluorescence has been an invaluable routine tool for the study of tissue structure and pathology. Fluorescent antibody techniques are also firmly established in diagnostic microbiology and were well described some years ago by Cherry and Moody (1965) and more recently in standard clinical laboratory manuals such as that from the US Center for Disease Control (Jones et al. 1977).

These techniques have established the usefulness and reliability of fluorescence methods in clinical laboratory diagnostic procedures. Classical immunofluorescence, however, is essentially a qualitative technique, dependent for its success upon the skills of the cytologist or histologist and use of the fluorescence microscope, as well as upon the specificity of the antigen : antibody interaction.

The exploitation of immunological reagents in quantitative analysis developed in parallel with immunofluorescence. The first, most widely applied, and for long the most sensitive of these techniques is the radioimmunoassay (Yalow and Berson 1960). The specificity of the antigen : antibody interaction, allied to the high detection sensitivity of radioactive labels, has enabled the quantitation of an immense variety of analyses even at very low concentrations in very complex biological sample matrixes, and has revolutionised many areas of clinical science, most notably endocrinology.

The present review will consider present-day and likely future developments in fluorescence technology in quantitative analysis, and potential applications in the clinical microbiology laboratory. The two concepts of specificity and detection sensitivity are stressed.

Early descriptions of quantitative microbiologic assay techniques employing fluorescent-labelled antibodies were based on membrane filtration to separate antigen : antibody complexes from unbound antibody (Neurath 1965) or centrifugation to separate immune precipitates containing labelled antibody (Head 1962). Direct fluorimetry of cell suspensions has also been applied (Gillis and Thompson 1978), and the potential applications of flow cytometers in quantitative analysis have been recognised (Lisi et al. 1982). Attempts to achieve quanti-

1 Efamol Limited, Woodbridge Meadows, Guildford, Surrey GU1 1BA, Great Britain
2 St. Bartholomew's Hospital, London, Great Britain

tation in fluorescence microscopy have included the use of micro-spheres (Deelder and Ploem 1974) and plane surfaces (Sedlacek et al. 1981) uniformly coated with antigen. Systems for the complete automation and quantitation of classical immunofluorescence detection have also been devised (Mansberg and Kusnetz 1966).

Further developments in quantitative fluorimetric techniques with applications in microbiology have been stimulated in large part by a desire to produce purpose-designed assay systems that provide optimum technical convenience and reliability in routine use. Several commercial companies have been active in this field and since the late 1970s a wide range of systems applicable to many aspects of clinical microbiology has become available. These include systems for the detection and quantitation of bacterial and viral antigens and antibodies; quantitative antibiotic assays with applications in therapeutic drug monitoring; and fluorimetric systems for antimicrobic susceptibility testing.

Such methods are often available with instrument control of critical steps in the procedure, and may incorporate dedicated data reduction and reporting systems.

The FIAX system (Harte 1981) is based on the use of sampler sticks carrying a small disc whose surface may be coated with specific antigen or antibody. The sticks facilitate reagent addition, washing and separation steps in a variety of diagnostic procedures involving fluorescein-labelled antibodies or antigens. Upon completion of the assay protocol, labelled reagent that is specifically bound to the coated disc is quantitated by insertion of the stick into a purpose-designed front-surface fluorimeter. FIAX methods have been applied to the detection of antibodies to a wide range of infectious agents, including rubella, toxoplasma, herpes, cytomegalovirus, schistosomiasis, *Candida*, gonorrhea, and legionella organisms (Harte 1981). In addition the technique has been adapted for the quantitative determination of aminoglycosides in serum, as an aid to the monitoring of antibacterial therapy with these potentially toxic drugs.

A wide variety of other approaches have been applied to the development of fluoroimmunoassays for therapeutic drug monitoring. Most rely on the use of fluorescent-labelled drug as a tracer reagent. The Immunostics systems is based on the use of antibodies coupled to magnetisable solid-phase particles which may be rapidly and reliably separated from assay incubation mixtures by the field of a simple permanent magnet (Johnson et al. 1982). Other methods avoid the need for a separation step ("homogeneous" assays). For example, the SLFIA system (Burd 1981) employs drug labelled with a fluorogenic enzyme substrate such that fluorescence signal is generated only upon addition of the appropriate cleaving enzyme; such activity is inhibited when the tracer is antibody bound. The TDX system (Jolley 1981) depends on the detection of the increase in the polarisation of fluorescence when a labelled drug molecule (a small species which undergoes rapid Brownian rotation in solution) is bound by antibody (where the much slower rotation of the large immunoglobulin molecule is experienced). These fluoroimmunoassays have been applied to all the clinically important aminoglycosides, and also recently to vancomycin (Filburn et al. 1983).

In addition to immunoassays, fluorimetric methods have wider applications in clinical microbiology. One example is the detection of bacterial enzymes by fluorescence release from fluorogenic substrates, a very sensitive measure of bacterial growth. This method has been applied to the quantitation of viable bacteria in the course of automated antimicrobic susceptibility testing in the Sensititre MIC system. This system is particularly elegant in that the fluorogenic substrate is included in the culture medium at the time of inoculation of the Sensititre MIC microtitre tray. These trays contain doubling dilutions of various antimicrobics which during incubation permit or inhibit bacterial growth according to the drug concentration and the susceptibility of the organism. Fluorogenic substrates are added to the Sensititre test broth at the time of inoculum preparation by addition to the broth tube of a reagent-impregnated paper strip. The substrates are rapidly released into the broth and are then sensitive to bacterial enzyme activity. Three enzyme functions are measured: phosphatase, esterase, and peptidase, carefully chosen to be generally representative of clinically significant non-fastidious bacteria. Release of fluorescence during the incubation period (5 h or 18 h depending on the reading time required) is derived entirely from bacterial surface enzyme activity, a direct measure of bacterial viability. The MIC result is obtained by reading the resulting fluorescence directly in the microtitre tray using a multireading fluorimeter, from which the raw fluorescence data are directly transcribed to a microcomputer for analysis and reporting. As would be expected from a method based on direct quantitation of viable bacteria, results from the Sensititre system are accurate and reliable, correlating at around the 95% level with accepted reference procedures.

Such application of fluorescence for bacterial quantitation in an MIC system suggests further possibilities in quantitative microbiology. The screening of urine samples, for example, is a major work load in routine bacteriology, and interpretation is heavily dependent on assessment of organism density. It is likely that organism quantitation as above would provide an accurate and sensitive screening method, with the immediate advantage of multi-sample reading on an instrument such as the Sensititre fluorimeter.

Fluorimetry in general has the advantage of relatively high sensitivity as compared, for example, with absorbance measurement. Those techniques that employ fluorogenic enzyme substrates may benefit from the concept of "amplification" of the signal, since one enzyme may cycle many substrate molecules and so liberate many fluorophores. Ultimately, however, sensitivity is limited not only by the magnitude of the specific signal (i.e. that produced by enzyme activity or from a specific fluorescent-labelled immunochemical reagent) but by the background or "noise" level. It is the signal-to-noise ratio that will determine overall sensitivity and thus control the scope for practical application of a particular technique (Ekins 1976). In the enzymatic methods, the nature of the substrate is crucial: it must be readily cleaved by the enzyme it detects but robust to all other attacks. Correct and careful design of such substrate molecules is the key to the exploitation of this technique in bacterial screening of urine samples and perhaps for the assessment of other enzymic changes taking place during the infectious disease process. Future advances in substrate design may well also lead to methods applicable to species identification of bacterial isolates and to the screening of blood cultures for bacteraemia.

In fluorimetry itself, background signal may be contributed by a variety of factors, including naturally fluorescent components of biological materials, impurities in assay reagents and light-scattering effects. It is these, rather than the intrinsic sensitivity of fluorescence detection, which control fluorimetric detection limits in practice (Matthews and Lytle 1979). In this respect, fluorimetry suffers in comparison with the detection of radioactive materials, where natural background levels are very low, and for this reason conventional fluoroimmunoassay methods have not attained the same sensitivity as radioimmunoassays employing such labels as the gamma emitter iodine 125.

Much enhanced fluoroimmunoassay sensitivity can now be achieved with the advent of the technique of time-resolved luminescence. This depends on the use of labels selected for their exceptionally long luminescence lifetimes. When a sample is illuminated with a short flash of excitation light (typically of a few microseconds duration), light scattering (which occurs essentially instantaneously) will follow the time course of the exciting pulse and fluorescent background will decay with a lifetime typically of up to a few tens of nanoseconds. If a label has a lifetime in the microsecond range, time-gating of the detection system permits measurement of the specific signal long after the exciting flash and background signals have decayed away. Particularly suitable as labels are chelates of certain rare-earth metal ions which may have fluorescence lifetimes of hundreds of microseconds (Wieder 1978; Soini and Kojola 1983).

Since background is eliminated, the sensitivity of fluorimetric detection is improved by several orders of magnitude, especially if photon-counting detection is applied: using simple purpose-designed instrumentation (Soini and Kojola 1983), rare-earth chelates may now be detected with sensitivity similar to that of iodine 125, but with much shorter measurement times (seconds rather than minutes). With the use of rare-earth chelate-labelled antibody reagents, sensitive time-resolved fluoroimmunoassays for rubella antibodies (Meurman et al. 1982) and for hepatitis B surface antigen (Siitari et al. 1983) have been developed. Other candidates as labels for time-resolved techniques are phosphorescent molecules, such as erythrosin (tetraiodofluorescein), whose emission lifetimes may also lie in the microsecond range or longer. Because phosphorescence in liquid solution is highly susceptible to oxygen quenching, the practical application of such "phosphoroimmunoassays" has depended on the development of a method for in situ chemical deoxygenation of the immunochemical assay mixtures (Sidki and Smith 1982).

A technique that is closely related to fluorescence is chemiluminescence: highly sensitive immunoassays employing chemiluminescent labels have been described, an application in microbiology being the detection of hepatitis B surface antigen using specific antibodies labelled with a luminol derivative (Schroeder et al. 1981). As compared with fluorescence, chemiluminescence detection requires the addition of extra chemical reagents to stimulate the light-emission process, which is time dependent; thus a chemiluminescence measurement in immunoassay is less convenient and precise than a fluorescence measurement, unless purpose-designed apparatus is employed (Schroeder et al. 1981). Also, each chemiluminescent label can emit only one photon (in the same way that a radioactive label can decay only once), while fluorescence may take advantage of the "amplification" concept since each fluorescent label can undergo many excitation/emission cycles and thus generate many individual "signals."

In the future development of fluoroimmunoassay technology, two important pathways can be foreseen. First, assays that involve a separation step ("heterogeneous") will increasingly be replaced by homogeneous methods, as is already occurring in therapeutic drug monitoring. Homogeneous assays are technically simpler to perform and hence liable to be more precise and robust, and more suited to automation. Second, labelled-antibody as opposed to labelled-antigen reagents will be more widely introduced. Labelled antibodies are typically employed in "reagent excess" (Ekins 1976), which in principle will lead to faster and more robust assays with maximal sensitivity that is not so critically dependent upon antigen-binding affinity as that of the "limited reagent" assays based on a labelled-antigen tracer reagent (such as conventional radioimmunoassay). The availability of monoclonal antibodies is likely to facilitate and stimulate the application of labelled-antibody techniques, in addition to the other analytical advantages conferred by these reagents.

The analytical advantages offered by monoclonal reagents derive from their true monospecificity. This is particularly applicable to the diagnosis of subgroups of viral agents, for example hepatitis A, B or "non-A/non-B." Reagents directed against different antigenic sites on the same molecule can also provide an approach to assays such as the Hybritech "Tandem" system. This method makes use of capture antibody and labelled antibody specific for different sites, and thus the assay can be carried out in a single step without interference between the two antibody binding stages. Research capability in generating monoclonal reagents is now widespread, and no doubt will soon provide a range of reagents offering new diagnostic specificities and improved assay performance.

However, reagent specificity is now not confined to immunological reagents. There is also broad interest in the development and use of DNA probes, defined strands of labelled DNA used as a detector for complementary sequences. As with antibody reagents, detection sensitivity is defined by the choice of label. Much of the early work made use of radio-isotope labels, but for routine use there is also great interest in enzyme or fluorescent labels because of their greater convenience in use.

All methodologies of course require instrumentation. Multi-sample reading fluorimeters are now available as well as automatic sample and reagent handling instruments. In general these employ well established principles of electronics, optics, and precision engineering. There are, however, various new approaches in development.

The first is the use of multichannel detectors which permit the rapid acquisition of complete spectral data. For example, a system for bacterial identification and differentiation has been developed based on the selective adsorption of several fluorescent dyes whose characteristic spectral profiles may be resolved from the three-dimensional excitation/emission matrixes measured in a video fluorimeter (Shelley et al. 1983).

The second is the use of fibre optics in fluorimetry (Sepaniak et al. 1983). The application reported involves the use of a pulsed laser to measure intrinsic molecular fluorescence of the anti-tumour drug doxorubicin, but the concept is relevant to any measurement of fluorescence in small volume samples, or perhaps in vivo.

The third development, still at a very early stage, is the use of "Biosensors." This term refers to devices capable of receiving and interpreting an electronic signal as the direct result of a biochemical event. Prospects for the use of chemical sensors based on fibre optics are already established in research use, and have recently been thoroughly reviewed (Seitz 1984) in both concept and application. In the not too distant future we may be testing "in vivo" or in micro samples, using biosensors to measure fluorescence in response to pulsed laser excitation taken to the investigation site by a fibre optic fluoroprobe.

Looking ahead to the near future, perhaps the most immediate question for the microbiologist is what is the role of the next generation of diagnostic technology in clinical microbiology. Advances in reagent specificity, coupled with increased detection sensitivity and signal data reduction, will offer faster testing of more complex mixtures. Sensitive detection of antigenic or enzymic activity in mixed populations offers the prospect for direct testing on patient specimens without intermediate culture. Developments firmly founded today in research terms hold the prospect of detection, identification, and susceptibility testing of bacteria and other micro-organisms in mixed culture. The diagnostic value of such tests and the clinical situations appropriate to their use are decisions for the clinical microbiologist.

References

Burd JF (1981) The homogeneous substrate-labelled fluorescent immunoassay. Methods Enzymol 74:79–87

Cherry WB, Moody MD (1965) Fluorescent-antibody techniques in diagnostic bacteriology. Bacteriol Rev 29:222–250

Coons AH, Kaplan MH (1950) Localisation of antigen in tissue cells. II. Improvements in a method for the detection of antigen by means of fluorescent antibody. J Exp Med 91:1–13

Coons AH, Creech HJ, Jones RN, Berliner E (1942) The demonstration of pneumococcal antigen in tissues by the use of fluorescent antibody. J Immunol 45:159–170

Deelder AM, Ploem JS (1974) An immunofluorescence reaction for *Schistosoma mansoni* using the defined antigen substrate spheres (DASS) system. J Immunol Methods 4:239–251

Ekins RP (1976) General principles of hormone assay. In: Loraine JA, Bell ET (eds) Hormone assays and their clinical application, 4th edn. Churchill Livingstone, Edinburgh, pp 1–72

Filburn BH, Shull VH, Tempera YM, Dick JD (1983) Evaluation of an automated fluorescence polarization immunoassay for vancomycin. Antimicrob Agents Chemother 24:216–220

Gillis TP, Thompson JJ (1978) Quantitative fluorescent immunoassay of antibodies to, and surface antigens of, *Actinomyces viscosus*. J Clin Microbiol 7:202–208

Harte RA (1981) Fluoroimmunometry and the FIAX system. In: Kaplan LA, Pesce AJ (eds) Nonisotopic alternatives to radioimmunoassay. Dekker, New York, pp 125–141

Head WF (1962) Immunochemical study of the diphtheria toxin-fluorescent antitoxin system. J Pharm Sci 51:662–665

Johnson C, Smith DS, Shaw EJ (1982) Fluoroimmunoassay of netilmicin. In: Richardson RG (ed) Netilmicin – a recent advance in aminoglycoside therapy (Royal Society of Medicine International Congress and Symposium Series, no 50), Royal Society of Medicine, Academic, London, pp 47–54

Jolley ME (1981) Fluorescence polarization immunoassay for the determination of therapeutic drug levels in human plasma. J Anal Toxicol 5:236–240

Jones GL, Herbert A, Cherry WB (1977) Fluorescent antibody techniques and bacterial applications. In: Centre for Disease Control laboratory manual. US Public Health Service, Atlanta, pp 1–11

Lisi PJ, Huang CW, Hoffman RA, Teipel JW (1982) A fluorescence immunoassay for soluble antigens employing flow cytometric detection. Clin Chem Acta 120:171–179

Mansberg HP, Kusnetz J (1966) Quantitative fluorescence microscopy: fluorescent antibody automatic scanning techniques. J Histochem Cytochem 14:260–273

Matthews TG, Lytle FE (1979) Blank limitations in laser excited solution luminescence. Anal Chem 51:583–585

Meurman OH, Hemmila IA, Lovgren TN, Halonen PE (1982) Time-resolved fluoroimmunoassay: a new test for rubella antibodies. J Clin Microbiol 16:920–925

Neurath AR (1965) Fluorometric estimation of antigens (antibodies). Z Naturforsch 20b:974–976

Schroeder HR, Hines CM, Osborn DD, Moore RP, Hurtle RL, Wogoman FF, Rogers RW, Vogelhut PO (1981) Immunochemiluminometric assay for hepatitis B surface antigen. Clin Chem 27:1378–1384

Sedlacek HH, Grigat H, Renk T, Seiler FR (1981). The fluorescence immunoassay using plane surface solid phases (FIAPS). Methods Enzymol 74:87–105

Seitz WR (1984) Chemical sensors based on fibre optics. Anal Chem 56:16A–34A

Sepaniak MJ, Tronberg BJ, Eastham JF (1983) Optical fiber fluoroprobes in clinical analysis. Clin Chem 29:1678–1682

Shelley DC, Warner IM, Quarles JM (1983) Characterization of bacteria by mixed-dye fluorometry. Clin Chem 29:290–296

Sidki AM, Smith DS (1982) Assay processes and materials therefor. U.K. Patent Application no 8227536

Siitari H, Hemmila I, Soini E, Lovgren T, Koistinen V (1983) Detection of hepatitis B surface antigen using time-resolved fluoroimmunoassay. Nature 301:258–260

Soini E, Kojola H (1983) Time-resolved fluorometer for lanthanide chelates – a new generation of nonisotopic immunoassays. Clin Chem 29:65–68

Wieder I (1978) Background rejection in fluorescence immunoassay. In: Knapp W, Holubar K, Wick G (eds) Immunofluorescence and related staining techniques. Elsevier/North-Holland, Amsterdam, pp 67–80

Yalow RS, Berson SA (1960) Immunoassay of endogenous plasma insulin in man. J Clin Invest 39:1157–1175

Detection of Viral Antigens by Time-Resolved Fluoroimmunoassay

P. Halonen [1], C. Bonfanti [1, 2], T. Lövgren [3], I. Hemmilä [3], and E. Soini [3]

Introduction

The principle of time-resolved fluorescence is reported in another chapter of this book and elsewhere by Soini (Soini and Hemmilä 1979; Soini and Kojola 1983). We have applied this technology to the detection of viral antigens directly in clinical specimens by indirect anti-species time-resolved fluoroimmunoassay (TR-FIA) and by direct one-incubation TR-FIA. The indirect TR-FIA was developed for the detection of rotavirus and adenovirus in stool specimens of patients with acute gastroenteritis and for seven respiratory viruses (influenza A and B, parainfluenza types 1, 2, and 3, adenovirus, and respiratory syncytial virus) in nasopharyngeal aspirates of patients with acute respiratory disease. More recently, a direct one-incubation TR-FIA has been developed, and the detection of rotavirus and adenovirus in stool and influenza A virus and adenovirus in nasopharyngeal aspirates have been used as model systems.

The results of the indirect and direct TR-FIAs and the comparison of these with indirect radioimmunoassays and enzyme immunoassays are described in this report.

Materials and Methods

Patients and Specimens

The specimens tested were obtained from the specimen library of our diagnostic unit. They had been sent mainly from hospitalized children in pediatric clinics and infectious disease hospitals but they also included specimens from young adults and elderly patients. Stool specimens were obtained from patients with acute gastroenteritis and diluted 1 : 10 in saline. After low-speed centrifugation, the supernatants were diluted 1 : 2 in RIA and EIA diluent (20% inactivated fetal bovine serum, 2% Tween 20, and 0.1% sodium azide in PBS, pH 7.2), tested in the diagnostic unit for rotavirus and adenovirus by indirect RIA or EIA. The original 1 : 10 diluted supernatant was stored at −20 °C. When tested by indirect TR-FIA they were again diluted 1 : 2 with the same diluent as for RIA and EIA; for the

1 Department of Virology, University of Turku, SF-20520 Turku 52
2 Institute of Microbiology, University of Brescia, I-25100 Brescia
3 Wallac Biochemical Laboratory, P.O. Box 10, SF-20520 Turku 10

direct one-incubation TR-FIA they were diluted 1:5 (final dilution 1:50) with the same diluent as above supplemented with 2% inactivated normal rabbit serum prescreened for rotavirus and adenovirus antibody.

Nasopharyngeal aspirate specimens from patients with acute respiratory disease were collected and treated as reported by Sarkkinen in this book. The original 1:5 or 1:10 diluted and sonicated specimen stored at −20 °C was tested without further dilution by indirect and direct TR-FIA.

Indirect Radioimmunoassays and Enzyme Immunoassays

The principle of these assays is the standard four-layer anti-species immunoassay as reported by Sarkkinen in this book. In the rotavirus and adenovirus EIAs and RIAs, each of the three incubations was for 1 h at 37 °C followed by 30 min with the substrate in EIA, whereas in the respiratory virus antigen assays the first incubation was overnight (about 16 h) at 37 °C and the other incubations as indicated above. Low positive specimens were checked by a blocking type of confirmatory test (Halonen et al. 1980).

Indirect Time-Resolved Fluoroimmunoassays

Principle. The immunological principle of the indirect TR-FIA was identical with that for the indirect RIA and EIA. The technique has been reported in detail elsewhere (Halonen et al. 1983). Briefly, polystyrene beads were used as the solid phase and the total immunoglobulin (Ig) fraction of guinea pig antivirus hyperimmune serum was employed as the primary capturing antibody on the solid phase. The secondary antibody was the Ig fraction of rabbit antivirus hyperimmune serum, and the indicator antibody was immunosorbent-purified sheep anti-rabbit IgG labelled with Eu chelate.

Eu Chelate Labels. Aminophenyl-EDTA-Eu was synthesized and conjugated to antibody by diazoreaction. More recently, the labelling was done using isothiocyanatephenyl-EDTA. Eu is actually in a nonfluorescent form in the label, but after the immunoreaction is completed and the label is bound on the solid phase, the Eu ion is dissociated with enhancement solution. A fluorescent Eu chelate is formed in solution which is quantitatively measured with a single photon counting time-resolved fluorometer.

Antiviral Antibodies. Viral hyperimmune sera were prepared by immunizing guinea pigs and rabbits with purified adenovirus hexon antigen (Halonen et al. 1980), purified Nebraska calf diarrhoea virus grown in cell culture (Sarkkinen 1981), or nucleocapsids and other structural proteins of respiratory syncytial virus (RSV), influenza A and B, and parainfluenza types 1, 2, and 3 (Sarkkinen et al. 1981a–c). Sodium sulfate precipitated Ig fractions of hyperimmune sera were used, and the optimal concentrations of guinea pig Ig on solid phase used as the catching antibody and rabbit Ig used as the secondary antibody were pretested by box titrations.

Procedures. Polystyrene beads (6.4 mm in diameter; Precision Plastic Ball Co., Chicago, Ill.) were coated with antiviral guinea pig Ig by incubating the beads overnight at room temperature in an optimal dilution of Ig in carbonate buffer, pH 9.6. The coated beads were stored in the antibody solution at 4 °C for 1–2 weeks. The 1 : 10 diluted specimen was pipetted as 200 µl aliquots into disposable polystyrene tubes, and the beads coated with antibody were then added. After 1 h (stool) or overnight (sonicated nasopharyngeal aspirate) incubation at 37 °C, the beads were washed, rabbit Ig was added, and the tubes were incubated again for 1 h at 37 °C, washed and the Eu-chelate-labelled anti-rabbit IgG antibody added. After the final incubation of 1 h at 37 °C and washing, the beads were transferred into clean tubes and the enhancement solution (0.1 *M* acetate buffer adjusted to pH 3.2 with potassium hydrogen phthalate and containing 15 µ*M* 2-naphthoyl-trifluoroacetone, 50 µ*M* tri-*n*-octylphosphine oxide and 0.1% Triton X-100) was added. The tubes were agitated gently and kept for 15 min at room temperature before measurement in the 1230 "Arcus" fluorometer equipped with a Xenon flash lamp. The length of the flash was 1 µs and the excitation wavelength 340 nm. After a delay time of 400 µs single-photon emission was counted for 500 µs at 613 nm. A new cycle was started after 1 ms from the previous flash. The cycle was repeated 1,000 times during the overall counting time of 1 s. The reference virus, usually in concentrations of 100, 10, and 1 ng/ml, and the diluent control (0 ng/ml) were included in triplicate in each assay.

Expression of Results. The results were expressed as counts per second (cps) values. In each test the background fluorescence of the enhancement solution was measured in tubes with noncoated beads and the mean value was automatically deducted from the cps value of each test sample. The ratios of reference and test specimens in duplicate were automatically calculated and printed by the Hewlett Packard units connected with the 1230 "Arcus" Fluorometer.

Direct One-Incubation Time-Resolved Fluoroimmunoassay

The principle of the direct one-incubation TR-FIA is shown in Fig. 1. The Ig or IgG fraction of antirotavirus, antiadenovirus (hexon antigen) or anti-influenza virus rabbit hyperimmune serum was used for coating the polystyrene microtitration strips at concentrations of 0.25 to 0.5 µg/well in carbonate buffer, pH 9.6. The coated strips were stored for as long as a few weeks at 4 °C. Before use they were washed twice with the washing solution (0.9% NaCl with 0.01% Tween 20). The diluent was the same as for indirect TR-FIA but 2% heat inactivated normal rabbit serum was added. Stool specimens at 1 : 50 dilution or nasopharyngeal aspirates at 1 : 5 dilution and the Eu-chelate labelled indicator antibody at concentrations of 50–100 ng/100 µl (the same rabbit Ig or IgG as on the solid phase for rotavirus and influenza A virus but immunosorbent purified rabbit antibody for adenovirus) were added to the coated wells and incubated for 1 h at 37 °C. After washing 6 times, the enhancement solution was added and the measurement was carried out in the same way as for the indirect TR-FIA. Using a noncoated strip, background fluorescence was measured from 12 wells and the mean (usually 600–700 cps) was automatically deducted from the cps values of the assay.

Layer	Immunoreagent		Incubation at 37°C
	Type	Concentration	
Anti-viral indicator antibody	Eu-chelate-labelled Ig or IgG fraction of rabbit anti-virus hyper-immune serum or immuno-sorbent purified antibody	25-100 ng/assay	1 hour
Specimen	Virus (structural protein)	1 ng - 1 mg/ml	
Catching antibody	Ig or IgG fraction of rabbit anti-virus hyper-immune serum	250-500 ng/well	
Solid phase	Polystyrene strip		

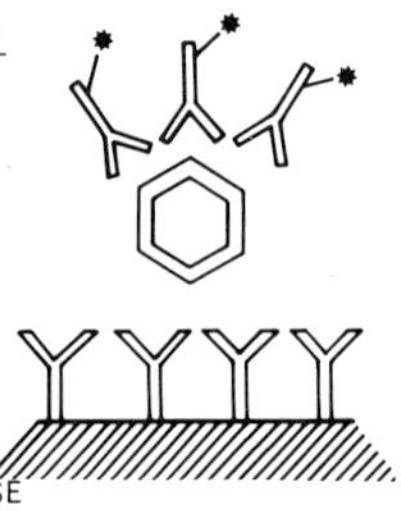

Fig. 1. Principle of direct one-incubation time-resolved fluoroimmunoassay for the detection of rotavirus and adenovirus in stool and influenza A and adenovirus in nasopharyngeal aspirates. Immunosorbent purified rabbit anti-hexon antibody was used as the label in the adenovirus assay

Results

Detection of Rotavirus and Adenovirus in Stool Specimens and Respiratory Viruses in Nasopharyngeal Aspirates by Indirect TR-FIA

When purified rotavirus and hexon antigen were tested by indirect EIA and indirect TR-FIA the sensitivities of each assay were almost identical. A comparison using rotavirus assays is shown in Table 1. The clinical sensitivities of the indirect rotavirus and adenovirus TR-FIAs were also the same and the agreement rates were higher than 97% (Table 2). In about 3% of the stool specimens, a nonspecific reaction was found according to confirmatory tests and the nonspecifically reacting specimens were the same in the indirect EIA/RIA and in the indirect TR-FIA.

Table 1. Comparison of indirect time-resolved fluoroimmunoassay (TR-FIA) and indirect enzyme immunoassay (EIA) in the assay of purified rotavirus reference antigen (NCDV)

NCDV (ng/ml)	Indirect EIA		Indirect TR-FIA	
	O.D.	Ratio[a]	cps ($\times 10^3$)	Ratio
100	1.511[b]	15.9	9.286[b]	18.2
10	0.360	3.8	1.085	2.1[c]
1	0.129	1.4	0.723	1.4
0	0.095		0.510	

[a] O.D. or cps (counts per second) value of the reference/O.D. or cps value of diluent control (0 ng/ml)

[b] Mean of triplicates

[c] Ratio value of 2.0 is used as cut-off for estimation of the assay sensitivity

Table 2. Comparison of indirect time-resolved fluoroimmunoassay (TR-FIA) and indirect radioimmunoassay (RIA) in the detection of rotavirus and adenovirus in selected stool specimens of hospitalized children with acute gastroenteritis

Indirect RIA	Specimens (*n*)		Indirect TR-FIA	
			Positive	Negative
Rotavirus positive	55		55	0
Rotavirus negative	138		2	136
		Positive rate	RIA 55/193 (28.5%)	
			TR-FIA 57/193 (29.5%)	
		Agreement	98.9%	
Adenovirus positive	68		67	1
Adenovirus negative	125		3	122
		Positive rate	RIA 68/193 (35.2%)	
			TR-FIA 70/193 (36.3%)	
		Agreement	97.9%	

Comparisons of indirect TR-FIA and indirect RIA in the titration of seven respiratory virus reference antigens (influenza virus A and B, parainfluenza virus types 1, 2, and 3, respiratory syncytial virus, and adenovirus) indicated similar sensitivities for these two assays. In sonicated aspirates of nasopharyngeal mucus specimens the clinical sensitivities of these two assays were almost identical. The total number of specimens tested was more than 300 with about 100 positive specimens.

Detection of Rotavirus and Adenovirus in Stool Specimens and Influenza A and Adenovirus in Nasopharyngeal Aspirates by Direct One-Incubation TR-FIA

Comparison of direct one-incubation (1 h) TR-FIA with indirect EIA (1 h + 1 h + 1 h + ½ h) in the titration of purified rotavirus and adenovirus hexon antigen indicated that the direct TR-FIA had the same or slightly higher sensitivity (between 1 and 10 ng/ml) than the indirect EIA (Table 3). In higher concentrations of reference viruses the TR-FIA ratios were considerably greater and the dose-response was more linear than in the indirect EIA. The TR-FIA also identified the same adenovirus and rotavirus positive stools as the EIA. However, the specimens reacting nonspecifically in EIA were negative in TR-FIA (Table 4).

The direct one-incubation TR-FIAs for the detection of rotavirus and adenovirus in stool specimens are now in day-to-day use in our diagnostic unit and a representative test is shown in Table 5.

Comparison of indirect EIA with incubations of 16 h + 1 h + 1 h + ½ h and direct one-incubation TR-FIA in detection of adenovirus and influenza A virus in nasopharyngeal aspirates indicated that the specimens positive for adenovirus in indirect EIA were positive in direct TR-FIA and that the sensitivities of these two assays were the same according to the reference virus titrations. A represen-

Table 3. Comparison of direct one-incubation time-resolved fluoroimmunoassay (TR-FIA) and indirect enzyme immunoassay (EIA) in the assay of purified adenovirus hexon antigen

Hexon (ng/ml)	Indirect EIA		Direct TR-FIA	
	O.D.	Ratio[a]	cps ($\times 10^3$)	Ratio
100	1.076[b]	16.1	10.458[b]	120
10	0.287	4.3	1.257	14.4
1	0.091	1.4	0.156	1.8
0	0.067		0.087	

[a] O.D. or cps (counts per second) value of the hexon/O.D. or cps value of diluent control (0 ng/ml)
[b] Mean value of triplicates

Table 4. Comparison of indirect enzyme immunoassay (EIA) and direct one-incubation time-resolved fluoroimmunoassay (TR-FIA) in detection of rotavirus and adenovirus in 142 stool specimens

Indirect RIA	Specimens (*n*)	Direct TR-FIA	
		Positive	Negative
Rotavirus positive	35	34	1[a]
Rotavirus negative	102	0	102
Nonspecifically reacting[b]	5	0	5
Adenovirus positive	5	5	0
Adenovirus negative	133	0	133
Nonspecifically reacting[b]	4	0	4

[a] The cps value was at the cut-off level
[b] According to confirmatory test

Table 5. A representative test of direct one-incubation TR-FIA in detection of rotavirus and adenovirus in stool specimens

Reference virus and specimens no.	Rotavirus		Adenovirus		Conclusion
	cps ($\times 10^3$)	Ratio[a]	cps ($\times 10^3$)	Ratio	
Reference 100 ng/µl	12.226	42	9.923	59	
10	1.432	4.9	1.032	6.1	
1	0.552	2.5	0.263	1.5	
0	0.294		0.170		
Specimen no. 1	50.630	172	0.170	1.0	Rotavirus
2	0.198	1.2	0.417	2.5	Negative
3	0.299	1.0	0.198	1.2	Negative
4	0.099	0.3	110.685	651	Adenovirus
5	0.238	0.8	0.372	2.2	Negative
6	0.306	1.0	0.128	0.8	Negative

[a] cps (counts per second) value of reference or specimen/cps of diluent control (0 ng/ml)

Table 6. A representative test of direct one-incubation TR-FIA in the detection of adenovirus in sonicated aspirates of nasopharyngeal mucus collected from hospitalized children with acute respiratory disease

References virus and specimen no.	TR-FIA		Conclusion
	cps ($\times 10^3$)	Ratio	
Hexon antigen 100 ng/ml	13.390	70	
10	1.372	7.2	
1	0.168	0.9	
0	0.191		
Specimen no. 1	0.249	1.3	Negative
2	0.449	2.4	Negative
3	7.721	41	Positive
4	57.722	303	Positive
5	0.209	1.1	Negative
6	0.354	1.9	Negative

Table 7. Comparison of indirect enzyme immunoassay (EIA) and direct one-incubation time-resolved fluoroimmunoassay (TR-FIA) in the detection of influenza A virus in sonicated nasopharyngeal mucus aspirates of hospitalized children with acute respiratory disease

Reference virus or specimen no.	Indirect EIA		Direct TR-FIA		Conclusion
	O.D.	Ratio	cps ($\times 10^3$)	Ratio	
Purified influenza virus					
100 ng/ml	2.268	119	16.953	137	
10	0.435	23	1.712	14	
1	0.163	8.6	0.233	1.9	
0	0.019[a]		0.124		
Specimen no. 1	1.106	61	1.627	13	Positive
2	0.070	3.9	0.095	0.8	Negative
3	0.012	0.7	0.172	1.4	Negative
4	1.184	66	1.240	10	Positive
5	0.926	51	0.763	6.2	Positive
6	0.110	6.1	0.259	2.1	Negative

[a] The unusually low O.D. value of the diluent control (0 ng/ml) resulted in exceptionally high ratios

tative test for the detection of adenovirus in nasopharyngeal specimens by direct TR-FIA is shown in Table 6. In the detection of influenza virus the sensitivity of direct TR-FIA was the same or slightly lower than the sensitivity of indirect EIA with overnight incubation (Table 7).

Conclusion

The principle and the immunoreagents used in indirect TR-FIA and in indirect EIA and RIA were exactly the same, the only difference between these three assays being the label of the anti-rabbit IgG antibody. Since the sensitivity and specificity of indirect RIA and EIA was demonstrated earlier (Sarkkinen 1981), it was not surprising that the indirect TR-FIA had the same sensitivity as RIA and EIA when tested by purified reference viruses and by clinical specimens. The specificity of these three assays was also identical.

The direct one-incubation TR-FIA was surprisingly sensitive within large detection range. In this assay the Eu-chelate-labelled anti-viral rabbit antibody is the same as on the solid phase and is added to the antibody-coated well immediately after test specimens or reference virus. The viral antigen should react first with the indicator antibody in solution and block the antigenic sites before the antigen is bound to solid-phase antibody. In addition, with higher antigen concentrations the label bound to excess antigen should be washed away resulting in decreased label activity on the solid phase. However, the experimental results do not support these theories and the findings may be explained by (a) ten times less antibody used for the labelled tracer than on the solid phase, and (b) stronger binding activity of the solid phase multiple antibodies than that of the labelled indicator antibody.

One-incubation TR-FIA is an ideal immunoassay. It has one incubation and one washing step only and it takes less than 2 h to carry out the test and to obtain printed results. Once the coated strips and the label are available the practical steps are simple: addition of the specimens and the label; incubation and washing; addition of enhancement solution and automatic reading. A particular advantage of the direct one-incubation TR-FIA is in the detection of respiratory viruses because in the indirect EIA and RIA the results are available in about 24 h. However, experience with direct TR-FIA is limited at present to only two respiratory viruses, and with adenovirus, immunosorbent purification of antibody for Eu-chelate labelling was a requirement for high sensitivity. If this is true with many other viruses then the preparation of the labeled antibody will be a problem unless monoclonal antibodies can be used (Siitari et al. 1983).

The general advantages of time-resolved fluoroimmunoassays are: the specific activity of the label exceeds that of radioisotopes thus making them ideal for assays in which high sensitivity is required; Eu-labelled preparations are stable for more than a year when properly handled; the measurement time is only one second; Eu has no known biological hazards. Our conclusion is that time-resolved fluorescence is one of the most promising alternatives available to nonisotopic immunoassays.

Acknowledgements. The excellent technical assistance of Ms. Kaija Johansson and Ms. Leena Soini is gratefully acknowledged. These studies were supported by grants from The Medical Research Council, The Academy of Finland, the Sigrid Juselius Foundation and C.N.R., Italy (Progetto „Controllo Malattie da Infezione").

References

Halonen P, Sarkkinen H, Arstila P, Hjertsson E, Torfason E (1980) Four-layer radioimmunoassay for detection of adenovirus in stool. J Clin Microbiol 10:614–617

Halonen P, Meurman O, Lövgren T, Hemmilä I, Soini E (1983) Detection of viral antigens by time-resolved fluoroimmunoassay. In: Bachmann PA (ed) New developments in diagnostic virology. Springer, Berlin Heidelberg New York, pp 133–146 (Current topics in microbiology and immunology, vol 104)

Sarkkinen HK (1981) Human rotavirus antigen detection by enzymeimmunoassay using antisera against Nebraska calf diarrhoea virus. J Clin Pathol 34:680–685

Sarkkinen HK, Halonen PE, Arstila PP, Salmi AA (1981 a) Detection of respiratory syncytial, parainfluenza type 2, and adenovirus antigens by radioimmunoassay and enzyme immunoassay on nasopharyngeal specimens from children with acute respiratory disease. J Clin Microbiol 13:258–265

Sarkkinen HK, Halonen PE, Salmi AA (1981 b) Detection of influenza A virus radioimmunoassay and enzyme-immunoassay from nasopharyngeal specimens. J Med Virol 7:213–220

Sarkkinen HK, Halonen PE, Salmi AA (1981 c) Type-specific detection of parainfluenza viruses by enzyme-immunoassay and radioimmunoassay in nasopharyngeal specimens of patients with acute respiratory disease. J Gen Virol 56:49–57

Siitari H, Hemmilä I, Soini E, Lövgren T, Koistinen V (1983) Detection of hepatitis B surface antigen using time-resolved fluoroimmunoassay. Nature 301:258–260

Soini E, Hemmilä I (1979) Fluoroimmunoassay: present status and key problems. Clin Chem 25:353–361

Soini E, Kojola H (1983) Time-resolved fluorometer for lanthanine chelates – a new generation of nonisotopic immunoassays. Clin Chem 29:65–68

Monitoring Microbial Growth by Bioluminescent ATP Assay

M.-R. Siro [1]

Introduction

Quantitative determination of total biomass carbon as a function of time is often important both in general and applied microbiology. The classic techniques for detecting and monitoring microbes include total visual or electronic cell counting, viable cell counting using plating techniques, estimating cell number by measuring the turbidity, and dry weight determination. Both viable and total cell counting are somewhat time-consuming and open to serious criticism [1, 2]. Although particle counting methods are faster, they suffer from inaccuracies caused by counting of any particulate matter, including air bubbles and dust particles. Moreover, direct counting includes both viable and nonviable cells. No rapid method for detecting or monitoring viable bacteria has thus far been available. Undoubtedly the determination of viable cells is important in all kinds of contamination studies and in processes that require viable cells. Because of the problems associated with conventional methods, there has been great interest in attempts to find a rapid routine procedure for determining the amount of microbes in different sources.

Adenosine 5′-triphosphate as an Indicator of Biomass

The role of the amount of adenosine 5′-triphosphate (ATP) as a parameter in estimating biomass carbon was proposed by Holm-Hansen [3] as early as 1966. Today ATP is a well-accepted indicator of microbial biomass and the method is based on four important assumptions [4]: (1) All living organisms contain ATP; (2) ATP is neither associated with dead cells nor adsorbed onto detrital material; (3) ATP is rapidly extracted from microbial material and can be precisely measured; (4) the ratio of ATP to cell carbon will be fairly constant for all microbial taxa independent of metabolic activity or environmental conditions.

Experimental evidence has shown that the first assumption is unequivocally acceptable and that the second one is valid if surface active material is not included in the microbial sample. The third assumption is valid under certain experimental conditions quite easily obtained using modern assay procedures. The most critical assumption is the fourth one. It has been shown that the ratio of ATP to cell carbon of microbial cells varies depending on the species and its met-

1 Department of Biochemistry and Pharmacy, Åbo Akademi, SF-20500 Åbo

abolic state as well as on the culture conditions and growth rate. Despite the fact that the accuracy of the biomass estimation based on ATP measurements has been questioned, the technique is still by far the most convenient and reliable method for estimating total microbiol biomass. High accuracy is obtained with a monoculture of a known species under controlled nutrient status of a population.

Assay of ATP with Firefly Luciferase

The assay involves the measurement of light produced as a result of an enzyme reaction as catalyzed by the firefly *Photinus pyralis* (American firefly). The overall reaction is as follows:

$$\mathrm{ATP} + \mathrm{D\text{-}luciferin} + \mathrm{O_2} \xrightarrow[\mathrm{Mg^{2+}}]{\text{luciferase}}$$

$$\mathrm{oxyluciferin} + \mathrm{AMP} + \mathrm{CO_2} + \mathrm{PP_i} + \text{light} \qquad \lambda\mathrm{max} = 562\ \mathrm{nm}\,.$$

The reaction is reversible and pyrophosphate is an inhibitor of the forward direction [5]. Extensive information on the mechanism of the bioluminescence reaction can be found in many reviews and books [5, 6].

In the firefly luciferase-based assay, the amount of light produced is directly proportional to the amount of ATP taking part in the reaction. Commercially available purified luciferin/luciferase preparations are optimized to give a constant light output [7]. The sensitivity and specificity of the ATP assay also increase considerably when purified enzymes are used, and concentrations as low as 10^{-11} *M* ATP can be measured in the assay mixture [8]. Moreover, other nucleotide triphosphates present in the sample do not take part in the luciferase-catalyzed reaction. Internal standardization of every assay improves accuracy and is used for converting relative light units to conentrations of ATP.

Extraction of Microbial ATP

The difficulty in determining ATP from different biological sources is the quantitative extraction of ATP from cells [9]. In general, the extraction procedures from plants, bacteria, and yeasts are more difficult because of their resistant cell envelopes compared with those of somatic cells. The susceptibility to different extractants varies even between species of microorganisms, and the extraction procedure has to be assessed separately in each system. The following aspects ought to be considered when extraction procedures are evaluated: (1) the time between sampling and inactivation of the nucleotide-converting enzymes in the sample must be short as the ATP pool may be turned over completely in a few seconds; (2) the method should give a reproducible and preferably high yield of ATP in biological material, and it should not dilute the sample excessively; (3) several different dilutions of the sample should be extracted to ensure that the yield of ATP is proportional to the amount of sample; (4) the stability of the extract should be good when stored under appropriate conditions.

Analytical Interference of the Assay Should be as Low as Possible

Analytical interference of the assay by the sample may be caused by the presence of specific inhibitors of the luciferase reaction or contaminating ATP-converting enzymes, protein-denaturating agents, and high concentrations of neutral salts. Several media used for growth or storage of cells as well as the agents used for extraction of ATP may have marked effects on the luciferase reaction [10–13]. To some extent, the interfering factors can be accounted for by the use of an internal standard in the assay. The inclusion of ethylenediaminetetraacetate (EDTA) in the extraction agents has been shown to improve the stability of the extracts by promoting irreversible inactivation of the ATP-converting enzymes in the sample [9]. Pretreatment including filtration and centrifugation of the sample may sometimes be necessary.

The procedures of ATP extraction described in the literature are many and vary according to the cell type present. The commonly used methods include chaotropic reagents such as perchloric acid (PCA) and trichloroacetic acid (TCA), organic solvents such as ethanol and butanol, boiling in buffer solutions, freezing, and the use of different detergents such as Triton X-100 and benzalkonium chloride [3, 14–20]. An evaluation of the extraction procedures used in different reports is difficult as the microbes and experimental conditions vary considerably. Moreover, different types of partially purified firefly luciferase enzymes have been used for the bioluminescence reaction.

In order to find the most effective ATP extraction conditions for yeast cells and to establish optimal assay conditions for the bioluminescence measurement, various extractants and their interference on the luciferase reaction were studied (H. Romar and M.-R. Siro, unpublished results). Reagents such as 4-hexylresorcinol, *N*-cetyl *N,N,N*-trimethylammonium bromide and benzalkonium chloride were tested (Fig. 1). The yields of ATP obtained from the cells were compared with extraction of 100% yield using 10% (w/v) TCA [19]. Only benzalkonium chloride proved to be as effective as TCA (Table 1).

Recently, Lundin and Thore [9] reported a study where ten different methods for extraction of bacterial ATP in five bacterial strains were compared. The methods selected for the study included extraction with acids, organic solvents, and boiling aqueous solutions. They compared the methods with regard to yield of adenosine nucleotides, interference of the luciferase assay, reproducibility of the method, and stability of the extracts. In all bacterial strains, extractions using TCA resulted in yields of ATP higher than those obtained by other extraction methods or very close to the highest ones. Furthermore, extraction with TCA resulted in complete recovery of added nucleotide.

Consequently, TCA or PCA have to be considered the only *general* extractants for bacteria and yeast cells today. They are strong inhibitors of the luciferase

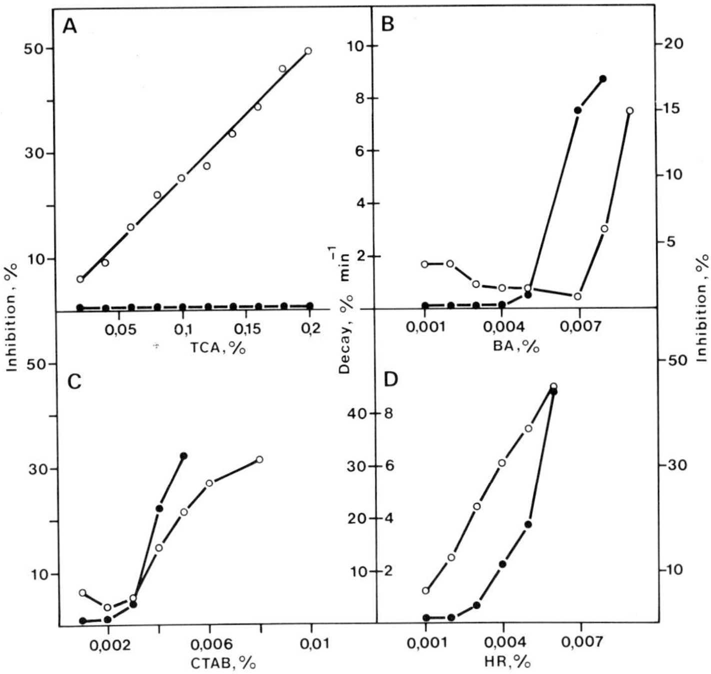

Fig. 1. The effect of varying concentrations (%; w/v) of trichloroacetic acid (TCA) **A**, benzalkonium chloride (BA) **B**, *N*-cetyl-*N,N,N*-trimethyl ammonium bromide (CTAB) **C**, and 4-hexylresorcinol (HR) **D** on the firefly reaction. The reduction of light emission as percentage inhibition (○) and the decay rate of light output as percentage decay per minute (●) are shown as functions of the concentrations of the extractants in the assay mixture. The experiments were performed by adding 5 µl varying concentrations of extractant to assay mixtures containing 10^{-7} *M* ATP, 0.1 *M* TRIS-acetate buffer, including 2 m*M* EDTA, pH 7.75, and ATP-Monitoring Reagent (LKB-Wallac)

Table 1. The amount of ATP extracted from 40 µg yeast (*Saccharomyces cerevisiae*) using different extracting agents

Extractant[a]	Concentration of extractant in extraction mixture (%; w/v)	Concentration of extractant in assay mixture (%; w/v)	Yield of ATP compared with TCA (%)
Benzalkonium chloride	0.02	0.004	100
N-cetyl-*N,N,N*,-trimethylammonium bromide	0.01	0.002	45
4-hexylresorcinol	0.015	0.003	20

[a] The extractant solutions were all 4 m*M* with respect to EDTA

[b] The extraction mixture contained 50 µl yeast suspension and 50 µl solution of extractant. The extraction mixture was rotated using a Vortex mixer for 1 min, then 300 µl 0.1 *M* TRIS-acetate buffer containing 2 m*M* EDTA pH 7.75 and 100 µl ATP Monitoring Reagent were added

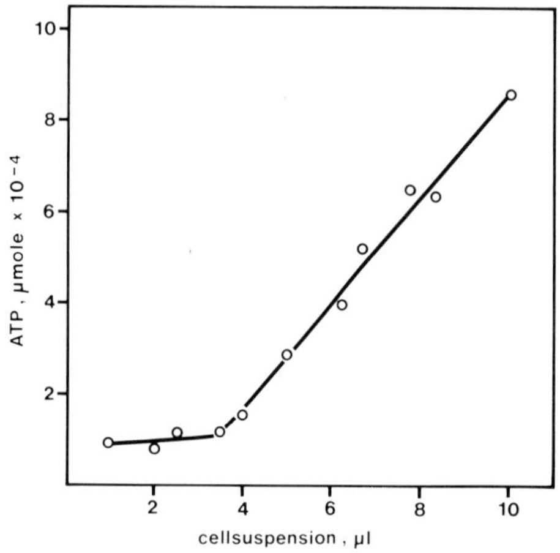

Fig. 2. ATP extracted from different amounts of cell suspensions of *Candida rugosa.* The cells were grown in a 7.5-l fermentor on a synthetic growth medium at 30 °C under vigorous aeration. ATP was extracted using 10% (w/v) TCA, including 4 m*M* EDTA. The extracts obtained were diluted before the bioluminescent assay of ATP

reaction but the analytical interference can be reduced in different ways. When PCA is used, the obtained extracts may be neutralized by KOH, and the sediment of perchlorate is easily separated from the sample before assay. If the amount of ATP in the sample to be measured is high enough, a dilution of the extract before the measurement of ATP gives good results [19]. In the presence of traces of TCA in the assay mixture, the decay of light emission is less than 1%/min and any interference can be corrected by internal calibration. However, the sensitivity of the assay decreases slightly when such inhibited assay conditions are used. Figure 2 shows the results obtained when *Candida rugosa* cells were extracted with 10% (w/v) TCA and ATP was measured in diluted extracts at 20% inhibition of luciferase activity. A linear correlation of the yield of ATP and the amount of cell suspension is obtained above 3.5 μl suspension. This sensitivity limit corresponds to 1.6×10^{-9} *M* ATP in the assay mixture, which is higher than the 10^{-11} *M* reported as the limit of the assay [8]. The inhibitory effect of TCA can also be avoided by removing the TCA from the extraction mixture before the ATP assay using diethyl ether.

Pretreatment of Sample when Measuring Microbial ATP

All living material contains ATP and interference from large amounts of nonmicrobial ATP in the sample may cause problems. Pretreatment of sample is thus often necessary in biomass determinations based on the ATP measurement technique. In general, somatic cells contain 10 to 100 times more ATP per cell than bacteria. The efficiency with which microbial ATP can be separated from ATP derived from the nonmicrobial part of the sample greatly affects the utility of the assay. There are at least two possible solutions to the problem. One is the selective extraction of the cells by exploiting the different sensitivity of each cell type to various chemical extractants. The method involves extraction of nonmicrobial ATP by incubation of the sample with a "mild" extractant, e.g., Triton X-100, followed by destruction of the released ATP by the ATP-hydrolyzing enzyme, apyrase. After pretreatment, apyrase is destroyed by heat or acid and the ATP of the microbial cells is extracted and measured as described previously. The procedure has been successfully used for rapid screening of bacteriuria [14, 25]. Another solution to the problem of interference from nonmicrobial ATP and other interfering components in the growth medium is separating the microbes from the rest of the material by centrifugation or filtration and then measuring the ATP in the separated organisms [21, 18]. Some interfering material might also be removed by sorption to a chromatographic material. Samples with low levels of microorganisms can be concentrated by micropore filtration.

Applications of Firefly Bioluminescence in Monitoring Microbial Growth

Reported applications dealing with the quantitation of ATP as a measurement of biomass can roughly be divided into two groups. In the first group, the principle has been used to estimate biomass in various natural environments, such as ocean water, and to estimate the bacterial content in clinical specimens, foods, and drinking water. The second group of applications is based on the use of ATP measurements as an index of growth in bacterial cultures and fermenting processes. Some of the applications are summarized in Table 2. The following applicatory examples illustrate the usefulness of the method in monitoring microbial growth or inhibition.

In the control of fermentation, ATP is both a parameter indicating the metabolic status of the organisms and a growth index. When baker's yeast was grown

Table 2. Applications of determination of biomass using the bioluminescent ATP assay

Application area	References
Estimation of microbes in different types of samples	
Marine water	Holm-Hansen and Booth [3]
Drinking water	Levin et al. [23]
Beer	Hysert et al. [20]
Meat	Stannard and Wood [21]
39 different foods	Sharpe and Woodrow [18]
Rumen	Forsberg and Lam [24]
Urine	Thore et al. [25]
Blood	Molin et al. [26]
Index of growth	
Fermenting processes	Hysert and Morrison [27]
Decomposition	Lehtokari et al. [28]
Susceptibility of bacteria to antibiotics	Thore et al. [29]
Antibiotica levels in biological fluids	Nilsson [30]

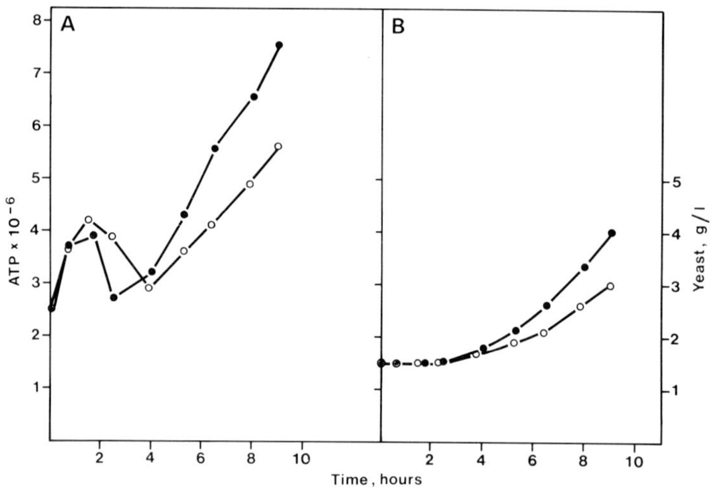

Fig. 3. The intracellular concentration of ATP **A** and growth **B** of yeast cells cultivated on a synthetic medium, containing either 0.05 *M* glucose (●) or 0.025 *M* maltose (○). The incubation was performed at room temperature without aeration [23]. Growth was monitored by measuring the absorbance at 554 nm and ATP extracted using 10% (w/v) TCA including 4 m*M* EDTA

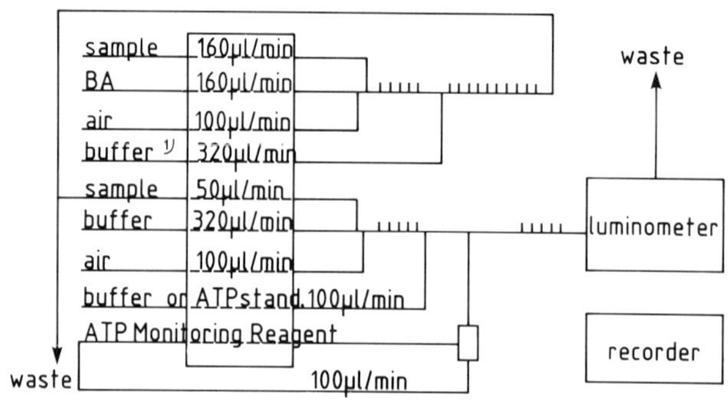

Fig. 4. Flow diagram of the continuous flow system using 0.16% (w/v) benzalkonium chloride solution (*BA*) as extractant for the bioluminescent measurement of yeast cell ATP. The system consists of two major components, the AKEA Automatic Chemical Analysis System and the LKB Wallac 1250 Luminometer with recorder. The pump has a constant speed of rotation and the amount of fluid pumped through each line is proportional to the inner cross-sectional area of the line. Flow rates of the lines and the arrangement of the equipment are shown. The flow in the lines was segmented by a bubble unit feeding air bubbles into the stream at regular intervals. [1)] 0.1 *M* TRIS-acetate buffer containing 2 m*M* EDTA, pH 7.75

in batch cultures on a synthetic growth medium, good correlation between the amount of cellular ATP and cell growth during the exponential growth phase was obtained (Fig. 3). During the lag phase of growth, changes in the ATP concentration were obtained, although no increase in the mass per culture was measured. On the basis of results obtained from manually performed experiments, two continuous flow systems were designed for measurement of ATP in the cells (Fig. 4). The experiments performed using the flow systems resulted in data which indicate a possible microbiological application of such systems for monitoring growth in fermenting processes.

When studying the ATP content of a selection of yeasts and bacteria, Hysert et al. [20] showed that a linear relation between the quantity of ATP and the number of yeast cells existed. The changes in intracellular yeast ATP pools during fermentation of wort by ale and lager yeasts were also determined. The results demonstrate that yeast ATP content/g dry weight does not change markedly during brewery type fermentations of wort. The assay procedure required only a few minutes per sample and it is applicable to microbiological quality control of unpasteurized bottled beer. Other possible applications of the method in the brewing industry were also suggested. These include monitoring biocide effectiveness in recirculating cleaning systems, evaluating the general microbiological state of brewing facilities and basic biochemical brewing research. Hysert et al. also suggested a further application of the assay [27] based on results reported in a separate study where the yeast and medium ATP contents and changes during growth on brewery wort of 36 different ale, lager, and miscellaneous yeasts at different temperatures were studied. Significant species to species differences were noted for these variables and for the influence of temperature on them. The patterns of changes in medium ATP levels for ale and lager yeasts, particularly at the lower temperatures, were suggested as a means of yeast classification and differentiation.

Most of the biochemical changes occurring during the decomposition process are direct results of microbial colonization, growth, and secession. ATP measurements have recently been used to monitor biomass changes during a laboratory-scale compost process [28]. Variation of the ATP content, temperature, and pH as a function of time were determined. Correlation of the ATP concentration with the temperature was obvious and the investigations showed that it is possible to estimate biomass changes occurring during the compost process by monitoring changes in ATP concentrations.

Some clinical applications of the ATP bioluminescence method involve the determination of the susceptibility of different bacteria to various antibiotics [29]. The ATP level of a susceptible bacterium is indirectly affected by the antibiotic by the decrease in growth, and directly by the disturbance in metabolism. The inhibition of growth of a suitable microorganism produced in vitro by an antibiotic may be quantified by measuring cellular ATP levels. The degree of inhibition is a measure of the susceptibility.

The amount of antibiotics in biological fluids has been determined by the bioluminescent ATP measurement by monitoring the depression of growth of a suitable microorganism [30, 31]. The dose-dependent effect of amikacin on accumulation of extracellular bacterial ATP was utilized to develop a rapid bioluminescent assay of serum amikacin. The accuracy of the assay over the therapeutic range expressed as a coefficient of variation varied between 1.7% and 4.6%, compared with 1.1% and 5.8% for an agar disc diffusion assay. The bioluminescent assay required only 25 µl serum and results were available within 90 min.

Conclusions

The bioluminescent technique for monitoring growth offers several advantages over traditional methods. The technique is highly specific and the results are obtained rapidly within a few minutes or hours. One major advantage is the fact that only viable cells are determined. Moreover, the method offers a high degree of sensitivity, accuracy, and reproducibility.

References

1. Holm-Hansen O, Karl DM (1978) Biomass and adenylate energy charge determination in microbial cell extracts and environmental samples. Methods Enzymol 57:73–85
2. Pringle JR, Mor JR (1975) Methods for monitoring the growth of yeast cultures and for dealing with the clumping problem. Methods Cell Biol XI:153–168
3. Holm-Hansen O, Booth CR (1966) The measurement of adenosine triphosphate in the ocean and its ecological significance. Limnol Oceanogr 11:510–519
4. Karl DM (1980) Cellular nucleotide measurements and applications in microbial ecology. Microbiol Rev 44:739–796
5. De Luca M (1976) Firefly luciferase. In: Meister A (ed) Adv Enzymol 44:37–68
6. Strehler BL (1968) Bioluminescence assay: principles and practice. Methods Biochem Anal 16:99–181
7. Lundin A, Rickardsson A, Thore A (1976) Continuous monitoring of ATP-converting reactions by purified firefly luciferase. Anal Biochem 75:611–620
8. Lundin A, Thore A (1975) Analytical information obtainable by evaluation of the time course of firefly luminescence in the assay of ATP. Anal Biochem 66:47–63
9. Lundin A, Thore A (1975) Comparison of methods for extraction of bacterial adenine nucleotides determined by firefly assay. Appl Microbiol 30:713–721
10. Holmsen H, Holmsen I, Bernhardsen A (1966) Microdetermination of adenosine diphosphate and adenosine triphosphate in plasma with the firefly luciferase system. Anal Biochem 17:456–473
11. Rasmussen H, Nielsen R (1968) An improved analysis of adenosine triphosphate by the luciferase method. Acta Chem Scand 22:1745–1756
12. Lin S, Cohen HP (1968) Measurement of adenosine triphosphate content of crayfish stretch receptor cell preparations. Anal Biochem 24:531–540
13. Thore A (1979) Technical aspects of the bioluminescent firefly luciferase assay of ATP. Sci Tools 26:30–34
14. Thore A, Ånsehn S, Lundin A, Bergman S (1975) Detection of bacteriuria by luciferase assay of adenosine triphosphate. J Clin Bacteriol 1:1–8
15. Chappelle EW, Levin GV (1968) Use of the firefly bioluminescent reaction for rapid detection and counting of bacteria. Biochem Med 2:41–52
16. Bagnara AS, Finch LR (1972) Quantitative extraction and estimation of intracellular nucleoside triphosphates of *Escherichia coli*. Anal Biochem 45:24–34
17. Guinn G, Eidenbock MP (1972) Extraction, purification and estimation of ATP from leaves, floral buds and immature fruits of cotton. Anal Biochem 50:89–97
18. Sharpe AN, Woodrow MN, Jackson AK (1970) Adenosine triphosphate (ATP) levels in foods contaminated by bacteria. J Appl Bacteriol 33:758–767
19. Siro M-R, Romar H, Lövgren T (1982) Continuous flow method for extraction and bioluminescence assay of ATP in baker's yeast. Eur J Appl Microbiol Biotechnol 15:258–264
20. Hysert DW, Kovecses F, Morrison NM (1976) A firefly bioluminescence ATP assay method for rapid detection and enumeration of brewery microorganisms. Am Soc Brew Chem J 34:145–150
21. Stannard CJ, Wood JM (1982) The estimation of viable bacteria in raw meat by measurement of microbial ATP. The British Food Manufacturing Industries Research Association (Research reports no 379)
22. Siro M-R, Lövgren T (1979) Influence of glucose on the α-glucoside permease activity of yeast. Eur J Appl Microbiol Biotechnol 7:59–66

23. Levin BV, Usdin E, Slonim AR (1968) Rapid detection of microorganisms in aerospace water systems. Aerosp Med 14–16
24. Forsberg CW, Lam K (1977) Use of adenosine 5′-triphosphate as an indicator of the microbiota biomass in rumen contents. Appl Environ Microbiol 33/3:528–537
25. Thore A, Lundin A, Ånsehn S (1983) Firefly luciferase ATP assay as a screening method for bacteriuria. J Clin Microbiol 17:218–224
26. Molin Ö, Nilsson L, Ånsehn S (1983) Rapid detection of bacterial growth in blood cultures by bioluminescent assays of bacterial ATP. J Clin Microbiol 18:521–525
27. Hysert DW, Morrison NM (1977) Studies on ATP, ADP and AMP concentrations in yeast and beer. Am Soc Brew Chem J 35:160–167
28. Lehtokari M, Nikkola P, Paatero J (1983) Determination of ATP from compost using the firefly bioluminescence technique. Eur J Appl Microbiol Biotechnol 17:187–190
29. Thore A, Nilsson L, Höjer H, Ånsehn S, Bröte L (1977) Effects of ampicillin on intracellular levels of adenosine triphosphate in bacterial cultures related to antibiotic susceptibility. Acta Pathol Microbiol Scand [B] 85:161
30. Nilsson L (1982) Rapid bioluminescent assay of serum amikacin. J Antimicrob Chemother 10:125–130
31. Nilsson L, Höjer H, Ånsehn S, Thore A (1977) A rapid semiautomated bioassay of gentamicin based on luciferase assay of bacterial adenosine triphosphate. Scand J Infect Dis 9:232–236

Bioluminescent Assay of Bacterial ATP as a Tool in Clinical Microbiology

L. NILSSON [1]

Introduction

The firefly bioluminescence assay of ATP can be used to quantitate bacteria [7, 8, 14]. In the assay of ATP, conditions are arranged to result in a light emission proportional to the ATP concentration. In clinical microbiology the technique has been applied to detection of bacteriuria [8, 14] and bacteremia [7], antibiotic susceptibility testing [2, 5, 6, 15], and antibiotic assays [1, 3, 4, 9–13].

Detection of Bacteremia

The presence of nonbacterial ATP from somatic cells is a major problem in the assay of bacterial ATP in body fluids. A selective method for distinguishing bacterial and nonbacterial ATP in urine specimens has been presented [14]. The method involved incubation of urine samples with detergent (triton X-100) and ATPase (apyrase). The detergent lysed various human cells, and the ATP released was destroyed by ATPase, but this treatment did not affect the ATP content in the bacteria [14]. After the elimination of somatic cell ATP, bacterial ATP was extracted with boiling Tris/EDTA buffer [14]. The extraction inhibits the action of the ATPase and disrupts the bacteria, thus releasing their ATP content.

When applied to blood cultures the detergent/ATPase treatment did not sufficiently eliminate the blood cell ATP. The remaining level of blood cell ATP (10^{-6} *M*) makes it impossible to detect less than 10^8 CFU of bacteria per milliliter [7].

Fractionated centrifugation ($250 \times g$) of samples from the blood cultures to settle most of the blood cells into a pellet, followed by treatment of the supernatant with detergent and ATPase, gave a further decrease of the background of blood cell ATP. This treatment did not release all blood cell ATP, since a background of ATP (10^{-8} *M*) reappeared after the extraction procedure with boiling Tris/EDTA buffer [7]. ATP probably remains bound to blood cell membranes and escapes the ATPase activity. Furthermore, the yield of bacteria in the supernatant fluid after centrifugation varied with different bacterial strains [7], probably due to their tendency to form aggregates.

A better separation of blood cells and bacteria was necessary to overcome this background of blood cell ATP (10^{-8} *M*). To achieve this, samples from the blood

1 Department of Clinical Bacteriology, University Hospital, S-581 85 Linköping

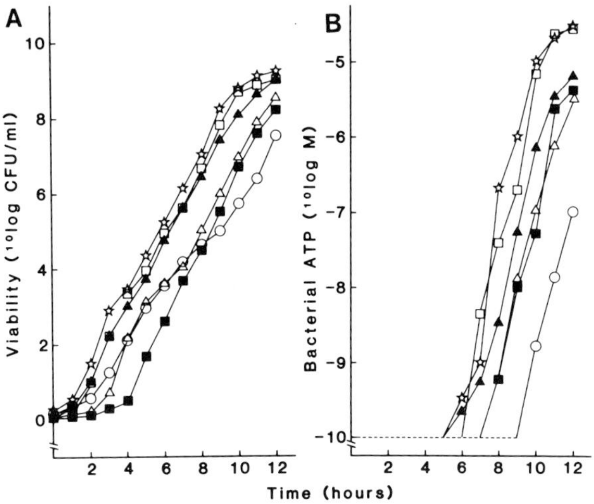

Fig. 1 A, B. Growth of *Escherichia coli* (□), *Klebsiella pneumonia* (☆), *Pseudomonas aeroginosa* (○), *Proteus mirabilis* (△), *Staphylococcus aureus* (■), and *Streptococcus faecalis* (▲) monitored by **A** viability (CFU/ml) and **B** bacterial ATP (*M*). The cultures contained 10% (vol/vol) blood

cultures were subjected to Percoll (Pharmacia Fine Chemicals) density gradient centrifugation. The small difference between densities of blood cells and bacteria necessitates lysing of blood cells with detergent before gradient centrifugation [7]. The released ATP was eliminated by ATPase.

Blood cell ATP which was bound to cell debris and escaped the ATPase activity was separated on the gradient. The position of the bacterial layer was determined with density marker beads (Pharmacia Fine Chemicals) or by gradient centrifugation of a bacterial culture with a bacterial count exceeding 10^6 CFU/ml, which gives a visible layer at a specific region of the gradient [7]. When samples without bacteria were subjected to gradient centrifugation, no blood cell ATP could be detected in samples from the specific region of the gradient in which the bacteria were known to be found [7]. The detection limit of ATP was 10^{-10} *M* due to the background luminescence from the luciferase reagent and the sensitivity of the equipment [7]. The yields of bacteria were equal to or greater than those in direct samples due to the zone sharpening effect of the centrifugation [7]. In simulated blood cultures inoculated with 10 CFU of bacteria per milliliter blood, bacterial ATP levels exceeded the detection limit (10^{-10} *M*) after 6–10 h of incubation (Fig. 1). At the time of detection the amount of bacteria just exceeded 10^4 CFU/ml. The time of detection with Gram's stain was 1–2 h later and the approximate detection limit was 5×10^5 CFU/ml [7].

The bioluminescent assay was evaluated on 512 clinical blood cultures. The results of macroscopic examination and subculture were the basis for establishing true positive and negative results. At the detection limit (10^{-10} *M*) the bioluminescent assay detected 94.2% of the true positive blood cultures within 6–18 h incubation, but the specificity was only 85.8%. The predictive values for positive and negative results were 29.5% and 80.3% respectively. By increasing the detection limit the sensitivity and specificity increased. The maximum correctly classified blood cultures were reached at a detection limit of 2×10^{-9} *M* ATP. At this limit of detection the bioluminescent assay detected 82.9% of the positive blood cultures and 99.6% of the negative blood cultures were correctly classified. The predictive values for positive and negative results also reached optimal figures, 93.6% and 99.6% respectively. Furthermore 52.9% of the positive cultures were detected 1 day earlier with the bioluminescent assay than with macroscopic examination and subculture.

These results indicate that the bioluminescent assay of bacterial ATP in blood cultures is an objective and sensitive method for rapid detection of bacteremia.

Antibiotic Susceptibility Testing

Rapid effects of antibiotics on levels of intracellular ATP in bacterial cultures have been studied [1–6, 9, 11, 15]. When bacteria are exposed to antibiotics the effect on viability (CFU/ml) is reflected in a parallel effect on intracellular ATP levels [2, 4, 9, 11, 15]. Leakage of ATP from bacteria exposed to membrane active agents interferes with the determination of intracellular ATP [10, 11, 15]. For example, in a culture exposed to ampicillin a discrepancy was found between ATP and viability (Fig. 2). The ATP level remained constant after 2–h incubation, while viability (CFU/ml) decreased gradually. If the culture was treated with ATPase (apyrase) before extraction of intracellular ATP, the remaining intracellular ATP parallelled with CFU (Fig. 2). This indicates accumulation of extracellular ATP. Extracellular ATP was determined by simply adding luciferase reagent to the bacterial culture. The amount of extracellular ATP in the culture closely corresponded to the difference between total ATP (determined after extraction)

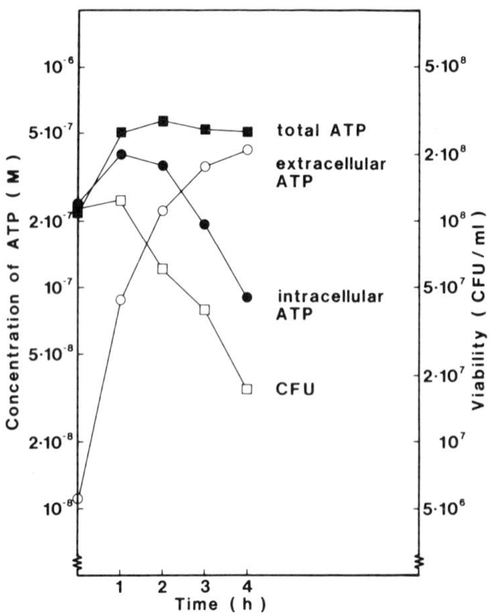

Fig. 2. Effects of ampicillin on viability (□) and on total (■), intracellular (●), and extracellular (○) ATP in a culture of *Sarcina lutea* exposed to 0.1 mg/l ampicillin

and intracellular ATP (determined after ATPase treatment and extraction (Fig. 2). The accumulation of extracellular ATP varies with the native ATPase activity of the test strain [11, 15]. To avoid interference from extracellular ATP all samples were exposed to ATPase before extraction and assay of intracellular ATP.

A close correlation was found between decrease in viability and decrease in intracellular ATP in bacterial cultures exposed to serial dilutions of antibiotics [2, 9–11, 15]. The decrease in viability and intracellular ATP occurred at a concentration of the antibiotic corresponding to the minimal inhibitory concentration (MIC) of the test strain. To utilize this phenomenon for practical purposes it is necessary to find a concentration which has the ability to discriminate bacteria with different antibiotic susceptibility.

Clinical isolates of bacteria were exposed to 10 mg/l ampicillin, and controls without ampicillin were incubated simultaneously. After 2-h incubation intracellular ATP levels were determined in all cultures. The ATP level in an exposed culture was expressed as percentage of its own control. The results were compared with MIC values obtained with broth dilution. The relationship between ampicillin effects on intracellular ATP and MIC values is shown in Fig. 3. Similar relationships between susceptibility and antibiotics effects on intracellular ATP have been presented for gentamicin [5], methicillin [6], nitrofurantoin [5], and doxycycline [2].

The results indicate that it might be possible to utilize the bioluminescent method for rapid grouping of bacteria according to antibiotic susceptibility.

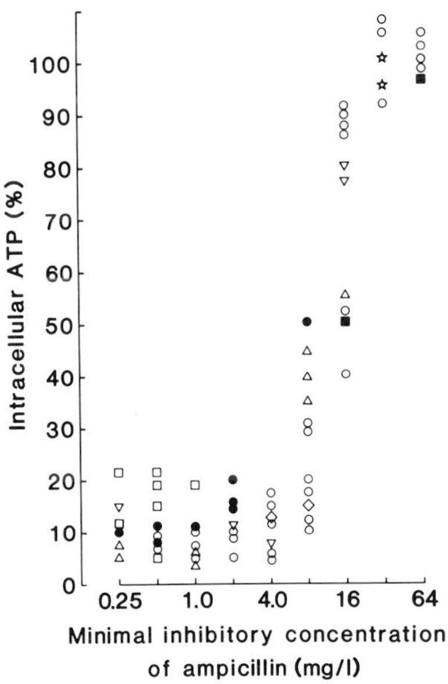

Fig. 3. Relationship between ampicillin effects on intracellular ATP levels and MIC values. Clinical isolates of 31 *Escherichia coli* (○), 2 *Klebsiella pneumonia* (◇), 2 *Proteus indole* – (■), 7 *Proteus indole* + (□), 2 *Pseudomonas aeruginosa* (☆), 6 *Staphylococcus aureus* (△), 5 *Staphylococcus epidermidis* (▽), and 8 *Streptococcus faecalis* (●) were used. After 2-h incubation intracellular ATP was determined in ampicillin-exposed cultures and control cultures. The ATP level in an exposed culture was expressed as a percentage of its own control

Antibiotic Assay

Rapid bioassays of antibiotics based on the determination of ATP levels in bacterial cultures exposed to standard antibiotic concentrations and to clinical serum specimens have been developed [1, 3, 4, 9, 11–13]. The dose-dependent effects of antibiotics on total ATP [1] and intracellular ATP [3, 4, 9] in bacterial cultures have been used. For quantitation of the antibiotics the linear part of the dose/response curves were used as calibration curves.

Outline of Intracellular Method

After a short (90–120 min) incubation at 37 °C of a control culture and cultures exposed to antibiotic standards or clinical specimens, elimination of extracellular ATP is performed by a short (10-min) incubation with ATPase (apyrase). After elimination of extracellular ATP intracellular ATP is extracted by heat [7] or by acid [15] and ATP is assayed [7]. The recorded light emission (arbitrary units) is plotted semilogarithmically against the concentration of the antibiotic. The concentrations of antibiotic in clinical serum specimens were calculated by comparing their light emissions with those of known concentrations plotted on the standard graph.

This assay procedure has been adapted for the determination of bactericidal antibiotics with different modes of action (ampicillin [3], cephaloridin [1], gentamicin [1, 9], and tobramycin [1]) and a bacteriostatic agent doxycycline [4].

The evaluation of the intracellular bioluminescent methods demonstrated that the accuracy was acceptable and that the method correlated well with the agar diffusion assay [1, 3, 4, 9].

The main advantage of the bioluminescent method is its sensitivity allowing use of microvolumes and its rapidity making results available within 2–4 h.

Agents causing damage to bacterial membranes lead to leakage of ATP into the medium [10, 11, 15]. Rapid effects of aminoglycosides on ATP levels in cultures of *Escherichia coli* LU 14 have been studied and a dose-dependent accumulation of extracellular ATP has been demonstrated [10–13]. This phenomenon has been used to develop a simple method for aminoglycoside determination [10–13] which does not require the extraction procedure needed in the intracellular method.

Outline of Extracellular Method

If cultures of *E. coli* LU 14 are exposed to standard concentrations of aminoglycosides for 45–60 min at 37 °C, the dose-dependent accumulation of extracellular ATP (recorded as light emission) shows a log-linear relation to aminoglycoside concentrations within the therapeutic range (Fig. 4). This relation was utilized as a calibration curve to determine aminoglycoside concentrations in clinical serum specimens. ATP leakage caused by aminoglycoside in patient serum samples incubated at the same time can be used to determine aminoglycoside concentration by referring to the calibration curve.

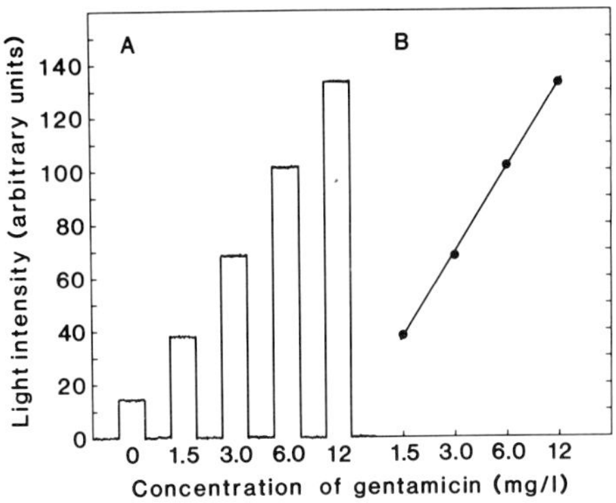

Fig. 4 A, B. Light emission obtained when mixing luciferase reagent and cultures of *E. coli*, LU 14, exposed for 45 min to gentamicin (12, 6, 3, 1.5, and 0 mg/l serum). The serum concentration in the cultures was 0.4%, **A** Chart speed: 0.5 mm/s. **B** Relation between recorded light emissions and log gentamicin concentrations utilized as calibration curve

Aminoglycosides are often encountered in combination with other antimicrobial agents, and thus it is important to use a test strain which is unaffected by the presence of other agents. *E. coli* LU 14 allowed for a specific assay, since therapeutic concentrations of other agents did not affect the bioluminescent assay of aminoglycosides [10, 12, 13].

The bioluminescent assay of aminoglycosides is rapid, results being available within 75–90 min and it has a level of accuracy similar to that of agar diffusion assay and enzyme immunoassay [10, 12, 13].

The bioluminescent assay correlated well with agar diffusion assay and enzyme immunoassay when evaluated on clinical serum specimens [10, 12, 13]. The required equipment and reagents are commercially available at a reasonable price. Thus, the simple, rapid bioluminescent assay offers an attractive alternative for laboratories performing aminoglycoside assays.

References

1. Harber MJ, Asscher AW (1977) A new method for antibiotic assay based on measurement of bacterial adenosine triphosphate using the firefly bioluminescence system. J Antimicrob Chemother 3:35–41
2. Höjer H, Nilsson L, Ånséhn S, Thore A (1976) In vitro effects of doxycycline on levels of adenosine triphosphate in bacterial cultures. Possible clinical applications. Scand J Infect Dis [Suppl] 9:58–61
3. Höjer H, Nilsson L, Ånséhn S, Thore A (1978) Evaluation of a rapid semiautomated bioassay of antibiotics. In: Siegenthaler W, Lüthy R (eds) Proceedings of the 10th international congress of chemotherapy. Current chemotherapy vol 1. American Society for Microbiology, Washington DC, pp 523–525
4. Höjer H, Nilsson L (1978) Rapid determination of doxycycline based on luciferase assay of bacterial adenosine triphosphate. J Antimicrob Chemother 4:503–508
5. Höjer H, Nilsson L, Ånséhn S, Thore A (1979) Possible application of luciferase assay of ATP to antibiotic susceptibility testing. In: Schram E, Stanley P (eds) Proceedings of the first international symposium on analytical applications of bioluminescence and chemiluminescence. State Printing and Publishing, California, pp 523–530
6. McWalter PW (1984) Determination of susceptibility of Staphylococcus aureus to methicillin by luciferin-luciferase assay of bacterial adenosine triphosphate. J Appl Bacteriol 56:145–150
7. Molin Ö, Nilsson L, Ånséhn S (1983) Rapid detection of bacterial growth in blood cultures by bioluminescent assay of bacterial ATP. J Clin Microbiol 18:521–525

8. Nichols WW, Curtis GDW, Johnston HH (1982) Analysis of the disagreement between automated bioluminescence-based and culture methods for detecting significant bacteriuria, with proposals for standardizing evaluations of bacteriuria detection methods. J Clin Microbiol 15:802–809
9. Nilsson L, Höjer H, Ånséhn S, Thore A (1977) A rapid semiautomated bioassay of gentamicin based on luciferase assay of bacterial adenosine triphosphate. Scand J Infect Dis 9:232–236
10. Nilsson L (1978) New rapid bioassay of gentamicin based on luciferase assay of extracellular ATP in bacterial cultures. Antimicrob Agents Chemother 14:812–816
11. Nilsson L (1981) Luciferase assay of bacterial ATP as a tool for rapid antibiotic assay. Thesis, Linköping University
12. Nilsson L (1981) Rapid bioluminescent assay of serum amikacin. J Antimicrob Chemother 10:125–130
13. Nilsson L (1983) Rapid bioluminescent assay for determining netilmicin and tobramycin concentrations in serum. Eur J Clin Microbiol 2:564–567
14. Thore A, Lundin A, Ånséhn S (1983) Firefly luciferase ATP assay as a screening method for bacteriuria. J Clin Microbiol 17:218–224
15. Thore A, Nilsson L, Höjer H, Ånséhn S, Bröte L (1977) Effect of ampicillin on intracellular levels of adenosine triphosphate in bacterial cultures related to antibiotic susceptibility. Acta Pathol Microbiol Scand [B] 85:161–166

Bacteriuria Detection in an Outpatient Setting: Comparison of Several Methods Including an Improved Assay of Bacterial ATP

A. LUNDIN [1], H. HALLANDER [2], A. KALLNER [3], U. KARNELL LUNDIN [1], and E. ÖSTERBERG [4]

Introduction

Bacteriuria testing is the most common laboratory investigation in primary health care. In the diagnosis of urinary tract infection (UTI) the concept of significant bacteriuria, i. e. the repeated finding of $\geqq 10^5$ CFU/ml, plays a major role [2]. However, in clinical routine the diagnosis is generally based on sediment analysis, chemical tests or simplified culture tests (e. g. dipslide tests) on a single urine specimen.

The possibility of using determination of bacterial ATP in bacteriuria testing has been investigated by several groups. As summarized in a previous paper [4] large variations in diagnostic performance have been obtained by different authors. Major reasons for these variations include: a) differences in classification of specimens due to the reference method used or its interpretation; b) difference in prevalence of UTI in the clinical material studied, and c) differences in method used for determination of bacterial ATP or interpretation of results, i. e. minimum ATP level required for a positive result (cut-off limit).

The present paper is part of a study aimed at finding an optimal method or combination of methods for bacteriuria testing in primary health care. Methods have been developed for determination of three pools of ATP: 1) Non-cellularly associated or free ATP, 2) ATP associated with somatic cells, and 3) ATP associated with bacterial cells. Three variations of the method for determination of bacterial ATP have been compared. For all methods effects of varying the cut-off limit for a positive test result have been studied. Results obtained by the ATP test are compared to those obtained by other bacteriuria tests. Finally the effect of follow-up testing of specimens with bacterial ATP levels around the cut-off limit has been studied.

1 Research Centre and Department of Medicine, Karolinska Institute, Huddinge Hospital, S-141 86 Huddinge
2 National Bacteriological Laboratory S-105 21 Stockholm
3 Karolinska University Hospital, S-104 01 Stockholm
4 Sollentuna Hospital, S-191 23 Sollentuna

Materials and Methods

Specimens

All 781 specimens were collected at Sollentuna health care centre. They were divided into 130 bacteriuric and 626 non-bacteriuric specimens by conventional culture methods performed by personnel from the National Bacteriological Laboratory. This method did not give a conclusive result in 25 specimens. These specimens were divided into 6 bacteriuric and 19 non-bacteriuric specimens by a combined method based on all laboratory data available. This method would have resulted in only 3.8% false results among those classified by conventional culture and correlated well with clinical findings in the 25 specimens classified with the combined method. Thus, in total there were 136 bacteriuric and 645 non-bacteriuric specimens.

Luminometric Tests

Luminescence was measured using a Luminometer 1250 and a slightly modified ATP Monitoring Reagent (LKB-Wallac, Turku, Finland). The ATP Monitoring Reagent contained a luciferase activity five times higher than the commercially available reagent. The buffer used for reconstituting the reagent (0.1 *M* Tris-acetate with 2 m*M* EDTA, pH 7.75) was prepared using freshly distilled water and thoroughly cleaned glassware. The reconstituted reagent was stored in the refrigerator overnight before use. In this way a reagent blank <1 p*M* ATP was achieved. All measurements were standardised by the final addition of a known amount of ATP Standard (LKB-Wallac). The following assays were performed (each in a separate tube):

a) Urine (10 µl) was added to ATP Monitoring Reagent (1 ml) and the luminescence corresponding to free ATP recorded. Triton X-100 (10 µl) was added in a final concentration of 0.1% and the luminescence corresponding to the sum of free and somatic cellular ATP was recorded. In each series of assays the slight inhibition of luciferase activity by triton X-100 was compensated for by a separate assay of luciferase activity with and without triton X-100.

b) Urine (10 µl) was added to an equal volume of 7.5% TCA (trichloroacetic acid) dissolved in distilled water. ATP Monitoring Reagent (1 ml) was added, and the luminescence corresponding to the total ATP concentration in urine was recorded.

c) Urine (1 ml) was mixed with an equal volume of a solution containing 0.2% purified apyrase (Sigma Chemical Co.) and 0.2% triton X-100 dissolved in distilled water (addition of calcium for stimulation of apyrase activity is not necessary since urine contains an adequate concentration of this ion). After 10 min incubation at 37 °C an aliquot (1 ml) was extracted by addition to four volumes of buffer (0.1 *M* Tris-acetate with 2 m*M* EDTA, pH 7.75) kept at a boiling water bath. After 5 min extracts were allow to cool and an aliquot (0.1 ml) was added to ATP Monitoring Reagent (0.9 ml). Finally the luminescence corresponding to bacterial ATP (37 °C) was recorded.

d) Urine (10 µl) was mixed with an equal volume of apyrase and triton X-100 (concentrations as in C). After 10 min incubation at room temperature an equal volume (10 µl) of 7.5% TCA was added. Finally ATP Monitoring Reagent (1 ml) was added and the luminescence corresponding to bacterial ATP (25 °C) recorded.

e) This assay was performed essentially as in d but the incubation of urine with apyrase and triton X-100 was done at 50 °C rather than room temperature. From the luminescence recorded bacterial ATP (50 °C) was calculated.

In preliminary experiments it was confirmed that a final concentration of 2.5% TCA is adequate for inactivating apyrase as well as extracting bacterial ATP. Calculations of ATP levels were based on the luminescence obtained from the known amount of ATP added as the last step in each assay. Corrections for reagent blanks were performed.

Other Bacteriuria Tests

Sediment analysis was performed on 10 ml urine using a phase-contrast microscope. Results were recorded as number of white cells per visual field and no, sparse, moderate or abundant amount of bacterial cells. Granulocyte esterase was estimated by a test strip dipped into the urine (Cytur test, Boehringer-Mannheim). Results were recorded as no, weak or strong colour. Nitrite was estimated by a test strip (Nitrite, Boehringer-Mannheim). Results were recorded as no colour (negative) or pink colour (positive). Simplified culture was performed by a dip-slide test (Uricult, Orion Diagnostica, Helsinki, Finland). Results were recorded with respect to amount of bacteria (no growth, $<10^5$ CFU/ml or $\geq 10^5$ CFU/ml) and type of flora (pure culture or mixed flora, with gram characteristics). With sediment analysis and the dip-slide test it was necessary in some specimens to record test results as uninterpretable. Particularly the type of flora on the dip-slide test resulted in a high rate of uninterpretables (6.3%). In calculations of sensitivity, specificity, and diagnostic efficiency uninterpretables are included in denominators to avoid overestimation of the diagnostic performance.

Results and Discussion

In a model experiment sterile filtered urine was inoculated with *Escherichia coli* and incubated overnight. Serial dilutions of this culture was prepared in sterile filtered urine. In these dilutions bacterial ATP (25 °C) was determined and shown to be proportional to viable count. Although the amount of ATP per bacterial cell may vary somewhat [3], it may be calculated that, with bacterial strains often found in bacteriuric specimens, 10^5 CFU/ml correspond to a concentration of bacterial ATP in urine of 0.1–0.4 n*M*.

In actual urine specimens clumping of bacterial cells may result in several cells being estimated as one CFU. There is also a risk for remaining non-bacterial ATP not degraded by the incubation with triton X-100 and apyrase. Furthermore urine specimens often contain low levels of non-pathogenic bacteria. Thus it is not surprising that the correlation between bacterial ATP and CFU is far from absolute [3]. A more practical approach than to use a cut-off limit based on model experiments would thus be to optimize the cut-off limit for a positive test with respect to the desired diagnostic performance in actual urine specimens. In bacteriuria testing this was actually the approach used in defining the cut-off limit 10^5 CFU/ml in conventional culture, i.e. the reference method, and also for sediment analysis and chemical bacteriuria tests.

In most groups of patients encountered in primary health care we feel that the risks of treating a patient without UTI with antibiotics are similar to not treating a patient with UTI, since the latter patient either will recover without treatment or will come back to the doctor. Thus a test for bacteriuria in primary health care should be optimized with respect to maximum ratio of true results, i.e. diagnostic efficiency.

It can be shown that at a given prevalence (P) a fixed value of diagnostic efficiency (E) is represented by a straight line in a diagram of sensitivity (SE) versus specificity (SP). Similarly, fixed values of predictive value of a positive result (PV_{pos}) or predictive value of a negative result (PV_{neg}) are represented by straight lines. Thus by including several sets of such lines in diagrams of sensitivity versus specificity all parameters describing the diagnostic performance (SE, SP, E, PV_{pos} and PV_{neg}) can be estimated. This way of presenting the data is used in the present paper ($P=0.17$).

In 32 bacteriuric and 160 non-bacteriuric specimens randomly selected from the entire clinical material all luminometric tests were performed. The effects on diagnostic performance of varying the cut-off limit (1–500 nM ATP) are shown in Fig. 1. It is obvious that the highest diagnostic efficiencies were obtained by methods estimating bacterial ATP, rather than free or somatic cellular ATP. Total ATP would be the second best alternative and would be somewhat easier to

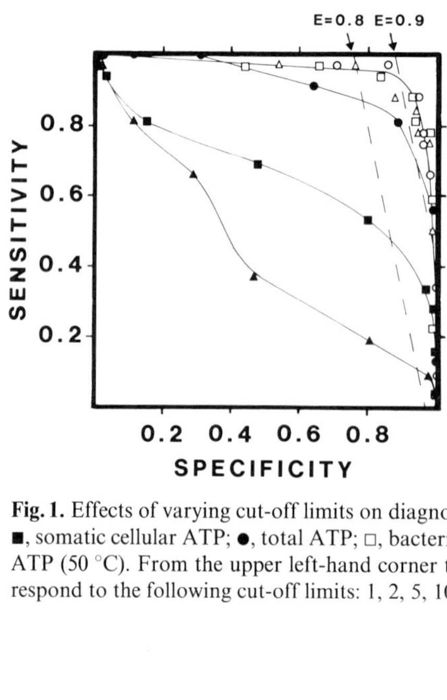

Fig. 1. Effects of varying cut-off limits on diagnostic performances of luminometric tests: ▲, free ATP; ■, somatic cellular ATP; ●, total ATP; □, bacterial ATP (25 °C); △, bacterial ATP (37 °C); ○, bacterial ATP (50 °C). From the upper left-hand corner to the lower right-hand corner the points plotted correspond to the following cut-off limits: 1, 2, 5, 10, 20, 50, 100, 200, and 500 nM ATP

determine. The three methods used for estimating bacterial ATP seem to give similar diagnostic efficiency. However, a lower cut-off limit (5 n*M* ATP) can be used for bacterial ATP (50 °C) than for the methods employing lower temperature at the incubation with triton X-100 and apyrase. Since model experiments with *E. coli* (the most frequently isolated bacterial strain) did not give a lower level of ATP at 50 °C than at lower temperatures, this may indicate a more complete degradation of non-bacterial ATP at high temperatures as suggested previously [1]. In the following, bacterial ATP (50 °C) was used with a cut-off limit of 5 n*M* ATP, i.e. considerably higher than the theoretically calculated limit. This means that commercially available ATP Monitoring Reagent can be used for bacteriuria testing without adding extra luciferase.

Optimization of cut-off limits for bacterial ATP (50 °C) and non-luminometric test was also done in all 781 specimens. The following limits for a positive test result were found to be optimal: a) moderate or abundant bacteria in sediment, b) >20 white cells/field in sediment, c) strong colour in Cytur test, d) pink colour in nitrite test, e) >5 n*M* in ATP test (a slightly higher diagnostic efficiency could be obtained at >15 n*M* ATP, but this would considerably reduce sensitivity), and f) $\geqq 10^5$ CFU/ml in dip-slide test (data on type of flora did not improve diagnostic efficiency due to too many uninterpretable results). The diagnostic performances of the tests using these cut-off limits are shown in Fig. 2. Diagnostic efficiencies were 0.94 with ATP test, 0.93 with bacteria in sediment, 0.92 with dip-slide, 0.92 with nitrite, 0.87 with white cells in sediment and 0.83 with the Cytur test. Only the ATP and dip-slide tests show high sensitivity and high specificity, and 95% confidence intervals show a considerable degree of overlapping.

All combinations of two tests were also evaluated. The highest diagnostic efficiency (0.96) was obtained by requiring that both ATP and dip-slide tests be positive for a positive combined result. Improved diagnostic efficiencies could

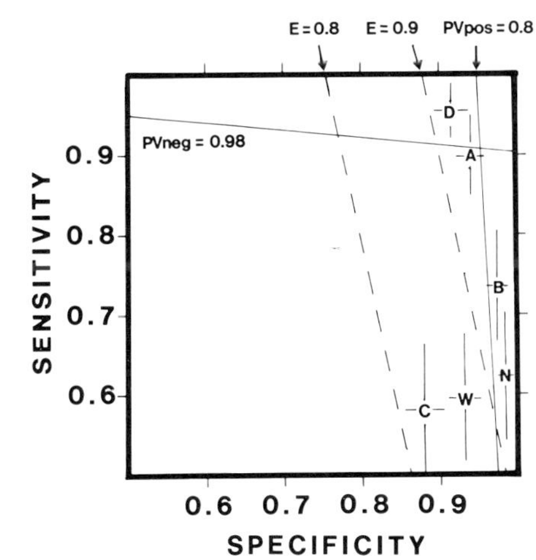

Fig. 2. Diagnostic performances of bacteriuria tests: *D*, dip-slide test; *A*, bacterial ATP (50 °C) test; *B*, bacteria in sediment; *N*, nitrite test; *W*, white cells in sediment; *C*, Cytur test. Confidence intervals (95%) are indicated on the figure

also be obtained by follow-up testing of specimens with uninterpretable or intermediate results in a primary test. The most promising results were obtained by follow-up testing of specimens with 3–25 n*M* ATP (12% of the specimens). Thus primary testing with ATP and follow-up testing with nitrite, dip-slide or conventional culture resulted in the following overall diagnostic efficiencies: 0.95, 0.96, and 0.98, respectively.

Conclusions

Among luminometric tests bacterial ATP resulted in the highest diagnostic efficiency. Among all bacteriuria tests similar diagnostic efficiencies were obtained by all methods based on bacteria (amount of bacterial cells or effect of bacterial metabolism on urine composition). Tests based on the presence of white cells resulted in low diagnostic efficiencies. Only the test for bacterial ATP has all the following properties: 1) The test provides an objective numerical result, which can be interpreted depending on clinical situation as confirming UTI (high PV_{pos} at high ATP level), as ruling out UTI (high PV_{neg}) at low ATP level) or as screening in uncertain situations (high E at optimized ATP cut-off limit). 2) The test can be performed with commercially available reagents and with modest laboratory facilities even while the patient is waiting. 3) The test can be used as a primary test with initiation of follow-up testing on the same day in specimens with intermediate test results or uncertain clinical situations. It is concluded that determination of bacterial ATP is a highly interesting alternative for bacteriuria testing in primary health care.

Acknowledgement. Many thanks are due to the personnel at the Sollentuna hospital and at the National Bacteriological Laboratory for excellent technical assistance. The study was partly sponsored by LKB-Wallac.

References

1. Johnston HH, Curtis GDW (1979) Detection of bacteria by bioluminescence – problems in removal of non-bacterial ATP. In: Schram E, Stanley P (eds) Analytical applications of bioluminescence and chemiluminescence, proceedings 1978. State Printing and Publishing, Westlake Village, pp 446–447
2. Kass EH, Finland M (1956) Asymptomatic infections of the urinary tract. Trans Assoc Am Physicians 69:56–64
3. Thore A, Ånséhn S, Lundin A, Bergman S (1975) Detection of bacteriuria by luciferase assay of adenosine triphosphate. J Clin Microbiol 1:1–8
4. Thore A, Lundin A, Ånséhn S (1983) Firefly luciferase ATP assay as a screening method for bacteriuria. J Clin Microbiol 17:218–224

Analytical Use of Luminous Bacteria in Veterinary and Medical Microbiology

S. Ulitzur, A. Naveh, and M. Barak [1]

Introduction

The use of intact luminous bacteria for analytical purposes has some clear advantages. Intact luminous bacteria are self-maintaining luminescent units and the in vivo luminescence of a single bacterium can be detected with the aid of a scintillation counter. In addition to naturally occurring species of luminous bacteria (*Photobacterium* and *Vibrio*), we cloned the luminescent system into different gram-negative bacteria, including *Escherichia coli* and *Salmonella typhimurium*. The luminous bacteria may be lyophilized and stored for many months in a cold room. Such lyophilized cultures promptly resume their full level of in vivo luminescence upon hydration.

The intensity of the in vivo luminescence emitted by luminous bacteria reflects both the cellular metabolic activity and the viability of the bacteria. Chemophysical and biological factors that affect the synthesis of the luminescence system, its activity, or the integrity of the luminous bacterial cell can be determined under appropriate conditions. In the field of medical microbiology, luminous bacteria offer a unique model and a powerful tool for studying the interaction between the bactericidal factors of the blood and bacteria. Another version of this system can be used to determine the activity and concentration of antimicrobial agents.

Determination of Antibiotic Activity with the Aid of Luminous Bacteria

The bioluminescence assays for antibiotics are based on three different principles. The first test, "the lysis test," uses a sensitive, highly luminescent culture of luminous bacteria and determines the activity of antibiotics than affect cell-wall or membrane integrity. Using this system, we were able to detect 0.01 μg/ml polymyxin B and E in a procedure that takes 20 min (Fig. 1); β-lactam antibiotics were determined with the aid of a supersensitive strain of luminescent (cloned) *E. coli*; 30 ng/ml pencillin G, ampicillin, and cephalosporin were determined in a procedure lasting 60 min. To test the antibiotics that act at the level of protein synthesis, two other tests were applied. The first one ("Induce-test"), recently described by us [1], is based on the ability of the tested antibiotic to inhibit the luciferase synthesis in the treated cells. For this test, we use a specific dark variant

1 Department of Food Engineering and Biotechnology, Israel Institute of Technology, Technion City, Haifa, 32000, Israel

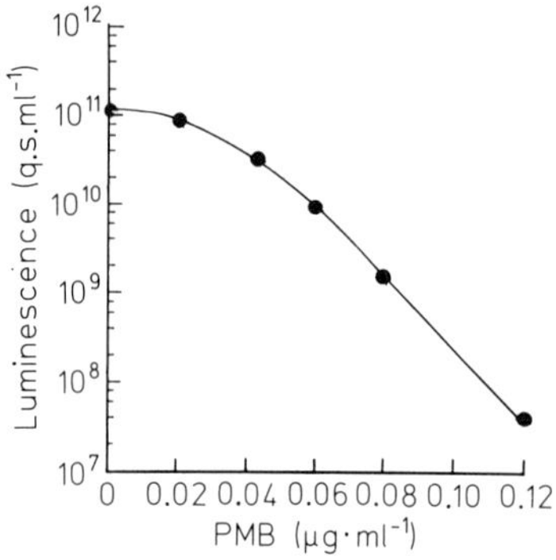

Fig. 1. 1-ml milk samples containing different concentrations of polymyxin B were incubated with *P. leiognathi* (10^7 cells/ml) for 10 min at 30 °C. 0.1-ml aliquots were diluted in 1 ml 3% NaCl containing 1% Difco peptone at pH 7.2. The luminescence was determined after 10 min at 30 °C

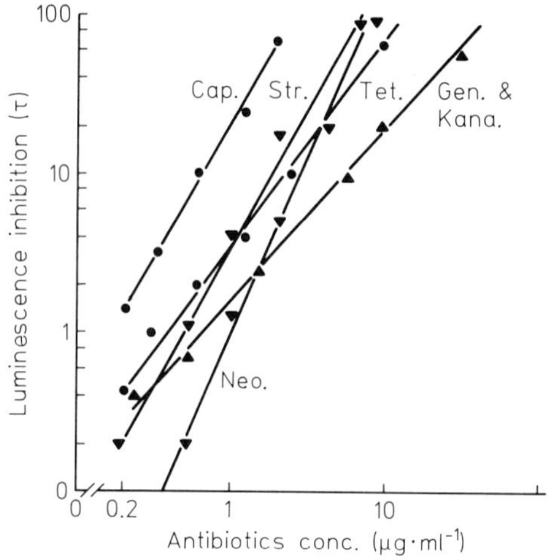

Fig. 2. Bioluminescence test for antibiotics in milk. Different concentrations of the tested antibiotics were added to 1 ml milk containing 2% NaCl and 50 m*M* MOPS buffer. The test for tetracycline was done at pH 6.0 in the presence of 25 μg/ml proflavin. The tests for the other antibiotics were done at pH 8.0 in the presence of 1.25 μg/ml proflavin. To each vial, 10^7 SD18 cells/ml were added. The figure shows the percentage of inhibition of luminescence after 45 min of incubation at 30 °C. Antibiotic symbols: *Cap*, chloramphenicol; *Str*, streptomycin; *Tet*, tetracycline; *Neo*, neomycin; *Gen*, gentamicin; *Kana*, kanamycin

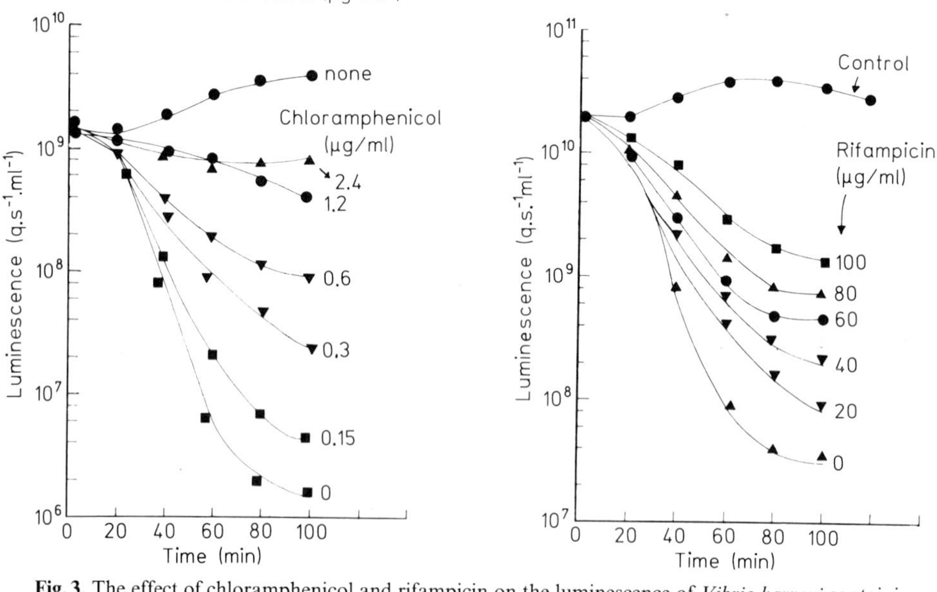

Fig. 3. The effect of chloramphenicol and rifampicin on the luminescence of *Vibrio harveyi* containing bacteriophage To. A culture of MAV, at maximal luminescence (2×10^8 cell/ml in ASWRP) was inoculated by 100 PFU/cell of the phage To and incubated for 5 min at 30 °C. 0.1-ml aliquots of the culture were then diluted in 1 ml ASWRP and different concentrations of chloramphenicol or rifampicin were added. The vials were incubated with shaking at 30 °C and the luminescence was determined with time

of luminous bacteria that undergoes prompt induction in the presence of certain DNA intercalating agents, such as norharman or acridine dyes. The in vivo luminescence of the acridine-treated cells increases more than 30-fold within 30 min. This increase in the in vivo luminescence is highly sensitive to the presence of very low concentrations of certain antibiotics. Using this system, we determined subtherapeutic concentrations of aminoglycosides, chloramphenicol, and tetracyclines in serum or milk (Fig. 2). The second test uses a biological coupled system in which highly luminescent bacteria are mixed with a specific bacteriophage (phage-test). In the absence of the antibiotic, the luminous bacteria are lysed within 45 min and their in vivo luminescence drops to almost zero. Antibiotics that inhibit DNA, RNA, or protein synthesis also inhibit the intracellular phage development and thus save the luminous bacteria (Fig. 3).

Use of Luminous Bacteria in the Determination of Serum Bactericidal Activity

Human serum bactericidal activity and the phagocytosis performed by polymorphonuclear granulocytes (PMN) were assessed by determination of the luminescence of the terrestrial luminous bacterium *Vibrio cholerae var albensis* [2–5]. The decrease in in vivo luminescence of *V. cholerae* cells in serum containing medium was directly proportional to the decrease in the viable count and was a function of serum concentration. Both immunoglobulins and the complement system are required to exert the serum bactericidal activity. Serum lacking immunoglobulins or certain complement components, especially C3, did not affect the luminescence (Fig. 4). The bactericidal effect of the serum on luminous bacteria was diminished by the presence of lipopolysaccharide or by pretreatment of the serum with different species of killed bacteria.

This system could easily distinguish between the classic complement pathway and the properdin pathway. Mg-EGTA, which inhibits only the classic complement pathway, did not inhibit the decrease in luminescence as did ethylenediaminetetraacetate (EDTA). Thus, it was possible to distinguish between deficiencies in complement components participating in both pathways and complement components that are involved only in the classic pathway. This system could also substitute the hemolytic system in complement-fixation tests (CFT) for syphilis and influenza A. With the aid of the luminescent system, a serum bactericidal activity at least tenfold higher was found in a patient with chronic benign idiopathic granulocytopenia. The activity was attributed to both elevated level (9.5 μg/ml) and activity of serum lysozyme.

Use of Luminous Bacteria in the Determination of Phagocytosis

The existing methods of phagocytosis evaluation are unable to determine all the real events occurring during phagocytosis, nor are they able to follow continuously the kinetics of the process. Phagocytosis of luminous bacteria offers an easy and simple method of determining the phagocytosis kinetics. The correlation

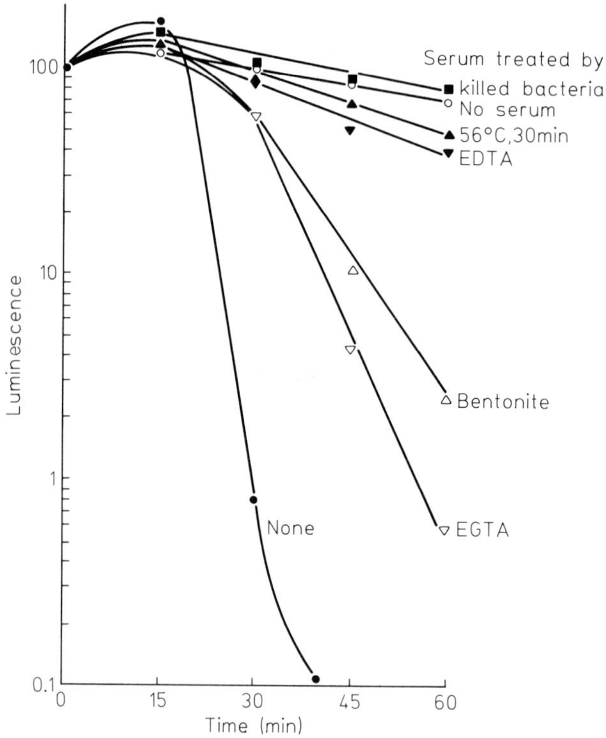

Fig. 4. Effect of various components of human serum on the in vivo luminescence of *Vibrio cholerae*. PBSC samples (1 ml) containing 30% pooled human serum were subjected to the following treatments: thermal inactivation of the complement system (56 °C, 30 min) (▲); immunoglobulin absorption by 10^8 heat-killed luminous bacteria (■); lysozyme absorption by bentonite (5 mg/ml) (△); EDTA (10 m*M*) (▼); EGTA 10 m*M*) (▽); bacteria alone (○); bacteria with untreated serum (●). To each sample, 0.1 ml of *Vibrio cholerae* cells (10^7 cells ml^{-1}) was added. The test was carried out at 37 °C, and the in vivo luminescence was determined with time

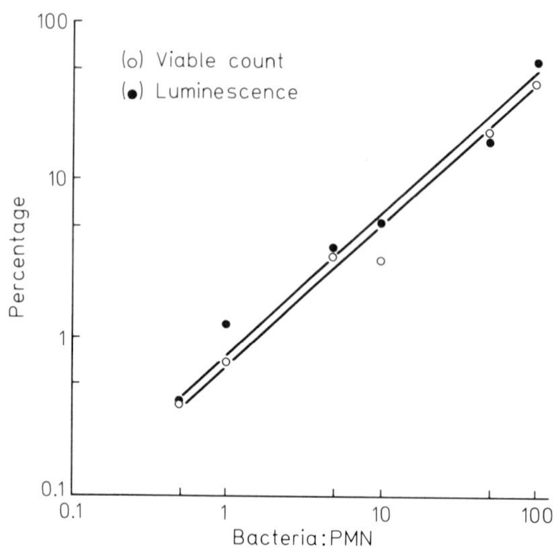

Fig. 5. Correlation between the viability and the luminescence of phagocytosed *Vibrio cholerae* cells at different bacteria to PMN ratios. *V. cholerae* cells (5×10^5–10^8 cells/ml) were suspended in 1 ml PBSC containing 10^6 PMN. The in vivo luminescence and the viable count were determined after 60 min of incubation. The decrease in luminescence (*solid circles*) and the viable count (*open circles*), both presented as percentage of the untreated bacteria, are shown as a function of the bacteria to PMN ratio

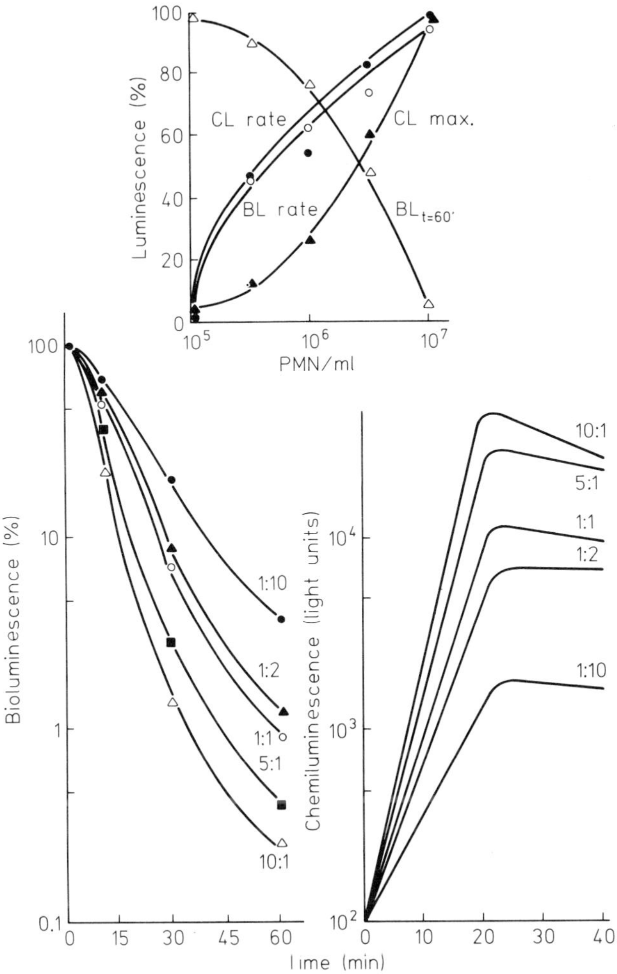

Fig. 6. The correlation between the bioluminescence and the chemiluminescence test for phagocytosis. The bioluminescence test was performed as described in Fig. 5 with 10^6 bacteria/ml and 10^5–10^7 PMN. The chemiluminescence test was performed with the same concentrations of dead bacteria in the presence of 10^{-5} *M* luminol. The correlation between the maximum bioluminescence and chemiluminescence as well as the rate of decrease in bioluminescence and the rate of increase in chemiluminescence during the process of phagocytosis were calculated as the ratio: observed value – minimal value: maximal value – minimal value

coefficient between the decrease in luminescence and the decrease in viable count is 0.999 (Fig. 5). The rate of the decrease in luminescence and the residual level of luminescence after 60 min of phagocytosis are directly proportional to the rate of increase in the phagocytosis-induced chemiluminescence and to its maximal level, respectively (Fig. 6). Opsonization requirements are comparable in both tests. Different inhibitors of the phagocytotic process caused similar changes in the rates of the bio- and chemiluminescence (correlation coefficient 0.974) and in the maximal luminescence level (correlation coefficient 0.804). The kinetics of the phagocytosis-associated decrease in luminescence can be described as a first-order reaction. The concentration of both the PMN and the luminous bacteria and the ratio between them affected the rate of the luminescence decay. The rate of phagocytosis, as expressed by the rate of luminescence decay, is a linear function of the bacterial concentration up to the saturation point 2×10^7 bacteria/10^6 PMN). Increasing the concentration of the serum used for opsonization increased the rate of the luminescence decrease, while extending the opsonization period shortened the lag in the luminescence decrease and increased the maximal decrease in luminescence. At high ratios of bacteria to PMN (100:1), the length of the opsonization period and the concentration of serum used were critical for the luminescence decrease. Under optimal conditions, up to 55 bacteria were killed by one PMN within 3 h of phagocytosis. As much as 10% of the luminous bacteria killing was found to be extracellular.

Anaerobic conditions or the presence of O_2 metabolite scavengers inhibit the luminescence decrease. The fact that there is no lag in the luminescence decrease after 90 min anaerobic phagocytosis and the direct microscopic test revealed that in anaerobic conditions the luminous bacteria are ingested, however the bactericidal systems are not activated, neither are the O_2-independent bactericidal systems effective in decreasing the level of luminescence. The reactive O_2 metabolites (hydrogen peroxide, superoxide anion, singlet oxygen, hydroxyl radical) that are formed during the active phagocytosis have been shown to induce mutations in bacteria. The test for the mutagenic activity was carried out as described [3]. Dark mutants of the luminous bacterium *Vibrio fischeri* reverted to hereditary stable luminescent forms when incubated with human PMN. The maximal mutagenic effect occurred during the first 15 min of phagocytosis and was dependent on the phagocyte to bacteria ratio as well as on the integrity of the PMN cells. Heat-killed phagocytes or disintegrated phagocytes did not show any mutagenic activity, while the supernatant of the phagocytosis reaction exerted mutagenic activity. Scavengers of the hydroxyl radical and singlet oxygen and the presence of superoxide dismutase prevented the mutations in the same way as they prevented the decrease in luminescence in bactericidal tests.

References

1. Naveh A, Potasman I, Basan M, Ulitzur S (1983) A new rapid and sensitive bioluminescence assay for antibiotics that inhibit protein synthesis. J Appl Bacteriol 56:457–462
2. Barak M, Merzbach D, Ulitzur S (1983) Determination of the serum bactericidal activity with the aid of luminous bacteria. J Clin Microbiol 18:248–253
3. Barak M, Ulitzur S, Merzbach D (1983) Phagocytosis induced mutagenesis in bacteria. Mutat Res 121:7–16
4. Barak M, Ulitzur S, Merzbach D (1983) The use of luminous bacteria for determination of phagocytosis. J Immunol Methods 64:353–363
5. Barak M, Ulitzur S, Merzbach D (1983) Elucidation of the phagocytosis mechanism with the aid of luminous bacteria. J Med Microbiol 18:65–72

Antimicrobial Susceptibility Testing Using Fluorogenic Enzyme Substrates

J. L. STANECK [1], S. D. ALLEN [2], and R. C. TILTON [3]

The past decade of antimicrobial susceptibility testing has witnessed the gradual departure from disc diffusion testing to technologies which offer quantitation, a greater spectrum of antimicrobials, semiautomation, computer assistance, and rapidity of data generation. Economic and clinical pressures will undoubtedly continue to encourage the development of rapid systems with low labor intensity. Historically, the endpoint in the majority of conventional assays has been direct recognition of growth and the inhibition thereof in a broth or agar matrix in the presence of varying concentrations of antimicrobial agents. This presentation will describe the application of a novel mechanism for the recognition of microbial growth using fluorogenic enzyme substrates to the flexibility and economy of microtiter plate methodology for quantitative antimicrobial susceptibility results.

Several substrates, which are nonfluorescent in their native state, have been studied by Seward Laboratories (London, England). When hydrolyzed by one or more bacterial enzymes, these substrates release chromophores which fluoresce upon excitation with appropriate wavelengths of light. Incubation of these substrates in the presence of an actively growing bacterial culture will produce detectable fluorescence roughly proportional to the increase in cell density. The presence of antimicrobial agents at inhibitory concentrations will not only prevent cell multiplication but will also markedly decrease the generation of fluorescence in comparison with antibiotic-free control cultures. It is, therefore, possible in principle to construct a format in which the fluorescent signal generated by a bacterial culture in the presence of stepwise dilutions of an antimicrobial agent will reflect the minimum inhibitory concentration (MIC) of the antibiotic for that organism.

The fluorogenic substrate principle was applied to the established Sensititre line of dehydrated microtiter susceptibility trays (Seward Laboratories, London, UK; Gibco Laboratories, Lawrence, MA, USA) for the determination of MICs. Primary objectives of Seward included: (1) the production of an instrument capable of automatically reading the standard Sensititre panels after the customary 18 h of incubation, and (2) the development of a same-day reading capability using this instrument.

1 Department of Pathology and Laboratory Medicine, University of Cincinnati Medical Center, Cincinnati, OH 45267-0714, USA

2 Department of Pathology, Indiana University Medical Center, Indianapolis, IN 46223, USA

3 Department of Laboratory Medicine, University of Connecticut School of Medicine, Farmington, CT 06032, USA

The conventional Sensititre system for susceptibility testing [1, 2] includes three separate microtiter plate configurations which provide for a total of 22 different dehydrated antimicrobial agents arranged in twofold dilution steps. An automatic inoculation device dispenses 50 μl broth with bacterial inoculum into each well of the plastic 96-well microtiter tray. The inoculation step, therefore, serves both to deliver the test organism to the plate and to hydrate the antibiotic dilutions with broth. An adhesive seal prevents desiccation during the ensuing 18-h incubation. The MIC is the lowest concentration of each antibiotic that prevents visible turbidity. Reading of the plate and reporting results may be computer-assisted in that manual reading of plates is guided by a mirrored electronic view box equipped with indicator lights that match an on-screen graphic display. Keyboard entry of inhibitory endpoints into a dual disc-drive microcomputer allows for data interpretation, report printing, and data storage for future retrieval and subsequent statistical analysis. This equipment was used to generate all control data with conventional Sensititre plates in the subsequent evaluation of the Sensititre Autoreader system.

Adaptation of the standard Sensititre plates for automated reading involves only the use of a modified cation-supplemented Mueller-Hinton broth formulation, to which predried fluorogenic substrates are delivered by means of a paper strip prior to the addition of bacterial inoculum. Inoculation and incubation of plates are done in the usual manner. Following approximately 18 h of incubation, automated reading is accomplished using a computer-driven automatic reader (Sensititre Autoreader). Upon keyboard command, a plate carriage transports the microdilution plate over an optical path that pulses the excitation wavelength of light and records the generation of fluorescence at the emission wavelength from each well of the tray. Raw signals from antibiotic-free control wells are compared with signals generated in the antibiotic test wells. A mathematical algorithm interprets the signals from the dilution series of each drug and assigns an MIC value that is subsequently displayed, printed, and stored.

A three-center collaborative study was established to validate the performance of the Autoreader in comparison with the manual reading of Sensititre susceptibility plates. Participants included the clinical microbiology laboratories at the University of Cincinnati (Cincinnati, OH), the University of Connecticut (Farmington, CT), and Indiana University (Indianapolis, IN). A wide spectrum of recent clinical laboratory isolates consisting predominantly of *Enterobacteriaceae*, *Pseudomonas aeruginosa*, *Staphylococcus aureus*, *Staphylococcus epidermidis*, and Group D streptococci were studied in numbers roughly proportional to their isolation rates in the clinical laboratory. Among the test isolates were representatives of 25 different species. Bacterial inoculum in all instances was nepholometrically adjusted by suspending several 18-h-old colonies in saline to match a MacFarland 0.5 turbidity standard. Appropriate dilution was made to provide an initial inoculum of 10^5 colony-forming units/ml in both the control broths and broths containing fluorogenic substrate. Three Sensititre plate formats were examined: APO1 contained antimicrobials active against gram-positive organisms; APO2A contained antimicrobials commonly used against gram-negative organisms, including second and third generation cephalosporins; APO3 contained antimicrobials commonly employed in the treatment of urinary tract pathogens. Each bacterial isolate was tested in parallel in a plate containing the fluorogenic substrate in buffered cation-supplemented Mueller-Hinton broth (test plate) as well as in a conventional plate containing standard cation-supplemented Mueller-Hinton broth without fluorogenic substrate (control plate). Both plates were incubated at 35 °C in ambient air for 18 h. MIC determinations were made from the control plate by visual reading using the computer-assisted viewing box; the automated reading of the test plate was done using the Autoreader instrument. In addition, a visual reading of the test plate was performed using the usual macroscopic growth criterion to determine MIC values.

All data were stored on floppy magnetic discs for later analysis. Each daily run of a particular plate type was accompanied by the testing of one or more ATCC quality control (QC) organisms having expected MIC values (suggested by the manufacturer for that plate type). This QC isolate testing was used both to monitor the daily test procedures and to provide day-to-day and interlaboratory reproducibility data. Furthermore, 15 clinical isolates were distributed to each center from Cincinnati to be run in triplicate for purposes of interlaboratory comparison.

Visually read control MIC values and those MIC values determined in the test plate by the Autoreader were analyzed utilizing a compilation microcomputer program that summarized the distribution of test MIC results relative to the results generated by the control plate. Each comparison between the control and Autoreader MIC for an individual organism drug combination will be subsequently referred to as a "data pair." A data pair was considered in agreement when the Autoreader MIC fell within plus or minus one well (or twofold dilution) of that of the control plate MIC value. A deviation of the test plate MIC value plus or minus two wells from that of the control was considered to be a minor discrepancy, while a major discrepancy occurred when the test MIC value exceeded the control value by greater than two wells in either direction within the twofold dilution series for a particular drug. A total of 11 247 drug/organism combinations (data pairs) was examined.

Table 1 contains the distribution among agreement categories for test MIC values relative to their respective control MIC values. The data are expressed as a percentage within each drug category for all gram-negative bacilli tested. Agreement was noted in 95.3% of all 9,619 data pairs. An additional 3.5% of all data pairs were categorized as minor discrepancies, while only 1.2% of all data pairs were considered to be major discrepancies. The percentage agreement analysis by antibiotic ranged from 91.9% for ampicillin to 98.4% for kanamycin. The largest percentage of major discrepancies was noted with cefoxitin (3.4% of all cefoxitin data pairs). Further analysis of major discrepancies revealed that approximately 60% of all major discrepant data pairs were associated with *Proteus*, in particular *Proteus mirabilis*.

Further examination of the *Proteus* discrepancy data revealed that all of the discrepant data pairs were associated with β-lactam drugs, notably cefoxitin, cefamandole, cephalothin, and ampicillin. It has been found that certain isolates of *Proteus* produce a number of enzymes with high activity toward the fluorogenic substrates utilized in the system. One can speculate that the longer interval necessary for cell-wall-active β-lactam antibiotics to exert an inhibitory effect com-

Table 1. Agreement of 18-h Sensititre Autoreader MIC values with control values 828 gram-negative bacilli

Drug	Agreement[a]	Minor[b] discrepancy	Major[c] discrepancy	No. of data pairs
Amikacin	96.5[d]	3.3	0.2	511
Gentamicin	97.9	1.7	0.4	828
Kanamycin	98.4	1.0	0.6	317
Tobramycin	97.1	2.7	0.2	511
Ampicillin	91.9	6.5	1.6	828
Carbenicillin	96.5	2.7	0.8	828
Piperacillin	96.7	2.3	1.0	511
Cefamandole	92.0	5.7	2.3	828
Cefotaxime	97.7	1.0	1.3	511
Cefoxitin	92.3	4.3	3.4	828
Cephalothin	92.4	5.6	2.0	828
Moxalactam	96.5	2.5	1.0	511
Chloramphenicol	96.9	2.9	0.2	511
Cotrimoxazole	96.5	1.9	1.6	317
Naladixic Acid	97.2	2.2	0.6	317
Nitrofurantoin	95.6	4.4	0	317
Tetracycline	95.6	3.1	1.3	317
Total	95.3	3.5	1.2	9619

[a] Autoreader MIC within ± 1 well of control MIC
[b] Autoreader MIC ± 2 wells from control MIC
[c] Autoreader MIC greater than ± 2 wells from control MIC
[d] Percentage within drug category

pared with other antimicrobials such as aminoglycosides might allow for the enzymatic release of background fluorochrome in wells containing levels of β-lactam drugs, which would subsequently cause inhibition of macroscopic growth upon 18-h incubation. Therefore, the Autoreader fails in the discrimination between background signal from these test wells and that of the control well and recognizes these particular test wells as positive. The Autoreader, therefore, reports the drug at a higher MIC value than would be reported from the visually read control plate. The observation that these discrepancies always involve a shift toward higher MIC values is consistent with this hypothesis. Also, no discrepancies were noted with *Proteus* and non-β-lactam antimicrobials. Of 140 *Proteus* isolates examined, 35 had major discrepancies usually with one to three β-lactam antibiotics each. In all, less than 15% of the total β-lactam/*Proteus* data pairs were found to be discrepant. One can further question the impact of these discrepant data pairs by examining the essential interpretation category shift, conventionally defined in terms of very major error (resistant to sensitive shift), major error (sensitive to resistant), and minor error (any shift involving an intermediate category). In 80% of the *Proteus* discrepant pairs, either no error or minor errors occurred, the remaining 20% were major errors, and no very major errors were noted.

Other than these data pair discrepancies involving *Proteus*, the distribution of major discrepancies is somewhat random with respect to both organism type and direction of deviation from the control well. It is useful in the context of procedural variation to examine other comparisons. The study design called for the Autoreader test values to be compared with control MIC values established visually in duplicate plates. To examine whether the addition of fluorogenic substrates to the broth would affect visually read MIC endpoints, comparisons were made between visually determined MIC values on the control plate and visually determined MIC values on the substrate containing test plate for all gram-negative organism-drug combinations. Agreement was greater than 97%; less than 0.6% of the data pairs were considered major discrepancies. Therefore, it was concluded that a visual determination of MIC endpoints in the presence of the fluorogenic substrates can be made with confidence should an automatic reading not be done or be questioned.

An additional comparison was made between the MIC values generated by the Autoreader and the values determined visually from that same substrate containing test plate. Category percentages for agreement, minor and major discrepancies were 97.5, 1.3, and 1.2, respectively. Agreement in this comparison was slightly higher than that noted in the earlier comparison of readings made from separate test and control plates, where small plate to plate variation might influence agreement. Therefore, readings made by the Autoreader, in a majority of the cases, accurately reflected growth patterns in the substrate containing plate.

Table 2 shows the distribution of data for 148 isolates of gram-positive cocci that were tested against 11 antimicrobials. Results were similar to those generated with gram-negative organisms in that 93.5% of the 1 628 data pairs agreed. Minor discrepancies were noted in 4.8% of the data pairs, while only 1.7% of the data pairs were considered major discrepancies. Agreements ranged from 86% among ampicillin and chloramphenicol data pairs to 98% for the gentamicin, kanamy-

Table 2. Agreement of Sensititre Autoreader MIC values with control values 148 gram-positive cocci

Drug	Agreement[a]	Minor[b] discrepancy	Major[c] discrepancy	No. of data pairs
Ampicillin	85.8[d]	11.5	2.7	148
Cephalothin	95.3	4.0	0.7	148
Chloramphenicol	85.8	10.8	3.4	148
Clindamycin	95.3	3.4	1.3	148
Erythromycin	94.6	2.0	3.4	148
Gentamicin	98.0	1.4	0.6	148
Kanamycin	98.0	1.4	0.6	148
Methicillin	93.9	5.4	0.7	148
Penicillin	90.6	4.7	4.7	148
Tetracycline	93.9	6.1	0	148
Vancomycin	98.0	2.0	0	148
Total	93.5	4.8	1.7	1628

[a] Autoreader MIC within ± 1 well of control MIC
[b] Autoreader MIC ± 2 wells from control MIC
[c] Autoreader MIC greater than ± 2 wells from control MIC
[d] Percentage within drug category

cin, and vancomycin data pairs. Examination of major discrepancies, however, showed that penicillin had the highest rate (4.7%), while tetracycline and vancomycin had no major discrepant pairs at all. Among the gram-positive cocci tested, the 99 isolates of *Staphylococcus aureus* showed the greatest agreement between Autoreader and control MIC values with 96%, 3%, and 1% of the data pairs in the agreement, minor and major discrepancy categories, respectively. It should be noted that approximately one-half of the *Staphylococcus epidermidis* tested either failed to grow in the control or test plates or failed to generate sufficient fluorescent signal to allow for an Autoreader MIC determination. Those 24 *Staphylococcus epidermidis* isolates that were run had 87.5% of data pairs in agreement with 10.5% and 2% of the data pairs showing minor and major discrepancies. Other than *Staphylococcus epidermidis*, rarely was an organism encountered, gram positive or gram negative, which failed to generate a readable fluorescent signal at 18 h.

Five quality control isolates – *Escherichia coli* ATCC 25922, *Streptococcus faecalis* ATCC 29212, *Staphylococcus aureus* ATCC 25923, *Pseudomonas aeruginosa* ATCC 25619, and *Flavobacterium meningosepticum* ATCC 13254 – were repeatedly tested in the three study centers. These studies indicated a high degree of both day to day as well as interlaboratory reproducibility. An analysis of variance for selected drugs for each QC isolate was carried out after converting the repetitive MIC values to Log_2 and adding the integer 9 in order to equalize numerically the distance between the wells of a twofold dilution series and keep all dilutions as positive values. The deviations could then be regarded as the distance from the geometric mean MIC. Virtually all QC organism drug combinations had standard deviation (SD) values ranging from 0 to 0.6, indicating less than a one-well deviation in repetitive readings made from the control plates or generated by the Autoreader. The largest SD in the visually read control values occurred with piperacillin and *Flavobacterium* (SD 1.28 for control, 1.0 for Autoreader). The only Autoreader SD above one well was noted with *Streptococcus faecalis* and clindamycin (SD 1.1); however, only eight repetitions were included in this analysis. In general, the Autoreader MIC values over a 4-month study period in three separate laboratories proved to be extremely reproducible on a day to day basis.

Fifteen randomly selected clinical isolates (three gram positive and 12 gram negative) were tested three times in each study center. The analysis of variance showed only small SDs from the geometric mean MIC as noted with the QC isolates. Most Autoreader SD values were between 0 and 0.6; rarely did a value exceed 1.0. Therefore, the reproducibility within both the control plate and the Autoreader plate groups was excellent. This analysis brought to light some interesting observations. Although deviations in repetitive determinations were small and virtually equal for visual readings of control plates between study centers, a large variance was noted between study centers with regard to *Proteus* and β-lactam antimicrobial combinations. Two isolates of *Proteus* included among the 15 circulated showed the same β-lactam major discrepancy between the Autoreader and the control MIC values noted in the clinical study; however, this discrepancy was noted only to occur in the repetitive runs conducted in Cincinnati and Connecticut. When the same two organisms were run in Indiana, the *Proteus*/β-lac-

Table 3. Autoreader/control comparisons: Distribution of agreement categories among three study centers[a]

	Agreement (%)	Minor discrepancy (%)	Major discrepancy (%)
Cincinnati	95.0	3.6	1.4
Connecticut	94.8	3.6	1.6
Indiana	96.5	2.8	0.7

[a] 828 gram-negative isolates, 9619 pairs, 18-h Autoreader system

tam MIC values were in agreement between the Autoreader and the control plates. Inspection of the raw fluorescent signal values from each center for these isolates shows the assignment of lower relative values in the Indiana testing for those β-lactam wells which were subsequently negative upon macroscopic reading 18 h than the corresponding signal assignments for those wells in the Cincinnati and/or Connecticut testing. It is not known whether this difference in values occurring only in certain *Proteus*/β-lactam combinations is due to slight differences in the electronics between instruments or to some unrecognized procedural variation between study centers. The enhanced discriminatory ability of the Indianapolis instrument was noted when percentage agreement distributions were compared between study centers for gram-negative isolates as indicated in Table 3. Major discrepancies occurred only half as many times in Indianapolis as they did in either Cincinnati or Connecticut. Further analysis of discrepancies indicated that 60% of the discrepancies in either Cincinnati or Connecticut occurred in the *Proteus*/β-lactam data pairs; however, no discrepancies involving *Proteus* and β-lactams were seen in the Indiana data. Elimination of the *Proteus* discrepancies would lower the major discrepancy rate among gram-negative bacilli to approximately 0.6% of all data pairs examined in the study.

A developmental phase has just been completed to establish a data base for a 5-h Autoreader system. Modification of susceptibility assay broth to encourage more rapid organism turnover, alteration of the concentration and nature of the fluorogenic enzyme substrates, and a change in the initial inoculum density from 10^5 to 2×10^6 CFU/ml, allows for the generation of detectable fluorescent signal in the majority of organisms tested after 5 h of incubation.

Approximately 7,000 organism/drug combinations among 600 clinical isolates were processed using the 5-h system in parallel with a duplicate control plate read visually at 18 h for the developmental stage of this project. The raw fluorescent signal at 5 h and the control MIC values have served as a basis for the construction of an interpretation algorithm capable of generating MIC values at 5 h. A nonoptimized 5-h algorithm was used to generate MIC data on 260 gram-negative isolates (3,048 data pairs) in Cincinnati and Connecticut. The 5-h Autoreader MIC values were compared with 18-h control values. Agreement was noted in 85% of the data pairs with 7.7% minor discrepancies and 8.4% major discrepancies. Aminoglycoside antimicrobials had agreement rates of approximately 93% with less than 2% major discrepancies. The β-lactam antimicrobials generally had lower agreements ranging from 78% with cefoxitin to 86% with cephalothin. Not surprisingly, the majority of discrepancies were associated with *Proteus* and occasionally *Morganella* against β-lactam antibiotics. A revised algorithm optimized for 5-h incubation has been developed which, when applied to the raw fluorescent signal data generated during the development study, increases agreements to control values in virtually all antibiotic categories to levels approaching the agreement noted in the 18-h Autoreader study. This new algorithm is expected to perform well with challenge organisms in a forthcoming 5-h validation study. Preliminary experience with gram-positive cocci in the 5-h system is very promising for those organisms that generate sufficient fluorescent signal at 5 h to allow MIC interpretation, notably *Staphylococcus aureus* and most group D enterococci. However, many *Staphylococcus epidermidis* and group D nonenterococcal isolates fail to produce signal at 5 h due to slow growth or poor enzymatic activity. Insufficient signal has been noted in approximately 15% of *Pseudomonas aeruginosa* tested, an observation consistent with the extended incubation times needed for *Pseudomonas* when using other presently available rapid susceptibility methods. It would appear that the primary application of the 5-h Autoreader system will be for *Enterobacteriaceae*, most *Pseudomonas*, and *Staphylococcus aureus*. The specific discrepancies with *Proteus*/β-lactam combinations that were noted during testing with nonoptimized interpretation algorithms hopefully will be resolved with further system modification. A validation study identical to that conducted for the 18-h Autoreader system will commence shortly at the three study centers.

In summary, the development of an instrument capable of automatically reading and interpreting quantitative antimicrobial susceptibility tests in microtiter trays using fluorogenic enzyme substrates for the monitoring of microbial growth has been highly successful. Sensititre dehydrated antimicrobial dilution trays containing fluorogenic substrate in a modified Mueller-Hinton broth can, after 18 h of incubation, be read automatically by the instrument both accurately and reproducibly. Modification of the broth, substrates, and interpretative algorithms promise similar results at 5 h of incubation for most commonly encountered clinical bacterial isolates. This innovative fluorogenic enzyme substrate technology should find application in other areas of clinical microbiology where the monitoring of growth parameters is integral to the testing modality.

Acknowledgment. The authors wish to acknowledge the excellent technical contributions made by Lana Weckbach and Anamaria West of Cincinnati, Ann Sedgwick of Connecticut, Janet Reynolds and Ed Harris of Indianapolis. Our gratitude also to Barb Boelscher for typing the manuscript.

References

1. Philips I, Warren C, Waterworth PM (1978) Determination of antibiotic sensitivities by the Sensititre system. J Clin Pathol 31:531–535
2. Pykett AH (1978) An assessment of the Sensititre method for determining minimum inhibitory and bactericidal concentrations. J Clin Pathol 31:536–538

Antimicrobial Susceptibility Tests: Automation and Mechanization

C. THORNSBERRY [1]

Introduction

Automation has come slowly to the clinical microbiology laboratory. In recent years, however, effort has been expended by microbiologists and by industry to develop automated or mechanized procedures for identifying bacteria and for testing antimicrobial susceptibility.

Methods using radiometry (DeBlanc et al. 1972), microcalorimetry (Bindford et al. 1973), bioluminescence (Gutekunst 1977), and electrical impedance (Colvin and Sherris 1979) have been tried, but systems utilizing these methods have not reached the stage where they have been studied extensively by collaborative groups. The procedures that have been fully evaluated and have significant use are generally based on optical determinations of bacterial growth by light transmission or light-scattering photometry.

Automated or mechanized systems for antimicrobial susceptibility testing yield either a category result (i.e., susceptible, intermediate, or resistant) or a minimal inhibitory concentration (MIC). In some cases, the same systems will yield either result, depending upon the user's selection of the appropriate test unit (cards, cuvettes, etc.). In the MS-2 system (Abbott Diagnostics, Irving, Texas), a computer program determines whether susceptible or resistant categories or an intermediate range of MIC results are reported.

In some systems, traditional susceptibility testing methods (Thornsberry and Baker 1981) have been automated or mechanized. In most of these, some part of the broth microdilution system has been automated but the reading and recording phases of both the broth microdilution and the agar dilution method have been automated. Some systems have been designed to yield results after only a few hours of incubation in an effort to have a same-day report.

The first automated, short-incubation system that was extensively studied was the TAAS system (Technicon Instruments Corp., Tarrytown, NY) (Isenberg et al. 1971). In this system, growth (number of bacteria) after 3 h in the presence of one concentration of an antimicrobial agent was compared with growth after 3 h in a control broth that contained no drug. A category judgment (susceptible or resistant) was made on the basis of an index calculated by comparing the number of bacteria in the presence of the drug with the number of bacteria in the control. The TAAS was automated except for inoculum preparation and contained its

1 Center for Infectious Diseases, Centers for Disease Control, Atlanta, GA 30333, USA

own computer system. Although this system is not available, many of the systems that were subsequently developed were based on similar principles.

Among the systems that were developed after the TAAS, the Autobac 1 and Autobac MIC systems (General Diagnostic, Morris Plains, NJ), the MS-2 system, and the AutoMicrobic System (AMS, Vitek Systems, Inc., St. Louis, MO) have been extensively and collaboratively evaluated by several investigators (Johnson et al. 1983; Schoenknecht et al. 1980; Thornsberry et al. 1980; Thornsberry and Baker 1981) and, for the United States, have been approved by the Food and Drug Administration. Although some similar systems are being developed and studied, activity on these types of systems seems to have slowed in comparison with the activity in the 1970s.

In the 1980s, there seems to be more commercial activity in developing automated or mechanized systems associated with the use of the microdilution system. These include dispensers for preparing the microdilution trays, inoculators, and automated and semiautomated reading devices.

In addition to susceptibility testing, many of the microdilution systems also offer bacterial identification on the same plate. For these combination systems that have an automated reading device, a microorganism is identified at the same time the susceptibility test is read. The identification and antibiogram are simultaneously printed on the report form.

Many systems also have data management capability, which, for the most part, consists of epidemiological data and summaries.

Automation in Agar and Broth Dilution Tests

Agar Dilution Tests

The replicating device described by Steers et al. (1959) has been used for more than 20 years to mechanize the inoculation of the agar plates used in agar dilution tests. With this device, plates can be rapidly inoculated with as many as 36 cultures. In the Repliscan System (Diagnostic Equipment, Inc., St. Paul, MN), identification and susceptibility plates are inoculated with a replicator similar to the Steers, and after overnight incubation results are semiautomatically read and recorded (Brown and Washington 1978). The susceptibility results are bases on growth or lack of growth in a single concentration of antimicrobial agent. Organisms are identified using the system's own data base and computer programs. The identification and susceptibility data can be viewed on a screen and are automatically printed on a special report form.

An agar dilution system (Mastascan) containing automated components has been developed in the Mast Laboratories in the United Kingdom. This system has been studied in Europe but has not been studied or approved for use in the United States.

Broth Dilution MIC Tests

The number of laboratories in which MICs are performed has markedly increased in the United States. The broth microdilution trays used in the tests can be prepared in the laboratory, or can be purchased in either frozen or lyophilized states.

For those who wish to prepare their own broth microdilution plates, automated dispensers are available which allow simultaneous dispensing of predetermined volumes of broth and antimicrobial agent into the cups of microdilution plates. The concentrations used are, therefore, not restricted to twofold dilutions, as was the case with earlier mechanized dilution systems. Three automated dispensers available commercially are the MIC-2000 system (Dynatech Laboratories, Inc., Alexandria, VA), the Anderson-Pasco system (Pasco, Inc., Wheatridge, CO), and the Quick Spense II system (Bellco Glass, Inc., Vineland, NJ). The MIC-2000 and Anderson-Pasco systems also provide automated or mechanized inoculators; the Quick Spense inoculator is not automated but does inoculate an entire plate at on time. In general, however, these plates are inoculated with disposable inoculators that are neither automated nor mechanized. Systems that automatically read the results of the broth microdilution tests are commercially available. The MR600 system (Dynatech Laboratories) can be programmed to use with different tray configurations (different antimicrobials in various wells). The MicroScan and API automated readers are made to be used with their respective systems. Other automated reading devices for microdilution tests are in various stages of development.

Autobac

The Autobac 1 is a system in which susceptibility tests are performed in broth and yield category results (susceptible, intermediate, or resistant) (Thornsberry et al. 1975). These results are based on comparisons of the amount of growth after approximately 3 h of incubation in the presence of the antimicrobial agent and the amount of growth after 3 h in the absence of the antimicrobial agent. The amount of growth in each chamber is determined using forward light scattering, and results are calculated with a built-in computer. The category of susceptibility is based on light scatter index (LSI) which is determined by comparing light scatter in the presence of antimicrobial agent and light scatter in the growth control (no antimicrobial agent).

The Autobac 1 system was extensively evaluated in collaborative investigation by seven laboratories located in various areas of the United States (Thornsberry et al. 1975). The results obtained by Autobac 1, by the standardized agar disc diffusion test (Bauer et al. 1966; NCCLS 1983), and by the agar dilution test were compared. The overall agreement of the three tests was similar and exceeded 90%. Most of the discrepancies occurred with certain organism-antimicrobic combinations: *Enterobacter* versus ampicillin and cephalothin; enterococci versus gentamicin, kanamycin, and tetracycline; indole-positive *Proteus* versus nitrofurantoin and tetracycline; *Pseudomonas aeruginosa* versus chloramphenicol, gentamicin, and kanamycin; *Serratia* versus polymyxins; and *Staphylococcus epidermidis* versus penicillin and tetracycline. In some cases, the cause of these discrepancies could be determined and remedied; e.g., in the collaborative study, a large number of discrepancies between nitrofurantoin and *Proteus* species were due to inappropriate concentrations of the drug in the elution disc (Butler and Gavan 1977). However, some drug-organism combinations have been contraindicated, e.g., cephalothin-enterococcus and kanamycin-*Pseudomonas aeruginosa*. Other published reports of Autobac 1 trials in clinical laboratories have shown general agreement with disc diffusion tests, but some additional problems with certain drug-organism combinations have been demonstrated, e.g., *Escherichia coli* and ampicillin (Stubbs and Wicher 1977; Waterworth 1976), *Staphylococcus aureus* and erythromycin (Lampe et al. 1975), and methicillin-resistant *S. aureus* and methicillin (Cleary and Mauer 1978).

The Autobac 1 system and method have been modified to permit determination of MICs. The Autobac MIC system was first developed to use three concentrations of antimicrobial agent, each contained in a disc as in the Autobac 1 test. Growth was measured by light scattering, and the MIC was reported as the least concentration that inhibited growth, as in a conventional broth dilution test.

This method was extensively evaluated in a collaborative study in which the results were compared with conventional microdilution MICs (Schoenknecht et al. 1980). During this study, it was determined that an MIC could be derived from two results (rather than three) using regression statistics done by the computer of the system. MICs obtained by the statistical calculations were generally equivalent to those obtained with the three concentrations and by microdilution.

MS-2 System

In the MS-2 system, susceptibility results are determined by comparing growth curves generated in the presence of one concentration of antimicrobial agent and in a control broth with no antimicrobial agent (Thornsberry et al. 1980). These curves are constructed by the system's computer using optical density readings made at short intervals (usually 5 min). When adequate growth has occurred, comparison of the appropriate portions of the two curves are made, and if the organism is resistant to the drug, the slope of the curve will be essentially the same as the slope of the control curve. If the organism is susceptible, the slope of the curve will be significantly different. The susceptibility result is printed out either as a category (susceptible or resistant) or as an MIC. The MIC is printed only when it is within the "intermediate" range and is determined by regression statistics. The category results are reported as susceptible, less than the lower limit of the intermediate range, or resistant and greater than the upper limit of the intermediate range, e.g., susceptible <4 µg/ml, resistant >12 µg/ml, or MIC $=$ 9 µg/ml.

The MS-2 system has been evaluated in a collaborative study by four laboratories in different areas of the United States (Thornsberry et al. 1980). In addition to evaluation with current clinical isolates, an additional set of selected challenge organisms was also tested to ensure that the system was challenged with as many drug-organism resistance patterns as possible. As one example, the challenge set contained inducible erythromycin-resistant *S. aureus* strains, which are seldom seen in the United States but are common in some parts of the world. In general, there was good correlation between MS-2, disc diffusion, and microdilution results. Discrepancies were usually associated with certain organism-drug combinations: enterococci versus cephalothin, penicillin, gentamicin, and kanamycin; *Enterobacter* versus β-lactam antibiotics; and *Serratia* versus colistin.

AMS System

The AMS is an outgrowth of systems that were developed for use in space. Its initial purpose was to identify microorganisms but has been extended to also do antimicrobial susceptibility tests.

The tests are performed in small plastic "cards" containing small wells or microcuvettes that are closed except for capillaries through which a broth suspension of the test bacteria passes to accomplish rehydration and inoculation. A carousel of inoculated kits is placed in a reader-incubator and subsequent manipulations are done automatically. An initial reading of light transmission serves as the basis on which future changes are based, i. e., the amount of growth that has occurred. Readings are made at intervals using a series of light-emitting diodes. Results are usually available in 3–6 h, and can be viewed on a screen and printed onto a special report form.

The susceptibility test card has 30 wells; one is a growth control and the others contain one or more concentrations of an antimicrobial agent. Category results, i. e., susceptible or resistant, are based on two concentrations of drugs, and MICs are based on more than two concentrations or, in some cases, statistical inference.

The system has been tested in collaborative studies and has been approved by the FDA. Recent reports indicate that agreement with the standard disc diffusion method generally varies from 85% to >95%, depending upon the organism-drug combination. Overall, however, the agreement with MIC tests is lower, though it is still, for the most part, >80% for most drug-organism combinations (Hansen and Freedy 1983; Johnson et al. 1983; Kelly et al. 1982).

ABAC and API 3600 System

In these two similar systems, tests are performed with two concentrations of each antimicrobial agent. If the test organism grows in the control well (no antimicrobial), categories of susceptibility are determined as follows: Susceptible, no growth in either concentration; resistant, growth in both concentrations; and intermediate, growth in the lower concentration but not the higher. In each case, the growth is read in a photometer, calculations are made internally, and category results are printed on a report form.

The ABAC system is not approved for use in the United States, and API has replaced the 3600 with their more conventional microdilution system.

Factors Affecting Results from Automated Systems

The methods and systems we have described are major technical advances in antimicrobial susceptibility testing, but they still have problems. Factors which affect the results of conventional susceptibility methods – type of media, pH of media, inoculum, time and temperature of incubation, and endpoint determination (Ericsson and Sherris 1971) – also can affect automated susceptibility tests. One factor that has caused problems in these systems is the time of incubation. Most of the automated systems, in contrast to the microdilution systems, have been developed as rapid tests to yield same-day results, and the actual growth time may be as little as 3 h. In some cases, factors which affect on overnight result have not been manifested at the time of the rapid-read. For example, a β-lactamase-positive staphylococcal strain may not produce adequate enzyme during a short incubation period to inactivate penicillin and thus would appear susceptible, whereas after longer incubation it would inactivate the penicillin and would be resistant. This problem may be partially resolved by manipulation of the amount of drug used in the test and by modification of the breakpoint parameters. Short incubation times may also contribute to problems of detecting methicillin-resistant staphyloccoci since these strains tend to grow very slowly (Cleary and Mauer 1978). Unfortunately, the literature tends to vary as to the efficacy of these systems for testing these organisms, so the extent of the problem is difficult to ascertain. Similar problems exist for drug-organism combinations in which the organism has high rates of mutation to resistance to the drug.

There may also be a more general disagreement between rapid-read results and overnight results. Lampe et al. reported that 28% of tests showed fourfold or greater increases in MICs after 18-h incubation as compared with 3-h readings. The correlation improved considerably when the inoculum for the 3-h test was 10^7 colony-forming units/ml and the inoculum for the overnight test was 10^5 colony-forming units/ml. They suggested that inoculum manipulation should be considered as a means to improve the correlation between rapid and overnight tests.

In the collaborative investigation of the Autobac 1 (Thornsberry et al. 1975), discrepancies were described as "very major" if the Autobac 1 system indicated that the organism was susceptible and the reference method indicated resistant, "major" if the reverse discrepancy occurred, and "minor" if an intermediate result was obtained for one system and not the other. Obviously, the "very major" discrepancies are the most serious and should occur very infrequently. An arbitrary rule of thumb has been that complete correlation should be over 90% and that the combined "major" and "very major" discrepancies attributable to the new system, should be less than 5%. Sherris and Ryan have recommended that the overall percentage of errors attributable to the new system should not exceed 5% and that very major errors should not exceed 1.5% (Sherris and Ryan 1982). Sometimes these levels of accurracy have not been achievable because of variability or inaccuracy in the reference method, but combinations with values less than 90% should always be examined carefully to determine whether changes in the procedure could be made to alleviate the problem or whether the combination should be contraindicated. It should be stressed again, however, that in a comparative study, no system can be better than the reference method, since with any discrepancy the reference method is assumed to be correct. If the reference method yields inaccurate results, the trial system that yields accurate results will be considered in error. Special attention should be paid to resistant strains whose resistance has been difficult to detect by other methods in other studies.

A correlation of 90% or better is not sufficient if the population tested did not contain any (or an adequate number of) strains that tend to show a disproportionate number of discrepancies and may occur if only sequential clinical isolates are used in the evaluation. Therefore, any new system should be tested with a set of selected strains to ensure that the system is challenged with all clinically relevant organism-antimicrobic combinations (Thornsberry et al. 1980).

Some problem drug-organism combinations may not have therapeutic significance, e.g., the kanamycin-*Pseudomonas aeruginosa* problem in the Autobac I system. Tests with combinations such as these can be contraindicated, and this has been done for certain combinations in both the Autobac 1 and MS-2 systems. With some combinations, it may be necessary to develop override systems in which the automated system is supplemented by another procedure known to yield the correct result.

Quality Control

Quality control of automated susceptibility tests is usually difficult because testing reference strains tends to yield little useful information since there is little chance to obtain an on-scale quantitative result. These strains are generally useful only to monitor gross changes that lead to changes in categories of susceptibility. The antimicrobial content of some elution discs can be monitored by disc diffusion.

Considerations for Users

If an automated system is being considered for routine clinical use, the prospective user should determine: (1) whether the system has been tested against an approved standard method in a collaborative study using an adequate sample of clinical isolates and a set of challenge strains containing selected resistance patterns, and whether the results of the study are available (preferably in a refereed publication); (2) whether specific problem organism-drug combinations are excluded from the system to be tested by another method; (3) whether the system is reproducible; (4) whether the system is mechanically reliable; (5) whether the system is flexible enough to be modified for newer developments or can be adapted for other uses; (6) the number of tests that can be performed relative to the cost of the instrument and the cost per test; (7) whether laboratory reports can be obtained within the "normal" working day; and (8) whether the system is approved by an agency such as the Food and Drug Administration or an equivalent agency, or whether it needs such approval.

It is also recommended that users read publications in which the systems are not only compared with standard methods, but also with each other, as has been reported by Hansen and Freedy (1983), Johnson et al. (1983), and Kelly et al. (1982). Users should discuss a system with other users, if possible. Furthermore, users should do their own comparative studies before adopting a new system. They should determine how the results compare with those obtained with the current method in use, how it fits into the work schedule, whether it saves labor or requires more than with the current method, and whether it is cost efficient. (See Sherris and Ryan 1982 for further discussion of this subject.) If the results do not compare favorably with those of the current method, the problem may be technical, but it could also be due to the presence of an unusually large number of isolates that are difficult to test accurately by the automated system.

References

Bauer AW, Kirby WM, Sherris JC, Turck M (1966) Antibiotic susceptibility by a standardized single disk method. Am J Clin Pathol 45:493–496

Bindford JS, Bindford LF, Adler P (1973) A semi-automated microcalorimetric method of antibiotic sensitivity testing. Am J Clin Pathol 59:86–94

Brown SD, Washington JA II (1978) Evaluation of the Repliscan system for identification of Enterobacteriaceae. J Clin Microbiol 8:86–94

Butler DA, Gavan TL (1977) Autobac 1: a rapid semi-automated antimicrobial susceptibility test system. In: Bondi A, Bartola JT, Prier JE (eds) The clinical laboratory as an aid in chemotherapy of infectious disease. University Park Press, Baltimore, pp 53–77

Cleary TJ, Maurer D (1978) Methicillin-resistant *Staphylococcus aureus* susceptibility testing by an automated system, Autobac 1. Antimicrob Agents Chemother 2:360–366

Colvin HJ, Sherris JC (1979) Electrical impedance measurements in the reading and monitoring of broth dilution susceptibility tests. Antimicrob Agents Chemother 12:61–66

DeBlanc JJ, Charache P, Wagner HN (1972) Automatic radiometric measurement of antibiotic effect on bacterial growth. Antimicrob Agents Chemother 2:360–366

DeGirolami PC, Eichelberger KA, Salfity LC, Rizzo MF (1983) Evaluation of AutoScan-3, a device for reading microdilution trays. J Clin Microbiol 18:1292–1295

Ericsson HM, Sherris JC (1971) Antibiotic sensitivity testing. Report of an international collaborative study. Acta Pathol Microbiol Scand [B] Suppl 217

Gutekunst RR (1977) Rapid procedures under development and evaluation: bioluminescence and impedance measurement. In: Bondi A, Bartola JT, Prier JE (eds) The clinical laboratory as an aid in chemotherapy of infectious disease. University Park Press, Baltimore, pp 85–100

Hansen SL, Freedy PK (1983) Concurrent comparability of automated systems and commercially prepared microdilution trays for susceptibility testing. J Clin Microbiol 17:878–886

Isenberg HD, Reichler A, Wiseman D (1971) Prototype of a fully automated device for determination of bacterial antibiotic susceptibility in the clinical laboratory. App Microbiol 22:980–986

Johnson JE, Jorgensen JH, Crawford SA, Redding JS, Pruneda RC (1983) Comparison of two automated instrument systems for rapid susceptibility testing of gram-negative bacilli. J Clin Microbiol 18:1301–1309

Kelly MT, Latimer JM, Balfour LC (1982) Comparison of three automated systems for antimicrobial susceptibility testing of gram-negative bacilli. J Clin Microbiol 15:902–905

Lampe MF, Aitken CL, Dennis PG, Forsythe PS, Patrick KE, Schoenknecht FD, Sherris JC (1975) Relationship of early readings of minimal inhibitory concentrations to the results of overnight tests. Antimicrob Agents Chemother 8:429–433

National Committee for Clinical Laboratory Standards (1983) M2–T3. Performance standards for antimicrobial disk susceptibility tests. National Committee for Clinical Laboratory Standards, Villanova, PA

Schoenknecht FD, Washington JA II, Gavan TL, Thornsberry C (1980) Rapid determination of minimum inhibitory concentrations of antimicrobial agents by the Autobac method: A collaborative study. Antimicrob Agents Chemother 17:824–833

Sherris JC, Ryan KJ (1982) Evaluation of automated and rapid methods. In: Tilton RC (ed) Rapid methods and automation in microbiology 1981. American Society for Microbiology, Washington, DC

Steers E, Foltz F, Graves BS, Riden J (1959) An inocula replicating apparatus for routine testing of bacterial susceptibility to antibiotics. Antibiot Chemother 9:307–311

Stubbs KG, Wicher K (1977) Laboratory evaluation of an automated antimicrobial susceptibility system. Am J Clin Pathol 68:769–777

Thornsberry C, Baker CN (1981) Antimicrobial susceptibility tests for bacteria. In: Balows A, Hausler WJ (eds) Bacterial mycotic and parasitic infections, 6th ed. American Public Health Association, Washington, DC

Thornsberry C, Gavan TL, Sherris JC, Balows A, Matsen JM, Thrupp LD, Washington JA II (1975) Laboratory evaluation of a rapid automated susceptibility testing system: report of a collaborative study. Antimicrob Agents Chemother 7:466–480

Thornsberry C, Anhalt JP, Washington JA II, McCarthy LR, Schoenknecht FD, Sherris JC, Spencer HJ (1980) Clinical laboratory evaluation of the Abbott MS-2 automated antimicrobial susceptibility testing system: report of a collaborative study. J Clin Microbiol 12:375–390

Turner A, Hawkey PM, Pedler SJ (1983) Computer-assisted determination of antibiotic susceptibility in photometer-read microtiter plates. J Clin Microbiol 18:996–998

Waterworth PM (1976) Automated sensitivity tests. J Antimicrob Chemother 2:104

Techniques for In Vitro Tests on New Antibiotics

S. W. B. Newsom [1]

Introduction

The large number of new antibiotics that have become available in the last decade have stimulated much "in vitro" testing in laboratories around the world. A typical paper will report the susceptibilities of perhaps 500–1,000 recent clinical isolates of pathogens against the test agent and usually include comparative data on one to seven relevant alternatives. A simple comparison of norfloxacin with nalidixic acid (see Table 1; Newsom et al. 1982) actually involved collection of 36,000 data points for susceptibility tests and a similar number for biotyping, but ended up as a single page of a journal.

The collection of so much data in a standardised way presents considerable problems, including the maintenance of a culture collection, preparation of inocula and handling the results. These have to be read, analysed, stored, and finally compared with others. All these steps are open to transcription errors and when done manually take a lot of time.

At the first Symposium on Rapid Methods and Automation in 1973 I reported on the use of a home-made multi-point inoculator, which in conjunction with 96-well trays allowed for tests on a culture collection (Newsom 1975). However one set of 96 organisms was all that could be handled in a week as the strains had to be taken from the freezer and grown up for 2 days to obtain a standard inoculum, and recording the data was tedious.

The introduction of cheap microcomputers and associated software has made a major impact on the scene, although systems such as the Sensititre (Seward), Sceptor (BBL), MicroID, and recently the Mastascan (Mast Laboratories) are aimed at routine susceptibility tests and might be expensive for a culture collection. This paper describes the development of our current system – which still maintains the original aims of "easy and economic" but has extended beyond the enterobacteria. Other topics to be discussed include the needs for worldwide standardisation of methods and for tailoring some tests to relevant clinical situations.

1 Department of Microbiology, Papworth Hospital, Papworth Everard, Cambridge, CB3 8RE, Great Britain

Table 1. Comparison of in vitro activities of norfloxacin and nalidixic acid

	Number of strains	Nalidixic acid		Norfloxacin	
		MIC 50%	MIC 90%	MIC 50%	MIC 90%
Enterobacter	250	4	16	<0.06	0.5
Escherichia coli	242	4	8	<0.06	0.5
Klebsiella	329	4	>32	0.12	2
Citro freundii	75	4	16	0.25	0.5
Proteus	200	4	8	<0.06	0.25
Pseudomonas aeruginosa	100	>32	>32	0.5	2
Staphylococcus aureus	79	>32	>32	2	4
Enterococci	48	>32	>32	4	8

Development of a Susceptibility Test System

Preparation of Inocula

The cultures are maintained at -70 °C in 3-ml screw-capped glass vials. Each strain is grown in brain-heart infusion for 18 h, then subcultured, and after 4 h an aliquot is placed in the vial and dimethylsulfoxide added to make a 20% solution (v/v). The vials are then placed in a polystyrene foam box and put into the freezer. The polystyrene retards the rate of freezing to approx 1 °C per minute, and after 24 h the vials are removed to storage racks. The slow cooling method is adapted from that originally used to preserve cells in liquid nitrogen (D. E. Pegg, personal communication).

The advantages of this method of storage are that the viable count remains unaffected by the process (unlike that in lyophilisation which decreases by at least 10^2) and a strain can be subcultured at any time by removing the stock vial and chipping off a piece of frozen culture. The viability of a stored culture is unaffected by this procedure.

To prepare a set of strains for an inoculum the relevant stock strains are subcultured on solid medium, checked for purity and then inoculated into brain-heart infusion broth and grown for 18 h at 36 °C. A 100-fold dilution is made in fresh broth, and dimethylsulfoxide is added as before. Aliquots of the dilution are then dispensed into a set of microtitre plates. The charged plates are sealed and frozen as before. Thus a set of identical inocula is prepared. The viable counts can be checked on one plate for the set. Individual plates are removed as required, thawed rapidly and used for direct inoculation of a set of media containing antibiotics either in broth (microtitre/trays) or agar (16-cm petridishes).

Preparation and Inoculation of Media

The simplest method is that of agar dilution. Iso-sensitest (or Mueller-Hinton) medium is used. A set of antibiotic-containing plates is made up, together with two control plates for purity tests on the inocula before and after the inoculation of test plates. The original home-made inoculator (Newsom 1975) has now been replaced by a simple hand multipoint inoculator (Dynatech) made with pins that can be sterilised by flaming. The inoculator can be standardised by measuring the carry-over of methylene blue from a concentrated solution into a 96-well plate filled with water. The amount of carry-over in each well can be measured in a photometer. Plates are inoculated over a template containing a standard pattern, and a centring mark is made on the rim of each plate. The plates are inoculated in sequence from that containing the most dilute antibiotic, and then incubated at 36 °C for 18 h. Antibiotic-free control plates are inoculated both before and after the test plates to ensure that the cultures are viable and not contaminated.

Plate Reading

The standard way to read plates has been by eye, recording the results of each on a pro forma as + or −, and discounting single colony growth. The pro formas are then studied in sequence and the MICs recorded in a book, in which each strain is recorded with its biotype and the results of previous antibiotic tests. As our collection of enterobacteria has grown a set of test plates evolved which had different species randomly distributed between them. Thus to add up all the members of a single strain, or to correlate different antibiotics required considerable concentration, and was prone to transcription errors.

An obvious way to avoid this problem was to employ a microcomputer. The low cost and ready availability of microcomputers has stimulated considerable interest in use of video cameras linked to processors for image analysis. One such set of equipment can be puchased in the UK from MAST laboratories under the name MASTASCAN, and is tailored for use with MAST Adatabs for agar dilution susceptibility tests. We have obtained a similar system based on a video camera with its own Z-80 processor linked to a BBC computer (Seescan, Cambridge). The software provided allows for the reading of a set of plates in sequence and computation of the MICs; in addition any discrepancies in the sequential results on a single strain are analysed and any "skip" values are reported.

Automated ELISA readers have been available for a few years, and are now provided with computer interfaces. As data collectors these are necessarily more complex and slower than video cameras, but may already be available in the laboratory. Dynatech provide software for the Apple computer to perform automatic processing of identification tests and antibiotic-susceptibility tests based on turbidity or colour readings from their plate reader. Control levels from a scan of the pre-inoculated plate can be used to subtract from those obtained after incubation. Similar equipment is becoming available for other readers, e. g. Artek and Flow. The growth of colonies on agar plates in a 96-well pattern produces enough change in optical density for automatic reading, and IQ-BIO (Cambridge) have provided us with a BBC computer programmed to process the data generated by the Dynatech reader from break-point susceptibility and biotyping tests. There is no reason why an agar-dilution MIC programme should not work equally well.

Data Analysis

Handling the mass of data generated by these tests can be a very time-consuming business. MIC 50% and MIC 90% must be worked out for each antibiotic. This is fairly simple, but comparison of the resistance of individual strains to different agents is more time consuming. However again the microcomputer can help. Data and record management systems are available for all microcomputers; the recently advertised Sinclair QL comes complete with relevant software for $ 600. We have used a "record management system" which is available in the UK for the Dragon computer at a cost of $ 100. The user specifies a number of parameters for each bacterial strain – e.g. microtitre tray number, identity, and susceptibility to antibiotics numbers 1 to *n*. The computer then prints an entry form, and the strains, their identity and tray numbers are entered. Susceptibility results can be added as they become available. The whole file of data can be indexed in different ways – thus the tray number or the MIC values for any individual antibiotic can be used. When the data has been entered any particular analysis can be done by writing a "report" routine. This instructs the computer to print any combination of data in line with an index. Thus one can print out all the "piperacillin-resistant" *E. coli* strains together with their MICs to other relevant antibiotics, or the MICs of enterobacteria to enoxacin in ascending order of resistance, in parallel with MICs to other quinolones. Such flexibility makes for easy analysis once the data has been entered. The file contents can be expanded at any time to include extra parameters or more strains. The last stage of this development will be to link the plate reader to the record system for automatic transfer of the data; this is feasible but will require more software.

Conclusion

The 96-colony agar dilution technique described above has been used satisfactorily on at least 18 different antibiotics. Provided care is taken with the inoculator, contamination has not been a significant problem, and contaminant colonies should be obvious as they would be away from the inoculum spots. There is a possibility that strains producing enzymes (e.g. β-lactamase) might interfere with the MICs of surrounding strains. This has not occurred in practise, although most of the β-lactam drugs tested by this method have been β-lactamase stable. The method has allowed our work to proceed without use of excessive technical time, but the recent addition of the automatic reader and record management system should reduce the latter significantly as well as obviating transcription errors.

Standardisation

A study of the world literature on the susceptibility tests of a new drug reveal a heterogenous pattern. Table 2 shows some results reported in the initial five papers published on norfloxacin (Ito et al. 1980; Newsom et al. 1982; Kahn et al. 1982; King et al. 1982; Downs et al. 1982). All the results are reported as MIC%s,

Table 2. Susceptibilities of pseudomonas to norfloxacin as reported in five papers

Organism	No.	MIC (%)			Authors
		50	75	90	
Ps. aeruginosa	50	2		4	Kahn et al. (1982)
	100	0.1		0.8	Downs et al. (1982)
	100	0.5		2	Newsom et al. (1982)
	175		0.38		King et al. (1982)
	50	0.2		2	Ito et al. (1980)

but four used MIC 50% and 90% and the fifth MIC 75%. The methods used also varied – two groups used broth cultures with inocula of 10^4 cells, two used agar dilution with an inoculum of 10^4 cells, and the fifth used agar dilution with 10^5 cells. In the case of norfloxacin the results were remarkably consistent; and quinolones in general do not have an "inoculum effect" unlike β-lactam antibiotics.

The dilutions used varied and some workers expressed results as an average value between dilutions. Thus it is very difficult to do more than observe from the data in Table 2 that norfloxacin is a very potent agent and note the need for some universal standardisation of tests and reports.

In the U.K. an unofficial agreement exists to test antibiotics in two-fold dilutions in a series including the value of 1 mg/l. The NCCLS have issued tentative standard methods for dilution antimicrobial susceptibility tests for bacteria which grow aerobically in the United States (draft – 1982) based largely on the information gathered by the International Collaborative Study (Ericsson and Sherris 1971); these include strict definition of inoculum sizes together with useful information on preparation of inocula and stock solutions and are to be welcomed. Hopefully such a system can be internationally accepted.

Special Tests

The standard laboratory test may not be a true guide to the clinical effectiveness of an antibiotic. Table 3 shows the results of tests done during a comparative trial of norfloxacin and cotrimoxazole for acute urinary tract infection in the community (Newsom et al. 1983). Both MICs of infecting bacteria and antimicrobial activity in the urine of patients receiving norfloxacin were repeated in Isosensitest agar, Isosensitest broth and pooled urine (pH 6). MICs in urine and the antibacterial activity of the patients urine are both much reduced, although less so for staphylococci than enterobacteria. Thus the much greater activity of norfloxacin than nalidixic acid in vitro may be much less in vivo, as Shah et al. (1983) showed the latter was much less affected by the presence of urine.

Table 3. The effect of urine on norfloxacin MICs

Strain	MICs		
	Agar	Broth	Urine
E. coli	< 0.06	0.03	2
E. coli	< 0.06	0.25	1
E. coli	< 0.06	0.06	8
E. coli	< 0.06	0.12	4
E. coli	< 0.06	0.06	8
St. saprophyticus	4		>16
St. saprophyticus	4		16
St. saprophyticus	16		16

Summary

This paper has underlined how the use of new reading devices attached to computers and the use of microcomputers for data storage and analysis has reduced the labour of testing new antibiotics at a cost that is unbelievably low compared to that at the time of the First International Symposium. Even so there is a need to ensure international standardisation of tests and to relate laboratory tests as much as possible to the clinical situation.

References

Downs J, Andriole V, Ryan JL (1982) In vitro activity of MK) 366 against clinical urinary pathogens including gentamicin-resistant Pseudomonas aeruginosa. Antimicrob Agents Chemother 21:670–672

Ericsson HM, Sherris JC (1971) Antibiotic susceptibility testing. Report of an international collaborative study. Acta Pathol Microbiol Scand [B] [Suppl] 217:1–90

Ito A, Hirai K, Inoue M, Koga H, Irikura T, Mitsuhashi S (1980) In vitro activity of AM 715 a new nalidixic acid analog. Antimicrob Agents Chemother 17:103–108

Kahn MY, Gruninger RP, Nelson S, Klicker RE (1982) Comparative in vitro activity of norfloxacin and ten other agents against urinary tract bacterial isolates. Antimicrob Agents Chemother 21:670–672

King A, Warren C, Phillips I (1982) In vitro activity of norfloxacin. Antimicrob Agents Chemother 21:604–607

Newsom SWB (1975) Easy economic typing of enterobacteria. In: Heden C-G, Illeni T (eds) New approaches to the identification of microorganisms. Wiley, New York, p 435–443

Newsom SWB, Matthews J, Amphlett M, Warren RE (1982) Norfloxacin and the g pyridone b carboxylic acids. J Antimicrob Chemother 10:25–30

Newsom SWB, Newsom RA, Toase D (1983) Norfloxacin vs cotrimoxazole for urinary tract infection in general practise. Abstracts of the 23rd interscience congress on antimicrobial agents and chemotherapy, no 519

Shah PM, Ottrad M, Stille W (1983) In vitro activity of norfloxacin in urine compared to that of cinoxacin, nalidixic acid and pipemidic acid. Eur J Clin Microbiol 2:272–274

MIC Reading with the Biology Laboratory Computer

M. KRESKEN and B. WIEDEMANN [1]

Introduction

In microbiological laboratories many techniques are used where the results are obtained by visual reading, especially if these methods are based on the turbidity of the cultures. Discrepancies between the different individuals cannot be avoided.

This is one reason for the development of several systems for automatic reading of results from microbiological tests. The Biology Laboratory Computer (BLC) has been constructed for use with different microbiological methods such as colony counting, MIC determination, and reading of inhibition zone sizes.

In this paper we want to evaluate MIC reading of the microdilution and agar dilution test by comparing the results obtained by visual and automatic reading.

Materials and Methods

The BLC has been described elsewhere (Förster et al. 1984). By the use of a television camera in connection with a personal computer, differences in turbidity are registered and calculated according to the user's needs.

Antimicrobials and bacterial strains are listed in Table 1. The microdilution was carried out on microtitre plates (Micur RST-System, Boehringer, Mannheim, Germany). The inoculum was 5.5×10^5 cfu/ml according to the guidelines of the NCCLS (NCCLS 1983). The nutrient medium was Isosensitest (Oxoid, Wesel, Germany). The agar dilution test was performed on Isosensitest agar (Oxoid, Wesel, Germany). The plates were inoculated with 5×10^4 cfu/spot using an automatic replicator (Mast Diagnostics, Hamburg, Germany). The swarming of *Proteus* strains was avoided by the addition of 0.025% of a detergent. Incubation for both methods was 16–20 h at 35 °C. For the agar dilution test the endpoint was defined as the antibiotic concentration which allows a growth of not more than three colonies or a density of the whole plaque with less than 35% of the density using the control as 100%. For the microdilution the BLC was asjusted in a way that the MIC was recorded where the optical density of the tubes was less than 35% using the control as 100%.

1 Institut für Medizinische Mikrobiologie und Immunologie der Universität Bonn, An der Immenburg 4, D-5300 Bonn 1

Table 1. Bacterial strains and antimicrobial agents

Species	n [a]	Antimicrobials for gram-negative rods	Antimicrobials for gram-negative rods / Antimicrobials for gram-positive cocci	Antimicrobials for gram-positive cocci
E. coli	10			
Citrobacter	10			
Klebsiella	10			
Enterobacter	10			
S. marcescens	10	Azlocillin	Ampicillin	Penicillin G
P. mirabilis	10	Cefsulodin	Mezlocillin	Oxacillin
P. vulgaris	10	Cefoperazone	Cefazolin	Methicillin
M. morganii	10	Ceftriaxone	Cefotaxime	Cefotiam
P. rettgeri	10	Latomoxef	Tetracycline	Erythromycin
Providencia	10	Ceftazidime	Gentamicin	Lincomycin
Acinetobacter	10	Cefoxitin	Cotrimoxazol	Vancomycin
P. aeruginosa	10	Azthreonam	Fosfomycin	Fusidic acid
S. aureus	9			
S. faecalis	8			
Control strains	5			

[a] Number of strains tested

Occurring skip tubes were defined as follows: If there was one skip tube the highest value was fixed as MIC, if there were more than one skip tube within the antibiotic dilution the lowest value was taken.

Results and Discussion

With regard to the data evaluation we asked the following questions:

1. How do the MIC values read by the BLC and the MIC values read with the naked eye correlate?
2. How often does BLC reading and reading with the naked eye give the same answer?
3. Is the difference in the MIC reading related to specific drugs and organisms?

Concerning the first question, Fig. 1 shows a scattergram for the agar dilution method with 2,140 antibiotic-germ combinations. The calculated correlation coefficient ($r = 0.989$) and the corresponding regression line ($y = 0.208 + 0.977\,x$) indicate a high interdependence. Although the correlation coefficient is close to one, there are some MICs that vary more than one dilution step up and down.

Figure 2 shows the scattergram for the microdilution method. Both the calculated correlation coefficient ($r = 0.970$) and the corresponding regression line ($y = -0.012 + 0.973\,x$) are almost identical with those of the agar dilution. A pronounced difference in the MIC readings is evident. There is a tendency to higher MICs if they are read with the naked eye.

If all MICs which have not been determined to the endpoint are excluded, the correlation coefficients do not change (Table 2). Therefore for the further evaluations all MICs can be taken into consideration.

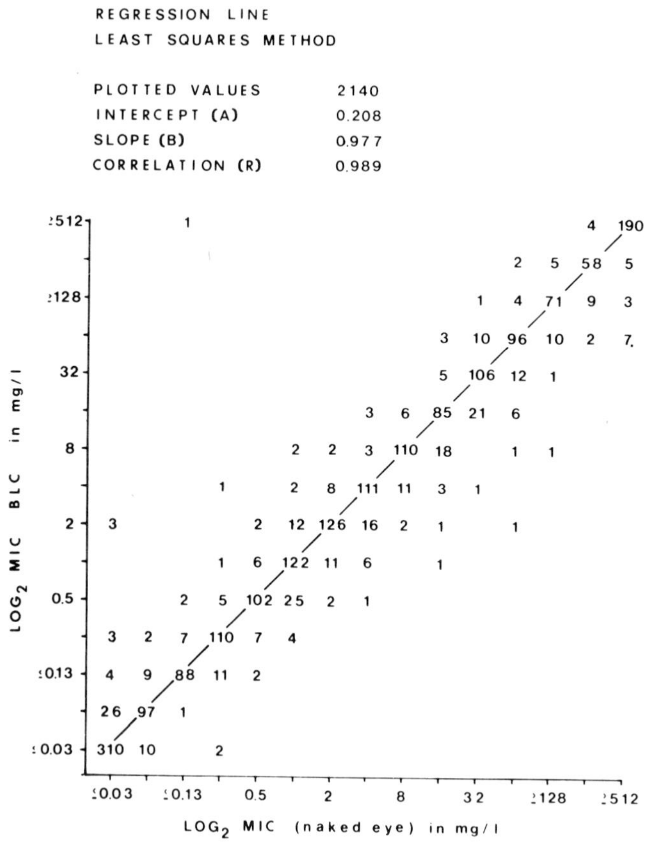

Fig. 1. Scattergram plot for the agar dilution method. The numerals indicate the numbers of data points at each location

Table 2. Correlation coefficients (r)

	Extreme values[a] included	Without extreme values[a]
Agar dilution	0.976	0.981
Microdilution	0.974	0.943

[a] MICs which have not been determined to the endpoint

A correlation coefficient gives more information on the intensity of two dependent variables than on their type of dependence. Therefore the seeming equivalence of the two methods was investigated by the statistical t-test. For this reason, in each of the 2,140 antibiotic germ combinations the difference:

MIC read by BLC − MIC read with naked eye was calculated (Figs. 3 and 4). A positive difference indicates a higher MIC for the BLC reading, a negative

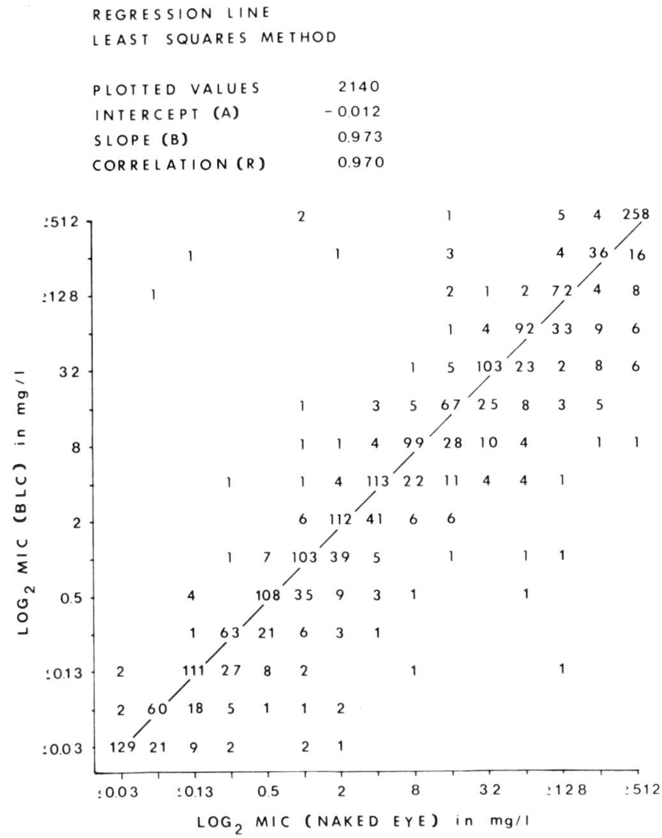

Fig. 2 Scattergram plot for the microdilution method. The numerals indicate the numbers of data points at each location

one a higher MIC for the naked eye reading. In the case of exact correspondence one would receive a difference of zero in all instances.

As described in Fig. 3, for the microdilution there are only 1,526 antibiotic-germ combinations fullfilling this praeposition. Within an acceptable range of one dilution step up and down there are 1,927 antibiotic-germ combinations, or 90%. 8.4% of all combinations show a negative difference of more than one dilution step, 1.6% a positive one. According to the statistical law of probability one would expect a nearly normally distributed population of difference values with a mean of zero. The t-test serves to test the hypothesis that the mean of the population is zero. Following the mean of -0.32 and the resulting t-value of -13.68, the hypothesis can be rejected at the 0.1% level. Hence it follows that there is a significant tendency to higher MICs if they are read with the naked eye.

Figure 4 shows the corresponding data for the agar dilution. Here 96.2% of all antibiotic-germ combinations are within the acceptable range of a one-step difference in each direction. The mean is close to zero ($\bar{X} = -0.03$) and the hypoth-

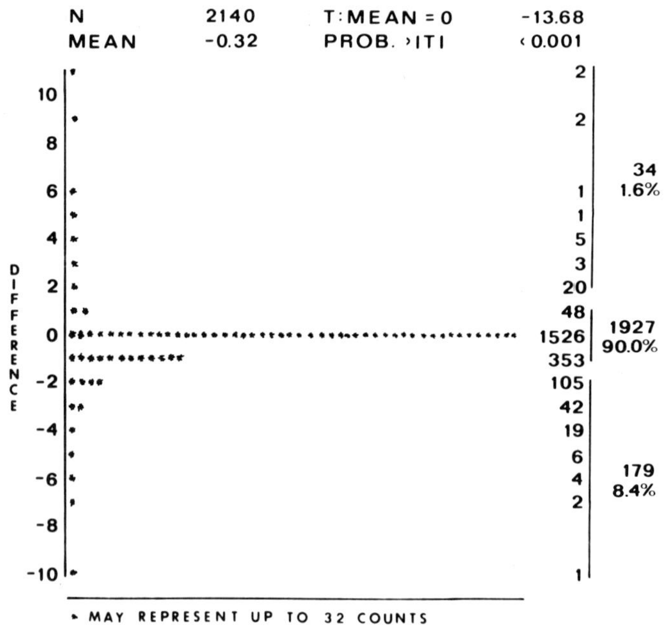

Fig. 3. The t-test for testing the hypothesis that the mean of the population: microdilution MIC read by the BLC – microdilution MIC read with the naked eye is zero

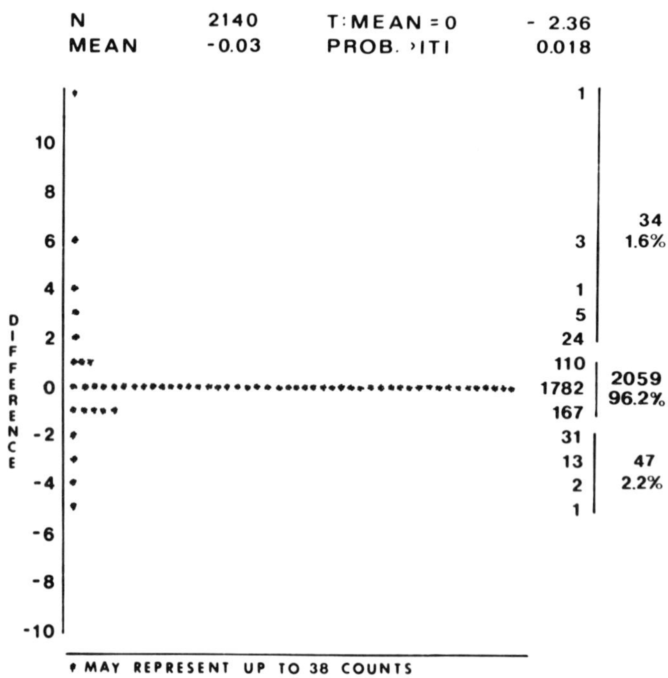

Fig. 4. The t-test for testing the hypothesis that the mean of the population: agar dilution MIC read by the BLC – agar dilution MIC read with the naked eye is zero

Table 3. Differencies of more than one dilution step between the two MIC reading methods with regard to single antibiotics and species

Antimicrobial agents/species	n [a]	BLC reading < naked eye reading	BLC reading > naked eye reading
Fosfomycin	37	37	0
Cefotaxime	25	21	4
Cotrimoxazole	21	16	5
Ampicillin	17	14	3
Tetracycline	17	11	6
Mezlocillin	14	10	4
Ceftazidime	14	13	1
Other	68	57	11
P. mirabilis	26	25	1
M. morganii	23	22	1
S. marcescens	20	18	2
Providencia	19	19	0
P. rettgeri	18	18	0
P. vulgaris	17	10	7
Acinetobacter	17	15	2
Other	73	52	21

[a] Number of cases

esis that the mean of the population is zero can only be rejected at the 2% level. Hence, there is only a slight tendency to higher MICs if they are read with the naked eye.

While for agar dilution the BLC reading of MICs does correspond to the reading with the naked eye, the results for microdilution have not yet been satisfactory. We thus sought the reasons of the noncorrespondence in the case of the microdilution.

Table 3 demonstrates the differences between the two MIC reading methods with regard to single antibiotics and species. It is obvious that some antibiotics and species have a special share in the general tendency to higher MICs when they are read with the naked eye. Among the antibiotics, above all fosfomycin, but also cefotaxime and cotrimoxazole, must be mentioned, and among the species the *Proteus* group and *Serratia marcescens* are worth noticing.

The following factors which cause a misreading by the BLC are evident: The main reason is a slight decrease in bacterial growth towards higher concentrations of the antibiotic. Owing to the definition of the threshold value, such a very slight growth of bacteria could not be detected by the BLC. Among the antibiotics this especially refers to fosfomycin, and among the species to the *Proteus* group. A better correspondence could be achieved if the threshold value were reduced. Furthermore, a misreading by the BLC was caused by a punctiform growth which could not be detected by the camera, an insufficient definition and interpretation of occurring skip tubes, a misreading by the naked eye, and a minor growth of streptococci and staphylococci at high concentrations of antibiotics which can lead to substantially higher MIC readings by the BLC.

Summarizing our experience with MIC reading by the BLC we were able to point out a very good correspondence to the MIC reading with the naked eye for agar dilution. In order to prevent the swarming of *Proteus* strains, a detergent must be added to the agar plates. This leads to an excellent reading of MICs by the BLC, but the MICs of some antibiotics, especially gentamicin, are reduced. The MIC reading for microdilution has to be further improved. By analysis of the different reasons for misreading by the BLC it will be possible to improve this automatic system.

Summary

The visual reading of MIC values from agar dilution and microdilution sensitivity tests was compared with the automatic reading by the Biology Laboratory Computer (BLC). Regarding 2,140 antibiotic-germ combinations, the difference in MIC reading with both methods was not more than +/− one dilution step in 96.2% with the agar dilution test, 90.0% with the microdilution test. In the reading of the microdilution tests there was a significant tendency towards higher MIC values with the naked eye. According to our experience the reading can be further improved by changing the threshold value.

Acknowledgement. The authors are grateful for technical help and fruitful discussions to Dr. D. Förster, Dr. H. Rost, H. Dedden, and B. Strauss.

References

Förster D, Dedden H, Grimm H, Rost H, Wiedemann B, Zembrod A (1984) Computer assisted reading system for microbiological tests. Fourth International Symposium on Rapid Methods and Automation in Microbiology and Immunology, seminar 59

National Committee for Clinical Laboratory Standards (1983) Tentative standard M7-T. Standard methods for dilution antimicrobial tests for bacterial which grow aerobically. Villanova, Pennsylvania

The Alteration of Microbial Growth Curves by Antibiotics

D. Greenwood [1]

Almost all the devices that have been described for the performance of rapid semi-automated susceptibility tests, including the Pfizer Autobac, the Abbott MS 2, and the Roche Cobas systems, use some form of turbidimetric measurement in order to determine the result.

It is perhaps surprising that the adoption of turbidimetry by these costly machines has passed seemingly unchallenged, since turbidimetric measurement as a means of assessing bacterial growth – and in particular the response of bacterial cultures to antimicrobial agents – has always been a subject of contention among microbiologists.

Opponents of turbidimetry point to the interpretive difficulties posed by measurement of microbial mass, which merely reflects the sum total of various changes that may have taken place in the culture, including any morphological alterations induced by antimicrobial compounds (Bergan and Carlsen 1980; Zanon 1977).

Those of us who have successfully used turbidimetry for many years to elucidate aspects of the response of bacteria to antibacterial agents acknowledge these difficulties but argue that with careful attention to experimental design and recourse to the microscope to observe the time course of morphological changes, important information can be gleaned by use of turbidimetric methods (Greenwood 1977; O'Grady and Greenwood 1978).

Many of the pitfalls and attractions of turbidimetry can be illustrated by reference to the effects of β-lactam antibiotics: since these agents produce concentration-dependent, morphological (and therefore turbidimetric) changes in microorganisms, and certain β-lactam antibiotics evoke anomalous morphological changes in gram-negative bacilli; furthermore, destruction of the antibiotics by bacterial enzymes may complicate interpretation of the results, and the responses of various bacterial species may differ fundamentally.

Concentration-Dependent Effects

Figure 1 shows the response of an exponentially growing culture of *Escherichia coli* to various concentrations of benzylpenicillin. Following exposure to 8, 16, or 32 mg penicillin per litre, growth continued normally for 1–2 h, after which a fall

1 Department of Microbiology and PHLS Laboratory, University Hospital, Queen's Medical Centre, Nottingham NG7 2UH, Great Britain

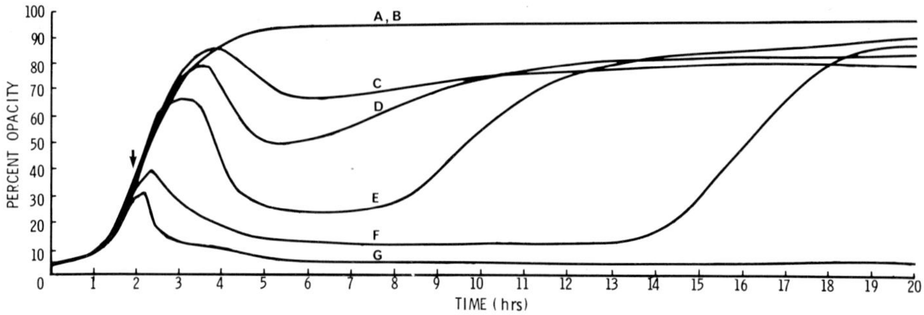

Fig. 1. Continuous turbidity records of *Escherichia coli*. Benzylpenicillin added at *arrow* to achieve the following concentrations (mg/l): *A*, 0; *B*, 4; *C*, 8; *D*, 16; *E*, 32; *F*, 64; *G*, 128. (Greenwood and O'Grady 1973 a)

in turbidity occurred, the extent of which increased with drug concentration. Later bacterial growth resumed, and the turbidity started to rise once more. Concentrations of benzylpenicillin above 32 mg/l caused an immediate deviation from normal growth and a rapid fall in turbidity. Recovery occurred during the overnight incubation period following exposure to 64 mg penicillin per litre, but not following exposure to 128 mg/l. The latter concentration, that at which no turbidity is visible after overnight incubation, corresponds to the conventionally determined broth minimal inhibitory concentration (MIC) for the high inoculum used.

Microscopy reveals that concentrations of penicillin allowing a period of normal growth before the turbidity falls inhibit septation, so that the bacteria grow as long filaments, some of which eventually lyse. Higher concentrations of antibiotic cause spheroplast formation leading to rapid lysis, unless osmotic protection is provided. Regrowth following continued incubation is associated with slow hydrolysis of the antibiotic, allowing persisting bacteria to resume growth.

This type of response is typical of the majority of gram-negative bacilli exposed to most β-lactam antibiotics, although there may be marked differences in the concentration at which the various morphological and turbidimetric changes occur (Greenwood and O'Grady 1973 a). However, some cephalosporins, notably the oral derivatives cephalexin and cephradine, induce only the filamentation response over a broad range of concentrations. Consequently, these agents cause a fall in turbidity only after exposure for several hours. Similarly, the amidinopenicillin, mecillinam, which induces unusual morphological changes in gram-negative bacilli (Greenwood and O'Grady 1973 b), allows bacterial growth to continue at a near-normal rate before any lysis occurs.

Some of the earlier cephalosporins, such as cephaloridine, cephalothin, and cephazolin, which evoke the typical range of concentration-dependent morphological effects in gram-negative bacilli, are susceptible to hydrolysis by the chromosomal β-lactamase possessed by all *E. coli* strains, so that regrowth of dense inocula is a common finding in turbidimetric tests. The dramatically different turbidimetric responses elicited by cephalosporins of both the cephalexin and the cephalothin types in *E. coli* are shown in Fig. 2. In conventional, low inoculum

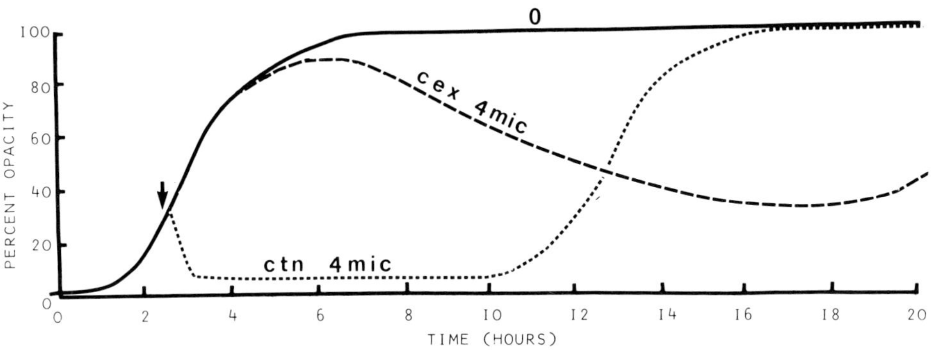

Fig. 2. Continuous turbidity records of *E. coli*. Cephalexin (*cex*) or cephalothin (*ctn*) added at *arrow* to achieve a concentration equivalent to 4 MIC

titrations, the MICs of cephalexin and cephalothin for this strain were identical, but the response of a dense inoculum to concentrations equivalent to 4 MIC was fundamentally different. The question of whether these different responses have any therapeutic implications has not been resolved.

Species Differences in Response to β-Lactam Antibiotics

These may manifest themselves in various ways.

Susceptibility to β-Lactamases

The problem of slow hydrolysis by chromosomal β-lactamases has already been referred to. Further difficulties may arise with gram-negative bacilli containing plasmids that mediate β-lactamase production. These enzymes, including the common TEM, SHV, and OXA types (Simpson et al. 1980), usually inactivate penicillins so efficiently that no response whatsoever is detected in turbidimetric systems, and the organisms appear unequivocally resistant. However, some cephalosporins, which penetrate the gram-negative cell more easily and are more slowly degraded by some plasmid-mediated enzymes, may cause profound lysis of the culture soon after antibiotic addition, with regrowth of persisters rapidly supervening as the drug is hydrolyzed (Greenwood and O'Grady 1973a, 1975). Such strains show very marked inoculum effects in conventional titrations and may appear susceptible in short-term turbidimetric tests, although the rapidity with which the cephalosporins are hydrolyzed raises serious doubts about their therapeutic efficacy.

Antipseudomonal β-Lactam Antibiotics

Other species differences may reflect idiosyncracies of bacterial response. For example, most *Pseudomonas aeruginosa* strains undergo filamentation as the pri-

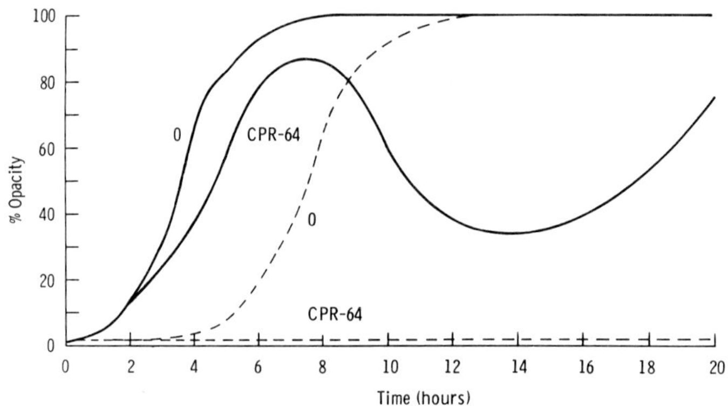

Fig. 3. Continuous turbidity records of *Pseudomonas aeruginosa* showing the effect of inoculum density on the turbidimetric response to an antipseudomonal β-lactam antibiotic. *Solid lines*, inoculum = 10^6 organisms/ml; *broken line*, inoculum = 10^4 organisms/ml. *CPR-64*, cefoperazone added at time 0 to achieve a concentration of 64 mg/l. (Greenwood and Eley 1982b)

mary response when exposed to antipseudomonal β-lactam antibiotics. With some *P. aeruginosa* strains exposed to certain antipseudomonal compounds, a fall in turbidity may not be apparent for 6 h or more, and such strains display very prominent inoculum effects (Greenwood and Eley 1982a, b). Such an example is illustrated in Fig. 3, which shows the turbidimetric response of two inocula of *P. aeruginosa* exposed to a concentration of cefoperazone equivalent to 16 times the conventionally determined MIC. Not only does growth proceed at a near-normal rate for about 7 h, but regrowth of the higher inoculum occurs during overnight incubation because of hydrolysis by the chromosomal β-lactamase of *P. aeruginosa*. These effects are masked when the lower inoculum is used.

Effect of Osmolality

A rather different situation arises with organisms like *Proteus mirabilis* and *Haemophilus influenzae*, which characteristically exhibit a low intracellular osmolality and hence resistance to lysis by β-lactam agents (Greenwood and O'Grady 1972; Roberts et al. 1974). Figure 4 shows the response to ampicillin (ca. 10 MIC) in broth of which the osmolality was 234 mosmol/kg, of strains of *E. coli* and *P. mirabilis* which, in conventional MIC titrations, displayed similar susceptibilities to ampicillin. In these conditions rapid, profuse lysis of *E. coli* occurs, whereas

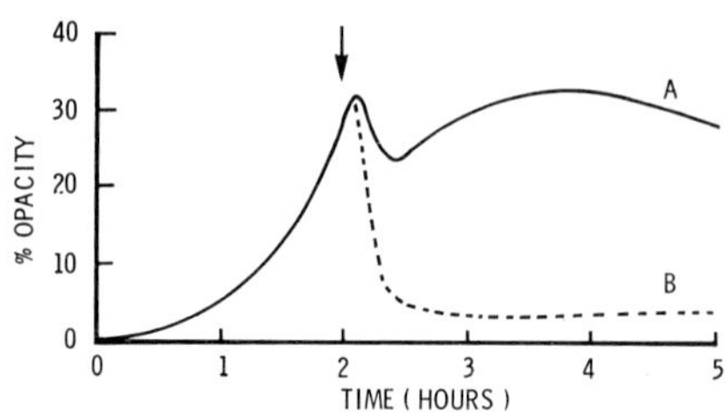

Fig. 4. Continuous turbidity records of *Proteus mirabilis* (*A*) and *Escherichia coli* (*B*) exposed (at *arrow*) to ampicillin (50 mg/l) in broth of which the osmolality was 234 mosmol/kg. (Greenwood and O'Grady 1972)

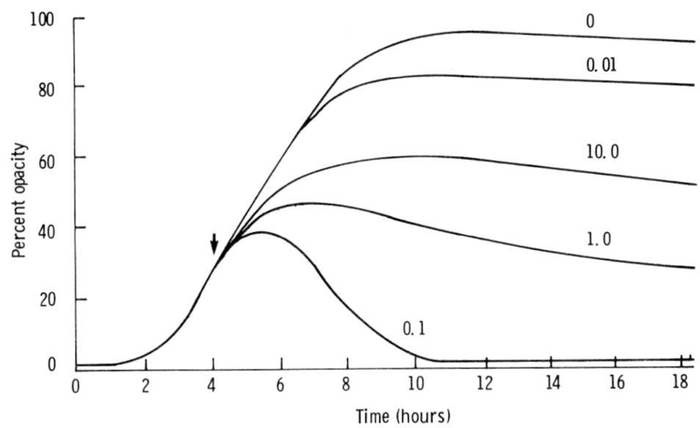

Fig. 5. Continuous turbidity records of *Staphylococcus aureus* showing Eagle's optimal dosage effect. Benzylpenicillin added at *arrow* to achieve the concentrations (mg/l) shown. (Elliott et al. 1979)

the majority of *P. mirabilis* cells transform into spheroplasts, which increase the turbidity before slow lysis ensues. Rapid lysis can be induced in *P. mirabilis* by lowering the osmolality of the culture medium, and conversely, lysis can be prevented in *E. coli* by raising the osmolality (Greenwood and O'Grady 1972).

Response of Gram-Positive Bacteria

Gram-positive cocci respond to β-lactam antibiotics in a rather different manner than gram-negative bacilli, probably because of basic differences in the nature of the lethal events occurring in the two types of bacteria (Greenwood 1982). Rapid lysis, such as is seen when gram-negative bacilli are exposed to high concentrations of β-lactam-antibiotics, is not a feature of the response of gram-positive cocci, and there may be considerable strain variation as well (Elliott et al. 1979). Moreover, concentration-dependent effects may take a different form in gram-positive cocci because of Eagle's optimum dosage effect (Eagle and Musselman 1948). This effect is illustrated in Fig. 5, which shows the response of a strain of *Staphylococcus aureus* to various concentrations of benzylpenicillin.

Interpretation of Turbidimetric Data

The above examples will serve to illustrate the wide variety of effects that might be encountered when the response of bacteria to antibiotics is monitored turbidimetrically. Many other examples could be cited involving other antimicrobial agents.

Clearly, the turbidimetric reading made after a few hours' exposure to antibiotic (i.e., the reading relevant to rapid, semiautomated susceptibility tests) merely reflects the peculiarities of response of a particular inoculum of a particular species (or strain) exposed to a particular agent. It may or may not bear a relationship to the conventionally determined MIC, which – at least in its broth titration form – represents the concentration that holds the turbidity of a moderate inoculum below the visible threshold after overnight incubation.

Use of turbidimetry to assess antimicrobial activity as a guide to therapy thus gives rise to two major difficulties:

1. The information generated by turbidimetry is qualitatively different from that provided by the MIC endpoint, which is an all-or-none effect that can be generalized to most bacteria and antibiotics. In contrast, the turbidimetric record represents the visible product of a response of the organism to the antimicrobial agent that may be unique to the particular conditions of the test.

2. Since the factors that determine the turbidimetric response and the MIC endpoint are fundamentally different, direct concordance between the two tests cannot be expected. Moreover, since the turbidimetric response must occur, by definition, in the visible range (normally $>10^6$ bacteria per ml), even the inocula used are not comparable.

Given these difficulties, it seems curious that the reliability of rapid, automated susceptibility methods has been assessed simply by trying to establish concordance with conventional MIC tests, often, even more curiously, through the intermediary of the Bauer-Kirby standardized disc diffusion test, which itself offers an imperfect correlation with the MIC.

Such concordance can be achieved within acceptable limits – at least for most organisms and antimicrobial agents – but only by manipulating the conditions of the turbidimetric tests so as to ensure concordance. If we were confident that MIC results provided the most useful indication of likely clinical benefit, this approach would be eminently sensible. However, there are few clinical situations in which this has been demonstrated.

An alternative approach, one which might be more useful, would be to accept the turbidimetric data irrespective of correlation with conventional laboratory methods and to attempt to establish the therapeutic relevance of the response obtained.

The attractions of this approach are twofold: the nature of the bacterial response can be taken into account and the response of the infecting organism to a therapeutically achievable antibiotic concentration can be followed over a period representing the normal interdose interval.

Turbidimetry is not an alternative means of measuring MICs of antibiotics; it is a means of generating information about bacterial response to antimicrobial agents that is not available from simple titrations. To force those devices that measure microbial susceptibility by turbidimetry to confirm to the MIC yardstick not only causes inevitable methodological difficulties in their use, but also results in a misconception of the kind of information generated by these devices and puts to ineffective use the information they are capable of providing.

References

Bergan T, Carlsen IB (1980) Bacterial kill rates of amoxycillin and ampicillin at exponentially diminishing concentrations simulating in vivo conditions. Infection 8 [Suppl 1]:S103–S108

Eagle H, Musselman AD (1948) The rate of bactericidal action of penicillin in vitro as a function of concentration, and its paradoxically reduced activity at high concentrations against certain organisms. J Exp Med 88:99–130

Elliott TSJ, Greenwood D, Rodgers FG, O'Grady F (1979) The response of *Staphylococcus aureus* to benzylpenicillin. Br J Exp Pathol 60:14–23

Greenwood D (1977) In defence of turbidimetry. J Antimicrob Chemother 3:286–287

Greenwood D (1982) Antibiotics of the beta-lactam group. Research Studies Press, Chichester

Greenwood D, Eley A (1982a) A turbidimetric study of the response of selected *Pseudomonas aeruginosa* strains to eight β-lactam antibiotics. J Infect Dis 145:110–117

Greenwood D, Eley A (1982b) Comparative antipseudomonal activity of some newer β-lactam agents. Antimicrob Agents Chemother 21:204–209

Greenwood D, O'Grady F (1972) The effect of osmolality on the response of *Escherichia coli* and *Proteus mirabilis* to penicillins. Br J Exp Pathol 53:457–464

Greenwood D, O'Grady F (1973a) Comparison of the responses of *Escherichia coli* and *Proteus mirabilis* to seven β-lactam antibiotics. J Infect Dis 128:211–222

Greenwood D, O'Grady F (1973b) FL 1060: a new beta-lactam antibiotic with novel properties. J Clin Pathol 26:1–6

Greenwood D, O'Grady F (1975) Resistance categories of enterobacteria to β-lactam antibiotics. J Infect Dis 132:233–240

O'Grady F, Greenwood D (1978) Estimation of antibacterial activity by turbidimetry. In: Siegenthaler W, Lüthy R (eds) Current chemotherapy. American Society for Microbiology, Washington, pp 471–472

Roberts DE, Ingold A, Want SV, May JR (1974) Osmotically stable L-forms of *Haemophilus influenzae* and their significance in testing sensitivity to penicillins. J Clin Pathol 27:560–564

Simpson IN, Harper PB, O'Callaghan CH (1980) Principal β-lactamases responsible for resistance to β-lactam antibiotics in urinary tract infections. Antimicrob Agents Chemother 17:929–936

Zanon U (1977) Sub-inhibitory levels of antibiotics. J Antimicrob Chemother 3:106–107

Clinical Needs for Determination of Antimicrobial Agents in Fluids and Tissues

S. Norrby [1] and C. E. Nord [2]

Introduction

Pharmacokinetic documentation of an antibiotic is needed at various stages of the development of a new agent before it can be used in clinical trials. Sometimes it is necessary to monitor dose size in routine clinical use of the drug. The first of these requirements is mandatory. A new antibiotic cannot be allowed to be used in the treatment of infections before its bioavailability and distribution in man is known from animal kinetic studies and phase I kinetic studies in healthy subjects. Animals studies alone will never suffice since many antibiotics have different kinetics in animals and man. Before a new antibiotic enters clinical studies, the plasma kinetics and excretion pattern must be known. However, due to ethical considerations, the distribution to peripheral compartments is only rarely studied in healthy subjects, and animal data will have to be extrapolated to the clinical situation until studies can be performed in patients. When the basic pharmacokinetic information is available and an antibiotic becomes a drug licenced for routine use, there may still be reasons to determine concentrations achieved in plasma, urine, tissues, or tissue fluids. Indications for such procedures, which are often costly, are lack of information on penetration of an antibiotic to a certain compartment and, above all, a narrow therapeutic spectrum, i.e., small differences between toxic and therapeutic levels. In the following review, the indications for pharmacokinetic analysis of antibiotic concentrations in the development of a new antibiotic and in its routine use will be discussed.

Bioavailability Studies

With drugs administered parenterally bioavailability is normally not a problem. It was not until chloramphenicol sodium succinate that serum concentrations have been reported which are lower after intramuscular (i.m.) than after oral administration [1]. The reason seems to be an insufficient hydrolysis of the ester to active chloramphenicol, and since the study indicated that chloramphenicol administered by the i.m. route may fail in typhoid treatment, the serum concentra-

1 Department of Infectious Diseases, University of Umeå, Umeå Regional Hospital, S-901 85 Umeå
2 Department of Microbiology, Karolinska Institute, Huddinge Hospital, S-141 86 Huddinge

tions must be monitored or, preferrably, the antibiotic should be administered by the intravenous (i. v.) or oral routes.

Orally administered drugs must be thoroughly studied in terms of bioavailability. The broader antibacterial spectrum that an antibiotic has, the better absorption from the gastrointestinal tract must be required. Insufficient absorption of a broad-spectrum drug will lead to increased risks of disturbances of the fecal flora with subsequent risks of diarrhea and/or selection of resistant strains. Many antibiotics, e.g., chloramphenicol, doxycycline, clindamycin, and cephalexin, are almost completely absorbed. Other antibiotics are dependent upon relation of the administration to the intake of food. Marked differences in the serum concentrations achieved with various preparations and in relation to food intake have been demonstrated with erythromycin. DiSanto and Chodos [2] found that, when the area under the serum concentration curve (AUC) was studied, following administration of unprotected or film-coated erythromycin base as single doses to healthy subjects, AUC was 47% larger with unprotected base if the subjects were kept fasting for a prolonged period before the intake. If, on the other hand, the preparations were taken 2 h after a meal, the AUC was 58% higher after intake of the film-coated tablets. A similar high degree of dependence on food intake has been demonstrated for the new axetil ester of cefuroxime which is considerably better absorbed if taken after food than in a fasting stage [3]. In all of these studies, healthy subjects were used. In the clinical situation other factors of importance should be considered; the patients may have deficient stomach function and it is difficult to guarantee patient compliance if a strict discipline is required in terms of a relation between food and drug intakes. Under clinical conditions, there are few indications for routine control of the bioavailability of antibiotics. However, assays of serum concentrations are indicated in patients with stomach dysfunctions or chronic enteric diseases, e.g., Crohn's disease, and as a mean to control a patient's compliance. With some well-absorbed drugs with a narrow therapeutic spectrum there may also be reasons to monitor serum levels in order to avoid toxic concentrations in patients who are at a risk to accumulate the drug during treatment. Examples of such antimicrobial agent are co-trimoxazole when used to patients with markedly impaired renal function and flucytosin. In the latter case, regular determinations of serum levels before and after administration should be performed to guarantee therapeutic levels, avoid development of resistance with the *Candida* spp. treated and maintain nontoxic levels [4, 5].

Routine Monitoring of Serum Concentrations

The use of aminoglycosides requires monitoring of serum concentrations to avoid toxic levels and achieve therapeutic concentrations. A clear relation has been showed between nephrotoxicity of aminoglycosides and serum concentrations before a new dose is administered. In a recent study by Moore et al. [6] the mean trough level of gentamicin or tobramycin was 2.6 ± 0.1 (mean $\pm$SD) mg/l in 184 patients who showed no signs of nephrotoxicity (defined as a 50% or more reduction of creatinine clearance) and 3.4 ± 0.3 mg/l in 34 patients with nephrotoxicity. The difference was statistically significant ($p < 0.04$). In another recent study [7]

15 of 27 patients with a serum creatinine increase of more than 30% during gentamicin treatment had trough levels above 2 mg/l while the corresponding figure in patients with no nephrotoxicity was 14 of 59 ($p < 0.025$). Surprisingly few investigators have discussed the dilemma that the increased serum levels of aminoglycosides may be the result of nephrotoxicity of the drugs rather than the cause of such reactions. Fee [8] studied factors of importance for development of ototoxic reactions to aminoglycosides and came to the conclusion that serum levels are not predictive. Similar results have been obtained by others [9] and may be explained by the fact that excessive serum concentrations are today avoided due to present monitoring routines. Since aminoglycosides do accumulate during treatment due to reduced glomerular filtration rate [10], monitoring of serum concentrations should continue at regular intervals during treatment, e.g., twice a week. Equally, if not more important as monitoring aminoglycoside concentrations to avoid toxicity is the determination of postdose levels to guarantee therapeutic levels. In many European countries a low dose of gentamicin or tobramycin, 1 mg/kg/dose, was routinely used for many years. At that dose, many of the patients did not reach 1 h postdose levels of 4 mg/l or higher and many of them were likely to have received inappropriate doses for treatment of systemic infections [10]. Even at a dose level of 1.5 mg/kg/dose, 9 of 86 patients in one study never reached a postdose level of 4 mg/l in repeated samples [6]. Provided that aminoglycosides are used on proper indications – severe infections – it seems more important to achieve therapeutic levels than to use inadequate doses in order to protect the patient from toxic reaction.

Other drugs which require routine monitoring of serum concentrations include flucytosin for the reasons mentioned above. With the cephalosporins, routine determinations of serum levels should not be necessary since these drugs have a wide therapeutic spectrum and excessive doses are required before signs of dose-related toxicity appear. An exception from that role may be ceftazidime, a new *Pseudomonas* active cephalosporin which is excreted unmetabolised by glomerular filtration only. In two studies, a marked reduction of the glomerular filtration rate has been demonstrated when moderately high doses were given to patients with mild or moderate preexisting renal impairment [11, 12]. Three patients who before treatment had marked renal impairment developed severe impairment during treatment with a high dose of ceftazidime; in two of them there were no other factors which could have explained the development [12]. Thus, it seems proper to suggest that in patients with preexisting renal impairment, e.g., glomerular filtration rate below 30 ml/min/1.73 m^2 body surface, ceftazidime levels should be monitored in a fashion similar to the one recommended for the aminoglycosides. A trough level of 40 mg/l should then not be exceeded. With cephalosporins which are metabolised, e.g., cefotaxime which undergoes hepatic desacetylation, the risk of accumulation is less obvious in patients with poor renal function and regular monitoring of serum concentration seems not to be indicated. Other antibiotics within that group have such a high degree of safety that monitoring is not meaningful. Examples of such drugs are cefuroxime, cefamandole, and cefazolin.

From a clinician's point of view, aminoglycoside concentrations should be performed as quickly as possible. The methods used should provide the results within a time which is sufficiently short to allow dosage corrections for the dose following the one from which the samples were obtained. In principle, that excludes the use of microbiological techniques and calls for methods which are more rapid. Such techniques are radioenzymatic assays [13], radioimmunoassays [14], and the enzyme multiplied immunoassay technique (EMIT [15]). All of these methods correlate well with the results obtained in conventional bioassays and fulfill the clinician's requirement for rapidity.

Due to increased problems with β-lactamase producing strains of *Hemophilus influenzae*, chloramphenicol has recently been more commonly used in newborns and neonates. Because of the potential increased toxicity of that antibiotic in this age group, regular monitoring of serum concentrations are required to avoid adverse reactions [16]. As demonstrated by Weber et al. [17], both radioimmunoassay and HPLC can be used with considerable precision, rapidity, and sensitivity.

Determination of Bactericidial Effect of Serum and Cerebrospinal Fluid

The value of antibiotic concentrations assays for the prediction of clinical efficacy can always be questioned. In order to give a true prediction, other parameters such as determinations of MIC, MBC, and routine antibiotic susceptibility must be true and enable a correlation between concentrations and in vitro antibacterial effect. As pointed out in several articles [18, 19] this is certainly not the case in many situations. Another approach to the problem is then to determine the bactericidal effect of serum, CSF, or other fluids on the causative pathogen. This technique is commonly used in patients with endocarditis and other serious systemic infections [20–23]. However, as discussed by Stratton et al. [24], the techniques are technically difficult and there are few, if any, convincing studies showing a direct relation between serum or CSF bactericidal effect and therapeutic efficacy of antibiotics. Such studies should be undertaken, although they would require considerable efforts and multicenter study designs.

Antibiotic Concentrations in Fluids and Tissues

Determinations of antibiotic concentrations in fluids other than urine or serum are rarely performed in routine clinical work. One reason is the obvious difficulty to achieve these materials, both for practical and ethical reasons. Moreover, tissues are difficult to handle and factors such as the binding of the drug to tissue components or protein are not easily evaluated. For documentation of the penetration of a new agent to peripheral compartments, such studies should, of course, be undertaken, but in routine clinical work they have a limited usefulness. An exception is CSF which, due to its composition, is easy to handle in the laboratory. With many new antibiotics only limited information about penetration over the blood-CSF barrier is available and therefore assays of CSF concentrations in clinical practice may be a guidance to dosage and prediction of clinical efficacy.

Conclusions

There is probably a considerable overestimation of the value of routine determinations of antibiotic concentrations in fluids and tissues. To be of clinical value such assays should be easy to correlate to other parameters of clinical efficacy or toxicity of the drug used. Since those parameters are only rarely at hand, the procedures used in routine work should be limited to the situations where we know that antibiotic concentrations will guide us in the treatment of the patient. Such situations, as reviewed here, arise then an aminoglycoside or other potentially toxic antibiotic is used and when adverse effects are related to concentrations achieved. Concentration determinations may also be of use to exclude deficiencies in gastrointestinal absorption of a drug and in controls of a patient's compliance. To be of value to the clinician, the results of concentration determinations are often required within a short period of time after sampling. Therefore, one should aim toward new and rapid techniques rather than microbiological ones. An indirect and still not fully evaluated technique is that determining the bactericidal activity of serum or CSF. These methods probably have a value in the choice of drugs and dosages in patients with certain serious systemic infections.

References

1. Du Pont HL, Hornick RB, Weiss CF, Snyder MJ, Woodward T (1970) Evaluation of chloramphenicol acid succinate therapy of induced typhoid fever and Rocky Mountain spotted fever. N Engl Med J 282:53–56
2. DiSanto AR, Chodos DJ (1981) Influence of study design in assessing food effects on absorption of erythromycin base and erythromycin stearate. Antimicrob Agents Chemother 20:190–196
3. Williams PEO, Harding SM (1984) The absolute bioavailability of cefuroxime axetil in male and female volunteers after fasting and food. J Antimicrob Chemother 13:191–196
4. Polak A, Eschenhof E, Fernex M, Scholer HJ (1976) Metabolic studies with 5-flurocytosin-6-^{14}C in mouse rabbit dog and man. Chemotherapy 22:137–142
5. Norrby R, Eilard T (1978) Treatment of opportunistic systemic mycoses. Scand J Infect Dis [Suppl] 16:59–64
6. Moore RD, Craig RS, Lipsky JJ, Mellits ED, Lietman PS (1984) Risk factors for nephrotoxicity in patients treated with aminoglycosides. Ann Intern Med 100:352–357
7. Scandinavian Study Group (1984) Imipenem/cilastatin versus gentamicin/clindamycin for treatment of serious bacterial infections. Lancet 1:868–871
8. Fee WE (1983) Gentamicin and tobramycin: comparison of ototoxicity. Rev Infect Dis 5 [Suppl 2]:S304–313
9. Barza M, Lauermann M (1978) Why monitor serum levels of gentamicin? Clin Pharmacokinet 3:202–215
10. Trollfors B, Alestig K, Krantz I, Norrby R (1980) Quantitative nephrotoxicity of gentamicin in nontoxic doses. J Infect Dis 141:306–309
11. Norrby SR, Burman LÅ, Linderholm H, Trollfors B (1982) Ceftazidime: pharmacokinetics in patients and effects on the renal function. J Antimicrob Chemother 10:199–206
12. Alestig K, Trollfors B, Andersson R, Olaison L, Suurküla M, Norrby R (1984) Ceftazidime and renal function. J Antimicrob Chemother 13:177–181
13. Holmes JK, Sanford JP (1974) Enzymatic assays for gentamicin and related aminoglycoside antibiotics. J Infect Dis 85:183–189
14. Minishew BH, Holmes RK, Baxter CR (1975) Comparison of a radio-immunoassay with an enzymatic assay for gentamicin. Antimicrob Agents Chemother 7:107–109

15. Phaneuf D, Francke E, Neu HC (1980) Rapid reproducible enzyme immunoassay for gentamicin. J Clin Microbiol 11:266–269
16. Lietman PS (1979) Chloramphenicol and the neonate – 1979 view. Clin Perinatol 6:151–161
17. Weber AF, Opheim KE, Koup JR, Smith AL (1981) Comparison of enzymatic and liquid chromatographic chloramphenicol assays. Antimicrob Agents Chemother 19:323–325
18. Greenwood D (1981) In vitro veritas? Antimicrobial susceptibility tests and their clinical relevance. J Infect Dis 144:380–385
19. Norrby SR (1982) Antibiotic treatment: relationship between pharmacokinetics, bacterial sensitivity and dose schedules. In: Bearn AG (ed) Antibiotics in the management of infections: outlook for the 1980s. Raven, New York, pp 61–78
20. Schlichter JG, MacLean HA (1947) A method for determining the effective therapeutic level in the treatment of subacute bacterial endocarditis with penicillin: a preliminary report. Am Heart J 34:209–211
21. Mandell GL, Sande MA (1975) Some new aspects of infective endocarditis. Geriatrics 30:97–102
22. Prober CG, Yeager AS (1979) Use of the serum bactericidal titer to assess the adequacy of oral antibiotic therapy in the treatment of acute hematogenous osteomyelitis. J Pediatr 95:131–135
23. Klastersky J, Meunier-Carpentier F, Provost JM, Staquet M (1976) Synergism between amikacin and cefazolin against Klebsiella: in vitro studies and effect on the bactericidal activity of serum. J Infect Dis 134:271–276
24. Stratton CW, Weinstein MP, Reller LB (1982) Correlation of serum bactericidal activity with antimicrobial agent level and minimal bactericidal concentration. J Infect Dis 145:160–168

Methods for Detection of Antimicrobial-Induced Changes in the Normal Human Oropharyngeal and Colon Microflora

C. E. NORD [1] and A. HEIMDAHL [2]

Introduction

The normal oropharyngeal and gastrointestinal microflora are stable ecosystems. Within a given person the microflora remains relatively constant over time. This ecosystem may however be disrupted. Irradiation of the head and neck causes changes in the oropharyngeal microflora and subsequently also in the gastrointestinal flora. Gastric surgery or pathologic conditions may affect normal peristalsis and can thus cause changes in the intestinal microflora. Hospitalization may also induce minor ecological changes. Ingestion of large bacterial inoculum may overhelm the resistance to new colonization and host defences (Nord et al. 1984). The most common and significant cause of disturbances in the oropharyngeal and gastrointestinal microflora is the administration of antimicrobial agents (Nord et al. 1984). The effect of antimicrobial agents on the patient's normal microflora can be separated into four general categories.

1. Antimicrobials may select for resistant microorganism in the normal flora and thereby cause overgrowth
2. Antimicrobials may induce microorganisms to produce toxins that can be deleterious for the patient like in *Clostridium difficile* associated colitis
3. Antimicrobials may induce colonization of the oropharynx and gastrointestinal tract with pathogenic organisms that may cause septic conditions
4. Antimicrobials may promote the emergence of resistant bacterial species that may reinfect the patient, e.g., surgical wounds.

Antimicrobials Capable of Inducing Ecological Disturbances

An increased risk for ecological disturbances of the oropharyngeal microflora is observed during administration of antimicrobial agents that are secreted in the saliva or from the gingival crevices and that have an antimicrobial effect on indigenous oropharyngeal microorganisms. In the gastrointestinal tract antimicrobials can affect the normal microflora after poor gastrointestinal absorption, by biliary or intestinal mucosal secretion (Nord et al. 1984). Suppression of the normal oropharyngeal and gastrointestinal anaerobic microflora seems to be associated with an increased risk for new colonization and with induction of toxin pro-

1 Department of Microbiology, Karolinska Institute, Huddinge Hospital, S-141 86 Huddinge
2 Department of Oral Surgery, National Bacteriological Laboratory, S-105 21 Stockholm

Table 1. Commonly used antimicrobial agents associated with different ecological disturbances of the oropharyngeal and gastrointestinal microflora

Bacterial overgrowth	New colonization	Toxin production	Antimicrobial resistance
Clindamycin	Clindamycin	Lincomycin	Tetracyclines
Erythromycin	Erythromycin	Clindamycin	Doxycycline
Fourth generation penicillins	Isoxazolylpenicillins	Ampicillin	
Third generation cephalosporins	Third generation cephalosporins	Third generation cephalosporins	

duction. Commonly used antimicrobial agents associated with ecological disturbances are presented in Table 1.

Induction of toxin production with subsequent diarrhoea and pseudomembranous colitis due to *Clostridium difficile* are most common after the administration of antimicrobial agents influencing the normal intestinal microflora. Antimicrobials mainly involved in this disorder are cephalosporins, clindamycin, and ampicillin (Aronsson et al. 1981).

Patients at Risk

Overgrowth with indigenous microorganisms may follow any antimicrobial therapy that interferes with the normal oropharyngeal and gastrointestinal microflora. Suppression of the anaerobic part of the microflora often permits aerobic strains to overgrow. Antimicrobials, like clindamycin and erythromycin present in saliva and in the gastrointestinal lumen, are often associated with overgrowth. Patients at risk for subsequent infection are those that are immunocompromised or otherwise severly debiliated by, e.g., tumor disease, aggressive treatment with cytotoxic agents, irradiation, or viral infection (Ringdén et al. 1984). Local infections, such as stomatitis or diarrhoea may, however, develop even in otherwise healthy persons after administration of antimicrobials present in high concentrations in the oropharyngeal and gastrointestinal tracts.

In healthy persons colonization with pathogenic microorganisms is generally of minor importance, while in patients with severe diseases, colonization is important for the emergence of severe infections with antimicrobial resistant, pathogenic microorganisms such as *Pseudomonas*, *Klebsiella*, and *Candida*.

Patients that are colonized with nonindigenous microorganisms in the oropharynx and are prone to aspiration are at risk for infection (Johanson et al. 1972). Thus not only patients with serious diseases, but also patients that are anaesthesized may develop pulmonary infection with gram-negative pathogens from the oropharynx. Introduction of endotracheal tubes probably increases the risk due to pressure damage of epithelial cells of the upper respiratory tract (Ramphal et al. 1980).

In patients with severe bone marrow failure, i.e., leukemia or granulocytopenia, infections with serious complications occur due to the low neutrophil count and impaired defense mechanisms. The microorganisms isolated from infections in such patients either belong to the patient's indigenous oropharyngeal or gastrointestinal microflora or are microorganisms acquired from the hospital environment. Gram-negative aerobic rods and fungi are the microorganisms most frequently isolated from these life-threatening infections.

Some antimicrobials, like tetracyclines, may promote rapid emergence of resistant strains among commensal microorganisms both in the oropharynx and gastrointestinal tract (Heimdahl and Nord 1983). Most purulent infections of the oropharyngeal tract as well as postsurgical infections of the oropharynx and gastrointestinal tract are caused by commensals of the normal flora. Thus the emergence of resistant commensal bacterial strains may render the antimicrobial agent useless in the prophylaxis and treatment of infections of the oropharynx and gastrointestinal tract.

Detection of Changes in the Normal Human Microflora

Microbiological Methods

The classical methods for analyzing changes in the normal human microflora are microbiological. Different selective media presented in Table 2 are often used to facilitate the identification of the different microorganisms (Sutter et al. 1980). However these studies are often very time consuming and expensive. Laboratory experience in handling and processing specimens from the normal flora are also required since the samples are mostly different from ordinary clinical samples obtained from infections.

Table 2. Media for normal microflora studies

Medium	Purpose	Dilutions	
		Saliva	Faeces
Blood agar	Total aerobic and anaerobic counts, predominant microorganisms	10^{-5}	10^{-8}
CLED agar	Enterobacteria. Enterococci	10^{-3}	10^{-7}
Esculin agar	Group D streptococci	10^{-1}	10^{-5}
Mitis-Salivarius agar	Alpha-streptococci	10^{-4}	
Hematin agar	Gram-negative cocci. Hemophilus	10^{-4}	
Sabouraud agar	Yeasts	10^{-1}	10^{-1}
Kanamycin-vancomycin blood agar	Bacteroides	10^{-3}	10^{-7}
Neomycin-vancomycin blood agar	Fusobacteria	10^{-3}	10^{-7}
Egg-yolk neomycin agar	Clostridia	–	10^{-3}
Veillonella agar	Gram-negative cocci	10^{-3}	10^{-4}
Lactobacillus selective agar	Lactobacilli	10^{-3}	10^{-5}

The changes in the flora can be followed microbiologically by three methods: (a) changes in the total number of all aerobic and anaerobic microorganisms, (b) changes in only the total number of anaerobes and (c) new colonization of aerobic gram-negative rods and yeasts.

Biochemical Methods

The methods that estimate the amounts of premetabolites or metabolites of the anaerobic flora may serve as presumptive methods for rapid detection of antimicrobial-induced changes in the normal microflora. Two methods have been described for following these changes: beta-aspartylglycine and volatile fatty acids.

Beta-aspartylglycine is normally metabolized by the anaerobic intestinal microflora and is normally not detected in fecal specimens. Increased amounts of beta-aspartylglycine indicates an impaired function of the anaerobic intestinal microflora (Welling 1978).

Volatile fatty acids produced by the anaerobic oropharyngeal and gastrointestinal microflora are involved in the inhibition of new microorganisms colonizing the oropharynx and gastrointestinal tract. Estimation of the amounts of free volatile fatty acids in saliva and feces may serve as an indicator of the anaerobic flora and the risk for colonization (Borthen et al. 1984). Table 3 shows the decrease of volatile fatty acids in patients receiving clindamycin for 7 days, when the numbers of anaerobic bacteria were significantly reduced.

Table 3. Mean concentrations of acetic, propionic, butyric, and isobutyric acids in feces of ten patients receiving clindamycin for 7 days

Day	Acid concentration (mmol/l)			
	Acetic	Propionic	Butyric	Isobutyric
0	112	29	21	2.8
1	153	31	29	2.8
2	93	11	20	<2.2
3	46	3.3	6.5	<0.2
4	33	1.4	3.8	<0.2
5	48	1.7	3.6	<0.2
6	48	1.5	3.4	<0.2
7	46	1.7	3.1	<0.2
9	43	3.0	6.9	<0.2
12	63	4.0	12	0.8
16	95	5.5	12	0.9
20	120	17.5	8.9	1.6
29	104	21	21	1.8

References

Aronsson B, Möllby R, Nord CE (1981) Occurrence of toxin-producing *Clostridium difficile* in antibiotic-associated diarrhoea in Sweden. Med Microbiol Immunol 170:27–35

Borthen L, Heimdahl A, Lindqvist L, Nord CE (1983) Determination of volatile fatty acids in saliva and faeces by gas-liquid chromatography: a rapid method for detection of antibiotic-induced changes in the oropharyngeal and gastrointestinal microflora. Int J Microbiol 1:35–45

Heimdahl A, Nord CE (1983) Influence of doxycycline on the normal human flora and colonization of the oral cavity and colon. Scand J Infect Dis 15:293–302

Johanson WG, Pierce AK, Sanford JP, Thomas GD (1972) Nosocomial respiratory infections with Gram-negative bacilli. Ann Intern Med 77:701–706

Nord CE, Kager L, Heimdahl A (1984a) Impact of antimicrobial agents on the gastrointestinal microflora and risk of infections. Am J Med 76:99–106

Nord CE, Heimdahl A, Kager L, Malmborg AS (1984b) Impact of different antimicrobial agents on the normal gastrointestinal microflora of humans. Rev Infect Dis [Suppl 1] 6:270–275

Ramphal R, Small PM, Shands JW, Fischlschweiger W, Small PE (1980) Adherence of *Pseudomonas aeruginosa* to tracheal cells injured by influenza infection or by endotracheal intubation. Infect Immun 27:614–619

Ringdén O, Heimdahl A, Lönnquist B, Malmborg AS, Wilczek H (1984) Decreased incidence of viridans streptococcal septicaemia in allogenic bone marrow transplant recipients after the introduction of acyclovir. Lancet I:744

Sutter VL, Citron DM, Finegold SM (1980) Wadsworth anaerobic bacteriology manual, 3rd edn. Mosby, St. Louis

Welling GW (1979) Beta-aspartyl-glycine, an indicator of decreased colonization resistance? In: van der Waaij D, Verhoef J (eds) New criteria for antimicrobial therapy: maintenance of digestive tract colonization resistance. Excerpta Medica, Amsterdam, pp 65–71

Chromatographic Methods for Determination of Antimicrobial Agents in Fluids and Tissues

J. A. WASHINGTON [1]

Introduction

Chromatography is a process which is used for separating complex mixtures based on their distribution between a mobile gas or liquid phase and a stationary liquid or solid phase. In liquid chromatography, a mobile liquid phase interacts with the sample molecules and with the stationary phase. "Reverse-phase" chromatography exists when a nonpolar stationary phase and a polar mobile phase are used. This type of liquid chromatography is especially useful in the clinical laboratory since a large variety of compounds of various weights and polarities in aqueous samples can be introduced directly into the chromatograph and separated on a single column by varying the composition of the mobile phase. The mobile phase is usually an aqueous buffer or a mixture of an organic solvent, such as methanol or acetonitrile, and water. An ion-pair reagent may be used with reverse-phase chromatography to modify the retention time of an ionic compound with an opposite charge. The most commonly used anionic ion-pair reagents are alkyl sulfonates ($RSO_{3}-$), while quarternary amines are usually used as cationic ionic-pair reagents (Gerson and Anhalt 1980).

HPLC has provided a means of separating virtually all antibiotics under development and in use today (Table 1) and is, therefore, a highly versatile tool for pharmacological studies and therapeutic drug monitoring. Its application in clinical laboratories has been predominantly for monitoring antibiotics with narrow therapeutic indexes, including, most frequently, the aminoglycosides and, to a lesser extent, vancomycin, chloramphenicol, and flucytosine. HPLC provides sensitivity, specificity, and speed (most assays can be completed within 30 or 60 min). Since most antibiotics have short half-lives, and hence short dosage intervals, and since many antibiotics, particularly the aminoglycosides, are given in combination with other antibiotics which may interfere with bioassays, the speed and specificity provided by HPLC represented a major advance in monitoring antimicrobial therapy. HPLC has also been useful in detecting and measuring antibiotic isomers and metabolites. Thus, HPLC provides the laboratory with the capability to assay antibiotics for which bioassay techniques are too slow and nonspecific or for which other techniques have not yet been developed.

1 Mayo Clinic and Mayo Foundation, Rochester, MN 55905, USA

Table 1. Liquid chromatographic assays of antimicrobial agents in serum, cerebrospinal fluid, or urine[a]

Aminocyclitols		
Amikacin	Spectinomycin	
Gentamicin	Tobramycin	
Netilmicin		
β-Lactams		
Cephalothin	Cefmenoxime[b]	Oxacillin[c]
Cephalexin	Cefmetazole[b]	Cloxacillin[d, e]
Cephradine	Cefoperazone[b]	Nafcillin[b]
Cephaloridine	Cefotaxime[b]	Amoxicillin
Cefazolin	Cefotetan[b]	Ampicillin
Cefaclor[b]	Cefsulodin[b]	Carbenicillin
Cefamandole	Ceftazidime[b]	Ticarcillin[f]
Cefoxitin	Ceftizoxime[b]	Azlocillin[b]
Cefuroxime	Ceftriaxone[b]	Mezlocillin[b]
	Moxalactam[b]	Piperacillin[b]
		Amdinocillin[b]
		Aztreonam[b]
		Clavulanic acid[b]
Other		
Adenine arabinoside	Nalidixic acid	
Amphotericin B	Niridazole	
Chloramphenicol	Nitrofurantoin	
Dapsone	Norfloxacin[h]	
Econazole	Rifampin	
Erythromycin	Sulfonamides	
Flucytosine	Tetracyclines	
Griseofulvin	Tinidazole	
Isoniazid	Trimethroprim	
Ketoconazole[g]	Vancomycin	
Metronidazole		

[a] Unless otherwise specified, references are contained in reviews by Anhalt and Moyer (1980), Gerson and Anhalt (1980), and Yoshikawa et al. (1980)
[b] References cited in Table 2
[c] Murai et al. (1981)
[d] Soldin et al. (1980)
[e] Teare et al. (1982)
[f] Kwan et al. (1982)
[g] Männisto et al. (1982)
[h] Paulikonis et al. (1984)

HPLC Assays

Aminoglycosides

Aminoglycosides are not measurable at clinically attainable concentrations with ultraviolet absorption and require derivatization of their amino groups with a fluorophore either before ("precolumn") or after ("postcolumn") HPLC separation. Removal of protein from the sample has been accomplished by precipitation, solvent extraction, silica gel adsorption, or ion exchange. Anhalt and Brown (1978) extracted the aminoglycoside from serum with a weak cation-exchange resin, a CM-Sephadex column, and then used reverse-phase, ion-pair chromatography and post-column derivatization with *o*-phthalaldehyde to measure amikacin, gentamicin, or tobramycin in serum. The methods were precise, and results correlated well with bioassays. Similar approaches can be used for other aminoglycosides, including neomycin, netilmicin, and sisomicin, but not streptomycin which lacks the primary amino groups required for derivatization.

β-Lactams

HPLC assays have been developed for relatively few penicillins (Table 1); however, there are numerous descriptions of HPLC assays of newer cephalosporins (cephems), cephamycins, oxacephems, and ureidopenicillins (Table 2). All of the *β*-lactam liquid chromatographic procedures listed in Table 2 were reverse phase, with the exception of one (Ascalone and Dal Bò 1983) for ceftriaxone in which an NH_2-bonded-phase column was used. Ascalone and Dal Bò (1983) cited rapid column conditioning, stability of column performance, and the absence of blocking of frits, phase shrinkage, or backpressure problems as advantages of normal (vs. reverse) phase liquid chromatography. All procedures cited used ultraviolet absorption for detection.

The speed, sensitivity, specificity, and precision of the HPLC assays cited in Table 2 have been reported to be excellent. In many instances, HPLC assays have been developed for pharmacological studies of a particular new cephalosporin so that their application to other cephalosporins is unknown. Brisson and Fourtillon (1981) described a method for the preparation and determination of cefazolin, cefoxitin, cefmenoxime, cefotaxime, cefamandole, and cefoperazone based on their extraction from acid-treated biological fluids into chloroform-1-pentanol and reverse-phase chromatography with varying contents of methanol or acetonitrile in the mobile phase. At a pH of 3, these cephalosporins are in their nonionized form and, therefore, soluble in the organic phase. This extraction could not be used for orally absorbed cephalosporins which contain an amino group that is ionized at pH 3. By varying the methanol content from 15% to 36% in a methanol-water-H_2SO_4 (1.8 *M*) mobile phase, Brisson et al. (1981) achieved good separation of cefotaxime, desacetylcefotaxime, cefroxadine, cefuroxime, cefazolin, cephalexin, cefamandole, cephalothin, and ceftizoxime, but not of cefsulodin, the most polar compound which was insufficiently retained to obtain a separation providing good sensitivity in plasma and urine samples. Danzer (1983) achieved good separation of cefotaxime, cefazolin, cefoxitin, cephapirin, and cefamandole by mixing the sample with acetonitrile solution containing 4′-nitroacetanilide as an internal standard and injecting the supernatant with a methanol/acetate buffer mobile phase (pH 5.5).

Although most studies of particular cephalosporins report lack of interference by other cephalosporins (which are unlikely to be administered concurrently) and aminoglycosides (which are not measurable with ultraviolet absorption), there are few, if any data on possible interferences by drugs other than antibiotics, such

Table 2. Analysis of newer cephalosporins and penicillins

Antibiotic	Extraction and/or deproteinization	Recovery (%)	Mobile phase	Column	Limit of detection (μg/ml)	Reference
Cefaclor	NS	NS	Methanol-acetate buffer (pH 5.2)	LiChrosorb C_8	0.2	Rofschafer et al. (1982)
Cefmenoxime	Centrifugal ultrafiltration of sodium dodecyl sulfate-treated plasma	NS	Acetonitrile-acetate buffer (pH 5.3)	μBondapak C_{18}	0.05	Granneman and Sennello (1982)
	Perchloric acid	73–76	Acetate buffer (pH 3.8)	μBondapak C_{18}	0.5	Reitberg and Schentag (1983)
Cefmetazole	Methanol-trichloroacetic acid	NS	Acetonitrile-citrate buffer (pH 5.4)	μBondapak C_{18}	0.4	Sekine et al. (1982)
Cefoperazone	Chloroform-pentanol	NS	Methanol-acetate buffer (pH 4.8)	μBondapak C_{18}	0.05	Brisson and Fourtillan[a] (1981)
	Methanol	97–101	Acetonitrile in tetrabutylammonium buffer	μBondapak phenyl	1	Muder et al. (1982)
Cefotaxime	Chloroform-acetone	83–115	Methanol-water-acetic acid or acetonitrile or	ODS Spherisorb	Plasma: 0.5–1 Urine: 5	Dell et al. (1981)
			Sulfonic acids	{ ODS Spherisorb, ODS Nucleosil, *or* μBondapak C_{18}		
	Trichloroacetic acid	NS	Methanol-water-sulfuric acid	LiChrosorb RP-18, 5 μm or μBondapak C_{18}	NS	Rouan et al. (1983)[c]
	Acetonitrile	NS	Methanol-acetate buffer (pH 5.5)	Radial Pak C_{18}	3	Danzer (1983)[d]
	DEAE-Sephadex anion-exchange	98	Acetonitrile-acetate buffer (pH 2.8)	C_{18}	1	Fasching and Peterson (1982)
Cefotetan	Trichloroacetic acid	NS	Acetonitrile-phosphate buffer (pH 3)	Nucleosil C_{18}	0.7	Nakagawa et al. (1982)
Cefsulodin	Centrifugal ultrafiltration of plasma-phosphate buffer mixture (pH 6)	99	Acetonitrile-acetate buffer (pH 4.2)	μBondapak C_{18}	0.2	Granneman and Sennello (1982)
Ceftazidime	Perchloric acid	99–100	Acetonitrile-formic acid-phosphate buffer	Hypersil ODS, 5 μm	0.6	Ayrton (1981)
	Acetonitrile	NS	Acetonitrile-acetate buffer (pH 4)	μBondapak C_{18}	0.8	Kemmerich et al. (1983)
	Methanol	96	Methanol-phosphate perchloric acid	MicroPak MCH 10	0.3	Myers and Blumer (1983)
Ceftizoxime	Trichloroacetic acid	89	Methanol-water-sulfuric acid buffer	LiChrosorb RP-18	Plasma: 0.2 Urine: 25	Rouan et al. (1983)[c]
	DEAE-Sephadex anion-	NS	Acetonitrile-acetic acid	μBondapak C_{18}	NS	Van Etta et al. (1983)

Ceftriaxone	Acetonitrile	95–102	Acetonitrile-water-ammonium-carbonate	LiChrosorb NH_2	Plasma: 0.3 Urine: 3 Saliva: 0.03	Ascalone and Dal Bò (1983)
	Acetonitrile	NS	Acetonitrile-phosphate buffer-ion pair	LiChrosorb RP-8	NS	Latif and Dajani (1983)
	Acetonitrile	100	Acetonitrile-phosphate-hexadecyltrimethyl-ammonium bromide (tetraoctylammonium bromide for urine)	Chromegabond C_{18}	NS	Salvador et al. (1983)
Moxalactam	Methanol-citrate	97	Plasma: acetonitrile-ammonium acetate (pH 6.5) Urine: methanol-heptylamine-phosphoric acid (pH 6)	Plasma: Chromegabond C_{18} Urine: Zorbax TMS	Plasma: 1.5 Urine: 7.5	Miner et al. (1981)
	Hydrochloric acid-acetate	87–98	Acetonitrile-phosphate buffer (pH 7)-calcium chloride-tetra-*n*-butylammonium hydroxide	μBondapak C_{18}	0.05	Brisson et al. (1982)
	Isopropanol-chloroform-isoamyl alcohol-acetate	90	Acetonitrile-water tetrabutylammonium	RAD-PACK A	0.5	Modai et al. (1982)
	Hydrochloride acid-acetate-tromethamine-octanol	75–81	Acetonitrile-ammonium acetate	μBondapak C_{18}	0.5	Ziemniak et al. (1982)
Azlocillin } Mezlocillin }	Sep-Pak C_{18}	92–102	Acetonitrile-phosphate buffer (pH 7)	μBondapak C_{18}	1.3–1.5	Hildebrandt and Gundert-Remy (1982)
Piperacillin	Methanol	85	Acetonitrile-acetate buffer (pH 4.6)	μBondapak C_{18}	NS	Aravind et al. (1982)
	Hydrochloric acid-chloroform-pentanol-phosphate buffer	84	Methanol-acetate buffer (pH 4.8)	μBondapak C_{18}	0.05	Brisson and Fourtillan (1982)
Amdinocillin (mecillinam)	Acetonitrile-diethyl ether	74	Methanol-phosphate buffer (pH 7)	μBondapak C_{18}	0.5	Lee and Brooks (1982)
Aztreonam	Acetonitrile	95–103	Acetonitrile-tetrabutylammoium	μBondapak C_{18}	Serum: 1 Urine: 5	Pilkiewicz et al. (1983)
Clavulanic acid	Ultrafiltration; derivatization with imidazole	112	Methanol-phosphate buffer (pH 3.2)	μBondapak C_{18}	0.1	Foulstone and Reading (1982)

[a] Method used for cefazolin, cephalothin, cefoxitin, cefotaxime, cefamandole, cefuroxime, and cefoperazone

[b] Method used for cefoxitin, cefoperazone, and moxalactam

[c] Method used for cefotaxime, desacetylcefotaxime, cefroxadin, cefuroxime, cefazolin, cephalexin, cefamandole, and ceftizoxime by varying methanol content in mobile phase and by using methanol-sulfuric acid for deproteinization of plasma for recovery of cephalothin

[d] Method used for cefotaxime, cefazolin, cefoxitin, cephapirin, and cefamandole

as diuretics and anti-inflammatory agents, which are frequently present in clinical samples.

Chloramphenicol

Gerson and Anhalt (1980) summarized five reverse-phase liquid chromatographic assays for chloramphenicol in their monograph on HPLC and therapeutic drug monitoring. Weber et al. (1981) found no significant differences in speed, sensitivity, specificity, and precision between an HPLC and a radioenzymatic assay for chloramphenicol and concluded that the selection of either method should be based on availability of equipment, reagents, and technical support.

Vancomycin

Uhl and Anhalt (1979) described an HPLC assay of vancomycin using a weak cation-exchange resin serum extraction procedure, an acetonitrile-phosphate buffer mobile phase, and reverse-phase chromatography with UV detection.

Other Antibiotics

Information regarding HPLC assays of other antibiotics is contained in Gerson and Anhalt's (1980) monograph on HPLC and therapeutic drug monitoring and in a review of the subject by Yoshikawa et al. (1980).

References

Anhalt JP, Moyer TP (1980) The role of gas-liquid chromatography and liquid chromatography in therapeutic drug monitoring. Lab Med 11:797–806

Aravind MK, Miceli JN, Kauffman RE (1982) Analysis of piperacillin using high-performance liquid chromatography. J Chromatogr 233:423–426

Ascalone V, Dal Bò L (1983) Determination of ceftriaxone, a novel cephalosporin, in plasma, urine and saliva by high-performance liquid chromatography on a NH_2 bonded-phase column. J Chromatogr 273:357–366

Ayrton J (1981) Assay of ceftazidime in biological fluids using high-pressure liquid chromatography. J Antimicrob Chemother [Suppl B] 8:277–231

Branneman GR, Sennello LT (1982) Precise high-performance liquid chromatographic procedure for the determination of cefsulodin, a new antipseudomonal cephalosporin antibiotic, in plasma. J Pharm Sci 71:1112–1115

Brisson AM, Fourtillan JB (1981) Determination of cephalosporins in biological material by reversed-phase liquid column chromatography. J Chromatogr 223:393–399

Brisson AM, Fourtillan JB (1982) High-performance liquid chromatographic determination of piperacillin in plasma. Antimicrob Agents Chemother 21:664–665

Brisson AM, Fourtillan JB, Berthon G (1982) Sensitive high performance liquid chromatographic analysis of moxalactam in biological fluids. J Chromatogr 233:386–391

Brückner O, Collmann H, Borner K (1983) Cefotaxime levels in ventricular cerebrospinal fluid, determined by bioassay and by high-performance liquid chromatography. Chemotherapy 29:237–243

Danzer LA (1983) Liquid-chromatographic determination of cephalosporins and chloramphenicol in serum. Clin Chem 29:856–858

Dell D, Chamberlain J, Coppin F (1981) Determination of cefotaxime and desacetylcefotaxime in plasma and urine by high-performance liquid chromatography. J Chromatogr 226:431–440

Fasching CE, Peterson LR (1982) Anion-exchange extraction of cephapirin, cefotaxime, and cefoxitin from serum for liquid chromatography. Antimicrob Agents Chemother 21:628–633

Foulstone M, Reading C (1982) Assay of amoxicilin and clavulanic acid, the components of Augmentin, in biological fluids with high-performance liquid chromatography. Antimicrob Agents Chemother 22:753–762

Gerson B, Anhalt JP (1980) High-pressure liquid chromatography and therapeutic drug monitoring. American Society of Clinical Pathologists, Chicago

Granneman GR, Sennello LT (1982) A very precise high-performance liquid chromatographic procedure for the determination of cefmenoxime, a new cephalosporin antibiotic, in plasma. J Chromatogr 229:149–157

Hildebrandt R, Gundert-Remy U (1982) Improved procedure for the determination of the ureidopenicillins azlocillin and mezlocillin in plasma by high-performance liquid chromatography. J Chromatogr 228:409–412

Kemmerich B, Warn H, Lode H, Borner K, Koeppe P, Knothe H (1983) Multiple-dose pharmacokinetics of ceftazidime and its influence on fecal flora. Antimicrob Agents Chemother 24:333–338

Kwan RH, MacLeod SM, Spino M, Teare FW (1982) High-pressure liquid chromatographic assays for ticarcillin in serum and urine. J Pharm Sci 71:1118–1121

Latif R, Dajani AS (1983) Ceftriaxone diffusion into cerebrospinal fluid of children with meningitis. Antimicrob Agents Chemother 23:46–48

Lecaillon JB, Rouan MC, Soupart C, Febune N, Juge F (1982) Determination of cefsulodine, cefotiam, cephalexin, cefotaxime, desacetyl-cefotaxime, cefuroxime and cefroxadin in plasma and urine by high-performance liquid chromatography. J Chromatogr 228:257–267

Lee TL, Brooks MA (1982) Determination of amdinocillin in plasma and urine by high-performance liquid chromatography. J Chromatogr 227:137–148

Männisto PT, Mäntylä R, Nykänen S, Lamminsivu U, Ottoila P (1982) Impairing effect of food on ketoconazole absorption. Antimicrob Agents Chemother 21:730–733

Miner DJ, Coleman DL, Shepherd AMM, Hardin TC (1981) Determination of moxalactam in human body fluids by liquid chromatographic and microbiological methods. Antimicrob Agents Chemother 20:252–257

Modai J, Wolff M, Lebas J, Meulemans A, Manuel C (1982) Moxalactam penetration into cerebrospinal fluid in patients with bacterial meningitis. Antimicrob Agents Chemother 21:551–553

Muder RR, Diven WF, Yu VL, Johnson J (1982) Determination of cefoperazone concentration in serum and muscle tissue with a versatile high-pressure liquid chromatographic method. Antimicrob Agents Chemother 22:1076–1078

Murai Y, Nakasawa T, Yamaoka K, Uno T (1981) High performance liquid chromatographic analysis and pharmacologic investigation of oxacillin and its metabolites in man. Chem Pharm Bull (Tokyo) 29:3290–3297

Myers CM, Blumer JL (1983) Determination of ceftazidime in biological fluids by using high-pressure liquid chromatography. Antimicrob Agents Chemother 24:343–346

Nakasawa K, Koyama M, Tachibana A, Komiya M, Kikuchi Y, Yano K (1982) Pharmacokinetics of cefotetan (YMO9330) in humans. Antimicrob Agents Chemother 22:935–941

Paulinkonis LT, Musson DG, Bayne WF (1984) Quantitation of norfloxacin, a new antibacterial agent in human plasma and urine by ion-pair reverse phase chromatography. J Pharm Sci 73:99–102

Pilkiewicz FG, Remsburg BJ, Fisher SM, Sykes RB (1983) High-pressure liquid chromatographic analysis of aztreonam in sera and urine. Antimicrob Agents Chemother 23:852–856

Reitberg DP, Schentag JJ (1983) Liquid-chromatographic assay of cefmenoxime in serum and urine. Clin Chem 29:1415–1418

Rotschafer JC, Crossley KB, Lesar TS, Zaske D, Miller K (1982) Cefaclor pharmacokinetic parameters: serum concentrations determined by a new high-performance liquid chromatographic technique. Antimicrob Agents Chemother 21:170–172

Rouan MC, Abadie F, Leclerc A, Juge F (1983) Systematic approach to the determination of cephalosporins in biological fluids by reversed-phase liquid chromatography. J Chromatogr 275:133–144

Salvador P, Smith RG, Weinfeld RE, Ellis DH, Bodey GP (1983) Clinical pharmacology of ceftriaxone in patients with neoplastic disease. Antimicrob Agents Chemother 23:583–588

Sekine M, Sasahara K, Kojima T, Morioka T (1982) High-performance liquid chromatographic method for determination of cefmetazole in human serum. Antimicrob Agents Chemother 21:740–743

Soldin SJ, Tesoro AM, MacLeod SM (1980) A rapid high performance liquid chromatographic procedure for the analysis of cloxacillin and/or nafcillin in serum. Ther Drug Monit 2:417–422

Teare FW, Kwan RH, Spino M, MacLeod SM (1982) High-pressure liquid chromatographic assay of cloxacillin in serum and urine. J Pharm Sci 71:938–941

Thirumoothi MC, Buckley JA, Aravind MK, Dajani AS (1981) Diffusion of moxalactam into the cerebrospinal fluid in children with bacterial meningitis. J Pediatr 99:975–979

Uhl JR, Anhalt JP (1979) High performance liquid chromatographic assay of vancomycin in serum. Ther Drug Monit 1:75–83

Van Etta LL, Fasching CE, Peterson LR, Gerding DN (1983) Comparison study of the kinetics of ceftizoxime penetration into extravascular spaces with known surface area/volume ratio in vitro and in vivo in rabbits. Antimicrob Agents Chemother 23:49–53

Weber A, Opheim KE (1983) High-performance liquid chromatographic quantitation of trimethoprim, sulfamethoxazole, and N^4-acetylsulfamethoxazole in body fluids. J Chromatogr 278:337–345

Weber AF, Opheim KE, Koup JR, Smith AR (1981) Comparison of enzymatic and liquid chromatographic chloramphenicol assays. Antimicrob Agents Chemother 19:323–325

Yoshikawa TT, Maitra SK, Schotz MC, Guze LB (1980) High-pressure liquid chromatography for quantitation of antimicrobial agents. Rev Infect Dis 2:169–181

Ziemniak JA, Chiarmonte DA, Miner DJ, Schentag JJ (1982) HPLC determination of D and L moxalactam in human serum and urine. J Pharm Sci 71:399–402

Microbiological and Immunological Assays of Antimicrobials

R. WISE [1]

Introduction

In antibiotic therapy, assays of serum (or other body fluids) from patients are performed for two principle reasons: to ensure that sufficient agent is present for adequate therapy and to monitor the patient for potentially toxic concentrations of the agent administered. Of the many methods of assay that are available, two, namely microbiological and immunological are mainly used in routine microbiological laboratories. In this paper I wish to examine the major advantages and drawbacks of the two methods.

Microbiological Assays

These may be agar diffusion assays (plate diffusion or vertical diffusion – the latter now little used), dilution assays (now very little used due to high inherent inaccuracy) and the urease method (again little used and will not be discussed further) (Noone et al 1971). It is the agar diffusion assays which have the widest applicability to antimicrobial therapy. The only agents causing difficulties are those which bind to the agar and hence diffuse poorly, namely the polymixins, however there is little call to assay them. Agar diffusion assays are also used in industrial quality control procedures and have attained the high degree of required accuracy. Their appeal is the cheapness of equipment necessary and their ability to assay a variety of agents with only minor modifications. They do though require a high degree of technical expertise as poor technique will lead to inaccurate results (Reeves and Bywater 1975).

In setting up a microbiological method the most important choice to be made is that of indicator organism. In the UK, where there has been almost 10 years of experience in quality control of antibiotic assays, there has been a steady move away from using a gram-positive organism (usually a *Bacillus subtilis*) to using a gram-negative, with a consequent improvement in results. In a recent survey (Reeves and Bywater 1983) 50% of those using a gram-positive indicator achieved poor results (mean +2 SD of mean of >50%) whereas only 15% of those using a gram-negative indicator achieved poor results.

Microbiological assays have certain inherent problems. They are by their nature slower, however methods taking only 2.5 h have been described (Shanson

1 Department of Microbiology, Dudley Road Hospital, Birmingham, B18 7QH, Great Britain

and Hince 1977). A major problem in seriously ill patients on combination antimicrobial chemotherapy is the effect a second antibiotic may have upon the one under assay. This can very often be overcome by the choice of an appropriate assay organism or by the addition of a compound capable of neutralizing unwanted antibiotics (Broughall 1978). Another approach is to measure the two or more antibiotics at once by the use of high-voltage electrophoresis – although this method does not lend itself to use in the routine laboratory.

In setting up a microbiological assay, it is important to alter the sensitivity of the procedure to the correct range by adjusting the conditions of the assay or by diluting the samples if so appropriate. The assay conditions which are variable are: (1) disc or well size, (2) pH or type of agar, (3) time of preincubation, (4) temperature of incubation. Factors which will affect the assay are many but included amongst these must be: (1) the availability of a magnifying zone reader – this in the UK experience has been found to be extremely important in increasing the accuracy of results from routine laboratories. Now, the use of callipers – or even worse, rulers – must be decried. (2) The use of large plates (25 cm × 25 cm) rather than a number of smaller plates is important so that all standards, controls, and tests are on the same plate and subject to the same conditions. (3) The use of levelling devices for the plates and high-grade glassware for making up standards is imperative. (4) The application of standards, controls, and tests to the large plate by using a random code (Broughall 1978) is also important as often wells at the edges of plates will give different zone sizes from those in the centre. (5) The use of a simple computer program to derandomize the tests and calculate a "line of best fit" and the answers is also desirable as this will reduce any errors which might otherwise occur in handling the data.

It is important when a new assay is being devised that special attention is paid to the choice of diluent for the standards. For the assay of human serum, pooled human serum should be used, though it is recognized that this adds to expense. It is possible to use cheaper animal serum but only if prior recovery studies have been performed and shown that the two fluids are equivalent. Unless this is done, problems may arise as many antimicrobials will have different protein-binding in different animal species. When fluids other than serum are being assayed it is important to have a fairly accurate knowledge of the protein content of that fluid. Inflammatory exudates such as pleural fluid from an empyema should probably be assayed against standards made up in 70% serum. Fluid from non-inflamed sites such as the peritoneum (in ascites) should have a lesser amount of protein present (as low as 20%). If there is doubt, the protein content should be measured first. It may well be important for certain agents such as aminoglycosides and 4-quinolones to ensure that the anion/cation content of sample and standards are similar when, for example, measuring levels of these agents in the prostate or urine.

Immunological Assays

Within the last few years a large number of new methods for assaying antibiotics have become available. These are all quicker and more specific than the microbiological methods and are available with different degrees of automation. In this review a brief description of the methods will be given.

EMIT Procedure

This was produced by the Syva Corporation and is now available with different degrees of sophisticated automation. Briefly the technique is as follows. The enzyme-labelled drug and drug in the sample compete for binding sites on the anti-aminoglycoside antibody. When the enzyme is bound, its activity is reduced and this reduction is proportional to the concentration of the drug in the sample. The enzyme label used is glucose-6-phosphate and it is detected by its reduction of NAD to NADH (Wills and Wise 1979), which is measured spectrophotometrically.

Polarization Fluoroimmunoassay

This has been developed by Abbott and marketed as the TDX system. It utilizes the property that when a fluorescein-labelled molecule (the aminoglycoside) is excited by polarized light, the polarization of the emitted light is dependent on molecular size. Free and labelled drug compete for binding sites on an anti-drug antibody. Only the free-labelled drug will scatter polarized light, and the scatter is proportional to the concentration of unlabelled drug (Weber 1953).

Substrate-Labelled Fluoroimmunoassay (SLFIA)

This method has been developed by Ames and is also available with different degrees of automation. The principle is that the drug labelled with an initially non-fluorescent compound (galactosyl-umbelliferone) competes with the antibody and unlabelled drug for the enzyme (β-galactosidase), which converts the substrate to a fluorescent product. The amount of antibiotic present will be proportional to the fluorescence detected.

Radioimmunoassays

These have now largely been replaced by non-radiometric methods due to cheaper detection systems, longer shelf lives and the removal of hazards associated with such methods.

Other Methods Requiring a Separation Step

Seward Limited have developed a fluoroimmunoassay which has the same principle as radioimmunoassay except that the antibody is attached to magnetic particles and the separation step can be performed by a magnet. The fluorescence is then "extracted" from the precipitate. Diagnostic Products have developed "SOPHEA" – a solid-phase enzyme immunoassay which also has a separation step.

Direct Antibody-Antigen Detection

Two methods are included here. Firstly, the automated nephelometric method of Beckman in which the direct antigen-antibody binding is detected. In a simpler system, "Macrovue" from Becton-Dickinson, a latex agglutination procedure is used. This latter system suffers from the fact that only discrete dilutional steps are possible.

Comparison of Methods

It is impossible to compare directly all the available methods. However, in a recent UK quality control survey (Reeves and Bywater 1983) an overall impression could be gained. In Table 1 the results are shown. It can be seen that the Syva EMIT and Abbott TDX performed particularly well.

We have evaluated many of these machines, in particular the TDX and EMIT, and this is presented in detail elsewhere (Andrews and Wise 1984). Briefly, both systems allowed the rapid, accurate and precise assay of gentamicin, other aminoglycosides and vancomycin. The presence of other antibiotics (cefoxitin 25 mg/l, fucidin 30 mg/l, penicillin G 30 mg/l, vancomycin 10 mg/l, chloramphenicol 25 mg/l), Intralipid (20 mg/l and 2 g/l) and gentamicin from various sources (with different component ratios) was assessed on the assay of gentamicin. The results obtained correlated well with the plate microbiological assay (TDX correlation coefficients, spiked samples 0.98, patient samples 0.95; EMIT correlation coefficients, spiked samples 0.99, patient samples 0.94). The coefficients of variation

Table 1. Analysis of methods and results of UK gentamicin quality control circulation, 1983 (Reeves and Bywater 1983)

Method	No.	Groups[a]		
		1+2	3+4	>5
FIA	36	16	15	5
RIA	2	1		1
EMIT	133	107	22	4
Latex agglutination	35	15	14	6
Beckman	3	1	1	1
Abbott TDX	10	8	2	
Microbiological gram-negative plate	93	54	26	13

FIA, fluoroimmunoassay; RIA, radioimmunoassay

[a] Group 1 = mean + 2 SD < 20, group 2 = mean + 2 SD 20–30, etc

Table 2. Assessment of intra-assay precision and accuracy

Assigned value (mg/l)	EMIT			TDX		
	Mean	CV	*P*	Mean	CV	*P*
2	2.01	5.6	0.8	2.12	4.4	<0.001[a]
6	5.99	6.2	>0.9	6.16	3.5	0.01–0.001
10	8.89	5.5	<0.001[a]	10.74	5.9	<0.001[a]

CV, coefficient of variation
[a] Significantly different from assigned value, TDX mean overestimation by 5.5%

Table 3. Assessment of inter-assay precision and accuracy[a]

Assigned value mg/l	CV	
	EMIT	TDX
1.5	14.8	5.5
5	3.2	4.8
10	6.1	5.7

CV, coefficient of variation
[a] Number of replicates = ten

(CV) for intra-assay studies were 3.5%–5.9% and 5.5%–6.2% for TDX and EMIT procedures, respectively (Table 2). The CVs for the interassay studies were 2.4%–5.9% and 3.2%–14.8% for TDX and EMIT, respectively (Table 3). Intralipid did affect the two assays (to give an overestimate of up to 14%), but the other antimicrobials did not. Gentamicin from differing sources may however affect results. Our general conclusions related to the two methods are as follows:

1) Both are suitable alternatives to the microbiological methods (UK groups 1 and 2)
2) Time: TDX – 5 min
EMIT – 2 min (longer reagent "assembly" time)
3) Cost: Abbott > Syva
4) No non-aminoglycoside antibiotic interference
5) Intralipid (up to 2 g/l) – ca. 12% overestimation by both
6) EMIT less accurate <2, >10 mg/l.

We have also had the oppurtunity to look at different systems and below are some comments.

1) EMIT method – Kem-o-stat machine – quicker than the Syva automated machine. Poor accuracy at <2 and >8 mg/l gentamicin. The standard curve could not be stored and the program could be improved.

2) Ames SLFIA – the manual system was rather time consuming and was, during the test period (when there was a very high ambient temperature), less ac-

curate. The automated Optimat performed well. When comparing it with Abbott TDX the correlation coefficient was 0.989. The intrabatch CV was 2.5%–11.7%, the higher figure being at the low end (ca. 2 mg/l); the interbatch CV was 1.1% –14.1% (again the poorer results were at ca. 2 mg/l).

3) Beckman nephelometric procedure – this was a relatively quick method, but inaccurate at <3 mg/l and >8 mg/l. The results were not as accurate as with TDX or ABBOTT.

There are, therefore, a number of methods available which have an acceptable degree of activity. Undoubtedly they are easier to perform than the microbiological procedures. However, it will be extremely unfortunate if the expertise that many have developed in microbiological methods is lost due to the convenience of automation.

References

Andrews JM, Wise R (1984) A comparison of the homogeneous enzyme immunoassay and polarisation fluoroimmunoassay of gentamicin. J Antimicrob Chemother 14:509–520

Broughall JM (1978) Aminoglycosides. In: Reeves DS, Phillips I, Williams JD, Wise R (eds) Laboratory methods in antimicrobial chemotherapy. Churchill Livingston, London, pp 194–207

Noone P, Pattison JR, Sampson D (1971) Simple rapid method for assay of aminoglycoside antibiotics. Lancet I:16–19

Reeves DS, Bywater MJ (1975) Quality control of serum gentamicin assays – experience of national surveys. J Antimicrob Chemother 1:103–106

Reeves DS, Bywater MJ (1983) Comments on antibiotic quality control. Public Health Laboratory Service Publication

Shanson DS, Hince C (1977) Serum gentamicin and assays of 100 clinical serum samples by a rapid 40 °C Klebsiella method compared with overnight plate diffusion and acetyltransferase assay. J Clin Pathol 30:521–525

Weber G (1953) Rotational brownian motion and polarisation of the fluorescence of solutions. Adv Protein Chemi 8:415–419

Wills PJ, Wise R (1979) Rapid simple enzyme immunoassay for gentamicin. Antimicrob Agents Chemother 16:40–42

Computerized Pharmacokinetic Evaluation of Antibiotics in Body Fluids and Tissues: An Overview

T. BERGAN [1]

Computerized pharmacokinetic evaluation in general aimes at a mathematical description of absorption, distribution, and metabolism of for instance antimicrobial agents. Pharmacokinetics uses established and objective criteria for the comparison of related drugs in health and disease. It can predict accumulation and the consequences of repetitive dosing and assist in generating suitable dosage regimens and steady state concentrations consistent with a probable therapeutic success and avoidance of side effects. In spite of very significant advances in the development of better antibacterial agents during the last decade, however, we need further information on the key factors governing penetration of antimicrobial agents to infected body sites in determining optimal dosage. In this presentation, I will deal with basic requirements for a pharmacokinetic assessment of antibiotics, including penetration of tissues and extravascular body sites.

In order to compare the properties of new agents, initial studies are usually performed on healthy volunteers. This is necessary for comparison within more standard populations, since the actual target group of the agents, infected persons, vary more in key characteristics and are more heterogenous pharmacokinetically. Computerized evaluations, though, must include patients to support relevant dosage predictions. Consequently, patients with for example a variable capacity for eliminating antibacterial agents, in particular those with reduced renal function, are studied. The effect of pregnancy is considered more rarely in view of uncertain teratogenicity of many new agents. However, a larger volume of distribution due to the fetus causes lower serum concentrations [10]. Pharmacokinetic evaluation in children, particularly newborns, is often performed during the later stages of drug development. Pediatric studies, though, are mandatory because of the complex and contradictory consequences in small children of a higher distribution volume. In neonates, this is counterbalanced by slower renal and hepatobiliary elimination. Consequently, a dosage on a kilogram body weight basis identical to that used for adults would commonly lead to underdosage in younger children. Cystic fibrosis (CF) patients, who have a higher clearance for β-lactam antibiotics than non-CF patients [5], comprise a special group.

Pharmacokinetic evaluation of a new agent must follow certain guidelines. Individual pharmacokinetic studies render slightly varying results within the range acceptable as statistically overlapping. This occurs because of an underestimated impact of chance selection of the individual participants. In order to generate re-

1 Departments of Microbiology, Institute of Pharmacy and Aker Hospital, University of Oslo, P.O. Box 1108 Blindern, 0317 Oslo 3, Norway

sults with a reasonably general applicability, the number of individuals must be adjusted in accordance with the degree of variation observed among the subjects. For instance, erythromycin shows such a considerable variation among individuals that as many as 20–25 volunteers are an absolute minimum for pharmacokinetic studies of this antibiotic. The number of necessary individuals should be adjusted according to statistical variance when the results of an initial minimum of 10–12 subjects are known.

The number of samples throughout the intervals studied must be so high that computers can generate statistically acceptable curve fittings. The number of samples should also be proportional to the degree of variations in serum levels. The number of samples per time unit should be high at about the time when curve inflections occur. Accordingly, the critical portions of the curves after oral or intramuscular doses are around the peaks and after intravenous doses during the initial, distribution phase and during the stage when a transition occurs between the α- and β-phases. Straight lines, e.g., in semilogarithmic plots, should be based on at least four points. Thus four or more observations are needed during the disposition phase.

For assay purposes, spiked samples of known concentrations should be stored together with the patient samples. This enables adjustment for possibly deteriorated activity during storage. An adequate number of samples is needed to differentiate between pharmacokinetic models.

For a complete input-output household, urine collection is necessary. Patients must be instructed to ensure recovery of the entire urine volume. Urine data allow calculation of disposition half-life as a control of serum $t^1/_2$. Differences between the rates of elimination from serum and urine data could follow inactivation in the kidneys. Such may be considered for imipenem and thienamycin, which are both inactivated by tubular cell dehydropeptidase-I.

In computerized assessment, penetration of infected body sites is crucial. Fibrin clots, abscess fluid, tissue chambers, and wound fluid have been particularly studied as experimental models. In man, extravascular penetration has been elucidated by skin blisters, elicited either by an irritating agent, e.g., cantharidine on the skin, or by suction on small circular areas [1]. The former tend to be inflamed, the suction blisters more noninflammatory. Although extensively studied [8], the relationship is somewhat obscure between fluid introduced by wicks into subcutaneous tissue and given tissue components. The threads pass through various blood capillaries and other vessels and elicit inflammation. Wicks are difficult to standardize with respect to the amount of fluid that escapes as they are withdrawn through the skin pore. Possible drug adsorption to wick fibers has to be considered. Another drawback of the assay is that the standards are prepared by soaking the wicks in sera of known concentrations. The amount of fluid adsorbed to the standard wicks prepared in vitro corresponds to saturation level, and this might differ from the amount adsorbed by wicks recovered from volunteer tissues. The type of wick material used is yet another consideration. This, perhaps, explains why wick fluid curves of the same agents differ considerably in relation to serum curves from study to study. My enthusiasm for the wick fluid approach is consequently somewhat guarded.

Other, possibly more preferable, approaches are peripheral human lymph [2, 3] or sterile filter paper discs, soaked while on the surface of the peritoneal cavity [12, 13]. The latter method has generated concentration profiles comparable with those of peripheral human lymph. These methods monitor tissue levels relatively faithfully since they represent penetration of nonspecialized tissues like connective tissue and muscle.

Organs involved in the elimination of the drugs, such as the didneys and liver, have to be studied separately as the concentrations here are higher than in other organs and may, according to tissue homogenates, even differ within the organs themselves. Otherwise, determination of penetration from tissue homogenates is extremely difficult to assess, because the contribution of tissue cells, extracellular fluid, and blood contents is rather variable and difficult to foresee.

Even with correction for blood contamination, the usefulness of information from tissue homogenates is now often considered to be limited. The best correction factors are based on either hematocrit plus hemoglobin in the tissue or counts of radioactive chrome-labeled erythrocytes. Correction factors are, e.g.: $c_t = c_{tot} - S$.

Where c_t is the tissue concentration representing the sum of interstitial fluid plus intracellular levels, c_{tot} the concentration determined in crude tissue homogenate, and S the amount per gram of tissue homogenate due to blood. $S = H_t/H_s \cdot c_s(1 - HCT)$, where H_t and H_s are the hemoglobin concentrations in tissue homogenate and whole blood, respectively, c_s is the serum concentration of the drug, and HCT the hematocrit. If a radioactive count (of chrome-labelled erythrocytes) is available, S is represented by $R_t/R_s \cdot c_s$, where R_t and R_s are the radioactive counts, respectively, per gram or milliliter of tissue and serum.

Tissue concentrations of drugs from homogenates, even after correction for blood contribution, may be difficult to interpret in physiological terms, since they represent a mixture of contributions from intracellular and intercellular fluid. The contribution from the former depends on the efficacy of cell crushing. Since the exact interpretation of tissue homogenate data is thus fraught with difficulties and the bacteria are exposed primarily to extracellular tissue fluid during their spread within tissues, methods for faithfull monitoring of the extracellular fluid levels are preferable.

Accordingly, we have recently assessed penetration of peripheral human lymph of the lower leg for several antibacterial agents, since lymph collects extracellular fluid (Table 1). The method has been described elsewhere [7].

Special limitations apply to the central nervous system. The concept of a blood-brain barrier is well established. It is due partly to the fact that the cerebrospinal fluid (CSF) represents a large volume compared with the surface area available for passage of drug molecules. In part, also, drugs like the β-lactam antibiotics are removed from the CSF by choroid plexus cells through active transport mechanisms similar to those responsible for renal tubular secretion. Both machanisms are blocked by probenecid.

Assessment of the ability of antimicrobial agents to pass into various parts of the body must take into account the hollow organs and their fluids, e.g., urine, bile, nasal or bronchial secretions, and vaginal discharge. The corresponding data assist in establishing suitable dosage intervals and dose size with regard to infections largely within the cavities of these organs.

Table 1. Penetration of antibiotics to peripheral human lymph [2–4]

Antibiotic	Serum protein binding (%)	AUC[a] (%)	Elimination half-life (h) Serum	Elimination half-life (h) Lymph
Mecillinam	5	97	1.2	1.8
Amoxicillin	15	88	0.8	1.1
Ampicillin	15	82	0.9	1.0
Clavulanic acid	20	79	0.9	1.1
Ticarcillin	40	58	1.1	1.1
Temocillin	85	56	4.4	4.9
Erythromycin	70	35	2.1	5.0
Gentamicin	0	100		
Sulfadiazine (SDZ)	56	63	8.0	10.7
Acetyl-SDZ	–	60	11.5	
Trimethroprim (TMP)	50	57	10.6	12.7
SDZ steady state		68	16.5	19.2
Acetyl-SDZ steady state		67	17.5	19.2
TMP steady state		59	9.4	8.9

[a] AUC, area under the concentration curve of lymph as percentage of serum AUC

From a pharmacokinetic point of view, optimal assessment of the data relating to extravascular body sites has to be made. It is necessary to follow concentrations over a long time period. The concentration curves of extravascular foci often have slopes different to those of serum. Therefore, spot tests of ratios have limited significance, although these are sometimes unavoidable with foci which cannot be monitored by repeated sampling, such as CSF and internal tissues in humans. Ratios between concentrations during the phase of invasion tend to be lower than ratios from the later part of the elimination phase, when extravascular concentrations are often relatively higher than serum concentrations, which may, indeed, surpass serum concentrations toward the end of normal intervals between doses [1].

Monitoring should be carried out with repeated samples over periods of up to five times the disposition half-life within a particular focus. Subsequent computerized curve fitting and pharmacokinetic evaluation is advisable. This enables the velocity of penetration of a particular focus, the peak concentrations, and disposition half-life ($t^1/_2$) to be determined and related to, e.g., the serum protein binding and basic chemical characteristics like molecular size, electrostatic charge, and lipid and aqueous solubility. The major criterion in demonstrating the penetrability of an antimicrobial agent (superior to using concentration ratios at particular points in time) is the ratio of the differentials between the concentration curves of an extravascular focus and the serum. The value corresponds to the total area under the concentration curves of a body fluid divided by the total area under the serum curve.

The choice of the correct methods in pharmacokinetic assessment of individual curves is another problem connected with computerized pharmacokinetic evaluation. Pharmacokinetic equations are now widely standardized and can be derived from standard texts [11]. However, two particular points should be discussed: the selection of the optimal method of curve fitting and which pharmacokinetic model to choose. A curve fit according to the one-compartment open model will generate different values of terminal $t^1/_2$, clearance, and apparent distribution volume than the curve fit resulting from the two-compartment model equations. Since the significance of curve fit improves the lesser the point scatter and pharmacokinetics fulfils a descriptive role, model-independent characteristics are useful but have their limits.

The particular computer method used for curve fitting decisively influences the assessment. The NONLIN program of Metzler [11] has become so widely distributed that it would appear to be a good candidate for an informal international standard. The use of this program does at least facilitate comparison between centers. The original version uses start parameters for eased curve fitting suggested to the computer from manual or desk calculation.

Sedman and Wagner have extended the NONLIN strategy by using a preceding program, AUTOAN [11], which through exponential curve stripping offers automatic selection of the most suitable pharmacokinetic model and start parameters, which are directly entered into NONLIN. The stripping considers the data as a sum of exponentials, $y = a_i - b_i t$, with $i = 1, 2, \ldots, n$, and n is the number of factors than can be determined by AUTOAN for the purpose of a suitable curve fit. a_i and b_i are characteristics determining the choice of pharmacokinetic model. AUTOAN objectivizes the process of model selection and this represents an improvement. AUTOAN/NONLIN has incorporated the possibility of a lag time and can differentiate between 12 different models. Of these, eight are based on first-order processes and four incorporate Michaelis-Menten elimination kinetics, i.e., of saturable processes.

A different strategy of curve fitting is used by a program developed by Engberg-Pedersen [6]. This is an iterative technique comparable with the polyexponential curve stripping of AUTOAN. However, it limits itself during curve fitting to a curve of the general form $y = ab/x + c \cdot x + d \cdot x^2$, in which y represents the natural logarithm of the concentration and x represents a time metameter; thus: $x = \sqrt{t - \phi}$. The letters ϕ, a, b, c, and d are unknowns to be estimated from the observations; ϕ is lag time.

The strategy fo the polyexponential curve fit program [6] involves five assumptions: (a) all are positive values; (b) at time $t = 0$, the concentration $c = 0$; (c) only one peak concentration is observed; (d) no minimum concentration is observed during the course of a curve; and (e) the serum curve proceeds asymptotically towards zero with time. All serum concentrations are converted to the corresponding natural logarithms before fitting. The polyexponential curve fit method deletes some of the data, notably concentrations of 0.1 mg/l or less, to avoid negative logarithms.

The Engberg-Pedersen program limits itself to determining the theoretical curve after fitting by multivariate regression without subsequent reference to pharmacokinetic theory. From the curve are generated the maximum concentration, the time of its occurrence, and the area under the fitted curve. In its published form, a weighting factor $w = 100 \cdot \sqrt{c/c_{\max}} \cdot (1 - (\phi/t))$ was used. In a later comparison on data analyzed in a challenge of the program with i.a. long-acting metronidazole, the weighting factor was changed to $w = (c/c_{\max})^p \cdot (1 - 0.9^{\phi}/t)$,

Table 2. Comparison between methods for pharmacokinetic analysis (mean values and SEM)

Study/ antibiotic[a]	Dose (mg)	n	Polyexponential curve fit c_{max}	Polyexponential curve fit t_{max}	Polyexponential curve fit AUC	Autoan/nonlin c_{max}	Autoan/nonlin t_{max}	Autoan/nonlin AUC	Actual values c_{max}	Actual values t_{max}	Actual values AUC
Study I											
Bacmecillinam-HCl	200[b]	11	2.72 (0.21)	0.65 (0.05)	3.94 (0.22)	2.70 (0.21)	0.64 (0.04)	4.34 (0.20)	2.73 (0.21)	0.64 (0.04)	4.20 (0.24)
Pivmecillinam-HCl	203[b]	11	1.55 (0.20)	1.01 (0.12)	2.96 (0.26)	1.59 (0.20)	1.02 (0.11)	3.38 (0.26)	1.63 (0.21)	1.0 (0.11)	3.20 (0.29)
P[c]			0.001	0.032	0.002	0.001	0.001	0.001	0.001	0.01	0.001
Study II											
Bacmecillinam-HCl[d]	400	10	7.27 (0.29)	0.75 (0.04)	12.73 (0.61)						
Pivmecillinam-HCl[d]	400	10	5.33 0.32)	0.98 (0.04)	11.12 (0.68)						
P			0.20	0.020	0.020						
Study III											
Amoxicillin	500	10	10.39 (0.76)	1.48 (0.11)	29.66 (2.33)	10.17 (2.39)	1.34 (0.20)	29.38 (2.21)	10.43 (0.76)	1.55 (0.54)	29.57 (2.21)
Pivampicillin	500	10	6.94 (0.58)	1.68 (0.09)	18.42 (1.18)	6.75 (2.07)	1.46 (0.19)	17.65 (1.83)	7.48 (0.70)	1.75 (0.51)	18.19 (1.18)
P			0.002	0.048	0.002	0.001	0.001	0.001	0.001	0.01	0.001
Study IV[e]											
Metronidazole A	800	11	17.80 (1.17)	1.64 (0.21)	204.83 (10.18)	17.96 (1.19)	1.65 (0.19)	192.38 (8.95)	18.23 (1.19)	1.50 (0.29)	201.1 (9.522)
Metronidazole B	800	11	18.33 (0.77)	1.21 (0.18)	192.29 5.78)	18.42 (0.77)	1.25 (0.19)	180.28 (8.08)	19.07 (0.66)	1.00 (0.19)	192.2 (5.63)
P			NS		NS	NS		NS	NS		NS
Study V											
Erythromycin[f]	500	8	3.55 (0.43)	2.37 (0.26)	13.54 (1.79)	3.15 (0.62)		14.2 (5.4)	3.19 (0.54)	2.5 (0.18)	12.14 (2.18)

[a] Studies I–IV were carried out according to a crosss-over design with all doses to all healthy subjects
[b] Doses corresponding to 136 and 138 mg mecillinam
[c] Wilcoxon's nonparametric test
[d] Study of Leo Pharmaceuticals, Copenhagen
[e] Comparison between two formulations
[f] Curve fitting not successful in none of the eight patients

where c is the concentration at time t, c_{max} the maximum measured concentration, and $p = {}^1/_2$ for i.v. and ${}^3/_2$ for oral doses. The choice of a weight function is arbitrary, but the fact that such widely differing weight factors have to be chosen to obtain reasonable results of curve fitting limits the objective application of this approach.

Results obtained using the Engberg-Pedersen strategy are given in Table 2 and compared the those obtained with AUTOAN/NONLIN and the actual laboratory assay results before computation. The purpose was to test the performance of the procedures on compounds with widely different disposition characteristics, both rapidly and more slowly eliminated compounds. In short, it is seen that the polyexponential curve fit strategy for substances with rapid invasion and elimination, such as mecillinam from the prodrugs bacmecillinam and pivmecillinam and ampicillin from prvampicillin and amoxicillin, may result in underestimation of the total serum concentrations as reflected by the areas under the curves (AUC). Similarly, the peak concentrations are seen to be below the values of both AUTOAN/NONLIN and the actual laboratory assays. The fact that the underestimates of the AUC are less apparent for preparations and doses giving higher serum concentrations follows from the strategy that concentrations below 0.1 mg/l were ignored. This demonstrates that data omission in general may have unforseeable consequences. This is a subjective element which many, therefore, would like to avoid. A method favoring the phase around the curve peaks tends to be more suitable for rapidly absorbed compounds, such as the prodrugs pivampicillin and pivmecillinam, which have been the special interest of the group that generated the polyexponential curve fitting method. In nine studies on amoxicillin and pivampicillin, the polyexponential procedure rendered mean peak values which were 93%–105% of the measured values (Engberg-Pedersen, personal communication 1980).

The differences observed with the polyexponential method in our study were not statistically significant in any of the cases using either AUTOAN/NONLIN or the actual preprocessed values. The differences thus appear more as trends and would need larger data pools for verification. Table 2 shows that the polyexponential curve fit handles metronidazole data better than it does the penicillin data. This is related to the circumstance that the changes in concentrations per time unit were more moderate with metronidazole and that slower elimination enabled monitoring over a long period. For a compound with a difficult and occasionally erratic time-course, such as erythromycin, the polyexponential strategy generated in AUC more consistent with the basic assay results, whereas the peak concentration was better estimated by AUTOAN/NONLIN.

It is a definite drawback that the polyexponential curve fitting procedure [6] does not incorporate any knowledge or awareness of the suitability of pharmacokinetic models. The mass of pharmacokinetic analyses which have generated strong support for the validity of first-order kinetics and first-, second-, or multicompartment kinetics should be incorporated into the strategies of computerized pharmacokinetic analyses, unless Michaelis-Menten kinetics can be demonstrated. It would be advantageous to incorporate this experience as part of a strategy to improve on curve fitting. A pharmacokinetic analysis carried out in this manner, as with AUTOAN/NONLIN, can determine characteristics like $t^1/_2$,

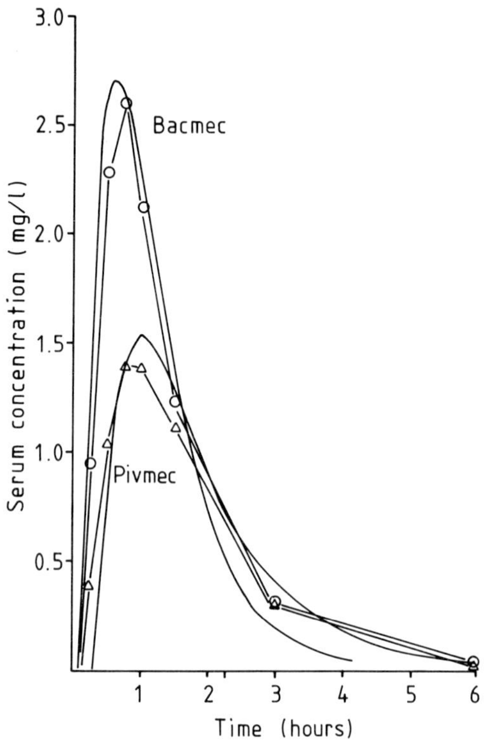

Fig. 1. Serum concentrations of mecillinam after the bacmecillinam and pivmecillinam prodrugs given as single doses of 200 and 203 mg, respectively, in cross-over fashion to 11 healthy volunteers (Study I). Polyexponential curve fitting: bacmecillinam, pivmecillinam. Mean curves from original assay data: bacmecillinam, pivmecillinam

clearance, and distribution volumes which are suitable for the sake of communication, objective comparison of different antibiotics, and calculation of optimal dosage regimes.

The finding that bacmecillinam is better absorbed than pivmecillinam was shown with a high degree of significance by all assessment procedures (Fig. 1). Both polyexponential procedures and AUTOAN/NONLIN showed higher serum peaks, earlier occurrence, and higher total serum concentrations (AUC) for bacmecillinam, which is thus suitable for oral administration of mecillinam. Mecillinam by itself is only available for parenteral administration. Commonly recommended are now 500-mg doses of amoxicillin and pivampicillin. It is thus significant that a previously published [9] cross-over study showed considerably higher ($p < 0.001$) concentrations of amoxicillin as reflected by both the total concentration (AUC) and the peak concentrations (Table 2), since these aminopenicillins have nearly the same antibacterial activity and tissue penetration [4].

Conclusions

Computerized pharmacokinetic analysis should employ methods which incorporate established pharmacokinetic model theory, such as AUTOAN/NONLIN, and not polyexponential curve fitting procedures independent of accepted pharmacokinetic laws. For a basic comparison between drugs, healthy volunteers represent the preferable cohort and the number of individuals should be ten, or more for agents with more erratic individual pharmacokinetic properties. Pharmacokinetic analyses in diseased patients should be performed more extensively, since these represent the very target group of the drugs and exhibit more variable disposition and distribution kinetics than the young healthy individuals often studied. This has been well established for, e.g., various renal and hepatic functions, cystic fibrosis, in neonates, gastrointestinal disease, and febrile conditions which affect the general physiology of the patients. Pharmacokinetic analyses of tissue penetration can be assessed by, e.g., peripheral lymph and peritoneal fluid.

References

1. Bergan T (1981) Pharmacokinetics of tissue penetration of antibiotics. Rev Infect Dis 3:45–66
2. Bergan T, Engeset A, Olszewski W, Solberg R (1979) Pharmacokinetics of bacampicillin and bacmecillinam in plasma and peripheral lymph. Lymphology 12:85–94
3. Bergan T, Engeset A, Olszewski W (1983) Temocillin in peripheral lymph. J Antimicrob Chemother 12:59–63
4. Bergan T, Engeset A, Olszewski W, Sjövall J (1983) Distribution to human peripheral lymph of amoxycillin and ampicillin from the oral prodrug bacampicillin. J Antimicrob Chemother 12:497–502
5. Bergan T, Michalsen H (1979) Pharmacokinetics of azlocillin in children with cystic fibrosis. Arzneimittelforsch 29:1955–1957
6. Engberg-Pedersen H (1974) Empirical equation for pharmacokinetic analysis of drug serum levels after oral application. Antimicrob Agents Chemother 6:554–562
7. Engeset A, Hager B, Nesheim A, Kolbenstvedt A (1973) Studies on human peripheral lymph. Lymphology 6:1–5
8. Hoffstedt B (1981) Penetration of antibiotics into subcutaneous tissue fluids with special reference to a new modified thread sampling technique. Scand J Infect Dis [Suppl] 30
9. Holten E, Solem J (1980) Konsentrasjon av amoxicillin og pivampicillin i plasma etter peroral medikasjon. T norske Lægeforen 100:181–182
10. Philipson A, Sabath LD, Charles D (1973) Transplacental passage of erythromycin and clindamycin. N Engl J Med 288:1219–1221
11. Wagner JG (1975) Fundamentals of pharmacokinetics. Drug Intelligence. Hamilton
12. Wise R, Donovan IA, Ambrose NS, Allcock JE (1981) The penetration of cefoxitin into peritoneal fluid. J Antimicrob Chemother 8:453–457
13. Wise R, Donovan IA, Drumm J, Dyas A, Cross C (1983) The intraperitoneal penetration of temocillin. J Antimicrob Chemother 12:93–96

Data Handling and Retrieval in Clinical Virology by Small Decentralized Computers *

K.-O. Habermehl [1]

Diagnosis of viral diseases requires an individually defined course of highly specialized laboratory procedures. In order to obtain relevant results it is necessary to get detailed data concerning the patient as well as clinical information about the type of disease and the date of its onset, including any special symptoms which may be present. On the basis of this information, the doctor in the laboratory has to decide what type of tests are to be performed and, depending on the preliminary results, which additional diagnostic procedures are necessary. Finally, the results of the laboratory tests together with the clinical symptomatology and the epidemiological or endemiological situation enable a report to be compiled.

The accumulation of large numbers of samples to be examined by different methods on the one hand, and the necessity to decide on a specific test for an individual patient on the other, is a contradiction in itself. This problem can be solved only by electronic data processing. Moreover, this information should be available for immediate access over a long period of time in case of later requests. In order to fulfill these requirements, it is necessary to computerize the available diagnostic data (Vroman 1979).

Requirements of a Computer System for Clinical Virology and Epidemiology

From the numerous proposals concerning the requirements for computer systems in diagnostic work in clinical microbiology (Krieg et al. 1971; Kobernick and Mandell 1974; Ridgway et al. 1980; Williams et al. 1978), it is now generally agreed that laboratory computers should be both easy to operate and accessible to laboratory personnel (Walter 1973). The programs should be designed in such a way that they can be adapted by laboratory personnel to specific needs without the assistance of a programmer. In addition, all data files should be available for immediate access. If diagnostic work is to be performed on a larger scale, even relatively short delays in the response to inquiries render the system impractical.

The combined requirements of immediate access and long-term storage of data lead necessarily to the application of decentralized minicomputers.

---

* Much of this chapter was originally published in New Developments in Diagnostic Virology (P. A. Bachmann, ed.), Current Topics in Microbiology and Immunology 104, Springer

1 Institut für Klinische und Experimentelle Virologie, Freie Universität Berlin, Hindenburgdamm 27, D-1000 Berlin 45

Moreover, it has been shown that the laboratory computer should be an independent unit under the supervision of the laboratory (Kobernick and Mandell 1974; Rappaport et al. 1969). Based on these considerations and on the experiences of other authors (Blair and Brown 1981; Farrar et al. 1975; Goodwin 1976; Goodwin and Smith 1976; Harvey et al. 1972; Mitchison et al. 1978; Vermeulen et al. 1974), we developed a concept for a laboratory computer for clinical virology (Habermehl 1980, 1981, 1983) that allows institute personnel to handle the system without any special knowledge or training.

Using a dialogue system, data input is performed in clear text or as abbreviations of the clear text of variable length, starting with the initial letters of selected words. The use of code numbers is avoided as much as possible. All data are stored for as long as possible, at least for several years. They remain available for immediate access and for linkage with the results from new diagnostic investigations. Thereafter, the data from previous tests are stored on tape, ensuring that the name and date of birth of the patient remain on the disc together with a reference to the location of his personal data file.

According to the individual request, working instructions are printed for specific diagnostic procedures, referring to the titers or results from previous tests of the patient, in order to facilitate diagnostic assays. After input of the test results, the system prints an interim diagnostic report for laboratory use only, containing all clinical data and results including linkage analysis, diagnosis, and information on previous requests. This interim report, printed out in mnemonics, enables the virologist to give a formatted interpretation of the results or a virological diagnosis, or to make additional remarks if necessary.

Upon request, the system gives a detailed report to the consulting doctor or a ward report in clear text without mnemonics, containing all data (present and previous) and laboratory results concerning the patient. The conclusion of this report is a virological diagnosis, and if necessary the report should contain additional information. Interim reports are also possible.

Programs for statistical or epidemiological procedures are also included in the dialogue system. The linkage system allows detailed statistical analyses of all data with any desired logical connection, for example, establishment of a serum bank with listings of the different antibody titers against several antigens. Further information such as an alphabetical listing of patients or data files is also available. In addition, the system performs all administrative work including the complete statement of accounts.

Functions of the Computer System

In order to obtain a comprehensive survey of the problem of data processing in clinical virology, it is necessary to demonstrate all steps of the diagnostic procedure.

Input of Clinical Data into the System

The software is designed for manual entry of information using terminals. It would be desirable to enter the data on prepunched or mark-sense cards, but this requires accurate and time-consuming work by the admitting service. Due to the variety of senders (several hospitals, general practitioners, pathologists, etc.) we have found that only the manual entry of information is practicable.

In order to obtain sufficient information in printed form which is easy to survey, we developed a single-sheet requisition form containing space for personal data of the patient and a listing of a selection of symptoms or syndromes. Three different symptoms can be marked by the sender. The type of test procedure can be proposed by the sender in clear text, but in the case of virological work it is advisable that the laboratory makes the final decision on the assay methods.

Patient admittance is the crucial problem of any computer system in clinical virology. We therefore developed time-saving procedures for the input of data concerning patients, senders, clinical diagnosis, and required investigations. Since most of the submitted samples are received from a limited number of senders, a data file exists containing their full addresses and any other pertinent information. For the admittance of a patient, it is necessary to answer the question concerning the identity of the sender with only the first two (or more) letters of the name. For example, the input of "ST" will be followed by the response "STEGLITZ KLINIKUM," and if the response is confirmed by the operator, the input of the complete information is fixed. If the sender's name is not stored in the data file, the input has to be performed immediately using a simple procedure regulated by the dialogue system. This time-saving principle has also been established for the input of clinical diagnoses, assay methods, specimen and virus designations, etc. The time-saving system of input, i.e., working with the initial letters of special terms instead of using code numbers, does not apply only to the admittance of patients but also to all other computerized procedures, including statistics and administration.

Requests for Testing and Input of the Patient's Data

The first step in this procedure is shown in Fig. 1. After input of the date of birth and the first two letters of the name of the patient, the system checks within several tenths of a second whether the patient's name is already stored in the file. If the patient's name is filed, the system indicates the number and date of the last admission, followed by the question "INFO" (Is more information desired?). If this is responded to with a "J" (yes), all previous data are presented immediately. Figure 2 illustrates the information called up onto the screen of the terminal, showing the data for the last admission, the number of the sample, personal data of the patient and clinical diagnoses, incluing the duration of the illness (2 days in this example). The results of the tests are given in mnemonics (only for internal use), followed by the previous virological diagnoses and the earlier information supplied by the institute. After the question: "Function (B/A)" (B, before; A, after) one can decide whether previous admissions should be recalled. If in Fig. 1

Date of birth	12. 09. 1953
Beginning of surname	MU
Name	Mueller Carola 12.09.53
Last entry	First entry of 05.04.81
Info	No
Pregnant	No
Sender	Steglitz Clinic
Ward	Ward 06
Insurance company	AOK
1 clin. diagnosis	Encephalitis
2 clin. diagnosis	Parotitis
3 clin. diagnosis	Bronchitis
Duration of illness	4
Investigation	5: Viruses causing diseases of the CNS
	7: Viruses causing diseases of the respiratory tract
Specimen	Venous blood
Method	Complement fixing reaction
Virus	Psittacosis
Subgroup	
Virus	
Method	
Specimen	

Correction (J/N/P/X)

Fig. 1. Admittance of patients into the computer. Progress of the input dialogue on the screen of the terminal

03.01.82	01091101/01	Schulz Heinz	M	01.08.52	2 days	
Steglitz clinic						
Ward 03		Pneumonia				
VBL-CFT-INA	4/1	VBL-CFT-INB	4/1		VBL-CFT-ADE	8/1
VBL-CFT-PAR-1	4/1	VBL-CFT-PAR-2	4/1		VBL-CFT-PAR-3	4/1
VBL-CFT-RES	4/1					
RAC-TCI:	INA-H3N2					
		Fresh infection with influenza A. Send a sample of venous blood please, in 8–14 days at the earliest				

Function (B/A)

Fig. 2. Information system on the screen of the terminal providing all data concerning a patient form one admission. Information from previous or following admissions can be requested by typing in the letter "B" or "A" (before or after), respectively, to the question "FUNCTION (B/A)"

the question “INFO” is answered with “N” (no) the dialogue system requests the name of the sender. As mentioned above, it is sufficient to type in the first letters of the name (for example “ST,” and if “ST” is on file, the answer will be “STEGLITZ KLINIKUM”). This same time-saving principle is used for input of the “STATION” (ward number, the “KLIN. DIAGNOSE” (clinical diagnosis), and so on.

Two different modes exist to request the designated laboratory tests. In the single mode the dialogue system asks for the type of specimen (“UNTERSUCHUNGSMATERIAL”), the method of testing (“UNTERSUCHUNGSMETHODE”), and the type of agent (“VIRUS”). Again, it is sufficient to type only the first two (or more) letters, which are then followed by confirmation of the whole term. Thereafter the system continues questioning for “VIRUS,” “UNTERSUCHUNGSMETHODE” (method of testing), and “UNTERSUCHUNGSMATERIAL” (specimen). If these last questions are not responded to, the system finishes the input procedure, gives the sample number for this patient, and prints it in sequential numbering order. Thereafter the computer begins the admission program for the next patient.

In the other mode of input, blocks of assays can be requested using code numbers. In this case the input mode for “UNTERSUCHUNGSGRUPPEN” (groups of tests) has to be responded to using a code number. Input of “5,” for example, is answered with “VIREN V. ERKR. DES ZNS” (viruses causing diseases of the central nervous system). Input of “7” is answered with “VIREN V. ERKR. DES RESPIRATIONSTRAKTES” (viruses causing respiratory tract infections). This latter group includes the information: complement fixation tests with venous blood for influenza virus types A and B, parainfluenza virus types and 1, 2, 3, respiratory syncytial virus, adeno viruses, and mycoplasm. Several groups of tests can be requested and combined in this mode. Apart from these group requests, it is possible to continue requests in the individual single mode system as described above. This time-saving principle allows for the admission of several hundred individual patients within a surprisingly short period of time. Further time reductions can be obtained by performing the input on several terminals simultaneously.

Printing of Worksheets

Upon requests, different working instructions are printed out for all working groups in the laboratory. The worksheets have different patterns according to the type of test to be performed. They remain as working protocols at the different work places in the laboratory. Figure 3 shows a working protocol for isolations from tissue cultures. Both the sample number and the name of the patient are always printed out simultaneously, in order to avoid confusion. The technician works with this form, recording the different types of tissue cultures, the dates of microscopic examinations, and the results. Another type of protocol is shown in Fig. 4, concerning hemagglutination-inhibition tests with rubella virus. At the top of the sheet the number of the type of protocol is given (in this example “0030”), followed by the identification of the protocol form. The number is necessary in

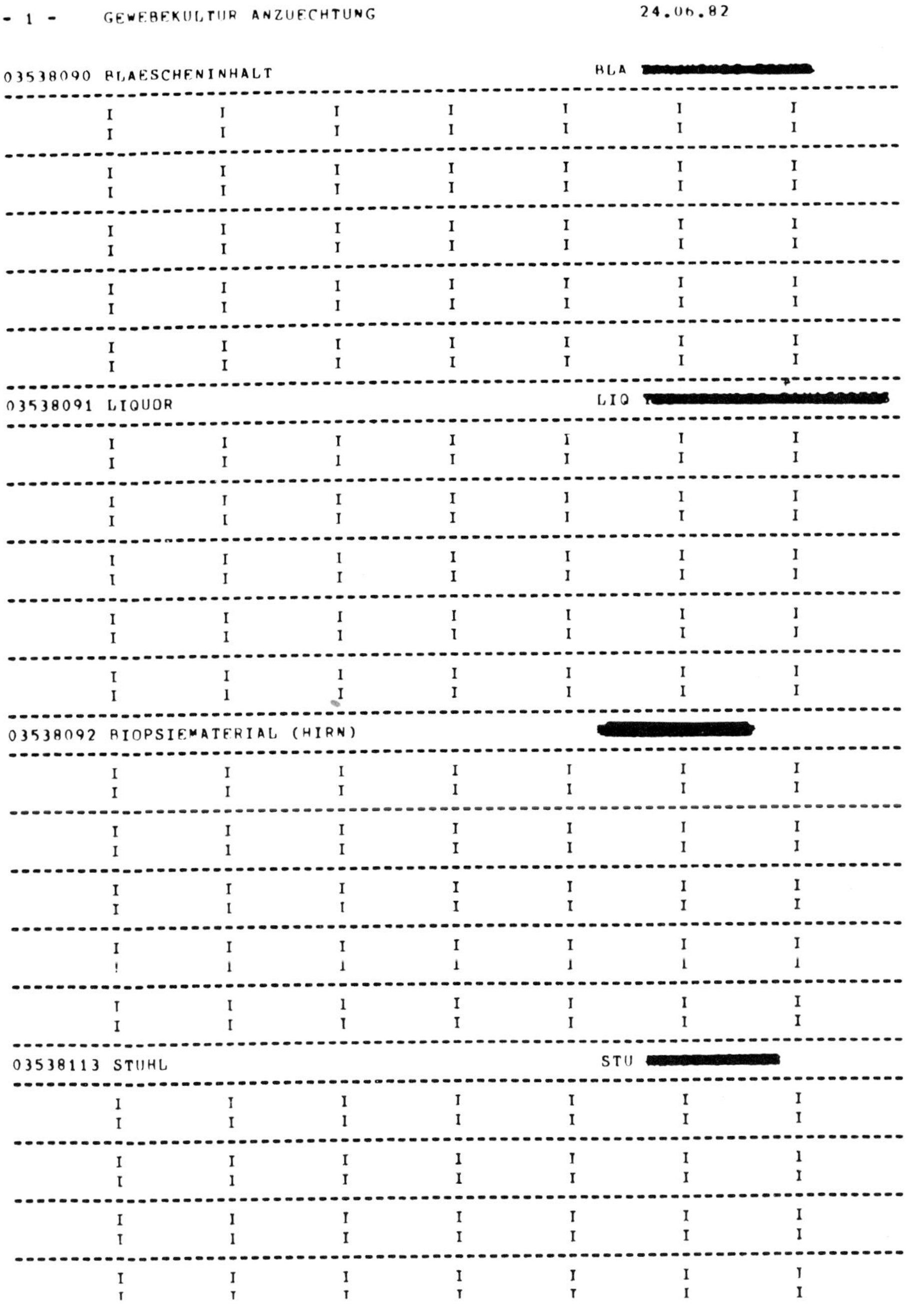

```
- 1 -     GEWEBEKULTUR ANZUECHTUNG                        24.06.82

03538090 BLAESCHENINHALT                            BLA

03538091 LIQUOR                                     LIQ

03538092 BIOPSIEMATERIAL (HIRN)

03538113 STUHL                                      STU
```

Fig. 3. Printout of a protocol for virus isolation from tissue culture

```
0030          HHT ROETELN              HAEMAGGLUTINATIONS HEMMTEST

                     I ROE I
                     I 000 I

03538094/01-VBL-I.....I

03537844/03-VBL      128I
03538097/04-VBL-I.....I

03538098/01-VBL-I.....I
SCHULZ ANNELIESE

03538101/01-VBL-I.....I

03538102/01-VBL-I.....I

03538106/01-VBL-I.....I

02348301/01-VBL      512I
03538116/02-VBL-I.....I

03538118/01-VBL-I.....I

03538121/01-VBL-I.....I

03519042/01-VBL      256I
03538128/02-VBL-I.....I
```

Fig. 4. Printout of a protocol for a rubella virus hemagglutination-inhibition test. Previous results are given in connection with the respective sample numbers. The requested tests are marked with *dotted lines*

order to retrieve the appropriate program for quick input of results. If previous data exist in the file concerning the same patient, the old sample numbers are listed together with the results. Samples and antigens are marked by the respective mnemonic terms, e.g. "VBL" (venous blood) or "ROE" (rubella). The requested examinations are marked with dotted lines. Listing the previous results is not only time-saving, since it enables the technician to start with the appropriate dilution, but it is also helpful for internal quality control, sometimes indicating unexpected developments in the antibody titer. The same principle is established in some of the other working protocols for serological examinations like complement fixation tests (Fig. 5), hepatitis radioimmunoassays (Fig. 6), or picorna-virus-neutralisation tests (Fig. 7). The requested assay procedures are marked with dotted lines.

Apart from the protocol sheets mentioned above an additional *master working protocol* is printed containing all the information and all tests requested for the

```
0010          VERSCHIEDENE VIREN                   KOMPLEMENTBINDUNGSREAKTION          24.06.82

                I ENT I CYT I MAS I MUM I VIZ I HSI I INA I INB I PIN I ADE I RES I MYK I TOX I RTA I ORN I ARB I LCM I QFI
                I 000 I 000 I 000 I 000 I 000 I 000 I 000 I 000 I 000 I 000 I 000 I 000 I 000 I 000 I 000 I 000 I 000 I 000

03538136/01-VBL-I     I.....I     I     I     I     I     I     I     I     I     I     I.....I     I     I     I     I

03530813/01-VBL       I    4I     I     I     I     I     I     I     I     I     I     I     I     I     I     I     I
03531315/02-VBL       I    4I     I     I     I     I     I     I     I     I     I     I     I     I    4I     I     I
03532058/03-VBL       I    4I     I     I     I     I     I     I     I     I     I     I     I     I    4I     I     I
03535351/04-VBL       I    4I     I     I     I     I     I     I     I     I     I     I     I     I    4I     I     I
03538138/05-VBL-I.....I.....I     I     I     I     I     I     I     I     I     I     I     I     I     I     I     I

03538139/01-VBL-I.....I.....I     I     I     I     I.....I.....I.....I.....I.....I.....I     I     I     I     I     I

03536945/01-VBL      4I   10I     I     I     I     I     I     I     I     I     I     I     I     I     I     I     I
03537336/02-VBL      4I   10I     I     I     I     I     I     I     I     I     I     I     I     I     I     I     I
03538140/03-VBL-I.....I.....I     I     I     I     I     I     I     I     I     I     I     I     I     I     I     I

03537658/01-VBL      4I   10I   40I   10I  160I   10I    4I    4I   80I    4I    4I    4I     I     I     I     I     I
03538141/02-VBL-I.....I.....I.....I.....I.....I.....I.....I.....I.....I.....I.....I.....I     I     I     I     I     I

03537620/01-VBL       I     I   20I     I   20I    4I     I     I     I     I     I     I     I     I     I     I     I
03538142/02-VBL-I     I     I.....I     I.....I.....I     I     I     I     I     I     I     I     I     I     I     I

02897501/01-VBL      4I    4I    4I   20I   10I    4I    4I    4I    4I   10I    4I    4I     I     I     I     I     I
02897502/02-VBL      4I    4I    4I   10I   10I    4I    4I    4I    4I   40I    4I    4I     I     I    4I     I     I
03538143/03-VBL-I     I     I.....I     I.....I.....I     I     I     I     I     I     I     I     I     I     I     I
```

Fig. 5. Printout of a protocol for complement-fixation tests with 18 different antigens marked by mnemonics. Format described in legend to Fig. 4

```
0050          HEPATITIS                            RADIO IMMUNASSAY

                I HBS I ABS I ABC I AAV I HBE I ABE I
                I 000 I 000 I 000 I 000 I 000 I 000 I

03538094/01-VBL-I.....I     I     I     I     I     I

03538097/04-VBL-I.....I     I     I.....I     I     I

03538098/01-VBL-I.....I     I     I     I     I     I

03538102/01-VBL-I.....I     I     I     I     I     I

03535838/01-VBL   3131I 2336I    4I   95I  876I  122I
03538112/02-VBL-I.....I.....I.....I.....I.....I.....I

03538114/01-VBL-I     I     I     I.....I     I     I

03538115/01-VBL-I.....I.....I.....I.....I     I     I

03538117/01-VBL-I.....I.....I.....I.....I.....I     I
```

Fig. 6. Printout of a protocol for hepatitis radioimmunoassays. The results of previous admissions are given by the counts of the gamma counter as well as by the significance, positive (+) or negative (−)

0020 COXSACKIE A,B, ECHO, POLIO NEUTRALISATIONSTEST 24.06.82

	CXA 009	CXA 021	CXA 023	CXB 001	CXB 002	CXB 003	CXB 004	CXB 005	CXB 006	ECO 003	ECO 006	ECO 008	ECO 011	ECO 019	ECO 025	ECO 030	POL 001	POL 002	POL 003
02402501/01-VBL	20	20	20	4	4	4	4	4	4										
02402502/02-VBL	20	4	4	4	4	4	4	4	4										
02402503/03-VBL		4	20	4	4	4	4	4	4										
03538095/04-VBL																			
02236102/02-VBL	4	4	4	4	4	4	4	4	4										
02236103/03-VBL	4	20	4	4	4	4	4	4	4										
03527034/06-VBL	4	20	4	4	4	4	4	4	4										
03538096/07-VBL																			
03538101/01-VBL																			
03538103/01-VBL																			
03538104/01-VBL																			
03538111/01-VBL																			
03538114/01-VBL																			
03530813/01-VBL	80	20	4	4	4	4	20	20	4										
03531315/02-VBL	160	20	4	4	4	4	20	4	4										
03532058/03-VBL	160	20	4	4	4	4	4	4	4										
03535351/04-VBL	80	20	4	4	4	4	4	4	4										
03538138/05-VBL																			
03538139/01-VBL																			

Fig. 7. Printout of a protocol for several picornavirus-neutralization tests. Format described in legend to Fig. 4

different work stations. This enables each technician to obtain general information on all the diagnostic steps to be performed with the same sample at different work stations in the laboratory.

Generation and Modification

The application of new methods or the introduction of new antigens in serology requires generation of new working protocols or modification of the old ones. For this reason special programs exist, operating in the dialogue mode, which enable the user to perform generation or modification of protocol forms without special knowledge of programming. To establish a new serology protocol, for example, the system proposes the number of the new form, asks for the name of the form, the type of sample (e.g., venous blood), and the different antigens (up to 20 in one protocol). After these questions have been responded to, the new form is automatically generated in the file, similar to those shown in Fig. 5. In addition, a special programm exists which is designed for changing previously established working protocols.

Input of Results

The system offers two modes for the input of results. In the mode concerning most of the information, the technician requests at the terminal the file for input of results and types in the number of the worksheet. The computer responds by listing on the screen the patient's name and the different tests carried out, in the same order as on the worksheet. The system stops automatically at the space designated for the input of the result. Thereafter, the technician has only to type in the antibody titer and after pressing a key the next space for the input of results appears automatically. The system has proven capable of allowing rapid input of results. It is easier and safer than marking the results on sense cards.

The second input mode for results is based on a program starting after the input of the sample number of the patient and listing the different tests in sequence. This mode permits input and correction of single results from individual patients. Both modes of input offer the possibility for correction at any desired time.

Printing of an Interim Diagnostic Survey

Upon request, usually once a day, the system prints a diagnostic survey, a "DIAGNOSE LISTE" (list of diagnose). This is a listing of all patients from whom test results have been obtained since the previous printout of a diagnostic survey. As shown in Fig. 8, this listing consists of a condensed survey of all information on the patients, including results, diagnoses, and a reference to previous admissions. This listing not only allows a survey on the course of the illness, which is the basis for a diagnosis, but it also enables the diagnostician in the laboratory to control the work routine and to request additional measures.

Most of the diagnoses and additional remarks made by the laboratory diagnostician are formatted and stored in a data file. For example, remark "24"

```
*** D I A G N O S E L I S T E  VOM 24.06.82 ***

 WILLY                                                    GESCHLECHT: M GEBR.:  8.07.12

PSYCHIATR.U.NEUROLOG. KLINIK               STATION

EINS-DATUM:        I 24.05.82 I
MAT-NR:            I  3536594 I
EINS-NR:           I       1. I

PARESEN (UNKLA     I        * I

LIQ                I        * I
VBL                I        * I

LIQ-GAZ            I  NEGATIV I
VBL-KBR-CYT-000I        10/1 I
VBL-KBR-ENT-000I         4/1 I
VBL-KBR-HSI-000I        20/1 I
VBL-KBR-MAS-000I         4/1 I
VBL-KBR-MUM-000I         5/1 I
VBL-KBR-VIZ-000I         4/1 I

24.05.82  3536594/01
ERGEB.D.AUSST.UNTERS.W.NACHGES.

-------------------------------------------------------------------------------------

 MARGARETA                                                GESCHLECHT: W GEBR.:  2.11.50

CHARLOTTENBURG KLINIKUM                   POLI

EINS-DATUM:        I  1.06.82 I 15.06.82 I
MAT-NR:            I  3537026 I  3537704 I
EINS-NR:           I       1. I       2. I

EXANTHEM (MAKU     I        * I        * I
ROETELN            I        * I          I

VBL                I        * I        * I

VBL-HAH-ROE-000I         4/0 I    512/3 I
VBL-KBR-HSI-000I         4/1 I      4/1 I
VBL-KBR-MAS-000I         4/1 I      4/1 I
VBL-KBR-VIZ-000I         4/1 I      4/1 I
VBL-MHH-ROE-000I             I     50/2 I
VBL-RME            I          I  POSITIV I

 1.06.82  3537026/01
IN 8 BIS 14 T.EINS.VON                  VBL


15.06.82  3537704/02
ERGEB.D.AUSST.UNTERS.W.NACHGES.

-------------------------------------------------------------------------------------

 SAEUGLING                                                GESCHLECHT: M GEBR.:         K

CHRISTOPHORUS-KINDERKRHS.                 STATION

EINS-DATUM:        I 10.06.82 I
MAT-NR:            I  3537494 I
EINS-NR:           I       1. I

DURCHFALL          I        * I

STU                I        * I

STU-ELM            I   RTA000 I
STU-GAZ            I    FEHLT I
```

Fig. 8. Printout of an interim diagnostic survey: a list of diagnoses. Given in a condensed form using mnemonics, it is designed for internal use only, in order to obtain an actual survey on the course of the diagnostic work. On this basis, diagnostic interpretations and additional remarks can be given for the medical report to the sender means: "PLEASE SEND ANOTHER SAMPLE OF ... WITHIN THE NEXT 8–14 DAYS." It is only necessary to type the number of the diagnosis or remark in the margin of the listing. These numbers will be introduced into the system in a specially shortened program run by a typist. This is the basis for the medical report to the sender.

Apart from the form text, free texts can also be generated. However, in order to make the diagnoses and additional remarks available for statistical evaluation, form texts are preferred. The data file contains approximately 100 form texts. In the same mode as mentioned above the text can be changed or new texts can be added into the dialogue system without the help of a programmer. In this respect, it should always be kept in mind that adjusting an old text should not change the meaning altogether, since otherwise the statistical evaluation of old and new texts would produce different results. Therefore, rather than adjusting old texts, new texts should be added to the data file.

One becomes accustomed to reading the diagnostic survey list very quickly. The data on more than 100 patients can be scanned in approximately 20 min. This survey is, in this author's opinion, the most important part of the system, as it fulfills the requirements mentioned in the introduction: to combine and link large numbers of test results and data with individual diagnostic procedures in virology.

Medical Report to the Sender

Upon request, data on all the patients given in the list of diagnose are printed out, including the diagnoses and further remarks, without any abbreviations or mnemonics (Fig. 9). Since especially in virology the diagnosis is characterized by at least two admissions, it is advisable to send a letter containing data on all the previous admissions. Some additional precautions have been introduced into this system: The operator can decide how many of the previous admissions should be printed in the letter. Furthermore, the precaution has been taken to print in the letter only the diagnoses of the last report. Remarks such as "PLEASE SEND ANOTHER SAMPLE OF BLOOD" are printed only once. If a second letter concerning the same admission is printed out, this remark does not appear in the letter, but it remains in the file. Finally, the printouts of the medical reports are sorted according to the sender, in order to facilitate delivery of the letters.

Since all information is stored in the integrated system, it is possible to generate a letter at any time. Interim reports can be given or a letter can be generated immediately after the arrival of a sample and before any results are obtained, giving only the message to "get more information" or to "get additional samples." These letters can be printed out immediately after the input of the patients' data. It is very helpful for establishing a successful diagnostic service to maintain good contact with the clinician.

PROF. DR. MED. K.-O. HABERMEHL
INSTITUT FUER KLINISCHE UND EXPERIMENTELLE VIROLOGIE (WE04)
FREIE UNIVERSITAET BERLIN

1000 BERLIN 45, HINDENBURGDAMM 27
TEL: (030) 7983696, 7983688

KLINIKUM CHARLOTTENBURG
POLI MED
SPANDAUER DAMM 130 DATUM: 24.06.82

1000 BERLIN 19

VERBINDLICHEN DANK FUER DIE UEBERWEISUNG DES UNTERSUCHUNGSMATERIALS
IHRER PATIENTIN:

GEB: 2.11.50

DATUM - EINSENDUNG:	1.06.82	15.06.82
EXANTHEM (MAKULOES)	*	*
ROETELN	*	
VENENBLUT	*	*
*** HAEMAGGLUTINATIONS HEMMTEST VENENBLUT		
ROETELN	1:4	1:512++
*** KOMPLEMENTBINDUNGSREAKTION VENENBLUT		
HERPES SIMPLEX	1:4-	1:4-
MASERN	1:4-	1:4-
VARIZELLEN ZOSTER	1:4-	1:4-
*** IGM HAEMAGGL.HEMMTEST VENENBLUT		
ROETELN		1:50+
*** ROETELN IGM ELISA VENENBLUT		POSITIV

VIROLOGISCHE DIAGNOSE:
ANSTIEG DER ANTIKOERPER GEGEN ROETELN.
DER NACHWEIS VON ROETELN - ANTIKOERPERN IM ISOLIERTEN IGM
SPRICHT FUER EINE FRISCHE INFEKTION.

MIT KOLLEGIALEN GRUESSEN

PROF. DR. MED. K.-O. HABERMEHL

- : TITER IM NORMBEREICH
+ : ERHOEHETER TITER
++ : SIGNIFIKANTER TITERANSTIEG

Fig. 9. Printout of a medical report to the sender in clear text without any abbreviations showing current and previous diagnostic results with interpretation and further remarks

Confidentiality

Access to the computer is gained by logging in. Each authorized user has a special log-in code which works in two steps. Different grades of the log-in code determine which parts of the system are accessible to the user. This concerns entering data concerning patients, printing out results, and correcting the data files.

A problem can arise if the visual display screen is left on after logging in. In this case, all information made available to the initial user could be obtained by other persons. This problem can be solved by two procedures. First, after finishing his or her work the user can stop the output of information onto the screen by typing in a code number. The second procedure is based on an automatic switch-off system. If the keybord of a terminal is not used for a given time (e.g., 2 min) access to the terminal is switched off automatically. In order to continue working on this terminal one has to log in again. This system of maintaining classified information has been criticized on the grounds that a consultant is able to examine another consultant's results. In agreement with Blair and Smith (1981) it should be pointed out, however, that this facility is generally considered beneficial rather than harmful.

Hardware

The system was started in 1975 as a pilot project using a small computer, the Wang 2200C: memory, 16 kilobytes; disc, fixed/removable, 4×5.5 megabytes; printer 180 character/s. After a thorough revision, the system was established in a larger computer, the Digital PDP L/34: multi-user operating system; memory, 256 kilobytes; disc removable, 2×30 megabytes, now extended to 300 megabytes. Printer, 300 lines/min. The software was developed by Dr. Buder Computer GmbH, Berlin.

Statistics

The linkage of clinical data, test results, and diagnostic methods allows detailed statistical evaluations to be carried out. Using the dialogue system, various statistical programs can be performed concerning epidemiological surveys, establishment of serum banks, or questions of administration.

At present, the epidemiological survey of viral diseases in the Federal Republic of Germany is performed using this system, based on the results of 49 virus diagnostic laboratories. In cooperation with the Deutsche Vereinigung zur Bekämpfung der Viruskrankheiten (DVV), the results from the various laboratories are submitted on optical mark-sense cards, on tapes or on discs. An epidemiological survey is made every 4 weeks and is the basis for the report of the German laboratories to the WHO.

After an initial pilot project in 1982, 30,000 cases with laboratory diagnosis of a virus infection have already been reported to our institute in 1983. These reports contain a large amount of different clinical and methodological informa-

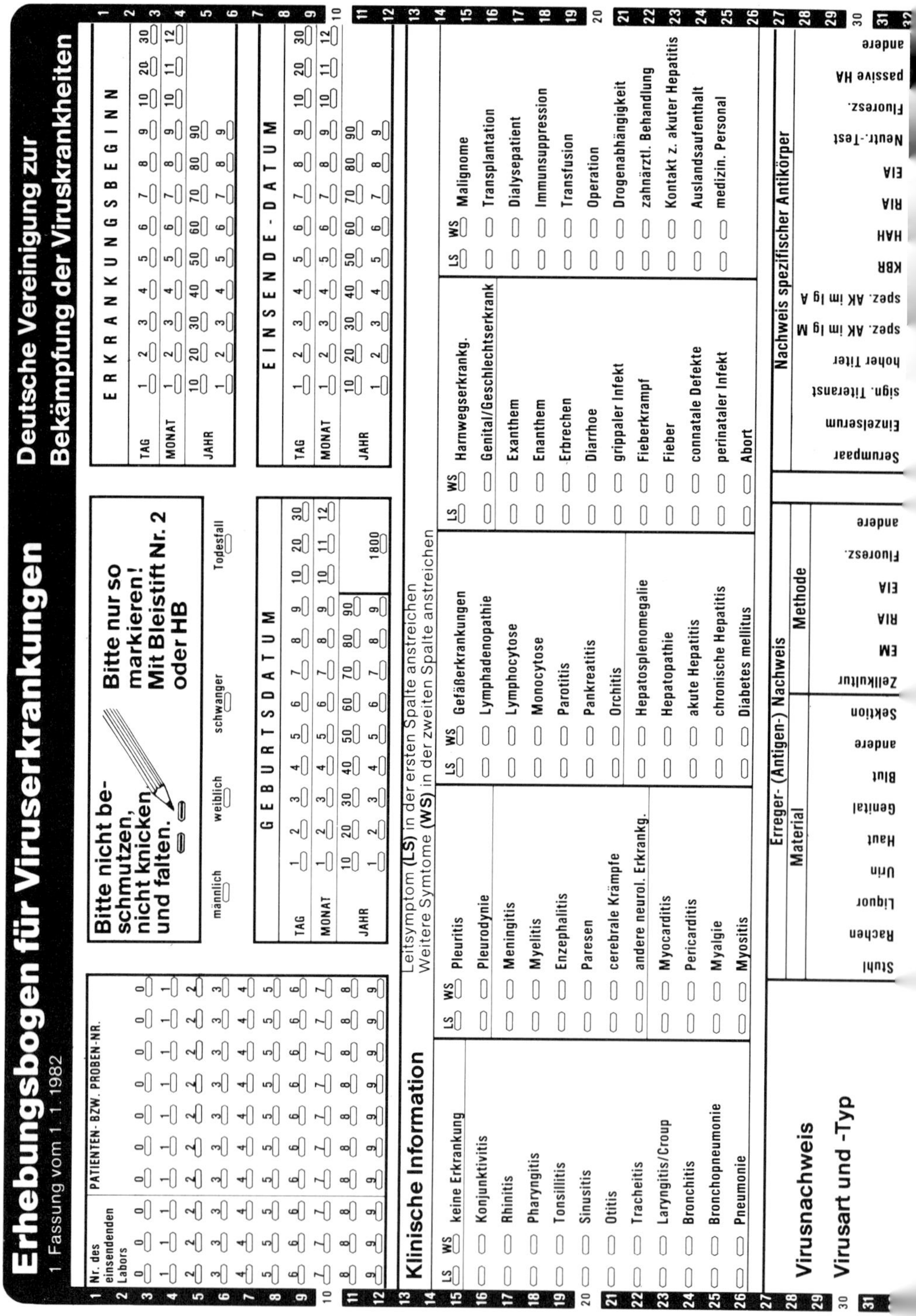

Fig. 10. Mark-sense card for submission of diagnosed virus infections from laboratories to the Institute of Clinical and Experimental Virology of the Free University of Berlin. The report includes number of the patient, date of birth, onset of illness, date of submission, clinical information, type of virus and

Nr.	Erreger	Typen
33	Adeno	n.typ. 10 20 30
34		1 2 3 4 5 6 7 8 9
35	Influenza	A B C
36	Parainfluenza	1 2 3 4
37	Resp. syncyt.	
38	Rhino	
39	* Mycoplasma pneumoniae	
40	* Chlamydia psittaci	
41	* Chlamydia trachomatis	
42	Coxsackie A	10 20 30
43		1 2 3 4 5 6 7 8 9
44	Coxsackie B	1 2 3 4 5 6
45	ECHO	10 20 30
46		1 2 3 4 5 6 7 8 9
47	Polio	Wild Impf. 1 2 3
48	Entero	n. typ. 70
49		1 2 3 4 5 6 7 8 9
50	Mumps	
51	Masern	
52	Röteln	
53	Epstein-Barr	
54	Herpes	1 2
55	Varicellen-Zoster	
56	Cytomegalie	
57	Rota	
58	FSME	
59	Corona	
60	* Coxiella burnetii	
61	Hepatitis A	
62	Hepatitis B	HB_s Ag
63		HB_c Ag
64		HB_e Ag
65	Für diesen Patienten wurde im Zusammenhang mit der gleichen Erkrankung bereits in einem früher eingesandten Formular eine positive Diagnose durch serologische Methoden gestellt:	
66	Für diesen Patienten wurde im Zusammenhang mit der gleichen Erkrankung bereits in einem früher eingesandten Formular eine positive Diagnose durch Nachweis des Erregers gestellt:	

method of detection. – Laboratories with an own dataprocessing system submit the data by tape, disk or on line

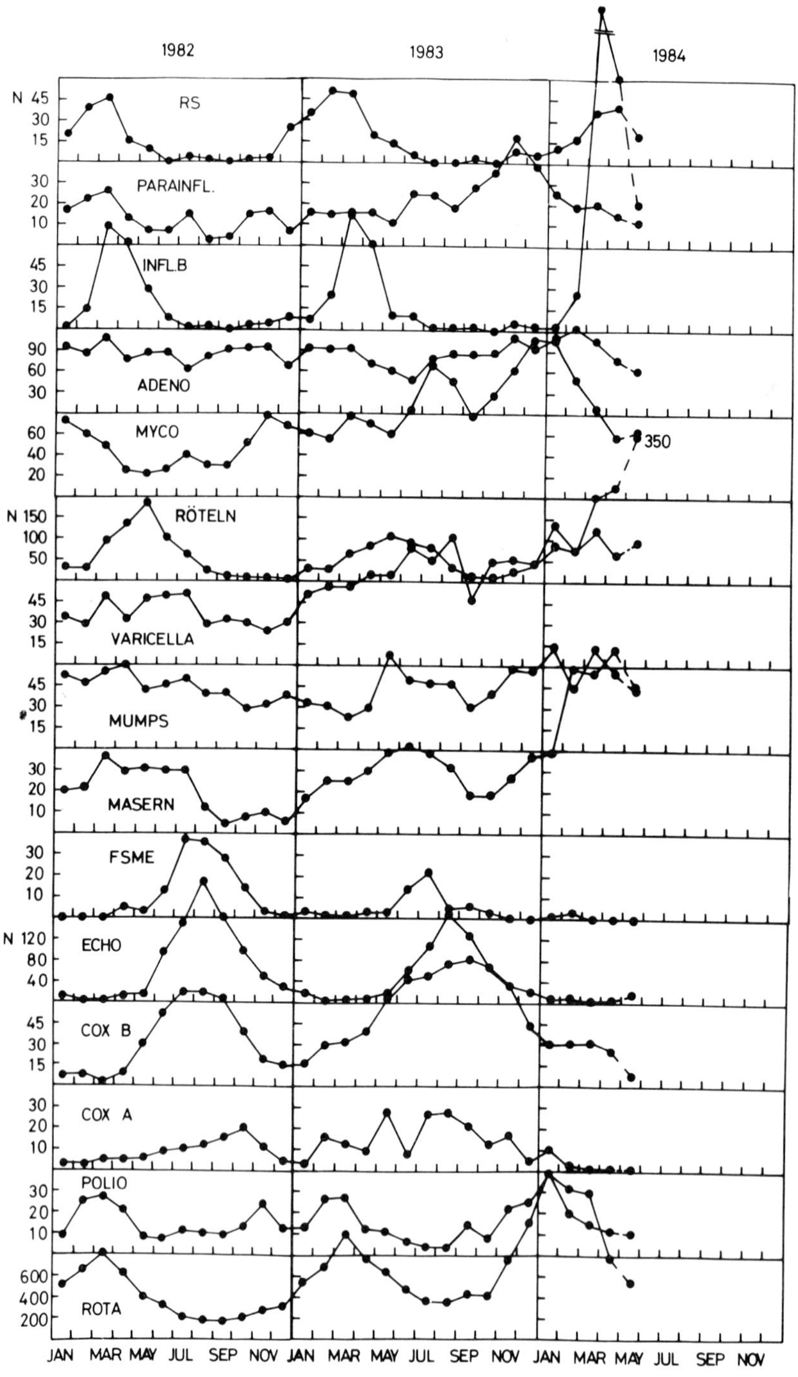

Fig. 11. Example of part of a monthly epidemiological survey demonstrating the total number of reported infections in the Federal Republic of Germany since 1982. In this report given in June 1984 the lines concerning May 1984 are dotted, indicating that the results from the last month are usually still incomplete and should be considered as preliminary

tion, allowing evaluation under different statistical aspects. Further details can be obtained by the mark-sense card developed in our institute (Fig. 10).

The laboratories report their results immediately, but there is a time lag due to the method of diagnosis, the reporting, and the evaluation. Concerning all these factors, the time difference between the onset of illness of the patient and the outgoing statistical report is approximately 2 months. The monthly reports are given under different aspects. One is shown in Fig. 11, demonstrating the total number of infections reported in the Federal Republic of Germany since 1982.

Conclusion

This integrated system for data processing and retrieval in clinical virology represents an advance in the use of computers in clinical work. The concept represents a fully developed, easy to handle system, applicable not only to virology or microbiology but also to other fields of clinical medicine. The principle and the software have been taken over by other laboratories in different countries on systems of variable size. If one pays some attention to the type of hardware, these laboratories can cooperate in integration without any additional expenditure.

References

Blair JN, Brown PP (1981) An on-line computer system for hospital bacteriology: description of its development and comments after five years' use. J Clin Pathol 34:1132–1137

Farrar JL, Brogan TD, Moulding RT, Page M (1975) Use of a computer for producing microbiological reports and for data storage and processing. J Clin Pathol 28:580–586

Goodwin CS (1976) Computer printing and filing of microbiology reports. 2. Evaluation and comparison with a manual system and comparison of two manual systems. J Clin Pathol 29:553–560

Goodwin CS, Smith BC (1976) Computer printing and filing of microbiology reports. 1. Description of the system. J Clin Pathol 29:543–552

Habermehl KO (1980) Einsatz der EDV in der Virusdiagnostik. In: Virusdiagnostik für Klinik und Praxis. Deutsches Grünes Kreuz, Marburg, pp 251–264

Habermehl KO (1981) Einsatz der Datenverarbeitung für die Diagnose von Viruserkrankungen. Hyg Med 6:82–87

Habermehl KO (1983) Data storage and retrieval in clinical virology. In: Waterson AP (ed) Recent advances in clinical virology. Churchill, Livingstone, Edinburgh London Melbourne New York

Harvey KJ, Were M, Heys W, Smith DD (1972) A time-shared computer system for data processing in bacteriology. Med J Aust 2:1076–1080

Kobernick SD, Mandell GA (1974) Implementing a laboratory computer system. Am J Clin Pathol 61:122–134

Krieg AF, Johnson TJ, McDonald C, Cotlove E (1971) Clinical laboratory computerization. University Park Press, Baltimore London Tokyo

Mitchison DA, Darrell JH, Mitchison R (1978) A computer-assisted bacteriology reporting and information system. J Clin Pathol 31:673–680

Rappaport AE, Gennaro WD, Constandse WI (1969) Should the laboratory have its own computer? Hosp Prog 50:114

Ridgway GL, Batchelor J, Luton A, Barnicoat M (1980) Data processing in microbiology; an integrated, simplified system. J Clin Pathol 33:744–749

Vermeulen GD, Gerster JW, Young VM, Hsieh RKC (1974) A computerized data storage and retrieval system for clinical microbiology. Am J Clin Pathol 61:209–215

Vroman L (1979) Observation: How, what and why. In: Reneman RS, Strackee J (eds) Data in medicine. Collection, processing and presentation. Nijhoff, The Hague, pp 1–10

Walter AR (1973) What to look for in a computerized information system. Lab Med 4:32–38

Williams KN, Davidson IMF, Lynn R, Rice E, Philipps I (1978) A computer system for clinical microbiology. J Clin Pathol 31:1193–1201

Application of Small Decentralized Computers in the Management of Bacteriologic Data in the Clinical Laboratory

K. J. RYAN [1]

Introduction

Hospitals and their clinical laboratories have been a major area for application of the advances in computer technology of the last decade. In particular, hospital information systems (HIS) and the numeric results produced by clinical chemistry and hematology laboratories have been well served by a wide variety of computer hardware and software systems. Fewer satisfactory systems have been developed for clinical microbiology laboratories primarily due to the increased complexity of both the workflow in the laboratory and the unique formatting requirements of patient reports. Adaptations of systems designed primarily to handle numeric data have generally not been successful in clinical microbiology.

As with all computer applications the size of the computer needed varies with the task. Many of the individual functions in the clinical microbiology laboratory could be adequately managed using any one of a number of microcomputers (Apple IIe, IBM-PC). For most laboratories microcomputers can store antimicrobial susceptibility data, perform epidemiologic sorts, calculate quality control results, and generate simple patient reports. Integrating these functions and generating the type of clinical report necessary in most hospitals requires a move up to the small computer or minicomputer range (Digital Equipment Corp. [DEC] PDP 11/70, Data General Nova). These machines offer the storage capacity and flexibility necessary for microbiology and other clinical laboratories at a price which makes their dedication to laboratory reasonable. While there is no fundamental reason the large mainframe computers (IBM-4341, Control Data Corp. Cyber 175) found in many hospitals could not do just as well, their use is often complicated by the billing or payroll needs for which they were purchased. Trying to meet the needs of the clinical laboratory as an afterthought can be an exercise in frustration no matter how large the computer. In general, the greatest success in this field has been through the dedication of an appropriate minicomputer to the clinical laboratory together with interfacing to other computers where appropriate. The interfaces may be with a mainframe computer acting as a HIS or with a microcomputer controlling the function of an instrument in the laboratory.

---

1 Department of Pathology, University of Arizona College of Medicine, Tucson, AZ 85724, USA

Microbiology Computer Systems

To be useful a computer need not take on all the information functions of the microbiology laboratory. As indicated above it is not necessary for patient reports to be computerized in order to make use of computers as an aid in work which utilizes patient data such as hospital epidemiology [1]. Creating a computer system which completely replaces the handwritten mode of information transmission in a clinical microbiology laboratory is more difficult because it must facilitate the work of both the technologist in the laboratory and the physician managing the patient. To do so the computer must adapt to the daily activities of its users rather than the reverse.

At the University of Arizona, we have developed a microbiology system as part of a comprehensive clinical laboratory system using a minicomputer dedicated to the laboratory. The microbiology project was a collaborative effort with James E. Peebles, who was then director of our Division of Computers and Biostatistics. Initial development was on a DEC PDP 11/70 with MUMPS as the programming language.

A major goal of the microbiology system was to simplify result entry to the extent it could be decentralized and linked to the actual performance of the work at the bench. This is needed due to the complexities of microbiologic results which are considerable when compared to the numeric results typical for chemistry and hematology laboratories. For example, in chemistry an accession number is answered with a single number (sodium: 127) or string of numbers (electrolytes). A comparable accession in microbiology (sputum culture) requires multiple result entries including microscopic observations (Gram smear), taxonomic terms, narrative text statements, and antimicrobial susceptibility data. Each of these may be in multiples linked to quantitative modifiers (4+; few), and susceptibility results which must be related to the organism and to a quantitation or interpretation ($\leqq$0.5 mcg/ml; resistant). In addition, these results become known at different times over hours to weeks and some will be sequentially modified as a matter of routine. For example, the observation of gram-negative rods in a blood culture bottle should be immediately entered into the computer although the result will have to be modified once or twice more before the organism is identified. These modifications are the rule rather than the exception in clinical microbiology, and any result entry system must make them possible without backtracking or unnecessary key strokes.

Our approach to this problem was to follow the entry style developed by Kunz and co-workers, which allowed keyboard entry of either codes or free text with new entries written over the old [2]. We added another level of entry by precoding the keys of the standard keyboard with the most common results. This allowed the entry of "*Proteus mirabilis*" or "no beta hemolytic streptococci isolated" with a single keystroke. Simple (dumb) terminals were placed at each workstation and provided with the ability to enter the majority of results using the "custom" keys and other results with codes or free text. All entries were immediately verified in full English text on the terminal cathode ray tube (CRT) screen regardless of the mode in which they were entered. This entry system allowed technologists to enter results as they performed the work with no noticeable loss of working time. This had the additional benefit of making microbiology results available for clinical use at the earliest possible time [3].

All verified results can be accessed by a variety of methods both within and outside the laboratory. The most important of these are remote terminals placed in busy hospital, clinic, and emergency room units, and the cumulative summary report printed by the laboratory computer under the control of laboratory data personnel. Just as the chemistry report organizes the patient's renal function tests in a fashion much more readable than individual paper slips, the microbiology cumulative summary groups blood cultures, urine cultures, parasitology, etc. with reports in chronological order within each category. The reports themselves use no abbreviations or codes and are flexible in format [3].

Automated Instrument Interfacing

The greatest potential for rapid and certain handling of laboratory data is by direct transfer form the microcomputer which controls the function of most automated instruments to the laboratory computer. If this is accomplished at the time the instrument generates the results with no manual entry, the instrument is considered to be "on-line". The interface process takes place in two phases. The first is the transmission of signals from the instrument computer to the laboratory computer via a standard interface boards. These signals, which are usually in ASCII code, must be translated into a form recognized by the laboratory system and matched to the patient, specimen, and test. The former requires documentation provided by the instrument manufacturer, and the latter a method to provide a firm link with the laboratory system's accession numbers. The second step is verification of these results followed by a decision to place them in the active patient file. This verification step presents unique problems with most automated microbiology instruments because the results are often generated outside the regular work day 4–12 h after the test run was started.

As part of our laboratory system we have created an on line interface with the Vitek AutoMicrobic System (AMS). The information flow goes from the AMS computer, a DEC PDP 11/04, through an RS 232C interface to a DEC PDP 11/03 in the laboratory. The PDP 11/03 is also linked to on-line instruments in the chemistry and hematology sections and acts as a sorting device prior to transfer of the data to the main laboratory computer recently upgraded to a DEC VAX/750. The link with the patient file is made by writing the laboratory-assigned accession number in the machine readable fields of the AMS cards. One of these fields was also used for an organism number to accommodate identifications and/or susceptibility tests on multiple organisms.

The second or verification/filing step was more difficult. In order to make maximum use of the speed of both the AMS instrument and the computer we elected to allow results to file on-line if they meet criteria programmed into the laboratory computer. For example, AMS EBC PLUS biochemical identifications of gram-negative bacilli were allowed to file automatically if the probability figure was >90% and the organism identified was not one excluded as unlikely such as *Vibrio cholerae*. In a study of the accuracy of on-line editing criteria for the AMS

EBC PLUS and URINE cards the frequency of identification errors filed was less than 2% [4].

On-line filing of antimicrobial susceptibility results is a more sensitive issue because an error is more likely to lead to mismanagement of the patient than is a report of *Enterobacter* when the organism was a *Klebsiella*. The benefits of interfacing are considerable, however, because of the large amount of data associated with each susceptibility test. In fact, much of the labor saved through automation will be lost if all the results must be manually entered from the continuous printout produced by the AMS terminal. A successful interface must take account of the weaknesses of the instrument's susceptibility method as well as preventing unusual results from entering the patient's file prior to verification by a microbiologist.

At the first level, establishing an interface for the AMS GSC PLUS card was done in the same manner as the other cards using the culture accession number and organism number on the card. Although the AMS program provides both a minimal inhibitory concentration (MIC) and an interpretation we preferred the flexibility of using our own interpretive criteria as we do for manual MICs. This required setting up a separate interpretive table for AMS results because the range of possible MICs is restricted with the GSC PLUS. The following example of such an interpretive table for *Enterobacter cloacae* and ampicillin is designed to bracket interpretive ranges which will include all possible AMS results. Since the highest possible ampcillin MIC is >16, only this result will be interpreted as resistant (R), 16 as intermediate (I), and all values between 16 and one as susceptible (S). The only possible results in this range are 8, 4, and 2 mcg/ml. The very susceptible (VS) category is reserved for MICs of 1 mcg/ml or less. Identical or different interpretive tables can be used for other gram-negative bacilli.

The use of an interpretive table in the computer also allows greater flexibility in designing exclusion criteria for on-line filing. Examples are shown in Table 1.

Table 1. Examples of on-line filing criteria for AutoMicrobic System susceptibility testing cards

Organism: *Enterobacter cloacae*
  Drug: Ampicillin
  Exclusion[a]: S, VS[b]

  Drug: Cephalothin
  Exclusion: S, VS

Organism: Klebsiella pneumoniae
  Drug: Ampicillin
  Exclusion: S, VS

  Drug: Carbenicillin
  Exclusion: S, VS

Organism: Staphylococcus aureus
  Drug: Methicillin
  Exclusion: I, R

[a] Results interface within the laboratory but are not placed in active patient file
[b] Interpretive codes as defined in the text

For *Enterobacter cloacae* responses of S or VS are not filed because they are not only unusual but have been shown to be weaknesses (false susceptibility) of the AMS and other automated susceptibility testing systems. This gives a degree of control not available in our routine system. Methicillin resistant strains *Staphylococcus aureus* which are rare in our laboratory can be excluded from filing since they are likely to be an error regardless of the method. As with the other on-line filing, the general principle is to allow the majority of data to enter the patient's file as soon as it tis generated holding back any questionable results for review prior to filing.

Epidemiology and Other Reports

In addition to routine patient reports, the microbiology system was designed to produce a number of management tools whose purpose is to improve efficiency in the laboratory. These include workload records, quality control values, antimicrobial susceptibility statistics, overdue tests, and a number of other reports which would be either difficult or impossible without the computer. All are accomplished without reentry of any patient or laboratory data and are available on demand by technologists, or by laboratory directors under the control of a password system.

Retrieval of microbiologic data for epidemiologic purposes is one of the most frequently cited benefits of laboratory computerization. For this purpose we elected to create a separate epidemiology file by programming the computer to select the epidemiologically relevant portions from each positive culture and store them in a file separate from the clinical report file. This allowed rapid searches and the ability to maintain immediate access to the data for 1 year or more which is not practical for the entire patient data file. Sorts for any combination of time period, organism(s), hospital unit(s), or specimen(s) are available within minutes and include the antimicrobial susceptibility pattern as well as the basic identifying data such as name, hospital number, dates, and accession numbers. These sorts are extremely useful for problem solving in hospital epidemiology and for a variety of other purposes.

In order to take the epidemiologic approach one step further we decided to attempt to have the computer aid in nosocomial infection surveillance by selecting the cases. This required a link with patient registration which included the date of hospital admission. We defined all cultures with positive findings received 4 days or more after hospital admission as potential nosocomial infections. Clinical interpretive criteria were applied to the reports selected, assigning them to negative (viridans streptococci in sputum), positive (> 100,000 *E. coli* in urine) or suspect (*Proteus* in wound drainage) categories. The category assignments were made according to written interpretive criteria but without reference to the patient's chart.

A comparison between this method and standard surveillance techniques which require chart review is shown in Table 2. The computer "positive" rates of 2.3% for urinary infection, 0.9% for pneumonia, 0.5% for primary bacteremia, and 0.5% for surgical wound infection are in reasonable agreement with their re-

Table 2. Comparison of routine[a] and computerized surveillance

Infection type	Nosocomial infection surveillance		
	476 Routine[b] cases[c]	575 Computer selected cases[d]	
		Positive	Suspect
Urinary	1.4[d]	2.3	1.9
Pneumonia	0.8	0.9	1.7
Bacteremia	0.7	0.5	0
Surgical wound	0.7	0.5	0.7
Other	0.7	0.9	0.7
Total	4.2	5.0	5.0

[a] Determined by chart review using Center for Disease Control (Atlanta, Ga.) criteria
[b] 13 months over a 4-year period (1978–1981)
[c] Percent of hospital discharges
[d] Computer cases, selected by criteria in text, over 6-month period in 1981

spective routine surveillance percentages of 1.4, 0.8, 0.7, and 0.7 collected over the previous 4 years. The suspect category added approximately an equal number of cases. In a 1-month direct comparison of routine and computer surveillance, 29 of 35 nosocomial infection cases determined by chart review appeared in either the positive or suspect computer list. Of the six cases missed by computer surveillance, two had the onset of their infection on the third hospital day and two had very short hospitalizations with detection of a nosocomial urinary infection as an outpatient. We now use the computer for routine surveillance of nosocomial infections reserving chart review for suspect outbreaks.

Summary

At the University of Arizona, we have found a minicomputer dedicated to laboratory use to fulfill all the needs of a busy clinical microbiology laboratory. A custom keyboard approach to data entry allowed the linking of result entry to the work-up of cultures at the bench. The efficiency of the laboratory and the speed of reporting were markedly enhanced by the installation of an on-line interface with the AutoMicrobic System. Additional reports and programs available from the computer without additional data entry have been labor-saving for those in and outside the laboratory. This includes a computer-based system for surveillance of nosocomial infections.

References

1. Ryan KJ, Paplanus SH (1975) Computer applications in hospital epidemiology. In: Schlessinger D (ed) Microbiology 1975. American Society for Microbiology, Washington DC, pp 101–110
2. Barnett O, Cameron M, Kissling WJ, Kunz LJ, Lazarus C, Moellering RC, Poitras WJ (1975) The role of the computer in microbiology. In: Prier J, Bartola J, Friedman H (eds) Modern methods in medical microbiology: systems and trends. University Park Press, Baltimore, pp 181–193
3. Peebles JE, Ryan KJ (1980) A microbiology information system. In: O'Neil JT (ed) Proceedings: The fourth annual symposium on computer applications in medical care. IEEE Computer Society, Long Beach, pp 534–538
4. Ryan KJ, Peebles JE (1982) On-line computer entry of routine and AutoMicrobic System bacteriology results. In: Tilton RC (ed) Rapid methods and automation in microbiology. American Society for Microbiology, Washington DC, pp 23–27

MicroLab and MicroSeek – A New Approach to Computing in Medical Microbiology

W. SHEPHERD, J. D. RANT, and R. K. A. FELTHAM [1]

Summary

MicroLab is a comprehensive software package for data-processing in bacteriology, virology, serology, mycology, and parasitology. It differs from other microbiology systems; its function was specified by a group of potential users from seven British medical microbiology laboratories selected for their variety and covering all aspects of hospital and community needs. The system serves three laboratories and is being installed currently in two more.

It is designed around a core, consisting of a data base and associated files which allow the generation, controlled maintenance and use of a range of coding tables. Most of the data entered into MicroLab are coded before entry: there are standard patterns which allow easy entry of common combinations of tests and results. The codes are readily generated by the application of a simple rule, which eliminates the need for reference tables, and makes data entry easy, quick and accurate. There is provision for some free text entry. Modules are provided from which selections of functions for data entry can be made to suit different requirements of individual sites. The layout of printed reports is specified by users, all coded data being translated to full textual form during printing.

MicroLab is a highly interactive system which uses a wide range of cues and prompts to guide the users, who identify themselves to the system with initials and a personal password, and this controls access to the various facilities of the system. The selections of transactions available to different groups of users are decided by the head of the laboratory, who has complete control of who has access to what.

There are two particularly valuable features in the system. First, there is the ease with which specimen information can be recalled, on line, using a variety of selection criteria, including Soundex matching of patient names. Second, there is the ability to analyse both coded and free text data using the specialized off-line enquiry system, MicroSeek.

The initial experience of users is presented.

1 The MicroLab Steering Group, Public Health Laboratory Service, London, Great Britain

Introduction

The Public Health Laboratory Service (PHLS) in England and Wales provides information for the diagnosis, treatment, prevention and control of communicable diseases to hospitals, practices, health centres, communities and the Communicable Diseases Surveillance Centre in London. The epidemiological aspects of prevention and control particularly distinguish medical microbiology from the rest of laboratory medicine. All these activities depend on accessible accurate records. Once a laboratory workload exceeds, say, 50,000 specimens/reports a year, conventional paper record systems begin to fail. Data recovery becomes slow, wasteful of time and effort and prone to error. Retrospective analysis, not defined in advance, may be difficult. We believe that a computer provides better recording and analysis of data, a means for improving the quality of service to patients, better control of hospital infection and more efficient administration of laboratories. Other reports to this effect exist (Aller and Elevitch 1983).

The Course of Development

In Norwich (1977–1979), this belief was explored by specifying requirements for such a system and buying a minicomputer with which a system (named CEDRIC) was devised and tested: it embodied all the essential ideas on which MicroLab is based. CEDRIC is a highly interactive "user-friendly" system employing a specimen-related filing system (which will allow the use of patient-oriented cumulative files) with configurable screen and report formats. It allows the entry of data and results, checking of data, authorisation, and printing of reports by unskilled staff. It depends on a unique powerful coding system derived from the application of a simple rule. Operators choose transactions from a series of menus and are guided through them by options and prompts offered by the system. Facilities exist for conducting previously unspecified retrieval of information from the database – a first version of MicroSeek.

In Leicester (1977–1980), methods for handling data by computer were explored (Feltham and Morrison 1980) in anticipation of the purchase of a computer. They emphasized that the system must be capable of a high throughput of data, reduce errors, facilitate reporting to the requester, produce accurate figures of work done and store data for retrieval. This should include antibiograms, antibiotic usage and information for the control of infection, among other administrative functions. Epidemiological matters should attract high priority. Improved patient care and knowledge of clinical microbiology would follow.

In 1980 the Public Health Laboratory Service, by forming the PHLS MicroLab Steering Group, combined the experiences of these two laboratories with the support of other representatives from the laboratories in Cambridge, Coventry, Exeter, Nottingham, Plymouth, and Southampton, Headquarters Computer Services and a manufacturer (then CTL). This group commissioned studies of laboratory needs in Birmingham, Ipswich, Manchester, Norwich, and Swansea by a consultant systems analyst and one member of the group (JDR) and, applying the results of the CEDRIC project and the studies in Leicester, wrote a functional specification for MicroLab, upon which a system specification was based and written and effected by CTL. It is intended to improve patient care, the collection and analysis of data – clinical, public health and epidemiological – to facilitate control of infection and to improve efficiency and administration. Minimally, it would be an electronic daybook which could support enquiries. A fully working system would support an extensive data base of coded and free text information capable of being updated and allowing the interrogation and editing of free text as well as changes and improvements, such as in screen and report layouts and dictionary contents. Specialist skills should not be required as full software support is called for. Existing staff should be able – with minimal training – to operate the system; extra staff should not be needed. The laboratory should be able to continue working and to cope with a failed system and its restoration. Installation must integrate the system into the laboratory without major change or constraint.

MicroLab

MicroLab does handle all aspects of medical microbiology and does operate in conjunction with an analytical facility – MicroSeek. MicroLab consists of a series of modules, illustrated in Fig. 1. Those employed by the user are available through a series of menus from which any user at any time and any terminal may select a transaction he wishes to use, without affecting any other operator's task

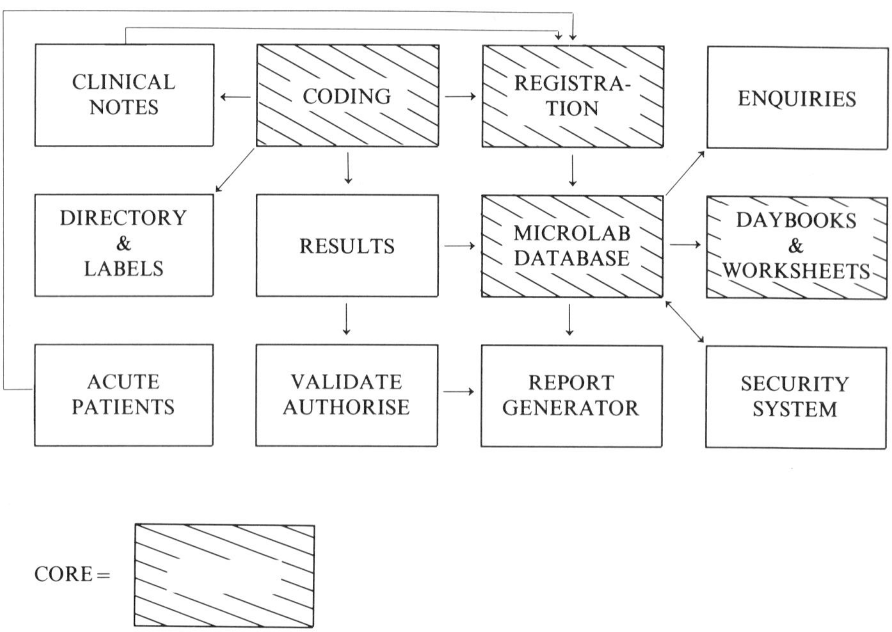

Fig. 1. MicroLab: The modules

or any system function. It is also possible to move from one transaction to another, and back, without losing control of either data or operation. Each module generates automatic cues and prompts to guide the user through the chosen transaction. This is a highly interactive system designed to give maximum help and to prevent the user, by signals, error messages and arresting transactions until correct responses are given, from making mistakes.

The "outward-facing" modules, such as specimen registration and report printing, can be configured to the particular needs of individual laboratories without affecting the essential operations of the system or the structure and function of the data base. Transferability, meaning that the system can be placed in and configured for any laboratory, is inherent in the system design.

Core. The core of the system – the electronic daybook – consists of the data base, specimen registration, coding and daybooks/worksheets/other lists.

The *data base* is an ordered array of all information about specimens received and processed, much of it in coded form: this requires interaction with a comprehensive coding system.

Specimen registration has one mandatory requirement: each specimen is assigned a unique laboratory accession number (LAN) which cannot be reused for this purpose and whose structure is defined by the individual laboratory. Each number identifies one particular record in the data base. The system will not operate without such a number being first assigned to a specimen and any accompanying document, after which all other entries are optional and the individual laboratory may specify its own selection. These may be changed as required. Both clinical and environmental specimens may be registered using different sets of options. Amendments may be made. A facility to search for a previous record and the option to copy details from it are provided.

The individual specimen-oriented record contains mostly non-numerical data and is not automatically one of a compiled time-ordered series. Epidemiological analysis must be possible. In patient-oriented compilations of records, one error corrupts the file; such files are not usually the object of epidemiological analysis, consisting mostly of numerical data (as in clinical chemistry). The option to generate such a cumulative record in MicroLab is not excluded, however.

Coding: The use of codes representing frequently used statements or standard patterns of information is central to the operation of MicroLab. Other systems of coding rely on extensive, sometimes complex tables – MicroLab does not. In principle, sequences of arbitrarily selected symbols lacking association with the content of statements represented cause difficulty. Sequences related in some way to the content of statements seem more logical, and a method of deriving codes from statements most logical of all.

Microbiological statements tend to be textual rather than numerical, as are phrases or assertions qualifying them. A simple rule (proposed by JDR) offers a method of deriving codes from statements or qualifiers: for one word, use the first three letters; for not less than two words separated by an embedded space, use the first letter of the first word, and the first and second letters of the second. There are limits to the number of three-letter codes which may be formed: there is no limit to the number of statements or qualifiers filed under such a code in order of frequency of use. These constitute a rank in the dictionary and are offered in turn in response to presenting the code, until one is accepted or the rank is exhausted. Ranks may be re-ordered and amended by adding new statements as required. Using an analagous method of "inspection," codes may be generated representing numbers such as cell counts and serological results, including those with paired sera, with the facility in the latter case of identifying fourfold changes in levels. Coding in MicroLab is easy to use, even for the unskilled operator, and is one of the great strengths of the system.

Daybooks etc. may be compiled according to local requirements.

The complete reporting system consists of the following modules added to the core: result entry, validation (data-checking)/authorization, report printing.

Result Entry: All data are entered easily, quickly, and accurately. Provision is also made for free text entry. All coded data recorded, whether for printed reports or presentation on screen, are automatically decoded to full textual form.

We distinguish two kinds of result, namely, observations of fact, and conclusions. This seems usual in most laboratories, and it is a waste of resources to use the computer for drawing conclusions the staff may wish to make themselves and are accustomed to doing. The use of codes for statements and standard pattern recall for result entry is particularly helpful.

The classes of result entered are culture, non-culture, and free text. "Culture" comprises culture for bacteria and fungi and tissue culture for viruses, including neutralization tests. "Non-culture" includes microscopy, cytology, serology, and immunology. "Free text" is for laboratory and clinical comment about results.

Validation (data-checking)/authorization confirm the accuracy and appropriateness of results before reporting them. In addition, authorization provides for the addition of medical or other comment by using free text, or the alteration or correction of the report content: it may include data-checking. There is a powerful facility for automatic authorization, where a user may specify individual rules for this operation.

Report Printing: It is better to have a standard format for all or most reports. Thus, preprinted stationary may be used, which is economical of resources, the alternative being to print the format *and* the contents of a report at the same time. It is also very convenient to arrange that all reports for one sender at a particular destination are printed as a batch, thus simplifying distribution. It is a further convenience to be able to print a second copy of a report for some other person to whom it may automatically be addressed.

Enquiries at three levels may be made on line, using up to five parameters in any combination, such as surname, sex, hospital number, sender, specimen type. The first level displays previous specimen and registration details, the second adds laboratory results that are to be reported (suppressed results are withheld). This allows telephone enquiries to be answered, with the opportunity to enter the enquirer's name in the record. The third level displays registration details, results and clinical notes for laboratory and medical staff. Internal audit details in relation to the record are also displayed.

Surveys of the data base are handled by MicroSeek, which will make previously unspecified searches defined at the time by the enquirer, allowing results and free text to be enquired upon. These surveys are best run at night since immediate answers are not possible. Help is available at all stages of this operation, eliminating the need for special training. The data fields to be examined are named, specifying matching criteria and relationships. Coded data fields are translated if necessary. Comparisons may be made using "greater than," "less than," "equal to," "not equal to," "between," and "contains." Free text may be analysed using "contains." Data items to appear in the report are also specified and output is in a form compatible with CTL's REPORTER software, whereby it may be reexamined in different ways.

User Experience

In Leicester, a very large laboratory serving a number of separate hospitals, representing the amalgamation of a group of laboratories with a workload in excess of 300,000 specimens/year and with university commitments has comfortably adapted to using MicroLab, initially as an electronic daybook, but latterly on complete system arrangements.

In Norwich, a district hospital laboratory with a workload of around 180,000 specimens/year has made a complete implementation on every aspect of its work, and the system can cope with printing more than 1,000 reports/day.

In Cambridge a successful implementation has been undertaken in a large university and teaching hospital laboratory with a large medical staff and a workload of the same order of magnitude. Great emphasis has been placed here on the clinical requirements which the system must meet.

MicroLab's performance may be analysed by using special software to monitor its working throughout a whole day or as may be required. This yields operational and management information, the latter describing the manner in which the system is being used by the staff, and the former about how it is performing.

MicroLab will accept microcomputers as terminals managing data or instrument systems on the bench. It is an evolutionary system with a large amount of latent design and inherent development. Further progress is under way.

References

1. Feltham RKA, Morrison JA (1980) Med Lab Sci 37:229
2. Aller RA, Elevitch FR (1983) Clinics in laboratory medicine, 3. Symposium – "Computers in the Clinical Laboratory"

Members of the PHLS Steering Group – 1984

A.D. Pearson (chairman), Southampton PHL; R.A. Bassett, Directorate, Headquarters; C. Clark, CTL, Hemel Hempstead; E.C. Coles, CDSC, Colindale; J.G. Cruickshank (secretary), Exeter PHL; R.K.A. Feltham, Leicester PHL; N.S. Galbraith, CDSC, Colindale; J.E. Griffin, CTL, Hemel Hempstead; M.J. Lewis, Nottingham PHL; C.J. Mitchell, Leicester PHL; P.R. Mortimer, Coventry PHL; J.D. Rant, Norwich PHL; W. Shepherd, Norwich PHL; C.E.D. Taylor, Cambridge PHL; R.E. Warren, Cambridge PHL; P.J. Wilkinson, Plymouth PHL

The Use of the Computer to Aid Decision Making by the Food Microbiologist

A. C. BAIRD-PARKER, D. C. KILSBY, and J. M. BROUGHALL [1]

No one can ignore the impact that the computer has had on society in the past decade. By computer we mean any electronic machine capable of storing data and of being programmed by the user to carry out predetermined tasks. In particular in scientific and technical areas the use of these machines is proving of ever-increasing value. As food microbiologists, we are also looking to these tools to aid us in our work.

One application of computer technology is the use of microprocessor controlled instrumentation to provide more reliable and versatile monitoring systems. Such systems are becoming more evident in the whole field of rapid methods and automation, the subject of this symposium. There is, however, another question: To what extent can the food microbiologist use computers to aid practical decision making?

Computers are at present used in several areas to help microbiologists in their decision making, notably in the area of microbial taxonomy and for data storage and sorting in clinical laboratories. This paper will explore some of the more advanced uses of computers in aiding decision making by the food microbiologist and demonstrate that this modern technology can be as useful to the practical microbiologist as his traditional tools, the wire loop and Petri dish.

A computer, whether it be a simple hand-held pocket computer or a mainframe unit, has certain characteristics which can be exploited by the food microbiologist. They can hold and store large amounts of information; they can sort and represent information in a different format; they can carry out complex calculations rapidly and reliably and they can repeat routine (boring) tasks reliably by scrupulously following rules given to them. This latter property is a positive asset but its usefulness is limited by the ability of the microbiologist to produce rules to define how he normally makes decisions.

In appraising how the computer can assist the food microbiologist, a categorisation of use functions is required. Here they are considered under the headings of their use for mathematical models, their application to analysing data according to predetermined rules, their use in laboratory management and their use in handling "knowledge." Each of these four categories will be considered with examples. It will then be necessary to consider (a) the discipline the computer imposes upon the microbiologist and (b) where future developments may lead the food microbiologist.

1 Unilever Research, Colworth Laboratory, Sharnbrook, Bedford MK44 1LQ, Great Britain

Mathematical Models

The simplest of computers can be used by microbiologists for the application of mathematical models. One of the simplest uses is in thermal process calculations, such as calculation of F_0 for determining a safe "botulinum cook." Models for this purpose are well known [1, 2]. Table 1 gives an example of a computer print-

Table 1. Calculation of cook time for a given F_0[a] (Vintners et al. [2])

RT	IT	f_h	j	z
120 °C	50 °C	40	2	10 °C

[a] Given F_0 value: 3 min
Calculated cook time: 57.7 min

Table 2. Calculation of thermal processes[a]

Initial product temperature = 40 °C

A:RT = 110	F_0		
	1	3	5
f_h = 20	43	70	96
40	74	106	132
60	101	137	166
B:RT = 115	F_0		
	1	3	5
f_h = 20	31	43	52
40	59	74	86
60	83	102	115
C:RT = 120	F_0		
	1	3	5
f_h = 20	26	32	36
40	51	60	66
60	72	84	92
D:RT = 125	F_0		
	1	3	5
f_h = 20	21	28	31
40	45	51	55
60	64	73	78

[a] All temperatures are in celsius, all figures in the tables are minutes cook time

out from such a model that is used by our own company. This allows for mathematical conversion of data on product heating characteristics into a predicted F_0 value. This print-out is from a micro-computer, and illustrates how any variable may be altered and its effect upon F_0 estimated. Table 2 shows how it might be used, for example, to compare various thermal process possibilities for a sterilised meat product.

This is of course a relatively simple model and can indeed be run on very basic pocket computers (we in fact run it on a Radio Shack PC-100). The principle is used in some commercially available F_0-integrators, and may be combined with graphical lethality output.

Other mathematical models may also be used in the same way. At Colworth House we have been developing modelling techniques which describe bacterial growth by means of a series of standard equations, in which different constants are applied for the various organisms under consideration (Figs. 1–3).

These models can describe the effect of up to three variables, for example temperature, pH and water activity, on the predicted pattern of bacterial growth [3, 4]. The models are all based on observed experimental data and offer a pragmatic data-fitting approach to forecast the likely extent of bacterial growth. A particularly valuable feature of this modelling technique is the ability to compile a "library" of fitted constants for the equations which can be interpreted through specially designed "user friendly" programs to give instant and easy access to predictions on the likely growth of a variety of organisms under various environmental constraints, in particular with fluctuating temperature conditions. Other workers have used a similar type of pragmatic approach in modelling observed data to predict bacterial growth or other kinetic values such as time to toxin production [5–7]. The basis of many models is the Arrhenius equation relating the effect of temperature to the observed kinetic pattern [8].

Table 3 shows an example of output from a hypothetical food process using one of the Colworth House models. Note that this model also includes a thermal death model as well as a growth model. As can be seen, it predicts no growth of

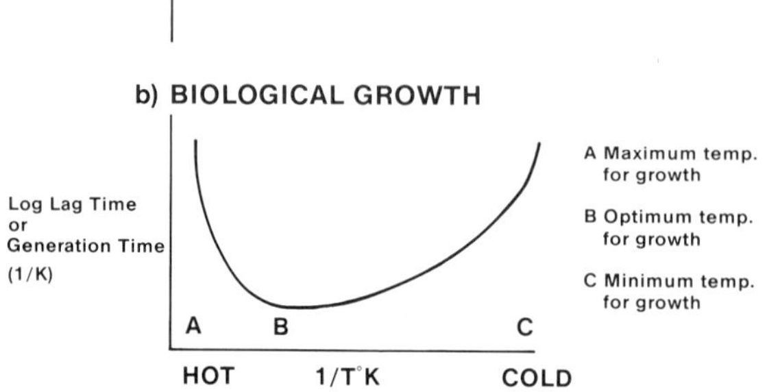

Fig. 1. Arrhenius plot (Broughall et al. [3])

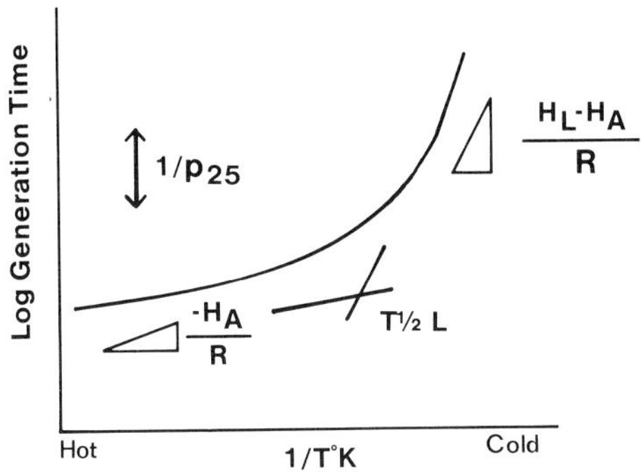

$$1/K = \frac{p(25) \times T/298 \times e^{\left|\frac{H_A}{R}\right|\left|\frac{1}{298} - \frac{1}{T}\right|}}{1 + e^{\left|\frac{H_L}{R}\right|\left|\frac{1}{T\,½\,L} - \frac{1}{T}\right|}}$$

Fig. 2. Low temperature portion of non-linear Arrhenius plot and equation

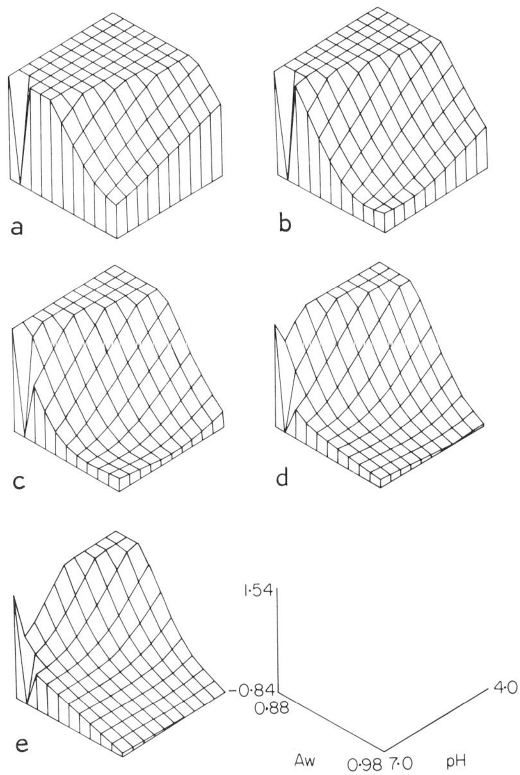

Fig. 3. *Staph. aureus* lag time; max. pH 7.0, Aw 0.98 (range of pH, 4.0–7.0; range of Aw, 0.88–0.98; max. lag time, 30 days). Surface **a** 10 °C; surface **b** 15 °C; surface **c** 20 °C; surface **d** 25 °C; surface **e** 30 °C; the same scale for the three axes of all five surfaces is given. The vertical axis represents the $\log_{10}$ of the lag time (Broughall and Brown [4]). With permission from Food Microbiology, vol 1. Copyright 1984 by Academic Press Inc (London) Limited

Table 3. Computer prediction of effect of temperature and time combinations on growth, survival and death of *Staph. aureus* and salmonellae through a hypothetical food operation

Process stage	Time (hours)	Temp. (°C)	*Staph. aureus* (numbers/g)	Salmonellae (numbers/g)
1	0.033	40	1.00 E+02	1.00 E+02
2	0.00	40	1.00 E+02	1.00 E+02
3	0.00	40	1.00 E+02	1.00 E+02
4	0.033	40	1.00 E+02	1.00 E+02
5	0.033	66	1.00 E+01	5.99 E−01
6	0.017	30	1.00 E+01	5.99 E−01
7	0.033	30	1.00 E+01	5.99 E−01
8	0.033	30	1.00 E+01	5.99 E−01
9	0.008	30	1.00 E+01	5.99 E−01
10	0.167	30	1.00 E+01	5.99 E−01
11	0.167	25	1.00 E+01	5.99 E−01
12	0.083	25	1.00 E+01	5.99 E−01
13	1.000	15	1.00 E+01	5.99 E−01
14	0.008	4	1.00 E+01	5.99 E−01
15	2.000	4	1.00 E+01	5.99 E−01
16	2.000	−20	1.00 E+01	5.99 E−01
17	0.017	12	1.00 E+01	5.99 E−01
18	0.167	12	1.00 E+01	5.99 E−01
19	0.167	4	1.00 E+01	5.99 E−01
20	0.083	− 2	1.00 E+01	5.99 E−01
21	2.00	4	1.00 E+01	5.99 E−01

Table 4. Computer prediction of effect of holding product at process stage 10 on growth of *Staph. aureus* and salmonellae

Process stage	Time (hours)	Temp. (° C)	*Staph. aureus* (numbers/g)	Salmonellae (numbers/g)
1	0.033	40	1.00 E+02	1.00 E+02
2	0.00	40	1.00 E+02	1.00 E+02
3	0.00	40	1.00 E+02	1.00 E+02
4	0.033	40	1.00 E+02	1.00 E+02
5	0.033	66	1.00 E+01	5.99 E+02
6	0.017	30	1.00 E+01	5.99 E−01
7	0.033	30	1.00 E+01	5.99 E−01
8	0.033	30	1.00 E+01	5.99 E−01
9	0.008	30	1.00 E+01	5.99 E−01
10	*8.000*	*30*	*9.38 E+05*	*7.18 E+02*
11	0.167	25	1.25 E+07	7.95 E+02
12	0.083	25	1.44 E+07	8.36 E+02
13	1.000	15	1.91 E+07	9.73 E+02
14	0.008	4	1.91 E+07	9.73 E+02
15	2.000	4	1.96 E+07	9.86 E+02
16	2.000	−20	1.96 E+07	9.86 E+02
17	0.017	12	1.96 E+07	9.87 E+02
18	0.167	12	2.01 E+07	1.00 E+03
19	0.167	4	2.01 E+07	1.00 E+03
20	0.083	− 2	2.01 E+07	1.00 E+03
21	2.000	4	2.07 E+07	1.01 E+03

salmonellae and *Staph. aureus* during the process and in fact some death occurs. If one of the time/temperature conditions in the process is changed (e.g. stage 10, Table 4) then it can be seen that the predicted level of salmonellae and *Staph. aureus* can change dramatically. Clearly such models can be very powerful tools for the microbiologist, and reference to Fig. 2, the model equation used, also shows why the computer is necessary to make such predictive modelling a practical proposition.

The whole area of mathematical modelling for risk assessment is a very exciting development and is made a practical proposition by the complexity of calculation possible with computers. Already at Colworth House, this type of predictive modelling is a standard procedure.

Analysis of Data

Although the food microbiologist may not recognise it as such, he often uses sets of rules to aid his decision making. The commonest example of this is in the interpretation of analytical data to decide whether a batch of a certain product passes a particular microbiological standard. Despite the development of such rules it is becoming increasingly clear that this is often a very inefficient way of making decisions on microbiological acceptability due to the imprecision inherent in many methods of sampling and in the subsequent methods used for microbiological examinations. The very reason for organising this symposium is to consider the state of the art of rapid methods of analysis. Advances in the rational interpretation of the resultant data are still very slow despite considerable interest in this area of analytical microbiology. The International Commission on Microbiological Specifications for Foods (ICMSF) [9] has done an excellent job in promoting this aspect of food microbiology, and has proposed practical, statistically based sampling plans to improve decision making. The use of computers allows far more sophisticated analyses of microbiological data, especially of concentration measurements. This is illustrated in Table 5, which is the print-out from one of our programs applying a standard variables approach [10] to the interpretation of analytical data. As can be seen, the decisions are made in an objective manner, dividing a batch of food into one of three categories; acceptable, outside Good

Table 5. Computer calculation (using standard variable statistics) of acceptance/rejection of product

```
Sample code: ASDF 2345
0 values above upper limit
0 values below lower limit
Mean=4.25337    Standard deviation=0.399771    No. of samples=10
GMP limit=4    Safety/quality limit=5
K1 value=2.98377    K2 value=1.02191
95% sure of rejection if>5% of lot exceeds S/Q limit
95% sure of acceptance if not more than 5% of lot exceeds GMP limit
Calculated S/Q level=5.4462    Reject
Calculated GMP level=4.6619    Reject
```

Manufacturing Practice standards but acceptable, and unacceptable (outside safety/quality standards).

The ability to analyse variables by use of computers enables the analyst to predict the result of further sampling. For example, it might be predicted that taking more sample units will increase the confidence of an analysis, allowing more precise prediction of a risk or compliance to a standard to be made in relation to the number of further sample units examined. Alternatively, analysis may show an unusual distribution of micro-organisms which make different (e.g. non-random) sampling plans more appropriate for carrying out microbiological examinations. More precise analysis of results may also allow microbial changes to be more clearly defined, such as the quantification of change in the level of a contaminant in a raw material through to a finished product. Our understanding of the relevant statistics dealing with the distribution of micro-organisms is often restricted by the analyses we can carry out. The computer offers a possible solution to this.

This is only one illustration of how computers can improve the quality of decisions made by food microbiologists. Data can often be more efficiently and reliably used by adopting trend analysis procedures, such as cusum and running average techniques [11], to measure performance and eliminate the effect of short-term variations in data relating to the microbiological quality and safety of products. Computers are ideal for such analyses.

Of course the computer will only work to the rules it is given. The food microbiologist must make the decision of what is, or is not, acceptable or significant.

Along with a higher quality of decision making, there may need to be a more objective and realistic assessment of the relevance of micro-organisms in foods by the food microbiologist. This cannot be achieved by the use of a computer, but requires microbiologists to make rational judgements based upon accurate microbiological data.

Laboratory Management

Computers can be excellent tools for assisting the food microbiologist in "managing" laboratory operations. The computer can also be programmed to select the microbiological tests needed for each type of sample received, to choose appropriate dilution ranges and to establish the number of sample units to be tested from each batch of material examined. The computer can use this information to produce a work sheet for the analyst. Figure 4 shows such a work sheet generated by one of our computerised quality assurance systems. The analyst carries out the appropriate tests, filling results onto the work sheet and then transferring the data to the computer. The computer can then interpret this data if appropriate decision making has been included in the programming. Figure 5 gives an example of a final results sheet in the quality assurance program referred to above. Apart from the improved decision making in the laboratory, it is our experience that such an approach to laboratory management can lead to real and significant cost savings.

		Work sheet	*Date: 30/03/84*
Product		Fish raw material	
Batch code:		QWERT1234	
Supplier:		FRIO	
Date received:		30/03/84	
Date of testing:		02/04/84	
		Product category = 3	
Tests	Dilutions	Results	
		Sample no.	
		1 2 3 4 5	
TVC	2, 3, 4		
Coliforms	1, 2, 3		
E. coli	1, 2		
S. aureus	1, 2		

Fig. 4. Example of computer-generated work sheet

Date: 05/04/84		Sample: Raw material – fish		Batch code: QWERT 1234
Organism	*N*	Mean	SD	Grade
TVC	5	5.43167	0.297446	A
Coliforms	5	3.93134	0.414578	B
E. coli	1	0	0	A
S. aureus	1	0	0	A

Fig. 5. Example of computer-generated results sheet

Knowledge Systems

It is interesting to speculate how much the "expert" is needed if his knowledge can be recorded elsewhere. Certainly the "expert" cannot be completely replaced, but some of the application of "expertise" is certainly merely the transfer of "knowledge." Furthermore, in making decisions the "expert" again exercise certain rules, although rarely consciously.

If a significant amount of this knowledge could be placed on the computer with basic rules on its use, a "knowledge" or "expert" computer system is produced. Such knowledge systems have been built for a variety of disciplines – and we are starting to build them for food microbiology at Colworth House.

Knowledge systems are only capable of being usefully constructed if the "expert" is willing to work out the basis on which he makes his decisions. This is difficult for two main reasons, firstly no "expert" likes to think his function can be largely carried out by a machine, and secondly we are not used to thinking in these terms. Nevertheless, this is likely to be one of the most significant developments in the use of computers by the working microbiologist now that software for this purpose is being produced.

Interfacing with the Computer

There are several levels at which the microbiologist can interact with computers. At the simplest, minimal level, the microbiologist merely asks the computer specialist to carry out the computing work for him. This may be satisfactory for building large standardised data bases or routine statistical analyses. There is no interaction, however, between the microbiologist and the computer.

A second level of involvement is when the computer specialist supplies the software for the microbiologist, and the microbiologist actually operates the computer. The success of this approach depends principally upon the skill with which the software has been produced. This is measured by how easily it can be used by the microbiologist and how much versatility it incorporates. Unfortunately there is no software specifically available for this purpose, although some packages, such as high level data base languages, may be useful. Good programs can only be produced if the microbiologist and the computer expert spend time working together and both get to understand the needs of each other.

The third level of involvement is when the food microbiologist produces his own specific programs. This requires a much higher level of involvement from the microbiologist. Once the necessary skills have been acquired then this is a very profitable way of exploiting computers. It has been our experience that many microbiologists can acquire the necessary skills and we have been very successful in following this route; thus the majority of our operational programs are now produced by working microbiologists.

Future Developments

The use of computers is still in its infancy. Some of the most exciting areas are in the development of knowledge systems and laboratory quality assurance programs incorporating objective decision making. As such, computers will enhance the role of the food microbiologist, increasing his contribution towards the safety and quality assessment of food products.

As new food processing units are built, microprocessor control will become an increasingly important part of the factory operation. The food microbiologist needs to develop with the new technology in order to exploit the opportunities it offers. With computer controlled food processes, the microbiological controls can be incorporated in the process. Control of processing times, operating temperatures and heating stages can occur automatically and reliably. Furthermore, and potentially more importantly, any deviation from normal can be clearly signalled and immediate, effective corrective action can be taken (if necessary by the computer). This not only increases microbiological safety but also leads to more cost-effective control of food processes. This trend can already be detected in the food industry and will increase as appropriate sensors for the measurement of chemical, physical, and biological changes that can be related to microbiological quality and safety become available.

Such a radical change in operations will give the food microbiologist great opportunities, but it will also demand significant changes in attitudes. The food microbiologist will no longer be required to apply his skills in the laboratory to the same extent, but will be using his knowledge to control processing operations.

Computers are useful tools which can be harnessed by the food microbiologist to improve the quality of the decisions he makes. Therefore if exploited effectively, computers will make a significant contribution to food safety and quality.

References

1. Stumbo CR (1973) Thermobacteriology in food processing. 2nd edn. Academic, New York
2. Vintners JE, Ratec RH, Halaby GA (1975) Thermal process evaluation by programmable computer calculator. Food Technology, p 42
3. Broughall JM, Anslow PA, Kilsby DC (1983) Hazard analysis applied to microbial growth in foods: development of mathematical models describing the effect of water activity. J Appl Bacteriol 55:101–110
4. Broughall JM, Brown C (1984) Hazard analysis applied to microbial growth in foods: development and application of 3 dimensional models to predict bacterial growth. Food Microbiology 1:13–22
5. Ratkowsky DA, Olley J, McMuckin TA, Ball A (1982) Relationship between temperature and growth rate of bacterial cultures. J Bacteriol 149:1–5
6. Roberts TA, Gibson AM, Robinson A (1981) Prediction of toxin production by *Clostridium botulinum* in pasteurized pork slurry. J Food Technol 16:337–355
7. Hauschild AHW (1982) Assessment of botulism hazards from cured meat products. Food Technol 36:95–104
8. Labuza TP, Riboh D (1982) Theory and application of Arrhenius kinetics to the prediction of nutrient losses in foods. Food Technol 36:66–74
9. International Commission on Microbiological Specifications for Foods (1974) Micro-organisms in foods. Sampling for microbiological analysis: principles and specific applications, vol 2. University of Toronto Press, Toronto
10. Kilsby DC, Aspinall LJ, Baird-Parker AC (1979) A system for setting numerical microbiological specifications for foods. J Appl Bacteriol 46:591–599
11. Massart DL, Dijkstra A, Kauffmann L (1978) Evaluation and optimisation of laboratory methods and analytical procedures. Elsevier, Amsterdam New York Cambridge

Computer-Assisted Reading System for Microbiological Tests

D. Förster [1], H. Dedden [1], H. Grimm [2], H. Rost [1], B. Wiedemann [3], and A. Zembrod [1]

The agar diffusion method is still the most common technique for testing the sensitivity of pathogenic bacteria in respect of antibiotics, and most therapeutic proposals are even today based upon this method.

The analysis of inhibition zones in the diffusion test by comparison with breakpoints fixed before hand is a central issue in the chemotherapy of bacterial infectious diseases. For years efforts have been aimed at standardizing this procedure. Up to now, however, all guidelines regulate the methodology: A measurement of inhibition zones is not prescribed. The decision about the antibiotic effect is made by optical impression (seldom with the aid of rulers, calipers, or templates) and depends directly on the experience of the microbiologists reading the tests.

Different attempts have been made to overcome this problem of uncertainty about the results of the disk diffusion technique:

1. The development of automatic reading devices for inhibition zones
2. The automation of susceptibility testing on the basis of other technological concepts

The common and ongoing trend in any case is to overcome the agar diffusion technique and to replace it with one of the dilution methods. This means using the well known tests, but giving them a new appearance.

John C. Sherris said about 10 years ago: "Many of us in the field at that time looked with some embarrassment at the developments in technology and performance control which were occurring in clinical chemistry, but consoled ourselves with the thought that our discipline was much more difficult (which it is), required the continuous application of informed judgement (which it does), and was, therefore, perhaps, inappropriate for the application of statistical standards of performance and the use of automated procedures (which it is not)!" (Prier et al. 1974).

The first approach to automatic reading and electronic data processing of the agar diffusion test using paper disks was published by Moellering et al. (1972). The measuring principle was based on a television camera, a video monitor, and a light pen with which the operator indicated the zone edge on the television screen. The information was then directly fed into the laboratory computer.

1 Pharma-Forschungszentrum, Bayer AG, Institut for Chemotherapie, Postfach 101709, D-5600 Wuppertal 1

2 Institut für Medizinische Mikrobiologie und Klinische Chemie, D-7987 Weingarten

3 Institut für Medizinische Mikrobiologie und Immunologie der Universität Bonn, An der Immenburg 4, D-5300 Bonn 1

There have been other semiautomated instruments for the purpose of inhibition zone readings, not only by image analyzers (Hallynck and Pijck 1979).

Several instruments, like the Autodata device (Hitchin, UK) or the Saxby Autobiocoder (Liverpool, UK), have been developed especially for the reading of bioassays with antibiotics on large plates. None of the instruments capable of only inhibition zone reading could be introduced on to the market.

As already mentioned, the second approach to automated and standardized antibiotic susceptibility testing of bacteria involved the attempt to replace the agar diffusion test by dilution techniques. This kind of automation split into two directions although the reading principle – turbidity measurement – was the same:

1. The development of completely new tests with special growth chambers, incubator units, and photometric reading
2. The automation of the microtiter technique beginning with prefilled plates, automatic inoculators, and finally automated photometric reading of the results

All the systems following this trend have another principle which is very often difficult to fulfill in the routine laboratory: They demand a complete change of the laboratory routine in accordance with the instruments. It is necessary to mention that every delay in the semiautomated system leads to a complete breakdown of the laboratory analyses if the old agar diffusion or another manual test is not carried out in parallel.

Apart from the early attempts already mentioned (Moellering et al. 1972), the image analysis systems did not enter the routine laboratory for clinical bacteriology. So far the only instruments based on video-technique used in the microbiological laboratory are colony-counting devices.

One additional application of every such image analysis instrument is the measurement of the area of inhibition zones of the agar diffusion tests. This is, of course, no technical problem, but it is a very time-consuming procedure, which limits its practical relevance.

McDonald and co-workers (McDonald et al. 1981) from Nottingham, UK, published an interesting and successful use of a television computer system for the automatic reading of the multipoint inoculated antibiotic sensitivity test. In combination with the Adatabs (Mast Laboratories Ltd.) and a multipoint inoculator, it represents a technology which might be adequate for the equipment brought onto the market by the microtiter system producers.

Material and Methods

The Biology Laboratory Computer (BLC) system presented here was originally introduced only for the reading of inhibition zones. The purpose was the automation of the reading of bioassays for the determination of antibiotic concentrations in body fluids and tissues, and electronic data processing of the results. In our laboratory the first experiences with the video-technique for the reading of inhibition zones were made with the Petriscop 720 P (Cambridge Instruments) connected to a HP 9825 B desk top calculator. While working with this machine the BLC concept was developed; it was planned as an universal measuring and data processing system for practical microbiological activities.

In brief it should be capable of:

1. Measuring the inhibiting areola in the agar diffusion test and bioassays
2. Reading of agar dilution tests (flexible adaptation to commercially available multiple inoculation devices; 21, 25, 48, 96, etc. stamps)
3. Reading of microdilution tests (perforated masks are available for commercial microdilution techniques)
4. Colony counts made in order to obtain killing curves, etc.

Plates are inserted manually in the device, whereas the reading operation and the subsequent data processing are automatic.

At present, the BLC system configuration comprises:

1. Video camera tripod housing with a control unit, a reflected light device, and an interchangeable microscope stage. At present there are three versions of the stage:
   a) Revolving stage for Petri dishes to be used in the agar diffusion test (reflected light method)
   b) Transmitted light illuminating stage with interchangeable templates for agar dilution and microdilution or dark field illumination
   c) x-y stage for epi-illumination of bioassay plates of various sizes
2. Video monitor
3. Print image and data entry keyboard
4. Gimix computer
5. Hewlett-Packard two-color plotter (HP 7440 A)
6. Epson Printer (MX80 FIT)

The following description gives more details of the system:

The television camera is interfaced to the microcomputer, which scans the sample over the interesting region and during each image frame digitizes a measuring line with 64 grey level conversions. After evaluation of the scan values the measuring line is moved to the next position of interest for detection of special characteristics of the sample. This may be searching for objects, detecting the edges of objects, or determination of position and qualities like diameters, areas, optical density, etc. It is done by the use of segmentation, smoothing, histogram, and planned edge detection methods to separate the relevant object information from the myriad of image data which are not needed.

The measuring line is mixed with the original television signal and the text lines of the computer display to allow the monitoring of the system operation. Guided by dialogues and by the visual display of the monitor, the operator easily controls the evaluation system.

The sample with the culture is illuminated by special lighting systems, dependent on the different test methods such as agar dilution, microdilution, inhibition zone measurements on Petri dishes or rectangular plates, or counting colonies. Enhancement of the optical contrast of the sample image for better signal to noise ratio is an important feature of the presented system. Dark field illuminators for each inoculated position in the agar dilution test plate for each well of the microdilution plate are one of the means for obtaining the necessary sensitivity, accuracy, and reproducibility of the measurements.

The motor control to move the Petri dishes or other plates, the image analysis for detecting the microbiological objects, and the data acquisition interfaces are executed by assembler routines written in the C language to establish the necessary speed for the measurements. Evaluation routines and all further programs are written in Pascal or Basic, to allow for easy changes or expansions.

This biology laboratory computer consists of a microcomputer GIMIX with 64 kB memory, expandable to 1 MB, with two floppy disk drives, a printer, and a plotter and the possibility of linking with other computers. The UNIX-like operating system OS9 with multitasking features not only supports the rapid routine measurements, but can also be used as an outstanding program development and word processing tool.

There is a software package of about 250 kB for data acquisition and image analysis especially designed to evaluate microbiological objects and to produce the different reports and graphical plots together with the required statistics. The modulator design of the soft- and hardware gives flexibility and ease of change for adaptation to the laboratory requirements and for further expansions.

Results of the Comparison of Visual and Automatic Reading with the BLC System

To evaluate the capability of the instruments, comparative studies were carried out in two different laboratories, one for the agar dilution and another for the microdilution. These preliminary experiments were followed by a multicenter study of the working group "Resistance" of the Paul-Ehrlich Society of Chemotherapy (PEG). The results of the latter will be presented partly here and at the seminar "Date Processing in Microbiology". The final results will be published elsewhere. The influence of different media, bacterial species, and antibiotics was of no special interest in the preliminary studies, as these studies served only to evaluate the hardware and software of the BLC.

In this paper only microdilution and agar dilution data are presented as the PEG study was restricted to the comparison of these two methods.

Preliminary Comparative Studies (Two Laboratories)

The first laboratory used Mueller-Hinton and Isosensitest agar successively in the agar dilution test. Various negative bacteria were included with no chosen prevalence of species (e.g., *Citrobacter, Morganella, Enterobacter, Serratia,* and nonfermenters). All strains were fresh clinical isolates. The inoculum was 10^4/CFU/ml.

Isosensitest broth was used in the microdilution test in the second laboratory. Bacterial strains and antibiotics varied, as described for the agar dilution test. The inoculum was 5×10^5 CFU/ml.

In both laboratories the two tests were carried out on different days. The test plates were first evaluated visually and then by the computer-assisted instrument

Table 1. Comparison of BLC and visual readings (preliminary results). △=absolute value of differences according to dilution

Method Study Nr.	Agar dilution 1	Agar dilution 2	Microdilution 1	Microdilution 2
Amount of measurem.	2003	819	1003	168
Differences:				
△=0	82%	88%	74%	66%
△=1	13%	11%	12%	27%
△≧2	5%	1%	14%	7%

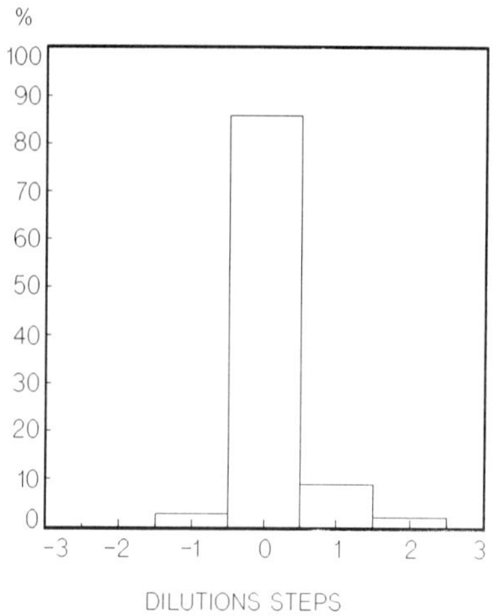

Fig. 1. Agar dilution test; BLC vs visual reading

through different persons. The results of these comparative studies are summarized in Table 1.

Combining test 1 and 2 a difference of two or more dilution steps was found in 3% and 10.5% with the agar dilution and the microdilution tests respectively. The overall agreement was 85% with the agar dilution and 78% with the microdilution test.

Figure 1 shows a diagram of the agar dilution test (tests 1 and 2). Complete accordance between visual and BLC reading is demonstrated by 85% of the measurements. The small columns on the right of the tall one indicate that the BLC reading gives slightly higher MIC values than the visual reading.

Multicenter Study (Seven Laboratories)

The data presented here are also preliminary; the final, complete statistical evaluation of all the data will be published elsewhere. This PEG study (see above) included 14 bacterial species, each representing ten strains. The MIC of these strains

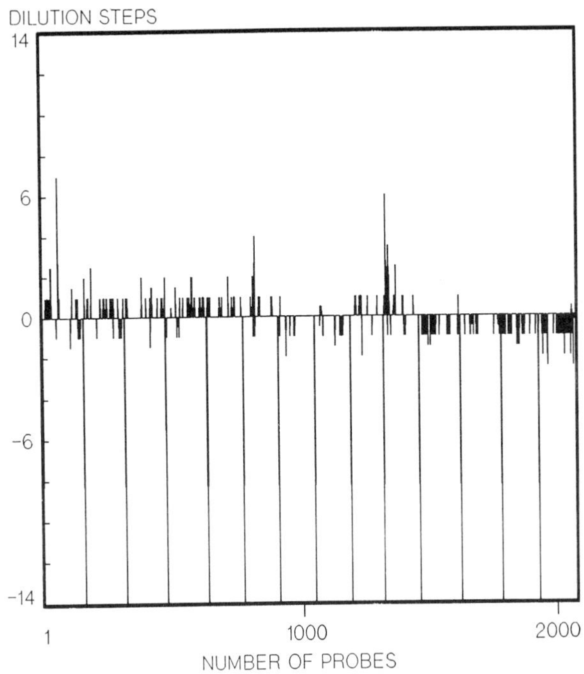

Fig. 2. Results from one laboratory in the PEG study comparing the agar dilution MICs as assessed by the BLC and visually (for explanation, see text)

was determined in relation to 24 antibiotics. For every strain–antibiotic combination, micro- and agar dilution MICs were determined according to the guidelines of the PEG working group (Dr. Grimm, Weingarten, 1984, personal communication).

The culture medium was Isosensitest broth. The inoculum contained 5×10^5 CFU/ml. As above, with the agar dilution test the microtiter plates were first analyzed visually and then by the computer-assisted reading system.

The results of one laboratory are summarized in Fig. 2.

The dilution steps are listed on the y-axis. The x-axis is divided into 14 columns, each referring to one species with ten bacterial strains. From the left to the right these are *E. coli, Citrobacter, Klebsiella, Enterobacter, Serratia, Prot. mirabilis, Prot. vulgaris, Prot. morganii, Prot. rettgeri, Providencia, Acinetobacter, Pseudom. aeruginosa, Staph. aureus,* and *Strept. faecalis.* Positive values (upper half of the figure) of the dilution steps again indicate higher MIC values of the BLC reading than of the visual reading; the negative values indicate the converse.

Figure 2 demonstrates that in the agar dilution test carried out in this laboratory the BLC reading results are often higher than the visually evaluated MICs. The largest differences reach up to seven dilution steps. The results of the last four species mentioned above (*Acinetobacter, Pseudom. aeruginosa, Staph. aureus,* and *Strept. faecalis*) are interesting insofar as here the BLC readings are lower than the visual readings. Nevertheless, the figure demonstrate that the agreement between the visual and the BLC readings is acceptable in this laboratory.

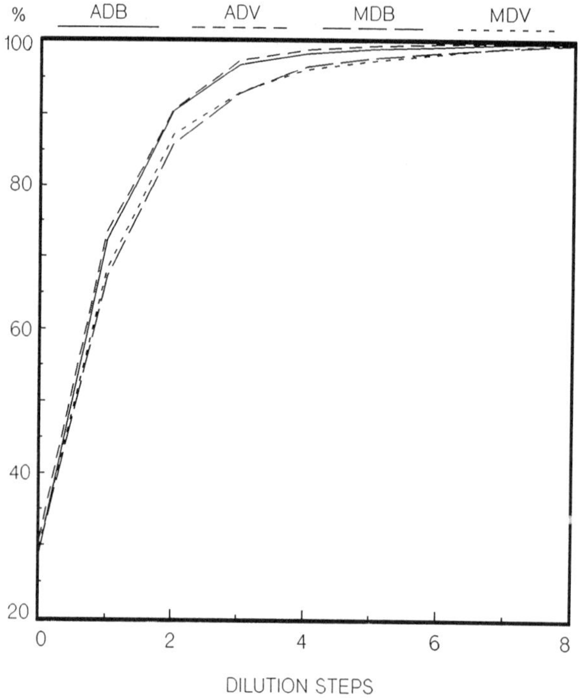

Fig. 3. Comparison of methods on the basis of the data obtained by the multicenter study (seven laboratories) (for explanation, see text)

A first general overview of the data obtained by the multicenter study is presented in Fig. 3. The comparison of the visual and BLC readings of seven laboratories was evaluated by the calculation and cumulative listing of the Q-ranges of the reading differences. The figure represents all values measured in this study.

The Q-range is the common method for the comparison of medians. It was chosen because:

a) There were seven values per group
b) Open classes have to be included
c) There was a double peak distribution

The percentage is written on the y-axis, while the dilution step differences are demonstrated on the x-axis.

Each of the four plotted curves consists of about 16,000 individual reading values. The two upper curves demonstrate the reading of the agar dilution test by means of the BLC system (ADB) and the visual method (ADV). The two curves represent the results of the microdilution method by BLC (MDB) and visual reading (MDV).

The uppermost curve shows the minimum deviation (straight line, agar dilution and BLC reading), but there is no significant difference to the next curve (dashed line, agar dilution and visual reading). The difference between the two lower curves is also nonsignificant (microdilution, BLC, and visual reading). But the difference between the two groups of curves is statistically significant at the

2% level. This evaluation therefore shows a better agreement between the seven laboratories regarding the reading of the agar dilution test than that of the microdilution test. Both BLC and visual measuring yield favorable results.

Discussion

Considerations concerning the development of the BLC were:

1. Without necessitating more personnel, and being well adapted to the otherwise unchanged laboratory activities, the unit was meant to provide easier work, faster reading operations, standardized data acquisition, and data protection (e.g., for statistical and resistance monitoring purposes).

2. The incorporation of existing standards (DIN, NCCLS) into the process of antibiotic evaluation was to counteract all the "playing around" with disk inoculations and the manipulation of breakpoints. It ought to be possible to relate in this manner the physicochemical properties of the various chemotherapeutic agents.

And in respect to other partly automated sensitivity testing devices:

3. The basic concept of other methods is the implementation of a new autonomous testing program featuring an inoculation mechanism, an incubation phase, and a facility for reading the results during and after incubation.

4. The principle of these other methods is the measurement of the turbidity of a liquid environment, which has been proved to lead to errors of evaluation in the case of β-lactam antibiotics.

To sum up: the other systems available on the market necessitate parallel testing in respect of the usual laboratory technique, as they are intended to replace the latter. The chosen measuring principle makes evaluating the effect of β-lactam antibiotics more difficult.

The variable BLC system will adapt itself to the practical requirements of laboratory work and is exclusively geared to the measurement and the evaluation of a common sensitivity testing procedure. Nevertheless, one should not overlook the fact that the value of such a device has to be proven in the clinical microbiological laboratory. The first results of laboratory evaluation of the BLC, although limited to the agar and microdilution techniques, were favorable in comparison with the visual reading.

There have been several technical problems with the BLC reading, especially regarding the preparation of the test plates. The addition of Pril and other substances to the agar caused reading problems because of the precipitation of the antibiotics. This leads to a turbidity in high concentrations of the antibiotic in the agar, where normally growth is unexpected.

The skip tube phenomenon and the hazy growth on agar dilution plates raised reading problems and called for adaptation of the software.

The final evaluation of the large data material of the PEG study will be helpful for the evaluation of BLC. So far the results seem to be encouraging.

Summary

A Biology Laboratory Computer system (BLC) is described which was originally introduced for use in measuring zones of inhibition for bioassays and single disk antimicrobial susceptibility testing. The additional concept behind this television computer system was to develop a measuring unit capable of reading not only inhibition zones but also agar dilution and microdilution plates as well as colony counts and other relevant tests in the microbiology laboratory.

The results of two preliminary experiments and one multicenter study which compared agar and microdilution readings obtained visually and with the BLC indicated that agreement and reproducibility are acceptable.

References

Hallynck Th, Pijck J (1979) New zone reading equipment in microbiological diffusion assays. J Antimicrobial Chemother 5:179–182

McDonald T, Pearson N, Johnson F, O'Grady FW (1981) Automated reading of the multipoint inoculated antibiotic sensitivity test. J Med Engineering Technol 5:243–245

Moellering RC Jr, Mercier BA, Kunz LJ, Poitras JW (1972) Evaluation of a computer-associated electronic zone analyzer in single-disc antimicrobial susceptibility testing. Antimicrob Agents Chemother 2:95–102

Prier JE, Bartola JT, Friedman H (1974) Modern methods in medical microbiology systems and trends. Proceedings of the 6th Annual of the Eastern Pennsylvania Branch of the ASM. University Park Press, Philadelphia

Problems in Operation and Data Transfer in a Multicomputer Environment

A. Schreiner [1]

Problems in a multicomputer environment arise when the different computers try to communicate with each other. Even if all computers are of the same type and are running the same operating system, provisions have to be made for common data bases, i.e., for sharing data between cooperating users. Thus care has to be taken with data formats, transfer protocols, data security, etc. If the computers are of different types, there are also problems of incompatibility.

The user has an increasing number of aids in his efforts to overcome incompatibility. Specialized hard- and software companies are using microprocessor techniques to overcome incompatibility. For a number of personal computers, chips are available allowing them to run with more than one operating system (for example, CP/M and MS-DOS). Large mainframes are already able to run two operating systems simultaneously, and the personal computers (PC) will certainly follow.

Incompatibility may come about for two reasons: one is the supplier, the other the user. Knowledge in the form of data and programs represents power. Therefore many users are, like the suppliers, not interested in having a system that is compatible with other computer systems, since they are afraid of losing control of the data.

To address the problem of incompatibility at its roots involves investigating an adequate system approach. We have to solve a long- not a short-term problem to avoid future incompatibility. The low price of microcomputers and the rapidly growing market of turn-key systems in specific working places make it difficult to analyze the whole field of operation and to consider an overall system integrating the microcomputers. Today we all know that because of costs software criteria are a higher priority in decisions than the hardware aspects. In the future the fit to the overall system will be of a still higher priority than those two factors combined, as it influences ultimately the efficiency of the whole organization.

Developing an overall system for an entire hospital often takes too much time, as the organizational propositions may have changed before planning has ended. But there is an economic optimum between long- and short-range planning. To find this optimum we have to answer three questions:

1. What part of an overall system is the most expensive? I believe it is the volume of stored data. Data are the result of expensive measurements, experiments, observations, conclusions, and decisions. Even the process of transmitting data from the outside world to computers in a machine-readable format is too expen-

1 Rechenzentrum, Universität Karlsruhe, Postfach 6380, D-7500 Karlsruhe 1

sive to do repeatedly; similarly, it is not profitable to neglect the use of stored data for other purposes.

2. What are the least durable parts in an overall system? These I think are the programs for evaluating data, which are used at each working place.

3. What is the main obstacle to designing an overall system in detail and realizing that plan? I believe this to be the permanent creation of new applications for computers.

To achieve an effective, economic, comprehensive inhouse computer system, we have to provide for mutual use all the data anyone has collected and prevent them from being limited to a present mode of application or to specific future one. To meet these requirements we need a dynamic system structure, i.e., an open-ended system whose development must not be limited to its last branch. What has to be defined are the rules according to which the system will grow, just as trees grow according to their genetic codes.

The only system structure which could meet this objective is a decentralized one. All owners of large or small computers can influence the development of such a decentralized system, for example, by becoming members on a committee for computerization. This kind of computerization need not necessarily be the job of a main computer center or administrative body.

Systems communicate with their environment via interfaces, the same applies to communication between systems. The overall inhouse system is normaly divided into subsystems. Initially, it may be expedient to set up just a few basic design rules for the overall system and gateways to the main system and only to develop local subsystems in more detail. The local systems may differ in character: Some may be more oriented to automatic data collection and experiment control, others to interactive work with data bases. Therefore, the structures, media, and protocols will also be different. For example, connecting a number of personal computers within a subsystem does not allow such expensive media as the transmission of graphic output to a screen.

So-called gateways for the transfer of data or messages from one network to another are gradually appearing on the market. Differences in the interfaces between devices and networks can be eliminated by so-called protocol-converters, of which a great variety are coming on to the market. It is necessary to be aware of the ever-increasing diversity in soft- and hardware. Beyond the overall system, the planning group also has to onsider the general requirements of communication in a hospital.

At present, two techniques are competing for the market of inhouse networks: the local area networks (LANs), promoted by computer companies, and the PBX digitized speech systems, promoted by the telecommunications industry. For port selection among mutually incompatible computers – not high-volume file transfer – a data PBX is preferable.

Within 2 or 3 years, the so-called ISDN, the data PBX of the German Federal Post Office, will be available in Western Germany; it will also cater for inhouse systems. This network will allow data and digitized speech to be transferred via telephone cables at a rate of 144 kb/s. This transfer rate is sufficient for a great many applications, for example a DIN-A 4 sheet of paper could be transmitted in less than 1 s, a floppy disc of 256 kb within 20 s.

LANs are primarily advantageous in resource sharing, i.e., the sharing of expensive equipment. Incompatible computers may share the same disc drive with LANs in many cases; only a few LAN products enable incompatible computers to read or update each others files, but this is mostly the case with popular personal computers. LANs are therefore preferable for connecting computers and devices in laboratories; they may, be linked by gateways between each other and to hospital-PBX system.

In choosing a LAN, four criteria have to be evaluated:

- Physical medium and transmission technique, e.g., copper wire, coaxial cable, or optical fiber
- Topology, i.e., bus, ring, or star structure
- Access method, i.e., the possibilities of transmission, which may be derived randomly as with Carrier Sense Multiple Access (CSMA) or in a deterministic way as with token passing
- Higher level services, i.e., protocol- or file-format conversion, network management.

Presently, the market is divided into three sectors:

- Broad-band networks with products such as INTERACTIVE SYSTEMS/3M, SYTEK, UNGERMANN-BASS
- General purpose baseband networks such as the ETHERNET-community and ARCnet
- Personal computer networks such as OMNINET, PLAN, etc.

The latter are the fastest growing group with the highest degree of sophistication, especially in protocol- and file-conversion.

Though the exchange of information between different computers is in its initial stages, now is the time to begin planning overall inhouse data communication: as system analysis, system development, organizing the procedure of proceeding will take longer than the development of certain technical features that are lacking at present.

The demand for these means of communication will grow rapidly. So-called expert systems,for example, those presently subject to intensive research in the field of artificial intelligence, can give immediate comprehensive information. They are used today in scientific libraries. The costs of such systems, especially the software, and their maintenance will be so high that it will be economical to share these systems, for example, via ISDN. A higher transmission speed will also be required in certain branches of the network, considering the rapid growth in the importance of graphic data-processing. This involves the possibility of communicating with computers by referring to three-dimensional models – perhaps of the body – displayed on screens, which can be enlarged, inspected internally or externally, rotated, colored in many ways, etc. There are also networks of broad-band technology, which employ television in addition to those using digital data transfer at high speed.

For ease of operation with such a multicomputer system and its network, some kind of data-base – on disc or paper – should keep track of hard- and software installations, for example, a description of each configuration, its operating system and standard software, data links and interfaces, the kind and structure of stored data, as well as the rights of access to them (if they are to be given outside a department), the structure of the inhouse network, the connection from the devices to the networks, the gateways between networks and to external networks and services. Such a data base is a necessary tool for developing a decentralized inhouse data-processing community. Provisions can also be made for easing compatibility if conversion aids are stored in such a data base.

Just as communication between individuals by means of the telephone created new methods of organizing work, so will communication between computers, especially with access to much larger amounts of data than those that can be managed locally, open up great new possibilities.

I will summarize here the main activities which have to be started to meet the challenges of low-priced software, hardware, and communication facilities:

1. Computerization of an organization should be guided by a committee consisting of all persons responsible for computers within the organization.

2. At least one expert should be on the committee to keep up with developments in the relevant computer hard- and software and data communication. If such an expert is not available within the organization, an external specialist should be regularly consulted. The variety of products is too large to be overviewed by nonprofessionals.

3. The guiding committee for computerization has to choose:

- One to three operating systems which will be standard within the inhouse multicomputer system. Provisions will have to be made for mutual data compatibility, training, and exchange of experience – anyone who chooses an operating system outside this setup will have to ensure that he can gain access to the overall inhouse system
- A set of hardware suppliers, as small as possible, to economize purchasing, maintenance, training, and operation
- The way of exchanging data, for example, the formats of floppy discs, the structure of data of general interest, the media, the rights of access to datas
- The structure and technical requirements for an inhouse network, e.g., transmission speeds, interfaces, protocols, provisions for maintenance and security, gateways for connections to other networks
- A chairman to promote the development of such an overall multicomputer system
- One or two people to take care of the operation and technical integrity of the network and the devices in general usage (for example a new task for the main computer center). This is not necessarily a full-time occupation.

It is a difficult task adapting traditional organization to a multicomputer environment, but those who meet the challenge and make use of the chances such a step creates will reap the benefits, as did those in the past when confronted with a major technological advance.

References

Datapro Research Cooperation. All about local area networks. Delran, NJ 08075 USA

Lockemann, Schreiner E (1983) Systemanalyse-DV-Einsatzplanung. Springer, Berlin Heidelberg New York Tokyo

A Philosophical Approach to Rapid Methods for Industrial Food Control

B. JARVIS [1]

Introduction

The annual cost of routine microbiological testing of foods in industry amounts to many millions of pounds. Within my own company, with 25 factories employing in the order of 300 factory laboratory staff together with the central laboratories, the annual budget cost for quality assurance is close to £5 million, of which more than £2 million is incurred specifically for microbiological testing. With such levels of expenditure, it is essential to question the raison d'être for undertaking routine microbiological testing.

Within a quality assurance operation, microbiological testing is used to monitor the quality of raw materials, of samples taken during processing, and of finished products; and to assess the hygiene of process plant which has been subjected to CIP or manual cleaning. In the dairy industry, tests are also performed on incoming raw milk for the presence of antibiotics and other inhibitors likely to have an adverse effect on the quality of fermented products. For product and process development, microbiological testing plays an integral part in the establishment of GMP through assessment of shelf life, development of product and process specifications, and so forth. Microbiological testing is used also to investigate retrospectively the causes of spoilage problems, reputed food poisoning outbreaks, etc., and has diverse roles in research, including the establishment of parameters inhibitory to spoilage and pathogenic organisms which are used subsequently in the development of new product and process formulations.

For many years microbiological quality testing in industry has been based upon traditional plate count methods together with the use of MPN and other tests for index and indicator organisms. In certain instances, simple, rapid test methods such as the 10-min resazurin test have been used in the U.K. dairy industry to assess the quality of milk prior to acceptance. In addition, process security has been based on enzyme analysis (e.g., the phosphatase test on milk and the α-amylase test on egg as indices of thermal pasteurization). Such indirect testing is more cost effective than extensive microbiological testing. However, if any of the data are not analyzed constructively to assess trends in ingredient, products, and process quality, then the testing is valueless.

1 Express Dairy UK Ltd., Victoria Road, South Ruislip, Middlesex HA4 OHF, Great Britain

The Needs of Industry

The methodological needs of industry can be divided into two groups consisting of general and specific needs. The general need is for methods which are simple to operate, with low running costs, having readily available stable reagents, requiring a minimal labor input and which are, ideally, automated or at least semi-automated. They should produce data which are compatible with those obtained by traditional methods, e.g., the colony count. While it will be possible in the long term to relate data from modern methods more directly to microbial quality and/or shelf life (see Wood and Gibbs 1982), I have no doubts that in the short term, it is essential for data derived by new methods to be capable of interpretation in relation to "traditionally accepted criteria" such as colony counts. Where automation of methods is possible, it is desirable that the data output be stored and analyzed by computer; it is also desirable that individual computer systems associated with different types of equipment be multicomputer-compatible, enabling computer networks to be developed. Capital cost, while important, is of less significance to laboratories in the developed world, provided the methodology actually works.

Industry also has specific needs for methods which will be more rapid than traditional ones. In some cases, exceptionally rapid methods (i.e., not more than 5 mins' delay time) are needed if incoming raw materials are to be off-loaded with the minimum hindrance to throughput and transport turn a round time. It is also essential that methods have high sensitivity and high specificity. In other words, methods are needed which are capable of detecting (relatively) low levels of organisms and which give only a low incidence of false-positive and false-negative results.

This is seemingly a formidable list of criteria for the development of new methodologies. To what extent are the methods which have been introduced in recent years capable of fulfilling these needs and of improving the efficiency and reliability of microbiological data obtained in the control of food manufacturing processes? At his stage, it must be admitted that within the industry undue emphasis is *still* placed upon retrospective product assessment, which can never properly be used to control food quality; it can be used only as an indication that problems have or have not occurred. Industry suffers considerably from the lack of proper education and training of many technicians; in many instances, they often appear to lack native wit and intelligence. In such a situation, it is clearly essential to have very simple, rapid feedback methods so that action may be taken if evidence for product contamination exists. How can we obtain such information?

Recent Method Developments for Enumeration

In the past 15 years, three groups of methods have been developed:

1. *Labor-saving methods,* based upon traditional techniques, for instance the Colworth Droplette technique (Sharpe and Kilsby 1971), the loop tile dilution technique (Ingram and Roberts 1974), automated colony count procedures by the

Table 1. Perceived characteristics of some rapid methods in relation to industrial needs

Need Criteria	Method							
	Impedance	Automated Colony Count	ATP	DEFT	Limulus	Dye Reduction	Instrumented Cell Counting	Automated Cell Counting
Simplicity	+	+	+/−	+	+	+	+/−	+
Capital cost	high	low-moderate	mod/high	low/high	very low	nil	high	?
Running costs	mod/high	as plate count	high	low	mod	very low	low	low
Stable reagents	NR	NR	+/−	+	+/−	+	+	+
Labor costs	low	mod/low	high	high	mod	mod	mod/high	low
Data compatibility	indirect	direct	indirect	direct	indirect	indirect	direct	direct
Computer interface	yes	possible	possible	possible	none	none	none	possible
Computer networkable	possible	possible	possible	possible	none	none	none	possible
Throughput (n/h)	120+	60+	10–30	10–15	10	100+	10–15	200+
Speed (elapsed time)	4 h+	days	5 min+	30 min	1–2 h	10 min–24 h	5 min	1 min
Sensitivity (CFU/g)	10^4+	10^3+	10^5	10^3–10^4	10^4	10^5+	10^3+	10^3+
Specificity	mod (potentially high)	variable	very low	low	low	mod/low	low	mod

NR, not relevant; Simplicity: +, simple; −, complex; Stable reagents: +, stable; −, unstable; ?, unknown, probability high (>£25,000)

plate-loop system (Thompson and Black 1967; Fleming and O'Connor 1975 b) and spiral plating (Gilchrist et al. 1973; Jarvis et al. 1977), automated counting of colonies on petri dishes (Fleming and O'Connor 1975 a; Jarvis et al. 1978), the hydrophobic grid system (Brodsky et al. 1982 a, b), and the use of dip slides (Mossel et al. 1976; Eustace 1974).

2. *Rapid methods* having an elapse time within the range of 4–12 h. This group includes techniques such as radiometry (Evancho et al. 1974; Rowley et al. 1976; Hatcher et al. 1977), microcalorimetry (Cliffe et al. 1973; Monk 1978), and impediometry (Hardy et al. 1977; Wood et al. 1977; Cady et al. 1978; Richards et al. 1978). Of these techniques, only impediometry has realistically been developed for use in the food and beverage industries.

3. *Very rapid methods* giving results in less than an hour. In the dairy industry the traditional method is the 10-min resazurin test. Newer methods include estimates of cell numbers by techniques such as the Bactoscan (Kielwein 1982) and the Direct Epifluorescent Test (DEFT) (Pettipher et al. 1980; Pettipher and Rodrigues 1982 a, b) and the estimation of cell metabolites and constituents such as endotoxin (limulus test) (Jay 1977; Südi et al. 1982; Hansen et al. 1982), ATP (D'Eustachio and Levine 1967; Sharpe et al. 1970; Stannard and Wood 1983), pyruvate (Marshall and Harmon 1978), etc.

Each of these methods has its supporters but to what extent do the methods really satisfy the perceived needs of industry? Table 1 summarizes the extent to which I consider these perceived needs to be satisfied.

Dye Reduction Tests

While dye reduction tests are simple to carry out, have exceptionally low running costs and stable reagents, and require only moderate labor input, the data obtained are only indirectly compatible with other methods and cannot be stored in a meaningful computerized data base. For routine purposes, dye reduction tests can be used for large numbers of samples (well in excess of 100 tests per hour), and for certain tests the elapsed time can be as short as 10 min. But the sensitivity is low, certainly in excess of 10^5 colony forming units (CFU) per gram, and the specificity is low. Furthermore, in recent years, the change to bulk storage of milk on farms, together with increased on-farm refrigeration, has led to a change in the predominant microflora of milk from gram-positive lactic acid bacteria to a predominantly proteolytic gram-negative bacterial flora. The latter are not readily detected by the traditional dye reduction tests, which are now of limited value. This poses very real problems in the rapid acceptability clearance of tanker loads of milk. One cannot expect a tanker driver to sit around and wait while elaborate microbiological tests are undertaken to check whether the quality of the product which he is delivering conforms with company- or industry-wide specifications! Similar situations occur in other industries which use large quantities of perishable raw materials (e.g., meat). Thus, alternative tests are required.

The Limulus Assay Test

The limulus assay (Jay 1977; Südi et al. 1982; Hansen et al. 1982), about which more will be heard in this symposium, is simple to perform, has only moderate running and labor costs and low elapsed time (60–120 min), but throughput is low. Furthermore, the specificity of the method is also relatively low, although sensitively is reasonably high (ca. 10^4 CFU/ml).

ATP Estimation

It might seem that detection of specific metabolites such as ATP would provide a very realistic alternative. The method for ATP assay is simple (Sharpe et al. 1970), but the running costs are high (certainly in excess of US $ 1 per test) when compared to those of the dye reduction test. The reagents are relatively stable, but because of low throughput, labor costs are relatively high, and for milk and other food materials which may contain somatic cells or intrinsic food ATP, separation of microorganisms from the food menstruum is needed prior to carrying out the analysis for ATP. Although a technique using separation processes has been developed for meats (Stannard and Wood 1983), the overall technique does not permit a high throughput, and it is of relatively low sensitivity, detecting the equivalent of 10^5 CFU/g or thereabouts. Other techniques not requiring separation of organisms have been developed in which nonmicrobial ATP is destroyed prior to analysis for microbial ATP (Bossuyt 1981, 1982). Although such techniques are reputed to give rapid estimations of microbial loadings on food materials, in our experience the methodology leaves much to be desired in terms of precision and reproducibility.

Cell Counting

Another approach is to estimate cell numbers directly using techniques such as the Coulter counter. Problems with such methods include relatively complicated operating systems with moderately high labor costs. Data are compatible with those derived from techniques such as the DEFT, the Breed smear count, or the plate count, but the throughput from manual equipment is relatively low (10–15 samples/h), although the elapsed time associated with such techniques is short (about 5 min). Sensitivity is high since the system will detect about 10^3 CFU per gram or per ml sample, provided of course that the level of background food particles does not interfere. Herein lies the problem because, except for certain beverages, differentiation of food particles from microbial cells is very difficult.

Exactly the same problem potentially applies to automated cell counting, although systems now being developed (Dow et al. 1984) will make it possible simultaneously to handle large numbers of samples and to get results very rapidly with very low labor costs and relatively low skill requirement from operators. Such techniques can undoubtedly revolutionize microbiological testing provided a solution can be found to the problem of quantitatively separating microorganisms from the food particles present in the test menstruum.

The Direct Epifluorescent Test

Developed by Pettipher and his colleagues (Pettipher et al. 1980; Pettipher and Rodrigues 1982a, b), the DEFT test is a special form of microscopic cell counting. It is simple to undertake, has low running costs and stable reagents, but because the throughput is relatively low (10–20 samples per operator per hour) labor costs are relatively high. Furthermore, the elapse time associated with the DEFT test is on the order of 30 min. Although much more sensitive than the dye reduction test, it still suffers from the disadvantage of low specificity and (reputedly) has low reproducibility.

Automated Colony Counts

The advantages of automated colony count procedures are that the throughput is increased and the media and diluent requirement reduced by comparison with manual plating techniques (Gilchrist et al. 1973; Jarvis et al. 1977), but the methodology still suffers the traditional drawback – that the elapse time is in the order of days rather than minutes. Nevertheless, the plate count is still the most sensitive method available because of its self-indicating nature, and it can of course be made highly specific by the use of appropriate selective agars.

Impediometry

Impediometry is a simple technique with moderately low running costs and low labor costs. In most cases, the data are compatible with traditional systems (i.e., detection times can be related to CFU), and it is possible to interface the computer output from impediometric instruments. Sample throughput is high, and elapsed time depends totally upon the level of contamination of the product – frequently in the order of 2–6 h for contamination levels within the order of 10^7–10^4 CFU/ml inoculum. Furthermore, the technique can be used at a range of temperatures and with a variety of culture media, thereby permitting a degree of selective testing to be undertaken (see Firstenberg-Eden and Klein 1983). Although several hours' elapse time is clearly an advantage over the several days required for colony counts, thereby enabling rapid testing of process and final product samples, the technique cannot be used for rapid screening of perishable raw materials.

However, impedance methods are not idiot-proof and require both intelligent application and careful calibration for each type of food system; even small changes in product formulation can adversely affect the calibration. For instance, we use the technique routinely for positive clearance tests on pasteurized fruit concentrates held in bulk containers and intended for use in yoghurts. After having worked satisfactorily for some time, the impedance system in our laboratory suddenly started at one point to reject batches of processed fruit for which it was not possible to obtain confirmatory evidence for contamination using traditional methods. Subsequent investigation revealed that the stabilizer incorporated in the

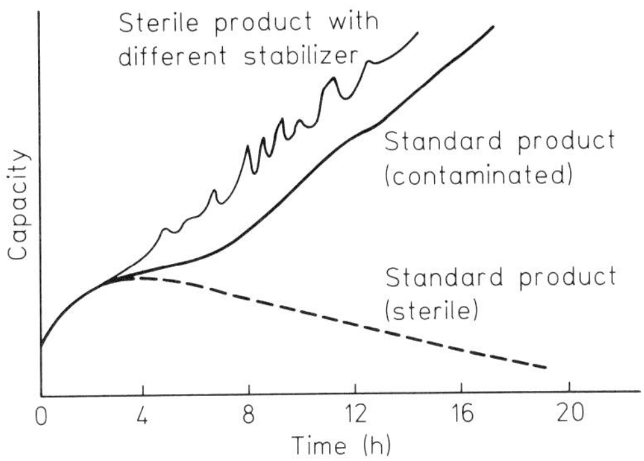

Fig. 1. Bactometer curves for fruit concentrate

fruit mix had been changed; this relatively minor compositional change adversely affected the ability of the system to detect yeast growth (see Fig. 1).

Detection and Identification Systems

It would be wrong not to make some reference to systems used for detection of specific organisms. Apart from routine tests for index and indicator organisms, the food manufacturing industry only rarely carries out tests for specific pathogens (such as salmonellae); even more rarely does industry seek to identify specifically organisms responsible for spoilage. Yet, if time could be saved on the more mundane quality testing, the opportunity to undertake a wider range of investigative testing could make the job of the industrial microbiologist both more interesting and worthwhile.

Table 2 summarizes some systems for specific detection of important microorganisms and/or their toxins. When fully validated, such methods will permit more rapid detection and, possibly, estimation of specific organisms than is possible at the present time. The major advantage will be to permit more rapid and therefore more cost-effective investigation of problems and more efficient food process safety evaluations than can be done at present. However, we must beware of being pressured into routinely testing for specific pathogens simply because rapid specific methods are available. Furthermore, the likelihood of detection of low levels of a specific organism will still be dependent upon the statistical vagaries of sampling and methodology. Unless we microbiologists recognize these constraints, industry will have increasingly unrealistic criteria for product "safety" imposed upon it by those who would seek "total absence of pathogens" in foods.

Over the past decade, many developments have been made to simplify identification of microbes. Miniaturized biochemical identification systems such as the API, the Enterotube, the Minitek, etc. (for comparison of these test systems with food isolates, see Cox and Mercuri 1979), are now commonplace. Other developments based upon biotechnology (e.g., DNA hybridization tests, immunological test systems, etc.) and sophisticated chemical technology (e.g., pyrolysis-GC-mass spectrometry; Shute et al. 1983) provide exciting opportunities for the

Table 2. Some methods for detection of specific organisms

Genus/Group	Method	Reference
E. coli	DNA-hybridization	Hill (1981)
Yersinia	DNA-hybridization	Hill et al. (1983)
Salmonellae	Fluorescent antibody	Barrell and Paton (1979)
	Enzyme immunoassays of flagellar extracts	Robison et al. (1983)
		Smith and Jones (1983)
	Solid-phase antibody adsorption immunofluorescence	Huang and Zhu (1983)
	Radioimmunoassays	Stewart et al. (1980)
	Bacteriophage + HPLC	Hirsch and Martin (1983)
	DNA-hybridization	Fitts et al. (1983)

future. However, while chemical techniques are undoubtedly valuable research tools, at this time the likelihood of their introduction into industrial control within the foreseeable future is remote. In contrast, the biotechnology-based test systems have enormous potential at the present time.

The Future Development of Rapid Methodology

It is my view that many of us have been obsessed with the concept of introducing high technology but have failed to identify the real day-to-day needs of industry. High technology methodologies which are useful now in research laboratories will lead to the routine methods of the future; but so far as industry is concerned, that is a long way ahead. Future development of appropriate biosensors should permit introduction of in-line microbiological analysis and provide data relevant to product shelf life and safety. Such developments will doubtless come from fundamental biotechnology initiatives. However, it is essential for us to recognize the intellectual credibility gap between those who develop methods and systems in academia and research institutes and those who would use the systems in industry. Greater educational initiatives are required for technical staff. Hardware sales based on hypothetical labor saving or the supposed feasibility of using technicians who are even leses well-qualified will backfire and further widen the credibility gap.

The food industry, faced constantly with increasing pressures on profit margins and with increasing labor costs, *needs alternative methodologies which it can employ now* in order to operate more cost-effective quality assurance programs. For many purposes, noncultural tests would be advantageous. Tests capable of detecting and enumerating relatively low numbers of specific microorganisms or their metabolites are needed *now*; but such tests must be both simple and exceptionally rapid. They must have low capital and running costs, and they must provide data which are compatible with traditional data systems. Ideally, they should be computer-interfaced for data analysis and storage. Some of industry's needs are being satisfied – but the credibility gap is wide.

Acknowledgement. I am indebted to Miss Sally Canty for bibliographic assistance and to past and present colleagues, especially Miss C. J. Stannard and Mrs. N. Shapton for helpful discussions and advice.

References

Barrell RAE, Paton AM (1979) A semi-automatic method for the detection of salmonellas in food products. J Appl Bacteriol 46:153–159

Bossuyt R (1981) Determination of bacteriological quality of raw milk by an ATP assay technique. Milchwissenschaft 36:257–260

Bossuyt R (1982) A 5-minute ATP platform test for judging the bacteriological quality of raw milk. Neth Milk Dairy J 36:355–364

Brodsky MH, Entis P, Sharpe AN, Jarvis GA (1982a) Enumeration of indicator organisms in foods using the automated hydrophobic and membrane filter technique. J Food Prot 45:292–296

Brodsky MH, Entis P, Entis MP, Sharpe AN, Jarvis GA (1982b) Determination of aerobic plate and yeast and mould counts in foods using an automated hydrophobic grid-membrane filter technique. J Food Prot 45:301–304

Cady P, Hardy D, Martins SD, Dufour SW, Kraeger SJ (1978) Automated impedance measurements for rapid screening of milk microbial content. J Food Prot 41:277–283

Cliffe AJ, McKinnon CH, Berridge NJ (1973) Microcalorimetric estimation of bacteria in milk. J Soc Dairy Technol 26:209–210

Cox NA, Mercuri AJ (1979) Rapid biochemical testing procedures for *Enterobacteriaceae* in foods. Food Technol (Chicago) 33:57–62

D'Aoust JY, Sewell A, Boville A (1983) Rapid cultural methods for detection of *Salmonella* in feeds and feed ingredients. J Food Prot 46:851–855

D'Eustachio AJ, Levine GV (1967) Levels of adenosinetriphosphate during bacterial growth. Bacteriol Proc 67:121

Dow CS, Khan MS, Scott NW, Johnsen T (1984) Automated particle volume distribution analysis for the diagnosis of bacteriuria. These proceeding W40

Eustace IJ (1974) The use of Orion dip slides in meat industry bacteriology quality control programmes. CSIRO Meat Res Report No 7/77

Evancho GM, Ashton DH, Zwarun AA (1974) Use of a radiometric technique for the rapid detection of growth of clostridial species. J Food Sci 39:77–79

Firstenberg-Eden R, Klein CS (1983) Evaluation of a rapid impediometric procedure for the quantitative estimation of coliforms. J Food Sci 48:1307–1311

Fitts R, Diamond M, Hamilton C, Neri M (1983) DNA-DNA hybridization assay for detection of Salmonella spp. in foods. Appl Environ Microbiol 46:1146–1151

Fleming MG, O'Connor F (1975a) An automated loop method for determining the total count of bacteria in milk. Ir J Agric Res 14:27–32

Fleming MG, O'Connor F (1975b) Evaluation of electronic colony counters. Ir J Agric Res 14:21–26

Gilchrist JE, Campbell JE, Donnelly BC, Peeler JJ, Delaney JM (1973) Spiral plate method for bacterial determination. Appl Microbiol 25:244–252

Hansen K, Mikkelsen T, Møller-Madsen A (1982) Use of the limulus test to determine the hygienic status of milk products as characterized by levels of Gram-negative bacterial lipopolysaccharide present. J Dairy Res 49:323–328

Hardy D, Kraeger SJ, Dufour SW, Cady P (1977) Rapid detection of microbial contamination in frozen vegetables by automated impedance measurements. Appl Environ Microbiol 34:14–17

Hatcher WS, DiBenedetto S, Taylor LE, Murdock DI (1977) Radiometric analysis of frozen concentrated orange juice for total viable microorganisms. J Food Sci 42:636–639

Hill WE (1981) DNA hybridization method for detecting enterotoxigenic *Escherichia coli* in human isolates and its possible application to food samples. J Food Saf 3:233–247

Hill WE, Payne WL, Aulisio CCG (1983) Detection and enumeration of virulent *Yersinia enterocolitica* in food by DNA colony hybridization. Appl Environ Microbiol 46:636–641

Hirsch DC, Martin LD (1983) Detection of *Salmonella* spp. in milk by using Felix-01 bacteriophage and high-pressure liquid chromatography. Appl Environ Microbiol 46:1243–1245

Huang C, Zhu G (1983) Immunofluorescence by slide solid-phase antibody adsorption (SPAIF) in food microbiology. Chin Med J 96:177–184

Ingram G, Roberts TA (1974) The loop-tile method of bacterial counting: a minimal apparatus method suitable for field work. Agric Res Counc Meat Res Inst Memo No 25

Jarvis B, Lach VH, Wood JM (1977) Evaluation of the spiral plate maker for the enumeration of microorganisms in foods. J Appl Bacteriol 43:149–157

Jarvis B, Lach VH, Wood JM (1978) Evaluation of the laser bacterial colony counter. Leatherhead Food RA Res Rep No 290

Jay JM (1977) The *Limulus* lysate endotoxin assay as a test of microbial quality of ground beef. J Appl Bacteriol 43:99–109

Kielwein G (1982) Experiences in enumeration of microorganisms in raw milk by Bactoscan. Kieler Milchwirtschaftliche Forschungsberichte 34:74

Marshall RT, Harmon CC (1978) The automated pyruvate method as a quality test for grade A milk. J Food Prot 41:168–177

Monk PR (1978) Microbial calorimetry as an analytical method. Process Biochem 13:4, 5, 8

Mossel DAA, Eelderink I, de Vor H, Keizer ED (1976) Use of agar immersion, plating and contact slides for the bacteriological monitoring of foods, meals and the food environment. Lab Pract (1976):393–395

Pettipher GL, Rodrigues UM (1982a) Rapid emumeration of microorganisms in foods by the direct epifluorescent filter technique. Appl Environ Microbiol 44:809–813

Pettipher GL, Rodrigues UM (1982b) Semi-automated counting of bacteria and somatic cells in milk using epifluorescence microscopy and television image analysis. J Appl Bacteriol 53:323–329

Pettipher GL, Mansell R, McKinnon CH, Cousins CM (1980) Rapid membrane filtration-epifluorescent microscopy technique for direct enumeration of bacteria in raw milk. Appl Environ Microbiol 39:423–429

Richards JCS, Jason AC, Hobbs G, Gibson DM, Christie RH (1978) Electronic measurements of bacterial growth. J Phys [E] 11:560–568

Robison BJ, Pretzman CI, Mattingley JA (1983) Enzyme immunoassay in which a myeloma protein is used for detection of salmonellae. Appl Environ Microbiol 45:1816–1821

Rowley DB, Previtte JJ, Srinivasa H (1976) A radiometric screening method for estimating the level of aerobic mesophilic bacteria in foods. In: Johnston MM, Newton SWB (eds) Rapid methods and automation in microbiology. Learned Information, Oxford

Sharpe AN, Kilsby DC (1971) A rapid, inexpensive bacterial count technique using agar droplets. J Appl Bacteriol 34:435–440

Sharpe AN, Woodrow MN, Jackson AK (1970) Adenosintriphosphate (ATP) levels in foods contaminated by bacteria. J Appl Bacteriol 33:758–767

Shute LA, Berkeley RCW, Norris JR, Gutteridge CS (1983) Pyrolysis mass-spectrometry in bacterial systematics. J Appl Bacteriol 55:xvi

Smith AM, Jones C (1983) Use of murine myeloma protein M467 for detecting *Salmonella* spp. in milk. Appl Environ Microbiol 46:826–831

Stannard CJ, Wood JM (1983) The rapid estimation of microbial contamination of raw meat by measurement of adenosine triphosphate (ATP). J Appl Bacteriol 55:429–438

Stewart BJ, Eyles MJ, Murrell WG (1980) Rapid radiometric method for detection of *Salmonella* in foods. Appl Environ Microbiol 40:223–230

Südi J, Suhren G, Heeschen W, Tolle A (1982) Die Anwendung des Limulus-Tests zur Untersuchung ultrahocherhitzter Milch und Ermittlung der bakteriologisch-hygienischen Wertigkeit des verwendeten Rohstoffes. Milchwissenschaft 37:341–346

Swaminathan B, Denner JM, Ayres JC (1978) Rapid detection of salmonellae in foods by membrane filter-disc immunoimmobilization techniques. J Food Sci 43:1444–1447

Thompson DI, Black LA (1967) The use of the 0.01 ml loop in the plate loop method for making viable counts of milk. J Milk Food Technol 30:273–276

Wood JM, Gibbs PA (1982) New developments in the rapid estimation of microbial populations in foods. In: Davies R (ed) Development in food microbiology I. Applied Science Publishers, London, pp 183–214

Wood JM, Lach VH, Jarvis B (1977) Detection of food-associated microbes using electrical impedance measurements. J Appl Bacteriol 43:xiv–xv

Essentials of Enumeration and Identification Methods for Microorganisms Which Interfere with the Safety of Foods with Particular Reference to Simplification Avoiding the Use of Nonconventional Instrumentation

D. A. A. Mossel, P. van Netten, and H. van der Zee [1]

The Rationale of Microbiological Monitoring of Foods

Despite a most marked reduction in the morbidity of infectious diseases in general, the incidence of diseases of microbial etiology transmitted by foods and of food losses due to activities of microorganisms has not been reduced worldwide, including the highly developed contries. This is most certainly due to the fact that it has only recently become generally accepted that control of microbiological safety and quality of foods cannot be achieved by the approach so far followed in most instances. This approach relies, in essence, on monitoring end-product samples and using the results in a feedback manner to improve manufacturing practices. What is really required of course is intervention (Kayser and Mossel 1984). This includes: (1) identification of hazard points (Bauman 1974) in manufacture, distribution, storage, and culinary preparation; followed by (2) design of measures leading to improved practices throughout. These should be elaborated and validated, guided by the examination of line samples rather than end products. Obviously immediate feedback to manufacture or preparation should follow whenever and wherever required (Mossel et al. 1984).

Even after the introduction of these so-called good manufacturing practices, frequent and regular tests are necessary to check that these practices are being followed and are effective, or that further modifications are required. Methods of examination to be used for this purpose should be: (1) as simple and rapid as possible; (2) economically feasible; (3) accurate, repeatable, and reproducible.

Not too much progress has been made in this respect until recently. This is partly due to the relatively late start of analytical food microbiology as a branch of science (Buttiaux and Mossel 1957). The incentive of using mechanized and even fully automated methods of microbiological examination of foods consequently did not always lead to an adequate response.

In this paper, an attempt will be to summarize the present state of microbiological monitoring of line and food samples, particularly in the light of more recent molecular-microbiological data on the attributes of the most important microbial groups as they occur in foods. This should enable microbiologists called to advise on, or introduce, new (including facilitated) techniques into the laboratories of industry and government inspection services to arrive at an unbiased

1 University of Utrecht, Faculty of Veterinary Medicine, P.O. Box 80175, NL-3508 TD Utrecht

evaluation of available technology, allowing a rational choice of procedures to be adopted.

Three general principles should always be adopted in this area of laboratory policy. First and foremost, the number of criteria to be used should be limited strictly to the minimum. The indispensable ones should be selected, guided by a careful study of the microbial ecology of every specific commodity, paying attention to health risks as well as to the food's so-called spoilage association (Mossel 1983). Furthermore, reference values should be available, against which the results of a given analysis can be gauged. This may sound trivial, but there exists a real problem here, i.e., in the assessment of such reference values. As in clinical medicine (Gräsbeck and Alström 1981), such reference values should be derived from surveys on specimens originating from production or catering lines previously inspected and noted for using correct practices, but modified before samples are drawn, if necessary (Mossel 1980). Finally and quite obviously, the methods used in assessing the conformity of production samples with reference values should be exactly the same as those used in determining reference values and rigorously standardized for that purpose. If this aspect of monitoring is neglected, the most embarrassing conflict of opinion can result between, e.g., production and quality assurance departments of the same factory, and worse, between manufacturer and buyer or government inspection services.

Analytical Essentials

Isolation and enumeration methods for specific groups of microorganisms in foods and drinking water invariably rely on the use of selective culture media. This entails two types of problem which require the permanent attention of the food microbiologist.

Limitations of Selective Media

In view of the intensive genetic flux observed in almost all niches of significance in food microbiology (Altherr and Kasweck 1982) no selective medium will exclusively grow the sought after microorganism for which it was designed. Even if it did initially, there is no guarantee that it will continue to do so (Dijkmann 1982). Consequently, most selective media are either insufficiently selective or else are inhibitory for the group of organisms they are supposed to enumerate. Every formula, whether made up in the laboratory or purchased in some form, therefore represents a compromise. This is in itself unavoidable and can in principle lead to workable situations provided the performance of a medium is "constant," i.e., does not vary too much from lot to lot.

This clearly calls for checking the functioning of media, both upon purchase and in the course of time when a given batch number is used for, say, more than 1 of 2 weeks. It goes without saying that such monitoring will soon be abandoned when testing methods become too complicated or time-consuming. Nonetheless, a medium should always be challenged by a selection of test strains that should

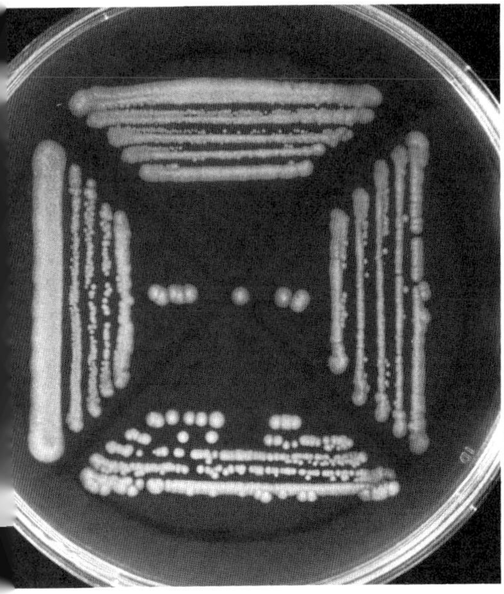

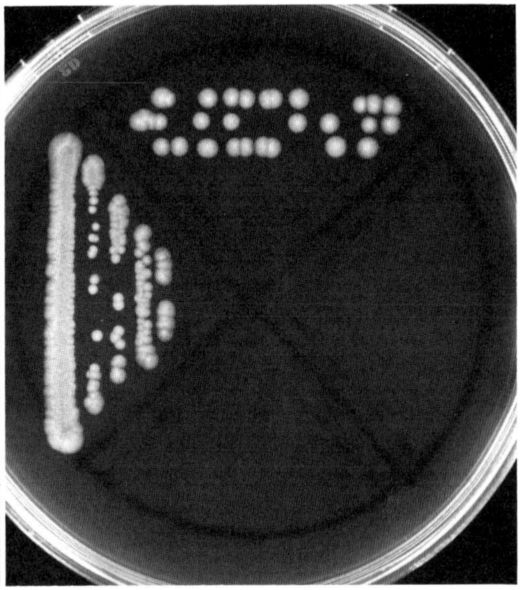

Fig. 1 **Fig. 2**

Fig. 1. Ecometric evaluation of a selective medium. Growth of a control strain

Fig. 2. Ecometric evaluation of a selective medium. Growth of an interfering strain: maximal acceptable level

be recovered quantitatively and another series that should be completely or almost totally inhibited.

Rather than using conventional fully quantitative techniques for this purpose, we found it useful to rely on "ecometric" monitoring, following the principle of constantly decreasing density of inoculation as practised in spiral plating (Masters and Palmer 1981; Reusse 1982). Disposable plastic 1-μl loops streaked sequentially over 21 lines on a standard Petri dish containing the medium under test (Figs. 1 and 2) appeared useful to this end (Mossel et al. 1983a). In the few instances where a higher accuracy is required, the same loops can be used for making conventional spread drop plates (Corry 1982).

The Effects of Sublethal Damage and Their Remedy

The use of selective media entails the risk that sublethally damaged cells among the population sought will not be included in counts. Virtually all foods in commerce will contain such damaged cells because these commodities are either intrinsically preserved by a reduced a_w, pH, or added antimicrobial substances (e.g., nitrite or sorbic acid), processed by heat, or else stored frozen or under refrigeration. These conditions will all induce sublethal lesions in organisms exposed (Mossel and Van Netten 1984); hence resuscitation steps (Allen et al. 1953)

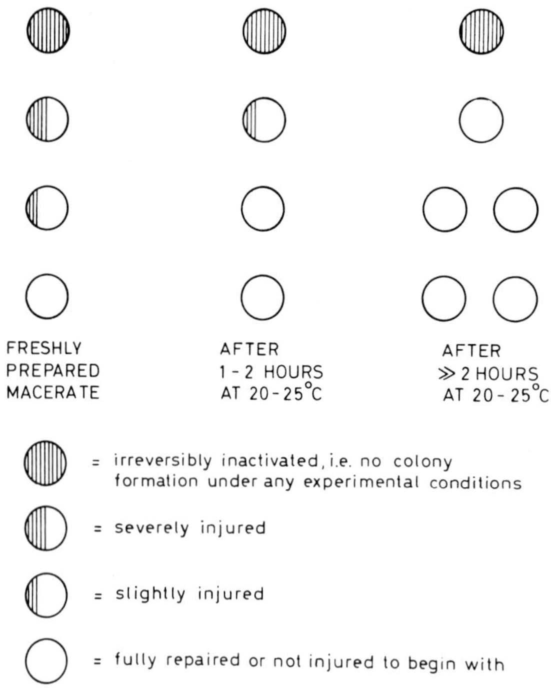

Fig. 3. Successive resuscitation steps in a microorganism following sublethal damage caused by exposure to intrinsic or extrinsic stress

should always precede the application of selective isolation and enumeration procedures.

Relying on fortuitous repair during preparation of food macerates and subsequent decimal dilutions is to be strongly discouraged, because this may lead to insufficient recovery of the more severely impaired cells in a given population (cf. Fig. 3). When macerates or dilutions are stored for longer periods of time, however, repaired cells may start to grow and consequently affect colony counts to the extent that falsely high results are obtained. Experimentally designed and validated, deliberate resuscitation steps should be used in all instances.

In presence-or-absence tests, dilution of samples and subsequent liquid medium repair ("nonselective pre-enrichment") has been found adequate (Van Doorne and Claushuis 1980). In selective colony counting procedures, liquid medium repair may also sometimes lead to complete revitalization of stressed cells (Mossel et al. 1980), though in many instances it may be inadequate. The solid-medium procedure of Speck et al. (1975), relying on recovery on a suitable solid medium, followed by overlayering with a selective medium, or replica plating onto such a medium, as with strictly aerobic taxa, is invariably effective. The use of special, injury-alleviating substances, such as catalase or pyruvate, in resuscitation media is sometimes essential to recover all viable colony-forming units (Van Netten et al. 1984). When such carefully devised resuscitation procedures are neglected, a substantial lack of reproducibility will inevitably occur.

For some selective media, there are indications that they will not inhibit stressed populations and hence can be used without a previous resuscitation step.

Table 1. Examination of frozen precooked shrimps at 42 °C versus 37 °C, with and without solid-medium repair on GTSYEP agar for 5 h at 23 °C ± 2 °C, using Baird-Parker's medium for ultimate assessment of CFUs

	Log_{10} Ng^{-1}				Log_{10} CFU Proportion	Log_{10} CFU Proportion
	37 °C		43 °C			
	With SMR on GTSYEP	Without SMR	With SMR on GTSYEP	Without SMR	GTSYEP 37 °C/GTSYEP 43 °C	BPM 37 °C/BPM 43 °C
	3.1	2.9	2.9	2.0	0.2	0.9
	3.3	3.1	3.0	2.6	0.3	0.5
	2.9	2.5	2.8	2.4	0.1	0.1
	2.8	2.7	2.8	2.0	0.0	0.7
	3.2	3.1	2.9	2.0	0.3	1.1
	3.2	3.1	3.0	2.4	0.2	0.7
Average	–	–	–	–	0.2	0.7

Abbreviations: GTSYEP, Buffered glucose tryptone soya peptone yeast extract agar with 100 ml egg yolk emulsion and 10 g sodium pyruvate per litre; SMR, solid-medium repair; BPM, Baird-Parker's medium

These media include: (1) Baird-Parker's glycine tellurite pyruvate egg yolk agar for the enumeration of *Staphylococcus aureus* (Idziak and Mossel 1980); (2) mannitol egg yolk polymyxin agar for colony counts of *Bacillus cereus* (Rappaport and Goepfert 1978); (3) Skirrow's highly selective blood agar, relying on the use of a cocktail of antibiotics for the enumeration of *Campylobacter jejeuni* (Waterman 1982; Hanninen 1982). However, omitting a resuscitation step when these media are employed entails a constant risk. Different stresses, e.g., heating versus freezing, may differ in their effects; moreover, the intensity of stress, and hence of damage, may vary widely, as illustrated in Fig. 3. This concern has recently been substantiated with respect to Baird-Parker's extensively tested medium. British workers demonstrated that cells of *Staph. aureus* stored for considerable periods of time in cured meats under anaerobic conditions might incur multifactorial damage leading to markedly reduced recovery on Baird-Parker's agar (Mansfield et al. 1983). We have observed similar phenomena in populations of *Staph. aureus* stored for many months in precooked shrimps at about −25 °C. As the data in Table 1 illustrate, such populations are not completely recovered on Baird-Parker agar at 37 °C, and even worse recovery on this medium occurs when incubated at 42.5 °C. The latter applies despite the fact that incubation at elevated temperatures had been found previously to lead to complete recovery of populations of *Staph. aureus* severely stressed by heating or osmotic shock (Van Doorne et al. 1982).

Securing Adequate Consistency of Data in Other Respects

Monitoring of selective media and the use of proper resuscitation procedures are essential parts of every analytical procedure to be used in food microbiology. Of

Table 2. Matrix to be followed when developing analytical methods in food microbiology and in the standardization of procedures to be used in trade and for intracorporate purposes

1. *Sampling plan*
   Randomization
   Numbers to be drawn per predefined lot
2. *Handling before examination*
   Transportation – time/temperature integral
   Challenge
      Incubation
      Inoculation in case of survival studies or assessment of resistance against colonization
3. *Preparation for examination*
   Defrosting – time/temperature integral; squeezing to release fluid
   Cleaning/disinfection of containers
4. *Drawing of subsamples ("aliquots")*
   Randomization
   Size of subsample
   Diminution and homogenization
5. *Preparation of macerate and dilutions*
   Size of second subsample, i.e., aliquot to be examined
   Preparation of macerate
      Compostion of maceration fluid
      Procedure, including time/temperature regimen
      Mode of dispersion
   Preparation of serial dilutions
      Composition of diluent
      Procedure, including time/temperature regimen
6. *Monitoring of culture media*
   Choice of test strains
   Selection of inoculation procedure
   Incubation, time and tolerance/temperature
   Reading
   Reference values to be used
7. *Resuscitation procedure*
   Composition of resuscitation medium
   Time/temperature program, including tolerance
   Mode of processing of resuscitated system
8. *Enumeration procedure*
   Composition of medium
   Preparation of medium
   Decontamination of medium
   Tempering of medium
   Aseptic precautions during inoculation
   Inoculation
      Procedure, qualitative
      Procedure, quantitative
      Holding or drying time/temperature
   Incubation
      Procedure, qualitative
      Circulation, ventilation and temperature tolerances
      Duration, including tolerances
   Reading
      Definition of target colonies
      Accuracy of colony counting
      "Emergency" handling of plates showing too few or too many colonies
      Reference values

Table 2 (continued)

Confirmation/identification
Extent of picking per given type of colony
Subculture for examination
Preliminary taxonomic grouping
Presumptive grouping
Complementary testing – intramural/extramural
Expression of results
9. *Essential annexes*
Justification and documentation of techniques
Laboratory precautions to be observed ("GLP")
Reference centers available for consultation
Recommended reporting form(s), storage and retrieval of data

course, they are by no means sufficient to secure adequate reproducibility of P-A tests or colony counts.

As emphasized before, all measures should always be taken to guarantee the highest degree of reproducibility. This is imperative in view of the relatively high coefficient of variation of colony counts (Mossel et al. 1980), but especially of P-A procedures (Wood 1950; Pretorius 1961; Lear 1962). This applies even when these determinations are carried out by highly qualified staff, taking all possible precautions and adhering most accurately to prescribed procedures. A few of the items requiring rigorous standardization are summarized in Table 2.

Tentative Taxonomic Grouping of Isolates

Ecological Needs for Identification

As indicated before, selective culture media in current use are reasonably effective in allowing sought after taxa to develop, while inhibiting most interfering organisms. For some purposes, gross typical colony counts will therefore suffice, particularly when adequate reference values have been generated and accurately standardized procedures have been laid down. For other purposes, identification of isolates, often termed "confirmation" of counts, cannot be omitted.

There is often also an ecological reason for tentatively grouping isolates obtained in colony counts. It has become common practice to examine foods processed for safety, either before or after hermetic packaging, for marker organisms belonging to the Enterobacteriaceae. The entire group of these well-defined bacteria can be used for this purpose, because their presence above a certain level in processed foods indicates either inadequate heat treatment or postprocess failure, i.e., reinfection of food and subsequent growth of this can occur (Mossel 1982). However, when using these organisms in the monitoring, if commodities that receive a last-minute addition of unprocessed condiments, such as grated cheese (Shelton 1961), or in monitoring products that are intensively handled after processing and therefore inevitably contain low numbers of Enterobacteriaceae (Beckers et al. 1981), hopefully of no health significance, the entire group cannot

Table 3. Recovery ($\log_{10}$ CFU ml^{-1}) of thermotrophic types of Enterobacteriaceae on MacConkey agar (VRBL) and infusion agar (IA) at 37 °C and at 42.5 °±0.5 °C

Strain	37 °C		42.5 °C	
	IA	VRBL	IA	VRBL
Enterobacter	9.2	8.9	9.2	8.9
Esch. coli 81	7.4	7.1	7.4	7.3
82	8.1	8.2	8.4	8.2
84	8.7	8.5	8.7	8.4
86	8.6	8.3	8.6	8.3
015	9.0	8.9	8.9	9.0
96	9.3	8.9	9.3	9.0
Klebsiella T 1	8.7	8.8	8.8	8.8
T 2	9.0	8.8	8.7	8.8
Kluyvera K	9.3	9.0	9.3	9.2
299	9.2	9.2	9.3	9.2
S 36	9.2	9.2	9.3	8.9
Salm. brandenburg	8.9	8.5	8.9	8.4
derby	9.0	9.2	9.0	9.2
eastbourne	8.8	8.6	8.8	8.7
enteritidis	9.3	9.2	9.3	9.3
hadar	9.1	8.5	9.1	8.4
heidelberg	8.9	8.8	8.9	8.6
indiana	7.2	7.3	7.2	7.2
london	8.9	9.2	9.1	9.1
montevideo	8.9	8.7	8.9	8.9
senftenberg	9.1	8.2	9.0	9.1
typhimurium	9.0	9.0	9.0	9.0
Shigella				
flexneri 2, H	7.9	7.9	8.0	7.7
flexneri 2, U	9.1	9.0	9.0	9.1
sonnei H	7.6	7.6	7.4	7.5
sonnei U	8.7	8.7	8.8	8.6

Table 4. Recovery ($\log_{10}$ CFU ml^{-1}) of psychrotrophic types of Enterobacteriaceae on MacConkey agar (VRBL) and infusion agar (IA) at 30 °, 37 °, and 42.5 °±0.5 °C

Strain	30 °C		37 °C		42.5 °C	
	IA	VRBL	IA	VRBL	IA	VRBL
Erwinia 96	8.7	7.2	8.7	7.9	<2	<2
Erwinia 240	–	–	8.8	8.9	<2	<2
Hafnia K	9.1	9.0	9.1	8.9	<2	<2
Hafnia 4	9.0	9.0	9.0	9.0	<2	<2
Hafnia 5	8.9	8.9	8.9	8.9	<2	<2
Hafnia 7	8.9	8.9	8.9	8.9	<2	<2
Yersinia enterocolitica						
serotype 0.3	9.0	8.9	9.0	8.9	<2	<2

Table 5. Diffferential characteristics of some *Clostridia*

	C. butyricum	*C. perfringens*	C. botulinum	C. difficile
Gelatin	–	+	+	+
H_2S	–	+	+	–
Indole	–	–	–	–
Nitrate	D	+	–	–
Lactose	+	+	–	–
Sucrose	+	+	–	–
Mannitol	D	–	–	+
Spores	S/S	S/N	S/S*	S/N
Motility	+	–	+	+

Abbreviations: S, subterminal; N, not swollen; S, swollen

be relied on as markers. We have suggested that the thermotrophic members of the group, which are generally the pathogenic genera (Stuart et al. 1942; Harvey and Thomson 1953; Kafel and Brian 1977), be used in addition to *Escherichia coli*, the classic marker, in these instances (Mossel et al. 1984). Table 3 illustrates how well these organisms are recovered on the customarily used MacConkey-type selective media at 42.5 °C, whereas Table 4 demonstrates the complete inhibition of psychrotrophic types of Enterobacteriaceae. Obviously, when obtaining substantial counts from certain products at this elevated temperature, taxonomic grouping is necessary in order to allow differentiation of bacteria like *Klebsiella* from the more relevant enteric types (Bagley and Seidler 1977; Uchino et al. 1984).

Another need for taxonomic grouping arose recently when studying the anaerobic microflora of dried weaning foods and their components (Mossel et al. 1973). Such products are as a rule examined by incubating decimal dilutions in sulfite agar in tubes that are oval in cross section (which reduce oxygen penetration compared with circular cross-sectional tubes), so as to attain anaerobic conditions (Mossel 1969). When virtually no black colonies are obtained from the first decimal dilution, products can be accepted. However, when substantial colony counts are observed, a need for roughly identifying the isolated organisms is presented. It is particularly important to differentiate the butyric acid-forming types from the pathogens, including *Clostridium botulinum, Cl. perfringens*, and *Cl. difficile*. The key, summarized in Table 5, has been found useful for this purpose.

Recommended Convenient Methods

Because the ingrained habit of "worship of numbers" (Wilson 1959) has long hindered food microbiology, rather few studies have been devoted to simple conventional systems for tentative taxonomic grouping of isolates obtained from foods.

Other workers (Bergan et al. 1982) and ourselves (Mossel et al. 1983b) have found the determination of the microscopic morphology of organisms linked to a catalase test most useful as the first step. In practice, it is quite surprising how many useful conclusions can be drawn from these two simple tests. They can be combined by subculturing pure cultures of isolates in small tubes containing brain heart infusion broth at 30 °C. The cultures are subsequently examined by a Gram stain, while the sediment is dispersed in a drop of freshly prepared 3% hydrogen peroxide, contained in wells of microtiter plates.

Further taxonomic grouping can be very conveniently carried out with the help of so-called polytropic tubes. These are single tubes allowing the determination of several taxonomic traits. The classic example here is of course Kligler's

Table 6. Simplified tentative taxonomic grouping of bacteria isolated from foods and examined for Gram stain and catalase reaction

Type of bacterium	Polytrophic tube		Supporting tests
	Composition	Characteristics	
Fermentative and nonfermentative Gram-negative rod-shaped bacteria	"Gram-negative diagnostic tube", i.e. Bottom: violet red bile, glucose agar Top: SIM agar	Oxidase, glucose dissimilation, Motility, formation of H_2S, indole, pigments	Open tube of glucose agar, multiwell plates with Simmons' citrate agar, urea agar; when required lysine decarboxylase, DNase, sorbitol
E. coli	Bottom: MacConkey lactose agar Top: SIM incubation at 44 °C ± 0.1 °C	Indole, attack on lactose	Multiwell plates with Simmons' citrate agar
Yersinia enterocolitica	Bottom: sorbitol agar Top: lysine broth	Dissimilation of sorbitol and lysine	Multiwell plates with Simmons' citrate agar
Vibrio parahaemolyticus	Bottom: 5% NaCl starch agar Top: MR-VP with 2% NaCl	Anaerobic dissimilation of starch; VP-reaction negative	Reactions in Gram-negative diagnostic tube (see above)
Gram-positive bacteria	Bottom: glucose agar Top: tryptone agar	Attack on glucose, catalase	Open glucose agar tube
Staph. aureus	Bottom: mannitol agar Top: rabbit plasma	Anaerobic dissimilation of mannitol, coagulase	–
Lancefield group D *Streptococci*	Bottom: kanamycin aesculin azide agar Top: 40% bile	Attack on aesculin, tolerance of azide, kanamycin and bile	–
Clostridia perfringens	Bottom: sulfite iron cycloserine agar Top: anaerobic motility medium, incubation at 46 °C	Tolerance for cycloserine Thermotrophic, nonmotile	–
Other *Clostridia*	Bottom: thiosulfate iron agar Top: sucrose agar, anaerobic incubation at 30 °	Formation of H_2S Attack on sucrose	Gelatin dissimilation Nitrate reduction Attack on mannitol cf. Table 5

glucose/lactose/iron agar (1918). Specific three-layer tubes have been developed for groups of organisms of special interest in food microbiology. They invariably contain: (1) a bottom layer allowing fermentative attack on a carbon source to be assessed (Hugh and Leifson 1953); (2) a layer of water agar to prevent interference between the biochemical processes of the bottom and top; and (3) a suitably designed top layer (Mossel et al. 1977).

In a few instances, additional tests must be carried out which cannot be combined with those determined in the polytropic master tube. These tests are done in microtiter plates or in square Petri dishes with 24 compartments. In Table 6, the various tests of this nature, which have been in use for over 10 years now, are summarized. It is clear that they save a tremendous amount of work and cost, because relatively small volumes of media and reagents are required throughout.

An Attempt at Noninstrumental Simplification of Approximate Assessment of the Colonization of Food

A Short Review of Available Methods

It is a general principle in biomedical monitoring that the use of a minimal number of criteria should be linked to every possible simplification of procedures. This allows many more specimens to be examined at a given laboratory and with given staff facilities. This is particularly important in food microbiology, where colonization may bear a markedly heterogeneous and sporadic character (Mossel 1982).

During the last 15 years, a good deal of progress has been made in developing mechanized or automated apparatus; often this has been concomitant with accelerated procedures. Procedures that also allow selective enumerations to be made, essential in food microbiology, include impedimetry (Bossuyt and Waes 1983; Firstenberg-Eden and Klein 1983; Firstenberg-Eden and Tricarico 1983; Weihe et al. 1984), ATP-related assessment of numbers of colony-forming units (Waes and Bossuyt 1981; Stannard and Wood 1983; Nichols et al. 1984), and the extremely elegant hydrophobic grid membrane, linked to enzymatic predigestion of food macerates (Entis 1983; Entis and Boleszczuk 1983). It cannot be denied, however, that these methods require rather expensive, though by no means uneconomic, equipment and also often use costly reagents. This makes these approaches somewhat impracticable in times of economic depression and essentially beyond the reach of developing areas and smaller food manufacturing industries.

The Nitrate Glucose Metabolism Test (NGT)

These considerations prompted investigations into very simple tests, relying on conventional modes of studying the velocity of attack on simple substrates, such as nitrates and glucose, which is of course related to numbers of colony-forming units (CFU). Attempts to use nitrate reduction tests in clinical microbiology (Sleigh 1965; Monte-Verde and Nosanchuk 1981) and dairy microbiology

(Sahlqvist et al. 1953; Klein 1955; Borneff 1957; Kandler 1961; Von Bockelmann 1962; Börger 1962; Peltola and Kylä-Siurola 1962; Crawley and Twomey 1969) date back many years, but have unfortunately never resulted in any practical application. This is probably because the older tests lack both precision, i.e., correlation between CFU numbers and response times, and selectivity (Mossel 1964).

Two essential developments might change this rather dim outlook for tests of this sort. Firstly, Popken and Bomar (1979) suggested that an additional test for glucose dissimilation be included in tests for nitrate reduction. This seems most appropriate because few, if any, bacteria of significance in food both lack a nitrate reductase and do not attack glucose by any pathway. Furthermore, reliable selective enrichment media are nowadays available for the determination of Enterobacteriaceae (Mossel et al. 1963; Mossel et al. 1974) and *Staph. aureus* (Van Doorne et al. 1981; Van Doorne et al. 1982; Terplan et al. 1983), the most important groups of bacteria looked for in foods. We have therefore studied the applicability of Popken and Bomar's principle of the NGT for the selective enumeration of these bacteria, particularly after Bartl et al. (1984) found it suitable for the approximate assessment of the total colonization ("aerobic plate count") of a variety of foods.

Development of a Procedure for Selective Enumeration

The NGT includes first of all a concentration/purification step, wherein the bacteria present in a food sample are concentrated by centrifugation and at the same time freed of any glucose or nitrate stemming from the food to be examined, which might interfere with the test. In order to avoid losses of viable cells exceeding some 10% by retention in the supernatant, it was found necessary to centrifuge for about 10 min at approximately $10^4 \times g$ (Mossel and Visser 1960).

Two approaches were followed to reduce the response times, i.e., time elapsing before the nitrite paper strip test becomes positive or the presence test for glucose, also carried out with test strips, becomes negative. The glucose concentration in the reaction system could be reduced without loss of consistency of the response to 0.25 g l^{-1}. In many instances, response times could be markedly reduced by increasing the incubation temperature to 43 °C – not surprising since many Enterobacteriaceae and *Staph. aureus* are thermotrophs (Mossel 1982b). We aimed at response times not exceeding 6 h, because this would allow results to be obtained within 1 working day, including the preparation of the samples by centrifugation.

With respect to accuracy, the tentative character of the NGT should always be kept in mind. The coefficient of variation of colony counts amounts to just under 10% (Mossel et al. 1980). Therefore, similar variations in response times in the NGT seem acceptable.

We determined the levels of detection of pure cultures of Enterobacteriaceae and *Staph. aureus* with this technique, initially using Enterobacteriaceae-enrichment (EE) broth (Mossel et al. 1963) and liquid Baird-Parker medium (Van Doorne et al. 1981) as culture media. Both were used with glucose content adapted to the required concentration. In the course of this investigation, it was found that EE broth can be advantageously replaced by a medium in which lauryl sulfate at a concentration of 0.1 g l^{-1} is added to Bartl's basal medium. Using positive nitrite tests as a criterion and a response within 5 h, the response became positive at levels of the order 10^2–10^3 CFU/g of food, for both taxa. The thresholds of detection relying on tests for glucose becoming negative, were one to two log cycles higher.

Experience with the Elaborated Technique

The newly developed technique is as follows:

1. Prepare a 1:10 macerate of the food in sterile peptone saline
2. Centrifuge the suspension for about 30 min at 10,000 × *g*
3. Resuspend the pellet in three media:
   a) 0.25 g/l glucose nitrate peptone broth (Bartl et al. 1984)
   b) Do. with sodium lauryl sulfate at 0.1/g/l (Mossel et al. 1974)
   c) Modified liquid Baird-Parker medium with 0.25 g/l glucose (Van Doorne et al. 1981)
4. Determine the time for the production of nitrite and/or depletion of glucose using dip sticks: in test 3a at 30 °C; in tests 3b and 3c at 43 °C
5. Compare with reference "scattergrams": relative response times for given foods with CFU counts and convert times to count.

It was applied to various types of foods, sampled from commercial sources. The results are presented in Figs. 4 and 5.

Generally, the levels of detection found in pure-culture studies could be substantiated. Within 6 h at 43 °C, levels of detection were of the order of 10^2–10^4 CFU g^{-1} when nitrite detection was the criterion used, but 10^4–10^5 CFU g^{-1} when glucose depletion was chosen as the parameter. These response times are certainly not higher than those observed in impedimetry (Firstenberg-Eden and Klein 1983) and do not require expensive apparatus.

From the scattergrams, a correlation coefficient between response time and log CFU g^{-1} of 0.8 and standard deviations from the straight line of about 0.5 can be calculated. Once more, this compares not at all unfavourably with the corresponding values observed in impedimetric measurements (Firstenberg-Eden and Tricarico 1983), which require expensive machines.

High levels of *Staph. aureus* appeared to reduce Enterobacteriaceae detection times. This is not surprising, because *Staph. aureus* can indeed grow weakly in media containing surface active agents (Mossel et al. 1974).

As shown by the scattergrams, response times were generally affected by the type of food under examination. Since in this procedure food constituents are eliminated by the centrifugation step, the food effect must result from physiological influences of the food on its associated microbial community structure (Mossel 1983).

Generally, therefore, results were most encouraging. They certainly warrant a further thorough evaluation of the NGT principle. Two aspects should be included in such an investigation: (1) determining whether glucose dissimilation tests are really required for the specific determinations which we elaborated, as

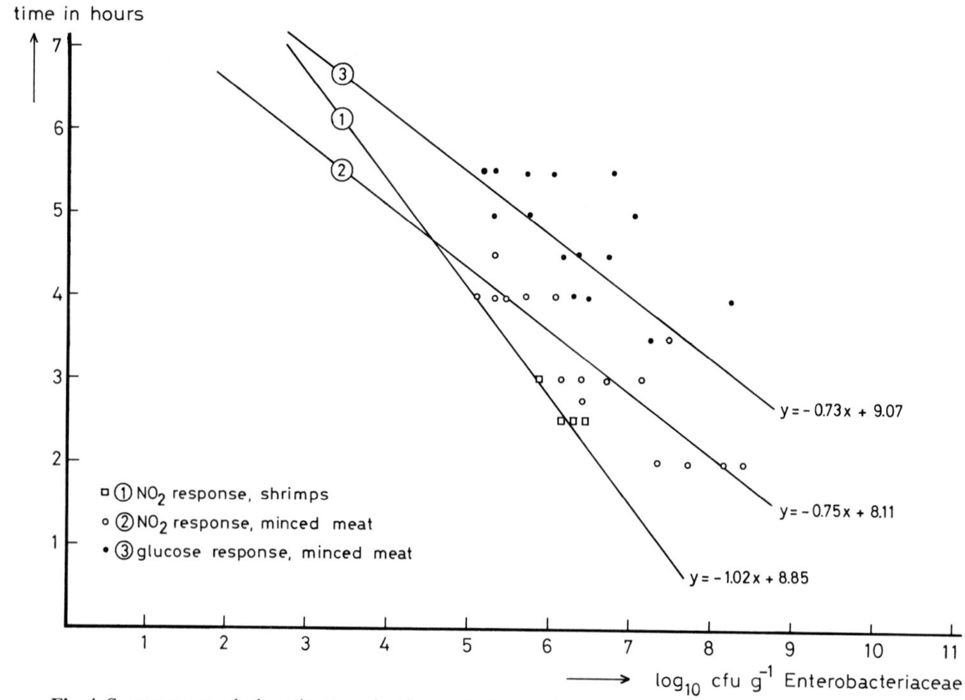

Fig. 4. Scattergram relating nitrate reduction and glucose depletion times with colony counts of *Enterobacteriaceae* in frozen minced meat and shrimps

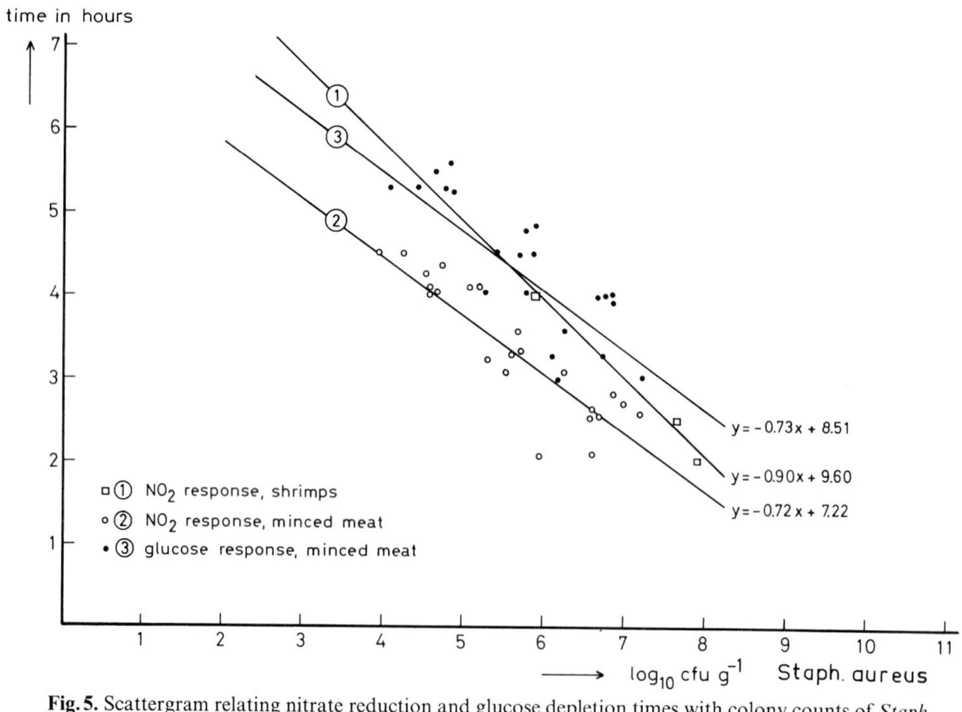

Fig. 5. Scattergram relating nitrate reduction and glucose depletion times with colony counts of *Staph. aureus* in frozen minced meat and shrimps

they are in the assessment of total colonization (Bartl et al. 1984); (2) attempts to eliminate the food effects, e.g., allowing a period of 1–2 h of liquid medium resuscitation (Mossel et al. 1980a).

We have also studied the possibilities of isolating broad taxonomic groups of bacteria, e.g., Enterobacteriaceae, or particular genera including *Salmonella, E. coli,* and *Yersinia,* from positive metabolism NGT by the use of selective subculture techniques developed earlier (Mossel et al. 1973; Mossel et al. 1974; Mossel et al. 1980b; Mossel et al. 1983a). These attempts were generally successful. This is not surprising because the response times of 6 h allow a marked enrichment of organisms sought without having to fear for overgrowth by competing bacteria (Mossel and Van Netten 1984).

Conclusions

The nitrate reduction response time technique, made specific for Enterobacteriaceae and *Staph. aureus* by the use of selective liquid media and incubation at 43 °C, was found promising for application in the monitoring of foods. Within 6 h, CFU levels of 10^3 g^{-1} could be detected. Selective subculturing procedures appeared quite successful in the isolation of particular bacteria or groups of bacteria of health significance from positive nitrate reduction tests. Previously developed identification procedures, relying on the use of various adapted polytropic tubes of the Kligler type, were used successfully for further taxonomic grouping of food isolates.

Acknowledgement. The authors are most grateful to Dr. N. V. Ashley, Research Division, Pedigree Petfoods, UK, for editorial advice and many useful scientific suggestions. They also wish to tender their gratitude to Ms. P. J. Noordijk, B. Vet. Med., and Messrs. A. van de Moosdijk, J. Moeys, and A. Valentijn for most skilful experimental assistance. It is a pleasure to thank Dr. F. Jaisli, Obipektin, Switzerland and the staff of the Pasteur Institute at Lille, France for their valuable assistance in the preparation of Table 3 and Mrs. K. A. van Hoeÿ, M. Sc., Nutricia, The Netherlands, for bringing the need for tentative classification of clostridial isolates to the authors' attention.

References

Allen LA, Pasley SM, Pierce MAF (1952) Conditions affecting the growth of Bacterium coli on bile salts media. Enumeration of this organism in polluted waters. J Gen Microbiol 7:257–267

Altherr MR, Kasweck KL (1982) In situ studies with membrane diffusion chambers of antibiotic resistance transfer in Escherichia coli. Appl Environ Microbiol 44:838–843

Bagley ST, Seidler RJ (1977) Significance of fecal coliform-positive Klebsiella. Appl Environ Microbiol 33:1141–1148

Bailey JS, Cox NA, Thomson JE (1983) Rapid procedure for biochemical characterization and serological confirmation of suspect Salmonella isolates. J Food Protection 46:764–766

Bartl V, Vanouskova S, Roubalova V (1984) Estimating the microbial quality of foods using glucose decomposition and nitrite formation tests. Acta Alimentaria 13 (to be published)

Bauman HE (1974) The HACCP concept and microbiological hazard categories. Food Technol 28:9:30–34, 74

Beckers HJ, Van Schothorst M, Van Spreekens KJA, Oosterhuis JJ (1981) Microbiological quality of frozen precooked and peeled shrimp from South-East Asia and from the North Sea. Zentralbl Bakteriol Abt I, Orig B 172:401–410

Bergan T, Hollum AB, Vangdal M (1982) Evaluation of four commercial biochemical test systems for identification of yeast. Eur J Clin Microbiol 1:217–222

Blickstad E, Molin G (1983) Carbon dioxide as a controller of the spoilage flora of pork, with special reference to temperature and sodium chloride. J Food Protection 46:756–763, 766

Börger K (1962) Untersuchungen über den Nitratreduktionstest. Milchwissenschaft 17:668–672

Borneff J (1957) Methoden und Ziele der Colitestung. Arch Hyg Bakteriol 141:505–531

Bossuyt RG, Waes GM (1983) Impedance measurements to detect postpasteurization contamination of pasteurized milk. J Food Protection 46:622–624

Buttiaux R, Mossel DAA (1957) L'analyse bactériologique des produits alimentaires périssables et conservés. Annls Inst Pasteur Lille 9:138–175

Chou SP, Dular R, Kasatiya S (1983) Effect of ferrous sulfate, sodium metabisulfite and sodium pyruvate on survival of Campylobacter jejuni. J Clin Microbiol 18:986–987

Corry JEL (1982) Quality assessment of culture media by the Miles-Misra method. In: Corry JEL (ed) Quality assurance and quality control of microbiological culture media. G-I-T, Darmstadt, pp 21–37

Cox NA, Bailey JS, Thomson JE (1983) Evaluation of five miniaturized systems for identifying Enterobacteriaceae from stock cultures and raw foods. J Food Protection 46:914–916

Cox NA, Fung DYC, Goldschmidt MC, Bailey JS, Thomson JE (1984) Selecting a miniaturized system for identification of Enterobacteriaceae. J Food Protection 47:74–77

Crawley WE, Twomey A (1969) The nitrate reductase test. NZ J Dairy Technol 4:61–66, 119–123

Dijkmann KE (1982) The optimal medium always the best choice? In: Corry JEL (ed) Quality Assurance and quality control of microbiological culture media. G-I-T, Darmstadt, pp 175–180

Edberg SC, Trepeta RW (1983) Rapid and economical identification and antimicrobial susceptibility test methodology for urinary tract pathogens. J Clin Microbiol 18:1287–1291

Edwards RA, Dainty RH, Hibbard CM (1983) The relationship of bacterial numbers and types to diamine concentration in fresh and aerobically stored beef, pork and lamb. J Food Technol 18:777–788

Entis P (1983) Enumeration of coliforms in nonfat dry milk and canned custard by hydrophobic grid membrane filter method: collaborative study. J Assoc Off Anal Chem 66:897–904

Entis P, Boleszczuk P (1983) Overnight enumeration of vibrio parahaemolyticus in seafood by hydrophobic grid membrane filtration. J Food Protection 46:783–786

Eijkman C (1904) Die Gärungsprobe bei 46° als Hilfsmittel bei der Trinkwasseruntersuchung. Centralbl Bakteriol 37:742–752

Firstenberg-Eden R, Klein CS (1983) Evaluation of a rapid impedimetric procedure for the quantitative estimation of coliforms. J Food Sci 48:1307–1311

Firstenberg-Eden R, Tricarico MK (1983) Impedimetric determination of total, mesophilic and psychrotrophic counts in raw milk. J Food Sci 48:1750–1754

Gräsbeck R, Alström T (eds) (1981) Reference values in laboratory medicine: the current state of the art. Wiley, Chichester

Hanninen M-L (1982) Effect of recovery medium on the isolation of Campylobacter jejuni before and after heat treatment. Acta Vet Scand 23:416–424

Harvey RWS, Thomson S (1953) Optimum temperature of incubation for isolation of Salmonellae. Monthl Bull Ministry of Health (London) 12:149–150

Hugh R, Leifson E (1953) The taxonomic significance of fermentative versus oxidative metabolism of carbohydrates by various Gram negative bacteria. J Bacteriol 66:24–66

Idziak ES, Mossel DAA (1980) Enumeration of vital and thermally stressed Staphylococcus aureus in foods using Baird-Parker pig plasma agar (BPP). J Appl Bacteriol 48:101–113

Kafel S, Bryan FL (1977) Effects of enrichment media and incubation conditions on isolating salmonellae from ground-meat filtrate. Appl Environ Microbiol 34:285–291

Kandler O (1961) Die Anwendung der Nitratreduktion zur Qualitätskontrolle der Rohmilch. Dtsch Molkerei Z 82:1283–1285

Kayser A, Mossel DAA (1984) Intervention sensu Wilson: The only valid approach to microbiological safety of food. Int J Food Microbiol 1:1–4

Kelly MT, Matsen JM, Morello JA, Smith PB, Tilton RC (1984) Collaborative clinical evaluation of the Autobac IDX system for identification of Gram-negative bacilli. J Clin Microbiol 19:529–533

Klein H (1955) Ein Beitrag zur Coli-Titer-Bestimmung in Milch. Berl Munch Tierarztl Wochenschr 68:58–60

Kligler IJ (1918) Modifications of culture media used in the isolation and differentiation of typhoid, dysentery and allied bacilli. J Exp Med 28:319–322

Lear DW (1962) Reproducibility of the most probable numbers technique for determining the sanitary quality of clams. Appl Microbiol 10:60–64

Leclerc H, Gavini F, Izard D, Trinel PA (1983) Les coliformes: mythe et réalité. In: Les bacilles à Gram négatif, d'intérêt médical et en santé publique: taxonomic – identification – applications. Coll Inst Natl Santé Recherch Méd 114:597–617

Lee BH, Simard RE, Laleye LC, Holley RA (1983) Microflora, sensory and exudate changes of vacuum- or nitrogen-packed veal chucks under different storage conditions. J Food Sci 48:1537–1542, 1563

Mansfield JM, Farkas G, Wieneke AA, Gilbert RJ (1983) Studies on the growth and survival of staphylococcus aureus in corned beef. J Hyg (Lond) 91:467–478

Masters PJ, Palmer GH (1981) A note on the use of the spiral plates to study the in vitro effects of amoxycillin, spectinomycin and chloramphenicol on a porcine Escherichia coli. J Appl Bacteriol 51:253–255

Monte-Verde D, Nosanchuk JS (1981) The sensitivity and specificity of nitrite testing for bacteriuria. Lab Med 12:755–757

Mossel DAA (1959) Enumeration of sulphite-reducing clostridia occurring in foods. J Sci Food Agric 10:662–669

Mossel DAA (1964) Platform tests for Enterobacteriaceae. In: Essentials of the assessment of the hygienic condition of food factories and their products. J Sci Food Agric 15:354

Mossel DAA (1980) Assessment and control of microbiological health risks presented by foods. In: Birch GG, Parker KJ (eds) Food and health. Applied Science, London, pp 129–166

Mossel DAA (1982a) Marker (index and indicator) organisms in food and drinking water. Semantics, ecology, taxonomy and enumeration. Antonie Van Leeuwenhoek 48:609–611

Mossel DAA (1982b) Microbiology of foods. The ecological essentials of assurance and assessment of safety and quality, 3rd edn. Utrecht University Press, Utrecht

Mossel DAA (1983) Essentials and perspectives of the microbial ecology of foods. In: Roberts TA, Skinner FA (eds) Food microbiology: advances and prospects. Academic Press, London, pp 1–45

Mossel DAA, Van Netten P (1984) Harmful effects of selective media on stressed micro-organisms – nature and remedies. In: Andrew MHE, Russell AD (eds) The revival of injured micro-organisms. Academic Press, London, pp 329–369

Mossel DAA, Visser M (1960) The estimation of small numbers of micro-organisms in opalescent non-alcoholic drinks by applying centrifugation. Ann Inst Pasteur Lille 11:193–202

Mossel DAA, Zwart H (1960) The rapid tentative recognition of psychrotrophic types among Enterobacteriaceae isolated from foods. J Appl Bacteriol 23:185–188

Mossel DAA, Visser M, Cornelissen AMR (1963) The examination of foods for Enterobacteriaceae using a test of the type generally adopted for the detection of Salmonellae. J Appl Bacteriol 26:444–452

Mossel DAA, Harrewijn GA, Van Sprang FJ (1973) Microbiological quality assurance for weaning formulae. In: Hobbs BC, Christian JHB (eds) The microbiological safety of food. Academic Press, London, pp 77–87

Mossel DAA, Harrewijn GA, Nesselrooy-Van Zadelhoff CFM (1974) Standardization of the selective inhibitory effect of surface active compounds used in media for the detection of Enterobacteriaceae in foods and water. Health Labor Sci 11:260–267

Mossel DAA, Eelderink I, Sutherland JP (1977) Development and use of single, polytropic diagnostic tubes for the approximate taxonomic grouping of bacteria isolated from foods, water and medicinal preparating. Zentralbl Bakteriol Abt I, Orig A 238:66–79

Mossel DAA, Veldman A, Eelderink I (1980a) Comparison of the effects of liquid medium repair and the incorporation of catalase in McConkey type media on the recovery of Enterobacteriaceae sublethally stressed by freezing. J Appl Bacteriol 49:405–419

Mossel DAA, Wijers B, Bouwer-Hertzberger S (1980b) The enumeration and identification of stressed Escherichia coli in foods using solid medium repair and Eijkman's thermotrophic criteria. J Appl Bacteriol 49:3, XVII–XVIII

Mossel DAA, Bonants-Van Laarhoven TMG, Ligtenberg-Merkus AMTh, Werdler MEB (1983a) Quality assurance of selective culture media for bacteria, moulds and yeasts: an attempt at standardisation at the international level. J Appl Bacteriol 54:313–327

Mossel DAA, Richard N, Gayral JP, Brissuel C (1983b) Etude comparative des tests préliminaires pour l'identification des bactéries isolées des matières alimentaires, des eaux et d'autres produits biologiques. Science Aliments 3:91–115

Mossel DAA, Van der Zee H, Corry JEL, Van Netten P (1984) Microbiological quality control. In: Herschdoerfer SM (ed) Quality control in the food industry. Academic Press, London, pp 79–168

Neilson AH (1978) The occurrence of aeromonads in activated sludge: isolation of Aeromonas sobria and its possible confusion with Escherichia coli. J Appl Bacteriol 44:259–264

Nichols WW, Curtis GDW, Johnston HH (1984) Detection of bacteriuria by bioluminescence effect of pre-analysis centrifugation of specimens. J Appl Bacteriol 56:247–257

Oblinger JL, Kennedy JE, Langston DM (1982) Microflora recovered from foods on violet red bile agar with and without glucose and incubated at different temperatures. J Food Protection 45:948–952

Overman TL, d'Amato RF, Tomfohrde KM (1979) Incidence of oxidase-variable strains of Aeromonas hydrophila. J Clin Microbiol 9:244–247

Park CE, Stankiewicz ZK, Lovett J, Hunt J, Francis DW (1983) Effect of temperature, duration of incubation and pH of enrichment culture on the recovery of Campylobacter jejuni from eviscerated market chickens. Can J Microbiol 29:803–806

Peltola E, Kylä-Siurola AL (1962) Nitrate reduction test in the control of milk hygiene, vol C. XVI. International Dairy Congress, København, pp 417–421

Popken AM, Bomar MT (1979) Schnellmethoden zur Bestimmung der bakteriellen Beschaffenheit von Speisen – Modelluntersuchungen. Alimenta 18:163–172

Pretorius WA (1961) Investigations on the use of the roll tube method for counting Escherichia coli I in water. J Appl Bacteriol 24:212–217

Rappaport H, Goepfert JM (1978) Thermal injury and recovery of Bacillus cereus. J Food Protection 41:533–537

Reusse U (1982) The use of the stomacher and spiral plate methods in Food Microbiology. In: Corry JEL (ed) Quality assurance and quality control of microbiological culture media. G-I-T, Darmstadt, pp 59–61

Sahlqvist B, Wallgreu K, Sjöström G (1953) Orienterande undersökningar over nitrat-reduktionsprovets anvendbarket för bestämning av coli-aerogenes-infecktion i mjölk. Svenska Mejeritidn 45:89–94

Shelton LR (1961) Frozen precooked foods – plant sanitation and microbiology. J Am Diet Assoc 38:132–134

Sheneman JM (1973) Survey of aerobic mesophilic bacteria in dehydrated onion products. J Food Sci 38:206–209

Sleigh JD (1965) Detection of bacteriuria by a modification of the nitrite test. Br Med J I:765–767

Speck ML, Ray B, Read RB (1975) Repair and enumeration of injured coliforms by a plating procedure. Appl Microbiol 29:549–550

Stannard CJ, Wood JM (1983) The rapid estimation of microbial contamination of raw meat by measurement of adenosine triphosphate (ATP). J Appl Bacteriol 55:429–438

Stuart CA, Zimmerman A, Baker M, Rustigiau R (1942) Eykman relationships of the coliform and related bacteria. J Bacteriol 43:557–572

Terplan G, Zaadhof KJ, Becker H (1982) Quality assurance of newer media for the enumeration of Staphylococcus aureus in food. Arch Lebensm Hyg 33:142–145

Uchino F, Hambali GG, Yatazawa M (1984) Nitrogen-fixing bacteria from warty lenticillate bark of a mangrove tree, Bruguiera gymnorrhiza (L.) Lamk. Applied Environ Microbiol 47:44–48

Van Doorne H, Claushuis EPM (1979) The quantitative determination of Enterobacteriaceae in pharmaceutical preparations. Int J Pharmaceutics 4:119–125

Van Doorne H, Baird RM, Hendriksz DT, Van der Kreek DM, Pauwels HP (1981) Liquid modification of Baird-Parker's medium for the selective enrichment of Staphylococcus aureus. Antonie Van Leeuwenhoek 47:267–278

Van Doorne H, Pauwels HP, Mossel DAA (1982) Selective isolation and enumeration of low numbers of Staphylococcus aureus by a procedure that relies on elevated-temperature culturing. Appl Environ Microbiol 44:1459–1462

Van Netten P, Van der Zee H, Mossel DAA (1984) A note on catalase enhanced recovery of acid injured cells of Gram negative bacteria and its consequences for the assessment of the lethality of l-lactic acid decontamination of raw meat surfaces. J Appl Bacteriol 57:169–173

Von Bockelmann I (1962) Jänoförelse mellan reduktasprovet och nitratprovet för bedömning av leverantörnjölkskvaliteten. Svenska Mejeritidn 54:587–592

Waes G, Bossuyt R (1981) A rapid method to detect postcontamination in pasteurized milk. Milchwissenschaft 36:548–552

Waterman SC (1982) The heat-sensitivity of Campylobacter jejuni in milk. J Hyg 88:529–533

Weihe JL, Seibt SL, Hatcher WS (1984) Estimation of microbial populations in frozen concentrated orange juice using automated impedance measurements. J Food Sci 49:243–245

Wilson GS (1959) Faults and fallacies in microbiology. J Gen Microbiol 21:1–15

Wood PC (1956) The enumeration of Escherichia coli I in sea water. A comparison of counts determined by the most probable number method in liquid medium with direct counts by the roll-bottle method. J Appl Bacteriol 19:26–30

Membrane Filter Techniques for Food Microbiological Analysis

A. N. SHARPE [1]

Summary

A combined membrane filter/enzyme-labeled antibody technique permits enterotoxin-producing *Staphylococcus aureus* and salmonellae to be rapidly and conveniently enumerated in foods. Brand to brand differences between membrane filters in the enumeration of fecal coliforms are eliminated by food debris deposited on them during filtration. A 1,600-point inoculator permits replication of colony patterns on hydrophobic grid-membrane filters for biochemical and immunological studies.

For many years, membrane filters (MF) have been essential to the bacteriological examination of water. Yet they have apparently received little attention by food microbiologists until recently. In one respect this is understandable: Most of us would almost certainly imagine that the average food suspension would immediately block the pores of MF. For example, I remember years ago siphoning out messy residues after trying to filter blended foods. In other respects, this lack of interest is remarkable, since MF would have many advantages over conventional agar-based food microbiology if only food suspensions were filterable. Surprisingly, it has transpired that almost all food suspensions can be made to filter nicely through MF, as numerous recent papers testify (Cousins et al. 1979; Sharpe et al. 1979; Peterkin and Sharpe 1980; Peterkin et al. 1982). The tricks are simple and few. The first step is to prepare the suspensions by the Stomacher method, which causes minimal tissue dispersal. Warming or chilling and employing a surfactant (Tween 80) are extremely effective in difficult cases, and (as a last resort) using on appropriate enzyme solution can be considered in very recalcitrant cases.

Now that the mechanics of handling food suspensions have been dealt with, we are in a position to explore how best to use the various attractions of MF. The obvious question is – what can MF do for the microbiology of foods? But there is also an intriguing inversion of this – what can foods do for microbiology on MF? I shall describe an example of progress in both of these aspects.

As far as the first is concerned, MF can:

1. Remove inhibitors derived from the food that might – in ordinary plate or MPN assays – cause low counts (e.g., acids, tannins, antibiotics). They can also remove food ingredients that might compromise a selective medium (e.g., sucrose introduced into coliform media).

1 Bureau of Microbial Hazards, Health Protection Branch, Health and Welfare Canada, Tunney's Pasture, Ottawa, Ontario, Canada

2. Allow damaged organisms to be resuscitated on a benign medium, transferred to selective medium, and still be counted.
3. Permit – in a remarkably large number of cases – the limit of detection to be improved by applying a larger inoculum than would normally be possible in plates or tubes.
4. Permit accurate counts from a single dilution over a wide concentration range. (This is only a property of the hydrophobic grid MF.)
5. Permit staining procedures on colonies (e.g., for indole) that are not feasible on plates.

These are powerful advantages. I want to concentrate on the demonstration of staphylococcal enterotoxin in colonies, both as an example of point 5 and to indicate the potential and problems of MF techniques in this area.

Notermans describes elsewhere in this volume the analysis of food for preformed staphylococcal enterotoxin (SE) by means of ELISA. We studied a different question – can MF yield a test for viable *Staphylococcus aureus,* superior in any way to the conventional plating/coagulase or thermostable nuclease (TNase) method?

For quality assurance, acceptable limits of *S. aureus* are currently finite but small, in expectation that microbial growth will not normally generate an unacceptable toxin level before the food is consumed. It is unlikely that direct ELISA on foods is sensitive enough to detect regulatory levels of a few organisms per gram.

While the threat from *S. aureus* comes from toxin production, not all strains of the organism are dangerous. This seems to have led to ambivalence among QA and regulatory microbiologists. On the one hand, the coagulase and TNase tests were developed as the best possible substitutes for the more difficult test of enterotoxigenesis. So we might expect that a simple MF test which directly confirms SE production by colonies and permits the organism to be enumerated would be most welcome. On the other hand, we have been puzzled by the cool reception of a test demonstrating SE producers as eminently countable purple spots. The reaction seems to be that any evidence for the presence of *S. aureus* (enterotoxigenic or otherwise) should be ample warning of lax manufacturing practice predisposing to a spread of the organism, i.e., negative results from a test detecting only SE producers may lead to a false sense of security. Yet this view does not seem to us to be in accordance with the widespread use of coagulase and TNase tests.

Perhaps one can have it both ways. For example, the following test (Peterkin and Sharpe 1984) was developed with SEB producers. We expect to confirm its value for other toxins when we receive cross-reacting antibody, since its target basically depends on the antibody employed. However, if nonenterotoxigenic *S. aureus* is also a concern and the mere ability to form typical colonies on Baird-Parker agar is not deemed evidence enough, our test can be easily adapted. Protein A is an excellent indicator of *S. aureus* and may be detected in this test simply by substituting an "incompetent" IgG for the anti-SE.

To test for enterotoxin production by colonies one needs suitable enzyme labeled antibody. Horseradish peroxidase (HRP) is an appropriate enzyme. Our early work was done with IgM monoclonal antibodies supplied by MA Gidney of the National Research Council of Canada. Though relatively free of interference from protein A, they gave weak reactions. Commercial polyclonal anti-SEB (Sigma) gave a stronger reaction. However, it is an IgG and its Fc portion causes it to react nonspecifically with protein A.

This is fine if one is interested simply in *S. aureus* but a decided problem if one is trying to enumerate only toxin producers, since SE negative colonies also stain. Digesting away the Fc portion of the antibody is one solution. However, a neater trick is to conjugate the enzyme, not to antibody, but to pure protein A (Dubois-Dalcq et al. 1977). This labeled protein A is a most useful reagent since it can be prepared in bulk and used to label almost any IgG quickly and simply. In so doing, it automatically prevents further nonspecific binding to protein A.

The test currently works best when we make a "blot" of the original MF onto the clean surface of a nitrocellulose membrane. The blot technique is essential when using polysulfone MF or HGMF because this material has no natural affinity for SE. However, this technique is preferable even for growth on mixed cellulose ester MF, which adsorb SE quite strongly. One simply lays the clean membrane on top of the colonies for a short while or rolls it over them using a cylindrical tool. Portions of the colonies adhere to the new membrane, in a mirror image of the original growth. We are not sure just why this is the better way, but we suspect that adsorption of products from colonies during the growth period may block adsorption of SE.

The remainder of the test involves reaction first with blocking agent (gelatin or serum albumin) to prevent further adsorption, then labeled antibody, and finally substrate (4-chloro-1-naphthol), these steps being interspersed by appropriate rinses. SE-containing colonies (or rather, areas of the membrane which were in contact with these) stain as insoluble purple spots.

I would like to be able to say that we carry out the various reactions and rinses using the most obviously efficient technique, which would combine the MF on a filter funnel and a vacuum. To date, this has given less well-defined images than the more tedious approach of swirling the MF gently in a Petri dish of fluid for several minutes at each stage. Again, we are not sure of the reason, but are investigating it.

If rows of various *S. aureus* colonies are grown on HGMF, the picture is typical of the response of the technique. One can see a strong response for SEB producers, no response for an atoxigenic but strong protein-A producer, and zero or very faint response for strains classified only as SEA, C, D, or E producers.

These were made using pure cultures. We have also used the technique successfully to recover (100%) an SEB producer inoculated into foods. Faced with massive competing microflora, one is forced to use selective growth media. Those based on salt allow *S. aureus* to grow, but also inhibit SE production. Baird-Parker (BP) medium allows good enterotoxin production but, of course, yields black colonies; their black residues on the MF tend to mask purple positives. Remarkably, SEB still expresses its antigenicity after dilute hypochlorite treatment. Thus, it seems that the black tellurite from BP medium can be oxidized away by an additional step without affecting the count of toxin producers.

Farber of the Health Protection Branch has been studying a similar test, using labeled Spicer-Edwards antisera to detect *Salmonella*. In cases where the suspension is sufficiently filterable, it seems possible to dispense with enrichment. For example, he obtained MF washings from poultry and incubated them directly overnight on bismuth sulfite and BGS agars. With this technique, it was possible within 1 day both to confirm the existence of salmonellae in birds and estimate the level of contamination.

It can be seen that we have been using HGMF for enumeration of *S. aureus*. In fact, the enterotoxin test was developed for the HGMF. I should like to digress slightly and describe a successful replicator system, which we expect will play a considerable role in the development of HGMF techniques. The main problem here was how to provide 1,600 clean and sterile points – one for each HGMF grid-cell – routinely, reliably, and inexpensively. Our device contains a roll of sterile film. A small section of this is molded temporarily into 1,600 points just before the replications are performed. Immediately after, the contamination is safely wound onto and confined by a take-up roll. The prototype machine is still being tested; the experimental model allowed us to pull at least 20 perfect replicates from a master, without cross- or aerial contamination.

The replicator has already proven invaluable for both the *S. aureus* and *Salmonella* studies. It is convenient, for example, to refrigerate a few dozen unincubated replicates. These may be put into the incubator the day before use, so that a predetermined pattern of known strains is at once available for experiment.

I believe that, for periods of a few weeks at least, this system is ideal for maintaining large numbers of cultures used in comparative studies. For example, with only every other row or column position filled (a very safe situation), 400 cultures could be maintained for daily use. Replicating such a plate takes seconds.

To illustrate the potential of the HGMF in this regard, the final pattern on the master HGMF could be built up from a number of contributing subpatterns. for example, each participant in a multilaboratory study could be assigned its own portion on the grid of an HGMF for isolates, leaving the remainder of the HGMF area blank. At the central laboratory, all these patterns could be combined onto one HGMF by the replicator, for further testing of biochemical reactions, etc.

The second question I want to cover is the way food debris trapped on the MF surface after filtration can modify its performance. Our study was prompted by criticism. A few introspective years of MF brought forth as many papers (Brodsky and Schiemann 1975; Dufour and Cabelli 1977; Green et al. 1975; Sladek et al. 1975; Standridge 1976; Tobin and Dutka 1977; Tobin et al. 1980), demonstrating that, at least where waterborne fecal coliforms are concerned, all MF are not equally effective.

While there is virtually no alternative to MF in water bacteriology, this is not true with foods. One can justifiably argue that plate count reproducibility is bad enough already, and that adding to this the possibility of an MF interbrand recovery variation of up to 1,000% is poor bedrock on which to develop a new food microbiology. To which we have always replied that: (1) We have never observed recovery variations with foodborne organisms attributable to MF brand, and (2) in any case, poor recovery by MF is thought to relate to surface pore structure and the degree of protection against desiccation and other stresses afforded to organisms on their surfaces. However, one would expect the food debris to surround the bacteria and insulate them from such surface stresses.

Eventually, such a means of defense must be tested. Purees of seven representative foods were inoculated with a "sensitive" strain of *Escherichia coli*, filtered through a less than optimal batch of MF, and plated. Food mats of varying thicknesses were used, and spread plates were simultaneously inoculated. Table 1

Table 1. Effect of food debris on counts of *E. coli* on Oxoid Nuflow MF[a, b]

Food	Weight of food (g)	Thickness of foot mat (μm)	*E. coli* count (%)[h]	Spread plate count (%)[b, h]
Carrots	0.0	0.0	100[a] (6.9)[e, f]	177
	0.001	0.06	135[b] (8.0)	
	0.01	0.6	167[c] (6.4)	
	0.1	5.5	185[d] (3.8)	
Cellulose	0.0	0.0	100[a] (8.9)	108
	0.001	1.2	102[a] (7.8)	
	0.01	12.3	–[g]	
	0.1	123.0	28[b] (18.1)	
Cheese	0.0	0.0	100[a] (10.6)	115
	0.001	0.14	115[a, b] (14.2)	
	0.01	1.4	122[b, c] (8.0)	
	0.1	14.0	136[c] (5.4)	
Green beans	0.0	0.0	100[a] (4.1)	108
	0.001	0.09	106[a, b] (6.7)	
	0.01	0.9	120[c, d] (1.9)	
	0.1	8.5	113[b, d] (2.9)	
Peach	0.0	0.0	100[a] (7.7)	129
	0.001	0.04	102[a] (13.8)	
	0.01	0.4	129[b] (7.5)	
	0.1	3.6	150[c] (3.6)	
Pear[c]	0.0	0.0	100[a] (3.8)	113
	0.001	0.06	109[a, b] (10.3)	
	0.01	0.6	122[b] (4.6)	
	0.1	5.6	115[b] (10.1)	
Tuna	0.0	0.0	100[a] (5.7)	144
	0.001	0.17	115[a] (11.1)	
	0.01	1.7	137[b] (8.5)	
	0.1	17.0	–[g]	
Turkey[d]	0.0	0.0	100[a] (6.5)	123
	0.001	0.09	98[a] (7.3)	
	0.01	0.9	135[b] (8.9)	
	0.1	9.0	–[g]	

[a] Diameter of filtration area = 76.9 mm
[b] m-FC agar (24 h; 35 °C)
[c] Too lumpy to measure without prefiltering
[d] Measured with a settled suspension
[e] Values for the same food bearing a common superscript letter are not different ($p > 0.05$)
[f] Coefficient of variation
[g] Plates uncountable as a result of colony-spreading
[h] Percentage of *E. coli* counts on MF containing no food mats

(Farber and Sharpe 1984) shows that all foods protected the organism, even in layers nominally thinner than single bacterial cells. A mat 1 μm or so thick (such as might be obtained by filtering 1–3 ml of the third decimal dilution onto a 47-mm MF) completely protected the organism, in that the MF count equalled or exceeded that of the spread plate. Scanning electron microscopy showed the bacteria nestled in crevices between fibers, just as if they were in the large protective pore openings of those MF tailored for fecal coliforms.

The MF needs to be retentive enough to trap bacteria, and it must be porous enough to supply nutrients from beneath. But I think it is safe to say that food microbiology using MF will be essentially free of the brand to brand variability we see in water bacteriology. For the average specimen, the microbes can be regarded as growing in a thin food matrix (which just happens to be on top of an MF), rather than growing on an MF.

This brings me to the topic of what foods can do for microbiology on MF. Whenever relatively clear fluids such as natural waters and carbonated beverages are filtered, bacteria may constitute the bulk of the material deposited on the MF. Such samples are, therefore, open to the vagaries of brand-dependent recovery. Of course, one may experiment to find the right brand and stick to it henceforth, hoping that it remains a good performer. But a simple treatment to equalize MF performances would be more satisfactory.

Table 2 shows how a thin layer (3–5 μm) of blended canned carrot protected fecal coliforms extracted from the Ottawa River and brought Sartorius and Millipore HA filters into line with Millipore HC counts (Farber et al. 1985). The procedure is very simple: 0.03–0.05 g of blended sterile carrot are added to the filtrant, for a 47-mm circle. Carrot is good in that it blends easily and is colorless

Table 2. Effect of carrot debris on counts of fecal coliforms on various MF[a]

Experiment number	Filter brand	Weight of carrot (g)	Thickness of food mat (μm)	Fecal coliform count[b, c]
1	Sartorius	0.0	0.0	33 ± 3.2
	Millipore HC	0.0	0.0	47 ± 7.3
	Sartorius	0.03	5.5	51 ± 8.2
	Millipore HC	0.03	5.5	55 ± 6.5
2	Millipore HA	0.0	0.0	27 ± 4.8
	Millipore HC	0.0	0.0	42 ± 7.2
	Millipore HA	0.03	5.5	36 ± 3.8
	Millipore HC	0.03	5.5	43 ± 7.5
3	Millipore HA	0.0	0.0	22 ± 5.4
	Millipore HC	0.0	0.0	36 ± 4.8
	Millipore HA	0.045	8.5	35 ± 6.1
	Millipore HC	0.045	8.5	37 ± 4.1

[a] Diameter of filtration area = 41.0 mm
[b] m-FC agar (24 h; 44.5 °C)
[c] Average ± standard deviations of ten replicates

in thin layers, but other foods would probably work as well. Workers with natural waters might find this a useful procedure.

Thus, we see that the overall process works both ways. MF can be useful in food microbiology, and foods can also be useful in MF microbiology.

References

Brodsky MH, Schiemann DA (1975) Influence of coliform source on evaluation of membrane filters. Appl Microbiol 30:727–730

Cousins CM, Pettipher GL, McKinnon CH, Mansell R (1979) A rapid method for counting bacteria in raw milk. Dairy Ind Int 44:27–39

Dubois-Dalcq M, McFarland H, McFarlin D (1977) Protein A-peroxidase: a valuable tool for the localization of antigens. J Histochem Cytochem 25:1201–1206

Dufour AP, Cabelli VJ (1977) Comparison of membrane filter brands for recovery of the coliform group. In: Bordner RH, Frith CF, Winter JA (eds) Proceedings of the symposium on the recovery of indicator organisms employing membrane filters, Report EPA-600/9-77-024. Environmental Protection Agency, Cincinnati, Ohio, pp 73–81

Farber JM, Sharpe AN (1984) Improved bacterial recovery by membrane filters in the presence of food debris. Appl Environ Microbiol 48:441–443

Farber JM, Sharpe AN, Kalab M (1985) Improved recovery of fecal coliforms from water by membrane filters in the presence of food debris. Can J Microbiol 31:16–18

Green BL, Clousen E, Litsky W (1975) Comparison of the Millipore HC with conventional membrane filters for the enumeration of fecal coliform bacteria. Appl Microbiol 30:697–699

Peterkin PI, Sharpe AN (1980) Membrane filtration of dairy products for microbiological analysis. Appl Environ Microbiol 39:1138–1143

Peterkin PI, Sharpe AN (1984) Rapid enumeration of *Staphylococcus aureus* in foods by direct demonstration of enterotoxigenic colonies on membrane filters using enzyme immunoassay. Appl Environ Microbial 40:1047–1053

Peterkin PI, Sharpe AN, Warburton DW (1982) Inexpensive treatment of frozen dairy products for membrane filtration. Appl Environ Microbiol 43:486–487

Sharpe AN, Peterkin PI, Dudas I (1979) Membrane filtration of food suspensions. Appl Environ Microbiol 37:21–35

Sladek KJ, Suslavich RV, Sohn BI, Dawson FW (1975) Optimum membrane structures for growth of coliform and fecal coliform organisms. Appl Microbiol 30:685–691

Tobin RS, Dutka BJ (1977) Comparison of the surface structure, metal binding and fecal coliform recoveries of nine membrane filters. Appl Microbiol 34:69–79

Tobin RS, Lomax P, Kushner DJ (1980) Comparison of nine brands of membrane filter and the most-probable-number methods for total coliform enumeration in sewage-contaminated drinking water. Appl Environ Microbiol 40:186–191

Status and Future of the Direct Epifluorescent Filter Technique

G. L. Pettipher [1]

The Direct Epifluorescent Filter Technique

General Principles

The direct epifluorescent filter technique (DEFT) was originally developed for counting bacteria in raw milk at the National Institute for Research in Dairying, Reading (Pettipher et al. 1980; Pettipher 1983). The method, which uses membrane filtration and epifluorescence microscopy, takes less than 30 min to complete and does not suffer from many of the disadvantages of other microscopic methods. Pretreatment of the milk with a proteolytic enzyme and surfactant at 50 °C lyses somatic cells and makes fat globules sufficiently fluid to enable 2 ml of milk to be filtered routinely through a 25-mm diameter membrane filter. Filtration concentrates and distributes the bacteria in a manner that makes counting easier and the techniques ca. 100 times more sensitive than the Breed smear. The use of a fluorescent stain and an epifluorescence microscope produces well-stained micro-organisms that are easily distinguishable from the small amounts of fluorescent debris. A polycarbonate membrane filter is used, the flat surface of which is better suited to microscopy than the uneven surface of cellulose acetate membranes. Micro-organisms on DEFT slides can be counted automatically, thereby considerably reducing operator fatigue. DEFT can be used rapidly to enumerate micro-organisms in urines, beverages and, with additional sample pretreatment, a variety of foods, in addition to milk.

Apparatus and Reagents

The main items of apparatus required for the DEFT include an epifluorescence microscope, filter manifold, filter towers, filters, vacuum pump, water bath, pipettes, containers for reagents, microscope slides and coverslips. The reagents needed for the DEFT include the enzyme trypsin, the surfactant Triton X-100, the fluorochrome acridine orange, citric acid-NaOH buffers, isopropanol and non-fluorescent immersion oil. Before use all reagents, except immersion oil, are prefiltered through 0.22-μm pore size cellulose acetate membrane filters to remove bacteria and particulate matter and dispensed into sterile containers. DEFT

1 Cadbury Schweppes PLC, Group Research, The Lord Zuckerman Research Centre, The University, Whiteknights, Reading, RG6 2LA, Great Britain

reagents are marketed in kit form by Foss Electric UK Ltd. (The Chantry, York, UK).

Method

Whole milk, enzyme and surfactant are incubated at 50 °C for 10 min in a sterile capped test tube. A polycarbonate membrane filter (Nuclepore, 0.6-µm pore size) is mounted, shiny side uppermost, in a membrane filtration unit of a filter manifold connected to a vacuum line. The filter is warmed by filtration of surfactant at 50 °C before the treated milk is filtered. Surfactant at 50 °C is used to rinse the test tube and filter tower after filtration of the treated milk. After filtration of the treated milk and rinse, the membrane is overlaid with stain. After 2 min the vacuum is reapplied, and the membrane is rinsed with buffer followed by isopropanol. The stained membrane is air-dried and then mounted in non-fluorescent immersion oil on a slide beneath a cover slip.

The mounted membrane filter is then examined by means of an epifluorescence microscope suitable for use with acridine orange. In the DEFT, because of the higher RNA content, metabolically active bacteria fluoresce orange whereas inactive bacteria fluoresce green (Fig. 1). Clumps of orange fluorescing bacteria are counted in microscope fields of view taken at random. Depending on the number of clumps per field, bacteria in 2–15 fields are counted (Pettipher et al. 1980). The DEFT count per millilitre is estimated from the mean clump count per field and the microscope factor (MF). The MF depends upon the amount of sample filtered through the given area of the membrane and the area of the microscope field. The MF is calculated as follows:

$$\mathrm{MF} = \frac{\text{Area of menbrane through which sample is filtered (mm}^2\text{)}}{\text{Microscope field area (mm}^2\text{)} \times \text{sample volume (ml)}} \,.$$

The MF can vary widely as it is influenced by sample volume, filter area, magnification of objective and eyepieces and type of objective used, e.g. standard or wide field. For a sample volume of 2 ml the MF will usually range from 2,000–11,000.

In addition to the count per millilitre of sample, a presumptive identification of the types of micro-organisms present on DEFT slides can be made by the operator. Generally the following types of bacteria are discernible by morphology; streptococci, micrococci, staphylococci, and some bacilli and coryneforms. Fungi, yeasts and spores can also be distinguished. A tentative identification of the types of micro-organisms present in the sample may be useful in locating their source, e.g. a high bacterial count in refrigerated farm milk where the majority are streptococci is usually indicative of mastitis in the herd. A full description of the method and staining reactions can be found in a monograph on the DEFT (Pettipher 1983).

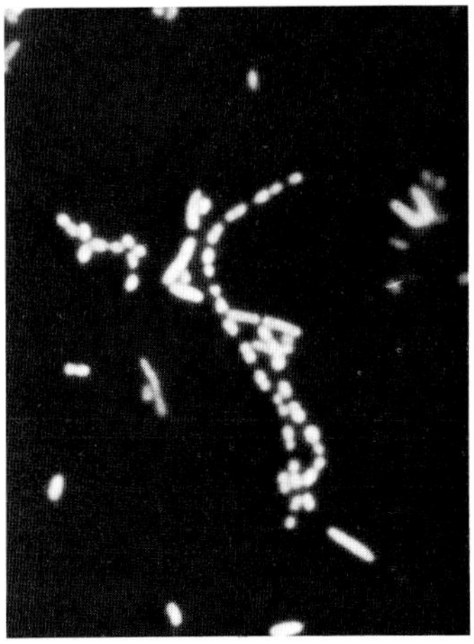

Fig. 1. Bacterial cells as seen in the DEFT

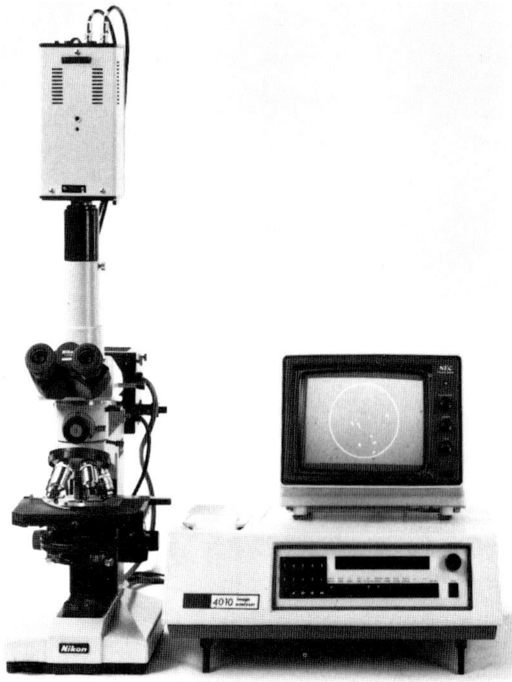

Fig. 2. Apparatus for the semi-automated counting of micro-organisms on DEFT slides

Automated Counting

The manual DEFT count is only suitable for 30–40 samples per day per operator as prolonged use of the microscope results in operator fatigue. Micro-organisms on DEFT slides can be counted automatically using the 40–10 image analyser (Analytical Measuring Systems Ltd., Shirehill, Saffron Walden, Essex, UK). A closed circuit TV camera (scanner) attached to the microscope camera port detects the microscopic image, and the resulting video signal is analysed to obtain quantitative data about the image (Fig. 2). Using the image analyser an operator can count ca. 50 slides per hour. The operator has to load the slide, focus the microscope and move the microscope stage manually. The image analyser can be controlled by a microcomputer which processes the generated data. The relevant information, e.g. sample code, count per millilitre and grade, is printed out and can be recorded simultaneously onto magnetic disc for storage purposes.

The semi-automated count of bacteria on DEFT slides agrees well with the corresponding visual DEFT count with a correlation of 0.94 (Pettipher and Rodrigues 1982a). For low count milks the semi-automated count exceeds the visual count but this difference becomes proportionally less as the count increases; there is close agreement over the range 1×10^5–5×10^6/ml. The higher semi-automated count is probably due to the inclusion of particles such as sediment and the few somatic cell nuclei which remain undigested by the pre-treatment. These particles are easily distinguishable from bacteria by eye and therefore would not be included in the visual DEFT count.

The counting system could be fully automated by including a cassette loading system for slides and focusing under microcomputer control. The technology for these modifications is already available (Ploem et al. 1979; Thomason et al. 1975). Automation of DEFT preparations (which would involved sampling, pre-treatment, filtration, staining, and rinsing stages) combined with automated counting would provide a system capable of analysing hundreds of samples per day.

Cost of Equipment and Disposable Items

The cost of equipping a laboratory for the DEFT is approximately £4,000 or $6,000. The major capital cost is the epifluorescence microscope which retails for ca. £3,200. It is possible that a dedicated epifluorescence microscope may be produced in the future which should cost only one-third of this amount. The other expensive items of equipment are the filter manifold and waterbath. The cost of the instrumentation for semi-automated counting is ca. £6,000. This includes the TV image analyser system and the microcomputer which, although not essential, greatly facilitates data logging, record keeping and report writing.

The current cost of disposable items used in the DEFT is ca. 50 pence or $0.70 per test. Of this, 30 pence is for the membrane filter and 10 pence for the enzyme. The microscope slide and coverslip cost 6 pence and the chemicals 4 pence. Savings may be possible by buying the membrane filters or reagents in bulk. For comparison, the cost of disposable items for a two dilution plate count is ca. 20 pence. Detailed costings for the DEFT based on labour, consumables and capital are given by Pettipher (1983).

Application of the DEFT

Raw Milk

The DEFT count correlates well with the plate colony count for farm, tanker, and silo milks, with an overall correlation coefficient of 0.91 (Table 1). Agreement between the two methods is close at the level of 10^4–10^5 bacteria per millilitre, but where counts exceed 10^6 per millilitre the plate colony count is higher than the DEFT count. This is possibly due to the increased break up of clumps during the

Table 1. Relationship between $\log_{10}$ DEFT count per ml (y) and $\log_{10}$ plate colony count per ml (x) for milk, meat and fish, and urine

Material	a	b	r
Raw milk	0.89	0.80	0.91
Fresh meat and fish	−0.41	1.00	0.91
Frozen meat and fish	1.31	0.74	0.87
Urine	0.02	0.88	0.85

$y = a + bx$; r, correlation coefficient

preparation of dilutions in plate colony counts for milks containing high numbers of bacteria.

Comparison of DEFT counts and plate colony counts obtained by different operators on fresh and stored milks shows that there is agreement between methods and between operators over a wide range of counts (Pettipher et al. 1980). For both methods the coefficient of variation (CV) between triplicate log counts within samples was $\pm 1.6\%$. The CV between mean log counts obtained by different operators was $\pm 2.2\%$ for the DEFT count and $\pm 2.6\%$ for the plate colony count. Repeatability of the DEFT count seems to be similar to that of the plate colony count.

The DEFT count, which takes less than 30 min to complete and is suitable for milks containing 5×10^3–10^8 bacteria per millilitre, is both sufficiently rapid for monitoring tanker and silo milk and sensitive enough for grading farm milk on the basis of bacteriological quality. Its use should permit better control and management of supplies, reducing the risk of poor quality products resulting from a high bacterial count in the raw material.

Foods

For stomachered suspensions of most foods tested, enumeration of micro-organisms in the DEFT proved difficult because of the presence of food debris. Pre-filtration of the suspension through nylon filters (mesh size 5 μm) removed most of the food debris and only slightly reduced the recovery of micro-organisms, as determined by the plate colony count (Pettipher and Rodrigues 1982b). The recovery rates fell as the microbial content of the food increased, e.g. approximately 100% at 10^4 per gram, 85% at 10^6 per gram, 70% at 10^8 per gram and 60% at 10^{10} per gram. The loss of micro-organisms is unlikely to substantially affect the count of micro-organisms in the food.

For milk, 0.6-μm pore size polycarbonate membrane filters are used in the DEFT but for some "cleaner" liquids, such as certain pre filtered food suspen sion, beverages and water, a proportion of the bacteria (up to 25%) may enter the pores of these membranes. If a sufficient volume of sample can be filtered to achieve the required sensitivity, it is recommended that a smaller pore size (0.4 or 0.2 μm) membrane filter is used for these products.

If Tween 80 is used as the surfactant for the pre-treatment of samples in the DEFT, the micro-organisms appear larger than if Triton X-100 is used. These larger organisms fluoresce more brightly and are more easily counted. As Tween 80 is relatively ineffective at lysing somatic cells, even in the presence of trypsin, Tween 80 should not be used as the surfactant for products containing large numbers of somatic cells. However, for other products where somatic cells are few or absent, e.g. butter and pre-filtered meat suspensions, the use of Tween 80 as the surfactant in the DEFT in place of Triton X-100 may result in better quality microscope preparations.

Depending upon the food, 4 to 15 ml of stomachered food suspensions (10 g food plus 90 ml diluent) could be filtered through a single nylon pre-filter and 3–10 ml of these pre-filtered suspensions could be filtered in the DEFT. A sample size of 2 ml was routinely used in the DEFT, together with a microscope factor of 57,500.

For fresh meat and fish, the pre-filtered DEFT count agreed well with the plate colony count over the range 10^4–10^{10} per gram, with a correlation coefficient of 0.91 (Table 1). For counts on frozen meat and frozen fish the prefiltered DEFT count also agreed well with the plate colony count over the range 5×10^4–5×10^7 per gram (Table 1). Of the other foods tested, there was good agreement between the pre-filtered DEFT count and plate colony count for cooked meats, cream doughnut and spices (Pettipher and Rodrigues 1982). The coefficient of variation between triplicate log DEFT counts on food suspension was $< \pm 1\%$, lower than that obtained for milk.

Recent results from a laboratory in Denmark show good agreement between the DEFT count and plate colony count for minced pork, beef, and bacon, various types of fish, mayonnaise salads, and milk (Boisen 1983). For minced beef, the manual DEFT count and automated DEFT counts had coefficients of variation of 10.2% and 7.9% respectively. The repeatability of automated counts on a single DEFT slide was $\pm 6.1\%$.

Urine

Most hospital pathology laboratories routinely culture all the urine specimens received for analysis. Generally $>80\%$ of these specimens are from patients without urinary tract infections and therefore bacteriologically "negative." There is a need for a rapid screening method for urines to reduce unnecessary microbiological analysis time.

In the DEFT, 2 ml of urine from non-infected patients will filter without pretreatment, but only 2 ml of a 1 in 100 dilution of heavily infected urine will filter routinely. As a general rule, if the urine specimen appears visually clear then it can be filtered undiluted, whereas if it appears cloudy it will only filter if diluted. Bacteria and somatic cells are clearly visible on DEFT slides prepared from infected urines. Early results (Pettipher 1983) show that there is a reasonable relationship between the DEFT count and plate colony count of bacteria in urine, with a correlation coefficient of 0.85 (Table 1).

The DEFT count on urine takes less than 10 min to complete as no pre-filtration or pre-treatment with enzyme and surfactant is required. It seems likely that the technique can be used rapidly to enumerate micro-organisms and somatic cells in urine, but to be of practical use in a hospital pathology laboratory the DEFT would first need to be fully automated. DEFT preparations from samples giving a high automated count would probably need to be visually assessed to see if the high count was due to micro-organisms or somatic cells; high counts of either are indicative of urinary tract infections.

Beverages

Methods using the same principles as the DEFT are used to estimate viable and non-viable yeast and bacteria in wines and on winery equipment surfaces (Cootes and Johnson 1980). The method for wine takes only 15 min to complete and has a sensitivity of less than 1 micro-organism per millilitre.

Water

The technique of membrane filtration and epifluorescence microscopy has frequently been used to enumerate bacteria in natural waters. Ignoring small variations, there is a consensus regarding the best method for enumerating bacteria in water. A 0.2-µm pore size polycarbonate membrane filter is used for filtration, the fluorochrome acridine orange is used for staining the bacteria and an epifluorescence microscope is used for enumerating the bacteria. Counts of bacteria in natural waters obtained by these techniques generally agree with counts obtained by other methods (Jones and Simons 1975; Watson et al. 1977).

Potential Applications

In addition to the applications of the DEFT previously described there are others which have not yet been fully evaluated. The method can be used rapidly to enumerate micro-organisms in broth cultures as a general laboratory counting technique. The method has been used successfully to enumerate bacteria in rumen and digesta samples, dental plaque, fruit mixes, shampoos, and machine cutting oils. Recently the DEFT has been used to predict the likely keeping quality of pasteurised milk and cream within 24 h (Rodrigues and Pettipher 1984; Griffiths et al. 1984).

Present State of Acceptance and Likely Future of the DEFT

For any new technique to gain acceptance it is advantageous if the results are similar to or correlate with those of a standard method. The DEFT count based on the number of orange fluorescing clumps of bacteria meets this requirement for raw milk, raw foods, and urines as it is in close agreement with the plate colony count. In the case of milk, these clumps may contain from 1 to 50 bacteria depending on the age of the milk. The bacteriological quality of milk and other foods is more closely related to the total number of viable bacteria rather than the number of viable clumps, as determined by the plate colony count. Once the technique has gained acceptance it is possible that a DEFT count based on the number of individual orange fluorescing cells may be preferred. This should be a more realistic measure of the bacteriological status of the product.

It is very important that any new technique should perform well in each laboratory that attempts it, otherwise its credibility may be adversely affected. The DEFT has now been introduced successfully into numerous laboratories. Where requested, assistance was given and training courses were held for prospective users. Only minor problems were encountered with the introduction of the technique, and these were quickly resolved. Operators in laboratories other than the

NIRD have obtained good correlations between the DEFT count and plate colony count. In a collaborative trial, laboratories generally achieved a correlation coefficient of >0.9 for raw milk (Pettipher et al. 1983). Independently, operators in Australia using minor modifications to the pre-treatment stages obtained a correlation coefficient of 0.95 between the DEFT count and plate colony count (Beck and Hehir 1982).

The DEFT is now operating satisfactorily in more than 40 laboratories which are situated in the UK, Eire, Germany, France, The Netherlands, Finland, Denmark, Sweden, and Australia. Of these laboratories, more than 25 use semi-automated counting. It has been estimated that approximately 50,000 DEFT counts are made per annum. The Jersey Milk Marketing Board was quick to see the advantages of the DEFT and has been using the semi-automated DEFT count to assess the bacteriological quality of farm milk for payment purposes since April 1982 (Anon 1982). Since its introduction, a considerable improvement in the bacteriological quality of the incoming milk has been achieved. In addition, the increased speed of testing has benefited the advisory service, and the use of microcomputers has greatly improved both the data storage and reporting. Over 20 laboratories at the larger creameries in the UK now have equipment for semi-automated counting. The DEFT has been included in the draft of the revised version of British Standard 4285 "Methods of Microbiological Analysis for Dairy Products" which should be published within the next year or two. This will give the method a "seal of approval" in the UK and it is hoped that other countries will include it in their standard methods. Currently the DEFT is being used mainly for raw milk but its use for other foods, most notably meat and fish, is increasing. No other method can give an accurate, sensitive count of bacteria in these foods in as short a time as the DEFT.

It is likely that, over the next few years, the DEFT will become progressively automated and miniaturised. Many potential users, e.g. centralised testing laboratories, cannot consider using the technique until it is fully automated. Preliminary results have shown that the volumes of sample and reagents and the membrane filter area utilised can be reduced by a factor of 10 without affecting the accuracy of the count. Miniaturisation will result in considerable savings in costs of reagents and membrane filters. The development and marketing of a reagent dispenser and, eventually, an automated filtration and staining device would be of considerable benefit to DEFT users. It will probably be a few years before an automated sampler and pre-treatment unit and a fully automated counting unit are available. The speed of these developments depends largely upon the potential market for such units and availability of the necessary funds. The recent application of DEFT to urines has considerably increased the potential market for automated units, and this should aid the raising of the money required to automate the method.

The acceptibility and the use of DEFT by microbiologists is increasing and the method is becoming widely used in the milk, food, and beverage industries. Automation of the technique is likely to speed its introduction, but it should not be forgotten that no automation is required if only a small number of samples (30–40) are analysed daily. A fully automated DEFT instrument has the potential to become the future method of choice for the bacteriological analysis of milk samples in central testing schemes and for screening urine samples in hospital pathology laboratories. At present the future of the DEFT looks very promising.

References

Anon (1982) Le "Jersey Milk Marketing Board" adopte la methode par epifluorescence pour l'analyse bacteriologique du lait en vue de paiment à la qualite. Tech Lait 967:29–31

Beck CG, Hehir AF (1982) The rapid fluorescence method for counting bacteria in milk. Aust J Dairy Technol 37:66–67

Boisen F (1983) DEFT – analyse af levnedsmidler. Dansk Vet Tidsskr 66:1030–1036

Cootes RL, Johnson R (1980) A fluorescent staining technique for determination of viable and non-viable yeasts and bacteria in wineries. Food Technol Aust 32:522–523

Griffiths MW, Phillips JD, Muir DD (1984) Methods for rapid detection of post-pasteurisation contamination in cream. J Soc Dairy Technol 37:22–26

Jones JG, Simon BM (1975) An investigation of errors in direct counts of aquatic bacteria by epifluorescence microscopy, with reference to a new method for dyeing membrane filters. J Appl Bacteriol 39:317–329

Pettipher GL (1983) The direct epifluorescent filter technique for the rapid enumeration of micro-organisms. Research Studies Press, Letchworth (Innovation in microbiology series, 1)

Pettipher GL, Rodrigues UM (1982a) Semi-automated counting of bacteria and somatic cells in milk using epifluorescence microscopy and television image analysis. J Appl Bacteriol 53:323–329

Pettipher GL, Rodrigues UM (1982b) Rapid enumeration of micro-organisms in foods by the Direct Epifluorescent Filter Technique. Appl Environ Microbiol 44:809–813

Pettipher GL, Mansell R, McKinnon CH, Cousins CM (1980) Rapid membrane filtration epifluorescent microscopy technique for direct enumeration of bacteria in raw milk. Appl Environ Microbiol 39:423–429

Pettipher GL, Fulford RJ, Mabbitt LA (1983) Collaborative trial of the direct epifluorescent filter technique (DEFT), a rapid method for counting bacteria in milk. J Appl Bacteriol 54:177–182

Ploem JS, Verwoerd N, Bonnet J, Koper G (1979) An automated microscope for quantitative cytology combining television image analysis and stage scanning microphotometry. J Histochem Cytochem 27:136–143

Rodrigues UM, Pettipher GL (1984) Use of the direct epifluorescent filter technique for predicting the keeping quality of pasteurised milk within 24 h. J Appl Bacteriol 57:125–130

Thomason BM, Herbert A, Cherry WB (1975) Evaluation of a semiautomated system for the direct fluorescent antibody detection of Salmonellae. Appl Microbiol 30:557–564

Watson SW, Novitsky TJ, Quinby HL, Valois FW (1977) Determination of bacterial number and biomass in the marine environment. Appl Environ Microbiol 33:940–946

Application of the Limulus Test for Detection of Gram-Negative Microorganisms in Milk and Dairy Products

W. HEESCHEN, J. SÜDI, and G. SUHREN [1]

Introduction

Quantitation of the contamination of milk with gram-negative saprophytic bacteria plays an important role in dairy bacteriology under the technological conditions currently prevailing in the dairy industry. The practical relevance of such determinations is seen from the following facts.

- Gram-negative saprophytic organisms are mainly responsible for the proteolytic and lipolytic degradation of milk and milk products, and these processes result in sensoric failures.
- A great number of gram-negative milk bacteria are psychrotrophic (Enterobacteriaceae, Pseudomonadaceae, Flavobacteriae).
- Gram-negative milk bacteria are killed by the legally requested pasteurization treatment. Since the shelf life of pasteurized milk products is critically dependent on the presence of gram-negative psychrotrophic organisms, their post-pasteurisation contamination has to be prevented.

None of the conventional methods is ideally suited for the determination of gram-negtive milk bacteria. When counting coliform bacteria or determining the coli titer one is dealing with a restricted specificity, while the opposite is true when counting psychrotrophic organisms. No selective medium is known for growing gram-negative, as opposed to gram-positive bacteria.

The possible use of the limulus test for determining the lipopolysaccharide content of dairy products in order to assess their contamination with gram-negative bacteria has been recently tested in a number of laboratories [1–4]. Following brief outlines of principles and methodology, we report here on the use of this test in dairy bacteriology.

Principles and Technique of the Limulus Test

Lipopolysaccharides (LPS) are an essential component of the gram-negative cell wall. The blood cells of the horseshoe crab (*Limulus polyphemus*) register the presence of LPS as denoting "alarm" and react through an as yet not fully understood chemical mechanism by the activation of a normally "silent" blood clotting enzyme. The activated proteolytic enzyme splits a clottable protein and triggers in this way a clotting reaction (Fig. 1). In the horseshoe crab, in vivo, this sequence

1 Institut für Hygiene, Bundesanstalt für Milchforschung, Hermann-Weigmann-Str. 1, D-2300 Kiel

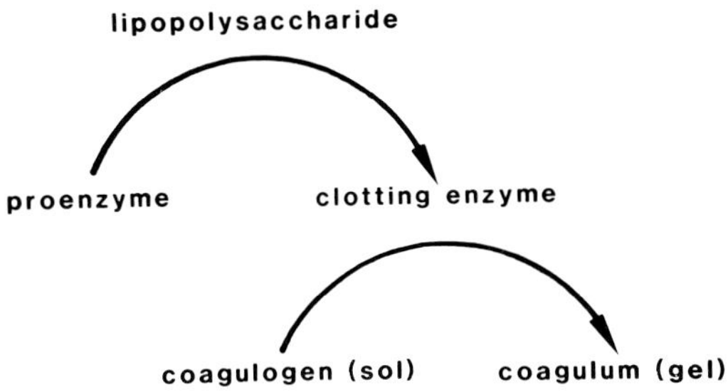

Fig. 1. Principle of the limulus test

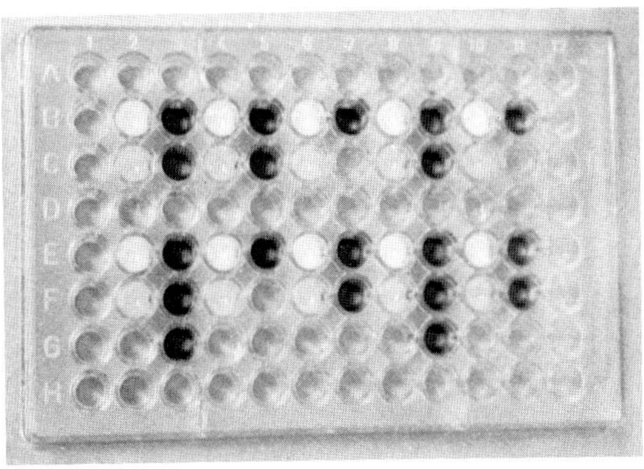

Fig. 2. Limulus test plate

of events initiates a blood clotting reaction which is apparently the only protective mechanism operating in this animal against bacterial infections.

The complete clotting system, excepting LPS, is obtained on a commercial scale from the blood cells (amebocytes) of the horseshoe crab and has been available for more than 10 years in the form of an easily soluble white powder as test reagent [5, 6].

The main area of application of the limulus test is in the pharmacy, more specifically in testing injection solutions and other parenteralia for the absence of pyrogenic contamination [7, 8].

More recently we have developed a microtitration method for applications of the limulus test in dairy bacteriology [3]. This microtitration test is already available as a commercial test kit [9]. The practicability of LPS determinations has been improved so significantly by these developments that, in addition to some new questions which can be answered with the limulus test, one should also consider some practicable biochemical alternatives to well-established cultural methods relying on the growth of viable organisms [9, 10].

The limulus test plates (Fig. 2) are regular microtitration plates made of polystyrene and having $8 \times 12 = 96$ cavities. In $5 \times 6 = 30$ positions the test kit plates contain freeze-dried limulus reagent of a specified sensitivity. The remaining 66 cavities are empty and can be used for serial dilution of the samples with pyrogen-free distilled water. Each of the plates can be used for testing, e.g., 3 dilutions of 10 samples.

After preparation of the required sample dilutions and dissolution of the reagent with 30 µl aliquots of the respective dilution, the test plate is incubated for

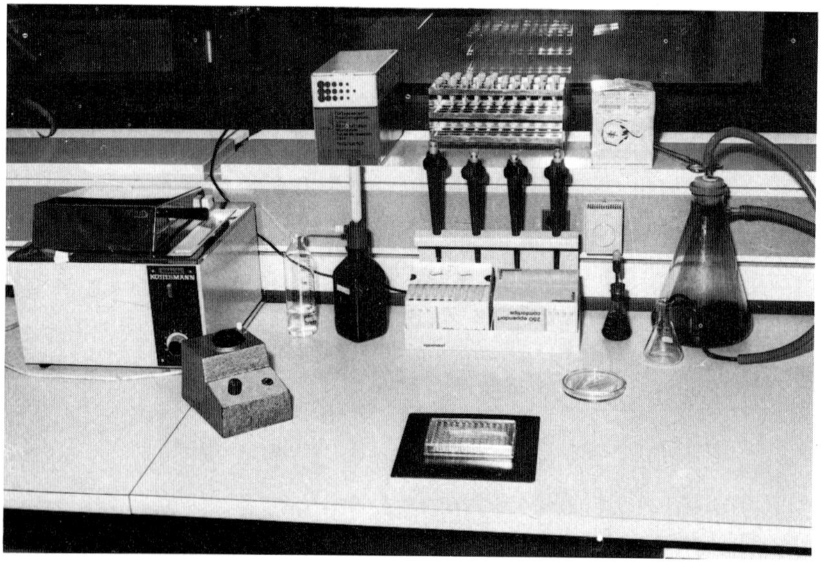

Fig. 3. Laboratory equipment for application of limulus test plates

1 h at 37 °C. In those cavities where the concentration of LPS attains a certain threshold value, there will be a firm gel. Reading the results is facilitated by adding one drop of a dye solution (1% toluidine blue) to each cavity. The final operation is then the application of a capillary tube connected to a water pump. If the reaction is negative, there is a dye solution in the cavity and this is instantly removed; in the case of a positive reaction, there is a colored gel, which is left in the cavity as a punched out disk.

The reagent in the test plate shows a predetermined sensitivity towards LPS. A purified LPS preparation is used as standard. Test plates of the commercial kit currently have a sensitivity of 0.032 ng/ml LPS. For determining the concentration of LPS in a sample one has to determine the greatest sample dilution showing a positive reaction. LPS concentration is calculated as the product of sample titer and test plate sensitivity.

The complete set of utilities and laboratory equipment which is required for application of limulus test plates is seen in Fig. 3.

Test Results in Pure Cultures

For interpreting LPS concentrations, it is necessary to determine the corresponding colony counts in pure cultures. From a sufficiently large material of this kind it is possible to calculate mean colony counts of gram-negative milk bacteria form LPS concentrations determined with the limulus test. It is to be noted that such tentative calculations always imply some "expected microflora."

Pure cultures of gram-negative isolates from milk and milk products were grown in UHT-treated low-count milk. The cultures were incubated at 9 °C and

Table 1. List of the strains tested

Strain	Origin	Strain	Origin
Enterobacteriaceae		Pseudomonadaceae	
Citrobacter freundii	Raw milk[a]	*P. putida*	Raw milk[b]
Enterobacter cloacae	Past milk[a]	*P. fluorescens*	Raw milk[b]
Erwinia herbicola	Past milk[a]		
Hafnia alvei	Raw milk[a]	Other	
Klebsiella aerogenes	Past milk[a]	*Flavobacterium* sp.	Raw milk[b]
Serratia liquefaciens	Raw milk[a]	*Acinetobacter* sp.	Raw milk[b]
Proteus sp.	Soft cheese[a]	*Alcaligenes* sp.	Raw milk[b]
Escherichia coli	Soft cheese[b]	*Aeromonas* sp.	Raw milk[b]

[a] Süddeutsche Versuchs- und Forschungsanstalt für Milchwirtschaft, Weihenstephan
[b] Bundesanstalt für Milchforschung, Kiel

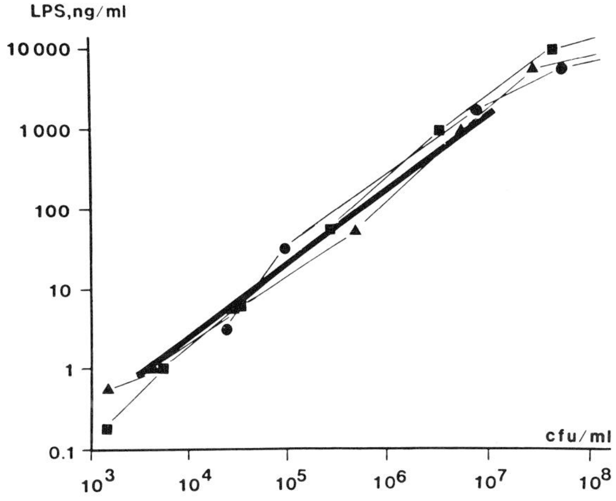

Fig. 4. Relationship between colony counts (plate count agar) and LPS concentration (*Proteus* sp., strain 16)

aliquots withdrawn at daily intervals for analysis until a colony count of ca. 10^7 cfu/ml was reached. A list of the strains involved is seen from Table 1. All strains were tested in at least two independent experiments. Representative results are shown in Fig. 4.

Table 2 shows the LPS concentrations found at 3 different cell count levels (10^5, 10^6, and 10^7 cfu/ml, respectively) as obtained from regression equations. The geometric means show a clear differentiation among the LPS concentrations found at 10^5, 10^6, and 10^7 cfu/ml. There are, however, marked differences among the strains with respect to the LPS concentrations corresponding to a given colony count.

Considering all available evidence we suggest that in raw milk 1 ng LPS corresponds to a mean colony count of 3,000–7,000 cfu/ml. One might transform LPS concentrations into colony counts with a corresponding factor for easier interpretation of the results [11]. Variation among individual strains may be of reduced significance when dealing with "field samples" since these will most probably represent a mixed microflora. For all practical purposes it appears to be satisfactory to regard 1 ng LPS as equivalent to about 5,000 cfu/ml [11].

Table 2. LPS concentration (ng/ml) in pure cultures at three cell count levels

	Cell counts cfu/ml		
	1×10^5	1×10^6	1×10^7
Citrobacter freundii (1)[a]	22	195	1,698
Enterobacter cloacae (5)	69	955	13,183
Erwinia herbicola (8)	20	182	1,622
Hafnia alvei (9)	7	70	661
Klebsiella aerogenes (10)	457	1,778	6,918
Serratia liquefaciens (12)	42	281	1,862
Proteus (16)	20	186	1,738
Escherichia coli (17)	20	166	1,349
Pseudomonas putida (21)	6	79	1,023
Pseudomonas fluorescens (24)	2	20	204
Flavobacterium (19)	0	0	6
Acinetobacter (20)	41	309	2,291
Aeromonas (30)	22	195	1,698
$\bar{x}_G$	15	126	1,113
s_g	7	7	6

[a] The numbers refer to the institutes strain collection

Analysis of Milk and Milk Products

Below we shall briefly discuss some typical applications of the limulus test. These results enable us to appraise the prospective applications of this method in scientific and technical investigations. An especially appealing aspect of LPS determinations is seen in the investigation of heat-treated products (like UHT milk or dry milk) in order to determine the bacteriological quality of the corresponding raw milk [3, 12–15].

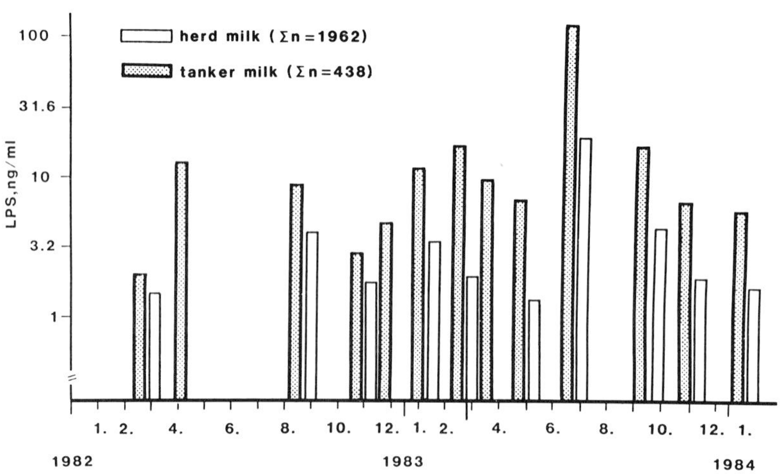

Fig. 5. LPS contents of representative herd milk and tanker milk samples delivered at a large North German dairy ($\bar{x}_G$)

Raw Milk

On 14 occasions between 1982 and 1984 we have determined the LPS concentration in farm milk and the corresponding tanker milks delivered at a large North German dairy with very high microbiological standards. Figure 5 shows an overview of the seasonal fluctuations of LPS concentrations. It is seen that the LPS level of the tanker milks is generally markedly higher than that of the herd milks. The corresponding frequency distribution diagrams are in Figs. 6 and 7. While

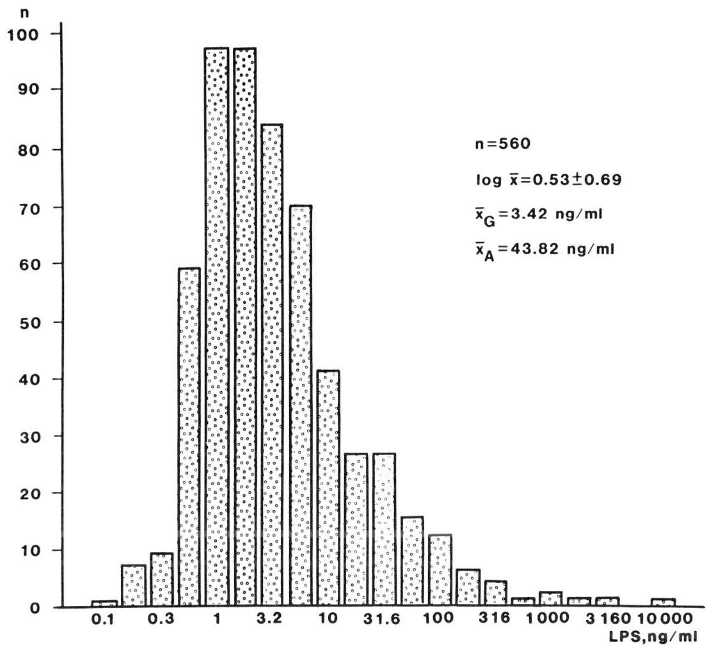

Fig. 6. LPS concentrations found in 560 herd milk samples collected by a North German dairy over a 2-year period

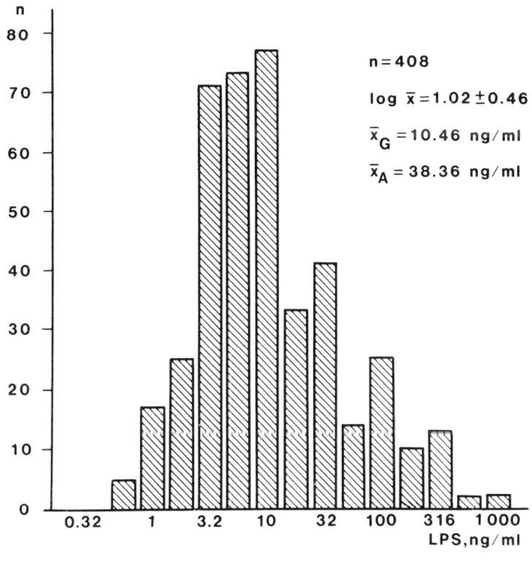

Fig. 7. LPS concentrations found in 408 tanker milks delivered at a North German dairy over a 2-year period

the geometrical mean of farm milks amounts to 3.42 ng LPS/ml, ca. 10.5 ng/ml is found in the tanker milks. Both sample populations show a broad spread implying a remarkably high "random" variation.

Heat-Treated Milk

In a North German state we obtained a representative collection of 415 freshly pasteurized consumer milk samples for analysis of LPS concentrations between the summer 1983 and the spring 1984. The results are shown in the form of a frequency distribution diagram in Fig. 8. The 415 samples were produced in 61 dairies. The means show a remarkable spread, implying well discernible differences from dairy to dairy (Fig. 9).

An especially interesting area of application can be seen in the analysis of UHT milk. Before application of the limulus test there were no reliable methods for determining the microbiological quality of the raw milk from an analysis of an UHT treated sample (heating the milk for a few seconds to 130°–150 °C).

This sort of information is quite important, since the microbiological prehistory of the milk might be decisive for the quality and shelf life of the UHT treated product [16, 17]. Of all bacterial metabolites which may lower the quality of heat-treated dairy products, thermoresistant microbial enzymes are to be mentioned first. Proteolytic and lipolytic enzymes of this type are likely to lead to sensoric failures during the prolonged storage of UHT milk. For this reason it is very op-

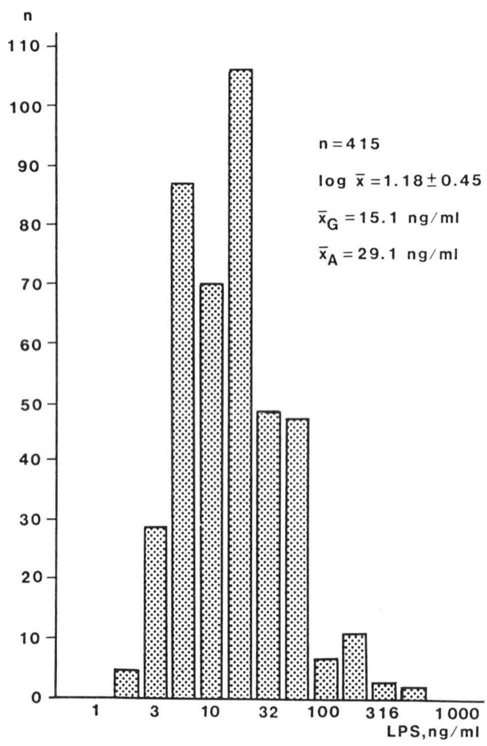

Fig. 8. LPS concentrations found in fresh pasteurized consume-milk produced in 61 dairies in a northern state of the F.R.G

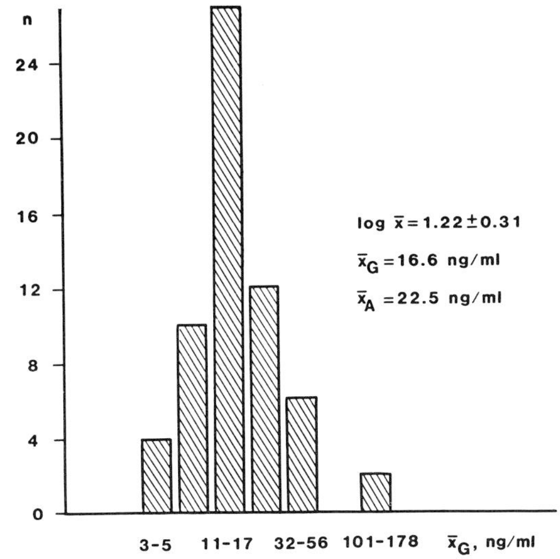

Fig. 9. Geometric means of the LPS concentration of pasteurized milk produced by 61 North German dairies (total number of samples is 422)

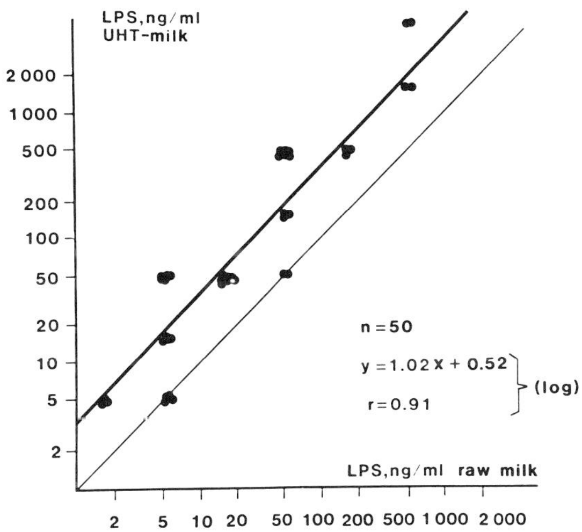

Fig. 10. Relationship between the analytical concentrations of LPS in UHT milk and the corresponding raw milk

portune to have a component of the bacterial cells available which survives undamaged all possible heat treatments, does not deteriorate during storage, and can be determined directly in the UHT milk.

There is a close correlation between the LPS concentration of raw milk and UHT milk (Fig. 10). However, it should be taken into account that the analytical LPS concentration is increased by an average factor of 3.55 through the UHT treatment. Accordingly, when "transforming" LPS concentrations into colony counts one should use different "transformation factors" (5,000 for raw milk and 1,500 for UHT milk, respectively). This effect on the analytical LPS concentration can be explained [10, 11] by the fact that UHT treatment leads to a very effective liberation of LPS from the cell wall structure of intact cells.

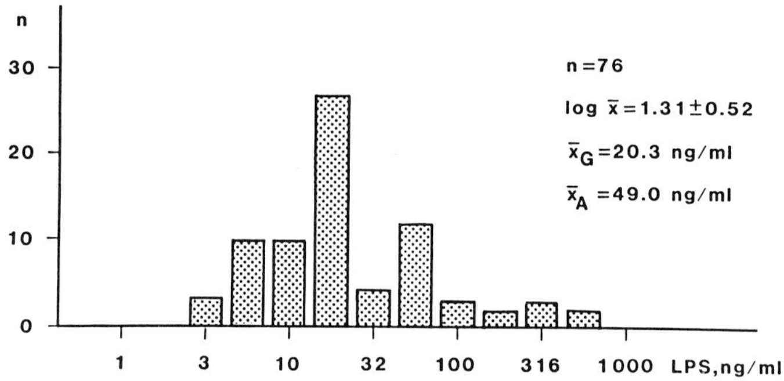

Fig. 11. LPS concentrations found in the UHT milk produced by five North German dairies over an 8-month period

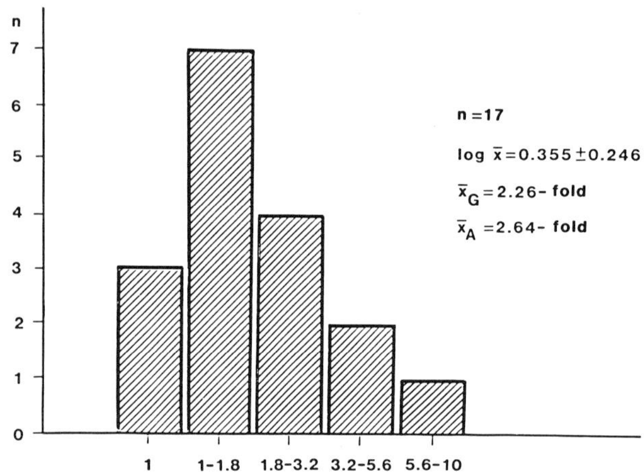

Fig. 12. Effect of the UHT treatment on the analytical LPS concentration in UHT milk (*abscissa*, ratio of LPS found in UHT milk and raw milk)

Figure 11 shows the results of a recent survey of UHT milk production in one of the North German states. It can be seen that there is a broad variation in the microbiological quality of the raw milk processed by the five dairies involved in the survey.

Results obtained in an earlier study [18] are shown in Fig. 12. Here we compare the analytical LPS concentrations found in dairy milk samples drawn just before and after the (commercial) UHT treatment. The figure shows an apparent "activation effect" of ca. 2.5 due to the UHT treatment ($\bar{x}_G = 2.26$; $\bar{x}_A = 2.64$). When comparing this result to the factor of 3.55 obtained in our own laboratory (see [11] Fig. 10), it is to be noted that the "raw milks" involved in these two investigations may not be strictly comparable. It is possible that lower "apparent activation factors" may be due to previous heat treatment(s) of the raw milk rather than to differential liberation of LPS by the laboratory UHT unit and its commercial counterparts.

Summary

The quantitation of gram-negative saprophytic organisms is of high practical relevance and is the aim of considerable scientific efforts in dairy bacteriology. Various gram-negative organisms can cause the proteolytic and lipolytic degradation of milk; a great number of gram-negative milk bacteria are psychrotrophic, forming the most important group of "recontaminants" which determine the shelf life of various milk products.

In the past few years the limulus test has found increasing interest as a sensitive and specific assay for measuring the concentration of LPS, an important cell wall component of gram-negative bacteria, as an index of the amount of gram-negative organisms in milk and milk products. The reagent is a preparation of the blood clotting system of the horseshoe crab which is purified from the blood cells (amebocytes) of the animal. The key component of this system is a proteolytic enzyme which is activated by LPS. Activation of this "clotting enzyme" triggers a chain of events which finally leads to gel formation from a "clottable protein."

We have developed a microtiter version of the limulus test for serial determinations in milk bacteriology, which is already available as a commercial test kit. The sensitivity of the test is about 0.03 ng/ml LPS, corresponding to ca. 300 cells/ml and to 10 cells/test (final vol. 30 µl). Gel formation is read visually. To aid visual reading a drop of a dye solution is added to each well and the suction of a water pump is applied through a capillary tube. If the result is negative, the well is momentarily emptied; if it is positive, stained gel is left behind with an empty "eye" marking the application site of the suction tube.

Pure cultures of milk bacteria yield very similar growth curves when followed by both cell counting and LPS determination. Nevertheless, wide variation is obtained if LPS–cell count ratios are calculated for such pure cultures. Concerning the natural mixed flora of raw milk, good approximation is obtained by using a conversion factor of 3,000–7,000 cfu/ng LPS.

The limulus test is a practicable method for assessing the bacteriological quality of raw milk used for UHT milk production. There is a close relationship between the LPS content of UHT milk and that of corresponding raw milk, even though UHT treatment results in an increased yield of detectable LPS. The correction factor has a mean value of 3.55. Accordingly, 1 ng LPS, when present in UHT milk, indicates 1,500 to 2,000 cfu of gram-negative bacteria per milliliter of corresponding raw milk.

References

1. Terplan G, Zaadhof KJ, Buchholtz-Berchtold S (1975) Zum Nachweis von Endotoxinen gramnegativer Keime in Milch mit dem Limulus-Test. Arch Lebensmittelhyg 26:217–221
2. Møller-Madsen AM, Mikkelsen T, Hansen K (1980) Gram-negativt bakterielt endotoxin (lipopolysaccharid) i maelk og mejeriprodukter. Statens Forsøgsmejeri, Hillerød, p 16 (240. beretning)
3. Südi J, Suhren G, Heeschen W, Tolle A (1981) Entwicklung eines miniaturisierten Limulus-Tests im Mikrotiter-System zum quantitativen Nachweis gram-negativer Bakterien in Milch und Milchprodukten. Milchwissenschaft 36(4):193–198

4. Mikolajczik EM, Brucker RB (1983) LAL assay. A rapid test for the assessment of raw and pasteurized milk quality. Dairy Food Sanitation 3(4):129–131
5. Cooper JF, Levin J, Wagner HN Jr (1971) Quantitative comparison of in vitro and in vivo methods for the detection of endotoxin. J Lab Clin Med 78:138–148
6. Cooper JF, Pearson SM (1977) Detection of endotoxin in biological products by the Limulus test. Dev Biol Stand 34:7–13
7. United States Pharmacopeia (1980) Bacterial endotoxins test, 20th rev. edn. Mack, Eaton, pp 888–889
8. Draft guideline for validation of LAL test an endproduct endotoxin test for human and animal parenteral drugs, biological products, and medical devices (1983) Pharmacopeial Forum 9(3):3012–3020
9. Südi J, Heeschen W (1984) Untersuchungen zur quantitativen Aussage des Limulus-Tests über die mikrobiologische Beschaffenheit von Milch und Milchprodukten. Arch Lebensmittelhyg 35:32–35
10. Südi J (1983) Les nouvelles methodes: le test limulus. La Technique Laitière 974:19–28
11. Südi J, Suhren G, Heeschen W, Tolle A (1982) Die Anwendung des Limulus-Tests zur Untersuchung ultrahocherhitzter Milch und Ermittlung der bakteriologisch-hygienischen Wertigkeit des verwendeten Rohstoffes. Milchwissenschaft 37(6):341–346
12. Zaadhof KJ, Terplan G (1981) Der Limulus-Test – ein Verfahren zur Beurteilung der mikrobiologischen Qualität von Milch und Milchprodukten. Dt Molkerei-Z 34:1094–1098
13. Südi J (1982) Limulus assay of bacterical lipopolysaccharides in UHT milk. Kieler Milchw Forschber 34(1):141–146
14. Hansen K, Mikkelsen T, Møller-Madsen A (1982) Use of the Limulus test to determine the hygienic status of milk products as characterized by levels of gram-negative LPS present. J Dairy Res 49(2):323–328
15. Hansen K, Møller-Madsen A (1983) Gram-genativt bakterielt lipopolysaccharid. Et mal for mejeriproducters hygieniske kvalitet. Statens Forsøgsmejeri, Hillerød, p 25 (253. beretning)
16. Law BL (1979) Review: enzymes of psychrotrophic bacteria and their effects on milk and milk products. J Dairy Res 46:573–588
17. Cousin MA (1982) Presence and activity of psychrotrophic microorganisms in milk and dairy products. J Food Prot 45(2):172–207
18. Südi J (1982) Rohstoffqualität für H-Milch – Ermittlung mit dem Limulus-Test. Dt Molkerei-Z 103(14):422 (abstract)

Enzyme-Linked Immunosorbent Assay of Staphylococcal Enterotoxins in Foods

S. NOTERMANS [1]

Introduction

Staphylococcus aureus is a major cause of bacterial food poisoning all over the world. Data from six countries attribute 25% of reported food-borne diseases of known etiology to *S. aureus* (Todd 1978; Smith et al. 1983). Staphylococcal food poisoning results from the ingestion of enterotoxins preformed in food by certain strains of *S. aureus* (Bergdoll 1979). Symptoms are usually observed within 2–6 h of eating toxin-containing food, and include nausea, vomiting, diarrhoea, and sometimes considerable weakness. A number of staphylococcal enterotoxins (SE) have been differentiated by serological techniques and are classified by the letter designation SEA through SEE (Minor and Marth 1976).

Demonstration of SE in leftovers of the suspect food provides good confirmatory evidence of staphylococcal food poisoning. The first methods applied for detection of SE were biological assays such as monkey and kitten feeding tests. Thereafter, immunological techniques were elaborated, based on the immunodiffusion technique of Ouchterlony (1958). Modifications of this technique include the optimal sensitivity plate method (OSP) as described by Robbins et al. (1974) and the microslide technique (Casman et al. 1969). Using these last techniques ca. 1 µg of SE/ml can be determined. When testing food for the presence of SE, the level of toxin necessary to cause symptoms must be taken into account. It is known that quantities of 0.5–5 µg of SE in a portion of food can cause an intoxication (Bergdoll 1970). For safety, therefore, it should be possible to detect at least 0.1 µg/100 g of food. Both the OSP method and the microslide technique fail to detect such small amounts of toxin. Therefore, other techniques are being developed with an increased sensitivity, and the method in vogue is the enzyme-linked immunosorbent assay (ELISA).

The ELISA for Staphylococcal Enterotoxins

Since the description of the ELISA by Engvall and Perlmann (1971) numerous modifications of the method have been developed. Most of these have been applied to the detection of SE, but the so-called sandwich method is preferred (Büning-Pfaue 1980; Berdal et al. 1981; Notermans et al. 1982a; Freed et al. 1983;

1 Laboratory for Water and Food Microbiology, National Institute of Public Health and Environmental Hygiene, P.O. Box 1, NL-3720 BA Bilthoven

Fey et al. 1983). The sandwich technique is carried out using polystyrene or polyvinyl as a solid phase, coated with anti-SE-IgG. After incubation with the test sample, the amount of adsorbed toxin is measured using enzyme-conjugated anti-SE-IgG. The enzyme activity is determined spectrophotometrically after the addition of a suitable substrate.

Sensitivity

The sensitivity of the sandwich ELISA depends primarily on the quality of the immunoglobulins used and the substrate for enzyme reaction. From different research papers it appears that the minimal detectable quantity of SE is approximately 1 ng/ml (Freed et al. 1982; Fey 1983; Berdal et al. 1981; Notermans et al. 1982a). For testing foods a simple extraction procedure is sufficient to detect SE in quantities below their intoxication level (Freed et al. 1982; Büning-Pfaue et al. 1980; Notermans et al. 1982a). Extraction can be carried out by homogenizing the food with an equal amount of distilled water. After centrifuging, the supernatant can be tested for the presence of SE. However, for testing foods with low water activity, such as dry sausages and cheese, a larger quantity of distilled water has to be used. In such cases concentration of the supernatant is recommended before toxin assay.

Cross-Reactivity

Nonspecific reactions which occur in the sandwich ELISA can be caused by cross-reacting antibodies and/or by nonimmunological binding of proteins. Cross-reacting antibodies may be present if the toxin fraction used for immunization is not completely free from other immunogenic substances. Since SE can be produced in a highly pure state (Shinagawa et al. 1975; Borja and Bergdoll 1967; Chu et al. 1966; Fey 1983), this type of cross-reaction can be eliminated. Furthermore, production of monoclonal antibodies against SE makes highly specific immunoglobulins available.

Another type of cross-reaction, so-called nonimmunological binding, is of more interest. Examples of nonimmunological binding observed by us in the sandwich ELISA are summarized in Table 1. Binding of staphylococcal protein (PA) to IgG is well documented (Mota et al 1978) and interaction of PA in the ELISA has been described (Notermans et al. 1982b). The binding of PA to IgG is prevented by using pepsin-digested antibodies, F(ab')2 fragments, instead of native IgG antibodies (Koper et al. 1980). It was demonstrated by Notermans et al. (1982b) that sheep IgG did not react with PA if conjugated with peroxidase according the method of Nakane and Kawaoi (1974). This is in contrast to the interaction of rabbit IgG with PA produced by nearly all *S. aureus*.

Nonimmunological binding may also be caused by certain enzymes with proteolytic activity. Of 14 different proteolytic enzymes tested (Notermans 1983, unpublished results) we found that papain (Sigma p1325) at concentrations above 10 μg/ml gives a positive reaction in the ELISA. Also, lysozyme gives a positive

Table 1. Examples of nonspecific reactions in the sandwich ELISA caused by nonimmunological binding of immunoglobulins

Substances	Reaction site	Concentration (μg/ml) giving positive ELISA results
Protein A	Binding to F_c fragment of IgG	0.01–0.1
Papain	Between F_{ab} and F_c fragments of IgG	1–10
Lysozyme	?	1–10
Certain extracts of shrimps	?	?
Certain extracts of mussels	?	?

reaction at comparable concentrations. It was observed that lysozyme of NBC (Nutritional Biochemical Corporation) reacted both before and after heat treatment (100 °C for 10 min), while lysozyme of Sigma (16876) and Serva (28260) reacted only after heat treatment. Lysozyme can be present in considerable amounts in food, especially food containing egg products. However, lysozyme present in eggs and other food did not react in the ELISA, even after heat treatment. Additionally, it was observed that lysozyme of NBC added to different food extracts did not react in the ELISA. This means that interference by lysozyme need not be feared in testing food extracts for the presence of SE. Other substances causing nonimmunological binding were found in extracts of certain lots of shrimps and acid-preserved mussels. Storing extracts at −20 °C resulted in a decrease of the nonimmunological binding.

Relation Between ELISA Results and Biological Activity

The ELISA for SE determines basically the immunological activity of the toxic molecule, which may differ from the biological activity. The biological activity in humans is attended by nausea, vomiting, abdominal cramping, and diarrhoea within 2–6 h after ingestion of the SE. Despite intensive research during the last three decades the mechanism of pathogenic action of the staphylococcal enterotoxins remains a mystery (Taylor et al. 1982). For testing biological activity, monkeys have been the preferred animal because peroral administration of SE elicits an emetic response from them (Sugiyama and Hayama 1965). Other animals (cats and dogs) display emetic responses to SE but not when enterotoxins are administered perorally (Bayliss 1940; Kocandrle et al. 1966).

The most important question is whether the results obtained with the ELISA are in good agreement with the biological activity of the toxin observed in human or monkeys. It may be proposed that polyclonal antibodies (PCA) used for detecting SE are directed not only to the biologically active site(s) of the toxic molecule. Therefore, false positive results may be obtained if only the biologically active site(s) of the molecule is damaged. Contrariwise, a sandwich ELISA using monoclonal antibodies (MCA) may yield a false negative result if only the matching antigenic site of the toxin is damaged. From this point of view, an ELISA performed with MCA active against the toxic site of the molecule is preferable. Notermans et al. (1984) tested different MCAs, which did not compete for the binding sites to SEA for neutralization capacities using the monkey feeding test. All MCAs tested, however, prevented vomiting in monkeys upon administration of SEA to which MCA was added. From these results it may be speculated that if SEA is coupled to an immunoglobulin the toxin does not reach its place of action in the body. If this is true, then the monkey feeding test is not suited for selecting MCAs only active against the biological active site(s) of SEA.

We made use of a more practical approach for testing the relation between ELISA results and the biological activity. SEA was added to food which was subsequently given common food processing treatment (heating, drying, freezing, storing, etc.). It was observed that, of the different treatments, heating caused immunological inactivation of the toxin. Especially, heating of SEA in minced meat extract, lettuce extract, and gelatine at pH 4.5 destroyed the immunological activity of the toxin as determined with the ELISA using either MCAs or PCA as coating antibodies. Using the monkey feeding test it was proven that the immunological inactivation was attended by biological inactivation. This result agrees with those obtained earlier by Humber et al. (1975) and Jamlang et al. (1971) comparing the immunological activity using the Ouchterlony gel diffusion technique and the monkey feeding test for detecting of heated SEA. In addition, Notermans et al. (1982a) found that the immunological inactivation of SEB upon heating in gelatine was also accompanied by biological inactivation. From these results it may be speculated that false negative ELISA results will not occur regardless of whether the test is carried out using MCA or PCA.

False positive results were traced by comparing ELISA results performed either with MCA or with PCA as coating antibodies. The same result was obtained after SEA-containing food was treated in different ways. This indicates that under normal food processing conditions the SEA molecule is not partially damaged. If this happened, the ELISA result using the different MCAs would become negative and the ELISA using PCA would remain positive. After heat treatment both ELISAs were negative coupled with a biological inactivation, suggesting that the whole SEA molecule was damaged.

Detection of Staphylococcal Enterotoxins in Food

During the last 5 years we have received leftovers of foods which were involved in food poisoning outbreaks and for which the symptoms resembled those caused by staphylococcal enterotoxin. Samples came to our laboratory from different regional Food Inspection Services. The leftovers were tested for the presence of SE; furthermore, *S. aureus* and other pathogenic microorganisms per gram of product were enumerated. Isolated strains of *S. aureus* were tested for SE production.

In general, two types of leftovers of food could be recognized: in one, *S. aureus* exceeded 10^6/g, while in the other it was absent or present at $<10^6$ g as strains which did not produce enterotoxin. In Table 2 some examples are presented.

Until now, SE could be detected in all leftovers containing enterotoxigenic *S. aureus* in excess of 10^6/g. There was one exception (Table 2, case No. 3). In this

Table 2. Analytical data for leftovers of food which caused symptoms resembling staphylococcal enterotoxicosis

Case no.	Type of food	Numbers of *S. aureus*/g	Enterotoxin production of isolates (type of toxin)	Amount of enterotoxin (ng/100 g)
1	Dry sausage	10^7–10^8	27/87[a] (SEA)	200
2	Chicken meat	10^8	2/2 (SEA)	4,200
3	Vanille custard	10^{6-7}	5/5 (SEB)	Neg.
4	"Osseworst"	10^{1-2}	0/2	Neg.
5	"Rookworst"	10^5	0/5	Neg.
6	Mussels	<10	n.t.	Neg.
7	Cheese	10^{5-6}	0/5	Neg.

[a] 27/87: of the 87 strains tested 27 produced enterotoxin

outbreak of food poisoning 40 cases of gastrointestinal illness occurred among 250 persons. They became ill with the following symptoms: vomiting (30) and diarrhoea (10), 4–5 h after eating vanilla custard. The vanilla custard contained a pure culture of SEB-producing *S. aureus*. No enterotoxin could be detected using the ELISA. Additional laboratory experiments indicated that, in spite of excellent growth of *S. aureus* in vanilla custard, SEB is not produced at 31 °C during the first 24 h of incubation. (These were the conditions used to store the vanilla custard which caused the disease.) Only after incubation for ≧48 h were small quantities of SEB produced. The same findings were obtained earlier by Olsvik (1982). Therefore, it should be considered that the symptoms caused by the vanilla custard were not caused by preformed SE. In all probability the symptoms were due to SEB produced in the intestinal tract of the patients. Such cases of enteritis have already been described by Surgella and Dack (1955). In the other outbreaks of food poisoning where low numbers of *S. aureus* were implicated (Table 2, nos. 4–6) no SE was detected. The *osseworst,* a fresh type of German sausage, was warmed up before consumption. The *rookworst* (cooked smoked ring sausage) was boiled before consumption. The mussels were acid preserved and eaten without heat treatment before. The cheese (Table 2, case no. 7), a home-made product, contained 10^{5-6} *S. aureus* per gram. However, isolated *S. aureus* strains did not produce enterotoxin as tested by ELISA and monkey feeding test.

These results obtained during the last 5 years have shown that detection of SE in leftovers of suspected food provides strong confirmatory evidence of staphylococcal enterotoxicosis. In leftovers of foods suspected of causing illness, in which no SE was detectable using the ELISA, only the symptoms of the patient suggested a staphylococcal enterotoxicosis. In these cases no further evidence, such as the presence of enterotoxigenic *S. aureus,* was obtained for a staphylococcal enterotoxicosis.

Summary

The enzyme-linked immunosorbent assay (ELISA) is being used increasingly in food quality control for detecting undesirable substances such as bacterial toxins.

For detecting staphylococcal enterotoxins (SE) the sandwich ELISA technique is the method in vogue. The technique is carried out using polystyrene or polyvinyl trays. Wells of these trays are coated with anti-SE-IgG, e.g., from sheep. After incubation with the test sample, the amount of adsorbed toxin is measured using enzyme-conjugated anti-SE-IgG. The enzyme activity is determined spectrophotometrically after the addition of a suitable substrate. For detection of SE in food a simple extraction procedure is sufficient to detect the toxins in quantities below their toxic level. In general, other proteins present in extracts do not interfere in the ELISA. However, interference may be expected if the food extract contains protein A (produced by *S. aureus*) and enzymes such as lysozyme and papain. These substances may result in false positive reactions. Although these reactions are observed only occasionally, they can be recognized by using appropriate controls.

To date an excellent correlation between the immunological activity of SE in food (as measured using the ELISA) and its biological activity (monkey feeding test) has been observed.

Evaluation of the ELISA for detecting SE in leftovers of food suspected of causing enterotoxicosis has shown that this assay provides strong confirmatory evidence of staphylococcal food-poisoning.

Acknowledgement. I am grateful to Dr. A. N. Sharpe, Health Protection Branch, Canada, for correcting the manuscript and for his critical review.

References

Bayliss M (1940) Studies on the mechanism of vomiting produced by Staphylococcus enterotoxin. J Exp Med 72:669–684

Berdal BP, Olsvik O, Omland T (1981) A sandwich ELISA method for detection of *Staphylococcus aureus* enterotins. Acta Pathol Microbiol Scand 89:411–415

Bergdoll MS (1970) Enterotoxins. In: Montie TC, Kadis S, Ajl SJ (eds) Microbiol toxins, vol 3. Academic, New York, p 265

Bergdoll MS (1979) Staphylococcal intoxications. In: Rieman H, Bryan FL (eds) Food-borne infections and intoxications, 2nd edn. Academic, New York, p 443

Borja CR, Bergdoll MS (1967) Purification and partial characterization of staphylococcal enterotoxin C produced by *Staphylococcus aureus* strain 137. Biochemistry 6:1467–1473

Büning-Pfaue H, Timmermans P, Notermans S (1980) Einfache Methode für den Nachweis von Staphylokokken-Enterotoxin B in Vanillepudding mittels ELISA-Test. Lebensm Unters Forsch 173:351–355

Casman EP, Bennet RW, Dorsey AE, Stone JE (1969) The microslide gel double diffusion test for the detection and assay of staphylococcal enterotoxins. Health Lab Sci 6:185–198

Chu FS, Thadhane K, Schantz EJ, Bergdoll MS (1966) Purification and characterization of staphylococcal enterotoxin A. Biochemistry 5:3281–3289

Engvall E, Perlmann P (1971) Enzyme-linked immunosorbent assay (ELISA): Quantitative assay of immunoglobulin G. Immunochemistry 8:871–974

Fey H, Pfister H, Rüegg O (1983) Comparative evaluation of different enzyme-linked immunosorbent assay systems for the detection of staphylococcal enterotoxins A, B, C, and D. J Clin Microbiol 19:34–38

Freed RC, Evenson ML, Reiser RF, Bergdoll MS (1982) Enzyme linked immunosorbent assay for detection of staphylococcal enterotoxins in foods. Appl Environ Microbiol 44:1349–1355

Humber JY, Denny CB, Bohrer CW (1975) Influence of pH on heat inactivation of staphylococcal enterotoxin A as determined by monkey feeding and serological assay. Appl Microbiol 30:755–758

Jamlang EM, Bartlett ML, Snyder HE (1971) Effect of pH, protein concentration and ionic strength on heat inactivation of staphylococcal enterotoxin B. Appl Microbiol 22:1034–1040

Kocandrle V, Houttuin E, Prohaska JV (1966) Acute hemodynamic and gastro-intestinal changes produced by staphylococcal exotoxin and enterotoxin in dogs. J Surg Res 6:50–57

Koper JW, Hagenaar AB, Notermans S (1980) Prevention of cross-reactions in the enzyme linked immunosorbent assay (ELISA) for the detection of *Staphylococcus aureus* enterotoxin type B in culture filtrates and foods. J Food Safety 2:35–45

Minor TE, Marth EH (1976) Staphylococci and their significance in foods. Elsevier, Amsterdam

Mota G, Ghettie V, Sjöquist J (1978) Characterization of the soluble complex formed by reacting rabbit IgG with protein A of *S. aureus*. Immunochemistry 15:639–642

Nakane PK, Kawaoi A (1974) Peroxidase-labelled antibody. A new method of conjugation. Histochem Cytochem 22:1084–1091

Notermans S, Boot R, Tips PD, De Nooy MP (1982a) Extraction of staphylococcal enterotoxins (SE) from minced meat and subsequent detection of SE with enzyme-linked immunosorbent assay (ELISA). J Food Prot 46:238–241

Notermans S, Timmermans P, Nagel J (1982b) Interactions of staphylococcal protein A in enzyme-linked immunosorbent assays for detecting staphylococcal antigens. J Immunol Methods 55:35–41

Notermans S, Boot R, Haas H (1984) Relation between immunological and biological activity of Staphylococcus aureus enterotoxin A in food upon different treatments (to be published)

Olsvik O, Myhre S, Berdal BP, Fossum K (1982) Detection of staphylococcal enterotoxin A, B, and C in milk by an ELISA procedure. Acta Vet Scand 23:204–210

Ouchterlony O (1958) Diffusion-in-gel methods for immunological analysis. Prog Allergy 5:1–77

Robbins R, Gould S, Bergdoll MS (1974) Detecting the enterotoxigenicity of *Staphylococcus aureus* strains. Appl Microbiol 28:946–950

Shinagawa K, Kunita N, Sakaguchi G (1975) Simplified methods for purification of staphylococcal enterotoxins A and C and preparation of anti-enterotoxin sera. Jpn J Bacteriol 30:683–692

Smith JL, Buchnan RL, Palumbo SA (1983) Effect of food environment on staphylococcal enterotoxin synthesis: a review. J Food Prot 46:545–555

Sugiyama H, Hayama T (1965) Abdominal viscera as site of emitic action for staphylococcal enterotoxin in the monkey. J Infect Dis 115:330–336

Surgella MJ, Dack GM (1955) Enterotoxin produced by micrococci from cases of enteritis after antibiotica therapy. JAMA 158:649–650

Taylor SI, Schlunz LR, Beery JT, Cliver DO, Bergdoll MS (1982) Emetic action of staphylococcal enterotoxin A on weanling pigs. Infect Immun 36:1263–1266

Todd ECD (1978) Food-borne diseases in six countries – a comparison. J Food Prot 41:559–565

Procedures and Methods for One-Day Analysis of Microbial Loads in Foods

D. Y. C. Fung[1]

Introduction

One of the goals of rapid methods in food microbiology is to be able to ascertain the microbial loads in foods within a working day. This time frame is too short to monitor samples with very low bacterial counts. However, it is possible to obtain important information for quality control purposes if microbiologists are willing to set "acceptable" and "unacceptable" levels for certain foods. Fung et al. (1980a) defined microbial load on meat surfaces as low, log 0–2 colony forming units (CFU)/cm^2; intermediate, log 3–4 CFU/cm^2; high, log 5–6 CFU/cm^2; and very high, log 7 CFU/cm^2. Using this guideline, samples with log 0–4 CFU/cm^2 would be considered acceptable food products whereas samples with log 5–6 CFU/cm^2 would be considered questionable and those with log 7 CFU/cm^2 unacceptable from a food spoilage standpoint. This of course does not include monitoring potential pathogens in foods such as *Salmonella*.

Using the above criteria one could estimate the microbial loads of food within a working day by several methods as 1-day analysis of microbial loads in foods. These methods can be roughly grouped as (1) microbial enzymes in foods (e.g., reductase, catalase, coagulase, thermoendonuclease, etc.), (2) microbial components and metabolites in foods (e.g., limulus lysate test, ATP levels, pyruvate, microbial toxins, etc.), (3) microbial activities in foods (e.g., radiometry, electrical impedance, microcalorimetry, etc.), and (4) direct count of microbial population in foods and their viability. In the Proceedings of the Fourth International Symposium on Rapid Methods and Automation in Microbiology and Immunology, Berlin, 7–10 June 1984, many methods within the above grouping have been discussed in detail by other authors. Review articles on many of these methods were presented by Goldschmidt and Fung (1979) and Fung (1984). The purpose of this paper is to review some of the works generated in the author's laboratory pertaining to possible 1-day analysis of microbial loads in foods.

Catalase Activities in Foods

Bacteria commonly associated with foods can be divided into catalase-positive bacteria and catalase-negative bacteria. Each group contains important food-

1 Food Science Graduate Program, Department of Animal Science and Industry, Kansas State University, Manhattan, KS 66506, USA

borne organisms. A bacterial culture is considered catalase positive when gas bubbles are generated when a drop of 3% H_2O_2 is placed on the colony. Catalase-negative culture will not generate gas bubbles. A semiquantitative scale using capillary tubes for catalase activities was described by Fung and Petrishko (1973). It is apparent that different bacterial species have varying degrees of catalase activities. It is possible, however, to use this concept to monitor catalase activities in certain foods. Some foods characteristically carry large populations of either catalase-positive or catalase-negative bacteria. A good example is the growth of psychrotrophs in aerobic, cold-stored food. Most psychrotrophs are catalase positive; therefore catalase activities in cold-stored food may be used to ascertain the microbial loads of these foods.

Gas Column Method

The author devised a simple catalase measuring unit. A Pasteur pipette is sealed at the narrow end by heat. From the wider end 0.05 ml of a liquid to be tested is introduced after which a 0.05 ml of 3% H_2O_2 (with 10^{-6} *M* EDTA) is added. The liquids are allowed to mix by rapidly moving the pipette in three circular motions. The entire liquid column is then "shaken" into the narrow portion of the pipette by a quick jerk of the wrist similar to shaking a mercury column in a mercury thermometer. After 5 s, the pipette is inverted. Surface tension holds the liquid column in the narrow part of the Pasteur pipette. Gas bubbles when generated will accumulate at the narrow end of the pipette (Fig. 1). After 15 min the length of the gas column and the total liquid and gas column are measured by use of an ocular micrometer. To minimize the effect of varying diameters of commercial Pasteur pipettes the gas column is expressed as percentage of total column, i.e., gas column/total column × 100 = percentage gas column.

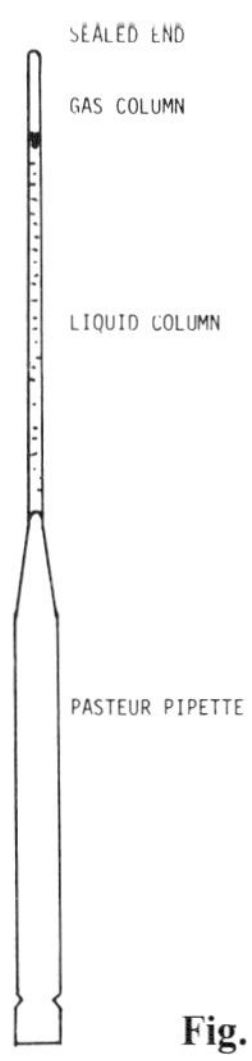

Fig. 1. Catalase tube

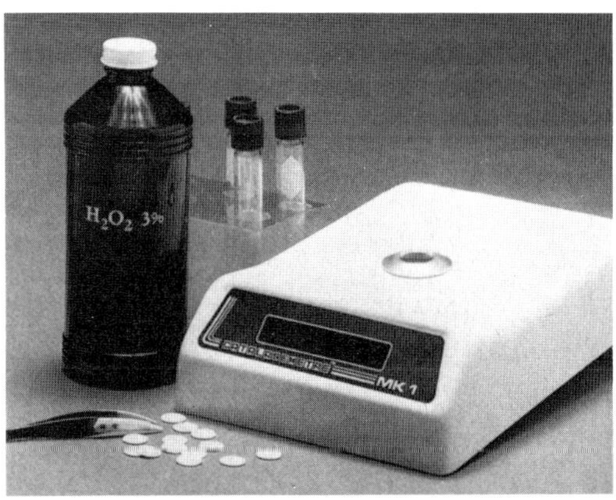

Fig. 2. Catalasemeter

Catalasemeter Method

An instrument called the Catalasemeter was recently introduced commercially by Bio-Engineering Group Ltd., (Fig. 2, Westport, Conn., USA). The principle is the disk-flotation method proposed by Gagnon et al. (1959). It is a simple instrument to operate. First, an optically clean glass tube containing 5 ml of 3% H_2O_2 (with 10^{-6} *M* EDTA) is introduced into a cavity in the instrument. A narrow beam of light passes through the liquid and is collected by a photoelectric cell which is attached to a digital timer. To perform the catalase test a paper assay test disk is saturated with the test liquid (0.12–0.15 ml with 0.02% ethanol added) and then dropped into the H_2O_2 solution. As the disk travels down the liquid by gravity the light path is disturbed and this activates the timer. In the event that the test liquid contains catalase, gas bubbles will be generated and trapped in the paper disk. This causes the disk to float. When the disk floats past the light beam the timer is deactivated. The time lapse is recorded in seconds and is the flotation time of the liquid sample. The flotation time is inversely proportional to the amount of catalase in the disk.

Results of a Comparative Study

Six catalase-positive bacteria and two catalase-negative bacteria were tested using the two catalase methods compared with the conventional viable cell count method. As expected, catalase-negative bacteria (*Lactobacillus brevis* and *Streptococcus faecalis*) showed no responses in the catalase methods. The correlation coefficient as well as the apparent limit of sensitivity of the Catalasemeter for the catalase-positive bacteria are as follows: *Staphylococcus aureus* (Fig. 3, −0.98, log 7.5), *Micrococcus luteus* (Fig. 4, −0.92, log 4.5), *Escherichia coli* (Fig. 5, −0.91, log 6.1), *Pseudomonas fluorescens* (Fig. 6, −0.84, log 4.8), *Bacillus cereus* (Fig. 7, −0.84, log 6.2), and *Salmonella typhimurium* (Fig. 8, −0.53, log 8). These

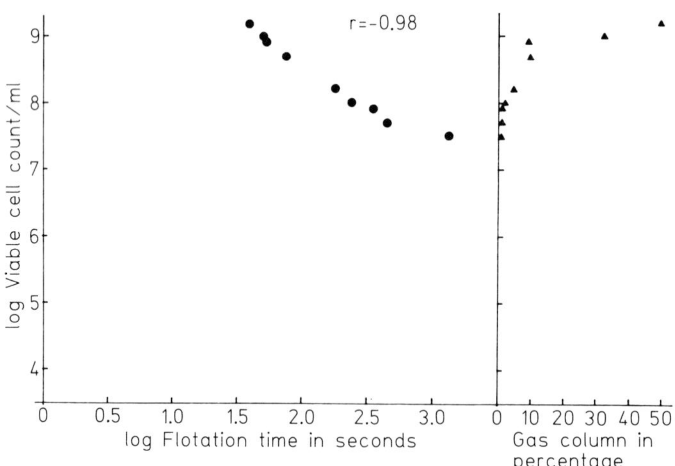

Fig. 3. Catalase activity of *Staphylococcus aureus*

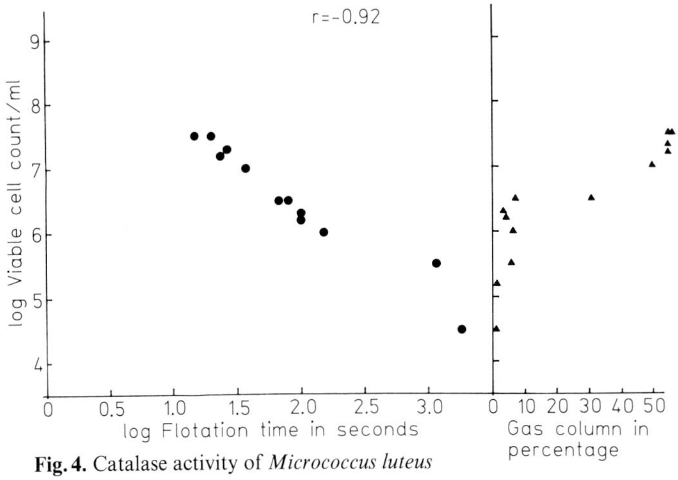

Fig. 4. Catalase activity of *Micrococcus luteus*

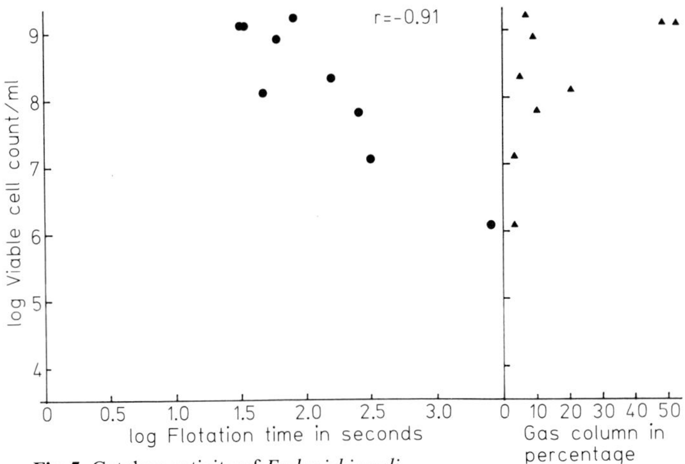

Fig. 5. Catalase activity of *Escherichia coli*

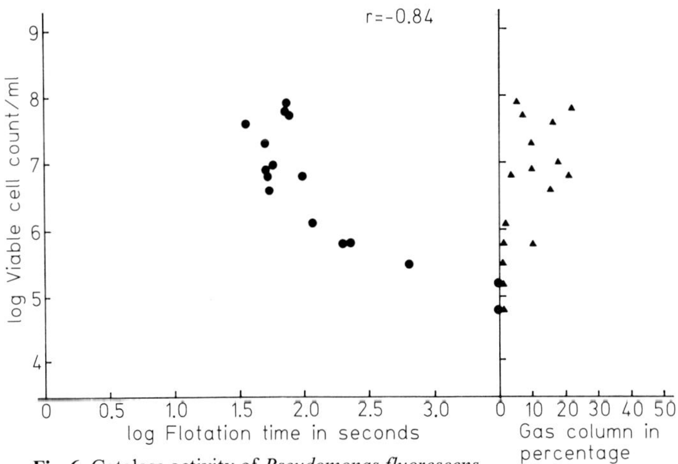

Fig. 6. Catalase activity of *Pseudomonas fluorescens*

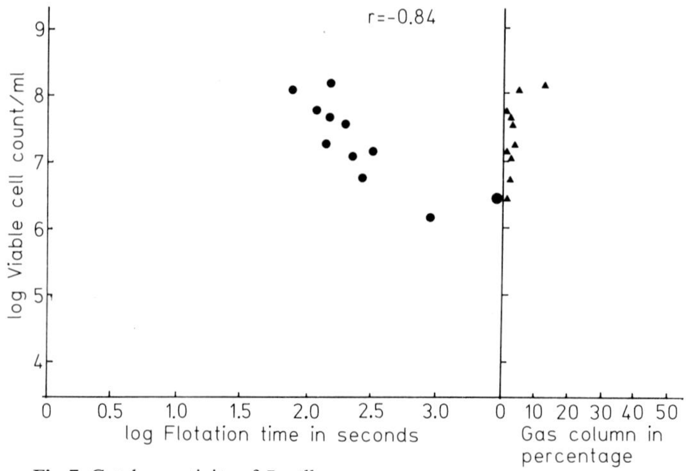

Fig. 7. Catalase activity of *Bacillus cereus*

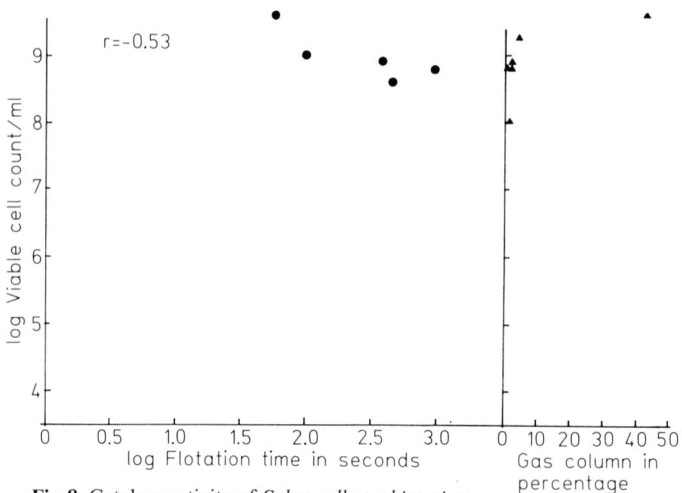

Fig. 8. Catalase activity of *Salmonella typhimurium*

data indicate that individual species produce different amounts of catalase and that the sensitivity of the Catalasemeter depends on the species. The correlation coefficient in general is good. The gas column method also showed that as the number of bacteria increased there was a concomitant increase in percentage of gas column. The curves were more sigmoidal compared with corresponding Catalasemeter data.

Practical application of these two catalase methods was tested on 24 samples of chicken wings (Fig. 9). The swab method was used to make viable cell counts of chicken wing surfaces. The correlation coefficient of viable cell counts and the Catalasemeter Data was −0.69, indicating that there is potential for this instrument to monitor load of cold-stored chicken. The data from the gas column also showed a positive correlation.

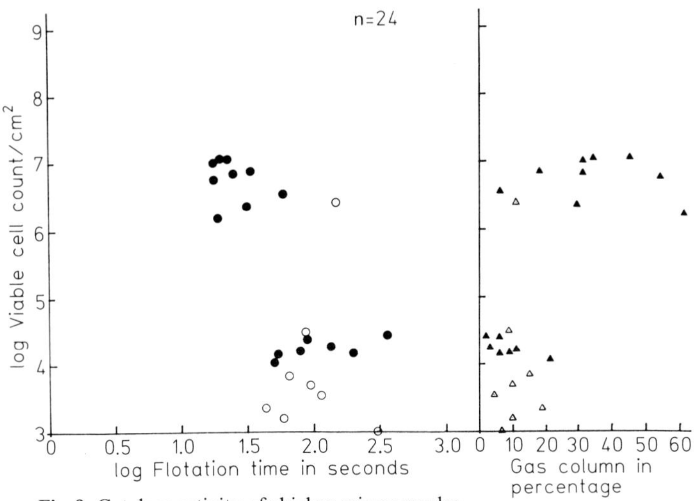

Fig. 9. Catalase activity of chicken wing samples

The concept of flotation time for measuring catalase activity has been tested on a variety of specimens. Recently Dodds et al. (1983) studying vacuum-packaged comminuted cooked turkey indicated that Catalasemeter results correlated well with the specific number of organisms but that it was not sensitive below 10^4 organisms/g of meat. The advantages of the catalase methods are simplicity of operation and speed of obtaining results (within minutes when high counts are encountered). The disadvantages include interference of nonbacterial catalase-like activities and the inability to detect catalase-negative organisms. Before routine use of either catalase methods the equipments must be standardized. The disks used to the Catalasemeter must be more uniform. The Pasteur pipettes currently on the market are not precision items. If the gas column method is to be used, more homogeneous tubes need to be employed. Ways must be found to differentiate bacterial and nonbacterial breakdown of H_2O_2 in the test systems for catalase measurements.

The conclusion of the preliminary study of catalase method for estimating microbial load in liquid cultures and on meat surfaces indicated that both the gas column method and the Catalasemeter method have potential as 1-day procedures for the analysis of microbial loads in foods. More data need to be collected before recommendation of either method for routine use.

Sequential Direct Count Method

Adhesive tape method is a simple and rapid method to estimate bacterial load on meat surfaces. Fung et al. (1980 b) reported that the mylar adhesive tape method provided highly reliable estimates of psychrotroph ($r = 0.95$) and mesophile counts ($r = 0.90$) on meat surfaces. The method involved placing an adhesive tape on meat surface to remove microorganisms and then transferring the microorganisms on the tape to the surface of nutrient agar. Contact time on the meat surface as well as on the agar was 1 min. After 24 h at 32 °C the colonies on the agar were counted and an estimate of microbial load on the meat surface could then be made.

Subsequent observations indicated that bacterial colonies started to appear on the agar in less than 12 h and sometimes as soon as 6 h. There were even instances when colonies started to appear in 4 h. It became apparent that with the aid of a stage microscope (15 ×) the appearance of bacterial colonies could be detected within 4 h after the tape had been impressed on the agar. However, the colonies were very small and difficult to distinguish from food particles or water droplets on the agar. To carry this process further we developed the "sequential direct count method."

The procedure is as follows: a 2×4 cm^2 adhesive tape is impressed on the surface of meat for 1 min. The tape is then transferred to nutrient agar surface and allowed to stay for 1 min before removal from the agar. The agar plate is then incubated at 32 °C. The *tape* is gram stained and then observed under oil immersion microscopy.

Ten microscope fields located at predesignated areas are observed and the number of colonies in each field counted. This is the "O-time" tape count. After 4 h of incubation of the agar plate at 32 °C a new 2×4 cm^2 tape is impressed on the area previously contacted with a tape containing microorganisms from the meat surface. After a contact time of 1 min, the new tape with bacteria from the agar surface is then removed, gram stained, and observed under oil immersion. This is the "4-h" tape count. The assumption is that the colonies on the agar are too small to be observed with the naked eye but when the bacteria are transferred to the adhesive tape, stained, and observed under oil immersion, a better estimation of the growth of the organisms can be made. Thus by comparing the O-time tape count and the 4-h tape count one can estimate not only the microbial load of meat surfaces at the time of sampling but also the viability of these microbes on agar surface in 4 h. An added advantage is the information on the proportion of gram-positive and gram-negative organisms on meat surfaces and their growth.

When the number of organisms on the meat surface is low, very few bacteria can be counted by this procedure. Conversely, when the microbial load is high all fields will be saturated, making accurate counting of bacteria impractical. Thus we have devised a semiquantitative scale as follows:

Very low count	(VL)	= All ten fields are countable and the total number is less than 20 bacteria
Low count	(L)	= All fields are countable and the total is between 20 and 100 bacteria
Intermediate count	(I)	= All fields are countable and total number of bacteria is ca. 300
High count	(H)	= Five fields are countable and five fields contain too many bacteria to count
Very high count	(VH)	= All fields contain too many bacteria to count

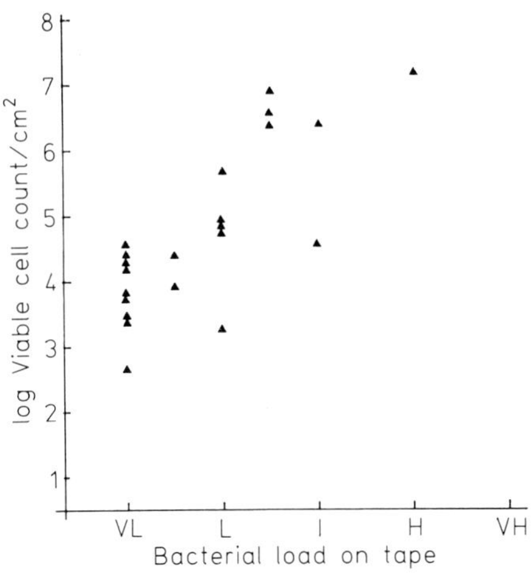

Fig. 10. O-time tape count of pork samples

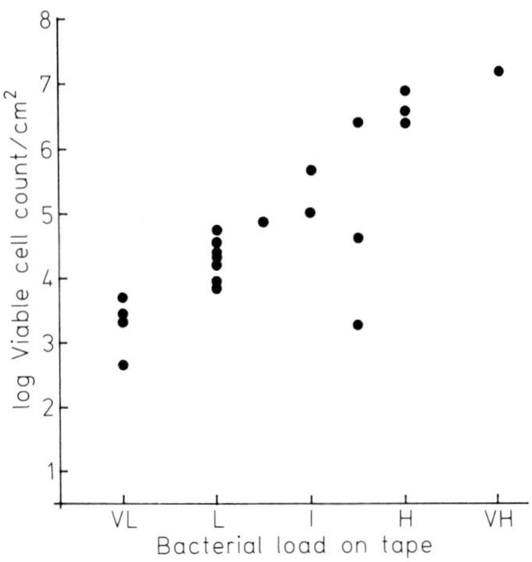

Fig. 11. 4-h tape count of pork samples

With this sequential direct count method the microbial loads of 22 pork samples were monitored. Figures 10 and 11 showed the O-time and 4-h tape counts, respectively, of pork samples compared with the swab method. The O-time tape count indicated that when pork samples had 10^5 CFU/cm^2 or below, the tape count method registered a VL to L count. When samples had 10^6 CFU/cm^2 or higher, the tape count registered I or H counts. The 4-h tape count data indicated that most of the sample below 10^5 CFU/cm^2 had an L or VL tape count whereas samples with 10^6 CFU/cm^2 or higher had an H or VH count. These data are not subject to statistical analysis due to the nature of the semiquantitative scale. The tape count method did provide useful information concerning the microbial load of the pork surfaces both at O-time and at 4 h. More work needs to be done on this procedure before recommendation for routine use. The procedure, when perfected, can be qualified as a 1-day procedure for analysis of microbial loads in meat surfaces.

It should be emphasized that the tape method is an *estimation* of the total population on meat surfaces. Every step in the sequential direct count method removes only a portion of the bacteria whether they are on the meat surface or on the agar. However, when performed in a consistent manner the data should be valid as an estimation of the total population.

Discussion and Conclusions

Many advances have been made in the field of rapid identification of bacterial colonies since the organisms were isolated on agar. Fung et al. (1984) reviewed most of the existing methods for rapid identification of isolates. Cox et al. (1984) discussed the advantages and disadvantages of many of the miniaturized diagnostic kits. Progress on rapid estimation of microbial loads in foods is lagging due to the requirement of growth of microbes in foods before they reach detectable levels. With the advances in genetic engineering methods and sophisticated instruments, minute quantities of enzymes, metabolites and the activities of certain groups of microorganisms can be monitored. The *value* of these measurements, however, is still the key issue in the field of applied microbiology. So far the reference point is still the viable cell count. Most systems are designed to match their activities or measurements against the number of viable cells per unit in certain foods. Until a better reference point is discovered or agreed upon by most microbiologists, a comparison of any proposed new methods for microbial loads in food must be made against the level of viable cells in the food under investigation.

The catalase methods described in this report under certain conditions provide reasonable estimation of the catalase-positive bacterial population in certain foods. The sequential direct count method can also provide valuable information on microbial load on meat surfaces. These methods are qualified as 1-day procedures when perfected.

Food microbiologists must continue to search for effective, simple, and reliable methods to ascertain the quality of food supply from the standpoint of food spoilage, food hygiene, and public health.

Acknowledgement. The author thanks George I. J. Wang and Jennifer Y. Lee for their technical assistance. The use of the Catalasemeter is appreciated through the courtesy of Bio-Engineering Group Ltd., Westport, Conn, USA.

References

Cox NA, Fung DYC, Goldschmidt MC, Bailey JS, Thomson JE (1984) Selecting a miniaturized system for identification of *Enterobacteriaceae*. J Food Protect 47:74–77

Dodds KL, Holley RA, Kempton AG (1983) Evaluation of the catalase and *Limulus* amoebocyte lysate tests for rapid determination of the microbial quality of vacuum-packed cooked turkey. Can Inst Food Sci Technol J 16:167–172

Fung DYC (1984) Rapid methods for determining the bacterial quality of red meats. J Environ Health 46:226–228

Fung DYC, Petrishko DT (1973) Capillary tube catalase test. Appl Microbiol 26:631–632

Fung DYC, Kastner CL, Hunt MC, Dikeman ME, Kropf D (1980a) Mesophilic and psychrotrophic populations on hot-boned and conventionally processed beef. J Food Protect 43:547–550

Fung DYC, Lee CY, Kastner CL (1980b) Adhesive tape method for estimating microbial load on meat surfaces. J Food Protect 43:295–297

Fung DYC, Goldschmidt MC, Cox NA (1984) Evaluation of bacterial diagnostic kits and systems at an instructional workshop. J Food Protect 47:68–73

Gagnon M, Hunting WM, Esselen B (1959) New method for catalase determination. Anal Chem 31:144–146

Goldschmidt MC, Fung DYC (1979) Automated and new instrumentation for microbiological analysis. Food Technol 32:63–70

The Potential of Chromatographic Techniques for the Manipulation of Viable Micro-Organisms *

P. D. PATEL and J. M. WOOD [1]

Introduction

In microbiology the conventional growth-amplification techniques (e.g. colony count) for the estimation of micro-organisms are both tedious and time consuming, requiring 24–48 h for a total viable count, while detection of specific organisms (e.g. *Escherichia coli*) may require a further 24- to 48-h incubation.

In recent years food manufacturers have sought novel microbiological methods that can produce quality control information more rapidly than the traditional cultural methods. This has resulted in the development of various instrumental techniques for the rapid estimation of total viable micro-organisms in foods. The techniques include the direct epifluorescent filter technique (DEFT) [20], luminometry [16, 26], impedimetric estimation [7] and radiometry [24]. In general, the instrumental techniques are less sensitive than the cultural methods and may also be subject to interference from non-microbial components in the milieu, e.g. ATP intrinsic to foods [25], and buffering constituents of the food can depress the impedimetric response of micro-organisms.

In these and future instrument-based techniques the ability to manipulate (i.e. enrich, separate or concentrate) micro-organisms by physical techniques may offer advantages in preparing samples for analysis. For instance, the separation and concentration of micro-organisms from other food constituents could result in a greater sensitivity of detection by instrumental techniques. Furthermore, separating specific groups of organisms from mixed microbial populations may lead to the more rapid selective estimation of the organisms.

Various techniques (e.g. chromatography, electrophoresis and density gradient centrifugation) have been used by clinical and research laboratories for separation and concentration of biomolecules. However, there has been relatively little work on the application of such techniques to microbiological analyses. The potential for separating micro-organisms using chromatographic materials such as anion exchangers [2, 6], cation exchangers and neutral gels [10] and hydrophilic-hydrophobic gels [9] has been demonstrated. However, the methods employed for estimating micro-organisms (i.e. microscopy and absorbance measurement) provided little information on the viability of micro-organisms recovered from the resins, and the chromatographic techniques were not suitable for routine

1 Applied Microbiology Section, Leatherhead Food Research Association, Randalls Road, Leatherhead, Surrey, KT22 7RY, Great Britain

microbiological analysis. Wood [33] has demonstrated the basis for resolving a mixture of viable *E. coli* and *Staphylococcus aureus* into individual components using the cation exchanger Bio-Rex 70 (Bio-Rad Labs.).

In the present study several chromatographic materials – cation exchange resin, Bio-Rex 70, hydrophilic-hydrophobic gel, Phenyl-sepharose CL-4B (Pharmacia Fine Chemicals) and an immunosorbent magnetic gel, Magnogel AcA 44 (LKB Ltd.) – have been investigated for the physical manipulation of viable micro-organisms from broth cultures or artificially contaminated foods. Further observations on the interaction between micro-organisms and the chromatographic materials have also been considered.

Interaction of Micro-Organisms with the Cation Exchange Resin Bio-Rex 70

Bio-Rex 70 resin (sodium form; 100/200 mesh) has a polyacrylic hydrocarbon matrix with carboxyl functional groups. In these studies two types of columns were used. The larger (1.6 × 6.5 cm), semi-automated system consisted of 12 g of resin and the small (0.6 × 6.5 cm), manual system consisted of ca. 2 g of the resin [17]. The adsorption or recovery of micro-organisms was calculated from colony counts of the column effluent.

The adsorptive capacity of Bio-Rex 70 for *Staph. aureus* was in the order of 10^9 cells/g [32] and the adsorption depended on the particle size of the resin (P. D. Patel, unpublished data, 1984). The affinity of various micro-organisms for the cation exchanger is shown in Fig. 1.

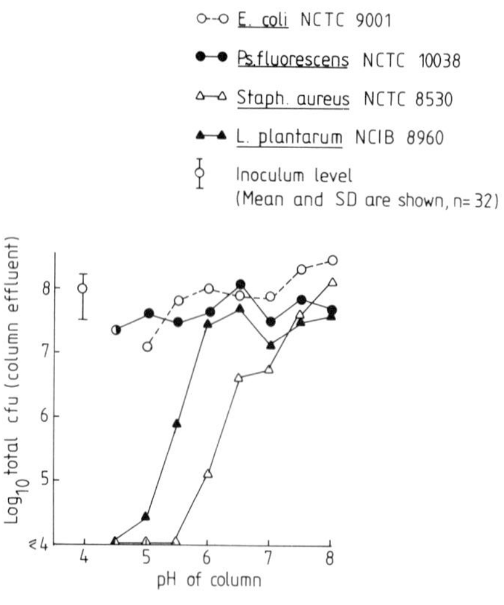

Fig. 1. Differences in the affinity of micro-organisms for the cation exchange resin Bio-Rex 70

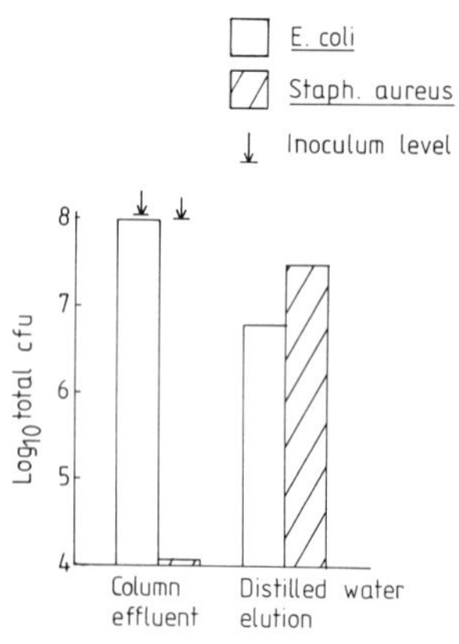

Fig. 2. Separation of *E. coli* from a mixture of *E. coli* and *Staph. aureus* (Bio-Rex 70, pH 5.5)

Separation of Microbial Species by Cation Exchange Chromatography

The differences in affinity of pure cultures of viable micro-organisms for the cation exchanger (Fig. 1) were exploited in order to enrich or separate binary mixtures of organisms into one or both of the components. When a mixed suspension of *E. coli* and *Staph. aureus* in 0.05 *M* phosphate-citrate buffer (pH 5.5) was applied to the column equilibrated to pH 5.5, approximately 80% of *E. coli* was recovered in the column effluent compared with $<1\%$ of *Staph. aureus* (Fig. 2). On elution of the column with distilled water, ca. 30% of the applied *Staph. aureus* was recovered compared with approximately 6% of *E. coli*. A similar separation was also achieved from a mixed suspension containing *E. coli* and *Lactobacillus plantarum* [17].

Application of Bio-Rex 70 Columns for Separation of Micro-Organisms Grown in Foods

Pure cultures of organisms were inoculated into 10% m/v food suspensions (in sterile 1/4-strength Ringer's solution containing 0.1% m/v Difco proteose peptone), homogenised and incubated at 37 °C for 16 h in order to simulate natural growth of organisms in foods. The food particles in the suspension were deposited at 2,000 g for 30 s and organisms were then concentrated by centrifugation at 2,000 g for 10 min and resuspended in the appropriate buffer. The subsequent separation of binary mixtures of organisms inoculated into the foods into one of the components using Bio-Rex 70 columns is shown in Table 1.

Table 1. Application of small Bio-Rex 70 columns for separation of organisms grown in foods

Food	Mixture of organisms	pH of column	Total $\log_{10}$ cfu applied to column	Total $\log_{10}$ cfu in column effluent
Raw pork (low count)	*E. coli* NCTC 9001 &	5.5	8.1	7.7
	Staph. aureus NCTC 8530		7.4	<3.6
Milk (pasteurised)	*E. coli* NCTC 9001 &	5.5	8.2	7.9
	Staph. aureus NCTC 8530		5.3	<2.6
Milk (pasteurised)	*L. plantarum* NCIB 8960	6.0	6.9	6.4
	Staph. aureus NCTC 8530		8.0	4.6

Involvement of Microbial Cell-Surface Proteins in the Interaction with Cation Exchange Resin

The involvement of a specific outer membrane protein 1 of *E. coli* in the interaction with Bio-Rex 70 was recently demonstrated by Wood [32]. In the present study, other specific cell-surface proteins were investigated for their influence on adsorption of the organisms to the cation exchanger. When *E. coli* strain

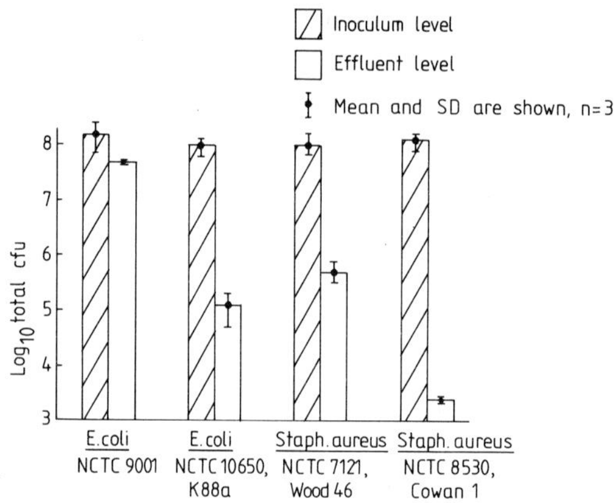

Fig. 3. Involvement of proteins in the interaction of micro-organisms with Bio-Rex 70 resin

NCTC 10650, possessing fibrillar K88a antigen [29], was applied to the Bio-Rex 70 column at pH 5.5, more than 99% of the cells were adsorbed compared with approximately 69% of the *E. coli* strain NCTC 9001, which does not possess the K88a antigen (Fig. 3). Figure 3 also demonstrates greater adsorption of the *Staph. aureus* strain Cowan 1, which possesses protein A on the cell surface, compared with the *Staph. aureus* strain Wood 46, which lacks surface protein A [8].

Mechanism of Interaction Between Micro-Organisms and the Cation Exchanger

There are several lines of evidence which strongly suggest that the interaction of micro-organisms with the cation exchanger may not involve to a significant extent the charge-charge interaction between the two bodies. It appears more probable that the organisms adsorb to the acrylic matrix of this resin by a mechanism involving hydrophobic interactions. There are three observations supporting this view.

Modification of the carboxyl functional groups of Bio-Rex 70 by various chemical treatments did not reduce significantly the adsorption of *Staph. aureus* in comparison with the non-treated resin (Table 2). Less than 40% of *Staph. aureus* which was adsorbed to methanol/HCl, carbodiimide-treated resin or the untreated resin were eluted with distilled water. However, less than 0.01% of *Staph. aureus* was eluted from the Alcian Blue-treated resins (Table 2). This may be due to adsorption of *Staph. aureus* to the strongly hydrophobic Alcian Blue dye.

Adsorption of both gram-positive and gram-negative organisms to large columns of the carboxyl-substituted hydrophilic matrix CM-sepharose gel was $\leq 60\%$ at pH 5.5 (Table 3) compared with $>99\%$ adsorption of gram-positive organisms to the carboxyl-substituted hydrophobic matrix Bio-Rex 70 resin (Fig. 1).

Finally, the *E. coli* strain NCTC 10650, which possesses fibrillar K88a antigen, consisting of a high proportion of non-polar amino acids [27], showed

Table 2. Effect of modification of carboxyl groups of Bio-Rex 70 on the adsorption of *Staph. aureus*

Treatment	Log_{10} total cfu		
	Applied to column	Column effluent[a]	Elution with distilled water[a]
Methanol/HCl	8.6	5.4	8.0
Alcian Blue[b]	8.9	2.7	3.6
Carbodiimide	8.8	5.7	8.2
Non-treated control[c]	8.8	4.3	8.4

[a] Represents a mean of 3–7 results
[b] In a separate experiment it was shown that viability of *Staph. aureus* in Alcian Blue was not reduced for up to 2 h. An experiment was normally completed in 1 h
[c] Represents a mean of 5 results

Table 3. Interaction of micro-organisms with large columns of CM-sepharose gel at pH 5.5

Organism	Log_{10} total cfu	
	Applied to column	Column effluent
E. coli NCTC 9001	8.2	8.2
E. coli NCTC 10650 (K88a)	8.0	7.9
Staph. aureus NCTC 8530	8.3	8.1
L. plantarum NCIB 8960	7.6	7.6

>99% adsorption to the Bio-Rex 70 resin (Fig. 3). However, the same strain showed only ca. 20% adsorption to the CM-sepharose gel (Table 3).

Overall, these results suggest that the carboxyl groups of the cation exchange resin play a minor role in adsorption and that the acrylic matrix is the major determinant of adsorption.

Interaction of Micro-Organisms with the Hydrophilic-Hydrophobic Gel Phenyl-Sepharose CL-4B

Micro-organisms also possess cell surface properties which confer hydrophobicity [29]. The approaches which have been used for examining the hydrophobicity of microbial cell surfaces include the adherence of bacteria to hydrocarbons [31] and plastics [23] and interaction of organisms with hydrophobic fluorescent probes [21]. However, the two techniques most commonly used for studying the cell-surface hydrophobicity are partition in aqueous polymer two-phase system [11] and hydrophobic interaction chromatography [9, 29].

Phenyl-sepharose CL-4B (Pharmacia Fine Chemicals) is a derivative of the cross-linked hydrophilic agarose gel Sepharose CL-4B and contains phenyl

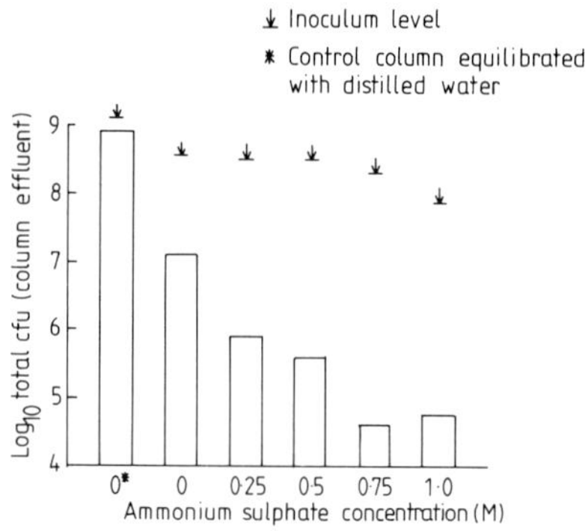

Fig. 4. Effect of various concentrations of ammonium sulphate on the adsorption of *Staph. aureus* NCTC 8530 to phenyl-sepharose gel

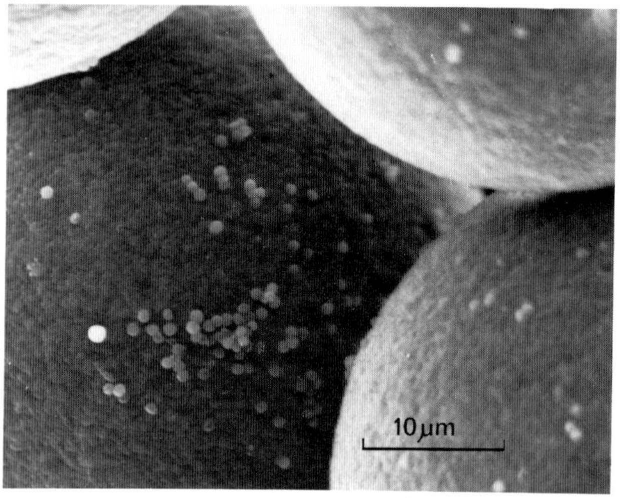

Fig. 5. Adsorption of *Staph. aureus* to phenyl-sepharose beads equilibrated with 1 *M* ammonium sulphate

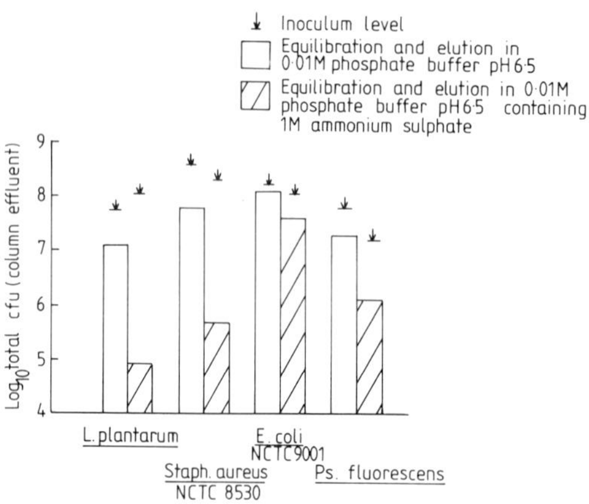

Fig. 6. Effect of ammonium sulphate (1 *M*) on hydrophobicity of micro-organisms as demonstrated by phenyl-sepharose chromatography

groups. The column chromatographic systems used in these studies were similar to that described in the previous section.

The adsorptive capacity of the phenyl-sepharose gel for *Staph. aureus* was in excess of 10^9 cfu/ml of the gel [9], and this permitted the study of interactions in small columns (ca. 10 ml) of the gel.

Within the time limit for completion of an experiment (i.e. <1 h) it was found that the viability of *E. coli, Staph. aureus* and *L. plantarum* was not reduced by 1 *M* ammonium sulphate although there was a slight reduction in the viability of *Pseudomonas fluorescens* [18].

The adsorption of *Staph. aureus* to phenyl-sepharose gel increased with an increase in ammonium sulphate concentration from 0 to 1 *M* (Figs. 4, 5). However, *Staph. aureus* did not adsorb substantially to the gel equilibrated with distilled water.

Figure 6 demonstrates the affinity of both gram-positive and gram-negative organisms for the phenyl-sepharose gel equilibrated in the presence and absence of 1 *M* ammonium sulphate.

Separation of Microbial Species by Hydrophobic Interaction Chromatography

When a mixed suspension of *E. coli* and *L. plantarum* was applied to a column of phenyl-sepharose gel equilibrated with 0.01 *M* phosphate buffer (pH 6.5) containing 1 *M* ammonium sulphate, 36% of *E. coli* was recovered in the column effluent compared with $<0.1\%$ of *L. plantarum* (Fig. 7). Although the recovery of *E. coli* was low, there was a considerable enrichment (ca. 1,000:1) over *L. plantarum* in the column effluent. On elution with distilled water, a further ca. 17% of *E. coli* and ca. 1% of *L. plantarum* were eluted (Fig. 7).

The experimental approach for separating binary mixtures of various microorganisms from milk by phenyl-sepharose chromatography was similar to that described using Bio-Rex 70 columns (see above). Table 4 summarises the various separations based on the differential affinity of micro-organisms for phenyl-sepharose gel (Fig. 6).

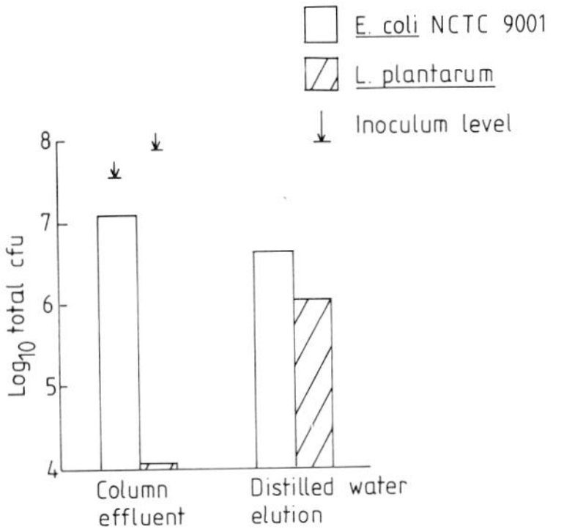

Fig. 7. Separation of *E. coli* from a mixed culture containing *L. plantarum* on phenyl-sepharose columns

Table 4. Application of small phenyl-sepharose columns (pH 6.5) for separation of organisms grown in milk

Mixture of organisms	Total $\log_{10}$ cfu applied to column	Total $\log_{10}$ cfu in column effluent
E. coli NCTC 9001 &	7.4	6.8
Staph. aureus NCTC 8530	6.7	3.6
E. coli NCTC 9001 &	8.2	7.1
Ps. fluorescens NCTC 10038	6.7	3.6
E. coli NCTC 9001 &	8.5	8.2
L. plantarum NCIB 8960	6.5	3.9

Mechanism of Interaction Between Micro-Organisms and the Phenyl-Sepharose Gel

There are two major lines of evidence which suggest that hydrophobic interactions are involved in the adsorption of micro-organisms with phenyl-sepharose gel. Firstly, the phenyl ligand of phenyl-sepharose gel, which has greater hydrophobicity at high salt concentrations (Pharmacia Fine Chemicals, hydrophobic interaction chromatography, applications book, 1979), was specifically involved in the interaction with micro-organisms at a high salt concentration (Fig. 8). Secondly, *E. coli* strain NCTC 10650, which possesses the K88a antigen consisting of a high proportion of non-polar amino acids [27], showed greater adsorption to the phenyl-sepharose gel than did the *E. coli* strain NCTC 9001, which does not possess the K88a antigen (Fig. 8).

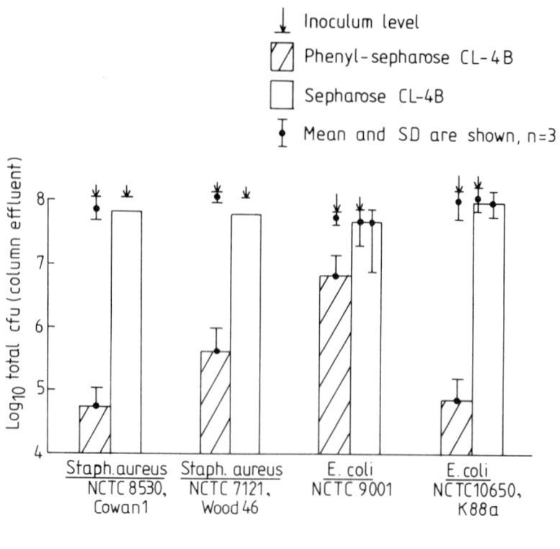

Fig. 8. Effect of phenyl groups on the adsorption of micro-organisms to phenyl-sepharose gel in the presence of 1 *M* ammonium sulphate

Interaction of Salmonellae and Other Enterobacteriaceae with an Immunosorbent Magnetic Gel

Act-Magnogel AcA-44 (LKB Ltd.) consists of 4% polyacrylamide and 4% agarose in a bead form with 7% iron oxide (Fe_3O_4) in the matrix, enabling it to be attracted by a magnetic field. The beads are activated with glutaraldehyde, and therefore free amino ligands (e.g. in proteins) can be immobilised to the gel.

Recently, magnetic enzyme immunoassay (MEIA) techniques have been developed for the rapid detection of staphylococcal enterotoxin B [15] and staphylococcal enterotoxin B IgG-type antibodies [19]. Magnogel was utilised in the MEIA techniques to effect a rapid magnetic separation of the bound enzyme conjugate from the unbound material.

In these studies Spicer-Edwards *Salmonella* flagellar antiserum or purified immunoglobulins from this antiserum were coupled to Magnogel. The immunosorbent gel was then used in a batch system to investigate the separation of *Salm. typhimurium* from mixed broth cultures and from food pre-enrichment broths.

The immunological specificity of adsorption of *Salm. typhimurium* to the immunosorbent gel is shown in Fig. 9. Maximum adsorption of *Salm. typhimurium* to the antiserum-coupled Magnogel occurred after 1.5 h of contact.

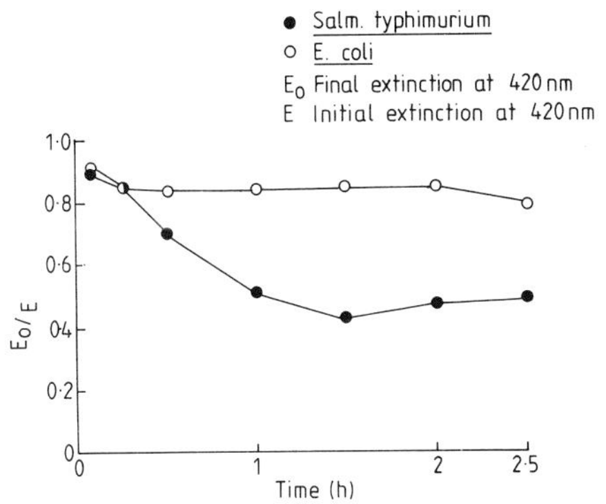

Fig. 9. Specific adsorption of *Salm. typhimurium* to Spicer-Edwards *Salmonella* antiserum coupled Magnogel

Physical Enrichment of *Salm. typhimurium* from Mixed Broth Cultures and from Food Pre-Enrichment Broths

When a mixed suspension of *E. coli* and *Salm. typhimurium* (ratio 11 : 1; Fig. 10a) was treated with the antiserum-coupled Magnogel, *Salm. typhimurium* was adsorbed to the gel as shown by the increase in ratio of *E. coli* to *Salm. typhimurium* in the supernatant (26 : 1; Table 5). There was a significant enrichment (11-fold) of *Salm. typhimurium* after elution from the Magnogel (Table 5; Fig. 10b). The enrichment of *Salm. typhimurium* from a mixture with *Cit. freundii* was 5-fold.

It was found that when organisms from *Salm. typhimurium*-spiked cocoa and chicken pre-enrichment broths were separated by centrifugation (2,000 g for 10 min), resuspended in 0.05 *M* phosphate buffer (pH 7.0) and treated with the antiserum-coupled Magnogel, high numbers (ca. total 10^7 cfu) of non-salmonel-

Table 5. Interaction of mixtures of Enterobacteriaceae with Spicer-Edwards *Salmonella* anti-serum-coupled Magnogel

Mixture of organisms[a]	Ratio of *E. coli* or *Cit. freundii* to *Salm. typhimurium*		
	Before Magnogel treatment	Supernatant after Magnogel treatment	After batch elution of Magnogel
E. coli & *Salm. typhimurium*	11:1	26:1	1:1
Cit. freundii & *Salm. typhimurium*	15:1	30:1	3:1

[a] Represents a mean of three separate experiments

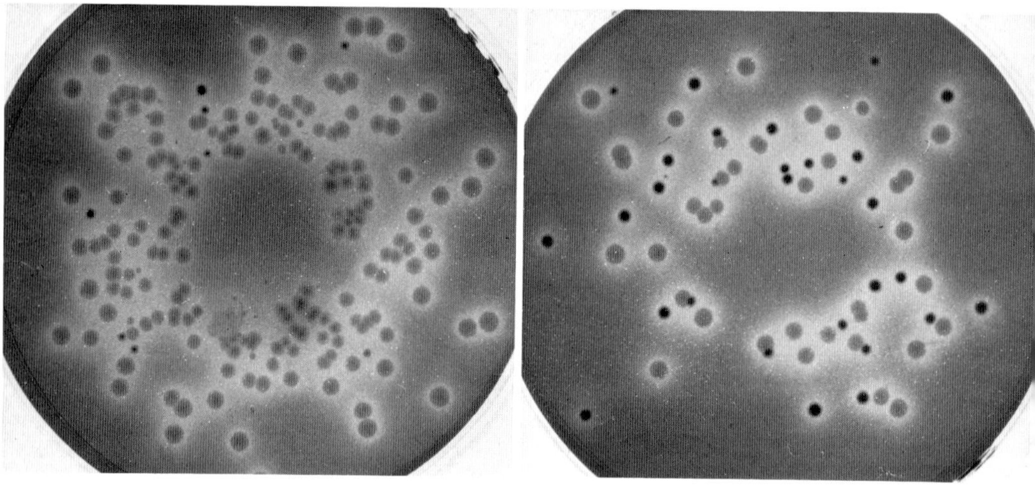

a

Fig. 10 a, b. Xylose lysine desoxycholate agar plates showing a mixture of *Salm. typhimurium* (*dark colonies*) and *E. coli* (*light colonies*). **a** before and **b** after the Magnogel treatment

lae were eluted from the gel (P. D. Patel, unpublished data, 1984). *Salm. typhimurium* was therefore not enriched from the food pre-enrichment broths. There may be at least two explanations for adhesion of high numbers of non-salmonellae to the Magnogel. Firstly, the non-salmonellae may adsorb non-specifically to the matrix of the Magnogel by way of a greater number of effective collisions between the organisms and the Magnogel. Secondly, the somatic "0" antibodies which cross-react strongly with other Enterobacteriaceae [1] may be present in the Spicer-Edwards *Salmonella* flagellar antiserum and therefore increase adsorption of non-salmonellae to the Magnogel.

Discussion

A vast range of techniques (see the Introduction) is used for the separation and purification of biological molecules on the basis of conformation, size, density, charge, hydrophobicity, and immunological specificity of the biomolecules. Since microbial cell surfaces are highly charged [12, 13] and may also possess hydrophobic [9] and immunological [22] properties, physical techniques may be used for the manipulation of micro-organisms.

In this limited study we have investigated the potential of three different chromatographic materials for manipulating viable micro-organisms. It demonstrates that physico-chemical and immunological characteristics of the cell surface may be exploited to separate or enrich considerably one species from binary mixtures of organisms in broth (e.g. Fig. 2) or artificially contaminated food (e.g. Table 1). However, application of the chromatographic materials for manipulating micro-organisms from naturally contaminated foods may present several problems.

In general, low numbers of organisms of significance (e.g. *E. coli*) in a heterogeneous population of organisms may be present in foods. In this case a growth-enrichment stage may be required prior to physical separation of micro-organisms by chromatography. On the other hand, food constituents (e.g. lipids, proteins, and inorganic salts) may interfere with the adsorption of organisms to chromatographic materials. However, the cation exchange resin has been used successfully in adsorbing out protein from raw meat homogenates prior to concentration and rapid estimation of the organisms [26].

Wood [33] has suggested that there are a large number of possible combination of resin types and ionic forms, conditions of pH and ionic strength, column size and flow rate that can be investigated for the separation of original microbial populations. However, this would invoke considerable empirical investigation whilst a better knowledge of the mechanisms involved may permit closer prediction of useful systems. It was demonstrated that microbial cell surface proteins (e.g. K88a antigen of *E. coli*) are involved in the specific interaction of the organisms with both the cation exchanger Bio-Rex 70 (Fig. 3) and the hydrophobic phenyl-sepharose gel (Fig. 8). The involvement of K88 antigen in *E. coli* [29], protein A in *Staph. aureus* [9] and protein M in *Strep. pyogenes* [28] in the interaction with hydrophilic-hydrophobic gels has been reported previously.

Several results regarding the adhesion of micro-organisms to both Bio-Rex 70 and the phenyl-sepharose gel suggest that hydrophobic interactions (i.e. van de Waal's forces [4]) could be involved in the adhesion process. For instance, Table 2 shows that *Staph. aureus* probably adsorbs to the hydrophobic acrylic matrix [5] of Bio-Rex 70, and Figs. 3 and 8 show >99% adsorption of the *E. coli* strain possessing hydrophobic K88a antigen [27] to both Bio-Rex 70 and phenyl-sepharose gel, respectively.

Daniels [2] reported that micro-organisms have low apparent isoelectric points (Ip $\leqq$ 5.0) and behave as "macroscopic zwitterions." At pH greater than Ip, micro-organisms bear an overall negative surface charge and interact with anion exchangers, while at a pH less than Ip they bear an overall positive charge and interact with cation exchangers. The results of this study showed that the gram-negative bacteria did not absorb significantly to the cation exchanger in the pH range 4.5 to 8.0 (Fig. 1). A similar result was obtained by Wood [33] on the interaction of some gram-negative bacteria with the Bio-Rex 70. This may be related to their low apparent Ip value. However, the gram-positive bacteria (e.g. *Staph. aureus*) showed a greater adsorption (i.e. >90%) even at pH 7.0, at which pH *Staph. aureus* would not be expected to adsorb.

A theory to explain the salt-dependent adhesion of *Staph. aureus* to Bio-Rex 70 resin at pH greater than Ip has been proposed by Wood [33]. From the data of Neihoff and Echols [13], who suggested that microbial cell surface carried a dominant exterior negative charge which overlay amino groups, Wood [33] proposed that these amino groups may be involved in the specific interaction with the Bio-Rex 70 and that the presence of salt accentuated the interaction. However, investigation of the interaction between *Staph. aureus* and both carboxyl-modified Bio-Rex 70 (Table 2) and the CM-sepharose gel (Table 3) suggests that negative charges on the cation exchangers are not involved significantly in the adhesion process.

The mechanisms involving attachment of a negatively charged bacterium to a negatively charged animal cell were recently proposed by Ofek and Beachey [14]. The energy barrier between the negatively charged surfaces was overcome by interaction of hydrophobic molecules on the bacterium and the hydrophobic portion of phospholipids on the animal cell membrane. Watt and Ward [30] suggested a similar mechanism in the attachment of gonococci to mammalian cells with gonococcal fimbriae acting as a bridge between both the negatively charged surfaces.

Although the Bio-Rex 70 resin represents a large and a highly rigid structure in comparison with an animal cell, a certain analogy may be drawn between the interaction of micro-organisms and the macroscopic surfaces. It is conceivable that hydrophobic constituents such as the K88a antigen on the surface of highly negatively charged *E. coli* [12] interact with the hydrophobic acrylic matrix of the negatively charged Bio-Rex 70 (Fig. 3), thus overcoming the energy barrier between the repulsive negative surface charges.

Salmonella typhimurium was not enriched from cocoa and chicken pre-enrichment broths using the Spicer-Edwards serum coupled Magnogel (see above). These results suggest that low levels (10^5–10^6 cfu/ml) of salmonellae normally present in pre-enrichment broths [3] of *Salmonella*-contaminated foods may not be separated using the antiserum-coupled Magnogel. However, the separation of salmonellae from other Enterobacteriaceae may be improved provided that adsorbents with a low non-specific binding and highly specific monoclonal antibodies with high avidity for salmonellae [22] become commercially available.

Overall, the results of these studies demonstrate the potential of chromatographic materials as a basis for sample preparation as new instrumental methods become established and signal/noise ratio becomes an important consideration in microbiological analyses.

At a more fundamental level, the study on the interaction between micro-organisms and both cation exchanger Bio-Rex 70 and hydrophobic phenyl-sepharose has demonstrated similarities in the mechanism of adhesion.

Acknowledgement. This work was supported by the Ministry of Agriculture, Fisheries and Food and is Crown Copyright. The advice and assistance of Dr. Paul A. Gibbs in the preparation of the manuscript is gratefully acknowledged.

References

1. Aleksićs (1982) Antigenic relationships between *Salmonella* and other Enterobacteriaceae: their significance with regard to the *Salmonella* diagnosis. Zentralbl Bakteriol Mikrobiol Hyg [A] 251:451
2. Daniels SL (1980) Mechanisms involved in sorption of micro-organisms to solid surfaces. In: Bitton G, Marshall KC (eds) Adsorption of micro-organisms to surfaces. Wiley Interscience, New York, p 8
3. D'Aoust JY (1981) Update on pre-enrichment and selective enrichment conditions for detection of *Salmonella* in foods. J Food Protect 44:369–374
4. Gerson DF, Akit J (1980) Cell surface energy, contact angles and phase partition. II. Bacterial cells in biphasic aqueous mixtures. Biochim Biophys Acta 602:281–284
5. Gerson DF, Scheer D (1980) Cell surface energy, contact angles and phase partition. III. Adhesion of bacterial cells to hydrophobic surfaces. Biochim Biophys Acta 602:506–510
6. Hall AN, Jafri SSA (1980) Gradient elution from a DEAE-cellulose column of capsulate and non-capsulate strains of *Klebsiella aerogenes*. J Gen Microbiol 117:263–265
7. Hardy D, Kraeger S, Dufour SW, Cady P (1977) Rapid detection of microbial contamination in frozen vegetables by automated impedance measurements. Appl Environ Microbiol 34:14–17
8. James AM, Brewer JE (1968) A protein component of the cell surface of *Staphylococcus aureus*. Biochem J 108:257–262
9. Jonsson P, Wadström T (1983) High surface hydrophobicity of *Staphylococcus aureus* as revealed by hydrophobic interaction chrmatography. Curr Microbiol 8:347–353
10. Kosinkiewicz B, Varanka M (1975) Behaviour of microbial cells on columns of some types of Sephadex gels. J Chromatogr 114:463–464
11. Magnusson KE, Kilkström E, Norlander LA, Norqvist J, Davies J, Normark S (1979) Effect of colony type and pH on surface charge and hydrophobicity of *Neisseria gonorrhoeae*. Infect Immun 26:397–401
12. Neihoff R, Echols WH (1973) Physico-chemical studies of microbial cell walls. I. Comparative electrophoretic behaviour of intact cells and isolated cell walls. Biochim Biophys Acta 318:23–32
13. Neihoff R, Echols WH (1978) Chemical modification of the surfaces of bacterial cell walls. Physiol Chem Phys 10:329–343
14. Ofek I, Beachey EH (1980) General concepts and principles of bacterial adherence in animals and man. In: Beachey EH (ed) Bacterial adherence, 3rd vol. Chapman and Hall, London, p 3
15. Patel PD, Gibbs PA (1984) Development of an inhibition Magnetic Enzyme Immunoassay (MEIA) technique for determination of staphylococcal enterotoxin B. Biochem Soc Trans 12:264–265
16. Patel PD, Wood JM (1983) The separation of micro-organisms from raw meats and their rapid estimation by measurement of microbial ATP. Leatherhead Fd RA Res Rep No 420
17. Patel PD, Wood JM, Gibbs PA (1983a) Physico-chemical interaction of micro-organisms with the cation exchange resin Bio-Rex 70. A preliminary application of the technique to foods. Leatherhead Fd RA Res Rep No 437
18. Patel PD, Wood JM, Gibbs PA (1983b) Interaction of micro-organisms with hydrophilic-hydrophobic gels (phenyl-sepharose Cl-4B and Octyl-sepharose Cl-4B): Preliminary application of the technique to foods. Leatherhead Fd RA Res Rep No 446
19. Patel PD, Wood JM, Gibbs PA (1984) Development of a Magnetic Enzyme Immunoassay (MEIA) technique for determination of anti-(staphylococcal enterotoxin) immunoglobulin G-type antibodies. Biochem Soc Trans 12:266–268
20. Pettipher GL (1981) Rapid methods for assessing bacterial numbers in milk. Dairy Ind Int 46:15–21
21. Radda GK (1975) Fluorescent probes in membrane studies. Philos Trans R Soc Lond Ser B 270:539–549
22. Robison BJ, Pretzman CI, Mattingly JA (1983) Enzyme immunoassay in which a myeloma protein is used for detection of salmonellae. Appl Environ Microbiol 45:1816–1821
23. Rosenberg M (1981) Bacterial adherence to polystyrene: a replica method of screening for bacterial hydrophobicity. Appl Environ Microbiol 42:375–377
24. Rowley DB, Previte JJ, Srinivasa H (1978) A radiometric method for rapid screening of cooked foods for microbial acceptability. J Food Sci 43:1720–1722

25. Sharpe AN, Woodrow MN, Jackson AK (1970) Adenosine triphosphate (ATP) levels in foods contaminated by bacteria. J Appl Bacteriol 33:758–767
26. Stannard CJ, Wood JM (1983) The rapid estimation of microbial contamination of raw meat by measurement of adenosine triphosphate (ATP). J Appl Bacteriol 55:429–438
27. Stirm S, Ørskov F, Ørskov I, Hansa B (1967) Episome-carried surface antigen K88 of *Escherichia coli*. II. Isolation and chemical analysis. J Bacteriol 93:731–739
28. Tylewska SK, Hjertén S, Wadström T (1979) Contribution of M protein to the hydrophobic surface properties of *streptococcus pyogenes*. FEMS Microbiol Lett 6:249–253
29. Wadström T, Faris A, Hjertén S (1980) Adhesion of enteropathogenic bacteria to hydrophobic surfaces. In: Berkeley RCW, Lynch JM, Melling J, Rutter PR, Vincent B (eds) Microbial adhesion to surfaces. Harwood, London, p 537
30. Watt PJ, Ward ME (1980) Adherence of *Neisseria gonorrhoeae* and other *Neisseria* species to mammalian cells. In: Beachey EH (ed) Bacterial adherence, vol 3. Chapman and Hall, London, p 253
31. Weiss E, Rosenberg M, Judes H, Rosenberg E (1982) Cell-surface hydrophobicity of adherent oral bacteria. Curr Microbiol 7:125–128
32. Wood JM (1979) The separation and estimation of microbial populations using biochemical and biophysical techniques. Thesis, University of Surrey
33. Wood JM (1980) The interaction of micro-organisms with ion exchange resins. In: Berkeley RCW, Lynch JM, Melling J, Rutter PR, Vincent B (eds) Microbial adhesion to surfaces. Harwood, London, p 163

Electrical Impedance Method for Determining Microbial Quality of Foods

R. FIRSTENBERG-EDEN [1]

Introduction

Impedance is the opposition to flow of an alternating electrical current in a conducting material. When monitoring the growth of microorganisms, the conducting material is a microbiological medium. Impedance changes occur in the medium as its chemical composition changes due to growth and metabolic activity of microorganisms. During growth, microorganisms convert larger molecules to smaller, more mobile metabolites, thus changing the impedance. When the microbial population reaches a threshold fo 10^6–10^7 cells/ml an exponential change in impedance signal is observed. The time required for this exponential change is inversely proportional to the initial cell concentration and is defined as impedance detection time (IDT).

The impedance method is highly versatile. It could be used to perform almost all microbiological tasks needed by the food industry, if proper protocols were to be developed. The method achieves faster identification of microbial problems (e.g., several hours vs. several days), is less labor intensive than traditional techniques, and provides automatic recording and presentation of data.

Considerations in the Development of Impedance Procedures

There are several fundamental differences between the impedance technique and standard methods for the enumeration of microorganisms. The impedance technique relies on the measurement of metabolic changes whereas traditional plate count procedures depend on the production of a visible biomass. Since the impedance technique depends on the rapid detection of metabolic change, factors such as time and temperature become critical parameters in the assay.

Media

Certain media are more useful in impedance testing than others. This may be due to the fact that organisms utilize different metabolic pathways in different media. Since some end products yield a stronger impedance signal than others, the selec-

1 Bactomatic Inc., P.O. Box 3103, Princeton, NJ 08540, USA

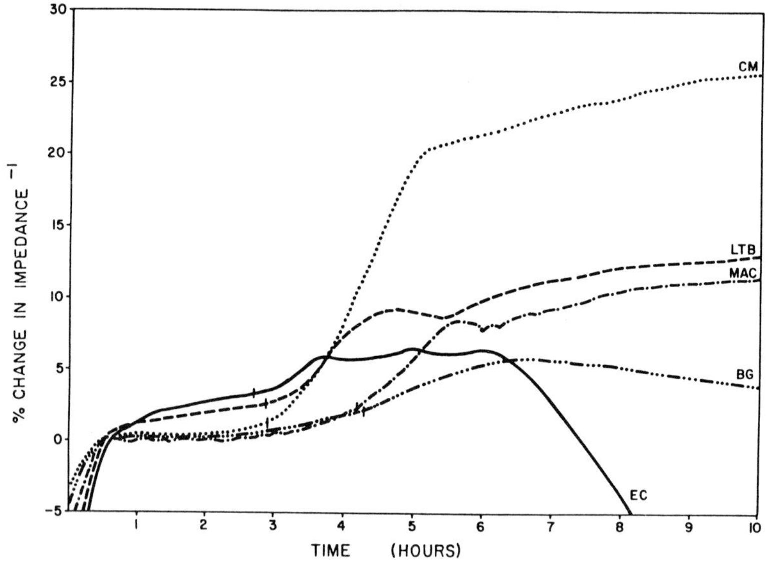

Fig. 1. Comparison of impedance curves obtained due to the growth of *E. coli* M1 in four traditional media, including MacConkey (*MAC*), Brilliant Green Bile (*BG*), Lauryl Tryptose Broth (*LTB*), E.C. Medium (*EC*), and in CM, the medium developed for impedance. (Firstenberg-Eden and Klein 1983)

tion of an appropriate medium is crucial. For example, Fig. 1 illustrates the impedance curves generated from four commonly used coliform media and CM broth (specifically designed for impedance testing). The curve produced in CM broth was associated with the steepest slope and the greatest total amplitude change. Therefore, the IDT in CM was the easiest to assess.

Minimizing Generation Time Differences

Although a successful impedance assay does not depend on the presence of identical groups of microorganisms in successive samples, the generation times for all isolates should be similar. Thus, in order to achieve the best possible correlation between standard methods and IDT, the generation times for all organisms should be brought as close together as possible under the experimental conditions employed.

Figure 2 compares the generation times for two psychrotrophs isolated from a single meat sample as a function of differing incubation temperatures. If such a mixture were to be incubated and tested at 30.0 °C, the *Pseudomonas* would replicate with a generation time roughly 4 times that of the *Coryneform*. With this difference, even at an initial ratio of 10^5 *Pseudomonas* to 1 *Coryneform*, the *Coryneform* population would reach the instrument's threshold first and thus be the organism detected. Consequently, in most cases only the *Coryneform* will be detected by the impedance method. While both organisms have the ability to form

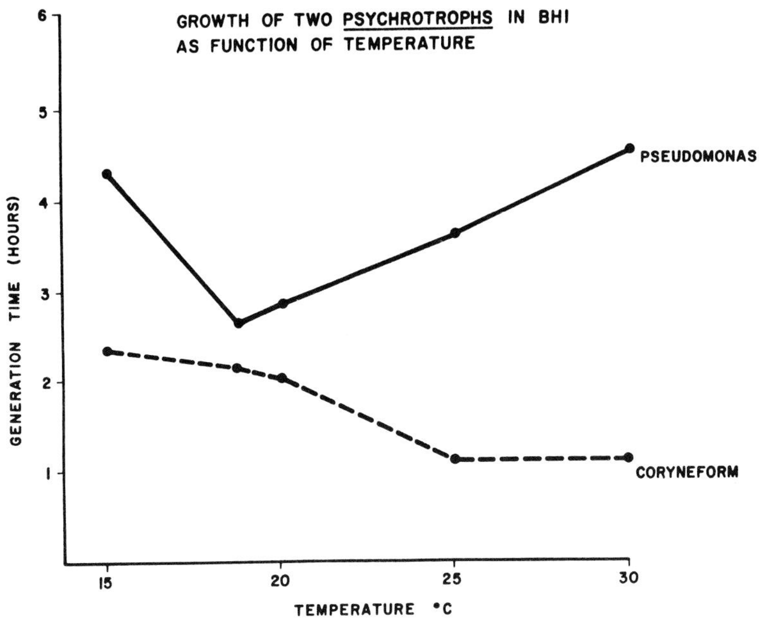

Fig. 2. Effect of temperature on the generation times of two psychrotrophs isolated from meat

colonies on plates incubated at 30 °C for 72 h, only the *Pseudomonas* would be countable on plates since the *Coryneform* will be diluted to extinction.

The above example clearly indicates that it would be impossible to create a meaningful relationship between IDTs and plate counts with samples containing a mixture of these two organisms if these are incubated at 30.0 °C. However, by merely lowering the incubator temperature in the impedance assay to 18.0 °C, the IDTs for samples containing pure cultures of either of these two isolates will correlate with plates incubated at 30.0 °C. This example demonstrates how the temperature used for performing the standard plate count might not be appropriate for use with the impedance method.

Selecting Between Impedance, Conductance, and Capacitance

Impedance is a complex entity composed of a resistive component and a reactive component. Impedance (Z) can be expressed in terms of conductance (G) and capacitance (C) as follows:

$$Z = \sqrt{\left(\frac{1}{G}\right)^2 + \left(\frac{1}{2\pi fC}\right)^2},$$

where f is the frequency in H_z. Both capacitance and conductance are frequency dependent when measured in an electrochemical well (Schwan 1963). At low frequencies the impedance is predominantly affected by the capacitance, while at high frequencies it is mostly affected by the conductance.

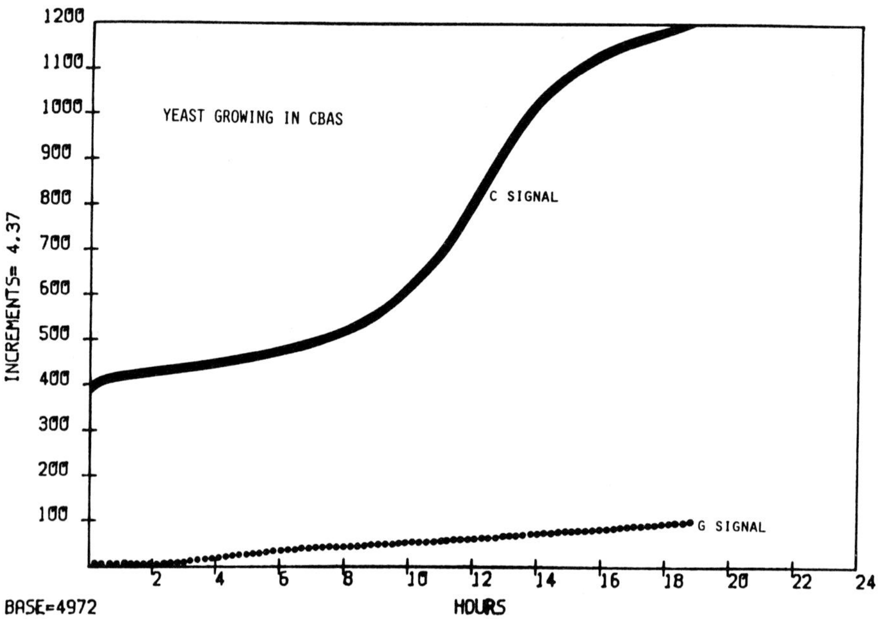

Fig. 3. Comparison of conductance (*G*) and capacitance (*C*) signal generated by a yeast isolated from orange juice

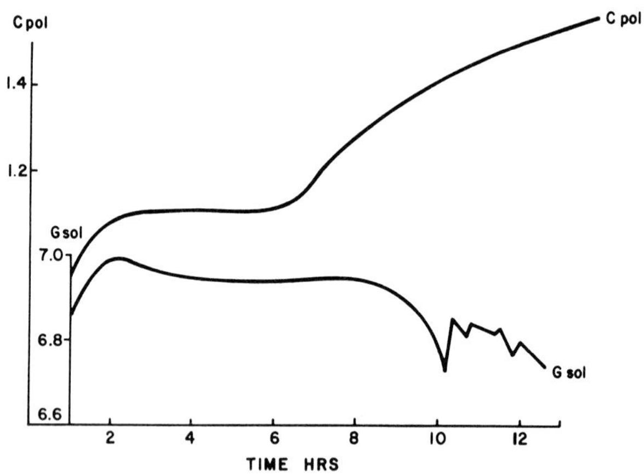

Fig. 4. The effect of growth of *C. utilis* in Orange Juice Serum Broth on capacitance (*C*) and conductance (*G*). C in μF and G in mS. (Adapted from Firstenberg-Eden and Zindulis 1984)

All three parameters (Z, G, and C) are potentially useful in monitoring microbial growth (Firstenberg-Eden and Zindulis 1984). In low conductivity media, such as Plate Count Broth, the change in G of the medium clearly correlated to bacterial growth. In more conductive media, such as Brain Heart Infusion (BHI), the relative changes in G were smaller, and measurements of changes in C were a better indication of bacterial growth. Yeast growth resulted in larger changes in C (Fig. 3) while the changes in G were very small. Furthermore, when the G signal was used to monitor the growth of yeasts, the direction of the conductance curves was unpredictable. While some yeasts resulted in an increase in G, others were associated with a decrease (Fig. 4).

Estimation of Microbial Levels with Impedance

Total Count

Currently the most commonly used application of impedance is to determine whether a sample contains above or below a predetermined concentration of organisms. This approach was used in conjunction with a calibration curve for such products as meats (Firstenberg-Eden 1983), raw milk (Firstenberg-Eden and Tricarico 1983), and frozen vegetables (Hadley et al. 1977). Prior to adopting an impedimetric method to estimate numbers of bacteria, a calibration curve should be generated. This curve defines the relationship between impedance and the standard method, usually the standard plate count (SPC) method.

In determining the number of samples required to construct the calibration curve a balance must be struck between the optimal number of data points required for reliable calculation of the curve, and the amount of time and effort required to obtain such data. Generally, as the numbers of samples analyzed increases, the calibration curve more reliably describes the performance of a particular product in the impedance assay. It has been determined experimentally that 100 samples are generally sufficient for the preparation of a reliable calibration curve having a significant correlation coefficient, and meaningful line parameters.

It is suggested that the 100 samples chosen should have microbial concentrations equally distributed over a four to five log cycle range. When microbial concentrations are restricted to a narrower range (e.g., one to two log cycles), the regression line is very sensitive to the addition of points in certain areas of the calibration curve. In such cases all statistical parameters can change markedly with the addition of data from one or two fresh samples.

The calibration curves described above may be prepared for most products. Therefore, the impedance method could be used to estimate numbers of organisms in a large variety of products. However, proper development of the analytical protocol is the key to success. The medium, temperature, pH, etc. routinely used in the standard method are rarely appropriate for the impedance method. Consequently, only the development of appropriate media that will optimize the impedance signal, choosing the appropriate temperature that minimizes generation times together with the appropriate signal selection, will yield accurate consistent impedimetric estimation of microbial counts. The importance of this can be demonstrated by examining the protocol developed for total count in raw milk.

The growth of typical milk microflora on the surface of modified plate count agar (MPCA) resulted in shorter IDTs and better curves than the growth in either diluted milk or BHI (Firstenberg-Eden and Tricarico 1983). Although at 30 °C

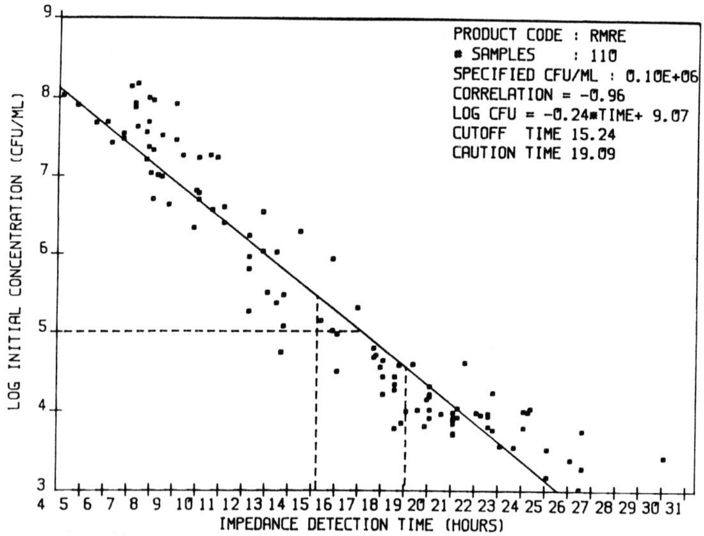

Fig. 5. The calibration curve relating impedance detection times to total plate count for raw milk samples. (Firstenberg-Eden and Tricarico 1983)

mesophiles often have generation times four to five times shorter than psychrotrophs, at 18 °C both populations have similar generation times. Therefore, 0.1 ml of raw milk was pipetted onto wells prefilled with MPCA and the wells were incubated at 18 °C. This procedure was tested with 110 milk samples from 14 different farms consisting of a varied ratio of mesophiles to psychrotrophs. The result (Fig. 5) shows a high correlation (0.96) between SPC at 32 °C and IDTs at 18 °C. This calibration curve can be used to classify the milk samples as above or below a certain specified level. If we choose a level of 10^5 CFU/ml, the corresponding IDT will be 17 h. By adding one standard deviation to each side of the IDT the time axis can be divided into three sectors: (i) 0–15.2 h, in which all the samples should have more than 10^5 CFU/ml; (ii) >19.1 h, where all the samples have less than 10^5 CFU/ml; (iii) 15.2–19.1 h, in which we have samples that cannot be classified with confidence. Using this scheme only one sample out of 110 was misclassified.

Since mesophiles are usually the predominant group of organisms present in farm fresh raw milk (Sogaard and Lund 1981; Zall et al. 1982), a method which estimates mesophilic load would be indicative of total count. Impedance monitoring of organisms grown at 35 °C on the agar surface (MPCA) correlated well to the number of mesophilic organisms present in the sample. For all samples containing a majority of mesophiles, the IDTs at 35 °C highly correlated ($r = -0.97$) to the SPC at 32 °C. This procedure allows for a rapid screen of raw milk samples. Any contamination greater than 10^5 mesophiles/ml can be detected within 4 h.

A collaborative study was carried out to establish the reproducibility of this impedance method in predicting counts of raw milk (Firstenberg-Eden 1984). Raw milk samples were sent to six laboratories to be examined by SPC and by the impedance method which produced bactometer predicted counts (BPC). The results of this study indicated that the suggested impedimetric procedure yielded accurate reproducible results, but required only a few hours to obtain results as compared with 48 h for the SPC.

Selective Media

A similar screening method can be used in conjunction with selective media to detect levels of a variety of groups of organisms of importance in foods, such as coliforms, yeast, lactic acid bacteria, and staphylococci. Again, the medium-temperature combination used in the conventional method might not be appropriate for the impedance method. Two examples of the use of selective media will be shown.

In Fig. 1 it was shown that CM broth, especially developed for the impedance method, resulted in better impedance curves than conventional media. The inhibitor level and the chosen temperature (35 °C) minimized the generation times of the coliforms tested (Firstenberg-Eden and Klein 1983). The IDTs from CM broth correlated better with the Violet Red Bile Agar (VRBA) counts confirmed in Brilliant Green than with the original VRBA data, suggesting that CM is more selective for coliforms than VRBA. This procedure was found effective for determining coliform levels in meat and a variety of dairy products (milk, cream, ice cream, yogurt, etc.).

It was not surprising that a medium especially developed for the impedimetric detection of yeasts (CBAS) resulted in superior impedance curves to all tested common mycological media (Zindulis 1984). The capacitance signal was found to provide stronger responses than the conductance signal (Figs. 3 and 4).

This medium was successfully applied for the detection of yeasts in orange juice and yogurt. A concentration of 10^2 yeast CFU/ml in a 1 : 10 dilution of orange juice in CBAS was detected within 25 h by impedance, in contrast to 3–5 days by plate count. The use of a similar procedure to detect yeast in yogurts resulted in detection times shorter than 20 h for 10^2 yeast CFU/ml.

Certain brands of yogurt or orange juice required a modification of CBAS. When a 1 : 10 dilution of the product in CBAS resulted in a pH below 4.5, addition of ammonium hydroxide to a final pH of 5.0–5.5 was found to improve the impedance signal generated by yeasts. The more favorable pH for yeast metabolism resulted in earlier detection times and better curves.

Utilization of Impedance for Noncounting Applications

In the previous section it was assumed that the impedance method is justified as long as it correlates with the traditional plate count method. This is indeed the situation, at least in food microbiology. However, it is important to understand that colony counts are not an end in themselves. Sharpe (1979) argues that plate count is a unique datum, nothing in the physical, chemical, biochemical, or im munological world corresponds to it. It provides very little information about the ability of food to deteriorate or to become toxic or infectious. The impedance method, which measures microbial activity and kinetics rather than numbers, can provide a better measure of acceptability and health hazards of foods than the plate count method.

Shelf-Life

Since the impedance method measures metabolic activity and, as a matter of course, integrates the effects of numbers of organisms and their metabolic activity, it would seem to be a most suitable tool for shelf-life prediction. It is therefore surprising that only a few studies are reported in the literature on the use of impedance for this purpose. Most of these reports are about prediction of shelf-life of pasteurized milk.

Recently Bishop et al. (1984) reported a new impedance test for the prediction of shelf-life of milk. This test involves the preincubation of diluted milk (1 : 1 in plate count broth) at 18 °C for 18 h, which allowed the quick multiplication of spoilage organisms. The preincubation step was followed by the addition of 0.5 ml of diluted milk onto the surface of MPCA and monitoring the impedance at 21 °C. Using this method the results are available in 25–38 h (including the preincubation time). The impedance method proved to have the more significant relationship with shelf-life ($r=0.88$) than either total count ($r=0.54$) or psychrotrophic count ($r=0.64$). Only the Moseley test, which required 7 days to complete, gave similar results ($r=0.84$).

Characterization and Identification

By measuring impedance changes associated with bacterial growth simultaneously in a number of different media, characteristic patterns can be obtained. Although little work has been done to date toward applying impedance measurements to bacterial identification, this field appears to have some potential because of the ease with which large numbers of channels (with different growth conditions) can be sampled. An extension of this concept is to use growth inhibitors to build a repertoire of different media so that the results of growth in such a set of media can be used to further characterize and even identify an unknown organism. Firstenberg-Eden et al. (1983) have successfully tried this approach to differentiate between salmonellae and non-salmonellae.

Microbial Growth Kinetics

In foods, various determinants of microbial growth such as temperature, water activity, pH, and concentrations of inhibitors affect the stability of the product. It would be advantageous to investigate the interactions between these various factors as they affect microbial growth, as we attempt to extend the shelf-life and safety of foods. In the development of starter cultures for the dairy, wine, or meat industry, for example, it is very important to determine in advance how active these cultures are going to be. These are only a few examples of the industry's need to assess growth rates and microbial kinetics. The impedance method provides a unique opportunity to study microbial growth and activity. The potential use of impedance to obtain a better understanding of microbial kinetics in complex systems is one of the most exciting areas for this technology. There is every possibility that it will provide new insight and understanding in many microbial systems.

Summary

The versatility of the impedance method makes it a valuable tool in food microbiology. Impedance can be used to estimate total number of bacteria or to detect levels of specific groups such as coliforms and yeasts. The impedance method provides a unique opportunity to determine potential shelf-life or microbial activity and kinetics. Impedance microbiology represents a different approach to qualitative and quantitative microbiology. Although it requires developmental work prior to its application, rewarding results can be obtained.

References

Bishop JR, White CH, Firstenberg-Eden R (1984) A rapid impedimetric method for determining the potential shelf-life of pasteurized whole milk. J Food Prot 47(6):471–475

Firstenberg-Eden R (1983) Rapid estimation of the number of microorganisms in raw meat by impedance measurements. Food Technol 37:64–70

Firstenberg-Eden R (1984) A collaborative study of the impedance method for examining raw milk samples. J Food Prot 47(9):707–712

Firstenberg-Eden R, Klein CS (1983) Evaluation of a rapid impedimetric procedure for the quantitative estimation of coliforms. J Food Sci 48:1307–1311

Firstenberg-Eden R, Tricarico MK (1983) Impedimetric determination of total, mesophilic and psychrotrophic counts in raw milk. J Food Sci 48:1307–1311

Firstenberg-Eden R, Zindulis J (1984) Electrochemical changes in media due to microbial growth. J Microbiol Methods 2:103–115

Firstenberg-Eden R, VanSise ML, Klein CS (1983) An impedimetric method for the presumptive identification of *Salmonella*. IFT 1983 Abstract 346, New Orleans

Hadley D, Kraeger SJ, Dufour SW, Cady P (1977) Rapid detection of microbial contamination in frozen vegetables by automated impedance measurements. Appl Environ Microbiol 34:14–17

Schwan HP (1963) Determination of biological impedances. In: Nastuk WL (ed) Physical techniques in biological research, vol VI. Electrophysiological methods, part B. Academic Press, New York, p 323

Sharpe AN (1979) Germ of a new food microbiology. New Scientist 13:860–862

Sogaard H, Lund R (1981) Psychrotrophic counts in raw milk and the keeping quality of pasteurized milk products. In: Psychrotrophic microorganisms in spoilage and pathogenicity. Academic, New York, p 91

Zall RR, Chen JH, Murphy SC (1982) The detection of psychrotrophic bacteria in milk using a dye test. Cult Dairy Prod J 17:7–12

Zindulis J (1984) A medium for the impedimetric detection of yeasts in foods. Food Microbiology 1:159–167

Are Rapid and Sensitive Microbiological Procedures Necessary for Monitoring the Environment?

W. J. Hausler Jr.[1]

Introduction

In today's society confusion reigns over the application, utilization, and interpretation of rapid methodologies. We have become accustomed to rapid travel by jets flying at speeds greater than the speed of sound or by manned space ships circumnavigating the globe in a little more than an hour. On the ground, we move from one place to another by rapid public transit or by automobile on restricted access highways or autobahns. In our homes, we have become particularly used to rapid methods in the kitchen, where convection or microwave ovens cook food in a fraction of the time utilized by conventional methods.

We walk rapidly, we talk rapidly, and we even communicate rapidly. Television satelite communications now permit an individual to view at first hand by electronic means an event that is occurring at the same instant halfway around the globe. In addition, television programs are presented so that even the most complicated mystery is readily solved within the program scheduling limits. Have you ever considered how audiences would shrink and how boring these complicated mystery stories would be if they were presented in a true-to-life span of time?

Please do not consider me antagonistic toward rapid diagnostic processes because of these preliminary remarks. Rapidity is necessary in a variety of situations, but one must assess the characteristics of the patient, client, or situation before rapid diagnostic methods can be usefully applied. Also, it is vitally important to recognize what it is you are looking for in order to decide whether a specific system or method should be deployed – rapidly or not.

In considering the subject of rapid and sensitive microbiological procedures for monitoring the environment, we must first define our patient, client, situation, or purpose. If this is not done, we run the risk of finding ourselves with an enormous stack of data in search of a problem. The environment includes everything that surrounds us, and we have all sorts of devices, methods, and senses to determine what is taking place in our surroundings. We can rapidly measure changes in temperature, atmospheric pressure, humidity, and wind velocity and direction. We can sense odors and see particulate-laden air. These forces effect changes in our well-being in one way or another. The quantity and quality of airborne particulates is usually measured over a short period, such as 24 h, without attempting to identify the discrete composition of the particulates. Oxides of nitrogen, sulfur

1 State Hygienic Laboratory, University of Iowa, Iowa City, IA 52242, USA

dioxide, and ozone are measured in much the same way. In those areas of the world where atmospheric dispersion of pollens and other allergens are a problem, there are systems enabling these to be measured and classified microscopically.

But how do we monitor specific microbes in the air? Unless it has been demonstrated that a particular infectious agent is causing a problem, specific monitoring for microbial agents is not generally practiced. Surrogate systems for monitoring the air environment are far more useful, practicable, and cost effective than rapid, sensitive, and specific microbiological monitoring. There are numerous devices for measuring the microbial load in a designated volume of air on either selective or nonselective culture or collection media. Unless required for some definite reason, rapid detection of specific microorganisms is unnecessary. I am not implying that we must invoke diagnostic systems only after a certain morbidity rate is reached. However, I am saying that effective surrogate monitoring systems are quite efficient and provide useful fundamental information on what one must then look for with more specific methods. It makes very little sense to monitor routinely our air for any and all of the microbes capable of causing disease in a biologic receptor that may be transported in our air environment.

Whatever its matrix, our environment is a slowly evolving system. It has a tremendous buffering capacity that adjusts to and accommodates for a large variety of unusual hazards and interferences; however, once a hazard or interference is suspected, there are a host of systems that can be deployed to identify the problem.

For the last 20 years, I have had the good fortune of directing a multidisciplinary environmental and public health laboratory, and during these two decades, I have had the opportunity of responding to a considerable number of environmental problems and disease outbreaks. After each episode, I ask myself and my staff if we could have been better prepared. Should we have had operational some system that would have provided a much earlier warning and greatly reduced the ultimate consequences? Over the years, this process has caused me to develop what I call the "Hausler theorem of $n+1$." If you establish n monitoring systems to detect any one of n hazards, the one that will wreck havoc on biologic receptors is $n+1$. It always seems to be one that was not anticipated. With so many different systems operating either independently or in concert, it is not difficult to conclude that a limited battery of rapid environmental monitoring methods may be useful, but in implementing them, caution must be exercised to prevent oneself from becoming too involved with detecting the specific while grossly overlooking the obvious.

Surrogate Systems in Environmental Monitoring

It should be apparent that I am a proponent of surrogate systems for monitoring the environment. In monitoring the microbiological quality of drinking water, for example, it is customary to use surrogate, after-the-fact methods. There are much simpler, nonmicrobiological means for determining whether or not water supplied for human consumption is potable. Primary treatment of raw water by coagulation, sedimentation, and filtration followed by chlorination is a standard treatment process. One can rapidly determine if the treatment process is functioning as expected by frequently assessing the turbidity and chlorine residuals of the finished water. Microbial monitoring is usually performed at various points in the distribution system, along with a determination of disinfectant residual. In those systems where no disinfectant is used, microbial monitoring is certainly of great importance. Regardless of the system, microbiological monitoring is usually performed by a surrogate system and in some situations may also include a total aerobic bacterial plate count. There are three principal surrogate systems for drinking water. One is a coliform or fecal streptococci determination, another is turbidity, and the third is chlorine. Although surrogate bacterial tests will provide us with some fundamental information about the supply, the really rapid and sensitive method which will alert us to a possible microbiological contamination is the test for chlorine. In a rather novel experimental procedure, Olivieri et al. (1984) recently evaluated the stability and effectiveness of residual disinfectants in a test water distribution system challenged by the introduction of sewage. Just as the presence of a residual disinfectant in the distribution system forms the primary barrier to disease transmission by post-treatment contamination, its absence is a warning or alert that chlorine-demanding substances have entered the distribution system. Thus, the presence of free chlorine in a chlorinated distribution system serves as a marker.

It is when these routine tests detect a failure somewhere in the system that we actually turn to rapid microbiological procedures. The need for such procedures is generated for instance when health authorities have installed a temporary replacement system that must be declared potable or when the distribution system itself must be closely and quickly monitored so that consumers can be provided with a safe drinking water source as quickly as possible.

Recently, Lippy and Waltrip (1984) reported a 35-year summary of waterborne disease outbreaks occurring from 1946 to 1980 in the United States. They observed that in 21.7% of the outbreaks, bacteria were the causative agent, 11.8% were due to virus etiology, and 7.1% were due to parasites – specifically *Entamoeba* and *Giardia*. Technological, methodological, and epidemiological changes during these 35 years provide a considerable bias to the data, and one can hope that recent advances in the recovery from environmental samples of microorganisms such as the *Campylobacter* (Mathewson et al. 1983) and *Yersinia* (Weagant and Kaysner 1983) will provide new and more complete information on waterborne diseases.

When discussing microbial pathogens in the environment, we tend to focus primarily on those agents causing human gastroenteritis. In a medical progress review of *Staphylococcus aureus,* Sheagren (1984) notes that, after *Escherichia coli,* this organism is the second most frequent cause of all types of hospital-acquired infections, as evident in recent data gathered by the Centers for Disease Control. Sheagren states that although the exact incidence of community-acquired *S. aureus* sepsis is not known, it continues to be the most common etiology of serious, progressive, skin, soft tissue, and post-traumatic infections in the community setting. Klapes (1983) believes that one way to recognize the significant presence of *S. aureus* in our environment is to acquire a better understanding of injury or stress physiology in staphylococci and to identify those materials and conditions which are favorable to organism recovery on selective media. Simply substituting Vogel-Johnson or Baird-Parker culture media in the membrane filter method is not sufficient.

An intriguing application of filter technology has recently been reported by Wallis and Melnick (1984). These authors claim that their semiautomated system can detect as few as 100 bacterial colony-forming units per ml in potable waters within 3 min. The procedure quantifies the number of bacteria in potable waters in this short interval of time and provides a permanent colorimetric record or results. Our other seminar presenters will elaborate on numerous other systems, so I need not take the time to be more specific. The important consideration is that rapid methods – even such simple devices as the field microscope – which will provide quick information on the situation at hand are readily available to be used in the field. If we recognize this information as primarily qualitative, it will aid greatly in rapidly circumscribing an environmental problem.

In searching for more rapid methods, we should not overlook the need to improve existing methods of monitoring potable waters. In the area of membrane filter (MF) technology, there is great need for a modification that will permit the ready detection of coliforms both in turbid waters as well as in the presence of large numbers of noncoliform bacteria (Burlingame et al. 1984). The MF technique provides a useful and rapid means for monitoring potability in a majority of drinking waters, but where the water is rather turbid, there is marked masking of coliforms. The same holds true for waters with a high microbial population in that, since the proportion of coliforms in relation to all other species present is relatively small, their presence is masked.

There is the additional problem that most microorganisms found in the environment which are of interest or importance to human health are, in all likelihood, stressed microorganisms. Existing methods as well as those still to be developed need to give more consideration to repair of these organisms before forcing them to perform rapidly in some new, fast-paced technology. LeChevallier et al. (1983) have suggested that verification of coliform density by standard procedures may result in significant underestimation. They reported that coliform density increased by 87% when they tested for presence of β-galactosidase and cytochrome oxidase rather than gas production in lauryl tryptose broth.

Another microbiological method of practical value recently reported by Littel and Hartman (1983) may have a very useful application in the determination of point-source pollution in surface waters. They incorporated a fluorogen (4-methylumbelliferone-α-D-glucuronide) and colorimetric starch into the gentamicin–thallous carbonate medium of Donnelly and Hartman (1978). Not only is this medium as sensitive as other media, but the authors claim its use makes it possible to differentiate fecal streptococci into phenotypic groups – hence its beneficial application to the determination of point-source pollution.

From the standpoint of drinking water potability tests, what is really needed is a system which is accurate, precise, sensitive, and reproducible, which will rapidly test a large number of specimens in a relatively short time with minimum human time contribution, and which can be connected to a computer for electronic processing and transmission of reports.

There are a number of systems on the market today that have the potential for providing this methodology someday, but their sensitivity of threshold of detection is not satisfactory at the moment for the low bacterial densities usually found in potable waters. Perhaps with continued encouragement, applications engineers will see the need for instrument modifications and produce a more sensitive instrument for water potability tests. It should not be difficult in this day of microsensitive detection elements coupled with electronic signal enhancement to meet the demands of high-volume environmental laboratories. Although a number of scientific publications have called attention to the potential application of impedance-measurement methodology to environmental microbiology (Munoz and Silverman 1979; Silverman and Munoz 1979; Cady et al. 1978), little actual progress has been made to date. The same holds true for the multichannel electrochemical detection method of Wilkins et al. (1978).

Another system that holds promise for adaptation to environmental microbiology is the potentiometric measurement of lipoic acid reduction in a very minimal lactose broth. Charriere et al. (1984) have reported on their modification of this basic system and have indicated their interest in pursuing its use in waters with very low bacterial concentrations.

In approximately the last decade, a new technology has evolved that has been increasingly applied. Its rapid development and wide-spread application have resulted in an expanding array of mini- and microcomputers, mini-mainframe computers, and personal and briefcase-type computers. I am referring to multivariate, computer-modeling systems.

Knowing the flow characteristics of a stream, its physical, chemical, and biological constituencies, as well as its many and varied uses, one can predict through computer modeling the effective impact of one or more new potential waste dischargers in order to determine whether these new discharges should be permitted. Kay and McDonald (1983) have reported development of a regression model designed to predict the concentration of *E. coli* and total coliforms in two British upland impoundments. The multivariate approach requires data on hydrological input to the reservoirs, physical and chemical water quality data, as well as many other parameters. This method permits a more accurate prediction of bacterial concentrations and provides useful information for those who must manage impoundments or reservoirs for recreational purposes. Application of this technology makes it possible to limit recreational activities at certain times or at specific sites in order to reduce unnecessary health risks.

Our environment is a highly complex, multi-variate system that cannot be reduced to rapid predictive episodes. Portions of it can be modeled and some specific parameters can be determined relatively rapidly, but these diagnostic efforts are usually of limited scope and application. As stated earlier, there are certain circumstances that can be predicted after acquiring limited or surrogate systems information. The best-known, of course, is the prediction of possible human fecal pollution of drinking water when coliforms are detected. One surrogate of very recent finding is that *Legionella pneumophila* has a greater chance of being present in institutional hot-water systems when they are maintained at 43°–45 °C rather than 58°–60 °C (Plouffe et al. 1983). Another is based on the evidence that normal free chlorine residuals found in water distribution systems will inactive hepatitis A virus, thereby establishing the surrogate that free chlorine residual monitoring will give added assurance that hepatitis A virus has little likelihood of being present (Grabow et al. 1983).

Various domestic and wild animal species can be excellent monitors of environmental quality. As with the coal miner's canary, Buck (1979) suggests that we closely monitor animals because they manifest metabolic activities comparable to man's. Levy et al. (1984) draw attention to the role macroinvertebrates play in causing tastes, odors, water discoloration, filter clogging, and high turbidity in potable water distribution systems, and they suggest that these biological forms may be the cause of coliform occurrences that are resistant to disinfection.

Arthropod surveillance as reported by Dorsey et al. (1978), Wong et al. (1978), and many other investigators provides an early warning system to the appearance of vectorborne diseases in the environment. Knowledge of vector presence and distribution as well as the prevalence of certain diseases is all that is necessary to provide rapid environmental monitoring.

In the study of nitrogen-fixing bacteria, Corbin (1984) has developed a rapid, 2-min determination of ammonia by a liquid chromatographic fluorescence method. This rapid method can be of considerable benefit to investigators who must run a great number of ammonia assays in a day's time.

Usefulness of Rapid and Sensitive Environmental Monitoring Procedures

This discussion has not ignored the need for sensitive environmental microbiological procedures – they are always needed, and they form the foundation for research in applied methodology. However, there must be a respectable balance between sensitivity and specificity. An environmental monitoring program should be sufficiently sensitive to detect minute changes that occur and yet, not be so specific that the balance of complex forces is overlooked. There is absolutely no doubt that the applied environmental sciences are in great need of more sensitive, specific, and rapid methodologies simply because the public has been led to expect results in the blink of an eye and to discuss data numbers without true regard for the units of measurement involved. The average lay person is not too concerned whether a certain methodology detects at the part-per-billion or part-per-trillion level but is greatly concerned if the results are reported as a "2" one time and perhaps "16" the next. We seem to be caught in the numbers game without fully realizing the impact of these contaminants on a highly fluid and complex environment.

Are rapid and sensitive microbiological procedures necessary for monitoring the environment? I believe they are, but not on the same order as rapid and sensitive methodologies applicable to clinical areas. In the environment, a rapid system may generate a result in a comfortable 48–72 h and still be of immediate utility, but in a clinical setting such a result is of little or no benefit. From the standpoint of sensitive environmental methods, laboratories are commonly expected to report chemical contaminants in the part-per-trillion range because of concern for environmental carcinogens. In the clinical area, such sensitive methods are not required to the same degree.

There is nonetheless a great deal that clinical and environmental investigators can learn from each other and, more importantly, what researchers engaged in developing or designing environmental monitoring methods can learn from other, more advanced rapid procedure activities. Little attention in environmental research has been given to the development of specific DNA probes for environmental hazards or the application of electronic and signal-amplified technologies so common in some rapid diagnostic areas. Direct transfer of current and/or future rapid technological systems is probably less important than the need for freer experimentation with new methods and the increased involvement of applications engineers in the microbiological system, including giving much more support and emphasis to microbiological engineering. We need only look at the multitude of benefits society has reaped since chemistry and engineering were combined many years ago into chemical engineering. The future for microbiological engineering is just as bright when we give full consideration to all of society's needs.

Summary

The necessity for rapid and sensitive or even routine microbiological procedures for monitoring the environment are generally predicated on making full use of the early acquisition of general information. In contrast to a biological system, an environment usually responds slowly to insults, and therefore its assessment mitigates against rapid procedures. Timeliness, cost-effectiveness, and utilization of derived information are fundamentally much more important than rapidity or sensitivity. The coordination of simple chemical, physical, and epidemiological information with either routine or rapid microbiological data is essential to a complete understanding of any environmental insult. More pressing than rapidity is uniformity of procedure. In this day of rapid technological advances, computer development, and immediate response capabilities, one is somewhat reluctant to admit that the most rapid and sensitive microbiological procedure for monitoring the environment is proper employment of the compound microscope. This sometimes forgotten instrument, coupled with an array of surrogate systems, should be utilized in the rapid monitoring of the environment of microbiological contaminants. It is my opinion that rapid and sensitive microbiological methods for monitoring the environment are necessary when meeting a specific need but that, in the normal course of events, surrogate systems are satisfactory. Far more urgent is the need for techniques to process a large number of environmental specimens having low bacterial density in a short period of time and with minimum labor.

References

Buck WS (1979) Animals as monitors of environmental quality. Vet Hum Toxicol 21:277–284

Burlingame GA, McElhaney J, Bennett M, Pipes WO (1984) Bacterial interference with colony sheen production on membrane filters. Appl Environ Microbiol 47:56–60

Cady P, Dufour SW, Lawless P, Nunke B, Kraeger SJ (1978) Impedimetric screening for bacteriuria. J Clin Microbiol 7:273–278

Charriere G, Jouenne T, Lemeland JF, Selegny E, Junter GA (1984) Bacteriological analysis of water by potentiometric measurement of lipoic acid reduction: Preliminary assays for selective detection of indicator organisms. Appl Environ Microbiol 47:160–166

Corbin JL (1984) Liquid chromatographic-fluorescence determination of ammonia from nitroginase reactions: A 2-min assay. Appl Environ Microbiol 47:1027–1030

Donnelly LS, Hartman PA (1978) Gentamicin-based medium for the isolation of group D streptococci and application of the medium to water analysis. Appl Environ Microbiol 35:576–581

Dorsey DC, Rowley WA, Wong YW, Brinker JP, Currier RW, Hausler WJ Jr (1978) Surveillance of arbovirus activity in Iowa, 1977. Mosq News 38:492–498

Grabow WOK, Gauss-Müller V, Prozesky OW, Deinhardt F (1983) Inactivation of hepatitis A virus and indicator organisms in water by free chlorine residuals. Appl Environ Microbiol 46:619–624

Kay D, McDonald A (1983) Predicting coliform concentration in upland impoundment: Design and calibration of a multivariate model. Appl Environ Microbiol 46:611–618

Klapes NA (1983) Comparison of Vogel-Johnson and Baird-Parker media for membrane filtration recovery of staphylococci in swimming pool water. Appl Environ Microbiol 46:1318–1322

LeChevallier MW, Cameron SC, McFeters GA (1983) Comparison of verification procedures for the membrane filter total coliform technique. Appl Environ Microbiol 45:1126–1128

Levy RV, Cheetham RD, Davis J, Winer G, Hart FL (1984) Novel method for studying the public health significance of macroinvertebrates occurring in potable water. Appl Environ Microbiol 47:889–894

Lippy EC, Waltrip SC (1984) Waterborne disease outbreaks – 1946–1980: A Thirty-five-year perspective. J Am Water Works Assoc 76:60–67

Littel KJ, Hartman PA (1983) Fluorogenic selective and differential medium for isolation of fecal streptococci. Appl Environ Microbiol 45:622–627

Mathewson JJ, Keswick BH, DuPont HL (1983) Evaluation of filters for recovery of *Campylobacter jejuni* from water. Appl Environ Microbiol 46:985–987

Munoz EF, Silverman MP (1979) Rapid, single-step most-probable-number method for enumerating fecal coliforms in effluents from sewage treatment plants. Appl Environ Microbiol 37:527–530

Olivieri VP, Snead MC, Kruse CW, Kawata K (1984) Stability and effectiveness of chlorine disinfectants in water distribution systems. US Environ Protect Agency MERL, Cincinnati, EPA-600/S2-84-011

Plouffe JF, Webster LR, Hackman B (1983) Relationship between colonization of hospital buildings with *Legionella pneumophila* and hot water temperatures. Appl Environ Microbiol 46:769–770

Sheagren JN (1984) *Staphylococcus aureus*. The persistent pathogen. N Engl J Med 310:1368–1373

Silverman MP, Munoz EF (1979) Automated electrical impedance technique for rapid enumeration of fecal coliforms in effluents from sewage treatment plants. Appl Environ Microbiol 37:521–526

Wallis C, Melnick JL (1984) An instrument for the immediate quantification of bacteria in potable waters. J Am Water Works Accos (to be published)

Weagant SD, Kaysner CA (1983) Modified enrichment broth for isolation of *Yersinia enterocolitica* from non-food sources. Appl Environ Microbiol 45:468–471

Wilkins JR, Young RN, Boykin EH (1978) Multichannel electrochemical microbial detection unit. Appl Environ Microbiol 35:214–215

Wong YW, Rowley WA, Dorsey DC, Hausler WJ Jr (1978) Surveillance of arbovirus activity in Iowa during 1972–1975. Mosq News 38:245–251

Searching for Rapid Methods in Environmental Bacteriology

E. E. Geldreich and D. J. Reasoner [1]

Introduction

With the evolution of treatment processes to control and improve the microbial quality of drinking water, wastewaters, and other infectious wastes, plant engineers have often requested operational monitoring methods that would provide a rapid characterization of the product water for determining treatment effectiveness. Furthermore, where downstream uses involve bathing or shellfish farming, application of a rapid test could provide a critical real time alert to potential public health problems in those designated areas. Unfortunately, monitoring programs today (Fig. 1) rely on standard test procedures that require 24 h or more sample processing time before results are available. As a consequence, this information is only a record of past treatment performance or of yesterday's water quality conditions to which the public was exposed.

Real time alert capability demands the application of rapid bacteriological characterization and a prompt action response. Once a rapid detection procedure is accepted for a specific indicator or pathogen, the logical progression is to integrate sample collection and analysis in an automated mode to provide more frequent information on water supply or wastewater treatment process effectiveness. Careful analysis of these data would clarify quality trends and provide more precise information on process control to improve product water. Such information would also be of singular value as an aid in restoring quality of drinking water after unexpected contamination caused by treatment failures, or introduced into distribution water through line breaks and cross connections.

Specifications for Rapid Methods

Primary considerations for selection of a rapid method must involve specificity, sensitivity, and precision. Test reliability should be capable of documentation with quality control procedures. Also key to test selection are the following considerations: (a) achievement of quantitative measurement within a few hours; (b) data that correlate with standard procedures; (c) methodology that is relatively uncomplicated for use by the average laboratory technician; and (d) cost effectiveness for benefits achieved. In general, nonspecific detection of microorgan-

1 Drinking Water Research Division, US Environmental Protection Agency, 26 West St. Clair Street, Cincinnati, OH 45268, USA

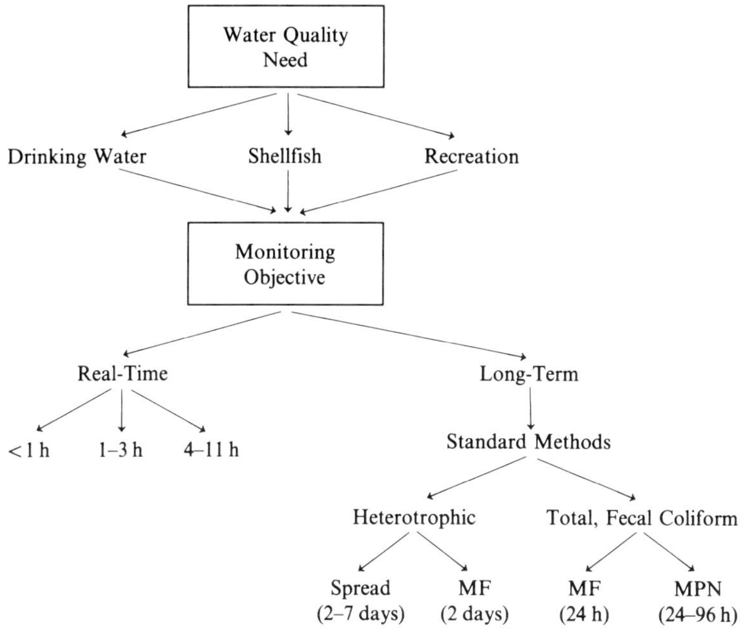

Fig. 1. Pathway for decision-making on monitoring objectives

isms has limited application in drinking water because many heterotrophic bacteria encountered are of no apparent health significance. However, if performance information on disinfection processes is desired, then nonspecific detection of general microbial activity is warranted.

Health significance of water quality is traditionally determined by testing water for organisms or groups of organisms common to the fecal flora of warm-blooded animals. The most common indicator groups include a broad spectrum of genera or species such as total coliform and fecal streptococci. Other indicators comprise a narrower grouping of related organisms typified by fecal coliform (fecal coli), enterococcus, *Bacteroides*, and sulfate-reducing clostridia. Individual species of bacteria, i.e., *Escherichia coli*, and *Streptococcus fecalis*, have also been recommended as more specific indicators of fecal contamination. Rapid detection of *Clostridium perfringens* might have application to disinfected water since its occurrence is suggestive of deficiencies in treatment. In practical terms, speciation of bacterial isolates using multiple biochemical test systems has not been applied to routine monitoring because the identifications are time consuming and labor intensive. Recent research in immunology and recombinant DNA techniques, however, may provide the means for rapid species identification.

Rapid tests for drinking water monitoring must be based on total coliform detection, not merely *E. coli* detection. The most frequently isolated total coliforms in situations of positive sample violations from contaminated drinking water that has been treated include seven species (*Enterobacter aerogenes, Enter. cloacae, Enter. agglomerans, Citrobacter freundii, Klebsiella pneumoniae, K. oxytoca,* and *E. coli*). Thus, rapid test concepts for the total coliform group are likely to incorporate a common biochemical response reaction (such as lactose fermentation) rather than serological techniques because of antigen antibody cross reactions among related organisms. Serological cross reactivity among the fecal streptococci may also restrict the application of rapid serological procedures for differentiation of this group, which includes *Streptococcus fecalis*, *S. faecium*, *S. bovis*, *S. equinus*, *S. avium*, and several subspecies and biotypes.

For water uses of greatest public health concern (drinking water, shellfish farming, and recreational waters), water quality standards are set at values thought to present a minimal risk of exposure to waterborne pathogens. Of these water uses, drinking water standards are the most stringent, requiring that no coliforms be detected in 100 ml water. For shellfish farming and recreational water, fecal coliform limits of 14, and 200–400 organisms/100 ml, respectively, have been established. The bacterial density limits suggested for source water used in treated public water systems are at the lower end of the water quality scale. For these sources, a maximum of 2,000 fecal coliform or 20,000 total coliform bacteria/100 ml is recommended to lessen the risk of microbial penetration of the treatment barriers. With these values in mind, and the susceptibility of coliform bacteria to reduction even in systems providing marginal treatment, it becomes apparent that until a breakthrough occurs in rapid test sensitivity, amplification of the original cell density through sample concentration or by enrichment will generally be necessary.

Cost Considerations

A major justification for a rapid bacteriological technique is its utility in operational monitoring. While public health is always of first concern, monitoring costs can be a very important consideration. What is a reasonable price for quicker results from bacteriological tests? Recently, the cost for routine coliform analysis performed by a city health laboratory was reported to be $ 3.32 per test by the membrane filter (MF) procedure, or $ 6.21 per test by the multiple tube procedure [7]. A survey of commercial laboratories showed that charges for these two procedures ranged from $ 8 to $ 12 for the MF test, and from $ 14.30 to $ 18.00 for the MPN test. The same cost factors would apply to analyses of samples from public water supplies, recreational waters, or sewage effluents.

Another approach to calculating monitoring costs uses information from a bacteriological monitoring program for several water plant operations serving cities with populations ranging from 60,000 to 900,000. The survey data revealed test costs of $ 15.00 per sample. Of course, monitoring cost will vary with the size of the system and the cost per test declines sharply for the laboratory involved in examining more than 60 samples per week. Part of this economy of scale is a function of the required number of samples examined per month based on population served. The price for meeting the bacteriological monitoring requirements mandated by US federal regulations has been estimated to be only 1 cent of the customer's cost per 3,785.4 l of public water supply [4]. With the preceding information in mind, it becomes apparent that a cost increase for a rapid monitoring procedure is justifiable and could approximate twice the cost of routine analyses for the benefits achieved.

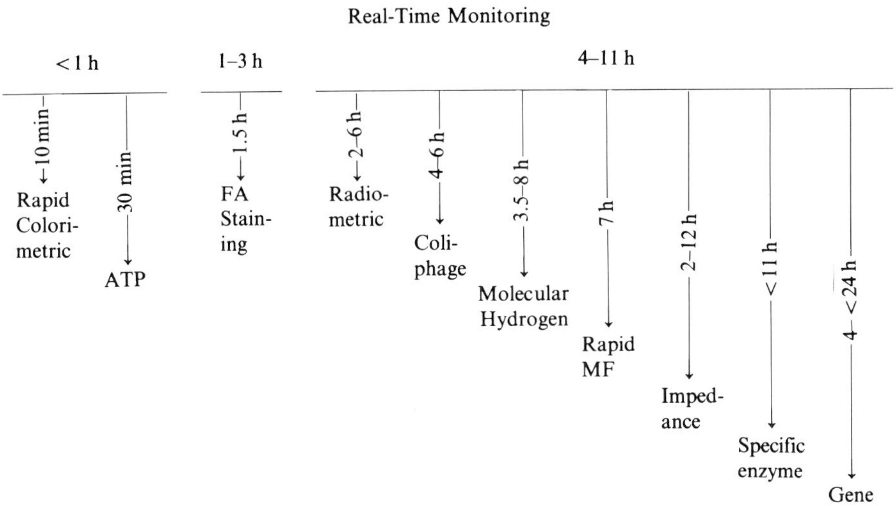

Fig. 2. Pathway for decision-making on real-time monitoring

The use of a rapid test in an effective monitoring program, including a quick action response when undesirable results are obtained, could be of substantial benefit. It would be particularly advantageous where the risk is due to: (a) fluctuations in source water quality that challenge existing treatment processes, (b) the direct reuse of high-quality wastewater, or (c) exposure to ambient natural waters in recreation waters or shellfish farms. While small water systems may have the greatest difficulties in delivering a continuous supply of safe drinking water, the application of a real time monitoring program is not realistic because of budget restrictions. In these situations, more frequent measurements of basic water quality criteria using simple, low-cost tests are more urgently needed to supplement annual sanitary surveys and operator training programs.

Searching for rapid methods in environmental bacteriology has revealed a variety of candidate methods that have potential in real time monitoring (Fig. 2). As might be anticipated, rapid tests that can be performed in less than 1 h have no specificity and may possibly include nonviable cells. Perhaps the more logical way to evaluate these procedures would be to examine their level of complexity with reference to suitability for use in the average laboratory or restriction to the laboratory with diversified staff expertise and in-depth instrumentation resources.

Rapid Test Candidates Using Conventional Procedures

The ideal rapid method candidate is one that would utilize methods and materials available to every bacteriological laboratory. Among the routine methods, the MF technique offers much promise, allowing concentration of organisms from large sample volumes. Furthermore, MFs permit visualization of developing microcolonies and use of or placement in contact with different media or biochemical substrates without disrupting their growth. Furthermore, colony isolation makes verification procedures easier.

Rapid Colorimetric Determination

The rapid colorimetric test is based on simple MF concentration of bacteria from a sample, followed by staining the trapped bacterial cells with Safranine 0 dye [20]. Excess dye is flushed away and an agent is applied which decolorizes only the dye on the MF, not dye absorbed by bacterial cells. Concentrations of bacterial cells $\geqq 10^5$ organisms/ml on the 10-mm diameter MF yield a pink to red color; bacterial densities below 10^5 organisms/ml yield a white or slightly yellow filter. The color of the membrane is then compared with a standard color guide, thus providing a semiquantitative indication of the bacterial density in the sample. Larger water sample portions (10 ml, 100 ml, or 1,000 ml) may be tested as necessary to achieve a detection level of 10^5 cells. Where water color interferes, the sample must be passed through an anionic exchange resin to remove color materials. If iron particulates occur, the sample is pretreated with a rust solvent (RoVer, Hach), so that metal oxides do not interfere with bacterial staining or the decolorization of the filter surface. Use of these sample clean-up procedures alone or in combination did not result in significant loss of the semiquantitative results. Obviously, this rapid test, which can be performed in a few minutes, is restricted to use in industrial cooling waters [19] or other process waters where a general measure of microbial density is adequate and the occurrence of nonviable organisms does not affect the interpretation of results.

Rapid Membrane Filter Approach

A 7-h fecal coliform test is possible using a lightly buffered lactose medium (M-7 h agar) and incubation at 41.5 °C [14]. Bromcresol purple and phenol red indicators in the medium change to a yellow color in the presence of developing fecal coliform colonies. The additon of 0.5% *d*-mannitol permits rapid differentiation of some fecal coliform strains which have an initial preference for utilizing mannitol. Use of 41.5 °C rather than 44.5 °C for incubation shortened the lag phase and reduced heat stress. Field results using this method correlated well with the 24-h standard MF fecal coliform test, although stressed indicator strains may produce poor differentiation with only 7-h incubation.

Coliphage Detection

The lysis of coliform bacteria by DNA or RNA coliphages to form plaques in a confluent growth lawn of a host culture offers a simple plating procedure that can produce results in 4–6 h. The test concept relies on a positive correlation between the density of coliform bacteria and coliphages in ambient waters and sewage effluents [9, 21]. Because coliphages appear to be more disinfectant resistant than coliforms, the quantitative relationship in treated drinking water may be distorted. However, coliphage detection in process waters has potential as a surrogate indicator of virus penetration of treatment barriers. Successful use of a rapid coliphage measurement is related to selection of a host *E. coli* culture that is susceptible to infection by the majority of coliphages present in the sample. *E. coli* C (ATCC no. 13706) appears to satisfy this requirement and is used at a standardized density of approximately 1×10^9 cells/ml for developing a lawn of host cells. Sensitivity of the test is one coliphage/100 ml.

Fluorescent Antibody Technique

The fluorescent antibody (FA) technique is a widely accepted procedure that is generally used in conjunction with an enrichment method for pathogens such as *Salmonella* and enterotoxigenic *E. coli*. However, this technique may be used for rapid direct indication of the presence of a specific organism in a water sample [2]. After membrane filtration to concentrate the organisms on the surface of a black (low autofluorescence) filter, entrapped cells are incubated for 1 h with a specific antibody fluorescein isothiocyanate conjugate in the humid atmosphere of a culture dish. Following staining, the filter is rinsed in phosphate-buffered saline, mounted in buffered glycerol under a cover slip and examined using a fluorescent microscope.

Limitations of the FA procedure involve the availability of specific FA conjugate for the organisms sought and laboratory capability to produce the specific FA conjugates needed. In addition, a fluorescent microscope is necessary for examination of the FA-stained preparation. Because of these requirements, application of the procedure may be limited to specialized laboratories. Furthermore, the fluorescence of nonviable cells in treatment processes may cause some concern in interpreting results.

Special Techniques Requiring Amplification of Bacterial Densities

Efforts to shorten the time required to detect either bacterial indicators or individual species in water sample flora to less than 5 h require the use of special instrumentation and materials. These procedures must be capable of quantitating viable organisms directly or after a brief period of amplification in a selective enrichment. Instrumentation and test materials for these methods are not readily available to the average laboratory and require some training of the laboratory staff plus an effective quality assurance plan to insure reliability of data. Due to the cost of the associated instrumentation and reagents, application of this category of rapid methods will be limited to large laboratories where resources in expertise and essential instrumentation are more likely to be available.

Bioluminescence-Chemiluminescence

The firefly bioluminescence system is the most studied of the known bioluminescence reactions for rapid detection and quantitation of microorganisms [11–13, 17]. Light generated in a living system or from biochemical components derived from an organism is termed bioluminescence, while light emitted from the reaction of chemicals that are not derived from a biological system is known as chemiluminescence. Using liquid scintillation spectrometry and the firefly luciferin-luciferase enzyme system, adenosine triphosphate (ATP) can be detected in standard solutions over a range of 10^{-9}–10^{-12} mol [17]. This system for measuring ATP has been used to measure total microbial biomass in lake sediments [11, 12] and a minimum of 100–300 bacteria/ml in spacecraft water supplies [13]. Recent studies to establish correlations between the heterotrophic bacterial population of drinking water and the ATP concentration have been unsuccessful because total viable biomass is more encompassing than a measure of the bacterial density in potable water. While this test is specific to viable cells, it cannot differentiate indicator organisms from other organisms in the water flora.

Luminescence measurements may also be based on the reaction of luminol with hydrogen peroxide when catalyzed by bacterial catalase or cytochromes. Using this chemiluminescence reaction, light emitted (which is directly proportional to cell density) is sensed by a photometer and recorded. Unfortunately, this approach may respond to soluble porphyrins and oxidizing agents from dead cells as well as viable bacteria.

Gene Probes for Specific Organisms

Developments in microbial genetics have led to methods that can detect minute quantities of genetic information from specific organisms. Rapid detection of enterotoxigenic *E. coli* (ETEC) from a mixed flora of a contaminated water is a promising example of gene probe application [6]. A water sample is processed by membrane filtration with incubation on M-FC medium (anaerobically at 35 °C rather than 44.5 °C to avoid loss of heat-sensitive plasmids coding for endotoxin) for 24 h. The bacterial colonies on the MF surface are then treated to remove non-DNA components of the cells. The residual cell constitutents are treated to denature the DNA, separating the two strands in the double helix. The single-stranded DNA is held in place on the MF and exposed to a large excess of specific probe DNA. Upon incubation of the probe DNA with the DNA released from bacterial cells on the MF, hybridization of probe DNA with complementary sample DNA may occur. Excess unbound probe DNA is washed away after the required incubation period. Labeling the probe DNA with a radioisotope, enzyme, fluorescent or chemiluminescent molecule permits detection of reaction. Except for radioisotopic labels requiring autoradiography, these assays can be completed in a few hours and are at least 1,000-fold more sensitive than serological tests.

Specific Enzyme Assays

Specific enzyme detection offers another approach to differential detection of viable bacteria from a sample, provided there are sufficient cells present to insure measurable enzyme concentrations. β-galactosidase and formic hydrogen lyase are inducible enzyme systems of coliforms that are associated with lactose fermentation [5]. Unfortunately, β-galactosidase is produced not only by *E. coli* but also by *Shigella, Proteus rettgeri, P. hauseri,* and *Clostridium* species [10]. For fecal coliform measurement, biosynthesis of glutamic acid decarboxylase can be used as a detection marker and is directly proportional to fecal coliform concentration. The glutamate decarboxylase enzyme measurement is coupled to use of a hollow fiber ultrafiltration system to concentrate the coliforms from a sample [18]. Based on the concentration of enzyme detected, the density of fecal coliforms per sample volume can be estimated. The sensitivity of this test requires 5×10^4 *E. coli*/ml, a concentration that would indicate grossly polluted water. For water samples with very low fecal coliform numbers, amplification by enrichment culture could achieve the minimum cell concentration needed, but would require an extended time period. For instance, one coliform in 100 ml of sample when enriched in lactose broth at 41 °C for 11 h would yield a positive glutamate decarboxylase test [10]. An automated system for glutamate decarboxylase detection has been developed which permits continuous or programmed sampling passage through a hollow fiber ultrafiltration system for amplification of fecal coliforms concentrated from 10 to 20 l, and analysis is completed within 12 to 13 h. While the automated aspects of the test are excellent, the time frame for analysis falls outside our definition of a rapid test.

Impedance Measurements

Microbial metabolism of carbohydrates in a nutrient medium converts these nonelectrolyte materials into a variety of electrically charged molecules and ions. The rapidity of the change in conductivity, measured as a change in electrical impedance to an alternating current passed through the culture, is directly related to the density of bacteria in a specific medium over a defined time period [3]. A minimum bacterial density, generally 10^6–10^7 cells/ml, is essential to achieving a detectable impedance signal [3, 16]. With high-quality water samples, a lag period will occur with no detectable change in the impedance signal until the target organism density reaches the threshold detection level. Since the length of the amplification period is inversely proportional to the original viable cell density in the water sample, a standard curve developed from a tenfold dilution series of a pure culture of a predominant indicator group organism is used to estimate the unknown bacterial density in the original sample. The baseline impedance value is determined by inclusion of a sterile medium control for comparison with the test sample.

Careful selection of the test medium and incubation temperature is important since many species of bacteria, yeast, and fungi growing in various media can produce a detectable impedance signal. For fecal coliform detection, specificity and minimal lag phase can be achieved by use of lactose fermentation in a selective medium at an elevated temperature of 41.5 °C. A selective lactose-based medium and anaerobic incubation at 35 °C to minimize interference should optimize recovery of total coliforms. Some success in rapid detection of heterotrophic bacteria has been achieved with the standard plate count medium; however, the potential for improved recoveries that may be achieved with the use of m-SPC and R2A media at 28 °C [1] needs to be explored.

Since impedance measurement is time-dependent and a minimum concentration of 10^6–10^7 cells/ml is needed to generate a change in impedance, application of this rapid test is limited to coliform monitoring of sewage effluent-laden receiving waters used as source water for public water supply, or recreational bathing areas. The use of a rapid impedance test for coliform detection in drinking water is difficult because detection sensitivity must be 1 coliform/100 ml. Such a sensitivity level would require a coliform amplification period of 18 h or longer. As an alternative, this test sensitivity might be achieved more quickly through MF concentration of low densities of coliforms. The MF with entrapped bacteria would then be submerged in a selective test medium for 35 °C incubation and periodic impedance measurement. The coliform density would still be correlated to detection time through the use of standard curves for predominant coliform species. Instrumentation is available which is adaptable to automation and multiple sample analyses to provide frequent monitoring of the quality of untreated water or sewage.

Radiometric Methods

The basis for all radiometric bacterial detection methods involves the use of a radioactively tagged substrate that is either incorporated or metabolized by bacteria with the release of a detectable labeled endproduct that can be measured, quantitated, and correlated with cell density. For coliform detection, the radioactive-labeled fermentable sugar, or other suitable substrate, is placed in the medium, a measured water sample is added, and the inoculated medium is incubated at a selective temperature to minimize interference from the noncoliform organisms and accelerate coliform growth [15]. Radioactive $^{14}CO_2$ from the fermentation is collected by precipitation as $Ba^{14}CO_3$ on filter paper moistened with a saturated solution of barium hydroxide placed in the lid of the culture vial. The activity (counts per minute, CPM) of precipitated $Ba^{14}CO_3$ is measured by a liquid scintillation spectrometer and compared with an uninoculated control for correction of background radioactivity. The net value obtained is then converted into an indicator density by interpolation of standard curves from pure culture strains. As few as ten bacteria can be detected in 1.5–8 h, depending on the organisms and whether stationary or agitated incubation is utilized. Detection times for *E. coli* prepared from pure cultures revealed that 1–10 cells, 100 cells, and 1,000 cells produced measurable activity in 6, 3, and 2 h, respectively. Results for detection of coliforms in natural samples are in general agreement with these standard curves. For drinking water samples, large sample volumes must be concentrated by membrane filtration to achieve detection sensitivity of less than 1 coliform/100 ml.

While significant progress has been made toward standardizing the radiometric rapid test protocol, precision and reproducibility need further refinement. The major problem appears to be the need to enhance the metabolic activity of coliforms stressed by characteristics of treated water supply and chlorinated sewage effluents. Extending the test time beyond 4 h defeats the objective of a rapid procedure. Improved results are more likely to come from the formulation of a medium that promotes the repair of damaged enzyme systems, the use of inducer compounds to stimulate earlier production or increased concentration of measurable metabolic products, and the inclusion of a supplemental radioisotope-labeled substrate that can be preferentially used by stressed coliforms early in their recovery effort [15]. However, control of the abiological release of $^{14}CO_2$ and improved control of the uniformity of radioactive labeling of substrates such as glucose, lactose, and mannitol may help improve the sensitivity of labeled metabolite-based rapid tests.

Molecular Hydrogen Detection

The production of molecular hydrogen by metabolizing bacteria can be measured electrochemically through a linear relationship established between bacterial density and length of lag phase before hydrogen ion release occurs [22, 23]. The procedure utilizes a platinum electrode and a reference electrode inserted into the test culture with incubation at a temperature appropriate for the organisms sought. Evolution of hydrogen ions during growth causes the development of changes in the electrical potential difference between the electrodes, with voltage differences being documented by a strip chart recorder. Incorporation of a multichannel monitor has permitted analysis of several samples every 80 s when desired.

This electrochemical detection concept has been used in a prototype in situ buoy system for monitoring fecal coliform densities at estuarine locations for early warning of pollution spills [8]. Data from the positioned buoy was sent by radio signals to a central monitoring location. The prototype design limited the number of sample analyses to ten before the unit was retrieved and resupplied with new medium. Deployment of the prototype was restricted to water temperatures above 8 °C to insure incubation temperature stability and the maximum operational time was limited to less than 10 days due to electrical power consumption.

Investigations with natural water samples indicated that detection of 10^2–10^3 total coliforms 100 ml of estuarine or fresh water was possible in 6.2–8.5 h. Fecal coliform concentrations of 10^2–10^3/100 ml of estuarine water were detected in 3.5–6.5 h and in 5.0–7.0 h in fresh water [22].

Summary

A survey of rapid methods for environmental microbiology reveals that candidate methods producing results in less than 2 h are nonspecific and fail to achieve the sensitivity needed to detect bacteria corresponding to a maximum acceptable density. Effective real time monitoring must include not only rapid test results but also a prompt action response within a total time of 2 h because of limitations in controlling process water for correction of treatment deficiency. Application of rapid tests for specific indicator species will not provide an adequate alert to incipient problems in treatment protection.

The most promising use of rapid tests in monitoring water quality appears to be in faster identification of specific pathogens. Efforts should be directed toward development of rapid detection methods for *Campylobacter, Yersinia,* viruses, and *Giardia,* because these pathogens demonstrate little apparent correlation with conventional indicators in specific situations. Among the rapid method concepts, radioisotope-labeled substrates, specifically labeled fluorescent antisera, organism-specific genetic probes, and molecular hydrogen detection offer the best approaches to realizing a breakthrough in water quality monitoring.

Finally, the increased cost of real time monitoring for conventional indicators or the more rapid identification of specific pathogens can be justified in terms of protection afforded to public health.

References

1. American Public Health Association (1985) Standard methods for the examination of water and wastewater, 16th ed. American Public Health Association, Washington DC
2. Brock TD (1983) Membrane filtration: a user's guide and reference manual. Science Tech, Madison
3. Cady P (1978) Progress in impedance measurements in microbiology. In: Sharpe AN, Clark DS (eds) Mechanizing microbiology. Thomas, Springfield, pp 199–239
4. Clark RM, Steve RG (1978) Meeting the drinking water standards: the price of regulation. In: Russell CS (ed) Safe drinking water: current and future problems. Resources for the Future, Washington DC, pp 271–324
5. Cundell AM (1981) Rapid counting methods for coliform bacteria. The bacteria, Vol 27. Academic Press, New York
6. Echeverria P, Seriwatana J, Chityothin O, Chaicumpa W, Tirapat C (1982) Detection of enterotoxigenic *Escherichia coli* in water by filter hybridization with three enterotoxin gene probes. Clin Microbiol 16:1086–1090
7. Geldreich EE, Kennedy H (1982) The cost of microbiological monitoring. In: Pipes WO (ed) Bacterial indicators of pollution. CRC Press, Boca Raton, pp 131–147
8. Grana C, Wilkins JR (1979) Description and field test results of an in situ coliform monitoring system. NASA Technical Paper 1334. Langley Research Center, Hampton
9. Kenard RP, Valentine RS (1974) Rapid determination of the presence of enteric bacteria in water. Appl Microbiol 27:484–487
10. Leclerc H, Catsaras M (1967) Un caractère constant chez l'*Escherichia coli* la presence d'une decarboxylase de l'acide glutamique. Bull Assoc Dipl Microbiol Nancy 107:39–49
11. Lee CC, Harris RF, Williams JDH, Armstrong DE, Syers JR (1971 a) Adenosine triphosphate in lake sediments I. Determination. Soil Sci Soc Am Proc 35:82–86
12. Lee CC, Harris RF, Williams JDH, Syers JK, Armstrong DE (1971 b) Adenosine triphosphate in lake sediments II. Origin and significance. Soil Sci Soc Am Proc 35:85–91
13. Levin GV, Usdin E, Slonim AR (1968) Rapid detection of microorganisms in aerospace water systems. Aerospace Medicine 39:14–16
14. Reasoner DJ, Blannon JC, Geldreich EE (1979) Rapid seven hour fecal coliform test. Appl Environ Microbiol 38:229–236
15. Reasoner DJ, Geldreich EE (1978) Rapid detection of water-borne fecal coliforms by $^{14}CO_2$ release. In: Sharpe AN, Clark DS (eds) Mechanizing microbiology. Thomas, Springfield, pp 120–139

16. Silverman MP, Munoz EF (1979) Automated electrical impedance technique for rapid enumeration of fecal coliforms in effluents from sewage treatment plants. Appl Environ Microbiol 37:521–526
17. Stanley PE, Williams SG (1969) Use of liquid scintillation spectrometer for determining adenosine triphosphate by the luciferase enzyme. Anal Biochem 29:381–392
18. Trinel PA, Leclerc H (1976) Essais de concentration de bacteries in suspension aqueuse par ultrafiltration. Ann Microbiol (Paris) 127B:201–212
19. Wallis C, Longoria C, Melnick JL, Tomlin JL (1980) Rapid determination of bacteria in cooling tower water with the bio-matic analzyer. Proceeding of the Cooling Tower Institute. Annual meeting paper no 222A. Cooling Tower Institute of America, Houston
20. Wallis C, Melnick JL, Longoria CJ (1981) Colorimetric method for rapid determination of bacteriuria. J Clin Microbiol 14(3):343–346
21. Wentzel RS, O'Neill PE, Kitchens JF (1982) Evaluation of coliphage detection as a rapid indicator of water quality. Appl Environ Microbiol 43:430–434
22. Wilkins JR, Boykin EH (1976) Analytical notes – electrochemical method for early detection and monitoring of coliforms. J Am Water Works Assoc 68:257–263
23. Wilkins JR, Stoner GE, Boykin EH (1974) Microbial detection method based on sensing molecular hydrogen. Appl Microbiol 27:949–952

Application of Automated and Rapid Microbiologic Assays in Various Health-Care Environments

M. S. FAVERO [1]

Introduction

In the past 15 years, there has been a virtual explosion of new techniques for the rapid detection and identification of microorganisms, as well as sophisticated procedures of automation (Goldschmidt 1980; Balows 1982). Currently, many of these procedures have progressed beyond the developmental and experimental phases and are used routinely in clinical microbiology and immunology, in food and water microbiology, and in the pharmaceutical industries. The purpose of this presentation is to discuss the application of some of these techniques to the field of environmental microbiology in a variety of health-care facilities such as hospitals and dialysis units.

Historically, procedures for environmental microbiologic sampling of intramural environments were derived from contamination problems associated with hospitals and spacecraft assembly areas (Favero and Puleo 1980). In the 1960s, in an attempt to understand the various modes of disease transmission, a variety of techniques were developed for sampling air and inanimate surfaces in hospitals. During this period, many hospitals in the United States instituted programs of routine microbiologic sampling of the hospital environment, which probably reached a peak in the early 1970s. Subsequently, it was the consensus among environmental microbiologists and hospital epidemiologists that a good deal of environmental microbiologic sampling in the hospital environment was not necessary. The science and art of infection control necessarily revolves around the patient – the patient who is hospitalized because of an infection, as well as the patient who does not have an infection but is at risk of acquiring one. Control measures are performed to prevent the spread of infection from any source to susceptible hospitalized patients or personnel. It is from this vantage point that microbiologic sampling in the hospital environment should be viewed.

In a program designed to control hospital-acquired infections, the highest priority is given to dynamic and efficient disease surveillance. Other important components of infection control programs include: (1) an efficient and effective infection control committee, (2) a competent clinical microbiologic laboratory, (3) effective patient care, (4) clearly understood and effective isolation policies and procedures, (5) staff education programs, (6) employee health service activities, and (7) a realistic environmental control program.

---

1 Hospital Infections Program, Center for Infectious Diseases, Centers for Disease Control, Atlanta GA 30333, USA

The most useful role of microbiologic sampling in an environmental control program is the investigation of specific problems within the hospital, and this type of sampling should be considered a necessary adjunct to the infection control program. Environmental microbiologic sampling, whether it be on a scheduled or a periodic basis, should be done only when it can provide meaningful and interpretable data that will aid in the identification of actual or potential contamination problems associated with specific procedures or instruments that have been incriminated epidemiologically in disease transmission. Thus, environmental microbiologic sampling may be an essential part of an investigation of an outbreak of hospital-acquired disease, or it may be an evaluation of environmental microbial contamination known to be associated with the transmission of infection.

The following guidelines should be kept in mind when contemplating the possibility of environmental microbiologic sampling. First, enforcement of rigid adherence to tested protocols of sterilization, disinfection, maintenance, and housekeeping practices is a much more efficient method for controlling hospital-acquired infections associated with patient-care equipment in critical environments than is scheduled microbiologic sampling. Second, environmental microbiologic sampling, whether in conjunction with an investigation of an outbreak of infections or in a study of the incriminated elements of the inanimate environment, should produce data that are relevant and possible to interpret. If this is not possible, environmental microbiologic sampling should not be performed. Third, microbiologic sampling which involves the intramural environment, such as air and inanimate surfaces, on a scheduled or routine basis often produces useless information and in some cases fosters a false sense of security and, worse, creates situations where environmental microbiologic sampling becomes independent from and even substituted for adequate overall programs of disease surveillance and control. Finally, if a particular procedure, medical device, or environment is considered by the epidemiologist and microbiologist important enough to sample microbiologically, it is extremely important that qualitative and/or quantitative microbiologic guidelines be objectively established. Without such guidelines, the microbiologist and epidemiologist have no idea of the meaning of data generated by microbiologic sampling. If such guidelines cannot be established or it is felt that densities of bacaterial contamination are not or cannot be related to the risk of hospital-acquired infections, these particular procedures or items should not be sampled as part of an infection control program.

Oftentimes, a particular part of the intramural environment of a hospital can play a major role in an outbreak of nosocomial infections. In these instances, the strategy that should be used for environmental microbiologic sampling depends on several things. First, the epidemiologist must determine whether certain procedures, medical devices, or other parts of the environment, such as water, are playing a direct or an indirect role in the outbreak. Second, if environmental microbiologic sampling is believed to be necessary, a microbiologist and an epidemiologist should coordinate the sampling scheme and determine which medical devices or other parts of the environment require microbiologic assay.

The application of a microbiologic guideline in this context differs from one associated with scheduled or periodic sampling in that the environmental microbiologic assays are usually directed towards the incriminated pathogen. Consequently, if the outbreak is due to *Pseudomonas aeruginosa,* this organism, and perhaps even specific serotypes, is sought in various environmental items that are sampled. In this respect, the guideline tends to be more qualitative than quantitative and is basically determined by the nature of the outbreak of infections.

The only health-care environment for which routine microbiologic sampling has been recommended is the dialysis unit. Gram-negative water bacteria can multiply relatively fast in water associated with hemodialysis systems, such as softened, deionized, reverse osmosis water, as well as the dialysis fluid itself. Although these fluids do not need to be sterile, excessive levels of gram-negative bacterial contamination pose a risk of pyrogenic reactions and septicemia (Favero et al. 1975). Quantitative microbiologic guidelines which are based on epidemiologic investigations have been recommended (Association for the Advancement of Medical Instrumentation 1981; Favero and Petersen 1977). It is suggested that dialysis fluid and water used to prepare dialysis fluids be microbiologically assayed at least once a month. Microbiologic guidelines for these procedures include sampling the water used to prepare dialysis fluid at that point where it is mixed with concentrated dialysis fluid. The level of bacterial contamination should not exceed 200 per milliliter. Dialysis fluid should be sampled at the end of a dialysis treatment, and the level of bacterial contamination should not exceed 2,000 per milliliter.

It is evident from these introductory remarks that the requirement for rapidity and automation in environmental microbiologic sampling in the hospital environment is not present ordinarily but could have application in the investigation of an outbreak of nosocomial infections where preliminary epidemiologic evaluations have incriminated specific parts or procedures associated with the environment. It is in this context that automated and rapid microbiologic assays which can be applied in certain health-care environments are discussed.

Limulus Amebocyte Assay for Detecting Lipopolysaccharide

Levin and Band (1968) first described the clotting properties associated with amebocytes from the horseshoe crab (*Limulidae*) when they were mixed with lipopolysaccharide or with gram-negative bacteria. This assay has been widely used to quantitate the amount of lipopolysaccharide (LPS) present in pharmaceuticals, waters, and other substances (Jorgensen and Alexander 1981). Gram-negative bacteria contain LPS (bacterial endotoxin). The specific assay can be performed rapidly, usually in less than 2 h, and is capable of detecting as little as 1 pg LPS.

This procedure has been used effectively in detecting bacterial endotoxins in fluids associated with hemodialysis systems (Carson et al. 1979; Carson and Petersen 1982) as well as intravenous solutions where patients had been experiencing endotoxemia. As mentioned above, gram-negative water bacteria can multiply relatively fast in fluids associated with hemodialysis systems and to such an extent that patients who are being dialyzed are at risk of acquiring endotoxemia and septicemia. In 1975, Hindman and co-workers (Hindman et al. 1975) investigated an outbreak of pyrogenic reactions among patients who were being hemodialyzed and determined that the source of bacterial endotoxin was not associated with a high level of gram-negative bacterial contamination in the hemodialysis system. Rather, it was determined by use of the *Limulus* amebocyte lysate (LAL) test that the source of the endotoxins was extramural and was associated with the community water supply. In this instance, the LAL test was invaluable not only as a rapid test for endotoxin but as a sole specific test to incriminate an extramural source of endotoxin in the field. In 1983, Kantor and colleagues Kantor et al. (1983) reported on an outbreak of pyrogenic reactions in a dialysis center where the source of bacterial endotoxin was traced to contaminated heparin. Analysis of heparinized saline solution used during the outbreak revealed a high density of *Acinetobacter calcoaceticus* var. *lwoffi* and a bacterial endotoxin level of 1,300 ng/ml. The LAL test was used effectively in this outbreak for rapidly screening a variety of solutions that were regarded as potential sources of endotoxin. These investigators also showed that the incriminated organism in concentrations found in solutions during the outbreak ordinarily produced relatively small amounts of endotoxins, but when grown in saline to which heparin had been added an unusually high amount of endotoxin was produced.

The LAL test has also been used as a rapid tool for detecting sources of bacterial endotoxin associated with parenteral fluids (Highsmith et al. 1979, 1982). In this context, the LAL test was used to detet bacterial endotoxins in parenteral fluids where no viable microorganisms were present, and patients who received these parenteral fluids experienced pyrogenic reactions.

Membrane Filtration and Epifluorescence Microscopy

Denyer and Ward (1983) in a series of laboratory studies, showed that bacterial cells recovered by membrane filtration and subsequently stained with acridine orange and directly examined by fluorescence microscopy provided a rapid method for assessing microbial contamination in intravenous fluids. The technique took less than 30 min for a result to be obtained. Contamination levels as low as 25 microorganisms per milliliter could be detected, and the sensitivity of the method could be increased by sampling larger volumes by using smaller filtration areas.

Although these studies were laboratory investigations on sensitivity and specificity and did not include field testing, the application as an epidemiologic tool seems reasonable. This technique has the advantage of being able to detect both gram-negative and gram-positive microorganisms whereas the LAL test is exclusively used to detect gram-negative bacteria or their endotoxins.

Luciferase Assay

The detection of adenosine triphosphate (ATP) indicates the presence of living microorganisms, and the quantity of ATP can indicate the number of viable organisms present. This assay has been used to detect viable bacteria in urine, blood specimens, drinking water, and marine waters, and there is a direct relationship between the quantity of measured ATP and cell numbers. Bopp and Wachsmuth (1981) described the use of the luciferase assay to detect bacterial contamination in intravenous fluids. They found that this test could detect rapidly cell concentrations of 10^4–10^5 per milliliter. The clear advantage of using the luciferase assay would be one of rapidity and not sensitivity, since levels less than 10^4 per milliliter cannot be detected but could cause infection. Nonetheless, the assay could be used as a screening test aid in an epidemiologic investigation of infections associated with intravenous fluids.

Hepatitis B Surface Antigen

There are a number of viral antigens that can be detected rapidly in bodily excretions or secretions. However, these have little application for environmental sampling in health-care institutions. A notable exception is hepatitis B surface antigen (HBsAg), which is associated with hepatitis B virus (HBV). A patient who is in the acute phase of infection with HBV or is chronically infected can have approximately 10^8 infectious HBV circulating in blood and approximately 10^{12} particles of HBsAg. Blood can be diluted to the point where it is no longer visible or detectable by chemical means and still contain sufficient levels of HBV to initiate infection in a susceptible human. It has been shown that the environment can play a major role in disease transmission in the health-care facilities where human blood contamination is extensive. These facilities would include hemodialysis units, oncology wards, dental operatories, and clinical laboratories where human blood is tested. Currently, HBV is unable to be cultivated by tissue culture techniques, and its presence can only be demonstrated by infectivity tests using a suitable test for animals such as chimpanzees. However, its presence can be inferred by the demonstration of HBsAg.

HBV has been shown to be a relatively stable virus in the environment (Bond et al. 1983) and can survive for at least 7 days under ambient conditions. HBsAg likewise is stable in the environment. HBsAg usually occurs in a concentration of 3 logs higher than HBV in circulating blood. Consequently, its detection cannot always be assumed to denote the presence of HBV, but it can be considered a valid marker of infectious blood. Favero and co-workers (Favero et al. 1974) showed that exposure of HBsAg-positive blood to 25 °C and 40% relative humidity for 14 days resulted in a loss of only 15%–20% antigenic activity. No reduction in HBsAg occurred on cotton swabs after a 7-day shipping cycle, and recovery of HBsAg from dried blood on stainless steel surfaces by the swab-rinse techniques was essentially 100% efficient. These results indicated that appropriate surface sampling in high-risk areas could be used as an epidemiologic tool to determine routes of hepatitis B transmission. Subsequently, Petersen and co-workers (Petersen et al. 1976, 1979) developed an air sampling technique for HBsAg. The procedure involved using a 0.45-μm membrane filter as a means of entrapping aerosolized suspensions of HBsAg and subsequent assay by radioimmunossay (RIA). The procedure had an efficiency of 22% and a coefficient variation of 11% and the capability of detecting as little as 5×10^{-5} ml of aerosolized HBsAg-positive serum in a single air sample. The technique was applied in hemodialysis centers and dental operatories and was instrumental in showing that the airborne route of HBV infection in these high-risk areas for hepatitis B infection was probably nonexistent and rather that the contamination of environmental surfaces was a much more important factor in the environmental transmission of HBV.

The procedures for assaying HBsAg by RIA and by enzyme immunoassay are rapid and specific. One major difference in performing these types of tests, however, is to include a set of environmental controls to establish a baseline rather than serum controls. This type of test can be used in the field, and when environmental surfaces associated with medical devices or inanimate objects or hospital-associated fluids are epidemiologically incriminated in HBV transmission, the procedure can be an excellent epidemiologic tool.

Airborne Microorganisms

Although there are a number of procedures to assay airborne microorganisms quantitatively, none of them would qualify as being rapid or automated. All require the use of culture media and the demonstration of visible colonies or bacterial growth in liquid culture medium. Furthermore, unless there is an outbreak of hospital-acquired infections that appear to be transmitted by the airborne route, there is virtually no reason to perform air sampling in health-care facilities. However, there bas been one report on the quantitative relationship between airborne viable and total particles (Oxborrow et al. 1975). These investigators measured the numbers of viable and total particles in a laboratory and in a class 100,000 cleanroom to determine if there was a predictable relationship that might aid in the monitoring of airborne microbial contamination. Six-stage Andersen samples were used to enumerate and size viable microbial particles and a modified Royco particle counter was used to enumerate total particles. Results showed that over 99% of the total particles present in both areas were less than 1 μm in size, but that only 1 in 10,000 of the particles in this size range were viable. On the other end of the particle size scale, it was found that less than a tenth of a percent of the total particles were greater than 5.4 μm in size and that only 4.5% of these particles were viable. There was a positive correlation in the combined data from both environmental areas between total and viable particles that exceeded 5.4 μm. It appeared that the amount of environmental control and personnel density and activity influenced the degree of correlation. Correlation between viable and total particulates might be increased in highly controlled environments such as laminar flow cleanrooms where total particle counts are dependent primarily on personnel density and activity. This type of estimation would not be able to be used in those areas where non-viable particles are generated by processes not directly related to personnel density and activity.

Although this technique would qualify as a rapid one, its applicability in health-care facilities is probably not realistic.

Summary

The requirement for rapid and automated environmental microbiologic assays in health-care facilities is significantly less than that of a clinical microbiology laboratory. The main application of these techniques is primarily as tools to aid in the epidemiologic investigation of outbreaks of nosocomial infections. Some of the more realistic and useful techniques include that LAL test that detects bacterial endotoxins associated with hemodialysis systems, medical devices, and intravenous solutions; techniques of epifluorescence and luciferase for IV solutions; and HBsAg for environmentally mediated HBV transmission.

References

Association for the Advancement of Medical Instrumentation (1981) American national standard for hemodialysis systems. AAMI, Arlington, Virginia

Balows A (1982) A decade of automation: new horizons in microbiolgy. In: Tilton RC (ed) Rapid methods and automation in microbilogy, American Society for Microbiology, Washington DC, pp 76–79

Bond WW, Favero MS, Petersen NJ, Ebert JW (1983) Inactivation of hepatitis B virus by intermediate-to-high-level disinfectant chemicals. J Clin Microbiol 28:535–538

Bopp CA, Wachsmuth IK (1981) Luciferase assay to detect bacterial contamination of intravenous fluids. Am J Hosp Pharm 38:1747–1750

Carson LA, Petersen NJ (1982) LAL-reactive material associated with hemodialysis membranes. In: Watson SW, Levin J, Novitsky TJ Endotoxins and their detection with the *Limulus* amebocyte lysate test. Liss, New York, pp 217–230

Carson LA, Petersen NJ, Favero MS (1979) Use of the *Limulus* amoebocyte lysate assay system for detection of bacterial endotoxin in fluids associated with hemodialysis procedures. In: Cohen E Biomedical applications of the Horseshoe Crab (*Limulidae*). Liss, New York, pp 453–464

Denyer SP, Ward KH (1983) A rapid method for the detection of bacterial contaminants in intravenous fluids using membrane filtration and epifluorescence microscopy. J Parenter Sci Technol 37:156–158

Favero MS, Petersen NJ (1977) Microbiologic guidelines for hemodialysis systems. Dialysis Transplant 6:34–36

Favero MS, Puleo JR (1980) Techniques used for sampling airborne microorganisms associated with industrial clean rooms and spacecraft assembly areas. Ann NY Acad Sci 353:241–254

Favero MS, Bond WW, Petersen NJ, Berquist KR, Maynard JE (1974) Detection methods for study of the stability of hepatitis B antigen on surfaces. J Infect Dis 129:210–212

Favero MS, Petersen NJ, Carson LA, Bond WW, Hindman SH (1975) Gram-negative water bacteria in hemodialysis systems. Health Lab Sci 12:321–334

Goldschmidt MC (1980) Instrumentation, automation, and miniaturization. In: Sonnenwith A, Jarett L (eds) Gradwohl's clinical laboratory methods and diagnosis, 8th edn. Mosby, New York, pp 1495–1553

Highsmith AK, Anderson RL, West CM, Dixon RE (1979) Characterization of an endotoxin-life substance recovered from an intravenous antibiotic solution involving pyrogenic reactions. In: Cohen E. Biomedical applications of the Horseshoe Crab (*Limulidae*). Liss, New York, pp 465–461

Highsmith AK, Anderson RL, Allen JR (1982) Application of the *Limulus* amebocyte lysate assay in outbreaks of pyrogenic reactions associated with parentral fluids and medical devices. In: Endotoxins and Their Detection with the *Limulus* Amebocyte Lysate Test. Liss, New York, pp 287–299

Hindman SH, Favero MS, Carson LA (1975) Pyrogenic reactions during haemodialysis caused by extramural endotoxin. Lancet 11:1

Jorgensen JA, Alexander GA (1981) Automation of the *Limulus* amoebocyte lysate test by using the Abbott MS-2 Microbiolgy system. Appl Environ Microbiol 41:1316–1320.

Kantor RJ, Carson LA, Graham DR, Petersen NJ, Favero MS (1983) Outbreak of pyrogenic reactions at a dialysis center. Am J Med 74:449–456

Levin ML, Bang FB (1968) Clottable protein in *Limulus:* its localization and kinetics of its coagulation by endotoxin. Thromb Diath Haemorrh 19:186–197

Oxborrow GS, Fields ND, Puleo JR, Herring CM (1975) Quantitative relationship between airborne viable and total particles. Health Lab Sci 12:47–51

Petersen NJ, Bond WW, Favero MS (1979) Air sampling for hepatitis B surface antigen in a dental operatory. J Am Dent Assoc 99:465–467

Petersen NJ, Bond WW, Marshall JH, Favero MS, Raij L (1976) An air sampling technique for hepatitis B surface antigen. Health Lab Sci 13:233–237

Problems in Detecting Water-Borne Viruses by Rapid Methods

N. GUTTMAN-BASS [1]

Introduction

The viral etiology of outbreaks of water-borne disease is epidemiologically difficult to prove, and evidence is limited to a small fraction of the many types of human pathogenic viruses found in water (IAWPRC Study Group on Water Virology 1983). In spite of this, there is a growing concern with the virological quality of water. Classical bacterial indicators of sewage pollution have not always been adequate, and infectious virus has been found in water samples meeting coliform standards (Deetz et al. 1984)

Routine monitoring of water for the presence of viruses is not generally performed, largely due to the current state of the methodology, which is difficult, expensive, and time-consuming. Steps in the assay of water for viruses include obtaining a representative sample of water, concentration of the viruses from the water, detection of the viruses, and identification. This paper is concerned with the use of rapid methods in one of these stages, that of virus detection. A useful method for detecting water-borne viruses should warn of the presence of viruses before the water is consumed. Rapid methods of virus detection, if adaptable to the requirements of the water environment, would present an attractive approach to the water virologist.

Requirements of a Water-Borne Virus Detection Method

The water environment presents difficulties in virus detection which differ in several ways from the problems of clinical isolation. Although speed of detection is important, a number of other points should also be considered in the choice of an appropriate method for the detection of water-borne viruses.

In the water environment, human pathogenic viruses are present in low concentrations due to dilution, adsorption to solids, and extinction. A primary requirement for a virus detection method is sensitivity. The sampling of large volumes of water is required, and in recent years techniques for concentrating viruses from water have been developed and tested, and are undergoing further improvement (Melnick et al. 1984). However, even one infectious virus in hundreds of

---

1 Environmental Health Laboratory, Hebrew University, Hadassah Medical School, P.O. Box 1172, Jerusalem, Israel

liters of drinking water is a cause for concern, and a virus detection method should be able to detect very small numbers of virus particles.

Pathogenic viruses enter water primarily through contamination by human or animal waste. Wastewater contains a variety of viruses which are present in the population and contribute in an additive way to the levels of pollution. A virus detection method should be specific in the detection of pathogens, yet be sensitive to a broad range of viruses. It is difficult to predict which viruses will be present in water at a given time, and variations may be hourly, seasonal, or annual. Since there is no single detection method for all water-borne viruses, a battery of tests would have to be used to cover the full complement of viruses. In practice, a single method is often used to detect a limited range of viruses. If a particular virus is of interest, then the choice of method is simplified to selecting the one which is most appropriate for the chosen virus.

The water environment contains contaminants, both chemical and biological, which may interfere with the virus detection method. This is particularly problematic with wastewater, which may contain high levels of toxic materials as well as large numbers of microorganisms. A detection method should be evaluated for various types of water to determine its usefulness with natural waters. In addition, the concentration of viruses from water may be accompanied by the concentration of interfering materials. A detection method should therefore be tested in conjunction with a concentration technique.

A virus detection method should differentiate between viable virus and viral remnants which are no longer infectious. Following treatment and disinfection, inactivated viral elements may remain in the water but no longer represent a health hazard.

Virus identification is desirable, but is not an absolute requirement. It is generally more important to know whether there are viruses in the water than to know which ones. However, it is often of interest to obtain a profile of the viral flora, and the detection method should provide for this option by directly identifying or enabling identification of the viruses.

Finally, a virus detection method for routine monitoring should be used frequently and be practicable for public health laboratories to perform. This is limited in part by cost, as well as simplicity of use. In order for the results of the virus assays to have validity, the method should be reliable and accurate.

If a method were found which met all of the above points, it would provide a practical approach to the reduction of water-borne disease. In reality, there is no such ideal method, and the choice depends of the aims of the user. For monitoring drinking water, the most sensitive and rapid method should be chosen. In this case, identification is of less importance than the knowledge that the water is contaminated by viable virus. For studies of sewage and the effects of treatment processes, the types and amounts of virus present are of interest. In this case, a less sensitive method may be acceptable, since the high virus concentrations in the wastewater are relatively easy to detect. Under certain circumstances, such as the investigation of a disease outbreak caused by a particular virus, water samples may be assayed for the suspected agent.

Water-Borne Viruses

Viruses which are water-borne share a number of characteristics. They are relatively stable in water and remain viable until they reach a new susceptible host. Most of the water-borne viruses are enteric and replicate in the intestinal tract of the host, although they are not necessarily restricted to that site.

The types of water-borne viruses are listed in Table 1. These include viruses which have been shown to cause water-borne outbreaks of disease (such as hepatitis A virus), and others which may have that potential. Potential water-borne viruses are those which are found or excreted into water, but have not yet been shown to transmit disease via water (such as reoviruses). In total, there are more than 120 types of human pathogenic viruses which may be found in the water environment. These viruses have been reviewed elsewhere (Banatvala 1981), and the following discussion emphasizes aspects of the viral groups relevant to their assay in water.

The enteroviruses are a group of 72 viruses containing single-stranded RNA, and are members of the picornavirus family. Although they frequently cause subclinical infections, they are also associated with more serious illnesses such as meningitis, paralysis, and hepatitis. Generally, antisera prepared against one type of enterovirus do not cross-react with the other types, and they are assigned to the group on the basis of morphology and composition.

Included in this group are the polioviruses, coxsackie B viruses, and echoviruses, which are relatively easily isolated and cultivated by tissue culture methods. Other members of this group are more problematic; for example, some types of coxsackie A viruses are difficult to cultivate in tissue culture. Hepatitis A virus, which has recently been designated enterovirus 72, is a virus of importance in the transmission of water-borne hepatitis. It is a virus which is cultivated with difficulty and has not been detected in water by tissue culture methods.

The enteroviruses as a group are frequently found in water, in part because of their abundance and in part because they are relatively easy to detect in cell

Table 1. Water-borne viruses

Virus group	Representative types	Group-specific antigen?
Enteroviruses	polioviruses coxsackie A viruses coxsackie B viruses echoviruses hepatitis A virus	
Reoviruses		Yes
Adenoviruses	enteric adenoviruses	Yes
Rotaviruses		Yes
Parvovirus-like	Norwalk agent	
Other fecal viruses	astroviruses coronaviruses caliciviruses	

cultures. This is particularly true of the polioviruses, which are common in areas where they are administered to the population in the form of attenuated live vaccines. The levels of specific enteroviruses in different water samples may vary widely (Morris and Sharp 1982), making it difficult to predict in advance which types will be present.

The reoviruses are a group of three virus types which contain double-stranded RNA. They replicate well in tissue culture and can be isolated directly from water. In some cases their abundance in water has been reported to exceed that of enteroviruses (Grabow and Nupen 1981). However, relative virus numbers are influenced by the tissue culture systems used in the specific assay (Sellwood and Dadswell 1981), and may not accurately reflect the absolute quantities of virus found in water.

The reoviruses share group-specific antigenic determinants so that an antiserum prepared against one should detect all three. Rapid immunoassays for these viruses are a feasible approach to their detection in water.

The adenoviruses are a group of DNA viruses which primarily cause upper respiratory tract infections, although they are also excreted into water. Recently new strains (40 and 41) of enteric adenoviruses associated with gastrointestinal disease which are difficult to cultivate in tissue culture have been found (Wigand et al. 1983). Although most work has indicated that adenoviruses are present in lower abundance in water than the enteroviruses or reoviruses (Irving and Smith 1981; Goddard and Sellwood 1982), this may be due to the inefficiency of the tissue culture methods used, which would not have detected the "fastidious" enteric adenoviruses. There is antigenic cross-reactivity among the adenoviruses, and a group-specific antibody might be a useful reagent in their detection in water.

The rotaviruses, which contain double-stranded RNA, are a major cause of infantile gastroenteritis. The human rotaviruses cross-react serologically and are currently typed into four subgroups (Wyatt et al. 1983). Although they have not been detected in water using standard tissue culture techniques, the recent application of rapid methods has allowed their detection, as will be described below.

The parvo-like viruses include the Norwalk agent, which has been shown to cause gastroenteritis and to be transmitted by water (Kaplan et al. 1982). These viruses have not yet been cultivated outside of a human host and have not yet been directly detected in water. Rapid methods may be the only useful ways to demonstrate their presence in the water environment.

Other viruses which are excreted into water may be transmitted through the environment; here all viruses are suspect. Some of the additional viruses listed in Table 1 may play a role in disease transmission via water, but their importance as water contaminants has yet to be investigated.

Although there are a large number of virus types which can be found in water, it may not be necessary to detect them all. Once their relative abundance and importance is better established, it may be possible to choose one or a few groups to indicate viral pollution and warn of possible contamination with other viruses as well.

Rapid Methods: Useful for Detecting Water-Borne Viruses?

Viruses may be detected by their effects on a host, whether an animal or cells in culture. These are considered to be standard methods of virus determination and detect replication, and tissue cultures are to be preferred over animals whenever possible. Rapid methods, which by definition detect viruses in a shorter time than the standard methods, usually detect components of the virus, such as the physical particle (electron microscopy) or its subunits (proteins by immunoassays and nucleic acid by hybridization).

The most common methods of virus isolation and enumeration in water are standard tissue culture techniques (cytopathic effect or plaque assay). However, not all of the water-borne viruses are detectable in cell culture, even when several lines from different sources are used. Water-borne viruses are not tissue culture adapted, making them more difficult to detect than laboratory strains. For those viruses which do grow well in tissue culture, such methods are sensitive to small numbers of viruses, and only viable viruses are detected. It is possible to choose a cell type sensitive to a broad range of viruses likely to be found in water, without having to preselect for specific viruses. In virus mixtures, however, more rapidly growing viruses may overgrow the slower viruses, preventing their detection. Tissue culture methods are well established in many laboratories, do not require expensive reagents, and are not technically difficult to perform. One major difficulty with tissue cultures is that they are often affected by toxic components in water samples, which could prevent detection of any viruses present. In addition, tissue culture techniques which can take several weeks to detect viruses may be too slow to enable the prevention of virus spread by the water route.

There is a continuing search to find more rapid methods for virus detection which are sensitive enough to be useful for water-borne viruses. The advantages and disadvantages of some general approaches to the rapid detection of viruses as applied to the water environment are discussed below. These methods are able to detect the presence of virus from within a few hours to a day. Although some require sophisticated equipment, once instituted they are not technically difficult to perform.

Electron Microscopy

Electron microscopy (EM) is a way to directly visualize virus particles in a sample. Viruses which have a distinctive morphology can be identified by this method, and all viruses present in sufficient concentrations in the sample should be detectable. A further refinement, the addition of antibodies to aggregate specific viruses (immunoelectron microscopy or IEM) allows more precise identification of the viruses, but limits the range of viruses detected. This would increase the sensitivity of the assay for the specific viruses chosen. The major drawbacks of EM and IEM are that they do not differentiate between viable and nonviable particles and are not sensitive to small numbers of virus particles (Elkana and Guttman-Bass 1983). Another problem is that environmental samples may contain large amounts of nonviral material which may obscure the presence of viruses. In addition, particles which resemble viruses may be falsely identified as human pathogens, particularly with viruses without distinguishing characteristics (apart from size and shape) such as the enteroviruses.

Immunoassays

Immunoassays utilize antibodies prepared against specific viruses to detect the presence of viral proteins in a sample. A preliminary selection must be made as to which viruses to detect, which in virus monitoring is a major drawback. A method for detecting water-borne viruses should be as general as possible, since the composition of the viral flora in water at any given time is not predictable. However, these approaches are useful for certain viruses difficult to detect by tissue culture methods, or for groups of viruses sharing a common antigen.

Immunoassays in common use include radioimmunoassays (RIA) and enzyme-linked immunsorbent assays (ELISA), in which the antigen is detected by antibodies labelled with a radioactive or enzymatic marker. These assays are rapid and easily automated, but are not as sensitive as tissue culture techniques. They are specific for viruses with antigenic components which can be detected by the antibody used, and are thus limited in range. Such immunoassays might prove to be useful for the detection of viruses with group-specific determinants in water. These immunoassays are probably less sensitive to interference by biological contaminants than tissue culture methods, but may be affected by chemical contaminants in the samples. Unless used in combination with a growth enrichment step, they do not differentiate between viable and noninfectious particles. One advantage of the ELISA is its portability; it can also be adapted to on-site virus detection more easily than the other methods.

Immunofluorescent (IF) methods, in which viral antigens are detected in infected cells by antibody tagged with a fluorescent probe, have a number of advantages over RIA and ELISA, although results may take longer to obtain (generally within 24 h). IF methods can be as sensitive as tissue culture techniques and can theoretically detect as little as one infectious virus unit. Viable virus is detected by IF more rapidly than in standard tissue culture assays. As with the other immunoassays, a specific antibody must be chosen in advance. It is possible, however, to combine antibodies to increase the range of virus detection (N Guttman-Bass and Y Tchorsh, unpublished results, 1983). There is a shorter time for microbial growth and contamination to interfere with the assay than with standard tissue culture methods, but since the assay uses cells, it is probably more sensitive to interference than the assays using only chemical reagents. The assay is limited not only by the antibodies selected, but will also only detect those viruses which produce antigen in the particular host cell used.

Nucleic Acid Hybridization

Nucleic acid sequences specific to viruses can be detected by hybridization to a labelled probe complementary to viral sequences. This method has not yet been used to detect viruses in water, but is discussed here as a possible direction for future investigation. The low virus concentrations in water would probably not allow the direct detection of viral sequences, but enrichment could be attained by growth in cell culture. Only partial growth would be necessary, to the stage where replication of the viral genome or accumulation of viral RNA would reach detectable concentrations. For virus groups with shared sequences, a probe complementary to that region could be used to detect the virus group. This assay would thus measure viable virus and might prove to be highly sensitive.

Current Use of Rapid Methods to Detect Water-Borne Viruses

The application of rapid methods to the detection of water-borne viruses is expanding as reagents become available. The groups of viruses which have been detected in water by rapid methods are listed in Table 2, along with the methods which were used.

A number of rapid methods have been used to detect enteroviruses in water. Immunofluorescence of infected cells, by a direct IF assay, has been used to detect poliviruses in sewage (Katzenelson and Kedmi 1979). Approximately half of the samples were negative for polioviruses by IF and positive for virus by plaque assay. This could be explained in part by the tenfold larger volume tested by plaque assay as well as by the absence of polioviruses from the samples. More recently, we have assayed sewage samples for virus by an indirect IF assay (Guttman-Bass et al. 1981) for coxsackievirus B 5, echovirus 7, and poliovirus 1 (N Guttman-Bass and Y Tchorsch, unpublished results, 1983). Virus was detected by plaque assay in unconcentrated and concentrated sewage samples at concentrations of up to 5,000 pfu/l. Fluorescent plaques of poliovirus were detected in one of the two sewage samples, but the concentrates were toxic for the cell layer. Neither of the other two viruses was detected.

These studies indicate some of the difficulties in the detection of enteroviruses by assays requiring a prior choice of antibody, limiting the range of detection. Unless the antiserum used has a broader specificity, this approach does not provide an attractive alternative to standard tissue culture methods for these viruses. Recently, antibodies which reacted with a number of enteroviruses were prepared using heat-disrupted virions as immunogens (Hasegawa and Inouye 1983). The use of such antisera, if reactive in an IF assay, might provide a less limited rapid assay for these viruses.

Table 2. Field studies using rapid methods

Virus group	Method used
Enteroviruses	IF, ELISA, RIA, EM, IEM
Rotaviruses	IF, ELISA, EM, IEM
Reoviruses	IF
Adenoviruses	EM

Hepatitis A virus (HAV), an enterovirus, is not easily isolated in cell cultures, although in combination with antibodies it can be detected. Due to the long incubation peroid of the disease, rapid detection of the virus in water followed by the introduction of public health measures might reduce the severity of a subsequent outbreak. Thus, a number of laboratories have chosen to look for this virus using rapid methods, since other methods of detection are too slow or not feasible.

HAV has been detected in sewage by a combination of affinity chromatography followed by IEM (Elkana et al. 1983). The one sample positive for HAV was from a broken sewer in a neighborhood with cases of infectious hepatitis, but the virus was not found in sewage from other nearby sites. The sample was also positive for virus by plaque assay. This method was also used to detect HAV in the Jerusalem sewage samples described above, but no HAV was found in 50 ml of sewage (N Guttman-Bass and Y Elkana unpublished results, 1984).

RIA was used to assay for HAV in samples of well water and sewage from a community undergoing an outbreak of gastroenteritis and hepatitis (Hejkal et al. 1982). HAV was found in three sewage samples and one well water sample taken a month prior to the outbreak of infectious hepatitis. Coxsackie B viruses were most frequently isolated from the water samples, and were implicated as the cause of the gastroenteritis outbreak.

ELISA has also been used to detect HAV in wastewater (Ryvarden 1982), but the relative levels of other viruses in the samples was not reported.

The importance of selectively assaying for HAV to the exclusion of a more general viral assay has yet to be demonstrated. It might be better viewed as a supplementary assessment of viral pollution and might aid in the timely immunization of the exposed population. It should be noted that the methods used for HAV detection in water did not determine viability. In all of the samples assayed by tissue culture methods as well, viable virus was found. The absence of HAV does not necessarily imply the absence of other pathogenic viruses, and HAV does not appear to be present consistently enough to warrant its use as a general virus indicator.

Rotaviruses have been the subject of increasing use of rapid methods for their detection. Since the strains cross-react antigenically, antiserum against all of the human types are relatively easy to obtain, and even cross-reacting animal strains can be used for antibody production. In addition, kits are available commercially for antigen detection. Although recently developed methods have made the direct isolation of rotaviruses from clinical specimens by tissue culture methods possible (Wyatt et al. 1983), the ease, sensitivity, and rapidity of immunoassays have made them the more common approach for isolation from water.

In particular, immunofluorescence of infected cells is a promising approach. Rotavirus has been detected in Texas sewage by idirect IF (Hejkal et al. 1982) and in drinking water samples in Mexico (Deetz et al. 1984). Using this method, Smith and Gerba (1982) detected rotavirus foci in seven secondary treated wastewater samples (0.4 ml tested), four of which were negative for viruses by plaque assay (6 ml tested). IF has also been used to quantitate the amount of rotavirus in Houston, Texas, sewage over the course of a year, with rotavirus concentrations ranging from 1 to 321 foci/l, while enteroviruses ranged from 7.5 to 800 pfu/l (Hejkal et al. 1984). The ELISA method was also tested, but did not correlate well with the IF results and resulted in a false positive rate of 23%.

Greater numbers of rotaviruses (detected by IF) than enteroviruses have been reported for polluted estuary waters in Texas, with rotavirus concentrations of up to 4,980 foci/378 l of water (Metcalf et al. 1984).

Monoclonal antibodies have been used in IF and immunoperoxidase procedures for rotavirus detection (SA Sattar and RA Raphael, personal communication, 1984). Sewage, effluent, and river water samples were successfully assayed for rotavirus. It would be of interest to compare the sensitivity of the method using monoclonal antibodies with the use of more general antisera.

ELISA and EM have also been used for the detection of rotavirus in water. Sewage samples in the German Democratic Republic which were positive for enteroviruses were assayed for rotavirus by two methods. Out of six sewage samples, three were positive by ELISA and one by IEM (Walter et al. 1982). In contrast, Steinmann (1981) compared the two methods for sewage samples from West Germany and found 6 out of 24 samples positive for rotavirus, with agreement between ELISA and EM methods. The EM was performed without antibody, and the virus identified by size and morphology. Viability was not determined. In England, Goddard and Sellwood (1982) detected rotavirus in sewage by ELISA, but less frequently than enteroviruses. Finally, in a West German study (C Epp, personal communication, 1984), rotavirus was detected in two out of eight samples of water from a sewage-polluted river. The samples also contained other viruses as detected by tissue culture methods.

In summary, rotaviruses are present in high enough concentrations in sewage-polluted waters to be detected by IF, ELISA, and EM methods. However, only the IF method detected rotaviruses in samples negative for enteroviruses, and this method would appear to be the most sensitive for rotavirus detection.

Other viruses have also been detected by rapid methods, although they have received less attention than the above viruses. Reoviruses, which in some cases are found in high concentrations in water, have been detected in sewage by IF (Ridinger et al. 1982). The IF method compared favorably to plaque assay, both in sensitivity and in rapidity of the assay. The plaque assay isolated nonreoviruses as well, and was less diagnostic for reoviruses. The detection of reoviruses by immunological methods would appear to be a promising approach.

In a study using EM to detect viral pollution of bore water samples on Norfolk Island, Australia, viruses were identified by their morphology (Murphy et al. 1983). Out of 32 samples, 1 was positive for rotavirus, 1 for adenovirus, and 2 contained "small round virus" particles. Tissue culture isolates were also made, and poliovirus type 1 and adenovirus (types 1 and 5) were isolated. Poliovirus was found in three samples which would have been acceptable by bacteriological standards. Although the EM did not test the viability of the viruses, it was broad enough to detect virus types which might not otherwise have been found.

Prospects

The use of rapid methods for virus detection in water has made possible the detection of important viruses such as HAV and rotavirus. Such methods are still problematic for the majority of the enteroviruses and will continue to be so until the introduction of rapid assays which can detect a wider range of such viruses. One direction might be in the area of nucleic acid hybridization. Recently, a cDNA probe to coxsackievirus B3 was used in a spot hybridization test to detect cells infected with other enteroviruses, including coxsackie A, echo, and polioviruses (Hyypia et al. 1984). Another approach is that suggested by the work of Deng and Cliver (1984), who used pooled human immune serum globulin in an ELISA to detect human enteric viruses in an additive fashion.

The development of rapid methods for adenovirus detection in water, either by immunossay or nucleic acid hybridization, is warranted to further explore their presence and distribution in the water environment.

The results of assays for rotavirus are accumulating, and this group of viruses appears to be present in high concentrations in water. Sensitive methods such as IF which are rapid and quantitative are in use to study these viruses in natural waters.

In summary, tissue culture methods, although in some cases still the method of choice for the detection of certain viruses, are being supplemented with rapid methods to detect a broader spectrum of viruses. At this point, rapid methods cannot completely replace more standard, if slower, methods of virus detection in water. However, as reagents and assays become available for the detection of more virus types by rapid methods, they will eventually be applied to the study of viruses in water.

References

Banatvala JE (1981) Viruses in faeces. In: Goddard M, Butler M (eds) Viruses and wastewater treatment. Pergamon, Oxford, pp 1–13

Deetz TR, Smith EM, Goyal SM, Gerba CP, Vollet III JJ, Tsai L, DuPont HL, Keswick BH (1984) Occurrence of rota- and enteroviruses in drinking and environmental water in a developing nation. Water Res. 18:567–571

Deng M, Cliver DO (1984) A broad-spectrum enzyme-linked immunosorbent assay for the detection of human enteric viruses. J Virol Methods 8:87–98

Elkana Y, Guttman-Bass N (1983) An evaluation of immune electron microscopy as a method for the quantitation of virus particles. J Virol Methods 6:53–58

Elkana Y, Manaa A, Marzouk Y, Manor J, Halmut T, Herscovici M, Barak S (1983) Detection of hepatitis A virus in sewage. J Virol Methods 7:259–262

Goddard MR, Sellwood J (1982) Isolation of enteric viruses from sewage. In: Butler M, Medlen AR, Morris R (eds) Viruses and disinfection of water and wastewater. Proceedings of the international symposium at the University of Surrey, Guildford, 1–4 Sept. University of Surrey Print Unit, Guildford, pp 1–6

Grabow WOK, Nupen EM (1981) Comparison of primary kidney cells with the BGM cell line for the enumeration of enteric viruses in water by means of a tube dilution technique. In: Goddard M, Butler M (eds) Viruses and wastewater treatment. Pergamon, Oxford, pp 253–256

Guttman-Bass N, Tchorsh Y, Nasser A, Katzenelson E (1981) Rapid detection of enteroviruses in water by a quantitative fluorescent antibody technique. In: Goddard M, Butler M (eds) Viruses and wastewater treatment. Pergamon Oxford, pp 247–251

Hasegawa A, Inouye S (1983) Production in guinea gips of antibodies to cross-reactive antigenic determinants of human enteroviruses. J Clin Microbiol 17:458–462

Hejkal TW, Keswick B, LaBelle RL, Gerba CP, Sanchez Y, Dressman G, Hafkin B, Melnick JL (1982) Viruses in a community water supply associated with an outbreak of gastroenteritis and infectious hepatitis. J Am Water Works Assoc 74:318–321

Hejkal TW, Smith EM, Gerba CP (1984) Seasonal occurrence of rotavirus in sewage. Appl Environ Microbiol 47:588–590

Hyypia T, Stalhandske P, Vainionpaa R, Petterson U (1984) Detection of enteroviruses by spot hybridization. J Clin Microbiol 19:436–438

IAWPRC Study Group on Water Virology (1983) The health significance of viruses in water. Water Res 17:121–132

Irving LG, Smith FA (1981) One-year survey of enteroviruses, adenoviruses and reoviruses isolated from effluent of an activated sludge purification plant. Appl Environ Microbiol 41:51–59

Kaplan JE, Feldman R, Campbell DS, Lookabaugh C, Gary W (1982) The frequency of a Norwalk-like pattern of illness in outbreaks of acute gastroenteritis. Am J Public Health 72:1329–1332

Katzenelson E, Kedmi S (1979) Unsuitability of polioviruses as indicators of virological quality of water. Apl Environ Microbiol 37:343–344

Melnick JL, Safferman R, Rao VC, Goyal S, Berg G, Dahling DR, Wright BA, Akin E, Stetler R, Sorber C, Moore B, Sobsey MD, Moore R, Lewis AL, Wellings FM (1984) Round robin investigation of methods for the recovery of poliovirus from drinking water. Appl Environ Microbiol 47:144–150

Metcalf TG, Rao VC, Melnick JL (1984) Solids-associated viruses in a polluted estuary. Monogr Virol 15:97–110

Morris R, Sharp DN (1982) Cytopathic enteric viruses in wastewater effluents and surface waters. In: Butler M, Medlen AR, Morris R (eds) Viruses and disinfection of water and wastewater. Proceeding of the international Symposium at the University of Surrey, Guildford, 1–4 Sept. University of Surrey Print Unit, Guildford, pp 32–41

Murphy AM, Grohmann GS, Sexton MFH (1983) Infectious gastroenteritis in Norfolk Island and recovery of viruses from drinking water. J Hyg Camb 91:139–146

Ridinger DN, Spendlove RS, Barnett BB, George DB, Roth JC (1982) Evaluation of cell lines and immunofluorescence and plaque assay procedures for quantifying reoviruses in sewage. Appl Environ Microbiol 43:740–746

Ryvarden G (1982) Serological screening procedure (ELISA) for the detection of water-borne transmission of hepatitis A virus. In: Butler M, Medlen AR, Morris R (eds) Viruses and disinfection of water and wastewater. Proceedings of the international symposium at the University of Surrey Guildford, 1–4 Sept. University of Surrey Print Unit, Guildford, pp 139–143

Sellwood J, Dadswell JV (1981) Viruses isolated from sewage 1974–1977. In: Goddard M, Butler M (eds) Viruses and wastewater treatment. Pergamon, Oxford, pp 27–30

Smith EM, Gerba CP (1982) Development of a method for detection of human rotavirus in water and sewage. Appl Environ Microbiol 43:1440–1450

Steinmann J (1981) Detection of rotavirus in sewage. Appl Environ Microbiol 41:1043–1045

Walter R, Dobberkau H-J, Durkop J (1982) A virological study of the health hazards associated with the indirect use of wastewater. In: Butler M, Medlen AR, Morris R (eds) Viruses and disinfection of water and wastewater. Proceedings of the international symposium at the University of Surrey, Guildford, 1–4 Sept. University of Surrey Print Unit, Guildford, pp 144–153

Wigand R, Baumeister HG, Maass G, Kuhn J, Hammer HJ (1983) Isolation and identification of enteric adenoviruses. J Med Virol 11:233–240

Wyatt RG, James HD Jr, Pittmann AL, Hoshino Y, Greenberg HB, Kalica AR, Flores J, Kapikian AZ (1983) Direct isolation in cell culture of human rotaviruses and their characterization into four serotypes. J Clin Microbiol 18:310–317

Detection and Identification of Viruses in Water by Immunoadsorbance and Energy Transfer

Y. ELKANA [1]

Introduction

Detection of viruses in water and wastewater is usually carried out in two steps: (1) concentration by nonspecific adsorption to various filters (Goyal and Gerba 1982; Gerba 1983); (2) detection, after elution of the adsorbed virus, in tissue culture (Smith and Gerba 1982). Lately, various immunoassays have been introduced, being used mainly for the detection of fastidious viruses. For example, Heijkal et al. (1982) describe the detection of hepatitis A virus (HAV) in sewage and well water by radioimmunoassay after concentration by filters.

The present work suggests another approach which involves immunoadsorbance and measurement of electronic excitation energy transfer for the detection of viral antigens, including those of fastidious viruses. The use of immunoadsorbance followed by immunoelectron microscopy for the detection of HAV is demonstrated in two cases: (1) recovery of HAV seeded into large volumes of tap water; (2) detection of HAV in the sewage of a town during a hepatitis outbreak.

Preliminary results show that the immunoelectron microscopic step can be replaced by fluorescence measurement of energy transfer.

The rate of electronic excitation energy transfer, which can occur between a fluorescent donor and a suitable acceptor, is proportional to R^{-6}, where R is the distance between donor and acceptor (Förster 1959). This phenomenon can be used for the identification of viruses in water in the following way: Antibodies against the virus are conjugated separately either to the donor or to the acceptor. When the two populations are mixed in solution at a low concentration no energy transfer can occur. Addition of the virus and the interaction of a single virus particle with the two kinds of antibodies brings the donor and acceptor into a close proximity which might enable measurable energy transfer.

Materials and Methods

Immunoglobulin G (IgG) was purified from a human serum having a high titre of antibodies against HAV, as tested by immunoelectron microscopy (IEM) and radioimmunoassay (HAVAB^{TM}, Abbot). IgG purification was performed on QAE-Sephadex according to Joustra and Lundgren (1969).

1 Division of Microbial and Molecular Ecology, Institute of Life Sciences, The Hebrew University of Jerusalem, 91904 Jerusalem, Israel

HAV was extracted from stools collected from siblings of children with clinically confirmed hepatitis, 1 week to 10 days before onset of symptoms, and confirmed by IEM.

Immunoadsorption was performed on columns of anti-HAV containing IgG coupled to Sepharose CL-4 B according to Elkana et al. (1983).

Fluorescent conjugates of IgG: Fluorescein and tetramethylrhodamine conjugates were prepared by mixing 10 μl of fluorescein- or tetramethylrhodamine-isothiocyanata, 10 mg/ml in dimethylsulfoxide, with the same amount of IgG at pH 9 (Forni 1979).

All reactions were carried out at 4 °C under gentle shaking overnight. Each reaction mixture was then passed through a Sephadex G-25 (PD-10, Pharmacia) column, equilibrated with 0.01 *M* phosphate buffer (Na^+), pH 7.5, to remove unreacted reagent.

Fractionation according to degree of conjugation was achieved by chromatography on DEAE–Sephacel. The sample was loaded on a 1-ml column equilibrated with 0.01 *M* phosphate buffer (Na^+), pH 7.5. Conjugates of increasing degree of conjugation were obtained by employing a gradient of 5 ml of the same buffer and 5 ml of the buffer containing 0.5 *M* NaCl.

Degree of conjugation was calculated from the absorption spectrum, using the formulae given by Forni (1979) for fluorescein and tetramethylrhodamine.

Fluorescence measurements were done on a Perkin-Elmer LS-5 spectrofluorometer.

Energy transfer was measured in a solution containing 0.06 mg/ml und 0.03 mg/ml of the fluorescein (3 dye molecules/IgG molecule) and rhodamine (1.5 dye molecules/IgG molecule) conjugates, respectively. A ten fold concentrated HAV-containing faecal extract was used as the source of HAV. It was concentrated by ultracentrifugation at $80{,}000 \times g$ for 3 h and resuspended in one tenth of the volume 0.01 *M* phosphate buffer, pH 7.5, containing 0.002% Tween-20. The reaction was carried out in the same buffer, with 0.1 ml of the concentrate in 3 ml reaction mixture.

Results and Discussion

Immunoadsorption can be an efficient method for recovery of HAV from water, as shown in Table 1. The virus was fully recovered after seeding into increasing volumes of dechlorinated tap water, up to 1 litre. When seeded into the original tap water, which is chlorinated, the efficiency of recovery decreased with increasing volume. This is probably due to prolonged exposure of the virus to chlorine in the larger volumes used. Our results agree with the finding of Grabow et al. (1983) concerning the time course of inactivation of HAV by chlorine.

Table 2 presents the results of sewage testing in a town during a hepatitis outbreak, using immunoadsorbance and IEM. HAV was detected only in a sewage sample collected from the affected neighbourhood. The negative results with the town's oxidation pond can stem either from dilution of the sample or from the specific sampling schedule used. As mentioned previously (Elkana et al. 1983), a 15-fold concentration of the sewage by conventional filtration methods resulted

Table 1. Recovery of HAV from seeded water by the immunoadsorbent method, tested by immuno-electron microscopy

Source	Seeded water	Volume of water seeded by a fixed amount of Fx (ml)[c]			
		0	10	100	1,000
Faecal extract (Fx)[a]	–	++++			
	Tap water		++	+	–
	Dechlorinated tap water[b]		++++	++++	++++

[a] Fx is the $6{,}000 \times g$ supernatant of HAV-containing stools suspended in phosphate-buffer saline to a concentration of 10% (w/v)

[b] Tap water was treated with $Na_2S_2O_3$ to reduce the chlorine

[c] 0.5 ml of Fx was mixed with the different water volumes and passed through a 1-ml column of IgG-Sepharose. The columns were then eluted with NaI (5 *M*) and the abundance of HAV (denoted by plus signs) was estimated by IEM as described (Elkana et al. 1983)

Table 2. IEM detection of HAV, recovered from sewage by the immunoadsorbent method

Sampling location[a]	Sample volume (ml)	Treatment[b]		
		Crude	After solid separation	Concentration ×15
Affected neighbourhood	50	++	++	+
Unaffected neighbourhood	50	–	–	–
Oxidation pond	50	–	–	–

[a] Sewage was sampled from various locations in a town and from the oxidation pond of this town, during a reported hepatitis outbreak (Elkana et al. 1983)

[b] Solids were separated with Freon (Elkana et al. 1983). Concentration was performed by a conventional filtration method

in a considerable loss of HAV. It is hoped that concentration by immunoadsorbance can circumvent this problem, as demonstrated in Table 1.

In the experiments described so far, IEM was the method used for the confirmation of the presence of HAV and for quantity estimation. This method requires skill and sophisticated equipment. Electronic excitation energy transfer offers an alternative method which may prove simpler to perform and more useful for quantification of the virus. A preliminary experiment in this alternative method is shown in Fig. 1. The fluorescence spectrum of a mixture of fluorescein-IgG and rhodamine-IgG (excited at 410 nm) was mainly that of fluorescein-IgG with a maximum at 519 nm. A ten fold concentrated HAV-containing faecal extract was added to the mixture. No change in the fluorescence spectrum was observed after 24 h. Several days later the distinct fluorescence of rhodamine-IgG, with a maximum at 575 nm, appeared. Its appearance is compatible with transfer of energy from fluorescein IgG to rhodamine-IgG as a result of addition of HAV-containing faecal extract. The presence of HAV in the extract was verified by IEM, as well as the antigenic activity of both fluorescent conjugates. The formation of

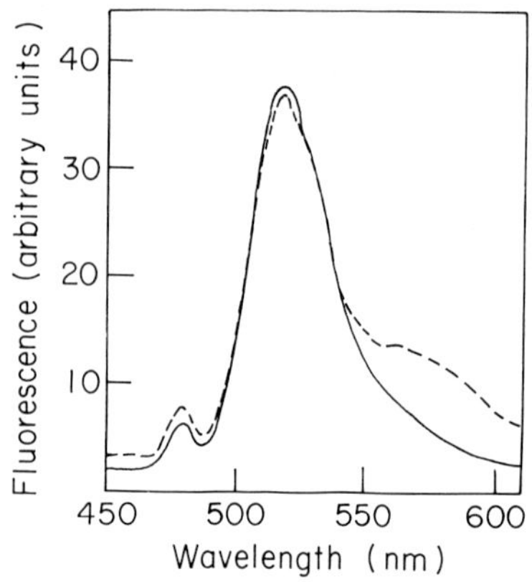

Fig. 1. Fluorescence emission spectrum of fluorescein-IgG ($\lambda_{max}=519$ nm) and rhodamine-IgG ($\lambda_{max}=575$ nm) with (---) and without (—) HAV-containing faecal extract ($\lambda_{ex}=410$ nm)

antibody–virus complexes is thus highly probable, although it was not checked in this case. This preliminary experiment does not rule out an additional contribution to energy transfer by non-specific aggregation of IgG with some component in the faecal extract.

Acknowledgement. I am grateful to Adnan Manaa', who carried out a large part of the work described. This research was supported by a grant from the National Council for Research and Development, Israel, and the K.F.K., Karlsruhe.

References

Elkana Y, Manaa' A, Marzouk Y, Manor J, Halmut T, Herscovici M, Barak S (1983) Detection of hepatitis A virus in sewage. J Virol Methods 7:259–262

Forni L (1979) Reagents for immunofluorescence and their use for studying lymphoid cell products. In: Lefkovits I, Pernis B (eds) Immunological methods. Academic Press, New York, pp 151–167

Förster T (1959) Transfer mechanisms of electronic excitation. Disc Faraday Soc 27:7–17

Gerba CP (1983) Methods for recovering viruses form the water environment. In: Berg G (ed) Viral pollution of the environment. CRC Press, Boca Raton, Florida, pp 20–35

Goyal SM, Gerba CP (1982) Concentration of viruses from water by membrane filters. In: Gerba CP, Goyal SM (eds) Methods in environmental virology. Dekker, New York, pp 59–116 (Microbiology series, vol 7)

Grabow WOK, Gauss-Müller V, Prozesky OW, Deinhardt F (1983) Inactivation of hepatitis A virus and indicator organisms in water by free chlorine residuals. Appl Environ Microbiol 46:619–624

Hejkal TW, Keswick B, LaBelle RL, Gerba CP, Sanchez Y, Dreesman G, Hafkin B, Melnick JL (1982) Viruses in a community water supply associated with an outbreak of gastroenteritis and infectious hepatitis. J Am Water Works Assoc 74:318–321

Joustra M, Lundgren H (1969) Preparation of freeze-dried, monomeric and immunochemically pure IgG by a rapid and reproducible chromatographic technique. Protides Biol Fluids 17:511–515

Smith EM, Gerba CP (1982) Laboratory methods for the growth and detection of animal viruses. In: Gerba CP, Goyal SM (eds) Methods in environmental virology. Dekker, New York, pp 15–47 (Microbiolgy series, vol 7)